Energy Saving in Buildings

Commission of the European Communities

Energy Saving in Buildings

Results of the European Communities Energy R & D and Demonstration Programmes (1979-1983)

Proceedings of the International Seminar, held in The Hague, The Netherlands, 14-16 November 1983

edited by

H. Ehringer

Commission of the European Communities, Directorate-General for Science, Research and Development, Brussels, Belgium

and

U. Zito

Commission of the European Communities, Directorate-General for Energy, Brussels, Belgium

D. Reidel Publishing Company

A MEMBER OF THE KLUWER ACADEMIC PUBLISHERS GROUP

Dordrecht / Boston / Lancaster

Library of Congress Cataloging in Publication Data
Main entry under title:

Energy saving in buildings.

At head of title: Commission of the European Communities.
The International Seminar: Energy Saving in Buildings was jointly organized by the Commission of the European Communities and the Netherlands Ministry of Economic Affairs.
Includes index.
1. Buildings–Energy conservation–Congresses. 2. Buildings–Energy conservation–Research–European Economic Community countries–Congresses. I. Ehringer, H. II. Zito, U. 1939- III. Commission of the European Communities. IV. International Seminar: Energy Saving in Buildings (1983 : The Hague, The Netherlands) V. The Netherlands. Ministerie van Economische Zaken.
TJ163.5.B84E575 1984 690 84-4902
ISBN 90-277-1770-2

Organization of the Conference by

- the Commission of the European Communities
  Directorate-General for Science, Research and Development and
  Directorate-General for Energy, Brussels
- the Netherlands Ministry of Economic Affairs

Publication arrangements by
Commission of the European Communities
Directorate-General Information Market and Innovation, Luxembourg

EUR 9150

Published by D. Reidel Publishing Company
P.O. Box 17, 3300 AA Dordrecht, Holland

Sold and distributed in the U.S.A. and Canada
by Kluwer Academic Publishers,
190 Old Derby Street, Hingham, MA 02043, U.S.A.

In all other countries, sold and distributed
by Kluwer Academic Publishers Group,
P.O. Box 322, 3300 AH Dordrecht, Holland

Printed in The Netherlands

P R E F A C E

The main objective of the Communities' energy policy consists in securing a sufficient energy supply to meet the present and future demand of its Member States, and in reducing the Communities' dependence on imported energy. In spite of the temporary fall in oil prices,energy saving measures will continue to remain imperative, as resources will continue to diminish rapidly. This is the reason why since 1975 the Communities have been stimulating universities, industries and national laboratories of the EC Member States to perform R and D work aiming at energy saving, and have rendered financial support to energy saving demonstration projects. More than 85 contracts concerning new techniques of saving energy in the building sector have been concluded. The work carried out under these contracts is coming to an end and results have become available.

The International Seminar ENERGY SAVING IN BUILDINGS jointly organized by the Commission of the European Communities and the Netherlands Ministry of Economic Affairs highlighted the results of energy saving projects carried out within the framework of the Communities Second Energy R and D Programme and the results of the Demonstration Programme (1979 - 1983). The seminar was also intended to convey these results to representatives of national authorities, public, industrial and financial organisations, architects, decision-makers and town-planners.

The seminar was openend by Dr. G. VAN AARDENNE, the Netherlands' Minister for Economic Affairs, and by Dr. A. STRUB, Director, Head of the Energy R and D Programme of the Commission of the European Communities.

In a plenary session on the first day, recognized energy experts illustrated the need for energy conservation research and commented on the potentials of various technologies in this field.

During the next two days results of research work and of demonstration projects carried out by contractors were presented in three sessions and three poster sessions on heat pumps, building envelope and improved conventional heating systems. During each session an overview of the work was given by a rapporteur and some contractors. At the end of the conference overall conclusions and recommendations were developed by Dr. M. DAVIS, Director at the Commission of the European Communities.

H. EHRINGER
Head of Division
Energy Conservation Research

U. ZITO
Head of Division
Energy Conservation
Demonstration Scheme

CONTENTS

B - DEMONSTRATION PROJECTS

B - DEMONSTRATION PROJECTS

OPENING

Introductory address by
Mr. G.M.V. VAN AARDENNE , Minister of Economic Affairs of the Netherlands

Opening address by
Dr. A. STRUB, Commission of the European Communities, Brussels

Podium at the opening session

Mr. G.M.V. VAN AARDENNE, Minister of Economic Affairs

INTRODUCTORY ADDRESS

by G.M.V. VAN AARDENNE
Minister of Economic Affairs

Ladies and Gentlemen,

I am extemely glad to have been invited here today to open this seminar on energy-saving in buildings. First of all because it gives me an opportunity to display just how much value we here attach to European Cooperation, particularly in such an important priority area as energy. In the second place, because of the importance of the large number of projects that will be presented at this seminar, which means that this meeting will focus extra attention both on energy saving itself and on the new ways to do so. Clearly, at the present time there is no reason to reduce our efforts in this field of energy conservation. If the expectation proves to be true that in the second half of the 1980s the demand for and the production capacity of oil will again come closer together, there will be the renewed possibility of fresh tension on the oil market and sharp price increases. This is a very important factor for buildings in particular, which in most cases represent capital investments with a long useful lifetime. Accordingly, in taking our decisions today we must allow for the energy situation of tomorrow. We will therefore have to find ways of reducing the future level of energy consumption still further. That requires international cooperation, especially to stimulate research in this field. The exchange of experiences about the design and results of the energy saving policy pursued in the European Member States is an important potential source of such stimulation. In addition, cooperation in research matters helps to enable optimum use to be made of the research capacity which is present in the various countries. Buildings, which account for about 40 % of total energy consumption, are one of the main priority areas for energy-saving policy. Indeed, in many countries this is the area where most of the energy conservation activities have developed.

Initially, energy conservation efforts in buildings focussed on a small number of measures such as insulation, but now there is a large number of energy-saving options in this sphere. To provide you as it were with a starting-signal for your discussion, I should like to take you on a short tour along these various options. In the first place, I would refer to the activities which precede the actual construction work, namely the physical planning stage. In this stage, energy-saving at all levels has to be integrated. Aspects to be considered here include the building density, the possibilities of utilizing waste heat, the position of houses and buildings in relation to the sun and wind direction, and their situation in relation to public transport facilities. During the construction stage itself, reducing heat requirements forms the second priority area. A good design is the first prerequisite in this respect. I see that your programme will also be devoting due attention to this matter. For new construction housing, energy requirements can be reduced in this way to 50 - 60% of the present level, and substantially further still if they incorporate the results of current research programmes. With this in mind, Holland is currently preparing to upgrade its programme for trial projects in this field. For the supply of the remaining heat requirements we can choose from among a wide range of energy-saving techniques, varying from local heating, instead of the individual central heating systems, which are still customary, to large-scale district heating systems, serving 50 000 houses and more. Important developments here are boilers with improved efficiency, large and small-scale combined generation plants, heat pumps, and technical modifications to the installations themselves. Beside these systems, we can also report energy-saving in lighting systems, especially in utility buildings. A good example of this is Nederlandse Gasunie's project, which is included in the European Demonstration Programme. The use of alternative energy sources such as solar and wind energy provide another way of achieving useful savings, although these techniques are not yet economically feasible.

It is not only by means of investments that energy consumption is being reduced, but the behaviour of occupants and users of buildings is also an important factor. Energy-conscious behaviour is capable of achieving large savings in itself, without necessarily leading to any loss of comfort. The introduction of new energy-saving techniques will also have to allow for the behaviour of these occupants and users, and provide for familiarizing them with these techniques.

When considering buildings, we should certainly not confine ourselves to new construction. In terms of quantity, the volume of existing buildings is far more important. And it is here that we find many of the inefficient energy situations stemming from the past. Accordingly, many countries have set up extensive schemes providing for insulation and other energy-saving measures for existing houses and buildings. In Holland, by the end of 1983 more than 1.6 million existing houses will have been insulated, with Government subsidies amounting to 1.2 billion guilders. This is more than 50% of the total number of existing houses to be insulated in this country. These energy-saving measures can very often be conveniently included in renovation projects for existing buildings. This brings us to the matter of how the Government, both at the national level and the international level, can respond to all these energy-saving options.

It is my opinion that the Government's efforts should be specifically directed at those areas where, otherwise, developments would take place insufficiently or only with excessive delay. Because of the present shortage of funds which exists in virtually all countries, we must aim for the maximum net effect from the allocation of Government money. Excessively detailed Government intervention, however, may lead to a rigidifying and impenetrable networking effect of regulations and stimulation measures.

I believe that a particularly important priority area for the Government is the setting-up of a well coordinated system for research and development and demonstraton projects. In this respect, the EEC has a very important role to play in Europe. Although many Member States have demonstration programmes of their own, a supplementary European programme is of great importance. One of the main arguments in favour of this is the broad application market for European demonstrations. In this way, using relatively few resources, we can achieve major energy-savings through the intended spin-off effects of demonstration projects. Moreover, a European programme would be particularly interesting to the producers of energy equipment, as this would offer them a convenient entrance to an international market.

After the demonstration project phase, the Government can also promote the introduction onto the market of new techniques, in cooperation with industry of course. I believe that the prime consideration in such a situation should always be the economic feasibility of energy-saving investments, so that in time the market is able to adopt these techniques itself, without Government subsidies. Because the real possibilities are often unknown in many quarters, it is important that the Government should strongly promote the building-up of an effective system for information and consultancy in energy-saving matters. Precisely in a situation where there is on the one hand uncertainty about developments on the energy market and on the other hand the availability of a wide range of energy-saving techniques, some even competitive with one another, selective information and consultancy are exceptionally important. In our society, however, we often still encounter obstacles to the introduction of worthwhile energy-saving measures, for example in the field of financing and organization. The removal of these obstacles is also, to some extent, a task for the Government, in many cases requiring a suitable tailored approach to the various target groups.

In conclusion, I should like to mention the importance of proper planning and organization of energy conservation activities at the regional and local levels. It is vital that local authorities, the owners and users of buildings, and the public utilities should cooperate towards this purpose, if we are to achieve tangible results.

Ladies and gentlemen, this brings me to the end of my introductory address. During the next few days you will hear about a large number of energy-saving techniques originating from all our Member States and financed with community resources. Techniques that are capable of making our society far less dependent on energy than hitherto. We must of course be prepared to provide sufficient creativeness and effort to put all the possible energy-saving techniques offered to us into actual practice. Only then will it be possible for research and demonstration projects to make our society truly energy-economical. I sincerely hope that this seminar will be able to make a valuable contribution towards that end. I should like to take the opportunity to express the wish that this meeting may contribute towards a continuation of fruitful European cooperation in the field of rational energy utilization.

Thank you very much.

O P E N I N G   A D D R E S S

Dr. A. STRUB
Commission of the European Communities
Brussels

I do not think it is necessary or opportune to remind this audience of the events and developments in the energy field, since 1973, which have led to "energy" becoming one of the major issues of preoccupation for any politician or economist.

The past three years have been marked by dramatic changes in the world oil market. By a combination of various factors, which I will not go into now, world oil demand went down by 20% and the price of crude oil has fallen correspondingly. What are the consequences of this on our behaviour? How should we react to this?

We can learn quite some lessons from this development, but we can also draw some wrong conclusions. One such would be that we can now relax and slow down our efforts to achieve further energy savings. While it is clear that the recent downward development in the oil market is due to a number of factors, not all of which were desired or welcome, it is also clear that without the coordinated and vigorous energy saving effort of the past few years, the fall would not have happened at all. Over the longer term, we cannot be certain that the trend will not reverse again. Therefore, a sustained effort of research, development and demonstration in the energy field is essential from the point of view of a long term Community energy strategy.

The Community's R, D and D programmes in the energy sector, some of which will be the subject of our conference today, are quite numerous. They cover the following fields:

- nuclear fission
- fossil fuels (coal, hydrocarbons)
- nuclear fusion
- renewable energy sources
- energy conservation.

All of these programmes are part of the Community's Framework Programme for scientific and technical activities. This Framework Programme, which was designed and adopted in June 1983, is aimed at representing the Community's research, development and demonstration activities in a coherent way. The ultimate goal of this Framework Programme is of course to continually examine the Community's activities and to adapt them so that they correspond to one of the seven declared Community R and D objectives, namely

- promotion of agricultural competitiveness
- promotion of industrial competitiveness
- improvement of the management of raw materials
- improvement of the management of energy resources
- reinforcement of development aid
- improvement of living and working conditions
- effective use of the Community's scientific and technical potential.

Up to now, the Community's R, D and D efforts have attributed a very high priority to the "energy" objective. In 1982, about 300 Mio ECU out of the Community's total expenditure for R, D and D of 450 Mio ECU, went into energy matters.

This high priority will remain. However, the elaboration of the Framework Programme has shown that important efforts will have to be undertaken immediately in several other strategically important fields, as otherwise the Community will be unable to keep or regain its place in the face of world competition, mainly from the USA and Japan.

The Commission has therefore proposed to the Council of Ministers and to the European Parliament not only a new R and D programme in the energy field, but also very important programmes in the fields of information technology, biotechnology, basic industrial technologies and others.

As already mentioned, the work presented today at this conference, relates to the Commission's demonstration projects and R and D programmes for energy conservation in buildings. Let me therefore make a few further remarks on this particular topic and on the trends in R and D.

The domestic and commercial sector accounts for the largest part (40%) of the primary energy consumption in the European Community. About two thirds of the primary energy consumption in this sector is used for the heating of buildings and one third for power consumption in lighting, domestic appliances etc. Over the long term, a reduction of at least 50% is possible in the energy requirements for the heating of buildings. Here, research and demonstration have their role to play. Basically, this reduction can be achieved in three ways:

- Energy savings of 20-50% by more efficient heat production. In the EC R&D programme, work is focussed on heat pumps and improvement of conventional heating systems. Heat pumps may in the long term reduce the primary energy need for domestic heating by 30-50% and this work is therefore strongly emphasized. Better control of heating systems for apartment buildings, office buildings and schools may lead to energy savings up to 20%.

- Energy savings of 20-30% through the reduction of transmission losses through walls, roofs and windows. Wall and roof insulation materials are already commercially available but problems still exist due to aging, condensation and inflammability. A considerable proportion of the heat losses in buildings (up to 30-40%) are caused by windows.

Window insulation is therefore very important. Infrared reflective window coatings, which are transparent for visible light, but which reflects radiant heat can double the level of heat insulation. This technique is being developed for new houses and for retrofitting in existing houses.

- Energy savings of 10-12% by reduction of ventilation losses. With increased insulation, the fraction of heat losses by ventilation (typically one air change per hour), which already now can be as high as 30-40%. Various possibilities are being explored to reduce these heat losses. The basic theme of all approaches is the construction of very tight houses with heat exchange between the outgoing and incoming air, or with ventilation which is controlled by CO_2 or H_2O vapour level detection.

Energy saving in the areas described above can be achieved through the development of specific energy saving techniques. However, the introduction of these techniques on a large scale also requires the development of energy saving building design concepts and the construction of buildings which incorporate these techniques and minimize their overall energy use. In such design concepts, computer models which simulate the thermal behaviour of buildings and their components, and playing an increasingly important role. Already, such models are being used by architects even though they are far from reliable and often yield large discrepancies. The study of computer models and their experimental validation was and remains an important subject in the EC R and D programme.

Another problem is human behaviour. The inhabitants of a building, by opening doors and windows, frequently interfere with heat saving measures, particularly in well insulated houses with controlled ventilation. You will hear, therefore, about different R and D projects dealing with the development of tracer techniques for measuring the heat exchange rate, particularly in large buildings.

Even if about two thirds of the primary energy consumption in the domestic sector is used for heating, electricity consumption (37% of the primary energy demand) is likely to yield further energy saving possibilities. In the EC R and D programme, in collaboration with industry, new refrigerators, washing machines and cooking equipment have been developed which consume considerably less of this high grade form of energy than the appliances which are currently used.

In order to demonstrate the economic feasibility of new energy saving technologies, a large number of full scale demonstration projects have been carried out by the EC for residential and other buildings (office buildings, hospitals, schools, sport centres). In a recent evaluation of these demonstration projects, it was concluded that the emphasis of EC demonstration support should shift towards projects undertaken by the large public and private investors, companies and cooperatives who manage housing complexes and estates. This would offer several advantages, including the rapid, widespread appreciation of successful demonstrations and the centralization of decision making. It can be said therefore that the following particular technologies deserve closer attention in future:

- building techniques which enable the heat requirements of large residential public and office buildings to be reduced;
- the use of micro-electronics to manage the energy requirements of existing or future residential and tertiary buildings.

To conclude, I want to stress the importance of technology transfer. If we want that the results of our R, D and D efforts are applied in practice, we must bring these results to the right people in the European Community. This conference is one way to bring about such a technology transfer. Here we will be discussing the outcome of the second EC Energy R and D Programme and the EC Demonstration Programme and we will review eighty projects in the building sector. I am sure that this will be a valuable experience.

I wish you a fruitful meeting and a pleasant stay in The Hague.

KEYNOTE SPEECHES

Building physics: indispensability or a vogue discipline?
A.C. VERHOEVEN, Delft University of Technology, Department Civil Engineering, Delft

The problem of the transfer of the results of research and innovations into the sector of energy conservation
G. ELIAS, Faculty of Engineering, University of Palermo

The low energy hospital
H. GOODMAN, Department of Health and Social Security, London

Le diagnostic: un préalable à la réhabilitation thermique des bâtiments existants
J.C. BONCORPS, Agence Française pour la Maîtrise de l'Energie, Paris

Development of modern heating technology
F. STEIMLE, Institut für angewandte Thermodynamik und Klimatechnik, Universität Essen

Ir. J.A. KNOBBOUT, presenting Prof. Verhoeven's paper

Prof. Ing. G. ELIAS

Mr. H. GOODMAN

Mr. J.C. BONCORPS

Mr. F. STEIMLE

BUILDING PHYSICS: INDISPENSABILITY OR A VOGUE DISCIPLINE?

Prof.Ir. A.C. VERHOEVEN
Delft University of Technology
Department Civil Engineering
Division Building Physics
Delft, The Netherlands

Introduction

In 1982 a Working Commission "Building Physics" of the International Association for Bridge and Structural Engineering was set up. They have formulated the following definition of Building Physics according to the present state of its field of knowledge: "Building Physics includes the phenomena of heat (energy), mosture, air, acoustics, fire and daylight which may occur in the interior of rooms, in building and structure components and in the environment of buildings and structures (building physics in urban environment)".

This definition shows, that Building Physics is closely connected to human comfort inside and outside of buildings in peoples daily lives. A short summary of the fields mentioned in the definition and the contribution of Building Physics to them will serve to elaborate on the existing interrelations.

Heat / Energy

The consequence of applying higher insulation levels in buildings are indeed a substantial saving of energy; the need of a rapidly responding heating system in order to prevent high temperatures in case of solar irradiation and in summer the necessity to use an effective sun shading device can be regarded as common knowledge.

In designing low-energy consuming buildings the non-stationary effects of varying outside temperatures and solar irradiation must be taken into account. Basically this can be done with the help of computers, whilst pocket calculators might be used for obtaining results that do not require high accuracy.
A Working Group of ISO TC 163 will shortly issue a technical report regarding this subject.
For office buildings it will be possible to assess whether an air conditioning is needed or whether an acceptable inner climate can be maintained by installing a simple ventilation system only.
Saving energy by putting in more insulation (a passive system) has proven to be very cost-effective and furthermore there are hardly any operating and maintenance costs involved. Using an active system the investment costs as well as the operating- and maintenance costs tend to be much higher. As a guideline for future developments it could be said, that aiming at energy savings in active solar systems should always be paired to developing maintenance-free equipment whereas the ration of energy-savings to (extra-) investment costs is small.
With Hfl 1000,- (Dutch guilders) invested in filling a cavity wall with insulating materials like foam plastic, fibre wool, etc. you can save roughly 400 m^3 natural gas per year; investing the same mount of money in solar collectors and their auxiliary equipment for space heating saves only about 40 m^3 natural gas per year!

Moisture

In the past buildings were ventilated with too large amounts of outside air. As a consequence of this low relative humidities were likely to occur in wintertime.
Nowadays windows, doors, etc. are sealed that tightly that the minimum fresh air amount (30 m^3/h per person) can barely come in.
This had led to an increase in the mean relative humidity inside houses and consequently to a growing danger of surface condensation on thermally weak points (thermal bridges) as well as internal condensation within constructions.
The protection against moisture deserves to be studied profoundly in order to prevent damages (cracking, deterioration, etc.) caused by hygric (and thermal) stresses. In the education of architects it is eminent therefore that the laws about hygric transports should get the emphasis they deserve.

Air

The wind-induced air infiltration of a building is depending on its orientation and its height as well as on the kind of the built area in front of the building. As wind velocities are changing continuously in direction, velocity, turbulence and so on, the airflow coming into every room of a building is basically non-stationary. A development in building industry that was to be expected is the application of balanced ventilation by which a controlled amount of fresh air is supplied to the building and the same amount of used air is exhausted meanwhile keeping the infiltration through windows at a minimum.
So the air change rate can be controlled and energy could be regained from the outlet-air.

Acoustics / noise

One is able, and indeed one should always try to protect people from high levels of environmental noise. Most often there is enough knowledge to predict the influence of traffic in a new town and what should be done when the noise level turns out to be too high.
In the earliest stage of a design it must be considered what rooms should be quiet (e.g. bed rooms) and locating them on the quieter side of the house should be considered. It can be stated, that generally one is capable to do a perfectly proper job, but mostly the money available is not hardly sufficient.

Fire

If the air flow pattern within a building and the kind of materials used are known, one is able to give judgement on it's fire sensitivity.
An adequate prevention for fire can only be reached by applying the knowledge and laws available.

Daylight

For reasons of energy savings it is favourable to design small windows on the east, north and west façades. This will also prevent a high temperature-rise in summer caused by solar irradiation. The consequence of such a statement is, that the daylight factor a a room

especially at the back can drop below 1%. Mostly this is not attractive (calling for artificial light, extra energy use, etc.).
So in a very early stage of the design already a compromise between great energy savings and small daylight factors should be reached.
With the aid of a computer the daylight factor and the energy losses can be calculated as function of the size of the windows to prepare these decisions regarding the window-sizes.

Building environment

If handicapped children or old people require an area in the open air where they can walk undisturbed it will be clear that such an area must be located in a calm part of the town where fast winds seldomly occur.
These parts can be pointed out before one stone of a new home for aged persons is placed on another. Sometimes a windtunnel is needed to designate the right area, but like in the case of the noise abatement, knowledge how to measure and how to predict is sufficiently available.

After these summarized examples it will be clear that Building Physics, being in most of the cases the initiator of the theoretical investigations that lead to today's knowledge is a must. So: Building Physics is indispensable for Architects, Designers and Consulting Engineers, who want to use modern materials, save energy and be sure that no damage will cause financial troubles in years to come.

MODIFICATION IN BUILDING PROCESSES

Description	Formerly	Nowaday
Functional demands on building structures	Age-long practically unchanged	Increasing requirements with respect to: - comfort standards - lighting and climatization of factories - specific new demands (TV-studios)
Materials and construction	Great continuity: use of heavy materials; traditional constructions	New, light and handy materials (important in connection with heat loss, impact and airborne sound insulation)
Production techniques	Great continuity; handwork using simple tools	New production-methods; prefabrication technology
Repetition	- no mass production - no disaster in case of errors	A high building production level: - tall building blocs - mass production of dwellings and schools - a single error demonstrates itself in a great number of buildings

THE PROBLEM OF THE TRANSFER OF THE RESULTS OF RESEARCH AND INNOVATIONS INTO THE SECTOR OF ENERGY CONSERVATION

Prof. Ing. Giacomo ELIAS

Full Professor of Technical Physics at the Faculty of Engineering of the University of Palermo

Director of the Finalized Energy Project, CNR, Rome

SUMMARY

The manifold and different "barriers" that must be overcome to let the innovations deriving from research become widely known by users and operators alike, are being made the object of consideration all over the industrialized world and by the European Community itself.

In the case of energy conservation, the solution of the problem is made all the more difficult by the interdisciplinary character of the research work and by the multiplicity and variety - at both the formative and cultural level - of the final users.

The aspects of the problem are analyzed and the indications resulting from the Italian experience of the Progetto Finalizzato Energetica of the National Council for Scientific Research are illustrated with special regard to energy savings in buildings.

INTRODUCTION

The European Commission does obviously follow, with interest, the problems related to the "transfer" of the products of scientific research, taking care to identify and overcome the so-called "barriers to innovation".

The latest report on this subject which we know is the work of Prof. A. Piatier: "Barriers to Innovation in the European Communities" (XII(81)04). The analysis contained therein is detailed and exhaustive; it represents a valid aid for tackling the problem in its entirety, quite aside from the sector under consideration.

It is our well-pondered belief, however, that even though the phenomenon has been manifesting itself in all fields with similar consequences, it has motivations and, hence, calls for remedies that are strongly dependent on the sector to which we refer.

This is all the more true when the point is not that of innovating a specific product or process, but rather that of applying measures involving most of the productive sectors and of the social behaviours, as required by any policy for the promotion of a rational use of energy.

It is perhaps for this reason that some of the measures adopted by the governments for incentivating energy conservation did not have the incisive quality and the profitability that was to be expected.

The definition itself of the term "innovation" that Prof. Piatier quoted as emerging from the Frascati meetings: "an idea turned into a saleable product or into an exploitable process", to be used in this case, should be interpreted in the widest sense. For example, the idea might even be transformed into a change of habits.

ENERGY CONSERVATION: AN INTERDISCIPLINARY RESEARCH

To conserve energy means, on the one hand, to eliminate waste and, on the other hand, to improve the use of sources, whenever this is possible.

Two lines of intervention thus emerge, that deeply differ from one another, even if they are not always kept clearly distinct either in the research programmes or in the corrective measures.

Waste can be described as the non-application or erroneous application, on the part of technicians, of knowledge previously acquired at the scientific level, or as the result of thoughtless behaviour on the part of users.

A clear-cut illustration of this point can be drawn from the sector to which this Seminar is devoted. Buildings constructed without taking into account the thermotechnic teachings and provided with systems with a low energy efficiency, bring about an energy waste in the same order of magnitude as the requirements corresponding to a correct planning.

But a considerable energy waste is also due to some behaviour of users, for which no plausible reason can be given, such as, for instance, an excess of natural ventilation with the heating system in operation.

It follows that, in tackling this type of problem, research, usually applied, has for its objective not so much "original" scientific contributions, at least in the conventional meaning of the term, but rather new projectual tools, strengthened by cognitive in-depth investigations (for example, on the behaviour of the installed construction materials). In this way, it will be easier for planners to apply in daily practice what science places at their disposal. At the same time, research must help achieving a "knowledge" of the user's behaviour, in order to identify the adverse aspects to be corrected.

The scarce scientific originality detracts nothing from the difficulties inherent in this type of research, especially for what concerns its most subjective aspects. This is the most intensely interdisciplinary area of the activity; it calls for solutions - original ones, this time - which must be innovative not merely in regard of the problem as such, but also as regards the choice, not always intuitive, and the detection on the data of the problem considered. For example, in the framework of the Progetto Finalizzato Energetica, in order to determine the behaviour of the existing "inhabited" buildings - the behaviour of the empty buildings not being regarded as significant - a special set of tools had to be evolved. This phase, introductory to the research proper, has called for electronic and not merely thermotechnical skills, as well as resources (men, money and time) in the same order of magnitude. Not only that: it has also produced as much innovation, since that set of tools is being used today for energy analyses.

Moreover, to think that the involved disciplines are only the strictly technical ones would be quite a limiting approach. The necessity of influencing the deeply-rooted behaviours of some components and/ or of the entire community, makes it indispensable to arrive at an approach, also sociological, to the scientifically-oriented problems, in support of normative, legislative or even merely promotional measures (if any). To exemplify: to impose a maximum temperature in rooms without considering the influence of the surface temperature of walls - and, hence, of the constructional techniques - on well-being conditions, is just as wrong as trying to obtain, through the application of a sanction, the compliance with regulations that, in the eyes of the public opinion, are not sufficiently grounded.

When the point is instead that of "consuming a lower amount of energy, the product or the service being equal", then the scientific originality of the contributions increases, while, as a rule, the interdisciplinary character of research decreases. Also in this case, the starting point is the acquired knowledge, but not with a view to utilizing them to better advantage, on the contrary, to overcoming them. A significant example of what precedes is represented by the concept and realization of "blue flame" combustors, or of those fed with "water-fuel" mixtures. In both cases, prior to realizing devices - nowadays sold on the market - with the indispensable reliability, it was necessary to study the basic aspects of the relevant combustion process, and to identify the stability fields of the flames; technological problems had also to be solved, having recourse to the use of non-conventional materials.

ENERGY CONSERVATION: MULTIPLICITY OF THE RECIPIENTS OF THE INNOVATIONS

Since they involve products, processes, planning methodologies, social behaviours and so forth and so on, the measures aimed at achieving energy conservation concern quite a range of final recipients.

At times, the findings emerging from the research work concern industrial sectors with a reduced number of operators, such as the automobile one. In this case, penetration is easier and faster, it being possible to involve in the research, from the beginning, all possible users. When the number of operators increases, as in the case of researches on window and door frames, the transfer process becomes proportionally more difficult, especially in regard of the smaller and less organized of them. And if the innovation concerns productive processes, it reaches the final users only after it has been adopted by the manufactures of the relevant systems and when the benefits it allows justify the replacement of the obsolete systems. It may take years before this occurs spontaneously, but if the convenience is made tangible by well-calibrated incentives, the process can be speeded up.

We have spoken about "well-calibrated" incentives on purpose. It would only be deceiving oneself to think that any innovations that are anything but convenient may assert themselves as a result of the granting of incentives. These are useful only if directed at breaking down the last resistances vis-à-vis those changes whose economic validity can be proved in the short or very short term.

In the case of energy conservation in buildings, the innovation may regard planning methods, construction materials, equipment and systems. The final user is he who lives or will live in the building;

he may enjoy the results of the research only through the work of planners, constructors, system installers and operators. In turn, each of these operators will influence the decisions of the other ones, in a game of interests that are often in conflict with one another.

The transfer process thus implies a broad range of actions, that can be only in part managed by those in charge of research, since the participation in it of a numerically significant component even of the main involved persons alone, cannot be realized. The collaboration becomes thus necessary of bodies, cultural and trade associations and, at times, of the legislator himself, who, by imposing and/or incentivating some given technical solutions, overcomes the conflicting interests on behalf of the community as a whole.

A major role in the transfer process must be played by the bodies in charge of the formulation of the technical standars, which, if in a position to acquire the innovations over a short period of time, can make their introduction easier by providing objective methods for the determination of services and of reliability, as well as criteria of comparison. Finally, of the utmost usfulness is the collaboration of the specialized and of the non-specialized press, depending on the object of the transfer. To ensure that the information is incisive and effective it is necessary, however, that the products of the research are adequately and synthetically illustrated. The need thus arises for a set of documents that may popularize the results in a scientifically correct way, a condition, this one, that is not always easy to obtain.

There emerges from the above a consideration that can be clearly taken for granted but that does not seem to be acquired, chiefly at the political level. While a "return" of investments may be expected in the applied scientific research, it is not correct, instead, to make the researches only answer for such a return. They can and must, in fact, be made liable for the results and useability of them, but it is the socio-economic system as a whole that must see to it that their work is actually exploited.

It is for this reason that the Italian Government has supplemented the National Energy Plan (NEP) with a National Research Plan on Energy (NRPE), conceived in such a way as provide the former with the cognitive support indispensable for its implementation. This is an absolutely innovative event in economic planning and, perhaps, not only for Italy.

THE EXPERIENCE OF THE PROGETTO FINALIZZATO ENERGETICA

The Progetto Finalizzato Energetica (PFE) of the National Council for Scientific Research (CNR) is a comprehensive research programme, whose object is the conservation of energy and the recourse to non-conventional sources, with the exclusion of the nuclear one.

The activities are carried out in a period of five years; the first five-year period is now over (1976-1981), while the second five-year period has just begun (1983-1987). The programme is nation-wide; the Government approves a feasibility plan and entrust CNR with the management of same. For the second five-year period, for a better coordination, a close financial and operational cooperation has been established with ENEA, a body in charge of industrial development in the non-conventional energy sector.

Since this is a special action to be completed over a fixed period of time, PFE has a budget of its own (70 billions the first, 270 the second) and has been provided with a managerial structure.

On the basis of the feasibility study, a working plan is prepared each year, which contains all the contracts required for the carrying-out of the research work envisaged for that year. The PFE operators come from the industrial sector and from universities, or are CNR workers. For example, the working plan for 1983 includes nearly 650 contracts, of which 47% to industries, 39% to universities and the remaining ones to CNR or ENEA laboratories.

Beside the administrative management and the scientific coordination of researches with one another and on the domestic and international plan, the PFE management is also entrusted with the task of promoting all necessary steps for the transfer of results. For this reason, a special service has been set up since the beginning, which provides for the organization of seminars, congresses, meetings with bodies and associations. It is also in charge of the publications which answer to criteria strictly connected with the nature of the recipients of them. There exist, therefore, the informative seminars and the final reports - of a more technical nature - while, in regard of the popularization of th PFE's activities, the so-called "white books" are issued, that are easy to read and extremely clear. The said service also produced short films that illustrate the activities carried out by PFE; they are translated into English and Spanish, in order that such information may reach foreign countries.

The experience acquired in the first five-year period has shown that all the steps undertaken, inclusive of those implemented through the specialized and daily press, are useful and necessary. Still something remains to be done. In particular, in the second five-year period the connections with the downstream structures will be further enhanced, so that the information items may be made to flow in a cascade towards the remotest users.

In particular, ENEA, thanks to its particular trend, shall certainly peform a major role in the transfer process. There were further inserted into the research groups, some components that are not essential for the latter's purposes, but that are certainly useful for spreading the results, providing that they are involved in due time.

Attempts will be concomitantly made to widen the number of experts to be placed at the disposal of the public administrations whenever they request them. The activity was finally strenghtened in the sector of the formulation of the technical standards, the importance of it having been recognized, as previously said, for the purpose of overcoming the barriers to innovation.

Summing up: what must be once again stated is that the transfer of results is not a supplement but rather an integral part of the research. The resources invested in it - some times in a consistent amount - realize the financement of the research.

THE LOW ENERGY HOSPITAL

GOODMAN H
CHIEF ARCHITECT DEPARTMENT OF HEALTH AND SOCIAL SECURITY
UNITED KINGDOM

A research and development exercise of the complexity of our Low Energy Hospital project can only be carried out by a multi-disciplinary team with a high level of technological support. Within each discipline must be those with an understanding of not only the energy demands and opportunities but also the clinical requirements of buildings they are researching. Whilst the results of the study will be presented as an amalgam of all these views synthesized and perhaps sanitized, each member of the team retains his individual identity. Research remains an essentially personal activity and despite the surroundings of both team and technology often the decisions about which avenue to follow relate to one individual in that team, and often then on his intuition rather than his scientific deduction. If one is therefore to draw conclusions and look forward, and that surely is the purpose of this keynote paper, then apart from technical and financial considerations, one is led by one's own prejudices.

The short-term results of our Low Energy Hospital study are published as one of the papers in this seminar and in any case are well known. We have succeeded in what appeared to be the nearly impossible task of reducing the energy consumption in a modern hospital by over 50%. But the success of the project will only be seen to be demonstrated if three interrelated aspects remain in balance during the lifetime of the project. Foremost of these is the ability to maintain a high-technology installation in a state of high performance. Secondly, the ability to continue the energy management function during the lifetime of the building. Thirdly, and most importantly, it is the 'will' to apply these first two. Our apparent success at this stage of the project is not 'a once and for all' achievement, indeed it is only a beginning. Our project will only finally be seen to be a success if it maintains its low level of energy consumption over the lifetime of the building. Although maintaining the technology in a functioning state and ensuring that energy management adapts to changing circumstances is important; the 'will' to do so is even more so. Whether we as Europeans or as nations or as individuals are determined to maintain this level of efficiency will ultimately depend on whether we as individuals can see it contributing to our own immediately perceived objectives and as a step on the road to even further and more important objectives. Any research and development programme is analogous to opening one door only to see another beyond, and the near certainty of the fact that there is another door beyond that.

When I come to draw conclusions then clearly they can only be the view of one member of this comprehensive team commenting on my own perceived objectives. Our low energy hospital is undoubtedly going to be efficient in energy terms; but more importantly it has emerged during our studies that we can design a hospital that is even more energy efficient than this first attempt and this we intend to do.

The remit for the first hospital was that we used 'off-the-shelf' technology although in very clever combinations. The remit for our second project is that all technologies new or yet to be developed might be incorporated. Already designs for this hospital aiming at 65% energy reduction are in hand.

However the fact that our project is a hospital is almost an irrelevancy. The real value of the project is that it is a large complex organisation with high energy demands and that it calls upon a wide range of energy conservation technologies to be used in combination. Whilst there is undoubted value in individual small scale studies of hardware it is unlikely that in real life that these will be used in isolation and that they will usually be used in combination; therefore there is a need for more projects which use combinations of technologies in a live situation. And particularly projects in which the client/user is part of the design team. But that places the emphasis of our work on energy rather than on efficiency. The issue raised is, that if we can achieve this level of efficiency in terms of energy, might we not be able to achieve this level of efficiency in other aspects of hospital organisation. In other words, the techniques that we have used to identify sources of demand and ways in which these could be satisfied at lower unit cost must be applicable to a whole range of other analogous activities within health buildings. We can think of the explosion in information technology and relate this to clinical activities, where the handling of increasing amounts of complex information regarding patients is critical to efficiency. We might consider materials handling where automation in industry enables us to store, handle and distribute ranges of goods at low cost and with minimal staff resources. We might think of management level where improved space-utilisation, improved efficiency in through-put and output and numerous other arenas - where in all of these, the same dedication would produce far greater returns. The received wisdom is that the cost of health care is becoming an intolerable burden on Western economies. Whether or not this is correct, or whether it suits the purposes of governments to say it is open to debate, but of what there is no doubt is that we can provide greater quantum of health care if we can improve our efficiency, and that can only be good.

But if we could return to place our emphasis on energy then it is in the national field that certainly the greatest returns and perhaps the greatest contention would arise. Reliable statistics are hard to come by in this field but there appears to be a general concensus view that in the United Kingdom buildings, and that is their environmental and process loads, appear to account for at least 50% of the total UK energy use. Hospitals are after all one of the most complex users of energy requiring generally high temperatures, high levels of environmental control, a twenty-four hour a day organisation and all within the context of a near fail-safe service. If within these constraints we can achieve energy savings of a magnitude that our research study shows, it is inconceivable that if the same techniques that we apply to the Low Energy Hospital were applied to a range of social, educational, commercial and industrial buildings then the same level of savings could not be achieved. The obstacles to such a programme would appear to be immense if one considers it at the bureaucratic level. But if one considers that the component parts of such an exercise would be probably individually smaller than our own study, and if one takes the point made earlier in my talk that the success of such projects depends on individuals then this series of studies would seem not only to be feasible but capable of producing results in a relatively short space of time.

We have been talking in terms of energy saved although I doubt whether there is any virtue in saving energy per se, certainly not in terms of incentives for the users. The real incentive must be the cash resource so saved and the ability of the users to re-invest this cash resource. The implications of this money so saved in terms of investment in industry, or in my own particular case to be devoted to patient care, are immense. Certainly this level of reduction is not without an effect on the energy industry and may well not be universally popular. In the case of my own country we have coal reserves which will last more than 300 years; this could be extended even further. Perhaps most importantly this reduction in demand might allow us to defer any further implementation of our nuclear generating programme until safeguards can be given against the health risks which seem to be emerging from this programme. I become increasingly concerned as evidence and counter-evidence concerning the dangers of radiation from the waste products emerge. If I apply the same scientific rigour to this evidence as I applied during the research stages of our project then I begin to become very frightened indeed.

But I will sum up on an optimistic note. I hope that in the short term the savings that we achieved in energy costs in hospitals will be invested to improve the service to our patients. In the longer term if we can mount this exercise to improve efficiency of our health buildings, again I hope that this will be invested into increasing the quantum of health provision. If I can make one aside here - of course in this conference we all look towards our work in energy conservation, being of benefit in a European context. But in my own field, that of Health, that it is the Third World who are most vulnerable. Too often we see European patterns for health care with their large expensive-to-run 'disease palaces' being repeated, without relevance to local, economic and social conditions. If only these efficiency studies allow us to influence development of Third World health services on the theme of 'efficiency and relevancy', then we will probably have done more good than any study of energy can possibly do.

I have said how we can apply these techniques to social building programmes but perhaps more importantly is their impact on commercial and industrial buildings. It is essential that savings achieved are re-invested and are devoted to improving the efficiency and competitiveness of those industries rather than being frittered away in other fields, or worst of all, to use as an excuse to increase the tax on fuel prices. And lastly, and perhaps where I need the greatest optimism we might achieve a reduction in nuclear power generation, which will improve the health of our nation by the removal of these risks from our environment.

LE DIAGNOSTIC : UN PREALABLE A LA REHABILITATION THERMIQUE DES BATIMENTS EXISTANTS

J.C. BONCORPS

Agence Française pour la Maîtrise de l'Energie

Le secteur du bâtiment est le secteur le plus important par la consommation d'énergie et le gisement d'économie d'énergie.

Dans ce secteur, les deux principales voies pour réduire les consommations d'énergie sont :

- la construction de bâtiments neufs très bien isolés et équipés de chauffage à haute performance énergétique, grâce à la réglementation ;
- la réhabilitation thermique des bâtiments existants, dont 90 % ont des caractéristiques thermiques très inférieures à celles des bâtiments récents, grâce à des mesures essentiellement incitatives.

Mais il est souvent difficile, pour les propriétaires de locaux existants, de déterminer les travaux les plus adaptés pour réduire leur consommation d'énergie ; d'où le diagnostic thermique, étude effectuée par des professionnels qualifiés, et destinée à donner aux décideurs toutes les informations nécessaires pour leur permettre de choisir les améliorations à apporter à leur bâtiment : caractéristiques thermiques, améliorations envisageables avec leur coût, les économies d'énergie et la durée d'amortissement, propositions d'un programme de travaux.

Le diagnostic thermique, qui doit être complet et objectif, est un préalable important pour toute opération de réhabilitation de bâtiments existants.

Le secteur du bâtiment est le secteur le plus consommateur en énergie ; plus de 250 millions de tonnes d'équivalent pétrole de consommation finale pour l'ensemble de la Communauté, soit 38 % de la consommation totale.

C'est également le secteur où le gisement d'économie d'énergie est le plus important : environ la moitié des économies d'énergie potentielles concernent ce secteur. Par ailleurs, les énergies nouvelles et renouvelables telles que l'énergie solaire, la géothermie ou la biomasse, peuvent être aisément utilisées dans les logements et les locaux à usage tertiaire.

Ce secteur est donc particulièrement important dans toute politique d'utilisation rationnelle de l'énergie.

Quelles sont les principales voies pour économiser l'énergie dans les bâtiments ? Elles sont au nombre de 4 :

- en premier, les modifications de comportement, telles que l'abaissement de la température de chauffage et la limitation des éclairages inutiles. Cette voie a déjà largement contribué aux économies d'énergie obtenues depuis 10 ans ; en France, les modifications de comportement représentent 60 % des résultats dans le secteur du bâtiment. Mais, dans l'avenir, leur contribution devrait être beaucoup plus faible, les gaspillages les plus importants et les plus faciles à supprimer ayant disparu ou presque disparu ;

- la deuxième voie, c'est l'amélioration, grâce à des actions de recherche et de développement, des performances énergétiques des appareils électro-ménagers, des appareils de cuisson, des téléviseurs et des matériels de bureau ainsi que des systèmes d'éclairage.

Ces deux premières voies ne peuvent plus apporter, dans l'avenir, qu'environ le quart des économies d'énergie potentielles du secteur. Les deux dernières voies visent la réduction des consommations d'énergie pour le chauffage des locaux et celui de l'eau chaude sanitaire, par une modification des bâtiments et de leurs équipements thermiques ;

- la troisième voie concerne la construction neuve de bâtiments d'habitation ou à usage tertiaire. Dans ce domaine, les actions entreprises depuis le premier choc pétrolier, c'est-à-dire depuis tout juste 10 ans, ont eu des résultats très importants. Grâce en particulier aux programmes de recherche-développement et de démonstration, il a été possible de maîtriser des techniques et procédés nouveaux et d'en assurer une large diffusion. Dans la plupart des pays de la Communauté, des réglementations ont été mises en oeuvre, avec des exigences de plus en plus rigoureuses en matière d'isolation et de permformance énergétique des systèmes de chauffage et de renouvellement d'air. En France, par exemple, les logements construits aujourd'hui ont des consommations d'énergie pour le chauffage réduites de 50 % par rapport à 1974 et un nouveau renforcement de la réglementation, prévu dans 2 ans, devrait conduire à une division par 3 des consommations de chauffage, en l'espace de 12 ans ;

- la quatrième voie, enfin, concerne la réhabilitation thermique des bâtiments existants. Cette voie représente un enjeu énergétique au moins aussi important que celui de la construction de bâtiments neufs à faible consommation d'énergie. Elle apparaît également fondamentale lorsque l'on considère que 90 % du patrimoine immobilier de la Communauté, soit 125 millions d'équivalents-logement, ont été construits conformément à des réglementations datant d'avant le premier choc pétrolier ; sauf exception, ces bâtiments ont des caractéristiques thermiques nettement moins bonnes que les bâtiments construits récemment. La réhabilitation thermique de cette très grande majorité des bâtiments existants permet, par ailleurs, de réduire les charges de chauffage qui atteignent souvent des niveaux difficilement supportables pour les ménages les moins favorisés.

125 millions d'équivalents-logement à réhabiliter du point de vue thermique dans les 10 pays de la Communauté, c'est un énorme programme qui nécessite un ensemble d'actions cohérent.

Evidemment, des travaux d'amélioration thermique sont entrepris de façon spontanée dans les bâtiments existants, soit à l'occasion du remplacement de matériels vétustes, soit dans le but de réduire les dépenses de chauffage. Mais le rythme de ces améliorations reste faible en général, si les Pouvoirs Publics ne mettent pas en oeuvre des actions spécifiques propres à favoriser ces travaux de réhabilitation thermique.

Plusieurs démarches peuvent être envisagées :

. Des mesures réglementaires ont été adoptées par quelques pays pour développer la réhabilitation thermique, au moins concernant certains points particuliers (régulation et comptage, par exemple) ; mais il apparaît qu'en général, la réglementation est difficile à élaborer, à mettre en oeuvre et à contrôler dans ce domaine, compte tenu de l'extrême diversité du patrimoine considéré, des impossibilités techniques fréquentes et des différences d'efficacité et de coût souvent très importantes, selon les bâtiments, pour un même type d'amélioration ;

. Ces difficultés, qui s'attachent à tout projet de réglementation dans ce domaine, ont conduit la plupart des pays membres à mettre en place des mesures incitatives en faveur des travaux d'amélioration thermique : primes ou subventions, avantages fiscaux ou financements privilégiés. Une large expérience de ces mesures incitatives depuis près de 10 ans a mis en évidence certains écueils à éviter :

- des modalités d'attribution des aides trop contraignantes conduisent en général à l'échec, car très peu de bénéficiaires acceptent de suivre des procédures complexes pour percevoir, en définitive, des aides d'un montant relativement modeste ;
- à l'inverse, des régimes d'aide simples et systématiques ne sont pas toujours utilisés en faveur des investissements les plus intéressants pour la collectivité ; ainsi, des crédits publics importants peuvent être dépensés pour des travaux d'amélioration qui n'ont qu'un impact très limité sur les consommations d'énergie.

Quelle est la situation, actuellement, dans les bâtiments existants par rapport aux objectifs de réhabilitation thermique ?

- d'abord, la plupart des gaspillages les plus importants ont été éliminés;
- ensuite, les travaux les plus simples et les plus rentables ont été réalisés en grande partie, notamment dans les maisons individuelles où les interventions sont les plus aisées ;
- enfin, les personnes susceptibles de décider des travaux sont très sollicitées par les entreprises, les industriels, les distributeurs d'énergie, ... qui vantent tel ou tel produit (qui l'isolation, qui le changement de chaudière, qui la régulation, qui la pompe à chaleur ou le chauffe-eau solaire). Ces personnes ne savent pas quoi choisir, d'autant qu'il n'est pas possible pour les Pouvoirs Publics, au-delà de recommandations générales, de conseiller les utilisateurs sur les meilleurs travaux à entreprendre ; en effet, à cause de la grande variété des bâtiments existants et de leurs conditions d'utilisation, il n'y a, en pratique, que des cas particuliers.

Ainsi, un frein important à la réalisation de travaux d'amélioration thermique dans les bâtiments existants réside dans la difficulté qu'ont les propriétaires et occupants pour déterminer les interventions les plus adaptées pour réduire leur consommation d'énergie.

Il convient donc de mettre en place un outil d'aide à la décision des maîtres d'ouvrage : collectivités locales, hôpitaux, organismes de logement social, gestionnaires de bureaux, d'hôtels ou de commerces, particuliers en maison individuelle ou en immeuble collectif.

Cet outil d'aide à la décision, c'est le diagnostic thermique ; il a pour objet de donner aux décideurs concernés toutes les informations nécessaires pour leur permettre de choisir les améliorations à apporter à leur bâtiment.

Le diagnostic est une méthode d'étude qui est un préalable nécessaire à toute décision d'entreprendre des travaux de maîtrise des consommations d'énergie dans un bâtiment existant. Le diagnostic doit être complet, objectif et indépendant de considérations commerciales. Il n'est donc pas une analyse sommaire de quelques améliorations possibles, ni un simple bilan énergétique, encore moins un devis de travaux.

Cette méthode d'étude se compose des trois phases suivantes :

<u>Phase 1</u> : <u>Description et examen précis</u> des locaux (utilisation, état du bâti et des installations, exploitation, usages spécifiques des énergies équipements particuliers, consommation, etc).

<u>Phase 2</u> : <u>Exploitation et traitement de données</u>, les calculs et leur interprétation permettant la mise en évidence et la description des améliorations à envisager, avec indication, pour chaque intervention, de son coût, des économies à en attendre et de la durée d'amortissement de l'investissement.

Phase 3 : Proposition d'un programme de travaux cohérent et adapté aux caractéristiques propres de chaque bâtiment étudié ; cette proposition est présentée dans le cadre d'un rapport de synthèse directement utilisable par le maître d'ouvrage et lui permettant d'orienter son choix de travaux dans les meilleures conditions de coût, de rentabilité et de délai.

Sur le contenu du diagnostic, il faut souligner quatre points essentiels :

a) la phase initiale du diagnostic description et examen ne doit pas se limiter au relevé de données sur le bâti : surface, caractéristiques des murs, et sur les installations thermiques : type et puissance du générateur, etc. Un examen détaillé (portant notamment sur l'évaluation du taux du renouvellement d'air, le fonctionnement du réseau de distribution et de la régulation, la détermination des rendements) est nécessaire pour identifier les anomalies et déficiences qu'il conviendra de corriger.

b) la phase centrale du diagnostic exploitation et traitement implique l'utilisation de méthodes de calcul adaptées au bâtiment considéré ; il s'agit, en particulier, de calculer les déperditions et les besoins thermiques et de tester toutes les améliorations envisageables : optimisation de l'isolation ; coût, résultat et rentabilité des interventions possibles, etc. Des méthodes utilisant la micro-informatique ou des ordinateurs plus puissants seront le plus souvent utilisées pour cette phase.

c) le diagnostic ne doit pas seulement indiquer les mesures correctrices qui s'imposent pour réduire la consommation d'énergie d'un bâtiment, il doit aussi suggérer des voies alternatives pour la satisfaction des besoins thermiques de ce bâtiment. Ces voies alternatives peuvent comprendre aussi bien des interventions sur le bâti visant l'accroissement des apports solaires que l'installation d'équipements nouveaux (pompes à chaleur, chauffage au charbon, chauffage au bois, capteurs solaires, etc).

d) pour certains types d'intervention (utilisation de techniques nouvelles, substitution d'énergie, ...), le diagnostic n'est pas suffisant pour permettre au maître d'ouvrage de décider les investissements à lancer. Dans ce cas, des études complémentaires sont nécessaires et elles doivent être mentionnées dans le rapport de synthèse du diagnostic.

Il apparaît bien que le diagnostic thermique revêt une importance fondamentale dans le processus qui doit conduire le décideur à prendre les choix les plus judicieux en matière d'utilisation rationnelle de l'énergie.

Pour emporter la conviction des propriétaires et maîtres d'ouvrage, il est nécessaire que ce diagnostic soit réalisé par un acteur indépendant de toutes considérations commerciales ou par un organisme qui, sous le contrôle des Pouvoirs Publics, offre toutes garanties sur la qualité du diagnostic thermique et la validité des travaux proposés.

Le diagnostic thermique, tel que je viens de le définir, est l'affaire de professionnels particulièrement qualifiés.

En France, où les Pouvoirs Publics soutiennent le développement des diagnostics depuis le début de l'année 1982, ils peuvent être effectués par les bureaux d'études, les ingénieurs-conseil, les bureaux de contrôle, les groupements d'entreprises avec garantie de résultat, les architectes, etc.

L'Agence Française pour la Maîtrise de l'Energie accorde des aides à une douzaine d'organisations professionnelles ; ces aides leur permettent d'engager des actions de formation, de mettre au point des logiciels utilisés dans les diagnostics, enfin, d'apporter une assistance technique décentralisée à l'ensemble des diagnostiqueurs opérant dans les différentes régions. Ces aides, et les contrôles qu'est amenée à réaliser l'Agence Française pour la Maîtrise de l'Energie, visent à assurer la meilleure qualité possible aux diagnostics.

Les diagnostics étant évidemment payants, les Pouvoirs Publics apportent une aide à tous les propriétaires et maîtres d'ouvrage qui en font la demande. D'une manière générale, les subventions couvrent 70 % du coût du diagnostic thermique, si bien que, toute aide déduite, un diagnostic ne représente qu'une faible dépense :

- 350 francs français (ou 50 ECUS) pour une maison individuelle
- 60 francs français (ou 10 ECUS) par logement en immeuble collectif
- 40 000 francs français (ou 6 000 ECUS) pour l'ensemble des bâtiments communaux d'une ville de 10 000 habitants, par exemple.

En dix huit mois, l'Agence Française pour la Maîtrise de l'Energie a mis en oeuvre des moyens importants pour soutenir le diagnostic thermique: 240 millions de francs français, soit 35 millions d'ECUS. Dans le même temps, 800 000 logements (ou équivalents-logement dans le secteur tertiaire) ont été diagnostiqués, et le rythme annuel de diagnostics devrait rapidement dépasser 1 million de logements.

Le diagnostic thermique, sous des formes sans doute un peu différentes, existe également dans plusieurs pays de la Communauté.

Au Danemark, par exemple, il y a un système qui s'apparente au diagnostic et qui a pour objectif principal, je crois, d'informer les futurs occupants des caractéristiques thermiques de leur logement et de la consommation d'énergie correspondante.

Par ailleurs, le Centre Commun de Recherche d'Ispra a entrepris deux actions : acquisition d'information sur le parc de bâtiments existants et la réalisation d'audits énergétiques visant en particulier l'amélioration des connaissances statistiques sur les caractéristiques thermiques de ces bâtiments.

En France, le diagnostic thermique est principalement considéré comme un outil d'aide à la décision : c'est une étude technique et économique complète permettant aux maîtres d'ouvrage, qu'ils soient simple particulier ou ville de 100 000 habitants, de prendre, en toute connaissance de cause, des décisions d'investissement en matière d'utilisation rationnelle de l'énergie.

Il est important de souligner que, si le diagnostic propose un programme de travaux d'amélioration, la décision d'entreprendre telle ou telle amélioration relève évidemment du maître d'ouvrage qui doit conserver toute liberté pour engager certains travaux, même s'ils sont moins rentables et moins intéressants pour la collectivité du point de vue de l'utilisation rationnelle de l'énergie. Mais, de toutes façons, avant de prendre sa décision, le maître d'ouvrage aura pu apprécier les différents aspects techniques, énergétiques et économiques des investissements envisagés.

Préalable important aux décisions d'investissement, le diagnostic thermique est maintenant, en France, obligatoire pour bénéficier de la plupart des aides aux travaux de réhabilitation thermique des bâtiments existants. C'est ainsi que, dans le cadre du Fonds Spécial de Grands Travaux, l'Agence Française pour la Maîtrise de l'Energie accorde des subventions aux investissements d'économie d'énergie, pour les collectivités locales, les hôpitaux, les logements sociaux, les réseaux de chaleur et, très bientôt, l'industrie : Ces aides sont attribuées au vu d'un diagnostic exigé systématiquement. Ce programme, représentant un montant d'aide annuel de 2 milliards de francs français, soit 300 millions d'ECUS, a permis de développer très rapidement les diagnostics thermiques. Par exemple, la moitié des hôpitaux publics français sont déjà diagnostiqués.

J'ai indiqué tout à l'heure que 800 000 logements (ou équivalents-logement) ont déjà bénéficié en France d'aide au diagnostic. Une grande partie de ces diagnostics sont actuellement en cours d'exploitation pour avoir une meilleure connaissance des caractéristiques des bâtiments et de leurs équipements, des consommations d'énergie, du gisement d'économie d'énergie, de l'efficacité et de la rentabilité des différentes interventions proposées, etc.

En particulier, l'Agence Française pour la Maîtrise de l'Energie a directement financé le diagnostic de tous les bâtiments (logements, écoles, commerces, bureaux, ...) de 3 villes de 40 000 habitants environ. Treize autres opérations similaires, pour des villes ou zones rurales de 10 à 20 000 habitants sont en cours de lancement.

Pour la ville de Meaux, entièrement diagnostiquée durant l'hiver dernier, nous disposons de rapports statistiques très intéressants portant sur 6 700 logements et 340 bâtiments à usage tertiaire ; on y examine l'influence de la réglementation, l'importance des comportements, le rendement des chaufferies, la nature et la rentabilité des travaux proposés. Il en ressort l'extrême diversité des bâtiments existants, même pour des bâtiments de même nature, datant de la même époque, utilisant la même énergie, dans une même ville.

Ceci est une justification de l'étude approfondie de chaque bâtiment par des spécialistes, que nous exigeons pour tout diagnostic thermique, au lieu d'indications fournies par des guides ou des banques de données dont on ne peut attendre que des valeurs moyennes.

Le diagnostic thermique est également un bon outil pour favoriser la diffusion des techniques nouvelles. En effet, l'examen, par des spécialistes, de l'ensemble des améliorations possibles permet que soit envisagée l'utilisation de ces techniques et procédés insuffisamment connus des maîtres d'ouvrage. Par ailleurs, l'expérience acquise à l'occasion des diagnostics permet de mettre en évidence certains problèmes spécifiques qui requièrent des actions de recherche et développement. Ainsi, les actions de recherche-développement et les projets de démonstration, dont un grand nombre seront présentés ici-même dans les deux prochains jours, contribuent largement, de façon complémentaire au diagnostic, à la réhabilitation thermique des bâtiments existants.

Enfin, je terminerai en soulignant que la réhabilitation thermique ne constitue pas seulement un volet important d'une politique d'utilisation rationnelle de l'énergie. Pour les occupants, elle constitue l'un des principaux moyens de réduire les charges d'habitation tout en améliorant le confort. Pour les propriétaires soucieux de la bonne conservation de leur patrimoine, elle constitue l'élément moteur de tout programme de réhabilitation de bâtiments anciens. Pour les industriels produisant des équipements thermiques, elle constitue un marché relativement nouveau et en développement susceptible de compenser la baisse, constatée dans plusieurs pays européens, du volume de construction neuve. Pour les professions du bâtiment enfin (architectes, bureaux d'études, entreprises et artisans), la réhabilitation thermique des bâtiments existants constitue un facteur d'évolution important : elle favorise l'adaptation à la situation nouvelle du marché du bâtiment, en développant leur qualification technique et en leur donnant de nouveaux atouts pour soutenir leur activité.

Development of Modern Heating Technology

F. Steimle, Essen

1. INTRODUCTION

By the end of 1973, it was realised in most of the industrial countries in Europe how important the energy question had become. During the last two decades mineral oil has played a dominant role among the other primary energies, mainly because it is an easy-to-handle fuel for domestic use and above all has been cheap.

The largest energy saving potential therefore lies in space heating. Besides the desired effect of energy saving, this is also the area where the great influence can be exercised to reduce the possible impact of politically uncertain oil supplies. However, the replacement of oil and the introduction of new technologies for space heating cannot be achieved quickly. Even optimistic estimates cannot hide the fact that it will take a long time before the success of new strategies for energy saving can have a noticeable effect. Economic considerations show, however, that even relatively small improvements in the efficiency of such processes which form a lage part of total energy consumption can lead to a considerable improvement of the whole energy consumption structure.

2. HEATING SYSTEMS

Figure 1 shows the systems for the several heat supply possibilities and the different heat distribution systems. They are possibly combined by heat recovery arrangements.

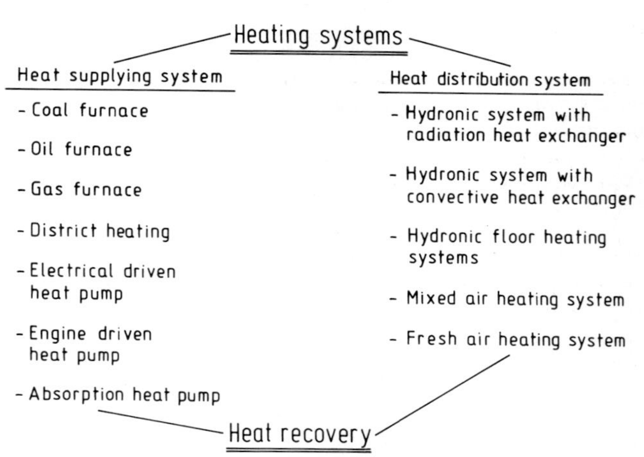

Figure 1: Heating systems

We have to face that a lot of work has been done to insulate buildings and to reduce the heat demand forspace heating. The next step of energy saving must be now to generate this heat demand with less primary energy. This question can only be answered by thermodynamical comparison of the different heat generating systems.

The second characteristic of a form of energy is its conversibility into other forms of energy. Whilst electrical energy can be converted at will into other forms of energy such as mechanical or thermal energy, thermal energy cannot be converted fully into electrical energy. The lower the temperature at which thermal energy is available then the lower the efficiency of energy conversion. In order to describe the proportion of convertible energy of a given total amount of thermal energy, the thermal energy is subdivided into a part which can be converted fully, called exergy, and a part which cannot be converted, calling anergy.

According the First Law of Thermodynamics can be restated as follows:

> In a closed system the sum of exergy and anergy remains constant.

It can, however, be seen that the ratio of exergy and anergy can change in the system. The possible changes are best indicated by the following formulation of the Second Law of Thermodynamics:

> For non-reversible processes in a closed system, exergy is converted into anergy. For reversible processes, the exergy remains constant. Anergy cannot be converted into exergy.

It can be seen that it is possible to produce anergy as heat from exergy whatever the ambient temperature will be, whilst it is not possible to produce exergy from anergy.

These results mean that we have to save exergy as our main aim.

3. HEAT SUPPLYING SYSTEMS

The changes of the different heating systems are shown in Fig. 2. This is a scenary for the Federal Republik of Germany, which shows a small decrease of the oil and a larger decrease for stove heating.

The development of furnaces have been done by better insulation to reduce the stand-by losses. An other possibility to reach this goal is the reduction of the heat supply temperature. This leads to so called low temperature furnaces. Using natural gas it is even possible to cool the exhaust gas to temperatures lower than dew-point to be able to use some latent heat, too. Two types of furnaces, which are used in the Netherland are shown in Fig. 3 and 4.

Many discussions have been rised about district heating with co-generation of heat and electricity. To use the heat of a large central power plant it is necessary to build up large networks of pipes, which cause fairly high investments. From the thermodynamic point of view it must be recognised that the condensing temperature of the Rankine-cycle of the power plant is rising from about 30°C up to about 130°C. This fact gives a decrease of electrical power output of about 20%. The only reasonable possibility of Co-generation of heat and electricity are small heating plants with electricity generation.

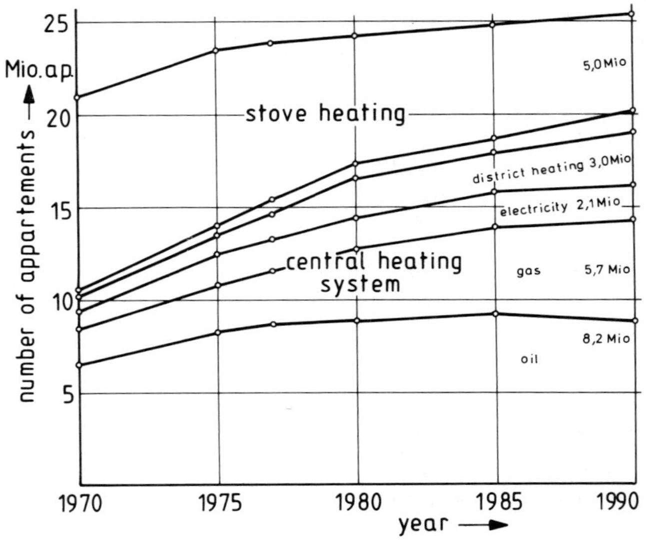

Figure 2: Heating development of dwellings

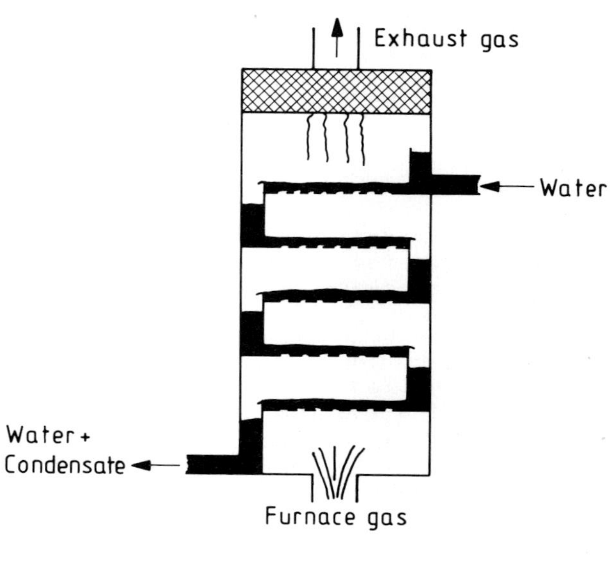

Figure 3: High efficiency furnace

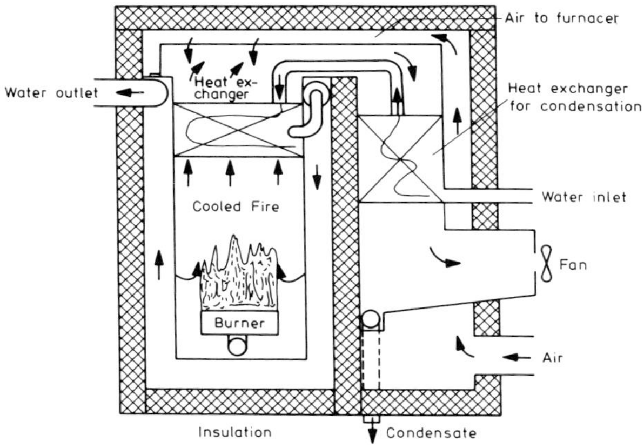

Figure 4: Scheme of a modern furnace

4. HEAT PUMPS WITH COMPRESSORS

The use of a heat pump is the only possibility to mix exergy coming from primary energy and anergy coming from the ambience or out of waste heat.

The mainly used type of heat pumps is in these days the electrical driven heat pump with ambient air as heat source. In the change of energy flow we have also to consider the energy losses in the power plant.

In the electrical power plant we havean efficiency on the primary energy of approximately 32%. The rest is waste heat which is given out normally to the environment. If we have a heat pump with a COP of $\varepsilon = 3$ we get 96% back as heating energy (Fig. 5).

Driving the pumps with combustion engines gives the possibility of regaining the waste heat from the engine and the exhaust gas at a high temperature level and using it for heating. We see in Fig. 5 that it is sufficient to spend 75% of the primary energy to obtain the required 100% heating energy. This is the main difference from an energetical point of view.

Another significant difference between drives is the heat capacity of the heat source. In combustion-engine heat-pumps this heat is much smaller and therefore the heat pumps can use smaller heat sources, too.

The heat from the cooling of the engine and the exhaust gas has a high temperature level. Therefore it is possible to store heat with a wide temperature range without any losses in COP of the heat pump itself.

Another advantage is the possibility of varying the speed of the engine within a wide range for capacity control. However, the fuel consumption of combustion engines depends strongly on the speed. Therefore it seems to be better to use an on-off operation with a small range of varying speed to fill up the heat storage tank (Fig. 6).

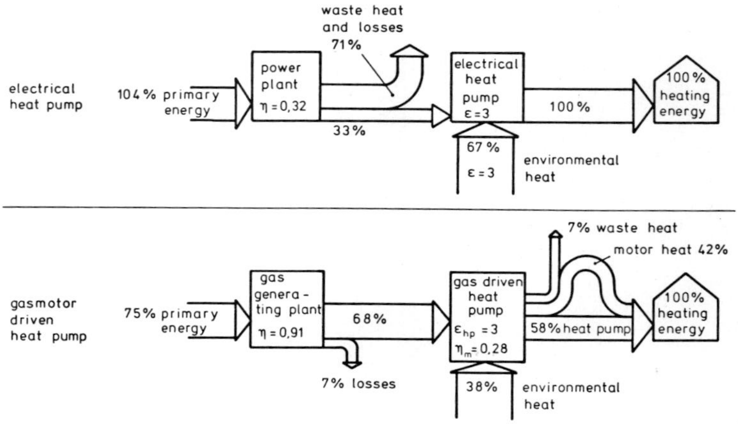

Figure 5: Primary energy demand for heat pumps

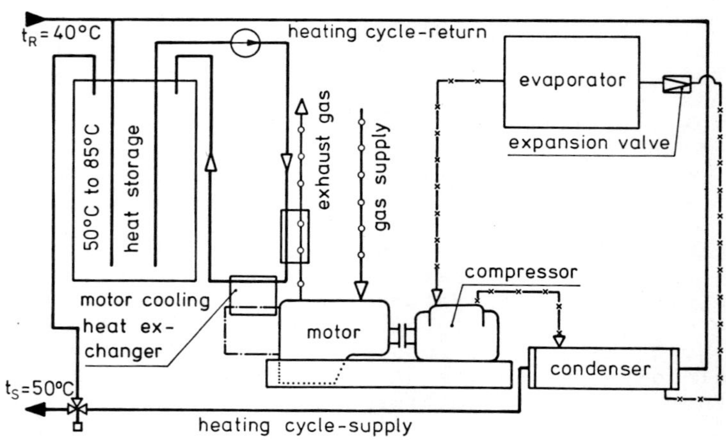

Figure 6: Gas-motor heat pump with storage tank

The different between an gas engine and a diesel engine it not so significant. The efficiency of the diesel motor is somewhat higher and it is not possible to cool the exhaust gas lower than the condensing point of vapour.

These types of heat pumps also have disadvantages. A great problem is the maintenance of the engine and also the noise level. However, there are ways to avoid these problems.

Quite another type of heat pump drive is the combination

of a Rankine cycle with the refrigerant cycle. The drive therefore is a small turbine. By using the same refrigerants as working fluid in both cycles, the expansion turbine and the compressor can be in the same shell.

5. ABSORPTION HEAT PUMPS

Comparing the circuit of a compression heat pump (Fig. 7) and an absorption heat pump (Fig. 8) the main difference is that the pressure rise is done in the first case in the gas phase whilst in the second case in the liquid phase which much less driving energy for the pump compared with the compressor.

Some more energy must be put in the absorption heat pump in the form of heat to the generator. Therefore, the theoretical absorption process can be shown in a T,s-diagram as two Carnot cycles one flowing clockwise and the other counter-clockwise (Fig. 9).

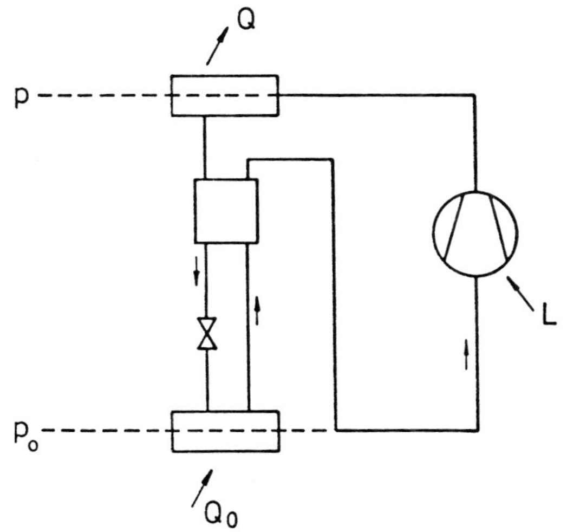

Figure 7: Compression heat pump

At present there are some pilot plants operating with NH_3/H_2O and H_2O/LiBr. Both pairs present some difficulties. NH_3/H_2O produce a high pressure in the condenser under heat pump conditions and needs rectification after the generator to ensure no water enters the evaporator. H_2O/LiBr has a narrow working range. It is limited in the lower part by freezing point and has an upper limit of the cristallization line.

Therefore, the main task is to discover other pairs and to evaluate their thermodynamic data.

Some criteria for choice are: a high difference in the boiling point between solvent and refrigerant; a high evaporation enthalpy of refrigerant; a low ratio between solution enthalpy and evaporation enthalpy; a low viscosity of the solvent and the mixture; good pressure range; no toxicity of all parts of the mixture; and no cristallization in the working range.

Another development in this field will be the 'heat transformer'. In this special type of absorption heat pump the heat of a middle temperature is transformed to a higher temperature. The exergy for this process comes out of a second process in which a cycle works between the mentioned middle temperature and the environmental temperature.

The search for new working couples with more advantages than NH_3/H_2O is one of the main aims of my laboratory at the university of Essen. This research is sponsored by the E.C. and therefore some results are presented at this conference, too.

Absorption heat pumps are built now in a small capacity range up to about 30 kW in one unit or built up like a chemical engineering plant with capacities above 1 MW.

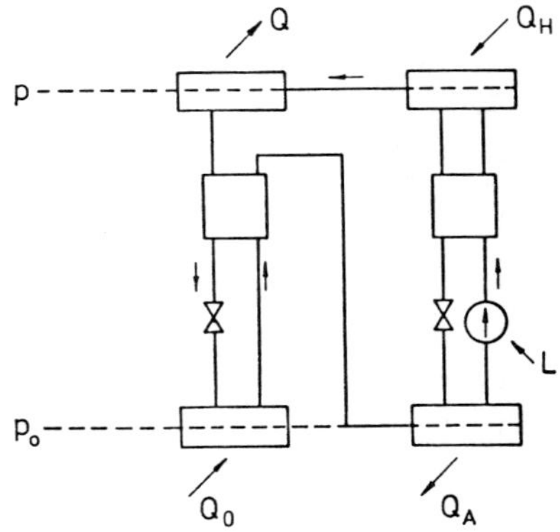

$$\zeta_{hp} = \frac{Q + Q_A}{Q_H}$$

Figure 8: Absorption heat pump

Figure 9: Scematic absorption process

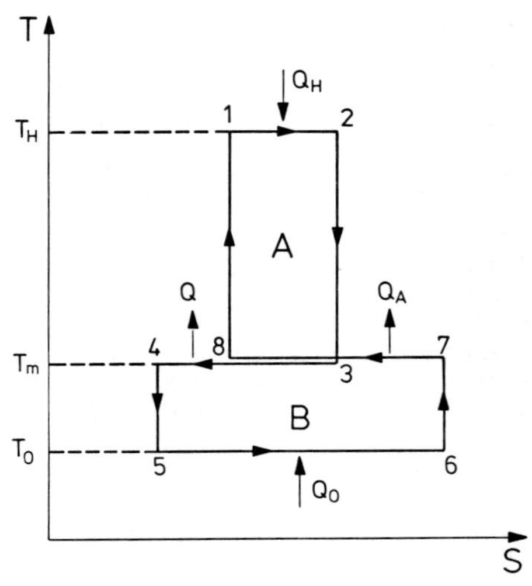

6. HYDRONIC HEAT WITH DISTRIBUTION SYSTEMS

In a lot of countries in Europe the hydronic systems are the most well-known for central heating plants (see Fig. 1). Very often the radiation heat exchangers are installed below the window areas. The reason is to avoid discomford caused by the low temperatures of the glass area of old and simple window construction (Fig. 10). With better windows and much lower overall heat transfer value the wall temperatures of the windows are rising into the comfortable range (above 17°C at 22°C room air temperature). Therefore also convective heat exchangers with much lower water temperatures are comfortable. Floor heating systems have the disadvantage that they have a long time constant in temperature control. In very well insulated buildings this control behaviour must be much quicker to avoid temperature rise during sunshine periods in winter.

$$\Delta \dot{q} = \dot{q}_{s1} - \dot{q}_{s2}$$
$$\Delta \dot{q} = \alpha_s \cdot \Delta t$$

$\Delta \dot{q}$ [W/m²]	physiological rating
< 23	comfortable
23 - 41	less comfortable
41 - 64	discomfortable
> 64	very discomfortable

with $\alpha_s = 7 \frac{W}{m^2K}$ and $\Delta \dot{q} = 35 \frac{W}{m^2}$; $\Delta t = t_1 - t_2 = 5\,K$

Figure 10: Heat fluxes for thermal comfort

7. AIR HEATING SYSTEMS

Combining the ventilation of the buildings with the heat distribution it could be very easy to use air as heat carrier from the central heating plant to the room. This could be done only with fresh air or also with return air (Fig. 11).

Air heating systems have the advantage that they have fairly low supply air temperatures, which must not be rised at low outdoor air temperatures. The capacity control can be done by changes of air volume (Fig. 12).

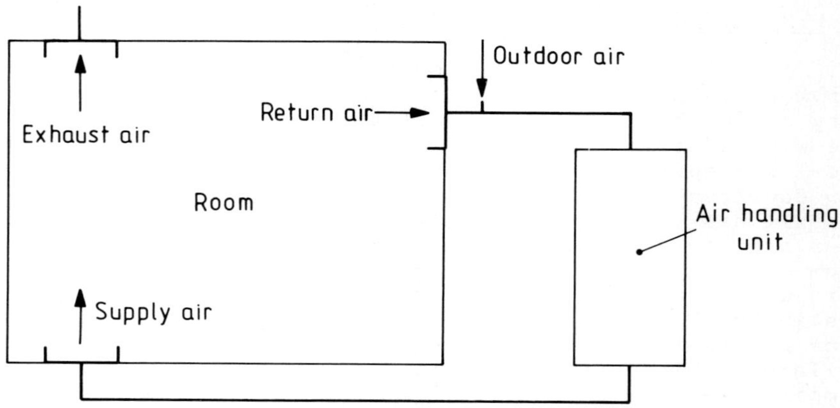

Figure 11: Air heating system

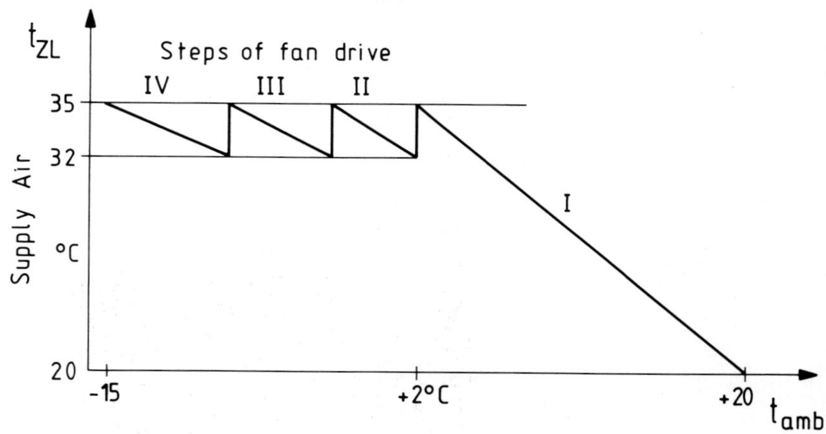

Figure 12: Control of air heating systems

8. HEAT RECOVERY

In well insulated buildings the ratio between the transmission heat losses and the heat losses by air changes are decreasing (Fig. 13). Recovery of heat out of the exhaust air gives therefore an important reduction of the total heat demand (Fig. 14).

The heat recovery systems can also be boosted by a heat pump with the exhaust air as heat source. Fig. 15 shows such an installation with an electrical heat pump and 0,8 air changes per hour. At the lowest outdoor air temperature about 20 W/m^2 are additionally necessary. Parts of this amount will be given by internal heat like lighting, equipment and the heat delivered by human beings.

Transmission heat demand: $\dot{Q}_T = A_w \cdot k_m \cdot \Delta T$

Heat losses by air changes: $\dot{Q}_L = \rho \cdot \dot{V} \cdot c_p \cdot \Delta T$

		Outdoor air changes in $m^3/h \cdot pers.$		
		20	40	60
Room length in m	4	1,13	0,56	0,38
	6	0,75	0,38	0,25
	8	0,56	0,28	0,19

Figure 13: Ratio between transmission heat losses and ventilation heat losses

Existing buildings	ca. 90 W/m^2
Well insulated buildings	ca. 50 W/m^2
Transmission heat losses	ca. 25 W/m^2
Heat losses by air change	ca. 25 W/m^2
Heat recovery by heat exchanger	ca. 15 W/m^2
Additional heat demand	ca. 35 W/m^2

Figure 14: Energy demand in dwellings

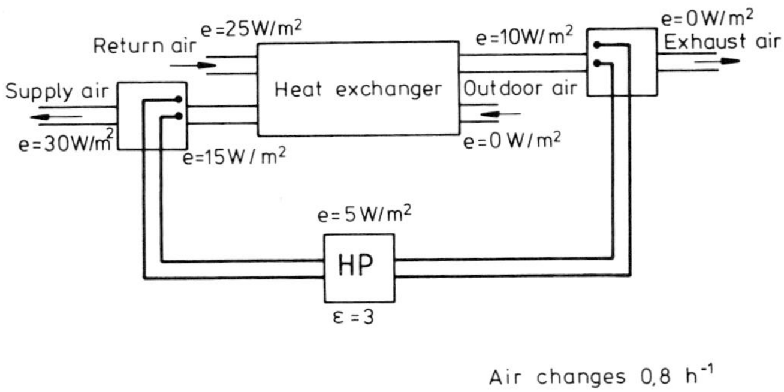

Figure 15: Heat recovery with electrical heat pump

If we use a fossil energized heat pump (gas engine driven heat pump or absorption heat pump) there is no demand of additional heating (Fig. 16). These figures show that a combination of heat recovery, heat pump and air heating system gives very good energy savings and also a possibility of a simple installation.

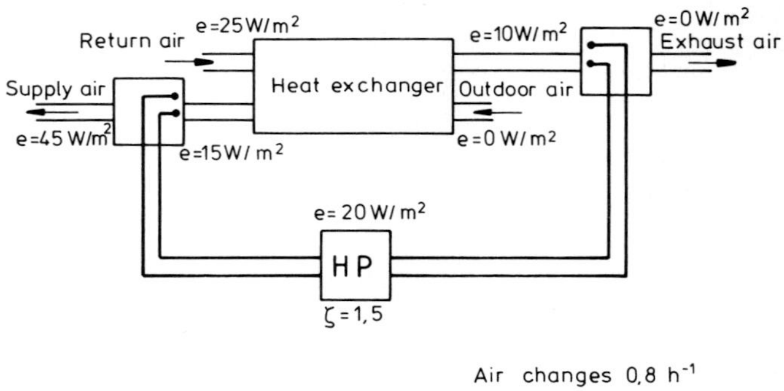

Figure 16: Heat recovery with fossil heat pump

The next step could be probably the use of Joule cycle heat pump with air as working fluid. Such a design is shown in Fig. 17. The coeffizient of performance COP is theoretically much higher than that of a compression heat pump, but it is necessary to get laboratory results and then a good machinery for it, before a final opinion can beformed.

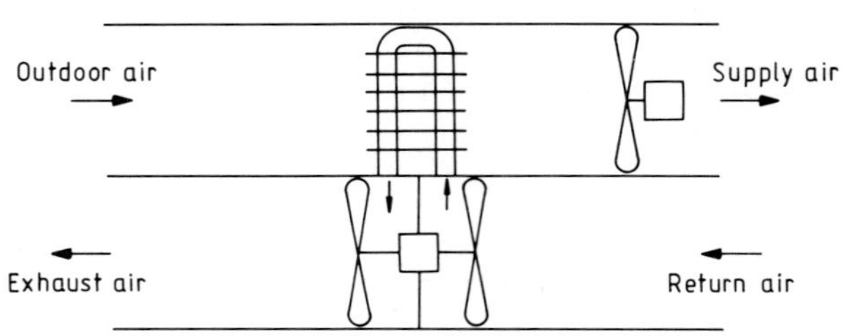

Figure 17: Joule-Process-Heat pump

9 CONCLUSION

The development of heating technology is going along in different areas. Not only the heat generation is of some interest but also the heat distribution systems are changing to low temperature installations.

All these things show that it is necessary to save energy by investment and brain storming, but, nobody wants to save energy but only money.

SESSION I - HEAT PUMP SYSTEMS

A - R & D PROJECTS

Chairman: Ir. J.A. KNOBBOUT
Rapporteur: Ir. P. ZEGERS

Personal impression of the session chairman

Influence of lubricating oil on heat pump performance - Data capture, analysis and control aspects

A single screw high performance heat pump

Investigation of commercial domestic heat pump performances

Optimized operation of a frozen air heat exchanger by using a micro-computer

Optimization of internal combustion engine driven heat pump (20 kW heating power)

Regulation and control of heating system using internal combustion engine heat pump

Monovalent electric air/water-heat pump plant for room heating in the houses for several families (social house building) to save energy - Heating system for low dwelling by means of heat pump

Energy roof and seasonal earth storage

Practical and theoretical investigation of a heating/cooling heat pump with ground storage and solar collector

Investigation into a complete earth-to-water heat pump system in a single-family dwelling focussing on the application of a vertical subsoil heat exchanger

Heating of a laboratory hall by means of a soil heat pump combined with unshielded solar absorbers

A high temperature lift absorption heat pump

Heat pumps with fluid mixtures applied to space heating

Practical applications of FRIMIP

Absorption heat pumps for space heating

Periodically operating absorption heat pump

Thermoelectric heat pumps for space heating

Systematische Konstruktion und Entwicklung einer kostengünstigen, verlustarmen und regelbaren Lösungspumpe für Absorptionswärmepumpen

Working fluids for sorption heat pumps

Untersuchung der Auswirkungen des Betriebes von Wärmepumpensystemen auf die Umwelt

Overview of heat pumps R & D for domestic heating carried out by the European Communities from 1979 to 1983 in the framework of the second EC Energy R & D Programme

PERSONAL IMPRESSION OF THE CHAIRMAN

J.A. KNOBBOUT

1. GENERAL

From the presentation of the reporter and the contractors and the general discussions it is clear that the heat pump has a high potential for energy conservation, but to fulfil the predictions made some years ago, a lot of work has still to be done.
In the presentation many aspects regarding the heat pump as a heating system have been presented, and we now have a clear view of where we stand and what can be expected in the future.

Before discussing the more technical points arising from the presentation it is necessary to review the mechanically driven heat pump and the absorption heat pump or more generally the sorption heat pump. Both types are suited for a very broad capacity ranging from the heating of individual dwellings, and buildings to district heating systems.

2. MECHANICALLY DRIVEN HEAT PUMP

The most developed heat pump is the mechanically driven heat pump, and it is clear that the success of such a heat pump depends strongly on climatic conditions and building structure. The acceptance of the heat pump depends heavily on fuel prices, which are rather diverse in the EC-countries for that matter.

The most attractive heat pump is the heat pump for heating buildings with a capacity of 100 to 300 kW, and very often driven by an internal combustion engine (mostly a gas engine) or an electric motor. The small heat pump for heating a house or a number of houses is frequently electrically driven and a large number have already been installed, but the number of heat pumps is not so high as would be desirable from an energy conservation point of view.

The heat source is an important sub-system for heat pumps, and a large number of heat sources have been tested, but the ground as a heat source seems to be particularly attractive. Sufficient data for design are available, also as regards cost and experience, for installing such a heat exchanger in the ground.

The performance of heat pumps is still far away from the theoretically possible performance and it is shown that an impressive increase in performance has already been reached. Further increase will be possible by step-by-step research, and the speed depending on the money spent in this area.

A number of general problems could be studied, among which is the oil circulation in heat pump systems, for example. The development of the oil-free compressor presented can solve this problem.
A detailed insight into the behaviour of components has been obtained, and the options for further research are clear. It has been proved that system control is very important, including heat storage in the system.

Also integration of the heat pump as part of the building structure is a requirement. From the present state of matters can be concluded that an electrically driven heat pump with a COP of 2.5 to 3 has already been realized, but that a further increase to 4 and even higher, is not to be excluded.
The use of non-azeotropic mixtures is a very successful means to increase performance and capacity. The commercial introduction of such a type of heat pump is being realised in France.

3. FUTURE RESEARCH ON MECHANICALLY DRIVEN HEAT PUMPS

The future in advanced heat pump research is directed to basic and general problems not excluding special topics which are crucial for the introduction of the heat pump. In general it is expected that the task of the European Commission will now be taken over by industry in the form of product development. Future developments are based on the general aim to lower the cost of heat pumps as an installed heating unit. This can be done by increasing the capacity pro unit or lowering the cost per unit of heat produced. With regard to this point attention should be given to advanced mounting and installing techniques, and lower servicing and maintenance costs. Consideration should further be given to the devélopment of components, in particular the evaporator and more in general to the two-phase flow in heat pump systems.

The possibilities of increasing performance and capacity should be further tested. The maximum temperature at which the heat pump performs should be increased. The noise problem is also a point to be considered. Attention should also be paid to more sophisticated heat pumps and the distribution of heat from small-capacity heat pumps for individual and from high-capacity heat pumps for district heating systems.

4. ABSORPTION HEAT PUMPS

The absorption heat pump is less advanced than the mechanically driven heat pump and this state of the art is reflected in the programme. In the demonstration programme, there is not yet a demonstration absorption heat pump for heating buildings.

The estimated potential of the absorption heat pump is high. The heat pump is heated directly by burning gas or fuel in general. The components differ greatly from the components of the mechanical heat pumps, and an other type of manufacturers can therefore construct the heat pump components.

The general impression is that an absorption heat pump will have more resemblance to a heating boiler, which will perhaps activate implementation. Also a gas distribution system which is available in a number of countries of the European Community will support the introduction of the absorption heat pump.

From the presented overview of the projects and the presentation it is clear that a choice of the ideal fluid pair is rather crucial for this pump, and required much attention. This includes not only thermodynamical data but also data on toxicity, stability of fluid pairs, freezing point, crystallisation, corrosion, etc., and it has to be said that everyone feels that an ideal combination has not yet been found. The choice of fluid pairs, especially for the smaller heat pumps, will also be limited by rules regarding pressure etc. The performance of components, and more in particular the absorber in which heat and mass transfer occurs, is rather crucial.

The design of more sophisticated systems, for instance resorption systems, to manage the pressure have resulted in an interesting combination of CO_2 and absorbent, which at first sight seemed not to be the best choice.

One of the projects has the aim to increase the outlet temperature of the heat pump so that for heating purposes the temperature is closer to the existing water temperature in a central heating system. This heat pump is particularly suitable for retrofitting.

5. FUTURE RESEARCH ON ABSORPTION HEAT PUMPS

It is expected that future research will be directed to basic and generally applicable rese'rch for the development and design of sorption heat pumps. This means, considering the state of the art, future activities will be in the line of the activities to-day. The work on the testing of new fluid pairs for the absorption or general sorption heat pumps in general will be continued.

The crucial components are to be developed further, and it seems that the range of priorities is as follows:

1. The absorber in which heat and mass transfer occurs.
2. The evaporator with a two-phase fluid system.
3. The generator.
4. The circulation pump should be cost-effective, and have controllable flow.

Attention will also be given to the periodic heat pump, which is an interesting system including a certain degree of thermal storage. The development of a control system for optimization is, also for an absorption heat pump, even more important than for a mechanically driven pump. The trend is to increase the maximum temperature of the generated heat and to increase the temperature difference between heat source and heat output.
Activities will also be directed to the advanced absorption heat pump to cope with pressure, performance, design, etc.

It should be pointed out that due consideration is to be given to rules for testing heat pumps under such conditions that the test results allow the practical performance to be predicted.
It is expected that in future demonstration programmes the presented results of the R&D programme will be reflected.

It is hoped and expected that in the next programme a large number of original proposals of high quality will be presented to activate the introduction of this interesting apparatus, which may lead to a high degree of energy conservation.

INFLUENCE OF LUBRICATING OIL ON HEAT PUMP PERFORMANCE
Data Capture, Analysis and Control Aspects

J.T.McMullan and D.W.Hughes

Energy Study Group,
New University of Ulster,
Coleraine, N.Ireland,
U.K. BT52 1SA

Tel: 0265 4141
Telex: 747597

Summary

A microprocessor based data logging and control system for the detailed examination of heat pump performance is described and some preliminary results are discussed of its application to the complex problem of determining the influence of lubricating oil on system performance. Results are presented which illustrate the powerful data analysis software which has been developed, and the relationship between data collection and analysis is demonstrated.

1. INTRODUCTION

The fundamental objective of this contract was to investigate the effects of varying the oil circulation in the evaporator and in the compressor of a heat pump system. The test rig therefore consists of a water-to-water heat pump which provides facilities for varying the oil concentrations as required. It also provides the essential facilities for recovering the oil and refrigerant from the liquid which has been removed from the heat pump. The design is an extension and development of a previous system which has been described elsewhere (1). The underlying scheme is a water-to-water heat pump, with an oil injection stage in the condenser liquid line, two liquid separation stages in the compressor suction and discharge lines, and two oil and refrigerant recovery systems. Figure 1 shows the layout of the system, though for simplicity, only the evaporator oil circuit is shown (the compressor oil circuit is virtually identical). The test rig is extensively instrumented and is microcomputer controlled to ensure the stability of the operating conditions. This of itself is not a trivial operation, as the system involves a number of dynamically interactive loops, which makes precision control of its behaviour difficult to achieve. Further complications arise when account is taken of the general requirement to make calorimetric measurements to acceptable accuracy (that is, to better than 5%), and of the particular need in this work to determine the fluid state points accurately, especially at the evaporator outlet. The reasons for this will be discussed later, but the result is that temperature and pressure measurements must be made with a precision which is not normally fully appreciated.

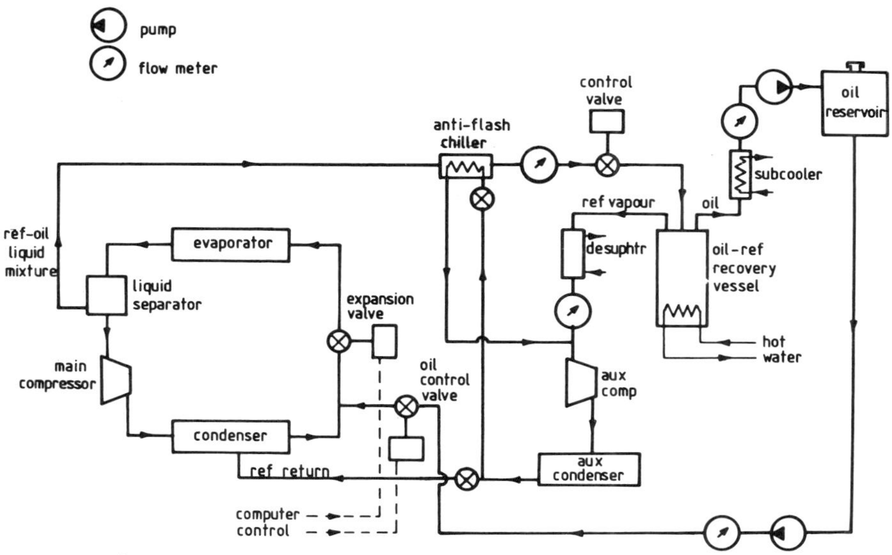

Fig. 1 Schematic diagram of test rig showing evaporator oil circuit.

2. DATA ACQUISITION AND INSTRUMENTATION

The acquisition and measurement of data are different in nature. Measurement means the production of electrical or other signals which quantitatively represent the properties being examined. That is, it is involved with the behaviour of the sensors and transducers. By contrast, data acquisition refers to the collection of these measurement signals and their recording in a form which is appropriate to the needs of the experiment. Neither problem can be considered in isolation because of the very strong interaction between the two.

Five types of measurement are being made: temperature, pressure, fluid flow rate, and electrical and mechanical power. For the temperature and pressure measurements, both absolute and differential measurements are required, to a degree of accuracy and resolution that depends critically on the purpose for which that reading is being made.

This point about the accuracy of the measurements has already been mentioned, and can be illustrated by two examples.

> Measurement of the performance of any heat pump system requires an accurate knowledge of the heat transferred at the evaporator and condenser. Indeed, any quoted performance figures will be useless even for commercial purposes unless they are accurate to much better than 4%. Since both the electrical power input and the water flow rates through the heat exchangers can be measured to within 1% under normal conditions, traditional error analysis shows that this accuracy can be achieved by measuring the temperature difference across the heat exchanger to within 3.5%. Since this temperature difference can be as low as 5 K, then either it must itself be measured to within 0.2 K, or the

absolute temperatures at the inlet and outlet must be measured to within 0.1 K. Thus, the best approach is to measure the temperature difference directly, for example using multiple junction differential thermocouples.

The second example is specifically related to the requirements of this project, for which it is important to obtain information about the quantity of refrigerant trapped in solution in the oil at various points in the system. This is very dependent on both temperature and pressure, so that good accuracy is required for both parameters. The achievement of a tolerably acceptable knowledge of this data implies measuring the temperature in the evaporator to within 0.1 K, and the pressure to within 1.6 kPa.

The data acquisition system has two distinct functions to perform. The first is the routine logging and conditioning of the data signals from the rig. This involves the handling of a relatively large volume of data at time intervals which are normally between 1 and 30 minutes, but which may be as low as 10 seconds as this is the default scan time of the microcomputer data logging and control program.

The second function is the control of the experimental conditions in the test rig. This involves a smaller number of parameters, but they must be sampled, conditioned, processed and acted upon as often and as quickly as is required to ensure the stability of the experimental conditions. Since there is likely to be competition between these two data logging functions in that both may demand servicing at the same time, the same acquisition system was used for both. Conditioning and processing could then be performed without interfering with the needs of the control system.

Thus the adopted approach was as follows. All the experimental data from the test rig were logged on every data acquisition cycle. Next, any necessary control functions were determined and executed. Finally, the data were conditioned, processed, and recorded. In recording the data, it was sent both to a floppy diskette storage medium and to a fast (9600 bits per second) serial data interface which could be accessed by the university mainframe computer or sent directly to a printer for reference purposes.

The total data logging burden was:-

- 31 temperatures using single or multiple thermocouples
- 3 differential temperatures using thermopiles
- 7 pressures using strain gauge bridge pressure transducers
- 14 fluid flows using axial turbine or positive displacement meters
- 3 electrical power measurements using modified eddy current kWh meters.

This represents a total of 41 analogue and 17 digital signals, of which, depending on the particular operating mode, up to 10 (6 analogue, and 4 digital) are needed to perform the control function.

Since 58 analogue and digital signals have to be monitored regularly, and ideally simultaneously, it was decided to use a micro-processor based data acquisition and processing system. Commercially available systems were examined, but were found to have three major disadvantages - insufficient accuracy in most cases, inflexibility, and lack of a suitably comprehensive control function. Thus, the system shown in Figure 2. was designed and constructed in our own laboratory.

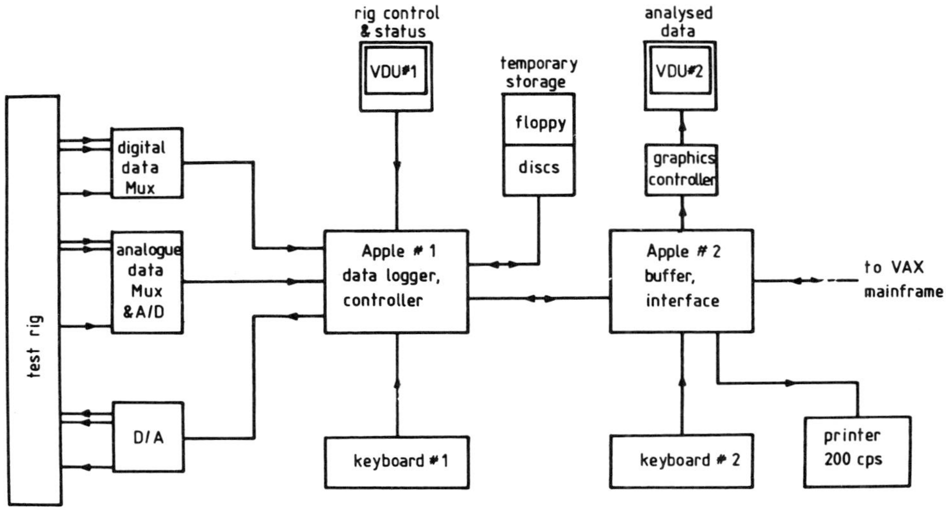

Fig. 2 Schematic diagram of control and data logger hardware.

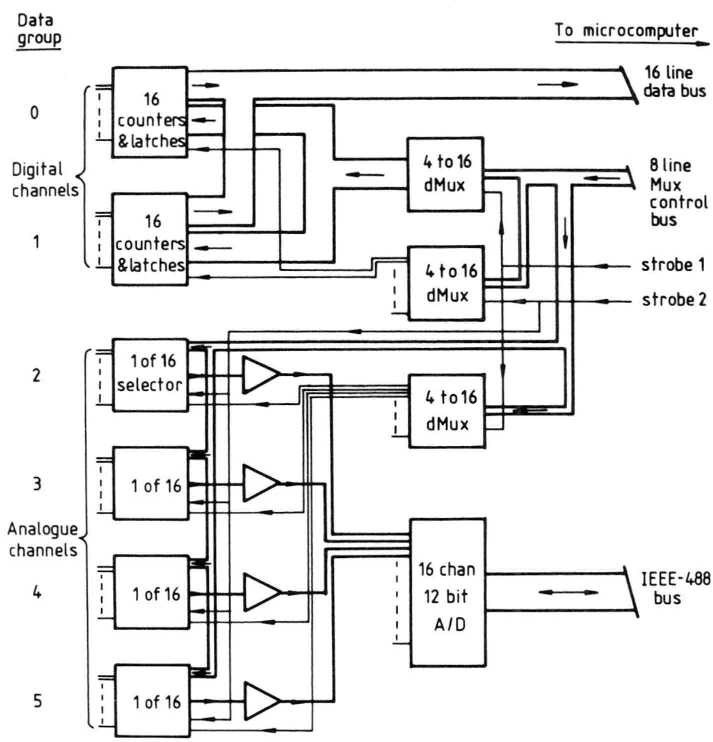

Fig. 3 Signal conditioning and multiplexing system.

An Apple][microcomputer was used as the basis of this system, primarily because of its flexibility and the ease with which custom designed hardware can be interfaced with it. The data acquisition system is shown in Figure 3, where it can be seen that the organisation is different for the digital and analogue data channels.

Each digital data channel consists of a frequency counter which monitors the pulse rate from a sensor, and latches it after a suitable interval. During the computer read cycle the output latches are scanned sequentially. The time required to scan all of these 16 bit channels is less than 1 ms.

The analogue signals are more complicated because they involve A-to-D conversion, and so must be preconditioned to a voltage range which suits the A-D converter. The amplification requirements vary from channel to channel. A particular complicating factor arises with the thermocouples, where the required absolute accuracies are in microvolts. This implies the use of high gain, very low drift amplifiers. Experience has shown that using a dedicated high precision amplifier for each channel is not the best approach. No matter how well they are matched, no two amplifiers are identical. It is preferable to multiplex all sensors with similar requirements into a common amplifier. This has the secondary advantage that calibration signals can be included with the data set, and sampled at every scan, thus overcoming any remaining amplifier drift problems.

The analogue channels are therefore arranged in groups of 16, with one precision instrumentation amplifier per group. The signals are multiplexed into the amplifier by relays with gold-plated contacts, and the amplifiers themselves are multiplexed into a single 12-bit dual slope A-D converter which is read by the microcomputer. In principle, the multiplexer control bus gives access to 65536 separate channels, however, in the present installation, only 256 of these are used.

After analysing this data, the microcomputer generates signals to position the control valves on the test rig. These signals are transmitted via a multichannel D-A system which produces command signals for the valve motor controllers. In this way, the function of valve positioning is off-lined from the computer, thus saving valuable computation time. The control philosophy adopted is very simple. The control signal for each channel is determined from the appropriate error signals for that channel using a proportional term to allow for the current deviation of the controlled parameter from its set point, and an integral term to cancel the droop from the set value that always appears with proportional control systems. The differential control term is also calculated but has been found to be unnecessary. The control signals for the various channels are evaluated independently and there is assumed to be no interaction between channels. There is actually considerable interaction between the various channels (for example, the condensing and evaporating temperatures), however, the system functions well and provides adequate control for present purposes.

The coefficients in the control algorithms for each channel were determined from measurements of the appropriate critical proportional gain and the system time constant. This process was conducted for each channel independently and the resulting coefficients were modified as necessary to compensate for the interactions between the channels. This approach is far from ideal but leads to the desired result.

Up to eight parameters may be controlled simultaneously, and at present evaporating temperature, condensing temperature and evaporator superheat have been selected. Other possibilities include the oil injection rates to the evaporator and compressor systems, the degree of refrigerant liquid subcooling at the expansion valve inlet, the water flow rates to the heat exchangers, the temperature and flow rate of liquid refrigerant injected into the compressor, and so forth.

A useful feature of the system is that selection of a particular control scenario for a given application is achieved at run time merely by choosing the appropriate software. In practice, this means selecting a particular floppy disk.

3. SOFTWARE DEVELOPMENT

The microcomputer software performs the following sequence:

1. Test keyboard for operator intervention.
2. Read the clock and delay until the next scan time (10 seconds).
3. Read the data from all channels.
4. Convert the raw binary data to the appropriate physical values, taking account of all current conversion and calibration factors.
5. Calculate new values for the control signals and send them to the output D-A converter for transmission to the valve controllers.
6. Display the currently selected set of parameters on the monitor screen. It is possible to select one of a number of different screen displays by pressing a key under 1. above.
7. Store the processed data for subsequent time averaging. This is used to eliminate the effects of random fluctuations.
8. At the required time, perform data averaging, store on diskette, transmit to the mainframe computer for analysis and archiving.
9. Repeat this sequence.

It is worth noting that interfacing to a timeshared mainframe computer is not a trivial problem, and has entailed an extensive hardware and software development programme to ensure that it is accomplished reliably and efficiently. We have developed a special microprocessor based system (shown as Apple #2 in Figure 2) which gives extensive data buffering, and provides detailed on-line analysis and graphics capabilities. One result of this is that, typically within 20 seconds, data can be collected from the test rig, analysed using the power of the mainframe computer, and displayed on the VDU screen or on a printer in the laboratory.

4. RESULTS

The results can be presented in three categories: control system behaviour, general behaviour of the heat pump system, and the study of specific aspects of system performance. Examples of each of these applications are presented below.

Figures 4. and 5. show the behaviour of the system under a typical startup sequence using a thermoelectric expansion valve under the control of the microcomputer. The chosen conditions were:

condensing temperature (parameter 86)	60 deg C
evaporator outlet temperature (parameter 83)	15 deg C
evaporator superheat (parameter 39)	2 deg C

Test of superheat control using thermoelectric expansion valve
9 : Main Ref Liquid Flow (L/S)
39: Ref Temp: Evap Outlet (C)
86: Sat Temp: Comp Discharge (C)
93: Superheat: Evap Outlet (C)
All Points Plotted
Data Sampling Interval: 10 Seconds

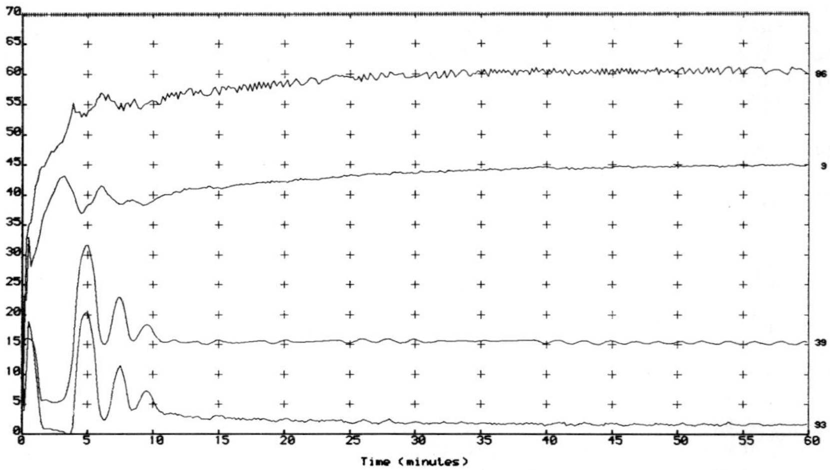

Figure 4. Time plot of heat pump startup showing ascillations in temperatures and flow rates

Test of superheat control using thermoelectric expansion valve
27: Compressor Power (O&S) (KW)
71: Evap Heat (Water) (KW)
73: COP (C) (Water)
All Points Plotted
Data Sampling Interval: 10 Seconds

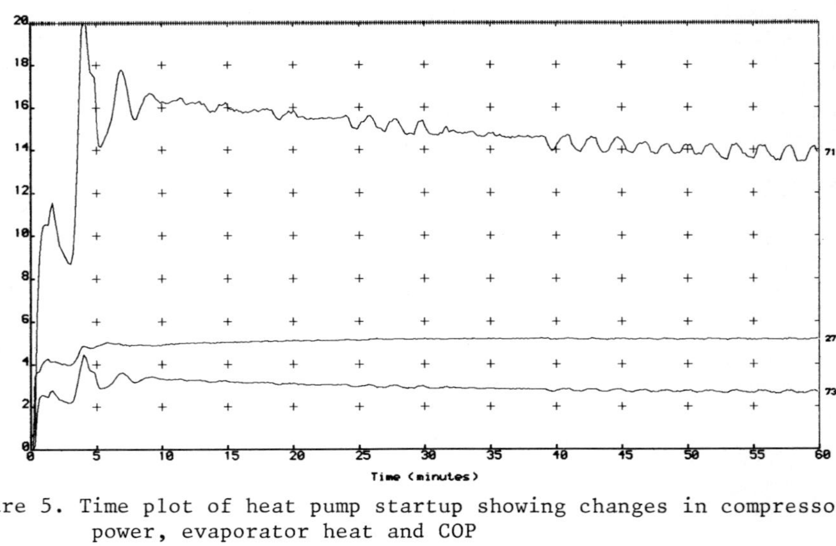

Figure 5. Time plot of heat pump startup showing changes in compressor power, evaporator heat and COP

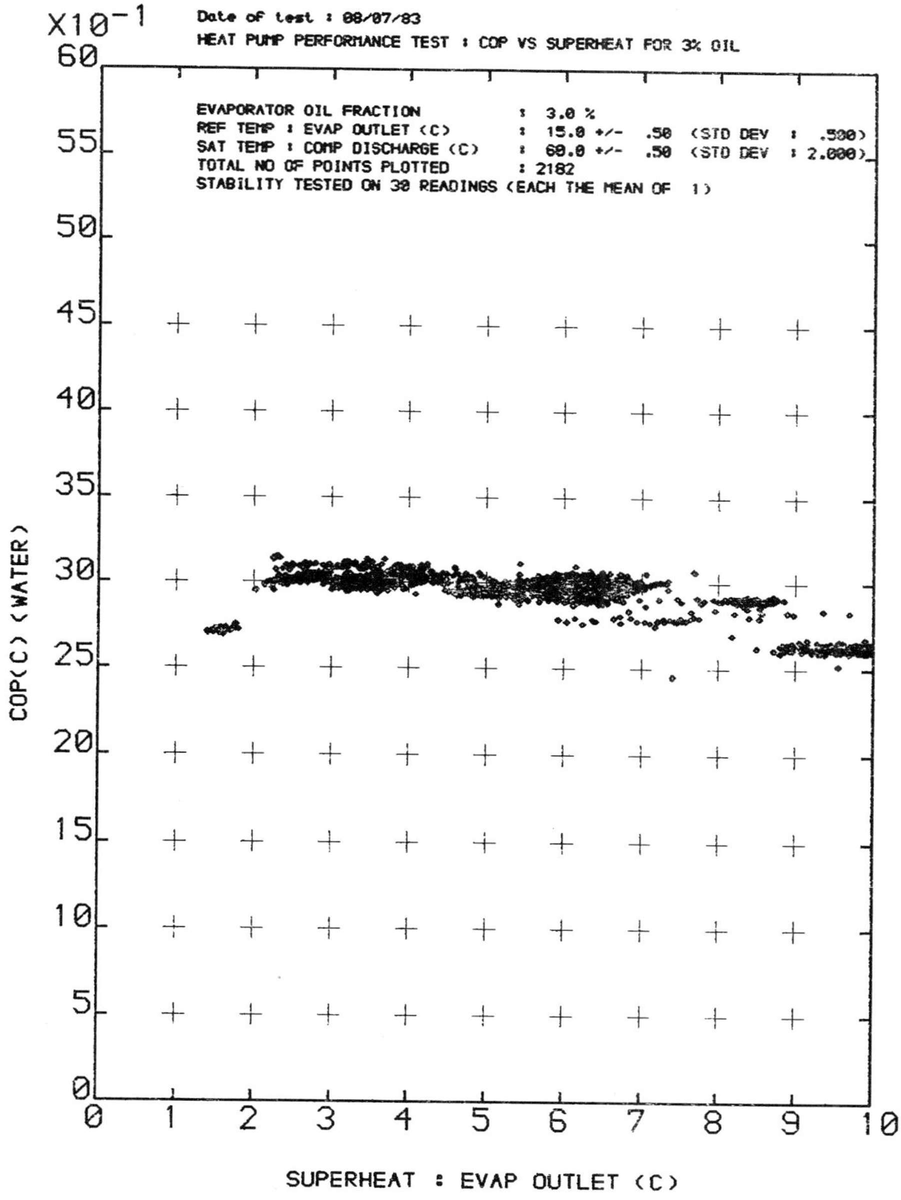

Figure 6. Example of data reduction and analysis.

It can be seen from Figure 4. that apparently acceptable operating conditions are achieved within about 30 minutes, but that the condensing temperature then overshoots by about 1 K, and is not steady at its set value until after 60 minutes. In fact, it takes between 90 and 120 minutes for full stability to be achieved. The early stages of the startup show characteristic oscillations in the evaporator superheat. This is due to an oscillation in the expansion valve control loop which in turn causes an oscillation in the refrigerant flow to the evaporator (shown on an aribitrary scale in the figure as parameter 9). (A different oscillation mechanism has been observed in other test runs in which the evaporator superheat fluctuates without a corresponding oscillation in the refrigerant flow rate. This is thought to be due to a flow instability within the evaporator tubes and will be discussed elsewhere).

Figure 5. shows some other performance data for the same startup cycle, and the oscillations in compressor power, evaporator heat transfer and COP can be clearly seen.

Figure 6. shows an example of the data reduction that is possible using this system. It represents a plot of the COP for cooling against the evaporator superheat setting, for selected evaporator and condenser temperatures using points drawn from 100 hours of data recorded at 10 second intervals. The oil circulation in the system was 3% by weight. The individual points were selected by choosing a moving time window (in this case 30 records or 5 minutes), and requiring that, within it,

(a) the mean values of the evaporator outlet and condensing temperatures were equal to particular selected values to within prescribed tolerances,
(b) the standard deviations of the evaporator and condenser data within the sample window was less than some prescribed value, and
(c) the standard deviation of the superheat value was less than another prescribed value.

In this way, both the stability of the system and the validity of the data under the specified experimental conditions have been ensured. The data reduction program is extremely flexible and allows any pair of parameters to be plotted, with the experimental conditions and stability being determined by tests on the mean and/or standard deviation of any desired combination of parameters. For example, the plot shown in Figure 6. could have included a further test of the liquid subcooling at the expansion valve inlet or of the stability of the evaporator heat transfer.

It is gratifying that this curve reflects the same trend in the effect of oil and evaporator superheat on the performance of the heat pump as was indicated in our earlier report (1). The difference this time is that we now have much greater confidence that the reported data is a true and quantitative reflection of the characteristics of the system. Figure 6. is a digest of 100 hours of continuous operation, and it contains 2182 data points which have been plotted only after a rigorous selection procedure.

5. REFERENCES

1. McMullan and Morgan, 1981, Development of Domestic Heat Pumps, CEC, EUR7098EN

A SINGLE SCREW HIGH PERFORMANCE HEAT PUMP

B. Zimmern & E. Kallmann

Omphale S.A.
F-92800 Puteaux Tel: (1) 775 0773

Summary: The scope of this research is a heat-only, air to water heat pump intended for a collective housing or business building, capable of a peak production between 100 and 300 kW. Can a single-screw compressor, operating with an injection of liquid refrigerant instead of oil, satisfy the operating conditions and the economic constraints? After initial problems, prototype compressors were designed and built that demonstrated the feasability of the project. An economic evaluation confirmed that they are competitive, especially if a semi-hermetic construction is adopted.

The asymmetric use of the capacity control device, a specific feature of single screw machines, permits using an economiser down to half load. Fringe benefits, arising from better use of the heat exchangers and from an original defrosting procedure, are expected.

Seasonal performance was calculated for the climats prevailing in Paris (oceanic) and Berlin (continental).

1. OBJECTIVES.

Heat-only, air-to-water heat pumps, as considered for the temperate zones encountered in most of the European Community, are presently serving with difficulty the collective housing and business buildings.

In the corresponding power range (from 100 to 300 kW thermal peak production) one can find commercially systems with a plurality of semi-hermetic reciprocating compressors, thus rather costly, whereas conventional oil-injected screw compressors are only competitive above this power range. They would indeed offer their users a number of well-known advantages: low noise, no vibrations, long lifetime, reduced maintenance,..

This research programme was initiated when Electricité de France, in its Research Laboratory in Chatou, found rather remarkable results while testing the 001 prototype of the PC single screw compressor. The efficiencies measured let us hope that, provided thorough development and several improvements, a heat pump with quite exceptional performances might be produced.

Without waiting for this Research Contract to be granted, Omphale, at its own cost built a fully instrumented new test rig, modified the initial prototype and built a second prototype.

The principle of the PC compressor is shown on Figure 1. Among all single screw structures, obtained by combining the geometric shape of the parts in mesh, the PC is the most elaborate. It uses a plane screw (hence the "P") co-operating with cylindrical pinion wheels (hence the "C"). It allows thus adjusting the screw-to-casing, screw-to-gaterotor and gaterotor-to-casing clearances to values as reduced as desired. By reducing in this way the leaks inside the compressor, one can hope dispensing of the oil as a sealant and evacuate the compression heat by injecting liquid refrigerant.

Moreover, the PC compressor permits easily to withdraw a gaterotor while running, thus creating an appreciable leak through which the gas can return to intake: this is an elegant and economic way to reduce the capacity by 50% increments, resulting in full-load, half-load, no load and each of these positions corresponding to an optimum in efficiency.

Therefore, following targets were set:

- build a prototype heat pump with a shaft power of approximately 30 kW, using a single screw compressor, in order to lower the limit of competitive application from the present value of 100 kW to 30 kW.
- by the same token, dispense of the injection of oil and of the associated circuits, and thus reduce system cost.
- obtain an efficient capacity control device, continuous or by steps, that could be implemented simply and at low cost.
- if possible, gain fringe benefits at the system level, e.g. by reducing the size of the heat exchangers, simplifying the defrosting devices, etc...

2. EXPERIMENTING WITH THE PC COMPRESSOR.

In a completely unexpected way, the first results obtained on the new test rig at OMPHALE have proven deceptive, as the efficiencies were some 10% below those measured by EDF.

First, these discrepancies were attributed to inadequate or inaccurate instrumentation or processing of the measured data. A thorough check of all instruments, the double checking of basic parameters, an exchange of instruments with EDF led to the conclusion that our measured values were quite correct.

In parallel, we explored whether the insufficient performances could not result from modifications made on the first prototype in view of improving it. A complete test campaign was scheduled to verify, point after point, all parameters, and this involved the construction of a third prototype.

At the end of November 1980 we had to acknowledge, without being able to supply an explanation, but only assumptions, that the technology of the PC compressor we had adopted was not able to meet the expected performance level, either because the initial values were erroneous, or because we had not been capable of reproducing an exceptional first machine.

To summarize the results obtained, let us note that we achieved an isentropic efficiency of 67 to 69% with an open compressor of 2500 l/min swept volume, running at 3000 rpm on Refrigerant 22, between 4 bars intake and 12 bars discharge pressure, while the objective value was 75%.

3. EXPERIMENTING THE CP COMPRESSOR.

As soon as mid-1980, envisioning that our tests with the PC machine might lead to a dead-end, we decided to risk an attempt with a more conventional type of machine, the CP type (cylindrical screw, plane pinions).

Up to then this compressor had only been manufactured as a refrigeration machine from 100 kW shaft power upwards. What we were aiming at was a machine 5 times smaller, thus requiring much reduced tolerances, and which took full advantage of all improvements that had been made during the work on the PC type. We were also in a position to start our tests without oil injection and could obtain very encouraging results in the very first days of 1981. An amendment to the programme of work was decided with the Commission, in which nevertheless all technical objectives of the initial contract were maintained.

3.1. Main features of the CP compressor.

The CP machines we tested (prototypes nr. 001, 002, 003) comprise a screw of 140 mm outside diameter, two plane "floating" gaterotors, i.e. each made of a cast-iron star supporting a fiber-reinforced plastic seal, driven by an electric 30 kW, two-speed (1500 and 3000 rpm) motor.

The cooling of the compressor and the sealing are provided by an injection of liquid Refrigerant 22, made in the initial stage of the compression, thus under a pressure differential available in the system without the need for a pump.

The sealing effect of liquid refrigerant is poor, due to its low viscosity; it is therefore necessary to reduce the play between the parts in relative motion (screw, casing, gaterotors) to achieve good efficiency. The CP structure, comparable to a pair of gears, permits no adjustment in operation, as did the PC. Thus the required accuracy must be built into the compressor and maintained during operation.

The main bearings and the end seal -in the open structure tested- are cooled and lubricated by a closed loop oil circulation, of which two designs were successfully tested. The corresponding circuit has small dimensions, because the load on the bearings is negligible and the required oil flow minimal. In a semi-hermetic version, there is no more end seal; as it is the main consumer of oil, the circuit can be further reduced.

3.2. Capacity control

The conventional capacity control device for large size refrigeration compressors, using two slides moving parallel to the shaft of the screw, was adopted. Its operation has nevertheless been modified in an unconventional manner: instead of moving both slides identically, a single slide is moved first, while the other is kept in full load position. So the range of 100% to 50% load is covered. At the latter value, the half compressor undergoing capacity reduction is fully unloaded, which provides operating conditions at optimum efficiency. Only below 50% load is the second slide used.

Nevertheless, unbalanced thrusts appear in the asymmetrical mode and, unless special provisions are made, they may increase the clearances in an irreversible way and thus reduce efficiencies again. We therefore decided to limit the amplitude of the asymmetric operation to 80% of the total range, thus moving one slide to reduce capacity from 100% to 60%, then moving the second down to 20%, and then possibly go to 0% using both slides.

We have evaluated the full asymmetric operation by computation and by experiment. It implies a certain redesign of the screw bearings, with no incidence on the cost of the machine.

3.3. Economiser.

This device, also named "superfeed", has been known for a long time. It simulates a double-stage compressor, thus increases system efficiency and, in the case of a heat pump, it increases the duty in proportion with the compression ratio. In other words it increases the heating effect inasmuch as the outside temperature goes further down, which is an appreciable bonus.

Presently, this device has not been systematically applied to screw compressors, because it produces its benefits only at full load, there being no compatibility between a good economiser and a capacity control device using slides. And a heat pump compressor operates most of the time at part load.

The interest of the asymmetric control procedure described is precisely to maintain one half-compressor at full load while the other half-compressor fulfills the capacity reduction down to 60% or even 50%. Therefore we experimented an asymmetric economiser with success, thus taking advantage of the increase in efficiency and in duty during the major part of the heating season.

In addition, we designed a compact device, replacing advantageously the conventional heat exchangers used for the economiser and their control devices, and providing an inexpensive way to implement the superfeed effect.

4. COMPRESSOR PERFORMANCES

The values obtained are coherent between the measures made at Omphale, those performed by Electricité de France (Direction des Etudes et Recherches, Laboratoire de Chatou) and by certain licensees of Omphale's.

4.1. Full load tests.

The basic parameter in a compressor is its isentropic efficiency. We have plotted in Figure 3 this efficiency versus pressure ratio. It shall be noted that efficiency decreases in a known way when pressure ratio increases, but the effect of the economiser is to reduce this drop in a large proportion.

4.2. Part load tests

It is well known that at part load, the shaft power decreases less than the duty, because certain losses remain constant when the load diminishes. Figure 4 shows the efficiencies measured when one slide is fully moved through, and the other one in full load position. It shall be noted that, at higher pressure ratio (PR=6), the economiser is quite effective in improving the performance.

4.3. Variation of operating conditions.

Tests have been made at 3600 rpm: as expected, performances are increased more than proportionnally, thus showing an increase in isentropic efficiency.
Tests have also been run at discharge pressures of 15 and 20 bars, corresponding to condensing temperatures of 40 and 50 degrees C. They represent a more realistic approach to heat pumps, but lower discharge pressures had been favoured initially for technical reasons (available drive power, vessel and instrument gages, etc..). Theoretical calculations had shown that an increase in discharge pressure should lead to an increase in efficiency. This was confirmed, as can be seen on Figure 5.

5. SYSTEM PERFORMANCE.

5.1. We have tried to figure the system performance over the whole heating season in two selected EC locations, i.e. Paris and Berlin. The basic parameter chosen is the C.O.P., whose annual mean value reaches 3.5 in Paris and 3.37 in Berlin, according to the system scaling adopted. This seems a good value for an air-to-water heat pump subject to large temperature differentials during the winter.

Various criteria for the scaling of the pump have been taken into consideration; they focus on similar optimum choices in the two cases selected here.

5.2. Noise is not a problem: from the beginning, though using a rather noisy motor and suffering from reflective walls in our lab, we measured 80 dBA at 1.5 metres at full load. It was decided that any action to reduce the noise further should be postponed until the system was production engineered.

5.3. With an air evaporator, defrosting is usually a problem. Conventionally, it implies inverting the heat flow, not only of the actual evaporator, but also of the other system components, in particular the condenser. The transient periods are non productive. During defrosting, gaseous and liquid refrigerant moves in the system to locations where it is not normally present, and secondary effects occur while it returns to its normal location.

We have therefore tested an original defrosting procedure, feeding the evaporator with hot liquid refrigerant. This has been made possible only because the single screw compressor accepts without harm significant liquid slugs at its intake. The tests have shown the feasability of this method, which remains to be enhanced during production engineering.

The advantage of this method is twofold: on one hand it does not invert the function of the condenser and of the evaporator, and thus reduces transients; on the other hand, it requires only minimal control devices, in fact a simple one-way bypass valve.

5.4. Economic evaluations, made on the basis of a semi-hermetic compressor unit -in order to compare with what exists in reciprocating machines of similar size- show a cost advantage, at system's level, of 24% approximately.

6. FURTHER STEPS AND PROJECTS

Omphale S.A., the company conducting this Research Contract, is a privately held invention development firm, i.e. its target is not to manufacture and sell products but to license major industrial firms who produce and sell its single screw compressors. Since 1981, the licensees of Omphale have been kept informed of the progress of this Contract, and one of them, APV Hall Products, Ltd., has already announced in 1982 the sales of oil injection free single screw compressors.

Improvements of the compressor itself and of the system have been found during the research period; as far as possible, they have been protected by patent applications, the major part of them being due for publication by the end of 1983. They relate essentially to

- the reduction of plays in the compressor
- the compact economiser
- the management of the lubricant
- the capacity control

Further work will be carried on in two main directions

- the theoretical and experimental improvement of the capacity control device (asymmetrical operation) in order to gain a few additional percent in efficiency

- the industrial development of a complete semi-hermetic system including the motor, compressor, economiser, lubrication device and capacity control, which could then be directly integrated in heat pump systems and probably also used in air-conditioning applications.

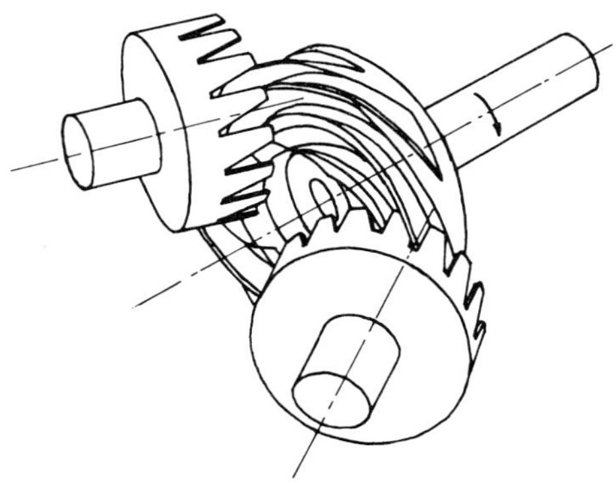

Fig. 1 - The PC Type

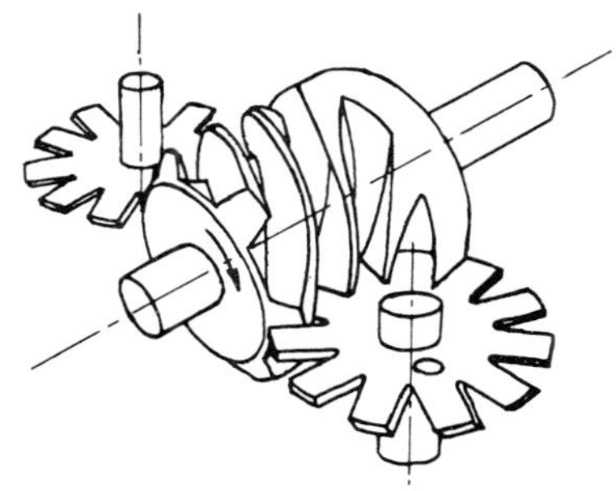

Fig. 2 - The CP Type

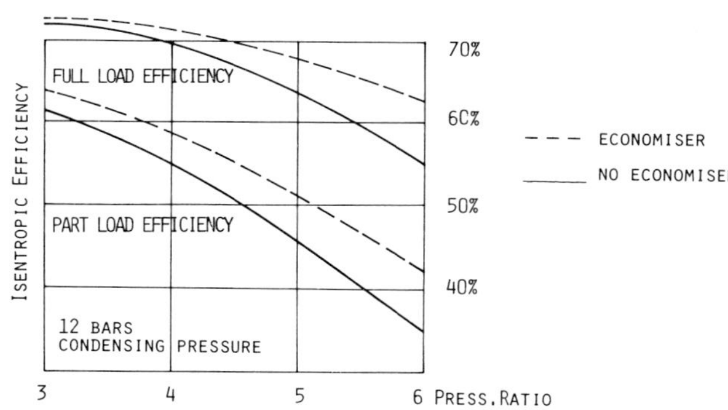

Fig. 3 (Full Load Efficiency) and Fig. 4 (Part Load Efficiency)

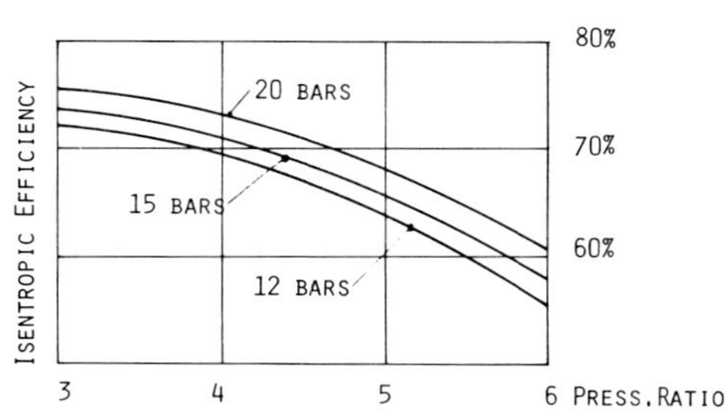

Fig. 5 - Full Load Efficiency vs. Condensing Pressure

INVESTIGATION OF COMMERCIAL DOMESTIC HEAT PUMP PERFORMANCES

H. MARTIN-NEUVILLE, J.P.PATUREAU and M. REYBILLET
Division Energétique - Société BERTIN & Cie

Summary

The research programe presented here has permitted BERTIN to set up both theoretical and experimental facilities with the objective of designing and testing ways of improving the performance of domestic heat pumps.

Apart from the technique for temperature control at the condenser, which can at the present time be applied to the commercial machines and permits a saving of up to 10 % of the annual consumption of electricity, the principal conclusion is the finding that the refrigerant circuit is unable to operate at a variable flow rate with acceptable efficiency.

This means that, first and foremost, it is advisable to reconsider the arrangement of the refrigerating circuit, particularly with regard to the following points :

- decoupling of the condenser from the evaporator, for example by means of a buffer capacity,
- adapting the evaporator to operate at a variable flow rate, by introducing, if need be, the notion of recirculation.

Apart from this approach, there is much scope for gain by improving the compressor, where a partial reassessment of the superheating (which should be replaced by an alternative safety device) would permit considerable reductions both in the energy consumption and in the size, and hence, the prime cost.

Such work can only be related to extensive programs, and only these would enable further progress to be made in heat pump technology.

1. BACKGROUND

The growth of the market for domestic heat pumps is at present restricted by two essential factors, namely the initial cost and the annual operating costs. While these two factors, in conjunction with the relative cost of energy (oil versus electricity) provide a measure, in terms of payback time, of the profitability of the operation, it appears that the initial investment itself represents a large obstacle in the consumer's decision.

Improvement of these conditions can be expected only as a result of farreaching measures, initiated by the public authorities, in the fields of vocational training, and of the reorganisation of the distribution networks.

Although they are assential to the development of domestic heat pumps, such measures will bear fruit only in the long term, provided that the will to succeed is demonstrated from now on.

In the short and medium term, the growth of the market of domestic heat pumps can rely only on the improvement in the technology, notably in two directions which are not incompatible.

1°) Reduction of manufacturing costs, the objective is to simplify the pump and its integration into the heating systems and to improve the optimization of the dimensions and the technology of the various components with value analysis considerations.

2°) Improvement in the performance of the domestic heat pump by paying attention both to the intrinsic performance and to the overall annual performance of the heat pump as installed. The latter, in particular, shows whether the heat pump is being used at the best of its capacity. Attemps will then be made to reduce the operating costs.

2. OBJECTIVES

With these technological improvements in mind, Société BERTIN has launched a research and development programme supported by the European Communities* and by the French Department of Industry.

In its initial phase, the work reported here has been aimed generally at the reduction in the operating costs, and therefore at the improvement in the performances. In this context, interest is limited exclusively to electric domestic heat pumps operating between outdoor air and heating water since, these represent the largest market, at least in Europe. The search for improved performance takes place at two levels :

- the level of performance at any given moment, reflecting the combined efficiencies of each of the components of the machine (compressor, evaporator, condenser),
- the level of actual annual performance which involves the manner in which the domestic heat pump is employed in the heating circuit.

The characteristic feature offered by the heat pump is, in fact, its adaptability to the sources between which it operates. While nothing can be done about outdoor air temperature, the temperature of the heating water at any one moment is, on the contrary, a decisive factor for the coefficient of performance of the machine. This temperature results from a balance between the power delivered by the heat pump, the heat losses of the accommodation in question, and the various circuit inertias, and it follows that the thermal control of the domestic acommodation plays a fundamental role in the annual electric consumption of the pump.

3. METHODS OF ANALYSIS

Both theoretical and experimental techniques are being employed.

3.1. Test stand

A test stand based on a commercial air/water system has been built at the Centre de Tarnos. An air circuit, with separate controls for temperature and humidity, enables the cold source to be simulated between - 10°C and 25°C. A water loop simulates the thermal behaviour of a domestic accommodation whose characteristics (inertia = water equivalent, transfer surfaces = radiators) can be modified according

----

* **Contracts n° EEA-4-048 F and EEA-4-064 F.**

to the case being simulated (the test stand is shown diagrammatically in Figure 6).

The entire equiment is instrumented for measuring temperature, humidity, flow rate, pressure and electric power consumption. The sensors (approximately 50) are managed by a microcomputer which allows the information to be acquired (up to 500 Hz), stored and processed. Specific software also enable the results to be processed by a more powerful computer. It is worth stressing here the extreme care with which the measurements must be taken (absolute accuracy, thermal drift, ...), particularly in relation to the refrigerant circuit temperatures, to enable the results to be meaningfully interpreted.

3.2. Mathematical models

Two mathematical programs have been developed.

3.2.1. Thermodynamic simulation of the system heat pump + accommodation

This mathematical model is intended to simulate the transfers of energy between the heat pump, accommodation and the environment and to calculate the annual electricity consumption of the pump (including that used for defrosting), taking account of :

- the type of regulation and thermostat temperature differential which is employed,
- the inertia involved,
- the characteristics of the components (heat pump, radiators, insulation),
- weather data.

3.2.2. Thermohydraulic simulation of the refrigerant circuit

The aim of this model is to analyse the complex phenomena governing the circulation of the refrigerant (two-phase flow, pressure drops, heat exchangers by boiling and condensation) in the heat pump circuit. The performances of the exchangers (evaporator and condenser) and of the expansion device are very closely linked to all these aspects, particularly during intermittent operation which is the case most of the time since the power control is of the on and off type.

4. CONTROL OF DOMESTIC HEAT PUMPS

When the external temperature increases relative to the nominal design point, the heat pump exhibits the distinctive characteristic of increasing the power delivered to the condenser, which is contrary to the thermal requirements of the accomodation which is to be heated.

In order to reduce the average power, two possible solutions exist :

- intermittent on and off operation,
- continuous operation with a variable speed compressor.

For each of these operations, it is of course necessary to try to maximise the COP at any given moment.

4.1. Temperature control at the condenser

In the case of an on and off operation, **Figure 2** illustrates the significance of the inertia of the heating circuit for the COP at any given time. In fact, if the inertia is low, the water temperature rises very rapidly and the COP remains low. If the inertia is high, on the other hand, the water temperature at the condenser remains low and the COP shows a much greater increase. To limit the influence of this inertia term, the choice which is made in this case is to regulate the water temperature at the condenser, through the intermediary of a buffer capacity, at a set point corresponding to the water temperature which suffices to provide for the needs of the accommodation. This set value naturally depends on the characteristics of the radiators. The control in the accommodation can then be refined by means of a water circulating pump or thermostat valves.

Figure 3 shows the results of a simulation (experimental and numerical) which compares, on an annual basis and for the Paris region climate, the conventional control ("free" condenser water temperature) and the control as proposed here, as a function of the inertia of the heating circuit. With the scale on the left showing the annual heat energy in kWh and that on the right the electric energy consumed, including defrosting periods, the results obtained make it possible to arrive at two findings :

- for the low inertias (typical of modern housing) the annual energy saving with the control at the condenser can reach 900 kWh,
- the heat consumption of the accommodation is also lower owing to a reduced temperature range in the accommodation. This also contributes to an increase in the comfort of the inhabitants.

The advantage of this system is that it can already at the present time be installed into the domestic heat pumps which are now on the market and at a moderate cost of investment (payback time estimated at about 3 years).

4.2. Control by variation of compressor speed

Speed variation experiments have been carried out with a commercial heat pump equipped with a frequency changer fitted to the terminals of the sealed compressor. Instantaneous performances have been measured between 3,000 and 600 r.p.m. for various source temperatures.

Figure 4 shows some of the results obtained during the test campaign. It is found to be possible, indeed, to adapt the thermal power Q_C to the condenser. The COP, on the other hand, shows a slight tendency to increase when the speed decreases (proportionately less as the temperature at the condenser is raised), but drops rapidly below 1,500 r.p.m.

Two reasons are put forward to account for these mediocre results :

- firstly, the efficiency of the compressor (especially that of the motor) probably diminishes with the speed,
- secondly and above all, the components of the refrigerant circuit (exchangers, expansion valve) are incapable of operating in an optimum manner when the flow rate of fluid varies.

The variable speed nevertheless remains a attractive concept, on condition that the two problems mentioned above are resolved first.

5. ANALYSIS OF THE REFRIGERANT CIRCUIT OF A DOMESTIC HEAT PUMP

Tests have been carried out on a commercial domestic heat pump to investigate the following two aspects :
- elimination or reduction of frosting-up of the evaporator,
- characterisation of the actual behaviour of the refrigerant at any given moment.

5.1. Elimination or reduction of frosting

This work was motivated by somewhat contradictory and, viewed retrospectively, exaggerated notions which existed several years ago concerning the actual effect of the defrosting on the annual running costs. Two processes were studied and tested.

5.1.1. Spraying the batteries with a low-freezing fluid

The aim in this case is to eliminate frosting altogether, and hence, to do without the cycle reversing system. After the problem of producing a uniform film on a large number of fins had been solved, two types of fluid were tested :

- Silicone oil with a low viscosity and a very low vapour pressure. The effect is purely mechanical and the frost is carried off by the oil. Quite a major drawback of this process, however, is the apparently unavoidable formation of an emulsion which needs to be separated at a later stage,
- absorbent brine (LiCl, NaCl of suitable molarity). In this case, the solution absorbs the moisture from the air, even beyond the dew point, which results in an improvement in the heat exchange. The solution, on the other hand, has to be regenerated either by boiling (with mechanical recompression of vapour) or by reverse osmosis.

The brines provide a completely satisfactory method of removing the frost, however, the problems associated with their use (corrosion, regeneration) restrict, a priori, their interest to very large refrigeration plants.

5.1.2. Preferential condensation with two evaporated in series

Finned batteries of the conventional type begin to frost up as soon as the air temperature drops below 7 to 8°C. This means that the temperature of evaporation is close to 0°C. The process consists in placing two evaporators in series with the air (in parallel with the refrigerant) and condensing as much water as possible on the first exchanger by keeping the evaporation stage slightly above 0°C. The elimination of the frost is not complete because the second exchanger is, of course, below 0°C. In addition, as soon as the air becomes too cold it is no longer possible to maintain the first evaporator above 0°C. Nevertheless, a reduction of the frosting can be expected, resulting in a wider separation or shortening of the defrosting cycles, or both.

Tests which bave been carried out on these lines have shown that, in the present state of the domestic heat pump techology (feeding the evaporator directly from the expansion stage), the total energy contribution of the defrosting of the commercially available heat pump used here does not exceed 5 to 6 % of the annual consumption. As a result, it is improbable that the defrosting system with two exchangers which was envisionned could have a reasonable pay-back time.

5.2. Characterisation of the actual behaviour of the refrigerant

The measurements carried out on the refrigerant determine the efficiency of the various components, in particular that of the compressor.

The main results are as follows :

5.2.1. Compressor

The overall efficiency as measured for the sealed unit varies between 30 and 50 % according to the operating conditions, the electrical motor efficiency being close to 80 %. For a while, oil carry-over was thought to be responsible for a large part in the efficiency loss. Measurements of oil entrainment, both by means of a separator and by evaluating the density of the liquid R22/oil mixture, indicate that the oil flow rate is usually below 1 % of the refrigerant flow rate for most regimes. In accordance, no noticeable effect was detected on the COP for operation with or without oil separation. Rather, it is now thought that the excessive internal superheating (probably of the order of 50 to 70°C and perhaps up to 120°C according to one european compressor manufacturer) accounts for a large part of the drop in efficiency. Another of its consequences, which is at least equally significant in terms of cost is the over-sizing of the compressor due to the increased volume/mass ratio on the suction side.

5.2.2. Exchangers, expansion valve

The most striking result, which can be seen in **Figures 5a and 5b**, is the extremely intermittent flow pattern of the refrigerant. The expansion valve shows considerable "cycling" which corresponds to temperature variations of 6 to 7°C and a period of 3 to 4 minutes at the evaporator tube level. This cycling naturally contributes to a lowering of the average evaporation pressure, and hence the average COP in operation. Furthermore, as demonstrated by **Figure 6**, the various evaporator tubes are not fed uniformly since, despite the considerable superheating which takes place in the instrumented tube, the fluid remains generally in two phases in the region of the collector. This means that the expansion valve may close slightly, thereby increasing further the "badly utilized" heat transfer area which produces the overheating. The loss to be made up which is due to this phenomenon is thus estimated at 15 or 20 %.

Another point to be noted is the very tight coupling existing between the high pressure section (condenser) and the low pressure section (evaporator) across the expansion valve. As a result, any disturbance or imbalance in the evaporator is rapidly reflected in a disturbance in the condenser. The pronounced and rapid changes in the flow rate can thus have quite an appreciable effect of the flow at the outlet of the condenser, leading to frequent carry-over of a two-phase mixture through the expansion valve, and consequently to a reduction in the efficiency of the refrigeration cycle.

Although it is knownat the present time how to adapt a heat pump circuit (condenser, expansion valve, evaporator and a circulating mass of refrigerant) for a nominal point with a good approximation and good performance, it appears that this circuit goes rapidly out of "tune" as soon as one leaves a narrow range of refrigerant flow rate around the nominal value, resulting in a considerable drop in performance. As a means of increasing the annual average COP is precisely to operate the heat pump at variable thermal power - or which is the same - at variable mass flow rate of freon, it is obvious that an evaporator capable of working at variable mass flow rate with high efficiency has to be developed.

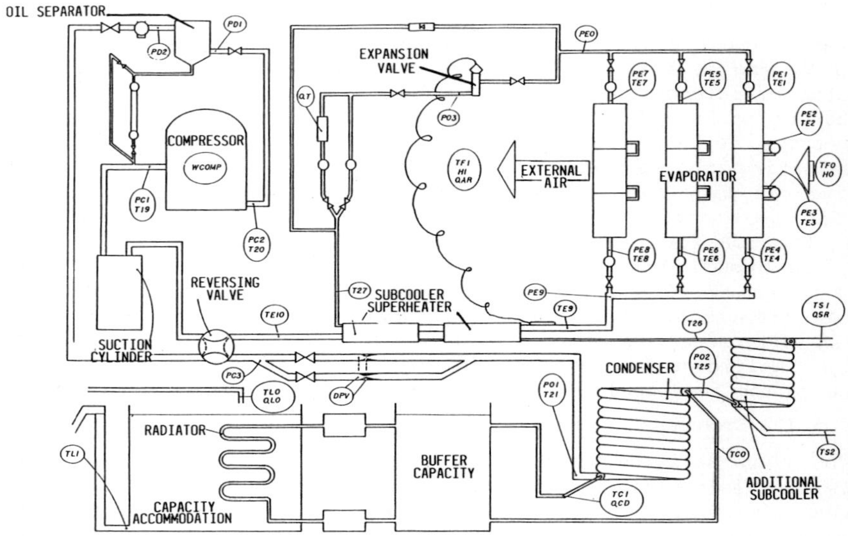

FIG. 1 - Diagram of the test stand and the location of measurements

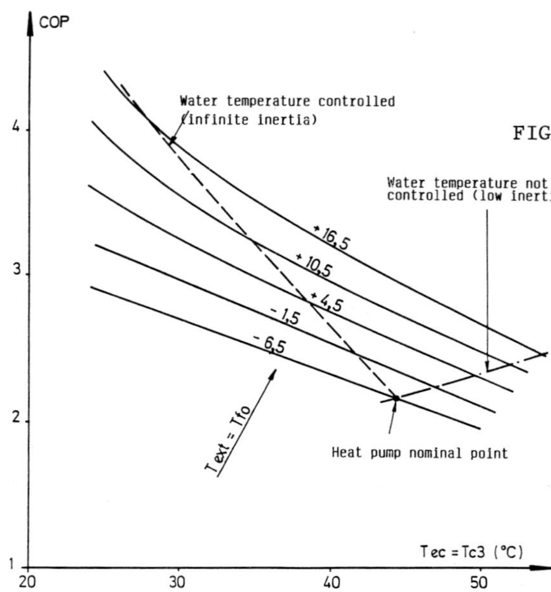

FIG. 2 - Adaptation of the domestic heat pump on part load

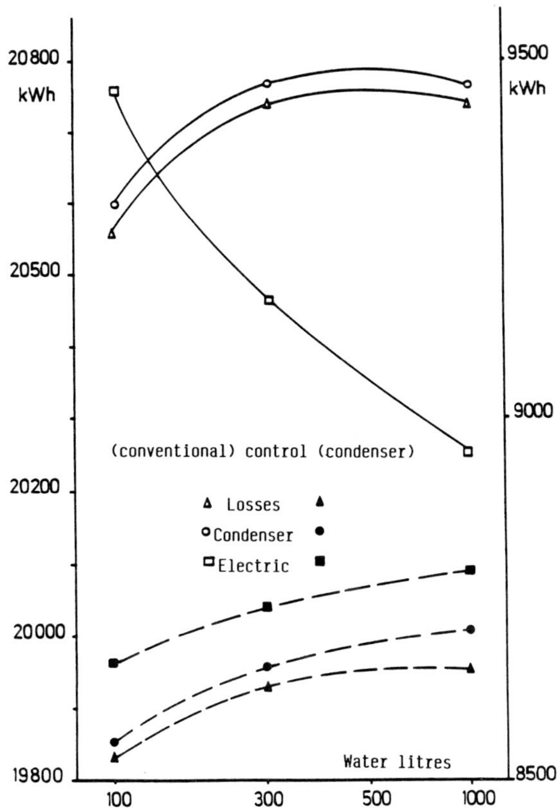

FIG. 3 - Annual consumption balance as a function of the control

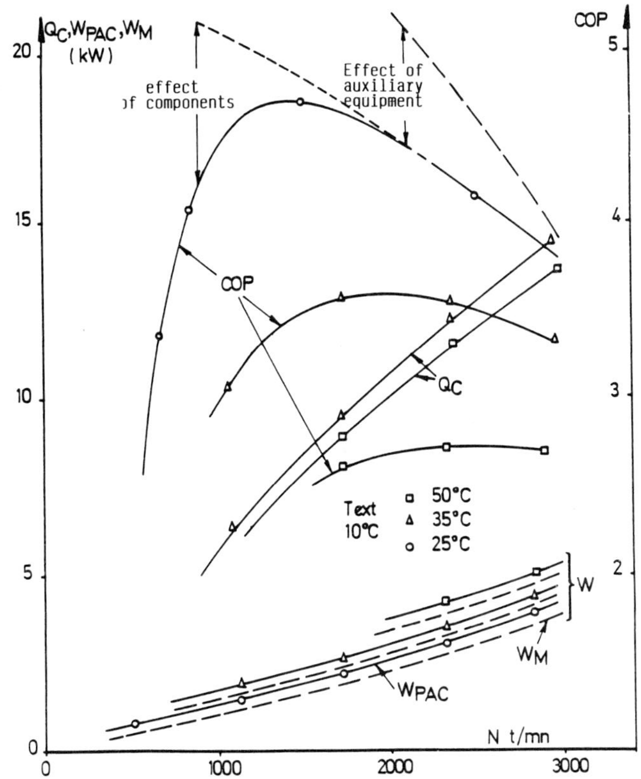

FIG. 4 - Results for variable speed

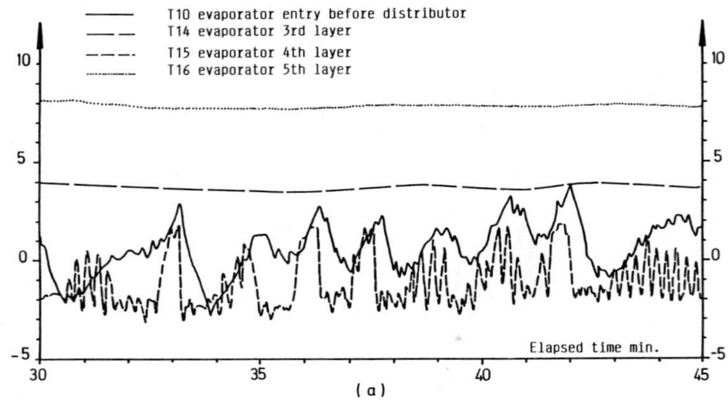

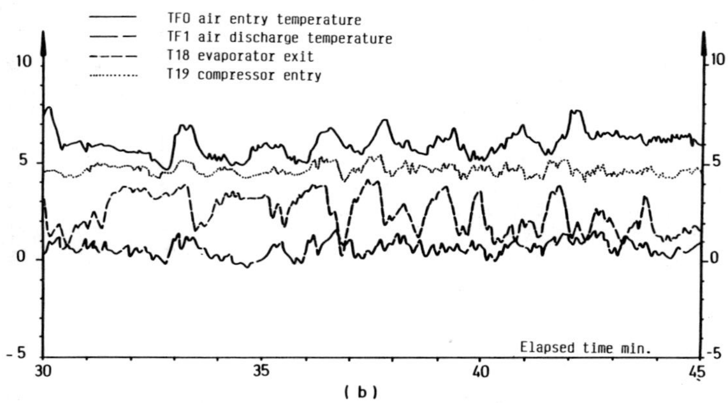

FIG. 5 (a + b)- Temperature variations during evaporation

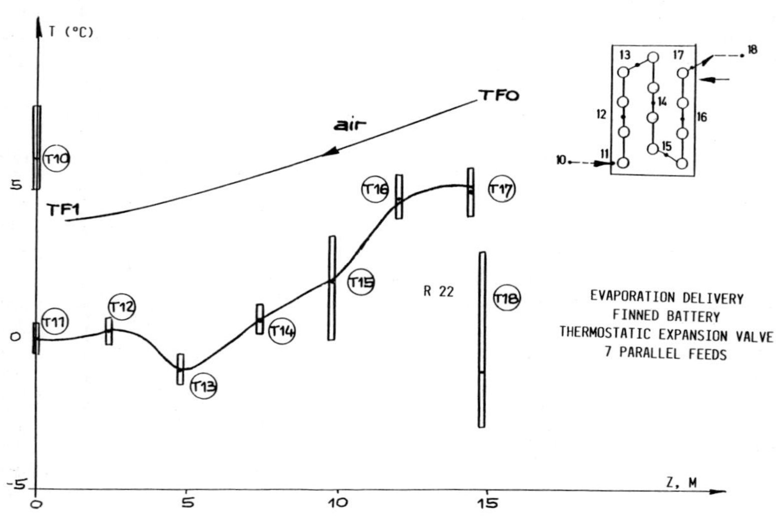

FIG. 6 - Evaporation profile in the 4th circuit

OPTIMIZED OPERATION OF A FROZEN AIR HEAT EXCHANGER BY USING A MICRO-COMPUTER

Robert ISERMANN

Summary

Herein the results of operating a frozen air heat exchanger within a monovalent heat pump system (air/water) controlled by a micro-computer are presented. We will be reporting about the operation of a micro-computer control, which allows the adjustment of any superheating temperature of the suction gas and a over all optimization. The most important parameters of the optimization of the evaporator operation were the fan operation/air flow rate and the whole defrosting procedure.
The various air flow rates (adjusted by varying the r.p.m. of the fan), have only little importance to the coefficient of performance. But a new criterion tested in practical operation for initiating the defrost procedure by using a calculated/estimated heat transfer coefficient is presented.

Monovalent and bivalent heat pump systems using 'outdoor air' as the heat source have to deal with a wide range of condition variation (e.g. air temperature and humidity). The difference between the possible and required heating power is especially significant for monovalent systems heating buildings. This must be removed by a fitting performance adaption.

Having a constant volumetric-flow the mass-flow of the refrigerant of a monovalent heat pump system can have a very different value due to the outdoor conditions. At the same time the heat- and mass-transfer conditions vary in a wide range.
For outdoor temperatures between +5 and −5 °C and high relative humidity values, there is a big frost build-up at the evaporator. An energy-saving adaption to these adaption to these circumstances causes the operation of a qualified control.

The presented research work had the goal to test the operation of a monovalent, electric-driven, heat pump system 'air/water' (low temperature heat distribution simulation) with a micro-computer for purposes of controlling the process and optimized performance. Therefore it was necessary to change the so-far used control-equipment to micro-computer technology, in order to operate the components of the system at optimized conditions and, at last, having costs for operation as low as possible.

In detail, the micro-computer had the following functions:

- Guaranty (control) of the system and operation safety, the check of the condensation, suction and oil pressure and the superheating temperature difference at the evaporator outlet,
- check of measuring positions,
- control of the evaporator heater for melting the defrosted frost build-up,
- control of the superheating temperature difference, of the inlet and outlet temperature of the heating water,
- adjustment of an optimized operation by controlling the level of performance of the compressor, the defrost procedure, the air flow rate, fixing the saddle-point for the superheating temperature difference and the inlet/outlet temperature of the heating system.

Therefore, the micro-computer opened the possibility to monitore certain characteristic values (to calculate the coefficient of performance C.O.P. etc.) or to set signals and control values to manipulate the process.

During this examination a micro-computer development system was used, which, with the exception of a safety pressure controller and the oil heater, managed the whole heat pump process control.

A new 'thermoelectric expansion valve' worked by the micro-computer was used to control the evaporator of the system. This arrangement allowed to set and control any superheating temperature differences of the evaporated refrigerant in all set points of performance.

After first tests we placed examinations on the different components of the heat pump system to find optimized performance for the compressor, the evaporator, the air flow-rate of the fan and the defrost procedure.

To reach the best performance we firstly found a set of parameters of the single components (e.g. the superheating temperature difference was examined by varying the evaporator-control). After that, the whole system was operated to minimize the annual operation costs and to limit the noise level of the fan to the allowed level. As the heating water temperature was conducted by the outdoor temperature, this temperature could be used as a characteristic value of the operation performance.

To optimize the evaporator operation the following control parameters were fixed and used:

* fan r.p.m/fan voltage supplyment for the parameter 'Air flow rate of the evaporator',
* Superheating temperature difference of the evaporating refrigerant at the evaporator outlet,
* the outdoor temperature to use the two alternative defrost-procedures -defrosting with the fan and with electric resistor heating-,
* evaporation pressure to finish the defrost procedure and
* heat-transfer-coefficient to initiate the defrost procedure

for the compressor operation

* a temperature limit of the outdoor air for usage of the alternative levels of performance ('half' and 'full')

The essential part of this work is represented by the examination of the heating operation, when the heat pump delivered energy to a heat storage, which was able to simulate a heating system by outdoor temperature lead heat consumption. During these tests especially the operation of the evaporator showed good results in optimization. The new criterion 'heat-transfer-coefficient' to initiate the defrost-procedure showed on the one hand a high reliability and on the other hand it allowed a well-significant correlation to the 'relative C.O.P. drop' as well as to the amount of frost build up. This allows to define the best point to initiate the defrost procedure.
This will be explained in detail in the following, while the start-up of the micro-computer controlled 'thermoelectric expansion valve was reported about on the contractor's meeting on April 28, 1982, held in Brussels.

The calculation of the heat-transfer-coefficient k_v (correlating on the difference between air inlet and evaporation temperature) to find the optimized point to initiate the defrost-procedure were examined that way: The evaporator was operated to very high levels of frost build-up, what resulted in a significant C.O.P. drop (see F i g . 1). Curve b is 'back-calculated' as the defrost procedure would have been initiated at the possible fines. It is evident, that energy-wa-

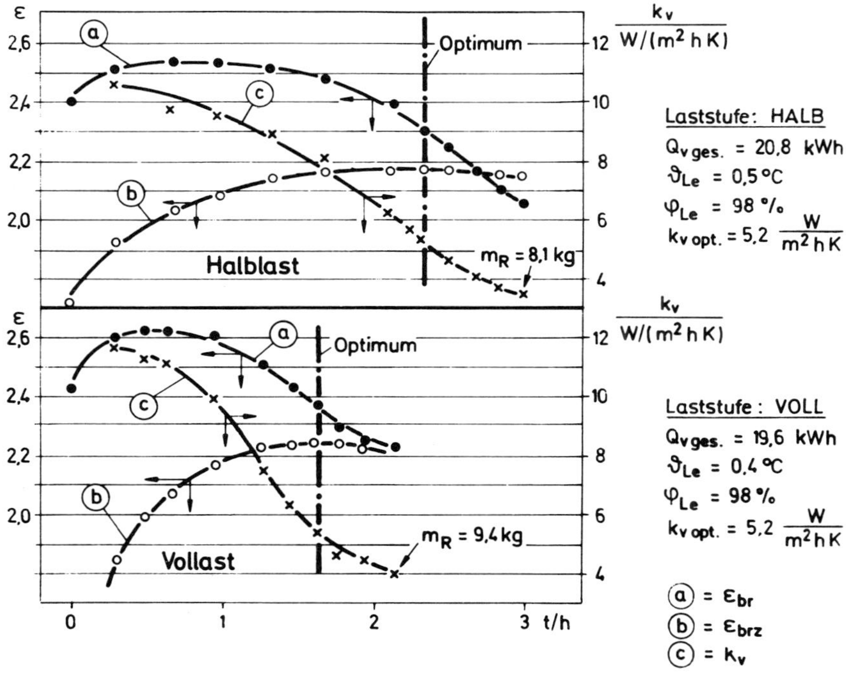

Figure 1: Overall C.O.P. vs. defrost initiation time for equal fan r.p.m. ('half' and 'full' performance of the compressor)

sting defrost procedures like electric resistors (used here) are more sensitive to initiating 'too early' than 'too late'.

The special qualification the of H.T.C. to optimize the initiation point of defrosting is shown in Fig. 2. The growing frost build-up causes increasing heat transfer resistance what is originated on the one hand by the build-up (conduction) and on the other hand by decreasing air flow V_L through the evaporator. The refrigerant side heat-transfer-coefficient α'_{vi}, related to the evaporator surface, is not decreasing too much compared to the frost side heat transfer coefficient k_{va}. From a certain amount of frost the 'frost-side' conditions are dominating the whole heat-transfer conditions and, therefore, the drop in C.O.P..

So, the heat transfer coefficient k_v represents both the parameters affecting the relative C.O.P. drop (related to the heating period): the energy consumption for defrosting and the present C.O.P. decrease.

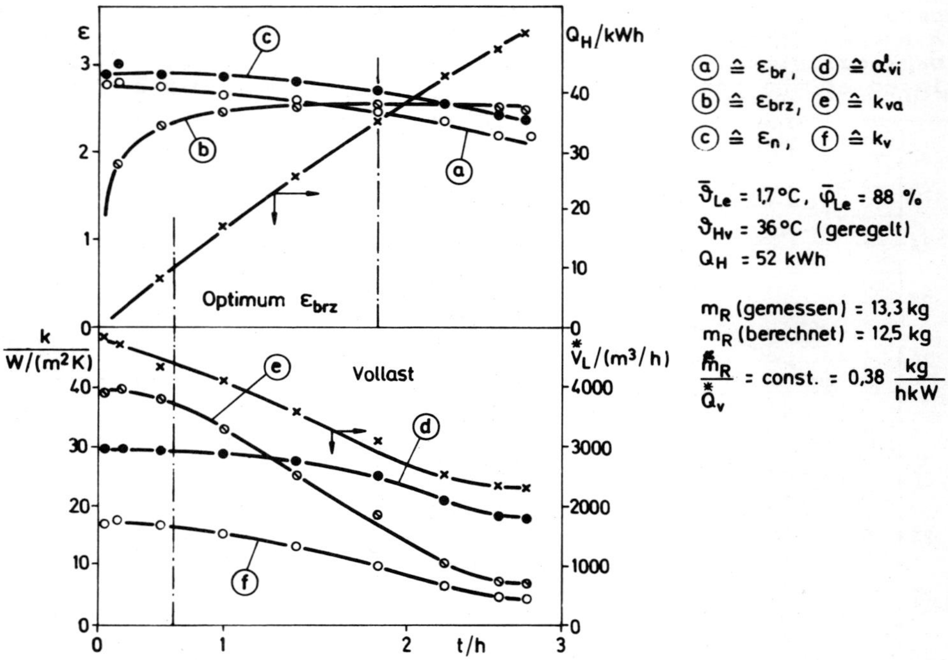

Figure 2: heat- and mass-transfer-conditions at an evaporator with increasing frost build-up

Fig. 3 shows two tests of frosting (examples), which represent the highest per-cent level of defrost-energy and the highest annual heating energy. The dependence of the C.O.P. on the over all heat-transfer-coefficient k_v is shown as a reference for a possible point of defrost initiation. The optimized

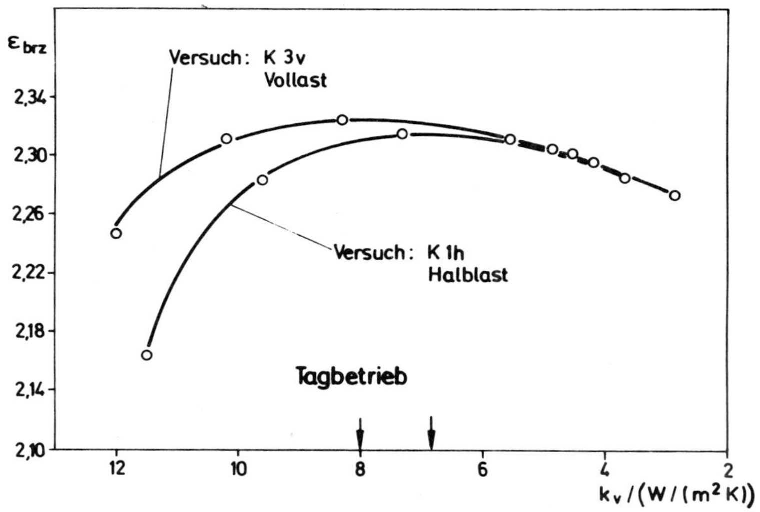

Figure 3: Average overall C.O.P. for a cycle vs. heat-transfer-coefficient as an indicator for defrost initiation

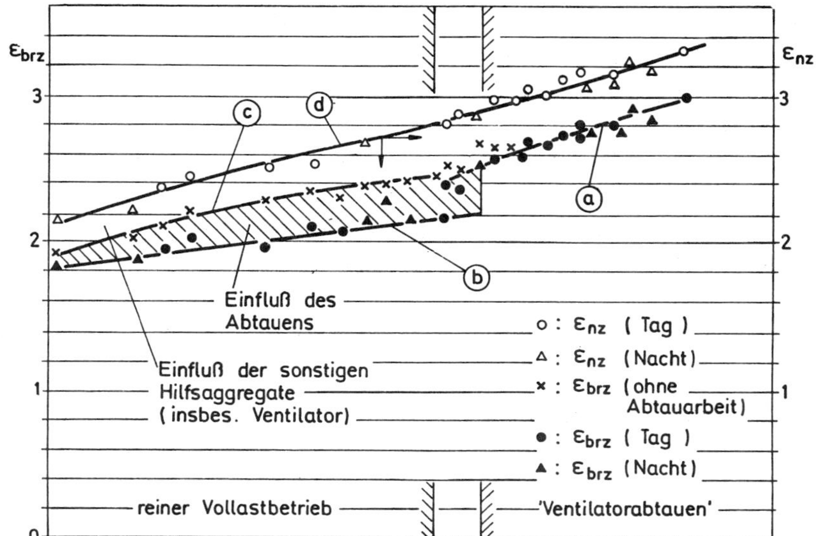

Figure 4: overall C.O.P. vs. outdoor temperature for optimized heating operation (simulation)

values are 8 resp. 7 W/m^2/K for full resp. half performance. These values also prooved within little deviations for all other outdoor conditions.

In F i g . 4 the dependence of the C.O.P. vs. the outdoor temperature for average-values is shown (rel. humidity 80 %). The hatched area represents the drop in C.O.P. caused by the defrost procedure (electr. resistors). Defrosting the evaporator by blowing the fan after stopping the heat pump, if the outdoor temperature was more than 3 °C prooved to be a very energy-saving procedure.

Basing on a representing sample test for the climate conditions of Dortmund, the energy used for defrosting was 3.5 % (related to the total energy consumption) for one heating period and thereby fairly low (see F i g . 5). For partly using the electrical resistor-heater thus is a very good result.
The energy consumption for all auxiliary components like fan, water pump, oil-heater and defroster was 13.5 % (of total energy consumption), the over all C.O.P. was $\beta_{br} = 2.54$ corresponding to a primary energy efficiency of $\xi = 0.9$.

The improvement in energy-savings using the optimized operating conditions can only be estimated, because there was not the same heat pump system running at the same time without using the optimizations. The improvement in C.O.P. is about 6 %.

The used micro-computer-system joint with the software was used to test and optimize the whole system. It offers a basis for a later usage in a commercial production e.g. a single-board-micro-computer of a monovalent heat pump system.

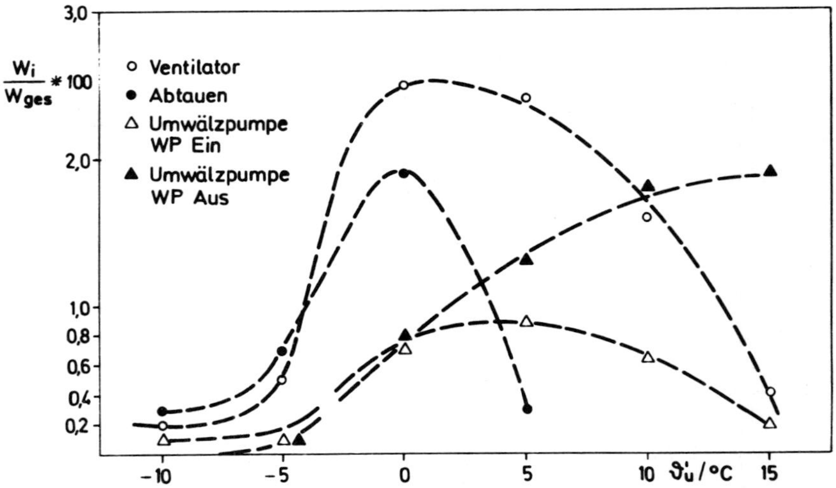

Figure 5: auxiliary energy (per cent usage) vs. outdoor temperature (for one heating period)

Formelzeichen

k_v	heat-transfer-coefficient of the evaporator	W/m^2/K
k_{va}	air side heat-transfer-coefficient of the evaporator (related to the temperature difference $\vartheta_{Le}-\vartheta_w$)	W/m^2/K
k'_{va}	air side heat-transfer-coefficient of the evaporator (related to the temperature difference $\Delta\vartheta_{effective}$)	W/m^2/K
k_{vopt}	k_v (optimal value)	W/(m^2 K)
m_R	mass of frost build-up	kg
$\overset{*}{m}_R$	frost build-up rate	kg/h
n_F	r.p.m. of the fan	1/min
Q_H	heat (produced)	kJ o. kWh
Q_v	evaporation heat	kJ o. kWh
$\overset{*}{Q}_v$	evaporation heat performance	kJ/s, kW
$W_i(\vartheta'_u)$	energy consumption type i for the temperature class ϑ'_u over one heating-period	kWh
W_{ges}	total energy consumption over one heating period	kWh
α	heat-transfer-coefficient	W/m^2/K
α_{yi}	refrigerant side heat-transfer-coefficient	W/m^2/K
α_{vi}	refrigerant side heat-transfer-coefficient related to the evaporator outer surface	W/m^2/K
β_{br}	overall C.O.P. for one heating-period	1
ε	C.O.P.	1
ε_{br}	overall C.O.P.	1
ε_{brz}	overall C.O.P. related to one cycle of operation	1
ϑ_{Hv}	temperature of heating water inlet	°C
ϑ_{Le}	air temperature at the evaporator inlet	°C
ϑ'_u	outdoor temperature of a class as an aveage value	°C

Optimization of Internal Combustion Engine Driven Heat Pump (20 kW heating power)

Dr. Klaus Lück
Fichtel & Sachs AG, D-8720 Schweinfurt

Summary

Since 1977 Fichtel & Sachs is concerned with the development of I.C. engine driven heat pumps for power ranges up to 20 kW. Using preferably the environmental energy source air, houses with a heat demand of up to 20 kW can be heated up to $-15^{o}C$ (monovalent); in those once with a heat demand of up to 50 kW the I.C. engine driven heat pump can be used on a bivalent basis.
We have developed an I.C. engine compressor unit particularly designed as a driving unit for the operation requirements of heat pumps. It can be operated either by means of natural gas or by fuel oil. In conjunction with the microprocessor control, designed for that very purpose, the heating capacity of the heat pump will be controlled depending on both the outside temperature and the heat demand of the house by speed control of the engine as well as by turning on and off the cylinders of the two-cylinder compressor. For the purpose of testing and optimization three heat pumps were installed in the R&D department and ten heat pumps were installed in private houses of members of our company. The results show that the expected data regarding power output, energy efficiency, and the ability to integrate the I.C. engine driven heat pump in an existing heating system can be achieved.

1. INTRODUCTION

Fichtel & Sachs has developed small I.C. engine driven heat pumps. They can be either run with gaseous fuels or light heating fuel oil. As a drive unit an engine compressor unit was specifically developed for this application.

The heat pumps have the following specifications:

- use in residential houses with a heat demand of 20 kW at $-15^{o}C$ outside temperature, monovalent, up to 50 kW, bivalent
- hot water temperatures up to $80^{o}C$, and thus particularly suitable for use in old buildings
- microprocessor control for the complete heating installation (burner/boiler system plus amount of heat)
- life span 15 years or approx. 40 000 hours of operation
- annual maintenance (about every 2 500 hours of operation
- 50 % saving in "primary energy" (oil/gas) for annual PER (Primary Energy Ratio) of min. 140 % (including defrosting and standstill losses, etc.)
- air is preferably used as an environmental energy source.

The present research study was prepared to optimize the individual features and to prove that such units can be used in residential houses without problems, that the advantageously achieve considerably higher energy savings than are obtained for example with largely improved conventional heating methods or with the electric and absorption heat pump.

2. BASIC FEATURES OF THE SMALL I.C. ENGINE DRIVEN HEAT PUMP

2.1 Design and Operation

Fig. 1 shows the design of the I.C. engine driven heat pump, using air as a heat source, in a split version, i. e. the heat source unit is set up in the heating room, in the attic, or in some other location outside the building and connected through refrigerant pipes to the engine compressor unit.

Areas:

- engine compressor unit with exhaust gas heat exchanger and separate sound absorbing hood for upward folding during servicing (upper part, Fig. 1)
- capacitor with refrigeration parts and safety devices (lower part, Fig. 1). This section is enveloped by simple covering sheets, because it is acoustically separated from the upper sound generating part.

  The refrigerant compressor, in this paper just shown as a schematic outline, is a 2-cylinder piston compressor with suction valves under pipe control.

  Our latest research results have shown that it is not absolutely necessary to have the engine compressor equipment set up in a separate location away from all the other refrigeration parts.

2.2 Basics of the heating control both for monovalent and bivalent operation

The developed microprocessor control equipment of the heat pumps is generally designed for monovalent and bivalent-alternative operation to ensure that the system can be easily integrated into existing heating networks. The overall control of the heating system is designed as a weather variable return flow temperature control, meeting all requirements of the modern energy saving control technology:

- sliding efficiency adaptation (modification)
- day and night drops to various programmes
- boiler priority circuit with efficiency adaptation
- vacation and anti-freeze mode
- malfunction and trouble diagnosis
- automatic defrosting function
- swichting to boiler/burner operation with 2- and 3-point-behaviour

The concept of a smooth efficiency adaptation by speed control and switching on and off the compressor cylinders is shown in Fig. 2.

The specific engine, compressor, and heat pump data obtained from extensive test series were used to simulate the heating power behaviour in a speed range of 1 200 min^{-1} to 3 000 min^{-1}, employing a complex simulation programme developed in our company. A monovalent heat requirement line (20 kW) for a non-limited heating operation is plotted. The heating power curves of the I.C. engine driven fuel oil heat pump are plotted for the lowest speed of 1 200 min^{-1} and the speed limit for bivalent operation of 2 400 min^{-1} each for a compressor swept volume of 200 cm^3 and 400 cm^3 and the maximum permissible speed of 3 000 min^{-1} for 400 cm^3 (peak demand on a monovalent basis for temperatures lower than $-10^{\circ}C$).

In this example we have purposely chosen 2 x 200 cm^3 for the compressor swept volume, to facilitate the explanation of the efficiency adaptation by the swept volume combination of 250 cm^3 and 150 cm^3 used in the field test.

The hatched section shows prohibited operation modes for an engine load of p_{me} longer than 5 bar (with regard to the service life, the mean pressure of the fuel oil engine was limited to 5 bar; if a gas engine is used, somewhat higher mean pressure values occur as the maximum gas amount cannot be set so exactly). This range can be shifted to higher values by optimization of engine and compressor (actual efficiency and output). The drawn in load lines of the heat pump (dot-and-dash-line = monovalent, continuous thick line = bivalent) distinctly show the advantage of a 2-cylinder compressor version: good efficiency adaptation to the required heat profile of buildings needing up to 40 kW without engine overload.

For the bivalent alternative operation an application is possible up to 0°C when limiting the speed to 2 400 min^{-1}, thus covering approximately 80 % of the annual heating power for full load heating requirements. For a limited heating operation - the actual heating requirements are lower than the calculated heating requirements of the building - the field of application is automatically enlarged up to outside temperatures that are obtained by the intersection of the limited heating requirement curve and the heat pump characteristic for 2 400 min^{-1} / 400 cm^3 (sliding bivalence temperature).

3. TEST PLANTS

A gas heat pump (air/water) and a fuel oil heat pump (air/water) were installed in the test field. The heat source units are integrated in separate closed air circulations to simulate all the climatic conditions occurring during the year (air temperature and humidity). Air temperatures between -20°C and +35°C can be achieved.

Complex measuring facilities were installed for monitoring, control and data acquisition. All measured values important to the heating power of the heat pump and for the distribution of the individual outputs of capacitor, cylinder, head cooling, engine and exhaust gas are collected by computer-controlled scanners.

Based on considerations of economic factors and installation methods, air/water heat pumps have proved to be preferred by the user. Thus we have concentrated our development work on the development of heat pumps using air as a heat source. The following passages will therefore be devoted to this system.

Fig. 3 shows the measured heating power (as a sum of capacitor output and waste heat output of the engine) over the outside temperature.

For an almost identical heating power of the heat pump procedure the deviations in the heating effect are due to the different outputs of the fuel oil engine (max. 36 %) and the gas engine max. 29 %). With the same mechanical driving power to the compressor, relatively more waste heat is available from the gas engine than from the fuel oil engine. Therefore the gas heat pump can achieve the same heating effect at lower speeds.

The characteristic of the gas heat pump was tested throughout with a compressor swept volume of 400 cm^3. The output of the fuel oil engine was limited on purpose, in order to remain within the favourable range for consumption and emission characteristics. For outside temperatures of 5°C the compressor swept volume must therefore be reduced to 250 cm^3, as the engine output is not sufficient for 400 cm^3. The fuel oil heat pump output is for -15° about 1.5 kW lower than the gas heat pump output.

Improvements in compressor and heat pump operation are possible to increase the heating power by 1.5 to 2 kW and to improve the PER. In Fig. 4 the PER's measured for both heat pump drives are plotted. The fuel oil unit values lie above the PER's of the gas system because of the better engine output.

On the basis of the PER's measured, annual PER's of up to 140 % can already now be obtained, taking into account all losses.

Based on the experience gained with the test plants, a different distribution of the compressor swept volume was adopted, i. e. 2 x 200 cm^3 (for the systems described here: 250 cm^3 / 150 cm^3).

This new concept (Fig. 2) is currently submitted to examinations on test stands and to a large-scale field test (EEC plan, contract No. EE-011/81).

4. FIELD TEST SYSTEMS

In the heating period 80/81 10 field test systems were installed in houses near the research station (7 fuel oil heat pumps, 3 gas heat pumps). One fuel oil and one gas heat pump each are continuously run to supply part of the heating energy to the research station. Particularly these two units shall prove the reliablity of the components used, as they are permanently in operation year round, taking into account that the heat demand of the research station is high compared to the available heat from the heat pumps.

Fig. 5 shows a survey of a wiring diagram for the integration of an I.C. engine driven heat pump into an existing heating system. In addition to the hydraulic integration, the total integration of the F&S control and the scanning is shown. The control of the conventional system is ensured by the heat pump control. In case of malfunction of the field test heat pump plants the existing control is automatically switched in.

This basic wiring diagram - here for burner/boiler operation without flow through the heat pump - was incorporated into the projection of the total system. Similar plans are also available for the electrical integration.

Prior to the integration and installation into the buildings all field test units were carefully tested for their function and performance on a field test stand particularly set up for this purpose.

No problems were encountered during the transport to the buildings, but the periods required for installation were 14 days on an average, because the heating builders were unexperienced with this kind of work, and the full range of measuring equipment involved a lot of installation work.

The results of the operation from the date of commissioning to December 31, 1982 are summarized in table 1 for all field test heat pumps. These are based on an evaluation of the detailed read-out records of the field test members via a desk computer with interactive mode.

Strikingly high is the number of operating hours of the plants WH 7 and WG 1. These were run to cover part of the main load of the research station and switched off at regular intervals solely for the purpose to determine the characteristic engine compressor value on engine test stands.

Despite high return flow temperatures the PER's are high compared to other plants and due mainly to low standstill losses during the long periods of permanent operation.

Run under equal conditions as WH 7, the gas heat pump WG 1 has a lower PER. This clearly shows the influence of the lower engine efficiency of the gas engine over the diesel engine.

Also comparably high are the PER's of the plants WH 1 and WH 6, resulting for the first plant from the connected swimming pool heating, and for the second one from the floor heating (low return flow temperatures).

The low values of the WG 2 can be explained partly by high cycle losses (small heat demands). In addition, the poor effective engine and compressor efficiencies defined from an analysis of the engine compressor data for this plant as well as for the WH 3, WH 4, and WH 5 plants must be considered. Generally we can say, that the range of variation of the characteristics of these first hand-made prototypes is very large.

As could be expected, it was noticed that the cycle frequency increases as the heat demand decreases. By the fine-tuned control and a better optimized efficiency adaptation in the future the switch-in frequencies can be maintained within a range of max. 4 000 a year. This is distinctly below the design value of 7 000/a.

The operating hours can be as high as 3 000 a year, particularly for important bivalent plants (30 to 50 kW), and even higher for an extreme additional swimming pool operation, this provides of course for increased absolute savings in primary energy.

Both experience and theoretical considerations have shown that particularly the deviations of the heating network temperatures towards higher values have their great influence on the annual PER, expecially for old buildings. From this we can conclude that allow temperature heating system using a heat pump is needed, but on no account should the forward flow temperature on the coldest day exceed 70°C. However, this is normally ensured by the largely dimensioned heating networks anyway.

Further factors that influence the PER are losses due to heat source defrosting, pressure losses in the refrigerant pipes and the combined sanitary water preparation, because high constant return flow temperatures of about 55°C have to be maintained.

To sum up, the annual PER's measured are between 1.15 and 1.35. Last but not least this is also due to largely improvable compressor and engine characteristics (efficiencies, supply rate).

To abtain information on the malfunction and trouble frequency of all plant and single parts used in the complete heat pump assembly, daily reports for each plant, broken down into main groups and typical components were evaluated.

The basic assemblies, such as engine, compressor and refrigeration circulation do not present any problems, but a bunch of difficulties has come up in the periphery of the total heat pump system, resulting mainly from the severe requirements for in-house operation, similar to conventional burner systems:

- fuel supply of the fuel heat pumps, preparation of the gas mixture of gas heat pumps (this should be the basis for completely new development work)
- waste gas
- noise absorption and engine suspension
- constructive layout of the evaporator
- tightness of fixture
- electric cabling

The results obtained and the experience gained in the operation of this research and development programme have brought about a constructive revision, especially of the heat pump main parts: engine with mixture preparation, compressor, heat source unit and microprocessor control.

We have consistently employed this know-how to make over 40 largely improved I.C. engine driven heat pumps that have been tested since 1982 according to an enlarged research and development programme. This research work is also financially backed by the EEC.

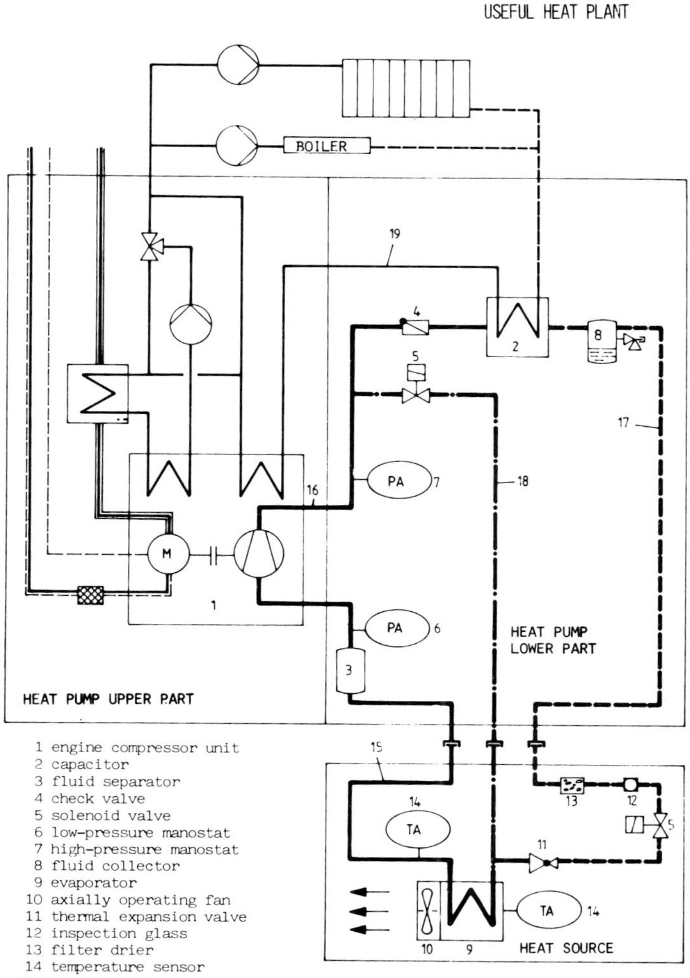

1 engine compressor unit
2 capacitor
3 fluid separator
4 check valve
5 solenoid valve
6 low-pressure manostat
7 high-pressure manostat
8 fluid collector
9 evaporator
10 axially operating fan
11 thermal expansion valve
12 inspection glass
13 filter drier
14 temperature sensor
15 suction pipe
16 pressure pipe
17 fluid pipe
18 deforster pipe
19 heating pipe

Fig. 1 - Schematic diagramm of an I.C. engine driven heat pump

PLANTS	WH 1*	WH 2	WH 3	WH 4	WH 5	WH 6	WH 7	WG 1**	WG 2	WG 3
heat demand in kW	55	25	15	30	30	25	basic load research station		12	25
heating network designed for °C	90/70	90/70	70/50	90/70	90/70	50/30	mean return flow temp. about 50°C		90/70	90/70
commissioning in 1981	03.03.	19.02.	12.02.	26.01.	19.01.	04.12.	15.03.	01.06.	05.09.	20.10.
operating hours	4743	4430	2880	4735	3776	4228	7170	7015	860	3640
cycle frequency	2649	7616	10410	15117	8300	5244	1273	892	1303	6492
running time per switch-in period	1,79	0,6	0,3	0,3	0,5	0,8	5,6	7,9	0,66	0,6
amount of heat generated in kWh	55100	47240	26680	44700	43570	44944	78550	108780	9192	39825
fuel consumption in l (m^3)	4173	3865	2255	3965	3822	3345	6137	8950	858	3538
PER incl. losses (defrosting, sanitary water, etc.)	1,32	1,22	1,18	1,13	1,14	1,34	1,28	1,18	1,04	1,17

* WH = fuel oil heat pump ** WG = gas heat pump

Table 1: Results obtained from commissioning to December 31, 1982 of the 10 field test plants

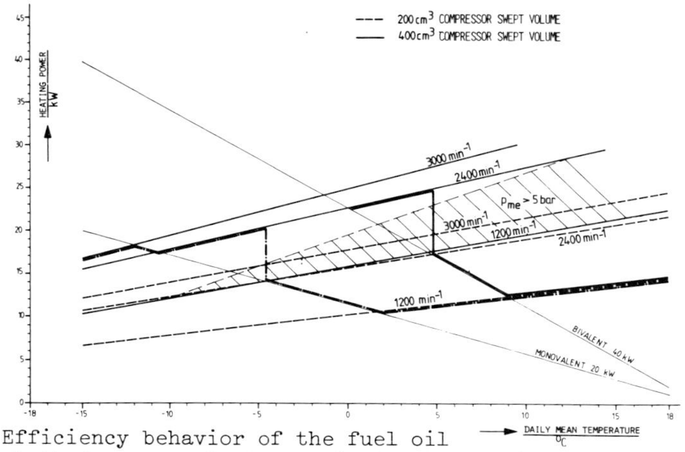

Fig. 2 - Efficiency behavior of the fuel oil heat pump both for monovalent and bivalent operation

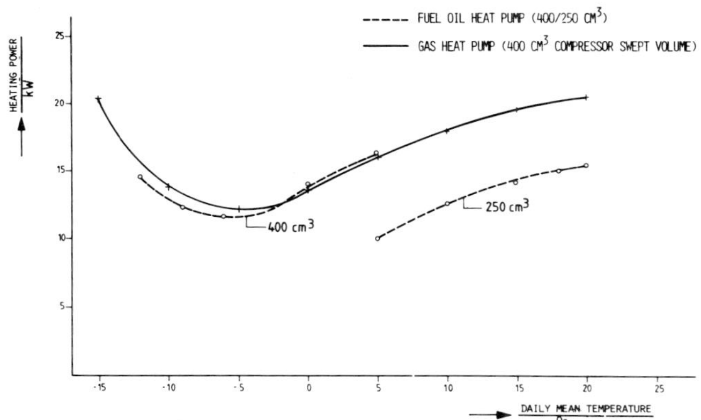

Fig. 3 - Measured heating power of the heat pumps with 400 cm^3 (250/150) compressor

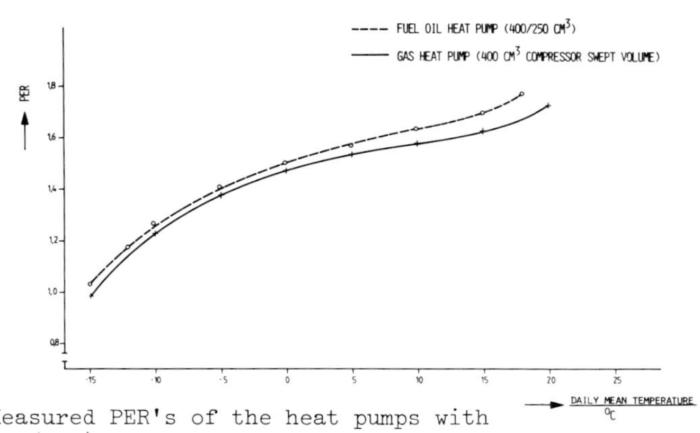

Fig. 4 - Measured PER's of the heat pumps with 400 cm^3 (250/150) compressor

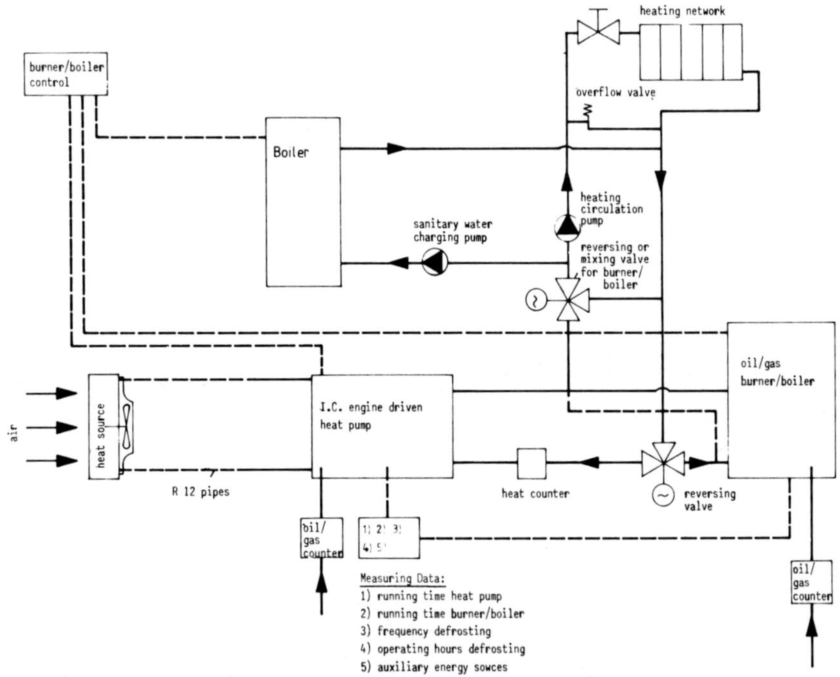

Fig. 5 - Integration plan for the field test units

REGULATION AND CONTROL OF HEATING SYSTEM USING INTERNAL COMBUSTION ENGINE HEAT PUMP

Y. MATHIEU, J.M. LEGRE, A. SARIGNAC
Renault Techniques Nouvelles Appliquées - Boulogne-Billancourt
Société Anonyme des Usines CHAUSSON - Asnières Cedex

SUMMARY

The study of a collective heating system by a heat pump with internal combustion engine was carried out during this contract.

A heat pump with internal combustion engine of automobile origin was constructed (thermal power: 190 kW) and the regulation of this apparatus by a microprocessor card was studied.

The study and our achievements have shown that the combustion of gas in Renault engines did not pose any technical problems and that the engines of automobile origin showed a sufficient reliability and life time for such an application.

The weak market for collective heat pumps in France does not, however, permit the industrialization of these pumps where components have already been developed.

1. INTERNAL COMBUSTION ENGINE FOR HEAT PUMPS

During the development project of heat pumps driven by internal combustion engine powered by natural gas, the excellent performances of the Renault engine, type A5M 245, equipped to take that type of combustible, have been confirmed.

The engine, which has a possible output of 35 kW to 3000 rpm at it's disposition, is fitted with an IMPCO carburator, a Barber Colman electronic speed regulator and totally electronic ignition.

The make-up of the lubricant is guaranteed by a simple and reliable automatic system.

A carefully studied synthetic oil for gas combustion of the ignition driven engines allows the spacing in between oil changes of several thousands of hours, eliminating certain maintenance constraints.

The mechanical energy developped for the engine shaft is used to drive a heat pump composed of an open type multicylinder compressor, a water/steam condensor and a water/steam evaporator, the freezing fluid used being the "Fréon 22".

Most of the heat from the engine is recovered by two thermal exchangers , water/water and water/fumes respectively.

As a result of this, the total thermal power provided by the pump is within the region of 150 kW, which gives the machine a performance coefficient of more than 1.8.

The heat pump comes in a compact unit, incorporated in a sound-proof covering mounted on a double chassis with elastic insulation.

2. PERFORMANCE OF THE HEAT PUMP

Tests carried out on the prototype give us the performances of the machine both at full capacity and at a reduced rate.

2.1 Parameters during operation

Calculated performances at full capacity:

- Overall power: 152 kW (available at the hot water source)
- Circuit hot water source: Temp. return : 35 °C
  Temp. departure : 45 °C
- Power required at the cold source: 52 kW
- Circuit cold water source: Temp. return : 8 °C
  Temp. departure: 1 °C
- Gas consumption: 83 kW PCI
- As a result the expected performance coefficient is 1.80

Performance at reduced rate:
- Overall power: 85.5 kW (available at the hot source)
- Circuit hot water source: Temp. return : 25 °C
  Temp. departure : 30 °C
- Power required at the cold source: 44 kW
- Circuit cold water source: Temp. return : 8 °C
  Temp. departure: 4 °C
- Gas consumption: 45.5 kW PCI
- As a result the expected coefficient is 1.88

3. REGULATION

Taking into account the power that it can produce (150 kW), the heat pump is intended for the heating of blocks of flats and for office blocks.

As a result of this, regulation is made at the return temperature of water at the condenser. The point at which the return temperature is fixed is the responsibility of the outside temperature and results from a code integrated into the regulation system.

All corrections are made by the internal combustion engine speed, in other words, by the intervention of the electronic regulator.

All the command, control and regulating functions are taken over by an electronic unit built around a micro-chip 6801 microprocessor.

The electronic card guarantees the acquisition of the data: temperature, speed, consumption; of command, fixed speed, safety permanent control of the maintenance of the probes and of the working order of the internal combustion engine (e.g. the exhaust temperature). From the card, an auto-control is also used in order to guarantee the correct operation of the programme.

In the regulation process, four different phases can be seen:
- the machine working at a reduced rate,
- the initiation phase,
- the regulation itself,
- the functioning at full capacity.

4. ELECTRONICS AND PROGRAMMES

The regulation unit is a microprogrammed system made up of several sub-units.

Everyone of the modules is independant from the others both from the electronics and programming points of view. This separation makes the adjustment and the maintenance of each unit easier.

What is more, this modular technique facilitates the adaptation of the system in case of development or change of the censors, the actuators or of the way in which they are run.

In addition to the microprocessor, the central unit is composed of an analogue to digital converter digitizer, multichannels for the acquisition of temperatures and a digital to analogue converter digitizer, which guarantees the speed adjustment in the form of a continual and variable tension.

The main system modules can be broken down into the following:

- power supply module
- control unit module
- speed motor module
- temperature modules (* hot water, outer
  (* motor exhaust
- control and safety module.

In order to adapt the reaction times of the actual slow phenomena and of the electronic reactions, the head of the control is made up of a programming pack which enables the system to adapt to the variation time of the temperatures.

It is this which guarantees an automatic adjustment to any change in the heating system during installation.

5. ON-SITE INSTALLATION AND TESTS

After the endurance tests on the heat pump, it is hoped that it will be installed on an actual site which has already been partly prepared for this very purpose.

This site, the Marcel BEC stadium, owned by "La Régie Nationale des Usines RENAULT", is a sports complex comprising:

- football and rugby pitches,
- tennis courts,
- indoor tennis courts and multi-sports halls,
- important changing rooms and toilet facilities.

The thermal production of the stadium is guaranteed by 870 kW boilers which provides low temperature heating and hot sanitary water.

The boiler room already consists of:

- a conventional gas boiler (700 kW)
- a condensation gas boiler (170 kW).

The annual energy consumption of the stadium is 1.1 10^6 kWh. The heat pump will be incorporated into this boiler room. The cold source is made up of a censor, 6000 metres long, burned beneath the surface of the stadium main pitch. The censor (a 3.2 mm diameter plastic tube), which is placed at the same depth as the drains will enable the extraction from the ground of about 80 kW.

In addition to the machine tests themselves, this operation should enable a saving in the order of 2 to 3 10^5 kWh a year, to be achieved.

CONCLUSION

This present document is the synthesis report marking the anticipated completion of the EEA-4-062 F contract.

GENERAL CONCLUSION

This contract, negotiated for the first time in 1979 showed, on one hand the willingness of Renault Techniques Nouvelles Appliquées (R.T.N.A.) and of the "Société des Usines Chausson" (S.U.C.) to develop their activity in the field of engine powered heat pumps, and on the other, the European Community Commission's wish to financially follow the development of technically evolved products.

Important studies have been carried out and several prototypes have been produced.

Research has enabled the following things to be carried out:

- a prototype heat pump with variable gears (theoretical capacity 150 kW),
- the adaptation of natural gas combustion and the modification of an internal combustion engine, adapted from a motorcar,
- the elaboration of a heat pump working system in a heating system,
- the study and design of a regulation system and of its interfaces by micro-processor with the heat pump and the environment,
- tests on the pump prototype,
- study of the setting up of the prototype on an actual site.

The industrial application of the micro-processor electronic card could not be carried out in the study proposed by RTNA.

However, RTNA and SUC, the contractors, have been forced to request the interruption of the contract in April 1983.

The French market for medium and high powered heat pumps has turned out to be a great deal weaker than it appeared 5 years ago. With openings for only several dozen heat pumps a year, the industrial profitability disappeared and from an economic point of view, production could not be maintained.

With the main objective for this type of pump being heating of blocks of flats, the tertiary sector and industry, the recession in this field as well as the depreciation time demanded by industrial firms, have reduced the market considerably.

Technical aspects

The development has brought out the following points:

- the heat engines of automobile origin have shown reliability for this type of machine. Renault engines using natural gas have been shown to have a lifespan of more than 10 000 hours.
- Noise from the heat engines could be easily kept within acceptable limits with use in boiler rooms.
- The problem of optimization of the dimensioning of diphasis heat exchangers remains only partially resolved.

Technical-economic aspects

Costs: the costs given below are budget costs towards which one must strive.

The "factory leaving" price of a heat pump with internal combustion engine within the 100 or 200 kW range is in the order of 100 000 to 130 000 French francs including VAT (per thermic kW).

If the heat pump with internal combustion engine had been correctly developped in France, the manufacturing cost per thermic kW installed would have been about the same as for an electric heat pump or a thermic engine heat pump.

Energy saving

The main characteristics of the heat pump with internal combustion engine is to obtain a seasonal performance coefficient (PC) on primary energy superior to the unit.

By comparison

- conventional boiler	PC	0.6 to 0.8
- condensation boiler	PC	0.9 to 1.05
- electric heat pump (with 1 kW = 2.5 primary units)	PC	1 to 1.3
- heat pump with internal combustion engine	PC	1.5 to 1.8

So the heat pump with internal combustion engine does allow important primary energy savings.

Used in blocks of flats and added to other boilers, such a machine would enable a saving of 0.35 TEP/year per flat, which, translated in terms of the machine, represents an annual primary energy saving of 25 TEP for a 150 kW thermic heat pump with internal combustion engine (in other words, more than a 25% reduction in consumption).

The depreciation of the investment is very dependant on the absolute energy cost and the variation compared with the costs of different forms of energy in relation to each other.

INTERNAL COMBUSTION ENGINE HEAT PUMP 843-20

1) engine
2) compressor
3) coupling
4) reducer
5) coupling
6) evaporator/air superheater
7) exchanger exhaust gases/water-glycol mixture
8) exchanger water-glycol mixture in engine/heating water
9) water-glycol mixture evaporator
10) heating water condenser
11) tank
12) reducing valve

Freezing fluid = R 22

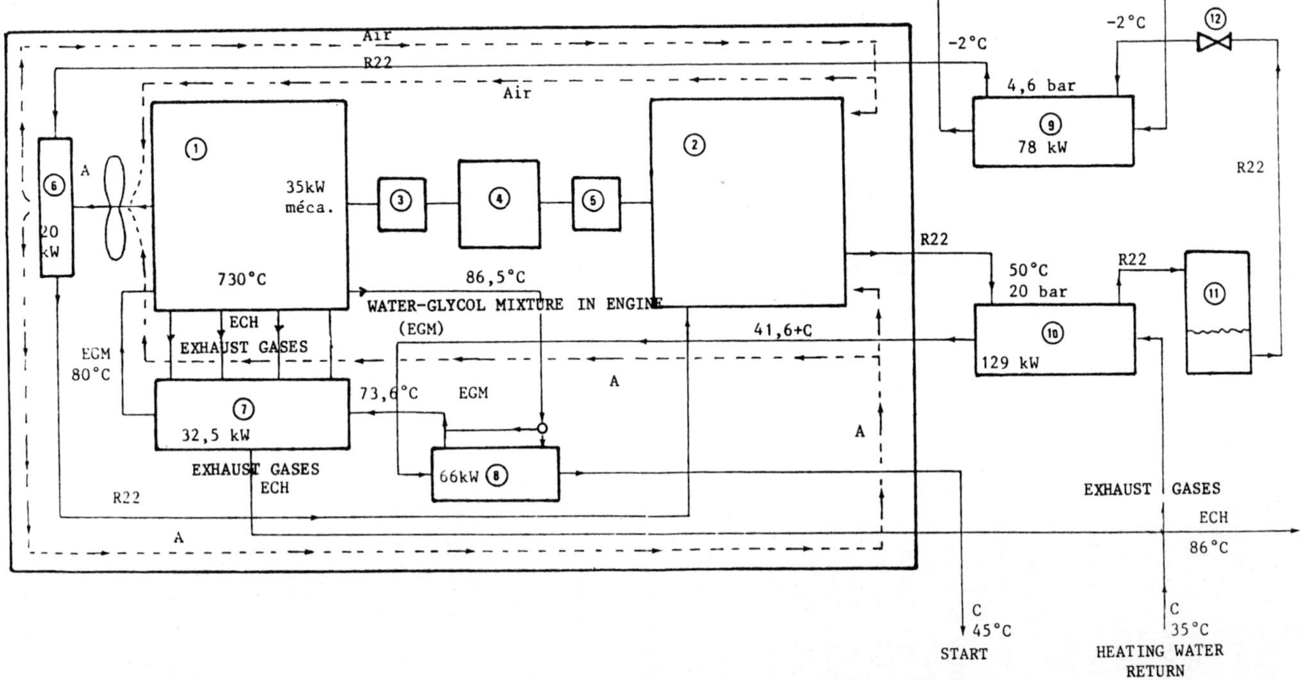

MONOVALENT ELECTRIC AIR/WATER-HEAT PUMP PLANT FOR ROOM HEATING IN THE HOUSES FOR SEVERAL FAMILIES (SOCIAL HOUSE BUILDING) TO SAVE ENERGY HEATING SYSTEM FOR LOW DWELLING BY MEANS OF HEAT PUMP

Dipl.-Ing. H. Klaus
Vereinigte Elektrizitätswerke Westfalen, Aktiengesellschaft, Dortmund

Zusammenfassung

Für 80 Wohnungen in Mehrfamilienhäusern wurde ein energiesparendes Heizsystem installiert bei dem nur geringe bauliche Anforderungen notwendig sind und somit auch in bestehende Gebäude eingebaut werden kann. Monovalente elektrische Luft/Wasser-Wärmepumpen mit einer Gesamtheizleistung von von 450 kW bei ϑ_a = - 12 °C sind in 2 Heizräumen untergebracht, die ehemals für Gasheizkessel vorgesehen waren. Auf den Flächdächern der Gebäude wurden die Verdampfer frei aufgestellt und geräuschdämmend verkleidet. Ein Warmwasserheizsystem mit gängigen Radiatoren übernimmt die Wärmeverteilung bei Heizwassertemperaturen von max. + 50 °C. Obwohl regelungstechnisch noch nicht alle Möglichkeiten zur optimalen Energieeinsparung ausgenutzt sind, war der spezifische Stromverbrauch für die Raumheizung während des Heizjahres 1982/83 mit ca. 75 kWh/m² Wohnfläche schon erfreulich niedrig.

Die dezentrale Warmwasserbereitung mit elektr. Durchlauferhitzern ist bei dem niedrigen Brauchwarmwasserverbrauch der 80 Wohnungsmieter die wirtschaftlich sinnvollste Lösung.

1. Beschreibung des Objektes

1.1. Ziel der Forschung

Nach dem derzeitigen Stand der Technik steht die Erprobung von Wärmepumpenheizungsanlagen zur Raumheizung im Mehrfamilienhaus erst am Anfang einer vielversprechenden Entwicklung. Die geringe Anzahl von Wärmepumpenheizungsanlagen dieser Größenordnung befinden sich in Neubauten, die mit Fußbodenheizungen zur Wärmeverteilung ausgerüstet sind. Demgegenüber besteht insbesondere beim Einbau von Wärmepumpenheizungsanlagen im Altbau die Notwendigkeit, vorhandene Wärmeverteilungssysteme mit Radiatoren weiter zu nutzen. Eine nachträgliche Umrüstung auf Fußbodenheizung ist nur in Sonderfällen möglich.

Aus diesem Grunde ist in diesem Objekt eine elektrische Wärmepumpenheizungsanlage zum Einsatz gekommen, der als Wärmeverteilungssystem eine Radiatorenheizung nachgeschaltet ist. Mit den in einem umfangreichen Meßprogramm zu erarbeitenden Betriebsergebnissen sollten die Möglichkeiten und Einsatzgrenzen einer elektrischen Wärmepumpenheizungsanlage mit Radiatorenheizung ermittelt werden. Mit den hieraus gewonnenen Erkenntnissen wird die breite Anwendung der Wärmepumpentechnik im Mehrfamilienhausbestand möglich.

Es ist zu beachten, daß die Entwicklung leistungsfähiger Wärmepumpenaggregate nicht Hauptpunkt des Vorhabens war. Diese Entwicklung kann teilweise bereits als abgeschlossen gelten. Nicht abgeschlossen sondern vielmehr am Anfang steht z.Z. noch der Kenntnisstand über die Wirkungsweise der gesamten Wärmepumpenheizungsanlage, bestehend aus Wärmepumpenanlage,

Wärmepumpenaggregat und Wärmeverteilungsanlage. Regelungstechnische Besonderheiten, die für Wärmepumpen sinnvolle Höhe des Temperaturniveaus des Heizwassers, Geräuschemission, Abtauoptimierung u.a. sind in ihren praktischen Auswirkungen auf die Betriebsergebnisse einer Wärmepumpenheizungsanlage bisher nicht oder nur ungenügend quantifiziert. Dieses Projekt soll in diesem Bereich Erkenntnisse liefern.

Die monovalente Betriebsweise der elektrischen Wärmepumpen mit der Wärmequelle Luft ist dabei insbesondere im Hinblick auf die Begrenztheit der verfügbaren Mengen und der erwarteten Preissteigerungen für Öl und Gas von besonderem Interesse.

Weiterhin sollte geklärt werden, wie sich in einem derartigen Objekt die Brauchwarmwasserversorgung energetisch dargestellt und wie sich die hier eingesetzte dezentrale elektrische Warmwasserversorgung (Durchlauferhitzer) im Vergleich zu einer zentralen Versorgung mit Wärmeerzeugung über monovalente elektrische Luft/Wasser Wärmepumpen in einem abgeschlossenen Forschungsobjekt mit 56 Wohnungen (EG-FB-310-77-9 EED) verhält.

1.2. Gebäudedaten

Die Durchführung des Projektes erfolgte in den Mehrfamilienwohnhausanlagen (Bild 1 und Tafel 1).

- Wohnblock I (Marsbruchstraße) in Dortmund-Aplerbeck mit 36 Wohnungseinheiten und einer Wohnfläche von rd. 2.620 m². Der Wärmebedarf des Gebäudes beträgt nach DIN 4701 rd. 190 kW ≙ 72,5 W/m². Bedingt durch die zeitweise Abtauung der Verdampferflächen steht die Heizleistung der Wärmepumpen nicht ständig voll zur Verfügung, aus diesem Grund ist die installierte Heizleistung 200 kW (2 x 50 kW) bei ϑ_a = - 12 °C.

- Wohnblock II (Eleonorenstraße) in Dortmund-Aplerbeck mit 44 Wohnungseinheiten und einer Wohnfläche von rd. 3.530 m². Der Wärmebedarf der Gebäude beträgt nach DIN 4701 rd. 240 kW ≙ 68,8 W/m². Die installierte Heizleistung beträgt 250 kW (5 x 50 kW) bei ϑ_a = - 12 °C.

1.3. Planung

Bei der Planung der Wärmepumpenheizungsanlagen waren hauptsächlich folgende Kriterien zu beachten:

- Die hohen spezifischen Investitionsausgaben für Wärmepumpen zwingen dazu, die installierte Heizleistung möglichst exakt dem Bedarf anzupassen.

- Die Wärmepumpe erreicht dann eine hohe Jahresarbeitszahl (cop), wenn die Differenz zwischen Heizwassertemperatur und Wärmequellentemperatur klein ist. Dies macht bei der durch die Außenluft vorgegebenen Verdampfungstemperatur der Wärmepumpe eine Optimierung der Größe der Heizflächen und eine dem jeweiligen Heizwärmebedarf angepaßte gleitende, außentemperaturabhängige Heizwasserregelung erforderlich. Dabei waren Kompromisse zwischen Raumbedarf für die Radiatoren, Investitionsaufwand für Wärmeverteilungssysteme und Wärmepumpenaggregat und optimaler Jahresarbeitszahl zu schließen.

- Aus elektrizitätswirtschaftlichen Gründen, d.h. Ausblendung der elektrischen Wärmepumpe zu Höchstlastzeiten der Stromversorgung und regelungstechnischen Gründen wurde zwischen Wärmepumpen und Wärmeverteilungssystem ein Speicher eingesetzt. Bei der Dimensionierung des Speichers mußten die o.g. Kriterien mit beachtet werden.

- Bei der Installation der Wärmequellenanlage war dafür Sorge zu tragen, daß der energetische Aufwand zur Abtauung der Verdampferflächen hauptsächlich im Bereich zwischen + 8 °C und - 5 °C Außenlufttemperatur minimal wurde, daß Körper- und Luftschallemission durch entsprechende schalltechnische Maßnahmen ausgeschlossen waren und daß die Luftfördermenge den o.g. Kriterien Rechnung trägt, damit für die Wärmepumpe eine möglichst hohe Verdampfungstemperatur erzielt wird.

- Der Kreisprozeß der Wärmepumpe verbindet konventionelle Fehler bei Wärmequellenanlagen, Wärmepumpenaggregat und Wärmeverteilungssystem. Mangelhafte Planungskenntnisse stellen derzeit mit die wichtigsten Hinderungsgründe für die breite Anwendung von Wärmepumpen insbesondere im Mehrfamilienhausbereich dar. Aus diesem Grunde sollen die Ausführung ein auf andere Fälle übertragbares Planungskonzept bestätigen.

1.4. Installation der Wärmepumpenpilotanlagen

Die Wärmepumpenpilotanlagen wurden gesplittet aufgestellt. Die Hauptaggregate mit Kompressoren und Kondensatoren sind in die Heizräme der ehemals geplanten Gasfeuerungsanlagen eingebaut, während die Verdampferteile mit Schalldämmkulissen auf den Flachdächern über dem letzten Wohngeschoss installiert sind. Wesentlicher Wert wurde bei diesen Aufstellungsvariante auf Vermeidung von Körper- und Luftschallübertragungen gelegt (Bild 2).

Die Auslegung der Wärmetauscherflächen und zugehörigen Anlagenteile ist mit dem Ziel erfolgt, bei der Nennheizleistung im Auslegungszustand eine möglichst hohe Leistungszahl (ε = 2) der Wärmepumpe zu erreichen. Damit ergeben sich günstige Verhältnisse bzgl. der aufgenommenen elektrischen Kompressorleistung für die Wärmepumpenbenutzer und für die Belastungsstruktur des Versorgungsnetzes.

Die Warmwasserverteilungsanlage ist in konventioneller 2-Rohr-Bauform mit örtlichen Heizflächen als Flachradiatoren, jedoch ausgelegt auf eine mittlere Heizwassertemperatur von + 45 °C, installiert und ausgerüstet mit thermostatischen Heizkörperventilen.

Für die Brauchwarmwasserversorgung in Bad und Küche wurde je Wohnungseinheit ein Durchlauferhitzer eingebaut, der über eine manuell zu betätigende Leistungsschaltung wahlweise mit einer Leistung von 14 bzw. 21 kW betrieben werden kann.

2. Finanzierung des Forschungsvorhabens

Die Anlagenkosten betrugen insgesamt DM 757.000,--. Von diesen Kosten haben die Wohnungsbaugesellschaften einen Anteil getragen, der an den Kosten für eine konventionelle Radiatorenheizung mit Gasheizkessel gemessen wurde DM 305.000,--
Die verbleibenden Anlagenkosten von DM 452.000,--
wurden laut Vertrag zu 50 % von der Kommission der Europäischen Gemeinschaften übernommen, die verbleibenden 50 % von VEW.

3. Betriebsergebnisse
3.1. Betriebserfahrungen

Erst nach Einbau langsam laufender Ventilatoren in die Verdampferteile auf den Flachdächern konnte die Geräuschsituation auf ein heute befriedigendes Maß gebracht werden. Es hat sich insgesamt herausgestellt, daß frei aufgestellte und der Witterung vollkommen ausgesetzte Verdampfer im monovalenten Wärmepumpenbetrieb sehr detailliert geplant und auch sorgfältig hergestellt werden müssen. Kleine Fehlerursachen führten anfangs mehrfach zum Ausfall der Wärmepumpen. Das sind auch die Gründe dafür, daß erst in der Heizperiode 1982/83 ein Heizbetrieb eingehalten werden konnte, bei dem die Wärmepumpen (von wenigen Tagen abgesehen) durchlaufend die Wärmeerzeugung übernommen haben. Die vergangene Heizperiode kann trotzdem nicht als repräsentativ angesehen werden, da die Außenlufttemperaturen durchschnittlich erheblich über denen eines Normaljahres lagen (Bild 3). Die tiefste Tagesmitteltemperatur lag nur bei - 6 °C.

Insgesamt konnten aus dem Heizbetrieb folgende grundsätzliche Erkenntnisse gewonnen werden:

. Die Heizwassertemperaturen sind in den Übergangszeiten so niedrig, daß die Nutzer ein Gefühl der Unterbeheizung ihrer Wohnräume äußern. Es kann nur eine gezielte und intensive Aufklärungsarbeit Abhilfe schaffen.

. Die Heizkörper-Thermostatventile werden in ihrer Funktion vom Nutzer häufig noch nicht richtig bewertet. Durch Wärmestau (unzweckmäßige Möblierung, überhängende Gardinen, Beeinflussung durch Fremdwärmequellen) treten oft fehlerhafte Regelungen auf, die dann zu einer mangelhaften Beheizung der Räume führen.

. Durch die geringe Temperaturdifferenz des Heizwassers zwischen minimaler Vorlauftemperatur und maximaler Vorlauftemperatur wird eine sehr exakte Regelung der Heizwasservorlauftemperatur erforderlich. Störgrößen wie Wind und hohe Luftfeuchtigkeit wirken sich stark negativ aus. Da diese Störgrößen hauptsächlich bei Außenlufttemperaturen über ± = 0 °C auftreten, kann die Veränderung der Vorlauftemperatur nicht mehr linear oder quasi-linear erfolgen. Da die zunächst eingesetzte außentemperaturabhängige Regelanlage den Ansprüchen nicht genügte, wurden während der Heizperiode 1983/84 neuartige Regelanlagen eingesetzt. Mit diesen ist es möglich, während der Zeiten mit relativ hohen Außentemperaturen eine Regelcharakteristik zu fahren, die von der Regelcharakteristik bei niedrigen Außenlufttemperaturen nach Bedarf abweichen kann. Die exakte Einregulierung der neuen Anlage war wegen der milden Witterung bisher nicht möglich.

. Der Betrieb einer Niedertemperaturheizung ist jedoch, wenn von den vorgenannten Schwierigkeiten abgesehen wird, auch im Mehrfamilienhausbereich und mit handelsüblichen Heizkörpern möglich. Trotz der regelungstechnisch bedingten leichten Streuung der Heizwasservorlauftemperaturen konnten grundsätzlich die vorausgeplanten niedrigen Heizwassertemperaturen eingehalten werden (Bild 4), sie können sich nach Einbau der neuen Regelanlage evtl. noch tiefer einstellen.

. Unerwartet schwierig verlief der Versuch, die Heizwassertemperaturen im Wärmepumpenheizkreislauf außentemperaturabhängig gleitend zu betreiben. Ziel war es, die Temperatur wie im Wärmeverteilungsnetz auch im Wärmeerzeugungsteil ähnlich streng an der Wärmebedarfslinie zu führen und dabei eine max. Abweichung von 5 K nach oben zuzulassen. Trotz aller bisher durchgeführten Veränderungen an der zuständigen Regelstrecke ist dieses für einen wirtschaftlichen Wärmepumpenbetrieb wichtige Ziel nicht erreicht worden (Bild 5). Gerade zu Zeiten mit hoher Außenlufttemperatur ist die Heizwassertemperatur im Wärmepumpenkreislauf bedeutend zu hoch und streut sehr extrem. Gründe dafür sind:

  - Leistungsunterschiede bei sich ändernden Bedingungen auf der Kondensator- und Verdampferseite allgemein.

- Sprunghafte und kurzfristige Veränderungen der Wettereinflüsse wirken sich direkt auf die Verdampfungstemperatur aus, da die Verdampfer frei auf den Gebäudedächern aufgestellt sind.
- Abtauvorgänge führen zu kurzfristigen Veränderungen im Kältemittelkreislauf.
- Alle bisher eingesetzten Regelsysteme waren nicht in der Lage, alle Veränderungen innerhalb eines zulässigen Zeitraumes entsprechend zu verarbeiten.

Grundlegende Änderungen am Regelsystem sind vorbereitet und lassen auf zufriedenstellende Ergebnisse in der kommenden Heizperiode hoffen.

3.2. Betriebsdaten

3.2.1. Betriebsdaten aus dem Heizbetrieb

Der Energieverbrauch für die Raumheizung war erfreulich niedrig. Nach Auswertung der ersten Meßergebnisse ist mit einem spezifischen Stromverbrauch von 70-80 kWh /m² Wohnfläche in einem Heizjahr zu rechnen. Gleichzeitig kann festgestellt werden, daß die Arbeitszahl der Wärmepumpen nicht ganz das erwartete Niveau erreicht hat, sie beläuft sich auf ß = 2,2. Auch die an anderer Stelle erzielte Leistungszahl von $\varepsilon = 2$ bei Außenlufttemperaturen von $\vartheta_a = -12$ °C wird hier nicht erreicht werden können. Grund dafür werden die langen Verbindungsleitungen zwischen Kompressor und Verdampfer sein und auch der Aufstellungsort der Verdampfer auf dem Flachdach, wo sie Wind, Schnee und anderen ungünstigen Witterungseinflüssen voll ausgesetzt sind. Letztgenannte Tatsache drückt sich auch aus in einem relativ hohen Anteil Abtauarbeit (Bild 6).

Weiterhin wirkt sich die zu hohe Heizwassertemperatur (siehe 3.1) stark mindernd auf die Arbeitszahl aus. Dieser Mangel soll behoben werden, so daß zukünftig doch noch mit einer Arbeitszahl von ß = 2,4 gerechnet werden kann.

Einen guten Überblick über den elektrischen Energieverbrauch der gesamten Wärmepumpenanlage bietet Bild 7 (Anlage Block III). Wird der theoretische Wärmebedarf dieser Gebäudegruppe mit 4.560 kWh/d im Auslegungszustand (4.275 kWh/d bei $\vartheta_a = -10$ °C) dem elektrischen Energieverbrauch mit zu erwartenden ca. 2.200 kWh/d bei $\vartheta_a = -10$ °C gegenübergestellt, so ergibt sich eine Leistungszahl bei $\vartheta_a = 10$ °C von knapp unter 2.
Im Bild 8 wird der theoretische Wärmebedarf durch den tatsächlichen Heizwärmeverbrauch bestätigt.

3.2.2. Betriebsdaten aus dem Brauchwarmwasserbetrieb

In der Tafel 2 sind die über 1 Jahr gemessenen Brauchwarmwasserverbräuche nach Haushaltgröße aufgelistet.
Insgesamt ergibt sich ein durchschnittlicher Verbrauch von 18,64 l/Person und Tag. Dabei liegt der niedrigste vorkommende Verbrauch bei 8,16 l/Pers. . d (1-Personenhaushalt) und der höchste bei 69,22 l/Pers.d (2-Personenhaushalt).
Die Verbräuche sind relativ gering und unter denen in der Literatur angegebenen. Eine wesentliche Rolle scheint die soziale Mieterstruktur zu spielen, da diese nach unserer Kenntnis relativ schwach ist. Andererseits wirkt sich erfahrungsgemäß eine dezentrale Warmwasserbereitung mit direkter Verbrauchsabrechnung gegenüber einer zentralen Versorgungsanlage schon verbrauchssparend aus.

In einem Vergleichsobjekt mit 56 Wohnungen mit einer zentralen Brauchwarmwasserversorgung und Wärmeerzeugung über monovalente elektrische Luft-Wasser-Wärmepumpen lag der gemessene Verbrauch, umgerechnet auf gleiche Kaltwasser- und Austrittstemperaturen, bei durchschnittlich 27,36 l/Person und Tag. Das bedeutet, daß der Verbrauch bei der zentralen Versorgung knapp 50 % größer ist als bei den dezentralen.
Bedingt durch Speicherverluste, ein weites Verteilungs- und Zirkulationsnetz, sowie hohe Kondensationstemperaturen der Wärmepumpen bei der Brauchwarmwassererwärmung war bei der zentralen Versorgungsanlage ein Verhältnis von Nutzenergie zu eingesetzter Energie von rd. 1 : 1 festgestellt worden. Wird bei der dezentralen Versorgung ein Wirkungsgrad von 98 % unterstellt, so ist in diesen Fällen aus energetischen Gründen allein die Versorgung mit Einzelgeräten einer zentralen Versorgung vorzusehen. Außerdem bringt die Einzelversorgung Investitionskostenvorteile. Erst bei überdurchschnittlich hohen Warmwasserverbräuchen könnte sich das Ergebnis umkehren.

4. Schlußbetrachtungen und Erwartungen

Insgesamt hat sich die Anlage gut bewährt, es mußten allerdings einige Anlaufschwierigkeiten behoben werden. Der niedrige Energieverbrauch bestätigt die Richtigkeit des Konzeptes, so daß auch zukünftig derartige Systeme bei der Nachrüstung im Mietwohnungsbestand Berücksichtigung finden können. Allerdings gibt es in der BRD z.Z. keine günstige Finanzierungsmöglichkeit, im mit öffentlichen Mitteln geförderten Wohnungsbau die Mehrinvestitionen aufzufangen. Außerdem muß noch viel Aufklärungsarbeit geleistet werden, um die Mieter mit den niedertemperaturigen Heizsystemen vertraut zu machen.

Eingehendere Daten aus dem Betrieb der Anlagen während der vergangenen Heizsaison stehen uns nach der z.Z. laufenden Auswertung der Bandaufzeichnungen zur Verfügung. Es werden noch Erkenntnisse erwartet, die zu weiteren Verbesserungen der Betriebsergebnisse führen und bei zukünftigen Planungen entsprechende Berücksichtigung finden können.

5. Wirtschaftlichkeitsbetrachtungen

Unter Zugrundelegung eines Energieverbrauchs von 75 kWh/m².a und eines Strompreises von brutto 0,17 DM/kWh ergeben sich Heizenergiekosten von 12,75 DM/m² . a. Dagegen entstehen bei einem durchschnittlichen Heizölverbrauch in der BRD von 30 l Heizöl/m² . a und brutto 0,70 DM/l Heizöl Energiekosten von 21,-- DM/m² . a. Der Differenzbetrag von 8,25 DM/m² . a dürfte ausreichen, die Mehraufwendungen für Wartung, Instandhaltung und Kapitaldienst abzudecken.

SUMMARY

An energy-saving heating system was installed to serve 80 flats. This system requires a minimum of structural alteration and can therefore be installed in existing buildings. Monovalent electric air/water heat pumps with a total heating capacity of 450 kW at ϑ_a = -12°C are housed in two boiler rooms which were formerly used for gas boilers. The evaporators were set up outside on the flat roofs of the buildings and soundproofed. A water heating system using normal radiators is used for heat distribution at a maximum water temperature of + 50°C. Although not every possible way of regulating the system so as to achieve optimum energy saving has been explored, the specific electricity consumption for heating in the heating year 1982/83 was, with approx. 75 kWh/m^2 of living area, very acceptable.

Since the hot water consumption of the 80 tenants is low, decentralized hot water production by means of electric flow heaters is the most economical solution.

TAFEL 1 - GEBAEUDEDATEN "MARSBRUCHSTRASSE"

Benennung	Wohnung St./Gr.	Wo.Fl. m²	Wärmeb. kW	Q_h WP ϑ_a = 12 °C kW
Block I	20/2-Raum 12/3-Raum 4/4-Raum	2.620	190	200
Block III	8/2-Raum 30/3-Raum 6/4-Raum	3.530	240	250
Gesamt	80	6.150	430	450
Spez.	1	76,90	5,38	5,63

TAFEL 2 - BRAUCHWARMWASSERVERBRAEUCHE IN 50 WOHNUNGEN MARSBRUCH-/ ELEONORENSTRASSE - Messzeitraum 1.7.1982 - 30.6.1983

Personen je Haushalt	Anzahl Haushalte	Durchschnittlicher l/m²· d	Verbrauch l/Pers.· d
1	1	0,13	8,16
2	9	0,65	26,43
3	19	0,72	19,62
4	15	0,76	15,90
5	3	0,67	11,55
6	1	0,80	12,75
7	2	1,00	13,64
Summe	50	0,72	18,64

Anmerkung: Die Verbrauchszahlen beziehen sich auf
. Kaltwassertemperatur = + 10 °C
. Austrittstemperatur = + 60 °C

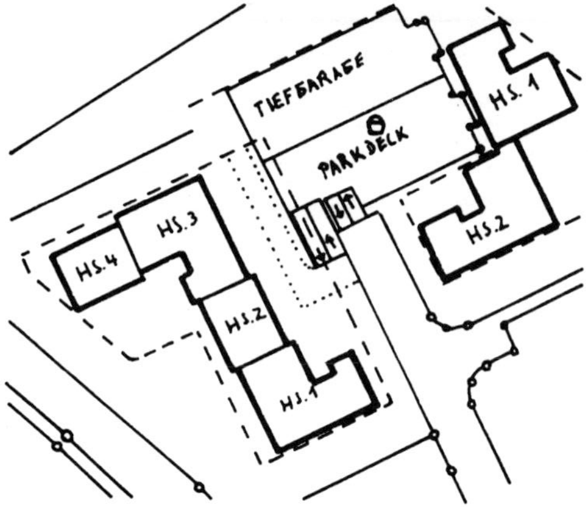

Bild 1

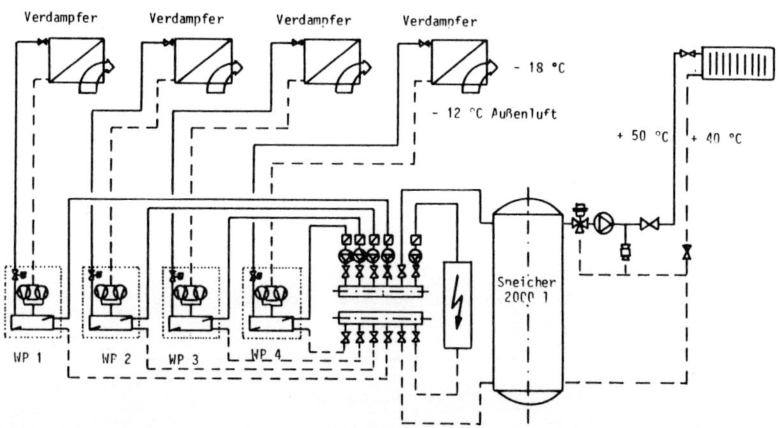

Bild 2
Schematischer Aufbau der Wärmepumpenheizungsanlage Dortm.Marsbruchstrasse mit 4 Luft/Wasser-Wärmepumpen mit einer Heizleistung von je ca 50 kW bei - 12°C

Ganglinie Aussenlufttemperatur 82/83

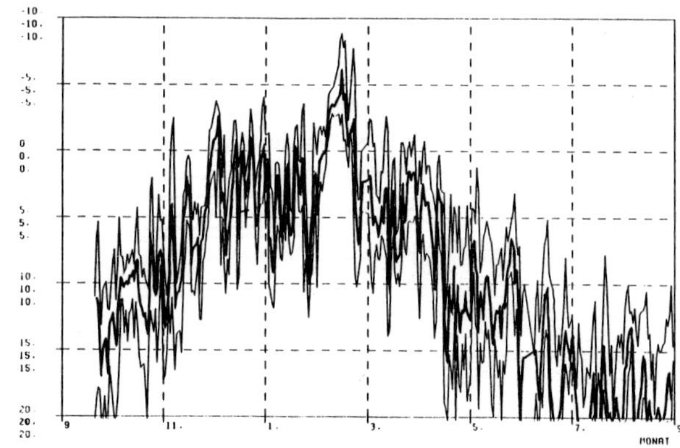

Grad C   Mittlere Aussenlufttemperatur (DIN)
Grad C   Minimum der Aussenlufttemperatur
Grad C   Maximum der Aussenlufttemperatur
- Tageswerte -

Bild 3

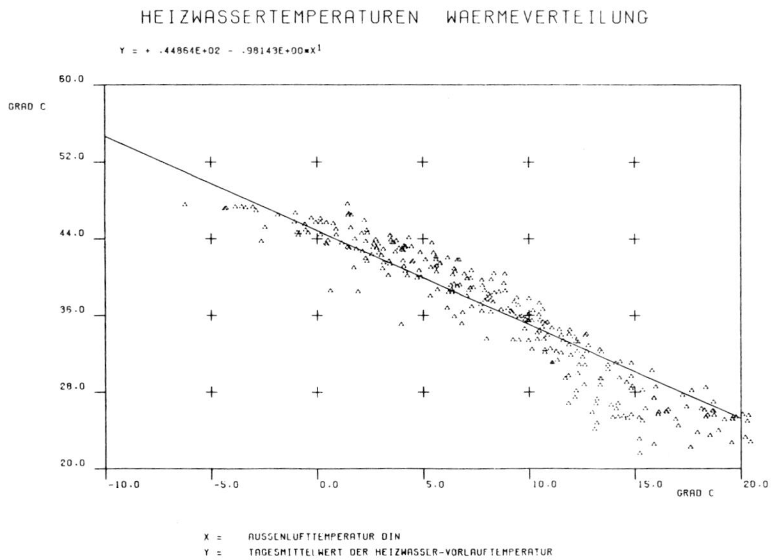
HEIZWASSERTEMPERATUREN WAERMEVERTEILUNG
Y = + .44864E+02 - .98143E+00*X1
GRAD C
60.0
52.0
44.0
36.0
28.0
20.0
-10.0
-5.0
0.0
5.0
10.0
15.0
20.0
GRAD C
X = AUSSENLUFTTEMPERATUR DIN
Y = TAGESMITTELWERT DER HEIZWASSER-VORLAUFTEMPERATUR

Bild 4

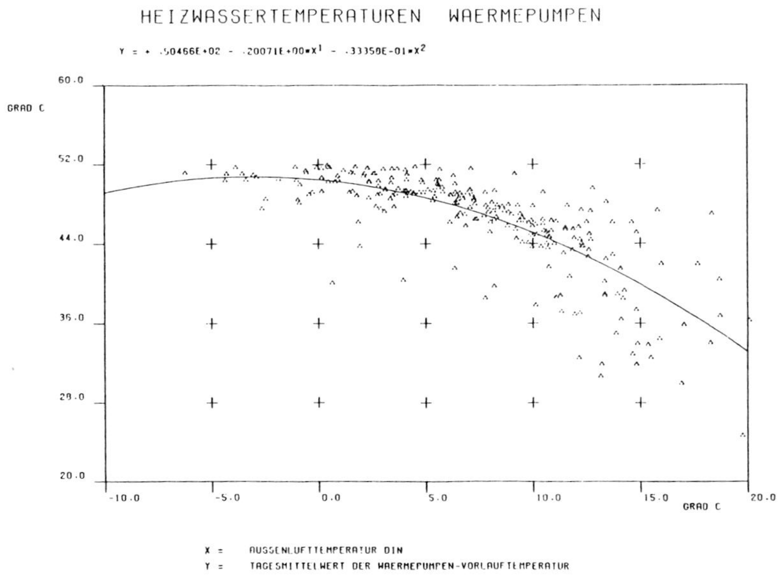
HEIZWASSERTEMPERATUREN WAERMEPUMPEN
Y = + .50466E+02 - .20071E+00*X1 - .33350E-01*X2
GRAD C
60.0
52.0
44.0
36.0
28.0
20.0
-10.0
-5.0
0.0
5.0
10.0
15.0
20.0
GRAD C
X = AUSSENLUFTTEMPERATUR DIN
Y = TAGESMITTELWERT DER WAERMEPUMPEN-VORLAUFTEMPERATUR

Bild 5

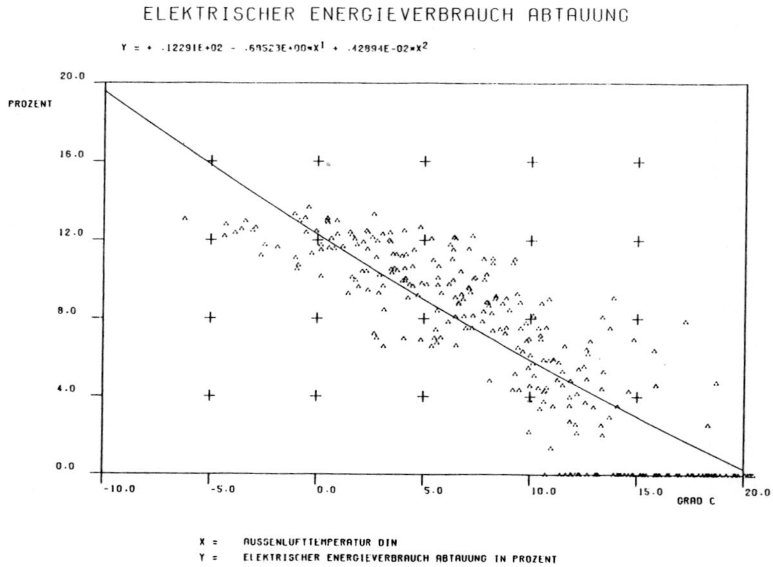
ELEKTRISCHER ENERGIEVERBRAUCH ABTAUUNG
Y = + .12291E+02 - .69523E+00*X1 + .42894E-02*X2
PROZENT
20.0
16.0
12.0
8.0
4.0
0.0
-10.0
-5.0
0.0
5.0
10.0
15.0
GRAD C
20.0
X = AUSSENLUFTTEMPERATUR DIN
Y = ELEKTRISCHER ENERGIEVERBRAUCH ABTAUUNG IN PROZENT

Bild 6

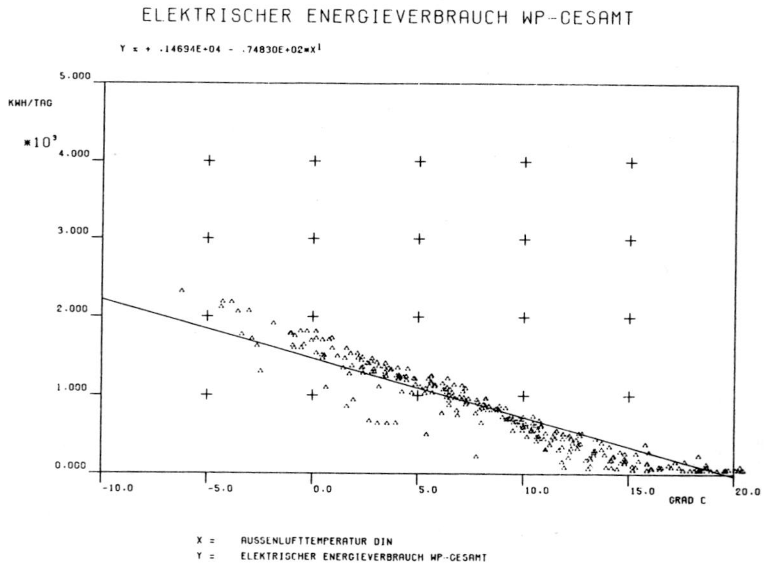
ELEKTRISCHER ENERGIEVERBRAUCH WP-GESAMT
Y = + .14694E+04 - .74830E+02*X1
KWH/TAG
*10³
5.000
4.000
3.000
2.000
1.000
0.000
-10.0
-5.0
0.0
5.0
10.0
15.0
GRAD C
20.0
X = AUSSENLUFTTEMPERATUR DIN
Y = ELEKTRISCHER ENERGIEVERBRAUCH WP-GESAMT

Bild 7

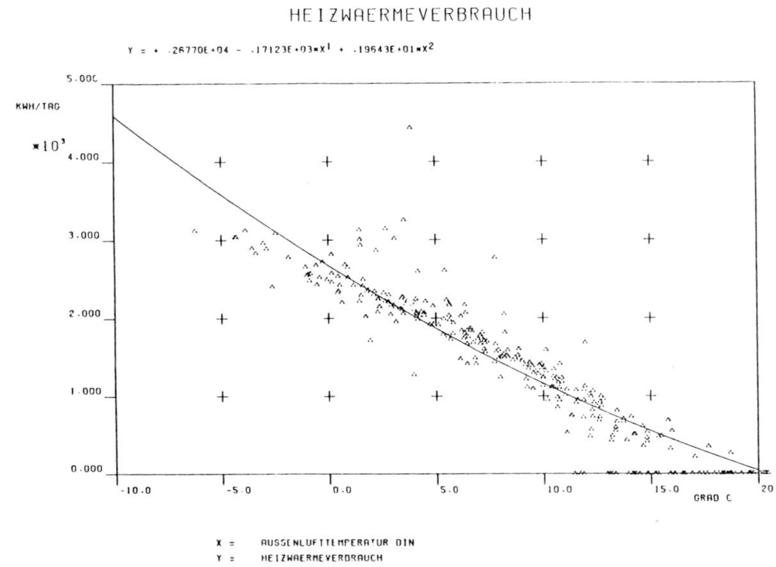
HEIZWAERMEVERBRAUCH
Y = + .26770E+04 - .17123E+03*X1 + .19643E+01*X2
KWH/TAG
*10³
5.000
4.000
3.000
2.000
1.000
0.000
-10.0
-5.0
0.0
5.0
10.0
15.0
20.0
GRAD C
X = AUSSENLUFTTEMPERATUR DIN
Y = HEIZWAERMEVERBRAUCH

<u>Bild 8</u>

ENERGY ROOF AND SEASONAL EARTH STORAGE

R.M. LAZZARIN (*)
Istituto di Fisica Tecnica della
Università di Padova, Padova, Italy

Summary

Low level heat can be collected by an energy roof and taken to a higher level by heat pumping. If the heat pump is utilized in building heating, the energy roof is useless in summer, i.e. when solar collection can be more efficient. Therefore the coupling of an energy roof with a seasonal storage can be useful. An experimental work is carried out at Musile di Piave near Venice concerning absorption heat pump heating of a building, using as cold source either an energy roof or an earth seasonal storage composed of 16 vertical tubes at a depth of 15 m. The earth storage is equipped with thermal resistances in order to know the temperatures in the soil both during summer charging and in the heating season. Details of the plant and the first experimental results in charging the storage by a 100 m^2 energy roof are reported.

1. INTRODUCTION

Absorption heat pumps are very attractive for countries where there is shortage of installed electric power. The possible PER (Primary Energy Ratio) can be about 1.1 ÷ 1.4 with a machinery which is silent, reliable and durable. Many researcheres are devoted to the study of alternative mixtures with more desirable properties than NH_3-H_2O or H_2O-LiBr, but it seems for now that these two mixtures will be predominant in the next years in commercial products. As regards H_2O-LiBr it is well known that the cold source cannot be at temperatures below 0°C even for modest temperature of the useful heat and probably the role of this mixture in absorption heat pumping will be in industrial applications.

In principle a larger range is allowable for ammonia-water with a cold source below -15°C: taking into account the limitations in generator temperature (160-180°C) and a useful heat available at 50-60°C, also ammonia-water wants a cold source possibly higher than 0°C in order to achieve an

(*) The author is Scientific Consultant of Metalux SpA, via Triestina, 74 30024 Musile di Piave (VE), Italy, Contract EEA-4-047-I(S).

acceptable PER. In effect a model on the market claims a minimum temperature for the cold source of 4°C.

It is easy to understand that in the absence of well or surface water a solar assistance is wellcome even for mild temperate climates (1,2,3). Heat pump solar assistance is affected by two main difficulties:

1) the cost of a sufficient collecting area;

2) the poor summer utilization of the solar section.

As regards the first point, absorption heat pump needs less collecting area than a compression one at the same PER and capacity (about half demand from the cold source), but the requested temperature is higher. The cost of the collecting area can be reduced adopting an energy roof, that is a simple system which substitutes the traditional roof and collects useful heat at a low level by suitable channels under the tiles.

As regards the second point, it is advisable to provide the plant of a seasonal storage so that useful solar heat can be collected all year long and can be supplemented during periods of poor or no insolation. Earth storage with vertical pipes is attractive to store energy, because the occupied surface area can be very limited increasing eventually the depth of the pipes and therefore it is almost always disposable near a building in contrast to horizontal pipes systems (4,5). Unfortunately poor information exist in literature on this rather new technique and it is difficult to predict the performance of a storage system in different types of ground and to correctly size the storage. Even the best techniques of inserting the pipes are questionable (6). Besides only experiments in heat extraction have been performed, but not yet in seasonal charging. It was claimed that whenever the storage operates below a 3m depth it is necessary to recharge it artificially, but no enough data exist on energy conservation and extraction in a summer charged storage at temperatures well higher than usual ground temperatures (7).

The aim of an experimental work carried out at Musile di Piave near Venice (Contract EEA-4-047-I(S)) is to study the performance of an ammonia water absorption heat pump in a building heating coupled either to a vertical tube earth storage or to an energy roof. During summer the energy roof charges the storage. In this paper the features of the experiment and the modes of operation of the system will be described; the experimental results obtained during August and the first half of September will be presented.

2. SYSTEM DESCRIPTION

The heated rooms were obtained by widening the building of the factory Metalux: they consist of a two floor body, 180 m^2 floor area, heated volume about 1080 m^3. The first floor is destinated to office, whereas the second one to a home. A section of the new building is sketched in fig. 1: one can easily see the two sheds oriented toward South, where the energy roofs have been installed. For now only one roof is working: it is made of copper tiles which are modular rhombed elements, drawn at the perimeter, diamond shaped at the centre, size 487x280 mm. The single tiles, even if they are tightly bound due to the deep drawing which allows a precise connection, can nevertheless bear the thermal expansions. The size of the roof is 25x4 m^2 (100 m^2). Copper channels under the tiles carry the heated fluid (a 30% mixture of glycol/water). The roof is set 11 cm apart the supporting surface in order to limit an air layer useful for air energy absorbing. The channels are connected to copper headers which take 5 channels each, through flexible connections so that thermal expansions are allowed. Soon a second similar aluminium energy roof will be built up.

Two different low temperature heating systems were installed in order to allow comparisons: fan-coil and floor heating coils. The heating machinery is an absorption heat pump (GAWP 25 Rekord) of water/water type with a nominal heating capacity of 25 kW. No auxiliary was provided as heat pump capacity was well in excess of the requirements (the smallest commercially available size). Instead a sensible storage is placed between heat pump and demand so that transient operations are minimized. Details on utilization circuit will be reported elsewhere: now only the circuit connecting energy roof, heat pump and earth storage will be examined.

The earth storage is formed of 16 reinforced poliethylene vertical tubes set in a square array 4x4 whose side is m 9.0. Each tube is composed by two concentric pipes. Details of the heading and size of the pipes are represented in fig. 2. The tubes are 15 m long and parallel connected in order to limit pressure drops. The whole length of tubes is therefore m 240, more about m 100 of distribution tubes at m 0.5 depth level with a total outside area of 80 m^2. Holes of enough diameter were obtained by drilling: the holes were filled up of water and the premounted tubes were inserted. According to the scarce data available in literature a specific power of 20-30 W/m can be expected (7,8,9), that is 4800-7200 W should be available, enough for the evaporator requirements of the absorption heat pump (about 7400 W at nominal rates).

Circuit connections are represented in fig. 3. Two couples of valves allow to isolate the four central tubes of the array, so that charging or discharging may be actuated only in the centre. Working operations of the system are multiple, as the energy roof is connected both to the evaporator of the absorption heat pump and to the earth storage (fig. 4). Thus the evaporator may be supplemented either by the energy roof directly (fig. 5) or by the earth storage (fig. 6). Moreover when the heat pump is not working and the roof can collect useful energy, the energy roof can be connected to the earth storage in order to charge it: this is the normal summer operation (fig. 7).

The controller functions are summarized in fig. 8. The first controlled variable is T_s, the temperature of the sensible storage after the heat pump: when this temperature is below 50°C the absorption heat pump is activated. In the presence of heating demand ($T_a < T_{set}$), this demand can be satisfied by the storage if its temperature is higher than 40°C. The heat pump evaporator is supplemented with priority by the energy roof whenever the useful temperature by it is higher than 10°C ($T_{ro} > 10$°C). If this is not the case the evaporator is connected to the earth storage until the cold source is higher than 5°C (in the case the heat pump is turned off).

In the absence of heating demand ($T_a > T_{set}$), the heat pump is eventually kept working if the cold source gives more than 5°C and the water storage is below 50°C. Otherwise the heat pump is turned off and the earth charging may be activated: this mode starts when the roof temperature overcomes 15°C which is presumably a useful temperature for charging. However the mode is mantained only if after a fixed time there is a sufficient increment between roof outlet and inlet. This somewhat involute strategy was necessary because no representative temperature can be attributed to the earth when the fluid does not circulate.

3. INSTRUMENTATION

Instrumentation can be subdivided into three main groups of measurements:

- meteoattinometric data;
- earth storage;
- thermal fluxes in and out the heat pump room.

As regards the first group, besides solar radiation and ambient temperature, one is interested in evaluation of energy roof performance as a function of wind velocity and air humidity (possible condensation): therefore the instrumentation is composed of sylicon pyranometer, resistance thermometer, paddle anemometer and a thermohygrometer.

The earth storage was equipped with 15 platfilm platinum resistance temperature detectors. They were disposed on two tubes a semidiagonal along and on three false not circulating tubes: in the array centre, in the middle of the considered semidiagonal and well outside the thermal storage in undisturbed soil. Each tube carries three sensors at a depth respectively of 1.5, 10 and 15 m (fig. 9). The resistance detectors were mounted on the tubes before insertion by tight fastening and careful sylicone protection. Stresses during insertion caused failure of the three detectors of the external tube (11,12,13). Subsequent failures of three other sensors (33,41,51) are attributed to short circuit. The three spoiled elements at m 1.5 depth will be probably replaced.

In the heat pump room the main interests lie on measuring of input/output of the absorption heat pump and in the input to the earth from the energy roof in summer. Also temperatures and flow rates are of interest. Temperatures are detected by immersion platinum resistance sensors; flow rates by rotameters; two heat counters are also provided. Unfortunately the counters gave no reliable results so that heat energy has been till now evaluated by temperature difference and flow rate measurements.

Separate measurements for the two energy roofs will be provided in order to compare the performances of the two roofs which differ not only in the tiles but particularly owing to the fact that one is ventilated whereas the other one is sealed at the top and at the bottom. In the heat pump room measures are easily readible through a digital thermometer and a 12 channels printing recorder.

4. THE FIRST EXPERIMENTAL RESULTS

Experimental surveys started on July 28th: the delay was due particularly to difficulties met in inserting the tubes into the soil and to a delay in masonry work of the new building.

All August long charging was operated only in the four central tubes of the array: the flow rate of the plant was unexpectedly low, i.e. only 500 l/h, owing to bad working of a threeway valve which could not be replaced in summer holyday period. The substitution took place on August 27th and the flow rate increased to 2500 l/h. Since September 1st the energy roof has been charging the earth storage through all the 16 tubes of the array. Experimental data are disposable till September 15th.

Though data are scarce and affected by the usual difficulties of plant running in (bad working of components, small liquid leaks, air in the circuit), some interesting observations can be reported.

(1) Initial temperature measurements of the soil revealed appreciable disuniformities which are easily detectable by Table I where temperatures at various dates and depths are reported the false tubes 2,4,5 along (see fig. 9). A soil stratigraphy showed a distribution of a first layer of clay (m4) and a second layer of humid sand (from m 4 to and below m 15). Perhaps the different distribution of water content affects temperature at various depths.

(2) At first the fluid from the roof reached easily temperatures as high as 40-50°C (daily average at about 40°C) owing to the low flow rate (as low as 5 l/m^{2}h). Nevertheless the temperature drop across the earth heat exchanger was never higher than 12°C with daily averages between 3 and 8°C, which means a maximum specific heat exchange of 50 W/m with mean values 12-35 W/m. Daily collected energy was between 20 and 40 kWh/day (72-144 MJ/day) with an average 5% roof efficiency.

(3) When the flow rate was increased to 2500 l/h the fluid temperature reduced to 35°C, the temperature drop across the earth heat exchanger was only 1-3°C, but the mean specific heat exchange rose to 20-60 W/m in spite of the lower fluid temperature at the inlet! Probably in the presence of a high temperature drop between inlet and outlet there is an important internal heat exchange of the fluid of the inner tube and the one flowing in the anulus which can be more important than the earth heat exchange. Also the effect of the long 1.5 m deep distribution tubes can be important when it is of the same order of the vertical tube length. With the higher flow rate the daily collected energy arrived at 70 kWh (250 MJ/day) with a roof efficiency as high as 12%.

(4) Finally the whole flow rate was sent to the 16 tubes after Septemberlst. There was a further increase in collected energy (lower return temperature to the roof, as low as 18°C, that is lower than average ambient temperature and near the soil first layer temeperature.The average daily collected energy was about 80 kWh in the less favourable September climate (shorter days and less insolation on the 16° tilt roof), arriving at 150 kWh in particularly favourable weather (energy roof efficiency at 20-30% for working temperatures at 30-35°C and ambient temperature at 20-25°C). The temperature drop across the earth heat exchanger was 3-6°C with a specific heat exchange of 20-40 W/m which might be increased for higher roof energy output. This will be experimented by operating the second not ventilated roof together with the actual one.

(5) The specific heat exchange is somewhat deceiving and should be investigated under different inlet temperatures and flow rates.

No conclusions can be drawn owing to the lack of experimental data. So-on extraction phase will start with absorption heat pump operation. Nevertheless some precise questions on the influence of temperature and flow rate for the performance of earth storage and energy roof have been posed. Next summer charging will take these interrelationships into account.

REFERENCES

(1) R.M. Lazzarin, L.Rossetto, Performance prediction of a NH_3-H_2O absorption heat pump, Preprints Commission E2, XVIth Int.Cong. of Refrg., Paris, 317-322, 1983.

(2) R.M. Lazzarin, L. Rossetto, L. Viero, Heating by absorption heat pump. Analysis through a simulation model, Preprints Commission E2, XVIth Int. Cong. of Refrg., Paris, 233-238, 1983.

(3) R.M. Lazzarin, Solar-assisted absorption heat pump feasibility, Solar Energy, 26, 3, 223-230, 1981.

(4) B. Bäckstöm, General principles for earth heat pumps, Nordic Symposium on Earth Heat Pump System, Göteborg, 24-30, 1979.

(5) G. Rosenblad, Earth heat pump system with vertical pipes for heat extraction and storage, Nordic Symposium on Earth Heat Pump System, Göteborg, 102-110, 1979.

(6) J.W.J. Bouma, A.D. Koppenol, Investigation into a complete earth to water heat pump system in a single family dwelling focussing on the application of a vertical subsoil heat exchanger, Proc. of Four Contractors Meeting on Heat Pumps, Brussels, 229-241, 1982.

(7) J.M. DeHoe, B. Geeraert, Heating of a laboratory by means of a soil heat pump combined with unshielded solar collectors, Proc. of Four Contractors Meeting on Heat Pumps, Brussels, 242-269, 1982.

(8) J.R. Goulborn, J. Fearon, Domestic heat pump with deep hole grounde source evaporator, Proc. of Four Contractors Meeting on Heat Pump, Brussels, 270-283, 1982.

(9) P. Nievergeld, A. Koppenol, J. van der Brugh, Experimental investigations about using the earth as a heat storage medium and as a heat source for heat pumps, Final Report, Contract 367-78-1-EE NL, Commission of the European Communities, 1981.

(10) M. Fordsman, A. Eggers-Lura, Analysis of the factors which determine the COP of a heat pump and a feasibility study on ways and means of increasing same, Final Report, Contract 370-78-EE DK, Commission of the European Communities, 1981.

depth	date	PIPE 2 +	4	5
1.5 m	start	19.5	-	-
	A27th	17.8	-	-
	S 1st	18.3	-	-
	S15th	18.2	-	-
8 m	start	12.6	10.7	12.7
	A27th	12.6	11.6	12.9
	S 1st	12.7	11.7	12.7
	S15th	13.0	12.3	12.9
15 m	start	12.0	10.6	12.7
	A27th	12.0	11.0	12.7
	S 1st	12.2	11.4	12.7
	S15th	12.6	12.1	12.9

TABLE I - Temperatures at different depths and dates the false pipes along; at 1.5 m depth temperature was dependent also on hour time: some tenths degree variation. Temperatures varied strongly the true pipes along according to circuit operation and fluid temperatures.
+ 1.5 m sensor at 8.30 o'clock

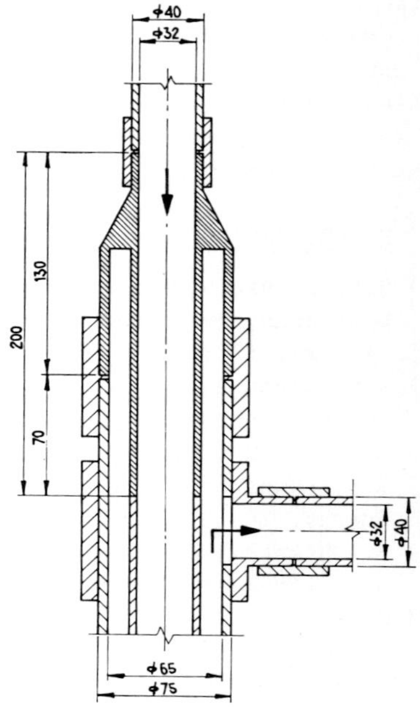

Fig. 1

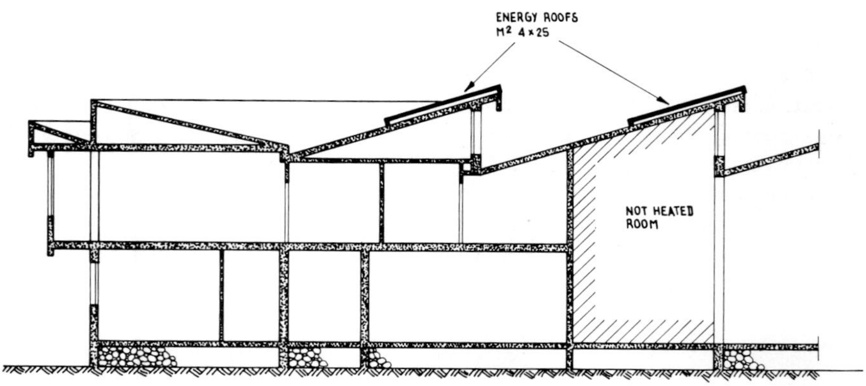

Fig. 2

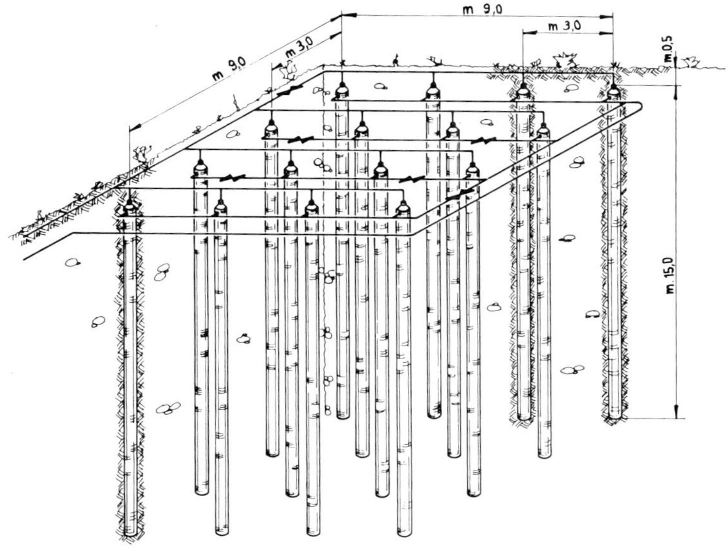
m 9,0
m 3,0
m 9,0
m 3,0
m 0,5
m 15,0

Fig. 3

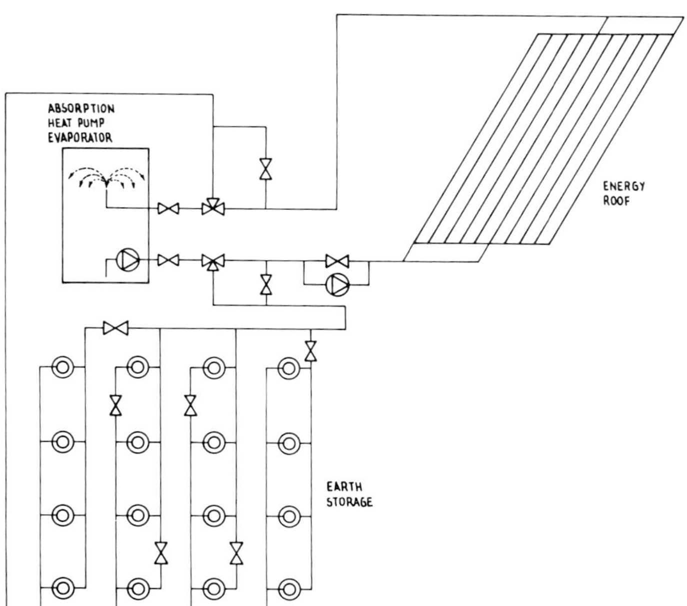
ABSORPTION
HEAT PUMP
EVAPORATOR
ENERGY
ROOF
EARTH
STORAGE

Fig. 4

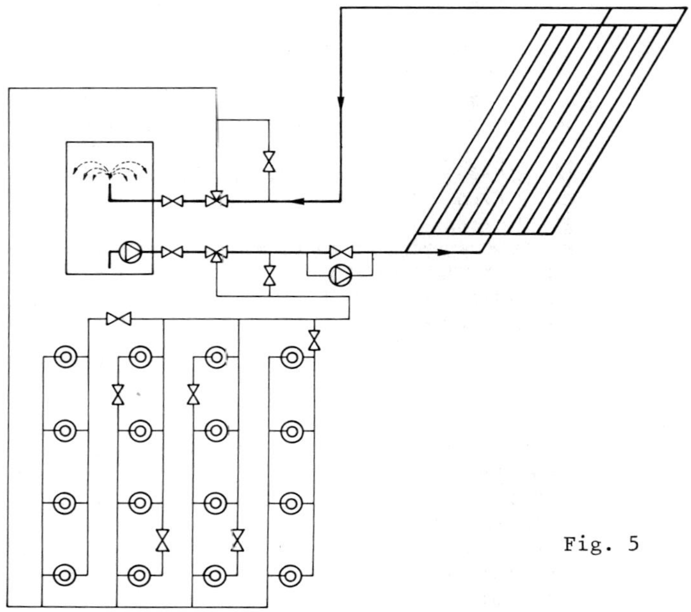

Fig. 5

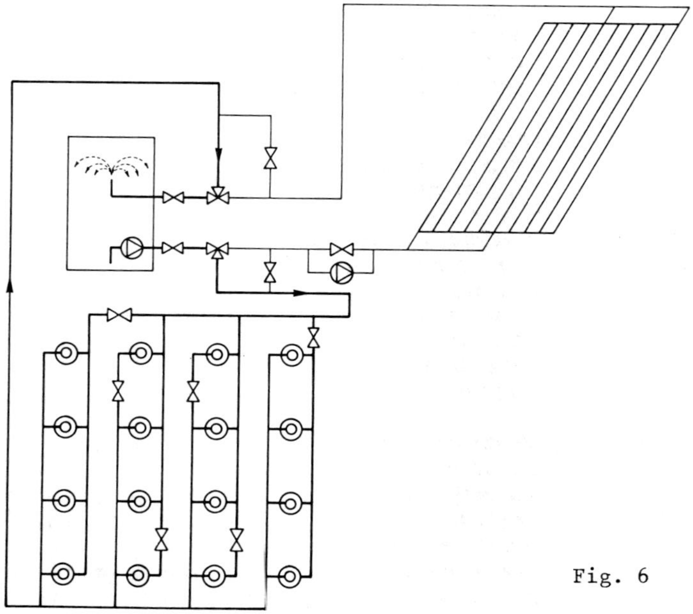

Fig. 6

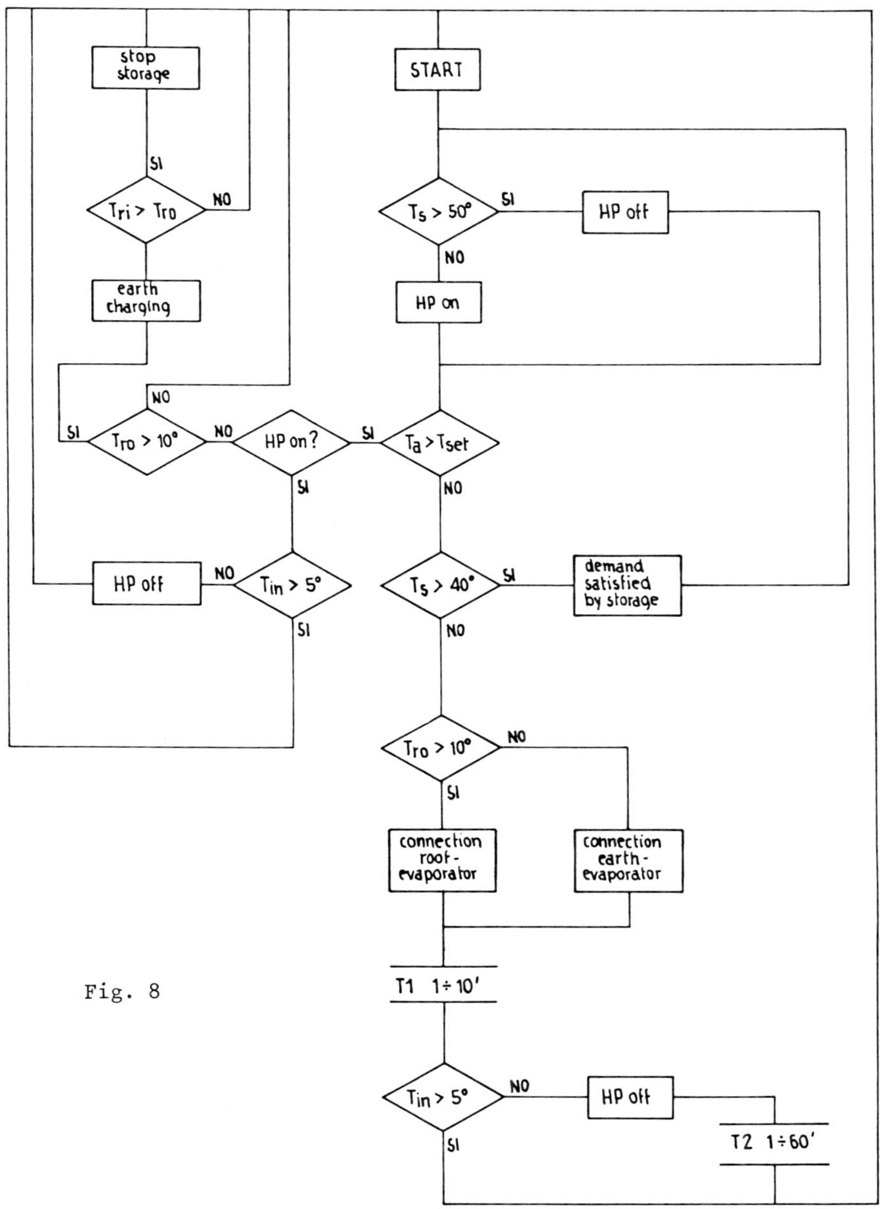
START
stop storage
SI
Tri > Tro
NO
earth charging
NO
SI
Tro > 10°
NO
HP on?
SI
Ta > Tset
Ts > 50°
SI
HP off
NO
HP on
SI
NO
HP off
NO
Tin > 5°
SI
Ts > 40°
SI
demand satisfied by storage
NO
Tro > 10°
NO
SI
connection roof - evaporator
connection earth - evaporator
T1 1÷10'
Tin > 5°
NO
HP off
T2 1÷60'
SI

Fig. 8

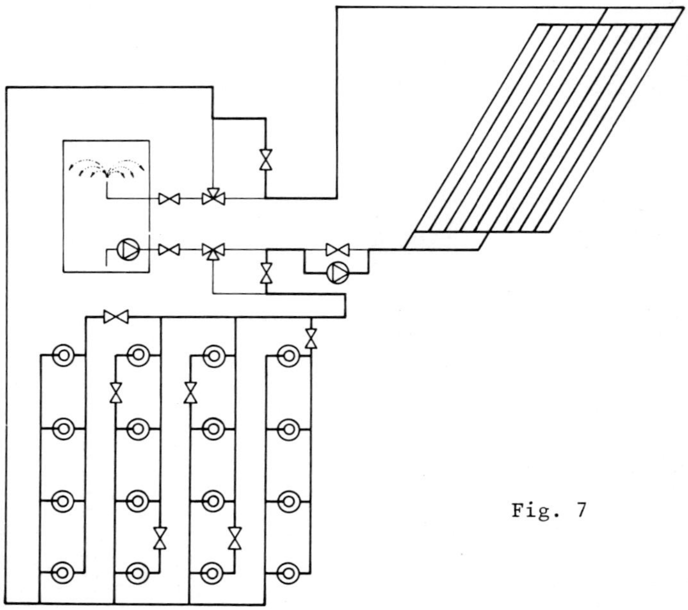

Fig. 7

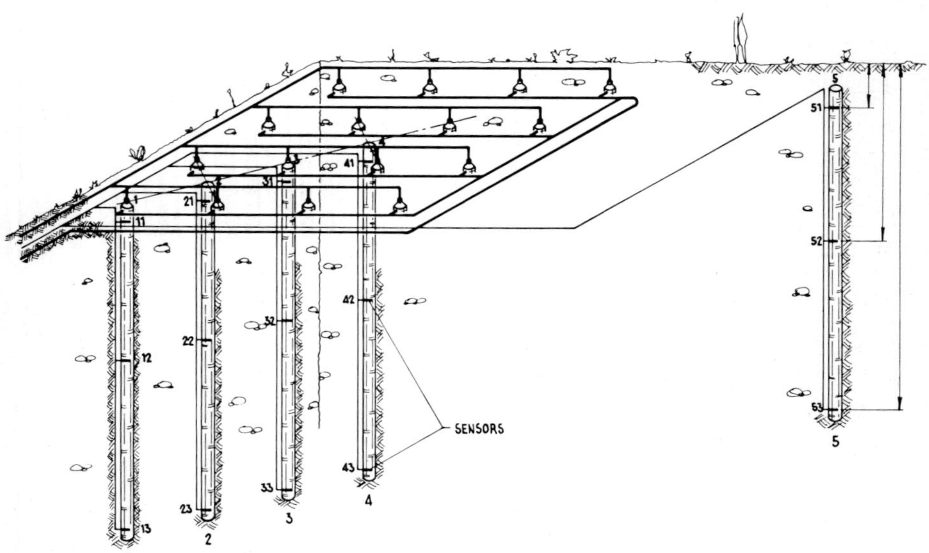

Fig. 9

PRACTICAL AND THEORETICAL INVESTIGATION OF A HEATING/COOLING HEAT PUMP WITH GROUND STORAGE AND SOLAR COLLECTOR

M. FORDSMAND
European Heatpump Consultors Ltd.
38, Skovshovedvej DK-2920 Charlottenlund

Summary

A heatpump installation for either heating or cooling of a single-family house is described. It consists of a ground body, a solar collector, a heat pump with subcooler, which is connected to a radiator system. The area requirements in the garden are only 32 m^2 and the average COP is 3.6.

1. INTRODUCTION

The experiences from heatpump installations with horizontal pipes show that the utilized energy is solar energy which is collected on the ground surface and stored in the ground from summer to winter. Therefore, it is interesting to investigate installations that utilizes more controlled ways to collect and store the solar energy. This investigation has also the aim to reduce the requirements for the areas for ground pipes.

2. SKANEKS

In 1972 at the firm Skaneks, Glostrup, Denmark, a practical installation of this kind was made. The brine temperature was 6-8° higher during the winter than on ordinary ground heat pumps and the COP was around 3.3.

The heatpump had a ground coil of about 500 m length, which was placed mostly down under the building and partly in the ground outside the building. The heat pump was of the water/water type with indirect evaporation and with fan-coil units for delivering the warm air to the rooms during the winter. During the summer the plant could be changed to air-conditioning by means of two 4-way valves, when cooled water was send through the fan-coil units and the hot water was send to the ground coils.

3. THEORETICAL INVESTIGATIONS

In 1978 with the aid of the EEC a theoretical study was undertaken to make a feasibility study on a combined summer cooling and winter heating system on three different destinations in Europe, namely Copenhagen, Paris and Marseille.

The aim of the project was to make a study on a solar augmented ground heat pump cooling and heating system in Europe.

The novelty of the concept is that in a combined heating /cooling heat pump plant, it is possible to store some of the condenser heat in the ground from summer to winter and also to collect quite a lot of solar heat. The result of this should be that the COP of the total installation is increased and that the energy consumption of the total installation is decreased. Fig. 1 shows the energy consumption on the three destinations.

A simplified computor model, which could be used on a TI-59 minicalculator was established. When comparing it with a finite element model, it was found that deviations in the heatsource were reasonable and therefore it was decided to use the simplified model for the study. It has not yet been proved by practical experiments how exact the model is.

The following main conclusions have been drawn:

A. It takes about three years for the ground to reach stable conditions.

B. It should be possible to achieve high COP values for a combined heating/cooling heatpump installation: Copenhagen 5.0, Paris 5.9, and Marseille 5.2.

C. For all three destinations a small pipe of only 54 m length is required.

D. For terrassehouses vertical pipes located below the house are suitable for this system and they have a sufficient storage capacity. The energy losses from the ground storage are very small, less than 10 % due to the moderate temperature level.

E. The primary energy saving by the heat pump sytem is considerable: Copenhagen 24 %, Paris 57 % and Marseille 29 %.

4. PRACTICAL INVESTIGATIONS

A logical continuation of the theoretical work was to verify the calculations with practical results. In 1981 with the aid from the EEC a project in using a heating/cooling heat pump system in a highly insulated single-family house was started.

4.1 DESCRIPTION

The system consists of a solar collector, a heat pump and a heat accumulator with vertical pipes in the ground. The function is to restore or rise the temperature in the ground during the summer with energy from cooling of the house and from the solar collector. The domestic hot water is produced by a heatpump that uses the exhaust air as a heatsource. Fig. 2 shows the lau-out of the system.

Firstly it was intended to use a new-built house but this was not possible. So for practical reasons it was necessary to place the ground body below the carport and to use the existing radiators.

4.2 THE HOUSE

The house is of a size of 137 m^2. Improvements in the insulation have been necessary so that 150 mm mineralwool has bee added in the attic one more layers of glass has been put in the windows so that there now are three layers of glass and all ventilation openings are now closed. 160 m^3 fresh air/h is fed into the house through the subcooler in the heat pump. The results of the improvements should be a transmission loss of 4.6 kW and the ventilation loss should mostly have been covered by the subcooler. However, the practical heat loss has shown to be higher and very dependent on the wind velocity. The load could increase with 50 % under severe wind. Therefore, the forward flow temperature was increased and the COP decreased more than it was expected.

4.3 HEATING INSTALLATION

The house has a radiator system constructed to deliver 9 kW at a temperature of 65° C. After the improvement in the insulation the forward flow temperature is 45° C.

In the cooling moad the radiators have proved to be unsuitable. Water and even ice were formed on the radiators. A better temperaturregulation system has improved this but radiators in general are not suited for this purpose.

4.4 GROUND BODY

The ground body is placed in the ground below the carport and it consists of 2x4 vertical pipes. There is an effective length of 116 m. See fig. 3 and 4.

The temperature and energy measurements seem to match fairly well with the theoretical calculations but it is too early to reach any definite conclusion.

4.5 SOLAR COLLECTORS

The solar collectors are made of polyethylen plastic. It can stand the effect of freezing water inside it is inexpensive and easy to install. The total area is 14.4 m^2 and consists of four collector elements. There must be used bellow-type connections to withstand the formation of ice.

The measurements show that the output is 35 % less than the theoretical calculations.

4.6 HEAT PUMPS

The heat pump for heating the house has worked satisfactorely. The condenser has a temperature differens of 2° K and the evaporator has a temperature differens of 4° K at average working conditions. The compressor is a 1.5 HP hermetic compressor with R 502.

The heat pump for the domestic hot water has also worked satisfactorely. The average energy consumption is 4.0 kWh/day including fan and pump. The average COP is about 3.0.

4.7 SYSTEMBEHAVIOR

The COP has been recorded daily. There are variations from 2.7-4.0 with an average about 3.6. The low values are connected to two factors:

A. The increase in roomtemperature requires higher forward flow temperature. This factor has risen the question whether or not night set-back is an advantage for heat pumps.
B. Some regulation methods, especially those with l ng running periods causes high forward flow temperatures which are unnecessary.

A new control system which uses running periods of one minute has been investigated. It has caused a lower temperature differens in the evaporator and the condenser, a higher COP and lower temperatures in the compressor.

It has not been possible to keep the outlet temperature from above 0° C. This is owing to the fact that the system was started in autum 1982. Such a system must either be started during spring or it must use water with antifreeze additives.

Tabel 1 shows the variations in the COP for the heatpump without a subcooler and with a subcooler and a cooling jacket around the compressor.

5. CONCLUSION

The experiment is not yet terminated but the preliminary results are as following:

A. It is possible to use a very small ground area, abt 4x8m for the energy storage.
B. The system results in a high COP, around 3.6.
C. The way in which the subcooler is used increases the COP at least 10 %.

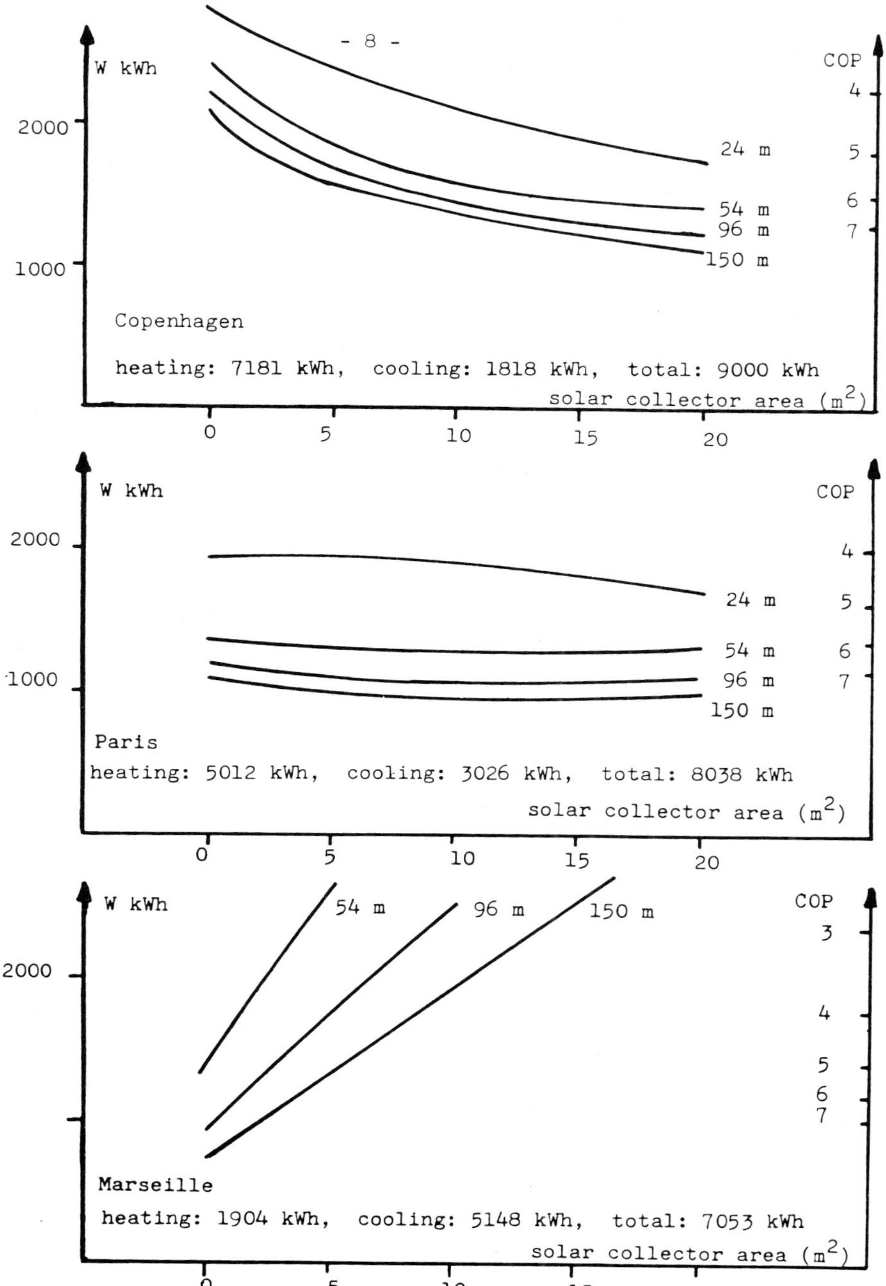
W kWh
2000
1000
COP
4
5
6
7
24 m
54 m
96 m
150 m
Copenhagen
heating: 7181 kWh, cooling: 1818 kWh, total: 9000 kWh
solar collector area (m^2)
0
5
10
15
20
W kWh
2000
1000
COP
4
5
6
7
24 m
54 m
96 m
150 m
Paris
heating: 5012 kWh, cooling: 3026 kWh, total: 8038 kWh
solar collector area (m^2)
0
5
10
15
20
W kWh
2000
54 m
96 m
150 m
COP
3
4
5
6
7
Marseille
heating: 1904 kWh, cooling: 5148 kWh, total: 7053 kWh
solar collector area (m^2)
0
5
10
15
20

Fig. 1

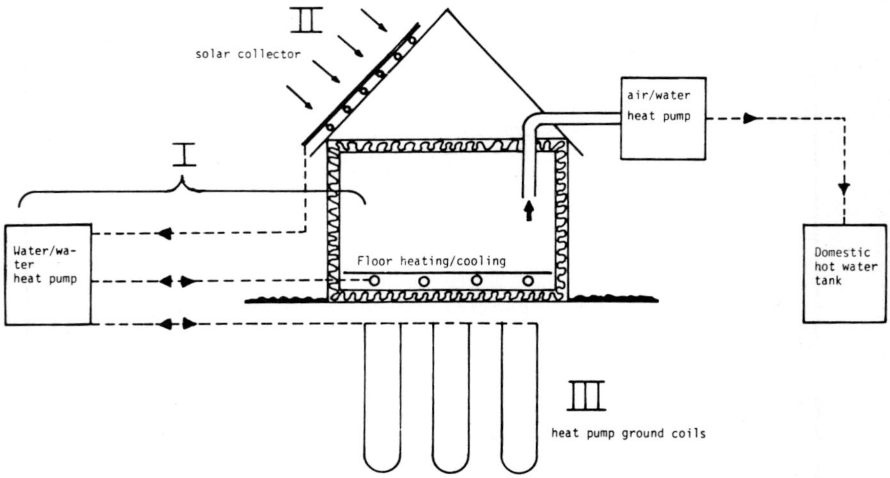

Fig. 2 Principle sketch of the solar/heat pump cooling/heating system

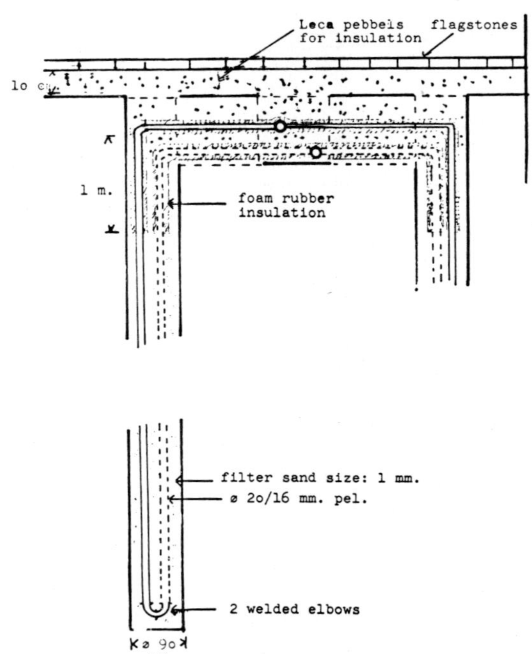

Fig. 3

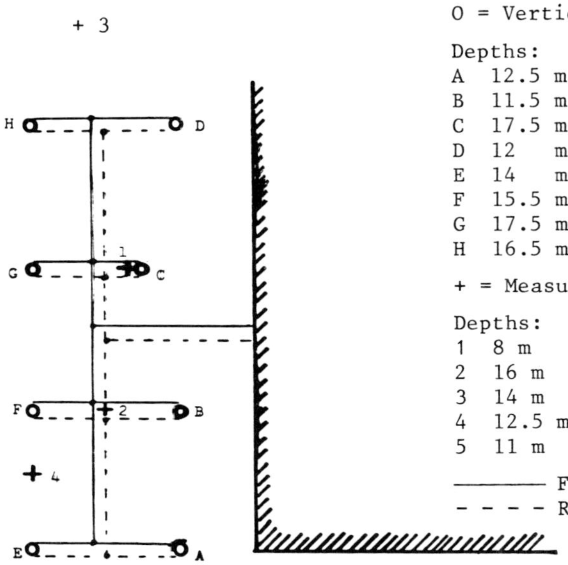

0 = Vertical pipes pel Ø 20 mm x 2

Depths:
A 12.5 m
B 11.5 m
C 17.5 m
D 12 m
E 14 m
F 15.5 m
G 17.5 m
H 16.5 m

+ = Measuring holes

Depths:
1 8 m
2 16 m
3 14 m
4 12.5 m
5 11 m

——— Forward flow tubes
- - - - Return flow Tubes

Fig. 4

Tabel for energy measurements in kWh/day during the winter 1982/1983

Months	dec.	jan.	feb.	april
Subcooler	0	X	X	X
Cooling jacket	0	0	0	X
Compress. input	24.06	30.74	27.76	13.53
Radiator output	75.2	103.4	95.1	48.4
Subcooler output	0	6.4	5.2	3.18
COP	3.1	3.6	3.6	3.8
Solar input	0	1.2	1.7	20.4

Note: The measuring period has been chosen, so that the temperature conditions are equal. i.e. The forward flow temperature is around 40°C, and the brine temperature is around 6°C.

The COP is calculated without circulationpumps.

Fig. 5

INVESTIGATION INTO A COMPLETE EARTH-TO-WATER HEAT PUMP SYSTEM IN A SINGLE-FAMILY DWELLING FOCUSSING ON THE APPLICATION OF A VERTICAL SUBSOIL HEAT EXCHANGER

J.W.J. BOUMA AND A.D. KOPPENOL
Organization for Applied Scientific Research TNO,
Division of Technology for Society

Summary

Vertical subsoil heat exchangers for heat pumps have been studied to establish the state of the art. As far as the design aspect is concerned a great variety of installations is possible. The types of heat exchanger most frequently used are concentric tubes and to a lesser extent parallel tubes. A review is given of the techniques of making the holes in the earth to accommodate the heat exchanger tubes. Water jetting by hand is favourable from an economic point of view, if only a limited number of tubes are to be installed. Computer simulations have been performed for seven tubes with a length of 10 m, 15 m and 20 m respectively. The present heat exchanger consists of seven tubes with a length of 14.5 m. They are positioned almost completely in ground water. The system comprises an electrical heat pump and a floor heating/radiator system. The system was monitored during the season 1982/1983. It has operated in a monovalent mode only, while 35% primary energy saving was achieved. Specific heat extraction rates are 40 to 65 W per m vertical tube length. Temperature recovery of the soil in a natural way during summer is almost complete.

1. INTRODUCTION

As a follow-up to earlier research work during which a computer model of a ground heat exchanger was developed (1) and experimentally verified for the horizontal concept (2), the attention has been focussed in this research programme on a complete electrical heat pump system with vertical subsoil heat exchanger, installed in an occupied residence. No heat storage in the ground has been considered. No domestic hot water is produced by the heat pump.

As far as the different types and the installing techniques are concerned, literature concerning vertical subsoil heat exchangers for heat pumps has been studied to determine the state of the art.

With the purpose of designing the subsoil heat exchanger the simulation model has been used to handle the characteristics of the system, comprising heat exchanger, residence, heat pump and climate.

The system has been in operation since October 22, 1982 and monitoring was completed by September 1983.

Before installation in the residence the characteristics of the heat pump have been measured in the laboratory under steady-state conditions.

In addition to the type selection and the design of the subsoil heat exchanger, the main objective of the research work is estimation of the extent to which the behaviour of vertical earth heat pump systems in a residence is predictable. Furthermore drawing up of design and application directives and instructions for such systems.

2. STATE OF THE ART

Vertical subsoil heat exchangers have been installed only on a limited scale and therefore experience is rather limited. The vertical systems dealt with here are of the closed type, i.e. no ground water is extracted. The types of heat exchanger mostly used are concentric tubes, the outer one of which is closed at the lower end, and to a lesser extent parallel tubes (U-shape). Also tubes with a diametric wall inside are used. The vertical tubes are mostly arranged in a network, divided in one or more groups. The groups are coupled together and connected to the heat pump by horizontal tubes. These are installed at a depth varying between 0.5 m and 1 m below the earth surface. In practice this means that heat exchange with the soil does not occur solely via the vertical sections but also via the horizontal lines, unless isolation is provided.

Plastic as well as metal tubes are used, often depending on the method of installing them. Common plastic used for the tubes is high-density polyethylene (HPE) and polyvinyl chloride (PVC). Metal tubes are made of steel and copper.

The presence of ground water flow is of great importance since heat extraction can be improved significantly in that case and temperature recovery of the soil in summer is improved. In The Netherlands the larger part of a vertical subsoil heat exchanger is situated in ground water throughout the year.

In practice a great variety in heat exchanger design exists for comparable installations. Especially the tube depth and the number of tubes installed differ strongly. Moreover, the tube distance, an important parameter with regard to heat extraction load and temperature recovery of the soil, is different (3). Depths vary between 10 m and 100 m. The majority of these systems, both experimental and semi-commercial, show a rather satisfying performance.

Two methods of heat extraction from the earth can be distinguished, i.e. the direct and indirect method. Most subsoil heat exchangers operate by the indirect method. In such systems a brine circulates through the heat exchanger and the evaporator. In some experimental installations the direct method has been applied. Here the subsoil heat exchanger acts as an evaporator. These installations were tested in Denmark, Belgium and Great Britain. In the majority of these experiments an improvement of performance, due to higher evaporation temperatures, did not occur.

In the case of vertical subsoil heat exchangers, extracting heat from depths between 10 to 100 m, recharging by the spring and summer climate is not ensured automatically. In Nordic countries it is considered necessary to regenerate artificially the soil temperature at these depths during the summer season. This is effected by means of solar collectors or passive air coils. However, in The Netherlands significant ground water flows may be expected locally in the vicinity of rivers, canals or lakes. In such cases recovery of earth temperatures is often complete. Model calculations and experiments with a 10 m single vertical tube of the U-shaped type in The Netherlands show, that temperature recovery by natural influences is nearly complete (2) (heat extraction continuously; maximum 40 $W.m^{-1}$ running length for a period of three months).

Sometimes the tubes are installed in the earth under an angle from a ground surface of approximately 1 m^2, which causes minor excavation work and site disturbance, an advantage in the case of existing residences with a garden.

An important factor in economic operation of vertical earth heat pump systems are the costs of installing the subsoil heat exchanger. These costs strongly depend on the length of the tubes, because the technique of making vertical holes is strongly related to the depth of the holes, the diameter of the borehole and the type of soil. Depending on depth the following techniques are used (3):

- Auger-technique: This method is only applicable up to the ground water table.
- Water jetting with a lance without casing (limited to roughly 15 m in the case of hand-made holes and 20 to 25 m with machine-made holes): The tubes are inserted after the lance has been withdrawn.
- Water jetting with a steel casing, preventing the hole from collapsing: The casing is withdrawn with a truck-crane after the plastic tubes have been inserted. Compared with the former method these holes do not necessarily have to be made in one single operation (realised for 10 m - tubes).
- Pulsing: This method is labour-intensive and attractive only when a few undeep (< 20 m) holes are needed. However, with this method high-quality soil samples can be obtained. Most other methods mentioned here only offer limited possibilities for accurate soil sampling.
- Pressing/vibrating: At least down to 10 m PVC-pipes can be pressed in clay-soil with the aid of a hand-winch. Another method is using a pneumatic device. Steel tubes, which are the outer tubes of concentric tubes, are pushed into the soil. They can be positioned vertically (locally) or under an angle from a central point (possible depth 20 to 30 m). In the case of installing the tubes under an angle, the upper parts of the tube are close together, which makes it easy to interconnect them. Contrary to other techniques the pneumatic machine can stay at one point. In The Netherlands a combination of vibrating and water jetting has succesfully been used to install U-type tubes with a length of about 21 m directly. This technique makes use of heavy machinery.
- Ramming: A steel liner is rammed into the soil with the aid of heavy machinery. At the lower part of the liner a plate is mounted which is lost when the liner is pulled up, after the heat exchanger tubes have been inserted. U-type heat exchangers of 35 m length have been installed easily this way in clay-soil in Sweden.
- Drilling (with washing): This method can be used for depths up to 150 m and more. In practice a distinction is made between the manners of circulating the fluid through the drilling tube, namely by sucking or pressing (mechanically or by air injection at the lower part in the mud). An advantage with sucking the mud through the drilling tube lies in the availability of representative soil samples. This technique is fast, especially in sandy soils. However, as in the case of ramming the heavy drilling equipment (truck-mounted) might be a constraint in applying this technique caused by the poor accessibility of existing premises. Moreover, sufficient water supply must be ensured.

Making holes in formations which are difficult to penetrate such as rock, hydraulic and conventional oil well drilling techniques can be used. These techniques however, are expensive and primarily used in the case of hard formations or deep holes, for instance with the purpose of exploring geothermal energy or applying heat storage on a large scale.

3. MODEL CALCULATIONS

Under a previous contract (1) a computerized predictive method for heat extraction from the soil and heat pump performance was developed. The calculation method applied is the finite-element method. The results achieved with the calculation model were in good agreement with the experimental results (2). This computerized method was also used in this investigation for the design of the subsoil heat exchanger and for the determination of the performance of the complete heat pump system (3). It is assumed that the tube has a uniform temperature throughout a vertical section, heat transfer in the soil is by conduction only, heat extraction is continuous and operation of the heat pump is in the monovalent mode. Calculations are restricted to one tube out of the group. Only heat extraction during the heating season is considered, heat recharging in spring and summer being governed only by natural processes. To simulate the heat pump and calculate the extracted heat relatively simple equations are used, i.e. a constant temperature difference has been assumed between evaporation - brine entrance temperature and condensation - water forward temperature, while the Carnot-efficiency is also constant. The calculation is carried out step-by-step in time. During each time step the COP, the extracted heat from the soil and the temperatures of the tube and soil are calculated.

Figure 1 shows the arrangement of the tubes making optimum use of the available ground area; seven tubes (diameter 0.05 m) in two rows and in a staggered arrangement. In this figure one symmetrical half of the upper side of the model geometry has been projected on one tube. Heat extraction takes place both from the garden and from the adjacent ground area. The site is located in an urban area in Apeldoorn.

The boundary conditions are as follows: Earth surface temperature at the upper side, an isothermal plane (10.2°C) 4 m beneath the tube and a uniform water/glycol mixture. The soil at the test site mainly consists of sand and gravel in ground water (thermal conductivity soil 3 $W.m^{-1}.K^{-1}$; volumetric heat capacity soil 3 x 10^6 $J.m^{-3}.K^{-1}$).

Calculations were performed for different tube lengths, i.e. 10 m, 15 m and 20 m, over a period of three years. Table I summarizes the calculated Seasonal Performance Factors (SPF) of the heat pump system (same heat demand dwelling; auxiliaries and thermal resistance tube wall and fluid film neglected). Obviously the tube temperature and the SPF are highest in the case of 20 m-tubes, but the quasi-stationary state of the soil is achieved quicker with 10 m-tubes, i.e. after two years. For the 15 m-tubes the simulation results have been presented in Figure 3.

4. SUBSOIL HEAT EXCHANGER

The type of heat exchanger tubes selected, on technical-economic grounds, are of the so-called hollow-wall type. This commercial tube system consists of two thin-walled (1.10^{-3} m) concentric PVC-tubes of which the annulus is divided into 11 flow channels, formed by radial walls inside (see Figure 2a). In addition to this type of tube also parallel tubes (U-type), concentric (loose) tubes and circular tubes with a diametric wall, PVC as well as PE, were considered (4). One of the factors which played an important role in the process of selection was the choice of drilling technique to be used, especially as far as the installation cost and the accessibility of the ground for drilling equipment was concerned.

The subsoil heat exchanger consists of seven vertical tubes of 14.5 m length, at 5 m distance apart (4), coupled with the heat pump via horizontal tubes. The net ground area of the garden is 110 m^2. In the central tube (diameter 36.10^{-3} m) the water/glycol mixture flows downwards and through the channels upwards. The tubes are placed in parallel, while technical measures were taken to ensure equal flow in each tube (see Figure 2b). The holes for the tubes were hand-made by water jetting with a lance. The horizontal connecting tubing is non-isolated and located at 0.7 m depth. Before operation air was vented thoroughly from the heat exchanger and a comprehensive leak-test was carried out to be sure that the subsoil heat exchanger is completely tight. The vertical tubes are located almost completely in ground water (91% of their length). Although about 40 m of trench had to be made, disturbance of the testsite and damage to existing overgrowth was no problem.

5. RESIDENCE - HEATING SYSTEM

The residence, built in 1975, is an attached single-family house with a floor heating system on the ground floor and radiators on the two upper floors. The residence is well-isolated; the maximum heat demand is about 7.5 kW at -12°C ambient temperature. The heat pump was installed in a detached garage at 14 m distance from the residence and connected with the heat distribution system via well-isolated underground piping.

The heating system is operated in a bivalent-alternative mode, using the present natural gas boiler in the dwelling (nominal capacity 23.3 kW). The small-capacity earth-to-water heat pump (5.5 kW heat output at 0°C brine temperature and 35°C water temperature) is electrically driven. In pursuing a higher COP the liquid leaving the condenser is subcooled, partially by heating surrounding air and by exchanging heat with the suction gas to the hermetic compressor.

6. DATA-ACQUISITION

The complete heat pump system was monitored with the aid of a data acquisition system, comprising a programmable data logger and a magnetic tape recorder with compact cartridge. In addition to soil temperatures at four positions and at several depths, water, air and brine temperatures as well as the energy flows in the system were measured, according to the data logger programme. Processing of measuring data took place with the institute's computer system.

7. MAIN RESULTS

Heat pump operation was started on October 22, 1982 and finished on May 31, 1983. Monitoring was continued until September 1983 in order to determine to what extent temperature recovery of the soil is obtained in a natural way. The winter 1982/1983 was rather mild. Therefore the heat pump was operated in the monovalent mode only.

In Table 2 the measuring results and the COP during the heating season have been summarized on a monthly basis (5). The SPF, including auxiliaries like brine pump, internal fan and heat pump controls, amounts to 3.23. The heat output to the air has not been included in these figures, because exact measurement was a problem and the heated air was not used for space heating in the residence (the COP will be 3 to 6% higher when air heating is included). In order to determine the Primary Energy Ratio (PER) a conversion from COP by a factor 0.35 has been made. In Table 3 PER-values have been given as a function of heat source and heat sink temperatures, on a daily basis. The PER based on the heating season amounts to 1.13. In Figure 4 a plot is presented of the results on a weekly basis, including the course of the average ambient temperature. The specific heat extraction has varied between a maximum of 65 W per meter running vertical tube length in October 1982 to a minimum of 41 $W.m^{-1}$ in February 1983 (operating period).

In Figure 5 the soil temperature at various measuring locations (see Figure 1) and the average brine temperature have been plotted versus time (weeks) for different depths. From the course of the difference between soil temperature at these depths and brine temperature it can be seen that during the first period of heat extraction the upper layers of the soil have played an important role, while halfway the heating season and in spring heat extraction from lower soil layers has dominated. The net positive contribution of the horizontal tubes to heat extraction, however, is limited.

The minimum daily averaged brine temperature at evaporator inlet was 3.2°C. Freezing of soil did not occur during the heating season under consideration. Due to the relatively mild winter 1982/1983 these results are slightly better than can be expected in a more rigorous winter.

The results of soil temperature recovery are also shown in Figure 5. Temperature recovery on August 21, 1983 was not yet complete for depths between 2 m and 15 m, however, the temperature difference was small, i.e. 0.5 to 1.5 K. A distinct effect of ground water movement on the recovery process has not been traced. It is expected that temperature recovery will be nearly complete at the start of the next heating season.

A comparison of field and calculation results has been made in Figure 6 on a weekly-average basis. The effect of the horizontal tubes has been included. A satisfying agreement has been obtained since the overall difference in extracted heat from the soil amounts to 7%, whereas the actual temperature behaviour of the soil at various depths has been simulated to a large extent.

8. ECONOMIC ASPECTS

The primary energy saving with the heat pump, in comparison with the present natural gas boiler, amounts to 35%, assuming a combi-boiler efficiency of 73% (lower heating value), in relation with the first heating season 1982/1983. Based on the 1983 energy prices, including VAT, the energy costs are by 7% lower than for the reference heating system. However, due to the relatively high investment costs for the heat pump system, especially the heat pump itself, the total system costs are still too high for economic application in The Netherlands for the present (pay-back time longer than life time). When a high-efficiency gas boiler is considered as a reference system the economy will become even worse for the heat pump.

The specific, commercial price of the subsoil heat exchanger, including the horizontal tubing, amounts to Hfl. 8.30 (3.0 ECU of 1981) per meter vertical tube and installing costs Hfl. 12.18 (4.3 ECU of 1981) per meter, which includes jetting the holes for the tubes and making the trenches. The costs for the heat source therefore form about 16% of the total cost of the installed heat pump system.

9. CONCLUSIONS

- There is great variety in vertical subsoil heat exchanger design that can be met with practical installations. Especially tube length, tube-to-tube distance and number of tubes strongly differ for comparable installations. The types of heat exchanger used most frequently are concentric and, to a lesser extent, parallel tubes made of plastic. The majority of these systems operate with indirect evaporation.

- Artificial recharging of the soil in the spring and summer period is practised especially in Nordic countries.

- Existing techniques for making holes for subsoil heat exchanger tubes greatly influence the installation cost. Water-jetting with a lance is a cheap technique, although practical depths attainable by hand are limited to roughly 15 m, provided no thick clay layers are present. In the case of existing premises the accessibility of the garden with heavy machinery may be a constraint.

- During operation 65 W per running meter vertical tube length was extracted at maximum and 41 W per m at minimum (coldest period).

- A primary energy saving (natural gas) of 35% was achieved in the heating season 1982/1983, in comparison with the present natural gas boiler.

- Temperature recovery of the soil in a natural way is almost complete for depths between 2 m and 15 m (on August 22, 1983). A significant effect of ground water movement on the recovery process has not been observed.

- Application of vertical earth heat pump systems, in new and existing residences, looks promising in countries with favourable conditions for heat pump application. Especially the vertical subsoil heat exchanger arrangement enlarges the possibilities of using the earth as a heat source at reasonable costs for installation and material, while only a small ground area is necessary.

10. REFERENCES

(1) Nievergeld, P. et al.
Investigations on using the earth as a natural heat source for heat pumps.
Final EC report 231-77 EEN. November 1978.

(2) Nievergeld, P. et al.
Experimental investigations about using the earth as a heat storage medium and as a heat source for heat pumps.
Final EC report 367-78-1 EENL. March 1981.

(3) Bouma, J.W.J., Koppenol, A.D.
Investigation into a complete earth-to-water heat pump system in a single-family dwelling focussing on the application of a vertical subsoil heat exchanger.
Progress report no. 1. Contract EE-A-4-033-NL. April 1982.

(4) Bouma, J.W.J., Koppenol, A.D.
Progress report no. 2. Contract EE-A-4-033-NL. September 1982.

(5) Bouma, J.W.J., Koppenol, A.D.
Progress report no. 4. Contract EE-A-4-033-NL. September 1983.

Table I: SPF VERTICAL BUNDLE; 7 TUBES

Tube length	20 m	15 m	10 m
1st year	3.39	3.29	3.12
2nd year	3.28	3.19	3.05
3rd year	3.23	3.15	3.03

Table II: MONITORING RESULTS; HEATING SEASON 1982/1983

Month	Running time (h)	Brine temp. (°C)	Water temp. (°C)	Electricity consumption compressor (kWh)	Electricity consumption auxiliaries (kWh)	Extracted heat from the soil (kWh)	Heat output to the water (kWh)	COP incl. aux. (-)	COP excl. aux. (-)
Oct.*	19.6	-	-	34.1	3.5	127.4	141.0	3.75	4.13
Nov.	90.3	9.4	30.0	156.4	15.0	505.1	609.5	3.56	3.90
Dec.	194.6	6.5	32.9	334.7	31.8	915.1	1208.0	3.30	3.61
Jan.	176.0	6.2	33.0	302.8	29.3	822.5	1078.8	3.25	3.56
Febr.	266.6	4.1	36.0	457.9	43.8	1111.3	1529.2	3.05	3.34
March	168.0	5.6	33.0	288.6	29.3	746.9	994.4	3.14	3.45
April	102.5	6.8	31.8	176.3	20.6	475.8	624.7	3.23	3.54
May	65.0	8.4	30.6	112.0	16.0	328.8	423.5	3.45	3.78
Total	1082.6			1862.8	189.3	5032.9	6609.1		

* from 22nd to 31st

Table III: PRIMARY ENERGY RATIO (PER); HEATING SEASON 1982/1983

Temp. of heat source (°C) \ Temp. of heat sink (°C)	26	27	28	29	30	31	32	33	34	35	36	37	38
3											1.06	1.06	1.05
4								1.09		1.07	1.06	1.07	1.05
5							1.11	1.08	1.09	1.08	1.10		
6					1.12	1.11	1.12	1.12	1.12	1.12	1.12		
7				1.18	1.17	1.16	1.16	1.16	1.17	1.17			
8				1.18	1.20	1.19	1.21	1.19					
9				1.23	1.23	1.23	1.20	1.26	1.24				
10		1.34	1.32		1.30		1.26						
11	1.37	1.35	1.35	1.37									

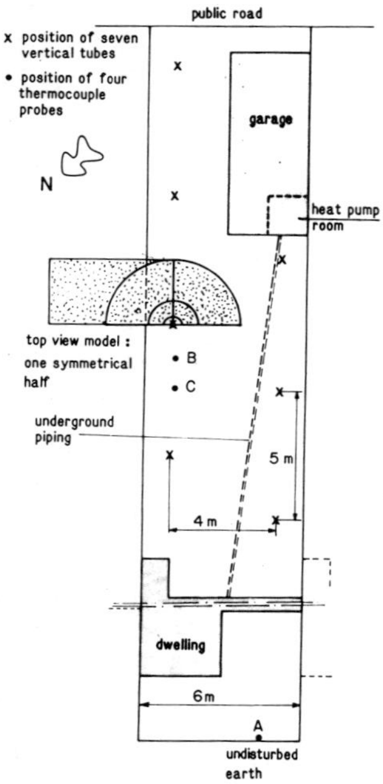

Fig. 1 - Map test site

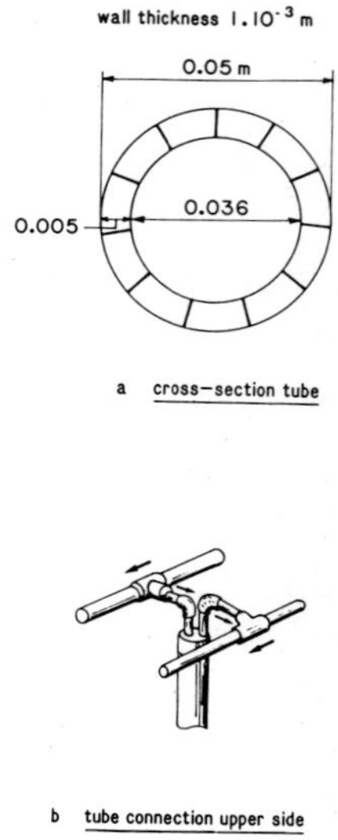

Fig. 2 - Details vertical tube

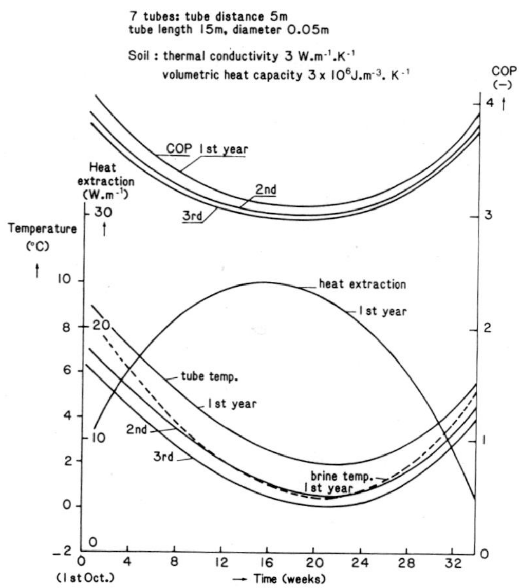

Fig. 3 - Calculated performance of vertical tube bundle

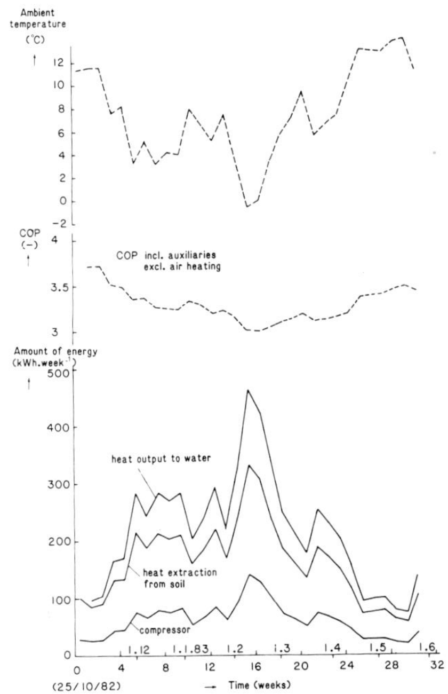

Fig. 4 - Heat pump performance in the field (weekly average)

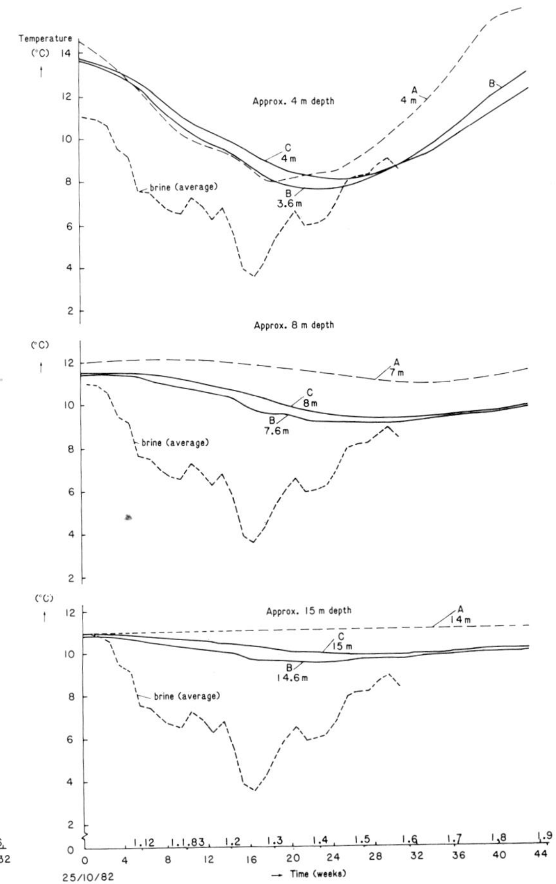

Fig. 5 - Course of soil and brine temperatures

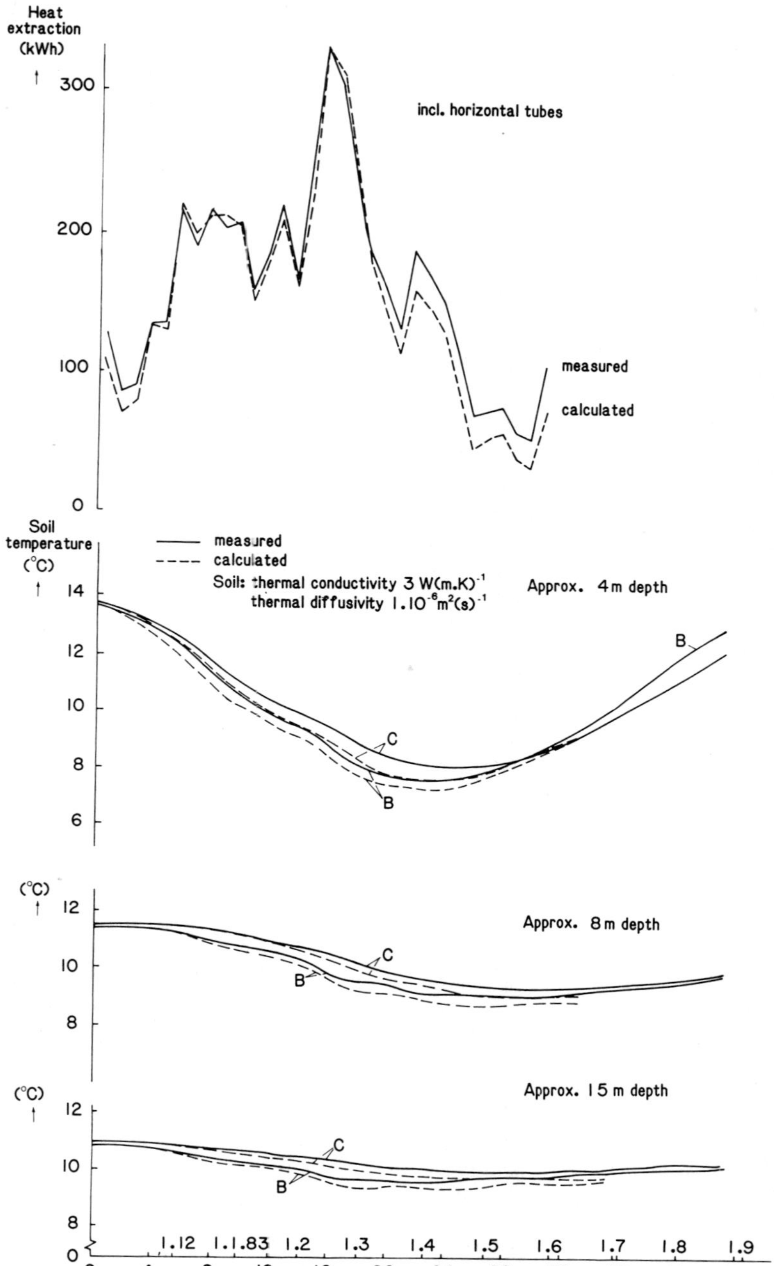

Fig. 6 - Comparison of monitoring and simulation results

HEATING OF A LABORATORY HALL BY MEANS OF A SOIL HEAT PUMP COMBINED WITH UNSHIELDED SOLAR ABSORBERS

B. GEERAERT and J.M. DE HOE
LABORELEC
Belgian Laboratory of the Electricity Supply industry
Department 8 : Electroheat and electrochemistry

Summary

A laboratory hall was heated by three heat pumps each of them using an horizontal ground heat exchanger as the only heat source during a first heating season. The installation performances were in good accordance with the results predicted by a simulation programme. Two of the existing soil heat exchangers were combined with energy absorbers during the two next heating seasons. No clear improvement of the yearly global COP followed from such a combination.
A short start-stop cycle time reduces the H.P. COP, the insertion of buffer tanks did not improve the yearly global COP. The stand-by consumption of the H.P. when the compressor is turned off is not negligible when the H.P. is running only a few hours a week.
The matching between H.P. without power control and energy absorbers is unfavourable because the H.P. are dimensioned for covering the nominal heat losses. Assistance of the soil heat exchanger by energy absorbers does not seem to be attractive but can bring a solution when the available soil surface is too small. A promising alternative for energy absorbers is their use as the only heat source of a bivalent H.P. system but cost reduction is a must.

1. OBJECTIVE OF THE WORK.

This research project has two main objectives :

a) an experimental verification of a theoretical computation model of an horizontal soil heat exchanger used as a heat source for a heat pump
b) and the study of the influence of coupling unshielded solar absorbers to an horizontal soil heat exchanger in order to improve the global installation COP or to reduce the needed soil surface.

2. DESCRIPTION OF THE EXPERIMENTAL INSTALLATION.

A laboratory hall (42 kW nominal heat losses - 457 m^2) was formerly equipped with direct resistance heating with fresh air intake through the walls (installed power = 57 kW) until the heating season 1979 - 1980.
Since the heating season 1980 - 1981, the laboratory is being heated by three soil heat pump circuits, each of which consists of a water to water heat pump (compressor power 5 kW) coupled with a soil heat exchanger on the evaporator side (4 loops of 100 m buried in two layers at 0,60 m and 1,20 m depth - 207 m^2 ground area) and 4 radiators or fan assisted convectors on the condensor side. Two of the existing soil heat exchangers have been combined with unshielded solar absorbers during the heating season 1981 - 1982.

52 m^2 of energy façade are coupled by means of a three-way valve with the first exchanger, working simultaneously with it.
A 40 m^2 energy enclosure is used alternately with the second heat exchanger. During the heating season 1982 - 1983, some modifications have been brought to the circuits, like automatic turn over from absorbers to soil exchanger or insertion of buffer tanks in the condensor circuit of the 2 heat pumps working with fan assisted convectors.
Figures 1 gives the schematic layout of the first heating circuit as it was during the season 1982 - 1983.

3. MAIN RESULTS.

3.1 Overall results.

A survey of the overall results is given in tabel 1. The heat demand has about been halved compared to the former resistance heating; this is due to the thermostat set-back during the closing hours and to the suppression of air intake through the walls at the rear of the electrical fan assisted radiators. The electrical consumption for heating purposes has been lowered from 1,41 kWh/W.season (with resistance heating) to 0,32 kWh/W.season; this is imputable partly to the heat demand reduction and partly to the use of solar assisted soil heat pumps.

3.2 Analysis of the heat sources behaviour.

- The influence on the COP of coupling heat absorbers with two soil heat pumps is not clear because other parameters are playing an important role (see point 3.3). Though, the reduction of the soil heat exchanger load due to the assistance of heat absorbers is evident when looking at the mean tube surface temperature of each heat exchanger (figure 2). The soil around the pipes of an unassisted soil heat exchanger gets frozen during the coldest winter conditions, whereas for solar assisted soil heat exchangers the temperature is always lying above the freezing point.
- Matching of the absorber surface, the heat pump power and the heat losses of the building.
  Classical commercial heat pumps are conceived for operation with a lowest glycol water outlet temperature of about - 7°C. The lowest air temperature below which such a heat pump can no longer run with the absorber as the only heat source depends on the installed heat pump power for a given absorber surface. The less the H.P. power, the less the temperature difference between ambient air and glycol water for a given permeability of the absorber. In the case of this experiment, the heat pumps have been choosen in order to meet the whole building heat demand at the lowest outside air temperature with a soil heat exchanger as a heat source. Some type of power control of the H.P. (double compressor speed e.g.) would allow to yield a lower turn-over temperature (0°C instead of 3°C in the case of the tested energy enclosure).
- Three energy absorber configurations have been tested : energy enclosure (33 W/m^2.K (*)), energy façade (20 W/m^2.K (*)) and energy stack at soil level (17 and 25 W/m^2.K (*))

  (*) : overall thermal permeability ambient air/glycol water of heat absorbers.

TABLE 1 : Heating of a laboratory hall by means of a soil heat pump combined with unshielded solar absorbers. Survey of the main characteristics.

<table>
<tr><th>Season</th><th>Electric heating system</th><th>Control system</th><th>Results</th></tr>
<tr><td>1979
-
1980</td><td>- resistances
- mechanical ventilation through the wall</td><td>- no heating at week-ends and holidays</td><td>- heat demand
= electric consumption
= 1,41 kWh/W.season (1)</td></tr>
<tr><td>1980
-
1981</td><td>- only 3 soil heat pumps
- no longer mechanical ventilation through the wall</td><td>- permanent heating at the same thermostat setting
- circulators synchronised to the operation of the heat pumps</td><td>- heat demand (1) :
1,17 kWh/W.season
- electric consumption (1)
0,52 kWh/W.season
<table>
<tr><th></th><th>COP HP</th><th>COP glob.</th><th>Oper. time h/seas.</th></tr>
<tr><td>HP 1</td><td>3,01</td><td>2,24</td><td>1 740</td></tr>
<tr><td>HP 2</td><td>2,86</td><td>2,41</td><td>642</td></tr>
<tr><td>HP 3</td><td>2,56</td><td>2,21</td><td>649</td></tr>
<tr><td>mean</td><td>2,85</td><td>2,27</td><td>-</td></tr>
</table>
- good accordance with predicted results</td></tr>
<tr><td>1981
-
1982</td><td>- 1st soil HP assisted by 52 m^2 of energy façade
- 2d soil HP assisted by 40 m^2 of energy enclosure (very partially)
- 1 buffer tank in the condensor circuit of the 3rd HP at the end of the heating season</td><td>- intermittent heating : thermostat set-back at night, week-ends and holidays
- aquastat in buffer tank controls the 3rd HP
- circulators running only when needed</td><td>- heat demand (1) :
0,78 kWh/W.season
- electric consumption (1)
0,32 kWh/W.season
<table>
<tr><th></th><th>COP HP</th><th>COP glob.</th><th>Oper. time h/seas</th></tr>
<tr><td>HP 1</td><td>2,41</td><td>1,99</td><td>467</td></tr>
<tr><td>HP 2</td><td>3,21</td><td>2,79</td><td>1 175</td></tr>
<tr><td>HP 3</td><td>-</td><td>2,03</td><td>316</td></tr>
<tr><td>mean</td><td>-</td><td>2,43</td><td>-</td></tr>
</table>
- strong influence of low power auxiliairies and running duration on the global COP.</td></tr>
<tr><td>1982
-
1983</td><td>- 2 soil HP fully assisted with energy absorbers
- buffer tank in the condensor circuit of the 1st and 3rd HP (working with fan assisted convectors)</td><td>- better repartition of the heat demand over the 3 HP by adjusting the set point of the wall thermostats.
- adjustment of the aquastat setpoint as a function of</td><td>- heat demand (1) :
0,72 kWh/W.season
- electric consumption (1)
0,32 kWh/W.season</td></tr>
</table>

TABLE 1 (continued)

Season	Electric heating system	Control system	Results			
				COP HP	COP glob.	Oper. time HP h/seas.
1982 - 1983		the outside temperature (from March 1983 on)	HP 1	2,58	2,02	487
			HP 2	2,92	2,55	532
			HP 3	2,49	2,15	624
			mean	2,67	2,23	-

(1) The heat demand and electricity consumption are reported to a type heating season of 2 087 degree-days 15°C/15°C (kWh/W nominal heat losses . season)

3.3 Analysis of the overall results.

3.31 General remarks.

- The steady state characteristics of the used heat pumps are those given in figure 3 except for the first circuit on which another heat pump with a better COP (+ 3 ... 5%) has been operating during the season 1980 - 1981.
- The auxiliaries (fans and circulation pumps at condensor and evaporator side) are working only when needed in this experimental installation.
- As shown in figure 4, a short start-stop cycle time considerably reduces the COP of the H.P. The duration of such a cycle is principally depending on the thermal inertia of the water content in the condensor circuit. In order to increase this start-stop cycle time in the circuits with low water content (circuits working with fan assisted convectors), buffer tanks were inserted in series.
  Problems occuring in this case are a decrease of the water flow rate and the best choice of the aquastat set point (the COP decreases if the set point is too high and the auxiliaries consumption increases if it is too low.
- Running hours of the heat pump.
  As shown in figure 5, the stand-by consumption of the H.P. when the compressor is turned off (... about 100 W for crankcase heater and control circuits) is not negligible especially during the period that the H.P. is running only a few hours a week. The H.P. and installation COP computed without stand-by consumption are given in table 2 in order to allow a comparison of performances independent of this parameter.

3.32 CIRCUIT 1 : energy façade (table 2)

The COP computed without taking into account the stand-by consumption is the best during the heating season 1980 - 1981 (soil only heat pump without energy façade); this can be explained by the better COP of the H.P. used during this first season.
The increase of the H.P. COP (+ 7%) between 1981/82 and 1982/83 is due to the insertion of a buffer tank that doubles the mean start-stop cycle time without affecting too much the water flow-rate (1 620 l/h instead of 1 800 l/h.

However, the installation COP remains quite constant, probably because the H.P. COP increase is counterbalanced by an increase of the auxiliaries consumption (1 162 working hours for the auxiliaries at the condensor side). There is no improvement of the installation COP due to the coupling with an energy façade working simultaneously with the soil heat exchanger (3,7 m^2 frontal surface/kW nominal heat losses - 20 $W/m^2.K$ thermal permeability of the energy façade).

3.33 CIRCUIT 2 : energy enclosure (table 2)

There is no significant difference in performances between 1981 - 1982 and 1982 - 1983 when looking at the H.P. and installation COP computed whithout stand-by consumption. The measured drop in H.P. and installation COP is only imputable to the greater importance of the stand-by consumption because the running hours of the heat pump itself have been halved (from 1 175 h to 532 h). The COP without stand-by consumption of the first heating season is lower than the one of the two others. Since the only modification brought to the installation between 1980/81 and 1981/82 is the coupling with the energy enclosure, one may suppose that the improvement of the global COP (+ 11%) is imputable to the coupling with the energy enclosure working alternately with the soil heat exchanger (2,9 m^2 frontal surface/kW nominal heat losses, 33 $W/m^2.K$).

3.34 CIRCUIT 3 : soil heat exchanger only (table 2)

The H.P. and installation COP without taking into account the stand-by consumption are quite constant over the three heating seasons. The only modification brought to the installation is the insertion of a buffer tank in series with the condensor circuit in the summer 1982.
Though the mean start-stop cycle time increases from 3,7 min to 14,3 min, there is no significant improvement of the H.P. COP without stand-by consumption; the reason is the drop of the water flow-rate over the condensor from 1 800 l/h to 1 300 l/h (nominal flow rate 2 000 l/h). The auxiliaries at the condensor side were in operation for 624 h during 1982 - 1983.

4. CONCLUSIONS

4.1 The heat demand for room heating of a 457 m^2 laboratory hall has been lowered down to about one half of the energy demand in the case of resistance heating. The main reasons are the suppression of mechanical ventilation through the wall and the thermostat set-back at night, week-ends and holidays.

4.2 The surface of the double layer soil heat exchanger of only 15 m^2 soil surface per kW nominal heat losses was fully sufficient for room heating in a climate with 2 087 degree-days 15°C/15°C for a building with 90 W/m^2 nominal heat losses.

4.3 When working without unshielded solar absorbers the installation performances are in good accordance with the results predicted by the simulation programme. The water contained in the clay ground around the pipes was froozen in the coldest weather conditions and the lowest pipe surface temperature of the non-assisted soil heat exchanger was - 2 ... - 2,5°C.

4.4 Due to the coupling of energy absorbers with the soil heat exchanger, the soil is no longer froozen around the pipes. The lowest mean value of the pipe surface temperature is + 2,5°C. This is clearly an improvement due to the assistance by the energy absorbers.

4.5 There is no clear improvement in the global COP due to the energy absorbers. Other parameters are playing an important role :
a) mean start-stop cycle time of the H.P.;
b) stand-by consumption of the heat pump when the compressor is turned off;
c) consumption of the auxiliaries.

4.6 When working with a linking between absorbers and soil heat exchangers in a monovalent heat pump system, some type of power control of the H.P. (double compressor speed e.g.) is necessary in order to yield a sufficiently low turn-over temperature with a limited absorber surface (3 ... 4 m^2/kW nominal heat losses).

4.7 This study yields a sufficient knowledge of the components of a solar assisted soil heat pump system (horizontal soil heat exchanger, solar absorbers, storage tanks, control system,...). Modelling of the interconnection between these components would be a useful follow-up of this study in order to allow consultant engineers to draw full profit from the information that has been acquired in the framework of this study. No attempt has been made for modelling the whole system because this requires a more fundamental study on modelling.

5. ECONOMIC ASSESSMENT.

5.1 Soil only heat pump system.

The investment for a direct resistance heating system at this moment is estimated to be equal to 57 kW x 5 000 BF/kW = 285 000 BF. The total investment (material and manpower) for the non assisted soil heat pump system amounts to 1 550 000 BF. Due to the experimental character of this installation, the investment has been higher than the one for a normal installation, which could be reduced to about 1 150 000 BF. The pay-back period of such a type of installation depends strongly on the heat demand pattern (thermostat set-back,...).
When taking into account :
- a yearly global COP of 2,3
- an electricity kWh price of 4 BF/kWh,
- an over-investment of 865 000 BF (1 150 000 - 285 000),

the pay-back period corresponding to a heat demand of 1,41 kWh/W.year, resp. 0,75 kWh/W.year, is equal to 6,5 years, resp. 12,2 years.

5.2 Solar assisted soil heat pump system.

No clear improvement of the yearly global COP follows from the assistance of the soil heat exchanger by energy absorbers so that such a combination does not seem to be attractive. However, the combination of energy absorbers and a soil heat exchanger can bring a solution when the available soil surface is too small and assistance of the soil heat exchanger being indispensable.

5.3 Energy absorber heat pump system.

Energy absorbers can act as the only heat source of a bivalent heat pump system with performances in the same order as these of classical air to water bivalent heat pump systems, but at a higher investment cost (+ 10 ... 20 %). Cost reduction of energy absorbers could be achieved by a better integration in the construction (energy roof, energy façade) or in the environment (energy enclosure). The use of energy stacks is a promising alternative allowing to reduce the installation costs.

TABLE 2 : Main results per circuit.

	Season	C.O.P. H.P.		Auxiliaries (fans, circulation pumps	C.O.P. instll.		Start - stop cycle time	Stand-by consumption	Running hours
		measured	stand-by consumption not included		measured	stand-by consumption not included			
	(-)	(-)	(-)	(W)	(-)	(-)	(min)	(W)	(h)
CIRCUIT 1	1980-81	3,01	3,10	1 200	2,24	2,30	-	61	1 740
Energy façade	1981-82	2,41	2,76	1 300	1,99	2,22	7,2	90	467
	1982-83	2,58	2,96	1 300	2,02	2,24	15,8	94	487
CIRCUIT 2	1980-81	2,86	3,18	950	2,41	2,63	-	90	642
Energy enclosure	1981-82	3,21	3,42	1 090	2,79	2,95	34,1	90	1 175
	1982-83	2,92	3,40	1 090	2,55	2,91	16,5	90	532
CIRCUIT 3	1980-81	2,56	2,73	850	2,21	2,37	-	70	649
Soil exchanger only	1981-82	2,29	2,72	890	2,03	2,35	3,7	70	316
	1982-83	2,49	2,75	890	2,15	2,34	14,3	70	624

CIRCUIT 1 (AND) Soil heat exchanger only
Soil heat exchanger and unshielded solar collectors

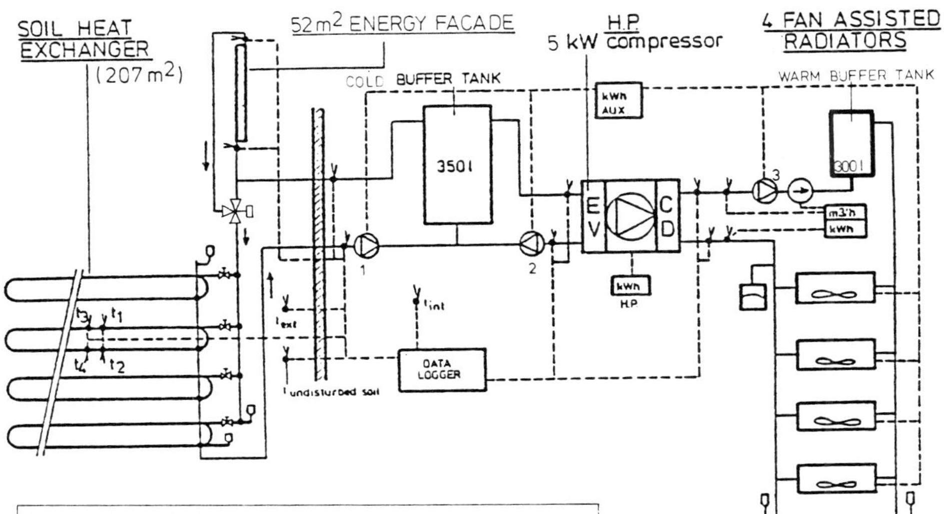

Control system
A thermostat on the wall controls fans and circulation pump 3.
An aquastat in the warm buffer tank controls heat pump and circulation pumps 1, 2 and 3.

Fig. 1

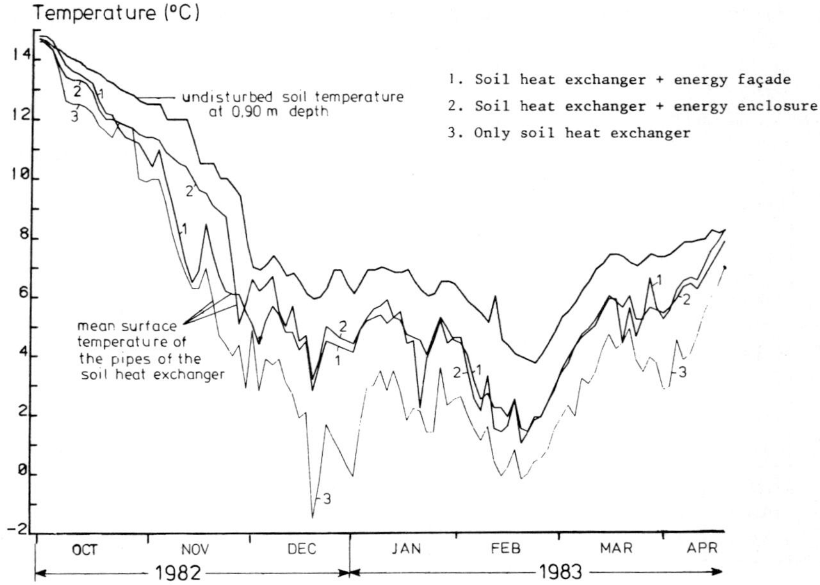

Fig. 2 Evolution of the soil temperature during the season 1982 -1983.

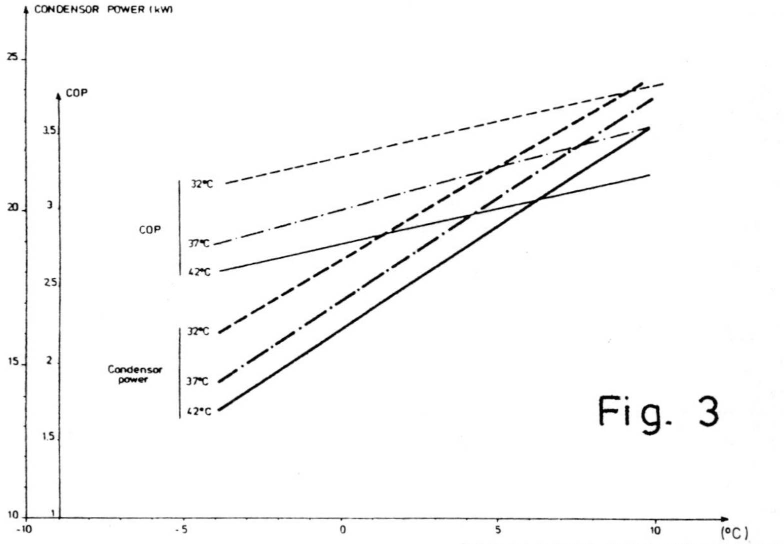

Fig. 3

Steady state characteristics of the heat pumps used on the soil heat exchangers.

Nominal compressor power : 5 kW
Water flow in condensor : ... 2,2 ... m^3/h
Glycol-water flow in evaporator : ... 4,2 ... m^3/h

32°C, 37°C, 42°C : hot water inlet temperature.

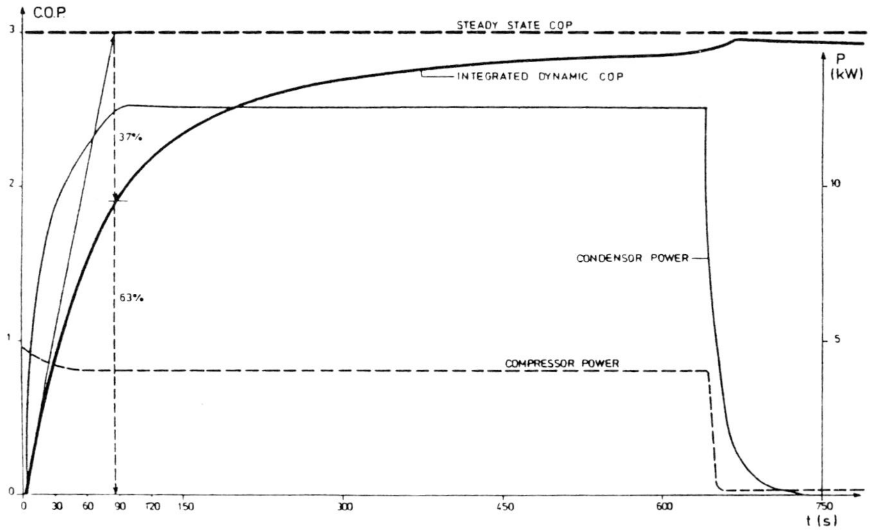

Fig. 4 Dynamic behaviour of a water to water heat pump. Evolution of the condensor power, the compressor power and the integrated dynamic COP as a function of the absolute operating time.

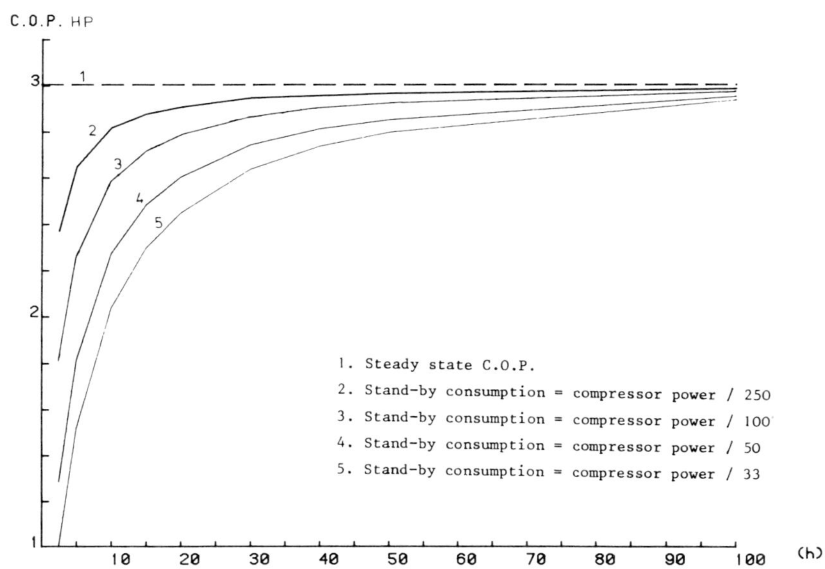

Fig. 5 Influence of the stand-by consumption (crankcase heater and control equipment) on the COP of the H.P. when the compressor is turned off, as a function of the weekly running hours of the H.P.

A HIGH TEMPERATURE LIFT ABSORPTION HEAT PUMP

C.O.B. CAREY, J.S. KHAHRA & I.E. SMITH
School of Mechanical Engineering
Cranfield Institute of Technology

Summary

Mixtures of alkali-metal hydroxides (principally sodium) with water have been identified as constituting a useful pair for use in an absorption heat pump capable of providing the high temperature lift that is necessary for hydronic domestic heating systems with ambient air as the heat source. A coefficient of performance in the region of 1.5 has been calculated for this combination using realistic input data. The evaporation of water at sub-zero temperatures has been studied, and appears to present no insuperable problems. Crystallisation problems, although not completely eliminated, have been reduced by the adoption of salt mixtures. An experimental heat pump has been constructed and preliminary performance results are listed and discussed.

1. INTRODUCTION

The absorption cycle has, in the past, been used for refrigeration and chilling purposes mainly with ammonia and water as the working fluids for the former and water and lithium bromide for the latter. Very little experience exists with working fluids other than these for heat pumping applications, although certain organic pairs have been proposed and are believed to have reached the prototype stage.

A potentially large market exists for absorption heat pumps as a boiler replacement in domestic heating situations, for in the UK alone 200,000 gas fired boilers are replaced annually. A heat pump with a heating COP of 1.5 would save one third of the fuel used by a conventional boiler.

However, to be compatible with existing hydronic heating systems and also to produce domestic hot water, such a heat pump must be capable of delivering hot water, at 60-65°C for a return temperature of 50-55°C, and of absorbing heat down at ambient air temperatures of zero and below.

The water-lithium bromide combination is incapable of a temperature lift greater than about 45°C at these temperatures, and although the ammonia-water cycle can achieve these figures it's COP is low as a result of the need for rectification; futhermore the condensing pressures at 70°C are of the order of 35 bar necessitating a high pressure solution pump. Toxicity may also be a problem in an ammonia system if the equipment is located within a building.

Whilst methanol with a mixture of bromides as the absorbent is capable of operating at -10°C in the evaporator and 75°C in the absorber, (SMITH et al (1)), the combination has been found to be unstable at the high generator temperatures (ca 180°C) that are necessary for its operation.

Apart from certain fluorocarbons, all organic compounds investigated have a questionable stability at these temperatures when a decade is considered as the required time scale, and fluorocarbons also have the disadvantage of a low latent heat.

In many, if not most, respects water is the ideal working fluid for the above application of heat pumps and this paper examines some problems associated with its use.

Absorbents for water that will produce a working temperature lift of about 80°C from normal environmental temperatures are not very numerous. A limited solubility eliminates most salts which exhibit a large negative deviation from Raults' Law and the most effective absorbent is undoubtedly sulphuric acid. However, the requirement of a high temperature lift is paralleled by the necessity for a high generator temperature (180-200°C) and at such temperatures highly concentrated sulphuric acid poses formidable problems in regard to corrosion. The only metallic material of construction capable of withstanding long term attack is tantalum, which is prohibitively expensive.

Ranking next to sulphuric acid is sodium hydroxide (potassium hydroxide has an inferior solubility). Sodium hydroxide is capable of providing an equilibrium temperature lift of from -10°C to 100°C at 70% salt concentration and up to 140°C at 80% concentration.

An obvious problem is the freezing of water at sub-zero temperatures which would seriously hinder heat transfer within the evaporator. Experiments have shown that water incorporating an anti-freezing component can evaporate effectively at temperatures considerably less than zero although the vapour pressure falls to rather low values, and these aspects will be dealt with later.

However, even supposing there is a lower limit to the evaporating temperature, and the heat pump becomes inoperative at temperatures below this, the consequences may not be too serious. In the UK climate the percentage of the annual heat load on premises heated between 06.00 and 23.00 hours is given approximately by $8.36\ T_A + 8$ where T_A is the ambient temperature (2). Thus, the heating demand when ambient temperatures are below zero, corresponding to an evaporator temperature of perhaps -10°C, is only 8% of the total. So,provided a heat pump can still deliver heat, even though not operating as a heat pump, it's viability may not be in jeopardy.

Such operation might be accomplished by shutting off the coolant supply to the condenser and permitting heat to pass directly from the generator to the coolant flow in the absorber.

2. SYSTEM CONSIDERATIONS

A typical absorption system employing an adiabatic absorber is shown in Fig. 1 from which the major components that will be referred to may be identified.

The equilibrium vapour pressure curves for pure sodium hydroxide solutions are illustrated in Fig. 2 and show that temperature lifts of over 100°C should be possible between evaporator and absorber (the shaded areas are where solidification occurs owing to hydration). Consequently quite large concentration changes in the absorbent may be tolerated whilst still leaving an ample driving potential for the absorption for temperature lifts of the order of 80°C, and this will result in low solution flow rates between absorber and generator. In addition these flow rates are also relatively low owing to the exceptionally high latent of water as compared with most other working fluids. The two factors combine to give a 'solution flow per kW evaporator heat' of the order of 2 g/s, in contrast to typical organic pairs which require flows of around 30-40 g/s. As a result a smaller solution heat exchanger is required and furthermore the COP becomes less sensitive to its performance. Fig. 3 shows the variation of COP with generator temperature T_G and solution heat exchanger effectiveness; the generator temperature is a measure of concentration change as well as giving the maximum operating temperature of the system.

One of the problems of using sodium hydroxide is that of crystallisation on cooling. At 65°C, sodium hydroxide forms a crystalline monohydrate containing 69% salt, as a result of which a solution of such a concentration solidifies at this temperature. Hence a system operating at these concentrations would always have to be maintained at temperatures above this level to avoid solidification of the solution accompanied by complete blockage of the equipment. Problems might not arise on short term cycling, but they most certainly would on complete shutdown. One way to avoid this might be by diverting water from the condenser to the generator and absorber and diluting the solution sufficiently to lower the crystallisation temperature. A typical dilution requirement would be 0.81 of H_2O per litre of strong solution. However, on starting this excess of water would have to be boiled off, during which period the absorber would not be operating at full temperature (although the condenser would). This would also lower the COP according to how long the unit had been operating following start-up. Investigations have shown, however, that mixtures of sodium and potassium hydroxides have appreciably lower solidification temperatures than either hydroxide alone. The p-T-x diagram for mixtures containing equal masses of sodium hydroxide and potassium hydroxide is shown in Fig. 4 from which it may be seen that a 70% solution of such a mixture only crystallises at 30°C in contrast to 65°C for the sodium salt alone although the equilibrium temperature lift (above 0°C) is reduced from 112°C to 82°C. This greatly reduces the quantity of pollution water necessary to prevent crystallisation on shutdown.

Other proportions of these two salts in water have been investigated, and although the results are not shown, there is a common trend of a reduction in the temperature lift accompanying a lowering of the crystallisation point. Work is in progress at present investigating other "additives" to sodium hydroxide that may have a beneficial effect. However, an equilibrium temperature lift of 80°C is still sufficiently large to enable a practical system to be designed and operated at, say, 70°C in the absorber. The temperature at which the condenser may be operated is, of course, limited only by temperatures that can be tolerated in the generator and solution heat exchanger, and these may well be determined by factors such as chemical corrosion, or the reduction in overall boiler efficiency.

Corrosion is an area that demands consideration and will dictate the material of which the heat pump is constructed, especially as regards the high temperature components. Although alkalis are less aggressive towards most metals than acids, they are nevertheless highly corrosive. Discussions with Imperial Chemical Industries (3) have indicated that nickel is extremely resistant to alkali corrosion at the temperatures in question, and although nickel is an expensive material in itself it is capable of being chemically deposited on less expensive base metals regardless of their surface geometry. Whilst this process is not at present cheap, it is not employed in any large scale production processes, and so the present-day figures may be unrealistically high. Another material that would also be satisfactorily corrosion resistant is nickel/cast-iron alloys, which have the advantage of being relatively inexpensive. Cast generators might well be not too dissimilar from existing boilers, although the solution heat exchanger would probably still have to be of fabricated construction.

For the lower temperature components (absorber and evaporator) existing grades of mild steel would probably be acceptable.

3. LOW PRESSURE WATER EVAPORATION

The development of a low temperature evaporator for an absorption heat pump with water as the working fluid poses two significant problems;

A) A low rate of heat transfer through the liquid to its surface and an unknown liquid-vapour interfacial temperature drop.

B) The freezing of water.

The liquid surface to hot wall temperature difference has been investigated by RABEN et al (4) in their experimental study of nucleate pool boiling at various pressures in the range of 13 to 1000 m bar. The results show temperature gradients which are almost an order of magnitude larger at a pressure of 13 m bar as compared with atmospheric. The heat transfer coefficients at this low pressure, for non-nucleate boiling are approximately 200 w/m^2K for a wall superheat of 3 to 10^oC.

This low rate of heat transfer at low pressures may be attributed to the diminishing contribution of convective heat transport and the disappearance of nucleate boiling in liquid pools. As a result, absorption chillers normally employ spray or falling film evaporators. In a falling film evaporator the heat is transferred by conduction through a thin water film and the heat transfer can accurately be predicted by using the Nusselt equation. However, the temperature drop owing to the interfacial resistance still cannot be predicted with certainty. The magnitude of this temperature drop may be estimated from kinetic theory by the equation;

$$\frac{\dot{m}}{A} = \left(\frac{M}{2\Pi R}\right)^{\frac{1}{2}} \sigma \left(\Gamma \frac{P_v}{T_v^{\frac{1}{2}}} - \frac{P_f}{T_f^{\frac{1}{2}}}\right)$$

This shows that for a given fluid and evaporating mass flux $(\frac{\dot{m}}{A})$ the interfacial temperature drop, $T_v - T_f$, increases with decreasing values of system pressure, P_v, and evaporation coefficient, σ. For a mass flux of 0.01 Kg/m^2-sec (25 kW/m^2) SUKHATME et al (5) have used this formula to calculate the interfacial temperature drop for water at decreasing pressures.

However, the correct value of σ to employ is a matter of debate, and is the largest source of uncertainty in the estimation of the interfacial temperature drop. For the generally accepted value of $\sigma = 0.04$ (JAMEISON (6)) for water, the calculated temperature drops would be 0.063^oC and 3^oC respectively for pressures of 1000 and 13 m bar at a heat flux of 25 kW/m^2. Although GALLAGHER-DAGGIT et al (7) reported a value of σ as low as 0.005 at 13 m bar and applied it with apparant success to their falling film evaporator, this should not be regarded as being conclusive.

An estimate of interfacial temperature drop on one commercial absorption chiller which employs water in a falling film evaporator, suggests that the value of σ lies between 0.014 to 0.03 depending on the assumed value of saturation vapour pressure inside the evaporator. Even at the lower value of 0.014 the overall heat transfer coefficient of the evaporator would still be of the order of 1.3 kW/m^2K. However, the value of σ at lower pressures, corresponding to the evaporation of water at sub-zero temperatures requires confirmation for this type of evaporator.

Experiments are currently under way to study the heat transfer and interfacial resistance phenomenon in falling film evaporators at low pressures and this equipment will also enable further studies concerning the second problem, namely the evaporation of water containing an anti-freeze component.

Preliminary tests have been carried out in a simple pool evaporator and have shown that water does evaporate satisfactorily at temperatures as low as -7°C from weak calcium chloride solutions (PRITCHARD (8)) but for reasons already discussed it was not possible to deduce reliable values of evaporation coefficient for application to falling film evaporators. The experiments also demonstrated that another problem, namely the re-freezing of pure vapour on cold surfaces was by no means as serious as the normal external frosting problem that occurs at higher air temperatures. An experimental heat pump (sect. 4) has been successfully operated with evaporation temperatures of -5.

It may be that certain types of evaporator might need to re-circulate the anti-freeze mixture and mix it with the incoming condensate. Rather than employ a pump for this process, the energy from the flash evaporation of hot condensate could be utilised to provide this circulation.

4. EXPERIMENTAL UNIT

An experimental heat pump has been designed and built to give an output of up to 3 kW. Since operating pressures are sub-atmospheric, much of the construction is in industrial glassware so that the behaviour of the system can be observed. For performance measurements, however, the equipment will be thermally insulated, but the facility to see what is taking place during initial tests is a valuable asset.

Electric heaters are employed in the generator and evaporator to enable power inputs to be accurately measured and controlled. An adiabatic absorber is used consisting of a packed column of ceramic Intalox saddles enabling the height of the column to be adjusted in order to optimise absorption performance : the maximum surface area obtainable is 3 m^2 with vapour pressure drops being less than 0.5 m bar.

Both absorber and solution heat exchangers are of the plate type which, with narrow plate spacings, favour heat transfer between viscous fluids with minimum pressure drop. The size of the heat exchangers may readily be adjusted by the addition or removal of plates. The high heat transfer rates during condensation allow a simple coiled pipe heat exchanger to be used for the condenser. With a surface area of 0.1 m^2, it is only a third of the area of the absorber heat exchanger.

For a concentration change of 5% the solution heat exchanger should be capable of exchanging 1.7 kW with solution flows of 6-7 g/s (ca 2.4 ℓ/min). In the absorber heat exchanger provision is made for circulating solution at up to four times these flow rates in order to permit operation at an almost constant temperature. The solution pump is of completely sealed construction with a magnetic drive to eliminate shaft sealing problems.

Because the system is sub-atmospheric it is imperative that no non-condensable gases are present, since this will prevent effective absorption and condensation. Although this should not be a problem for a commercial system (e.g. of welded construction and with palladium cells for hydrogen removal), it is difficult to make an experimental unit sufficiently leak tight when subject to periodic dismantling. The smallest achievable leaks are of the order of 10^{-2} - 10^{-3} mbarls^{-1} and so provision is made continuously to evacuate the system with a pump via a throttle valve. The leak is balanced with the throttle, and any water vapour lost is held in a LN_2 trap.

The unit has not yet operated with fully concentrated solutions

(65/70% salt) although temperature lifts of 60°C have been achieved with potassium and sodium hydroxide mixtures. The principal results obtained are summarised in table 1.

The generator has successfully run at 150°C and above for many hours with no apparent damage to the glass vessel, the nickel plated electric heater or the plate heat exchangers, and crystallisation has not been found to be a problem.

5. CONCLUSIONS

The main objective behind this work has been to evaluate the potential of sodium hydroxide and water as a working pair for absorption heat pumps capable of serving a boiler replacement unit in small scale (domestic) applications. Experience has been gained in using sodium hydroxide in an experimental heat pump and of the problems associated with the use of water as a working fluid at low pressures. At the time of writing the full potential of the high temperature lift of which sodium hydroxide is capable has yet to be realised, but there seems little doubt that it will be achieved in due course.

6. REFERENCES

1. I.E. SMITH & S.E. EL-SHAMARKA : Absorption Heat Pumps for Space Heating, Proc. 3rd Int. Conf. Future Energy Concepts, London (1981) pp. 232-237.
2. British Gas Corporation. Private Communication.
3. I.C.I. Ltd. Private Communication.
4. I.A. RABEN, T.B. RICHARD and C. GEORGE : A Study of Nucleat Pool Boiling of Water at Low Pressure. Sixt National Heat Transfer Conference, A.I.CH.E.-A.S.M.E. August 11-14 (1963).
5. S.P. SUKHATME and W.M. ROHSENOW : Heat Transfer during Film Condensation of a Liquid Metal Vapour. Journal of Heat Transfer, Transactions of A.S.M.E. Feb. (1966) pp. 20-28.
6. D.T. JAMEISON : The Condensation Coefficient of Water. National Engineering Laboratory, Report No. 186, May (1965).
7. G. GALLAGHER-DAGGIT, R. Hopes and J. Brown : A Thermochemical Heat Pump/ Energy Storage Scheme Based on The Sulphuric Acid-Water System. Rutherford Laboratory, Report No. RL-79-041, April (1979).
8. M. PRITCHARD : Development of a Low-Temperature Evaporator for Absorption Heat Pumps. M.Sc. Thesis, Cranfield Institute of Technology U.K., Sept. (1982).

	generator	condenser
maximum power transferred	1·7 kW	1·2 kW
maximum temperature (vapour)	165 °C	80 °C
	absorber	**evaporator**
maximum power transferred	1·1 kW	0·7 kW
maximum temperature t_e 0 °C	60 °C	
minimum t_e		-5 °C

Table 1 - Test rig performance

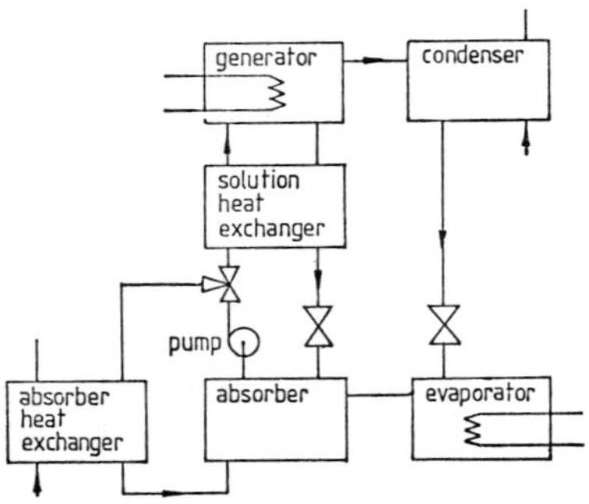

Fig. 1 - Diagram of an absorption system

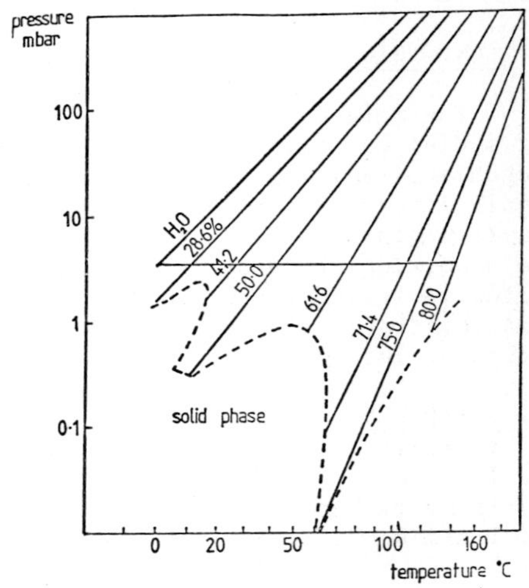

Fig. 2 - p-t-x NaOH:H_2O system

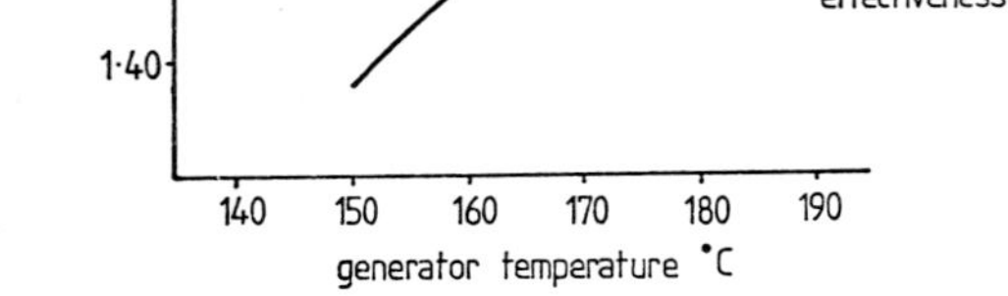

Fig. 3 - NaOH:H_2O heat pump performance

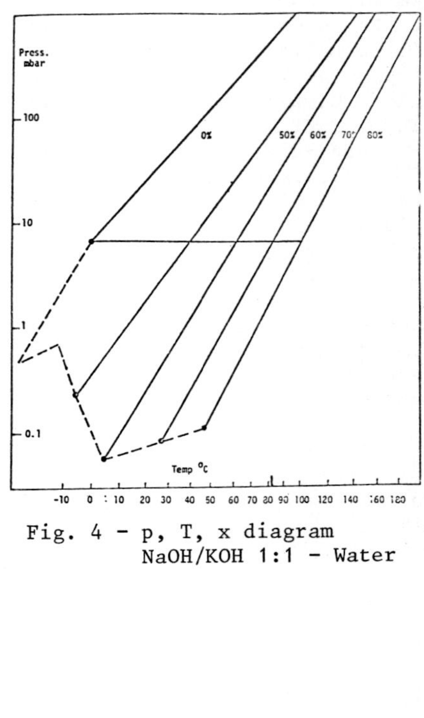

Fig. 4 - p, T, x diagram
NaOH/KOH 1:1 - Water

HEAT PUMPS WITH FLUID MIXTURES APPLIED TO SPACE HEATING

CL. RAMET - A. ROJEY & J. DURANDET

Institut Français du Pétrole
1 et 4 Avenue de Bois Préau
92502 RUEIL-MALMAISON (France)

AIM : Research and development program of compression heat pumps operating with fluid mixtures. This work is being carried out by IFP and LEROY-SOMER.

A. - GENERAL BACKGROUND

1. Interest of non azeotropic fluid mixtures

In the field of compression heat pumps for space heating, the use of non-azeotropic fluid mixtures makes it possible :

- to improve the coefficient of performance,
- to increase the thermal capacity.

Each way can be followed by involving specific optimized mixtures.

1.1 COP improvement

The performance of compression heat pumps can be improved through a better reversibility of heat exchanges at the evaporator and the condenser.

When heat is exchanged, the outside fluid temperatures generally vary, whereas in conventional heat pumps the temperature of the working fluid remains virtually constant during the evaporation and the condensation stages.

The mixture is selected in order to have a temperature variation in the evaporator and the condenser as close as possible to that of the outside fluid during heat exchange. It is essential that the heat exchangers operate under counter-flow.

In the applications of the heat pumps for space heating, the interval of heat recovery is usually lower than the interval of heat supply. Therefore a specific flowsheet involving an additional heat exchanger makes it possible to obtain the optimal conditions of reversibility with an appropriate fluid mixture.

Under such conditions, a COP improvement as high as about 10 % can be expected over the performances obtained with conventional refrigerants used in water-to-water or air-to-water heat pumps for space heating.

1.2 Increase in thermal capacity

The thermal capacity of a compression heat pump can be increased through the use of a selected non-azeotropic fluid mixture, which has a higher volumetric heat output, defined as the ratio of the condenser effect above the volumic flowrate at the compressor intake. A low specific volume of the vapor and a large latent heat of vaporization are required. An optimized fluid mixture used instead of a pure refrigerant makes possible the thermal output to be substantially increased, whereas the COP is conserved.

The increase in capacity which can be obtained is much higher than that obtained through the use of an azeotropic mixture, having a lower boiling point than the pure refrigerant, i.e. over 20 % without any change in the lay-out of the heat pump. Therefore, for a given heating requirement, this advantage leads to a lower investment cost.

2. Selection of the fluid mixtures

When selecting a mixed refrigerant, a lot of properties have to be considered, among them : the vaporization and condensation temperature ranges needed ; the latent heat of vaporization ; the specific volume of the vapor phase ; the ratio of heat capacities C_p/C_V ; the thermal stability ; the products cost and safety (toxicity ; inflammability).

For optimizing the fluid mixture composition, a thermodynamical model based on the Soave equation is used. This model represents all the fluid mixture properties, namely vapor-liquid equilibrium and enthalpic characteristics. For predicting the behavior of either a pure component or a mixed refrigerant, different computer simulation programs are used.

B. - IFP EXPERIMENTAL WORK

1. Description of the test loop

An experimental water-to-water test loop was used for testing and comparing the performances of fluid mixtures and pure components. The heat pump was composed of a lubricated piston compressor associated with an oil separator and of standard double-pipe finned exchangers.

The experimental COP was the ratio of the condenser thermal effect over the compression power on the shaft.

2. Example of experimental result

The operating conditions correspond to a case of radiator heating for a low delivery temperature :

- Evaporator : T_{in} = 12° C T_{out} = 5° C
- Condenser : T_{in} = 37° C T_{out} = 45° C

Working fluid	R22	Mixture FRIMIP
Thermal output (W)..........	2010	2493
Compression power (W).......	424	520
COP.........................	4.74	4.79
Compression ratio...........	3.38	3.44

This test, in agreement with the calculated results, shows that an important increase in capacity (here 24 %) and simultaneously a conserved COP can be achieved by using a selected fluid mixture, without any change in the flow-sheet.

C. - APPLICATION TO COMMERCIAL HEAT PUMPS FOR SPACE HEATING

1. Field of investigations

The most promising application selected consisted in selecting a mixed refrigerant, making it possible to increase the thermal capacity of standard Leroy-Somer heat pumps. The economic interest is the reduction of the investment cost of the installations. The development program covers the application of mixtures to water-to-water and air-to-water heat pumps in a range of thermal capacities from 5 kW up to 60 kW ;it is made up of the following steps :

- selection of the mixture and optimization of the compositions on the basis of calculations and experimental tests,

- study od adaptations required (refrigerant-flow control device ; recommended operating conditions : safety and command regulations),

- compatibility of the operating conditions and the fluid nature with the existing equipment (reliability ; range of utilization),

- economic assessment.

2. Water-to-water heat pumps

a) Performances

A gain in thermal output between 20 % and 25 % through an optimized mixture compared to R22 was demonstrated with different machines, the coefficient of performance being practically conserved. This improvment was realized through the use of FRIMIP, a new mixture patented and developped by IFP.

The comparison of performances between the mixed refrigerant and the pure component depends mainly upon the temperature lift between heat source and heat sink and also the water temperature increase on the condenser side.

The refrigerant-flow control is ensured through a conventional thermostatic expansion valve adapted to the mixture properties.

b) Reliability

A long-term tet run has been conducted with three units filled of mixture during 6 months under different operating conditions ; no particular failure due to the fluid nature was encountered.

Comparative tests of swelling of elastomers carried out by IFP have shown the same behavior for both R22 and the mixture.

c) Economic assessment

An economic evaluation conducted by Leroy-Somer over the whole range of the available units has determined a reduction in investment cost between 10 % and 12 %, by assuming a rate of increase in capacity of 20 %. A marketing development of Leroy-Somer water-to-water heat pumps with IFP mixture is planned at the end of the year.

3. Air-to-water units

A gain in thermal capacity of 18 % with a decrease of 4 % in COP was measured with the unit PALMYRE 120 of 11 kW_{th}, under the design test conditions of the program PERCHE, the used mixture making it possible to reach a maximum supply temperature of 55° C.

More generally, as far as the heat transfer resistance of the evaporator is located on the air side, the optimal performances of COP and capacity which are to be expected by using the selected mixture, require either an oversizing of the finned area of the coil or an increased air flow.

These adaptations are now being investigated with the unit PALMYRE 75 (7 kW_{th} with R22 in the design test) ; technical and economic conclusions should be drawn shortly, concerning this application.

D. - CONCLUSION

The possibility of increasing the capacity of heat pumps applied to space heating without any extracost by using optimized mixed refrigerants was shown. A new generation of water-to-water heat pumps manufactured by Leroy-Somer and filled with the fluid mixture FRIMIP developped by IFP will be commercialized at the end of the year 1983. The better competitivity of this new system should extend the market of such heat pumps in France and in Europe.

PRACTICAL APPLICATIONS OF FRIMIP

H. BAGLIONE DIRECTOR HEAT PUMPS

DEPARTMENT LEROY SOMER
ANGOULEME FRANCE

The heat pumps department was created en 1976 in our company. In application of our policy which is to create employement in the country rather than inside the town, our organisation is based on three geographical centers.
HASPARREN near BAYONNE we have a factory of 12000 m2 where are manufactured heat pumps for the domestic market. The actual production is of 1200 heat pumps per month.
MARENNES on the atlantic coast near LA ROCHELLE we have a second factory of 4000 m2 where are manufactured heat pumps for the commercial, industrial and agricultural markets.The actual production is of 120 heat pumps per month.
ANGOULEME is the marketing center of this activity. We carry research and development on the heat pumps in our laboratories.
We have a training center for heating services and we coordinate the sales force.
Actually our organisation covers FRANCE with 150 distributors, 4000 installators and 200 service teams.

Thanks to the Energy policy which is applied in FRANCE since 1975 the heat pump market knows a constant development. The objective is to reach 100000 heat pumps per year around 1985. In order to reach this target we have to find solutions reducing the cost of heat pumps. This cannot be achieved by mass production, the quantities being too small. We must find technological improvements reducing cost. This is the objective that we are trying to achieved with the new refrigerant FRIMIP.

Our first application concerns WATER/WATER heat pumps. Our actual range of heat pumps of this type is as follows

TYPE	GT 45	GT 75	GT 120	GT 200	GT 400	GT 600
Heating capacity KW	4,6	7,5	12	19,5	40	60
Cost F.F.	14.750	19.560	26.440	32.400	42.350	62.000
Cost/KW	3206	2608	2203	1660	1058	1033

The use of FRIMIP gives us an increase of 24% in capacity with practically no change on the technological design and a slight cost increase coming from the fact that FRIMIP is more expensive than the usual fluid R22 and that the way of manipulating is changed.

By using FRIMIP we are able to present a wider range of heat pumps without increasing the number of components. This enables us to meet more closely the thermal requirements of the houses where heat pumps are installed.

Therefore we have the following range

TYPE	GT 45	GT 75	GT 90	GT 120	GT 150
Heating capacity KW	4,6	7,5	9	12	14,4
Cost F.F.	14.750	19.560	20.730	26.440	28.000
Cost/KW	3206	2608	2300	2203	1940

TYPE	GT 200	GT 250	GT 400	GT 500	GT 600	GT 750
Heating capacity KW	19,5	23,4	40	48	60	72
Cost F.F.	32.400	34.350	42.350	44.900	62.000	65.750
Cost/KW	1660	1460	1058	935	1033	913

We see that we have an increase in heating capacity of 20 % and an increase in cost of only 6%. Futhermore with the components of 6 types we come out with a range of 11 types.

COMPLEMENTARY TESTING

FIELD TESTING

20 heat pumps have been tested on standard installation giving all satisfactions.

LABORATORY TESTING

Long run test have been and are carried to measure the variation of characteristics. Here again the results are satisfactory.

APPLICATIONS TESTED

The FRIMIP has been used on other products such as water/air heat pumps and dehumidifiers for swimming pools. Here again the results are satisfactory.

MANUFACTURING PROCEDURES

Manufacturing procedures have been developped in order to insure a proper filling of heat pumps which insures the right proportion.

CONCLUSION

FRIMIP is today used on WATER/WATER heat pumps. It will be shortly applied to WATER/AIR and DEHUMIDIFIERS. Leroy-Somer and IFP are still investigating the interest of fluid mixtures applied to AIR-TO-WATER heat pumps.

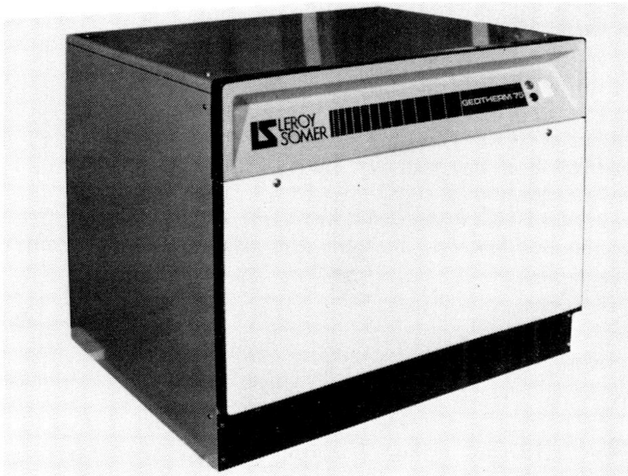

Fig. 1 - Leroy-Somer water-to-water unit GT75

Fig. 2 - Leroy-Somer air-to-water installed unit PA120E

ABSORPTION HEAT PUMPS FOR SPACE HEATING

J. CHERON - A. ROJEY
Institut Français du Pétrole
1 et 4 Avenue du Bois Préau
92502 RUEIL MALMAISON (France)

SUMMARY

Absorption resorption cycles with non toxic, inexpensive and non flamable fluids are investigated to be used in heat pumps for space heating. CO_2 is a good fluid because of its properties. An exhaustive list of solvents has been drawn up. A test loop has been built and experiments are running.

Economic studies have been carried out with an absorption heat pump compare to condensing burner and classical burners.

A. - INTRODUCTION

One device which can make a significant contribution to energy saving for space heating is the heat pump.

Heat pumps are now available in many shapes, sizes and types but those operating on the vapour compression cycle are the most common, especially for individual domestic applications.

For space heating applications in Europe some problems still exist :

- in the low temperature of the most common cold source (outside air),
- in the use of electricity mainly during winter season leading to a costly peak power problem.

Thus in recent years, absorption heat pumps have received considerable interest (1-4) due to some features :

- high primary energy efficiency,
- lack of compressor giving a long life expectancy,
- reduced noise problems (no compressor),
- possibilities for direct firing of various sources (gas, fuel oil, coal, wood...).

Research work has been undertaken at IFP in the field of absorption heat pumps (and heat transformers) within a more general framework of a program on processes for energy conservation.

The objectives for absorption research are the following :

- to limit the working pressure in order to decrease the materiel cost by using absorption resorption cycles,

- to improve the coefficient of performance by the use of an auxiliary fluid,

- to avoid the problems of toxicity of NH_3 by using new working fluids.

This paper presents some results of these works. The new working fluids research has been undertaken in the framework of the EEC contract N° EE-A-4-042 F.

B. - LIMITATION OF PRESSURE (Absorption-resorption cycles)

The general principle of absorption-resorption cycles goes back to Altenkirch (5).

Its main advantage is that the maximum pressure in the system can be adjusted by adapting the solute concentration in a re(ab)sorption loop.
It is possible to reach or surpass the critical temperature of the solute.

The IFP is developing heat pumps using a cycle with ammonia and water, derived from absorption resorption principle.

Figure 1 illustrates this combination of absorption and resorption cycles. High level heat is supplied to the generator and low-level heat is sent to the desorber.

These heats are transformed into medium-level heat which is transmitted to the hot source in the resorber and absorber.

The absorption-resorption cycle is thus a good way for reducing cost of equipments by decreasing the internal pressure of ammonia-water pair working at high temperature.

This process is mainly attractive in order to produce heat at temperatures above 80 C (the vapor pressure of pure ammonia being about 32 bars at 70° C). In France such a temperature level is uncommon for space heating needs.

A pilot plant with an output around 300 kW will be soon tested to produce heat up to about 120° C. With such an absorption-resorption cycle it is now possible to use a well known fluid pair ammonia-water either for producing heat at high level (industrial applications) or for producing heat at low level (domestic heating applications) without problems of cristallization such as for the water-lithium bromide working pair.

C. - COLLECTIVE SPACE HEATING APPLICATION (Auxiliary fluid)

1. Improvement of the coefficient of performance

In France, 3 million flats in collective housing are heated by oil.

For a new technique like absorption heat pump a pay back time of 6 years is the limit not to be exceeded.

Figure 2 shows the investment cost per thermal kW, the coefficient of performance COP and the pay back time in years.

The main objective of IFP for absorption heat pump processes is then to reach a COP of about 1.8 leading to a pay back time of 4 at 6 years.

2. Multistage process

A well known way to improve the COP is to use a multistage process (6).

In a two stages heat pump the working fluid is first evaporated at medium pressure, absorbed by a resorption loop and then evaporated a second time at low pressure (Fig. 3). It is thus possible to improve the thermal power from the outside source for a given thermal power of generation.

We did not choose this way because of the need of extra investments and the complexity of this system.

3. Auxiliary fluid

IFP has developped a heat pump with an auxiliary condensable fluid.

Fig. 4 shows the general diagramm of such heat pump.

The effects are very similar to those achieved by a multi-stage process but without its complexity. Certain thermodynamical properties are required for this auxiliary fluid. IFP has developed a complete computer simulation of such cycles and carried out experiments which confirm the predicted results.

Another interest of the auxiliary fluid is to have good performances for heat pumps even with temperature variations of the "cold" and "hot" sources.

A coefficient of performance up to 2.0 is then possible and for collective domestic heating uses a very high COP is practically feasible.

D. NEW WORKING PAIRS. INDIVIDUAL SPACE HEATING

Many solute-solvent pairs can be used in absorption but the most frequently used pairs have some disadvantages for individual space heating applications :

- for the water-lithium-bromide system there are risks of corrosion and the "cold" source must be at temperature above 0°C. For the latter reason the Japanese water-lithium-bromide cooling systems are not easily used for domestic heating applications,

- for the ammonia-water system, precautions must be taken when used for individual home heating because of the toxic nature of ammonia.

To find new pairs useful in individual domestic heating certain physical and chemical criteria have to be defined and require large research and development efforts 4.

In the framework of this contract new solute solvent pairs are investigated particularly with resorption cycles which widen the range of fluids to be used in heat pumps.

Non toxic, inexpensive, non-flamable and easily availabe fluids have been investigated and among them, IFP is considering the possibilities of use of CO_2.

1. Advantages of CO_2

Figure 5 gives some heats of vaporization at boiling temperature versus boiling temperature of the fluid.

It is quite easy to conclude that :

- H_2O is the best fluid but as its boiling temperature is 100° C, a heat pump using it will be at very low pressure and quite impossible to use below 0°C,

- NH_3 is the second best fluid but it is toxic and its pressure is too high,

- alcohols (CH_3OH, C_2H_5OH) have good heats of vaporization but they are flamable and (for CH_3OH) toxic,

- methylamine ethylamine is toxic.

So we proposed CO_2 which is a fluid with the particular properties :

- great availability,
- non toxic
- easily soluble.

CO_2 is a well known refrigerant fluid in compression cycle. But its critical point (T_C = 31° C) is too low for heating uses.

We developped an absorption-resorption cycle with which it is possible to use any fluid above its critical point.

An exhaustive list of solvents has been drawn up (Fig. 6), solid or liquid, chemical physical or mixed types. Alkanolamines such as MEA (monoethanol amine) DEA (diethanolamine) are the most used solvents for the removal of CO_2 and H_2S in oil and gas industry.

Figure 5 also shows the values of heat of reaction between CO_2 and aqueous amine (lean and rich solutions). This heat of reaction is high enough (between H_2O and NH_3 heats of vaporization) to select CO_2 as a fluid in heat pump process.

2. Experimental work

Equilibrium data

Many experimental data are available (equilibrium data, corrosion and thermal stability data) but for use in heat pump cycles supplementary data, especially at low temperatures, are required. An equilibrium pick up device has been built for this purpose and we have collected solubility data of aqueous and non aqueous solvents.

Test loop

A test loop has been built for working pressures up to 30 atmospheres and temperature up to 200° C (Photo N° 1).

It includes precise flow-meters and platinum 4 wire resistances to reach the energy balance.

At the present time we are running experiments with the resorption loop.

3. Computer simulation work

From the IFP knowledge in thermodynamics simulation some computer programs have been developed. But as there are complexation reactions between CO_2 and solvents, the classical models did not give very accurate correlation.

Special modifications have been introduced in our models and more experimental data an required to obtain as good a computer simulation as we have for compression fluids.

4. Expected performances

A COP of about 1.5 is expected as up to near 200° C it is possible to find solvents which don't decompose.

Internal pressure is around 20 bars and the two circulation rates are acceptable.

E. ECONOMY

Studies have been carried out for individual and collective housing.

Collective housing (ammonia-water + auxiliary fluid)

With a basic COP of 1.7 the absorption heat pump is the better converter for primary energy.

A pilot plant with an output arond 100 kW will soon be tested with a gas burner to confirm technical and economical data.

For another study a 133 kW absorption module has been designed.

POWER	133 kW
OUTSIDE TEMPERATURE	5° C
WATER INLET TEMPERATURE	33° C
WATER OULET TEMPERATURE	44° C
COP (Low heating value)	1.69

INVESTMENT	(KF)
MATERIEL	153
LABOUR COST	33
OUERHEADS	14
COST PRICE	200
SALE PRICE	250
INSTALLATION	70
TOTAL	320

A pay back time of about 5 years is possible.

Individual housing

For a large house (85 MWh annual energy consumption) studies have been carried out with an absorption heat pump (COP 1,5) and condensing burner (0.9) compare to classical burners (0.75 and 0.7 efficiency)

	High cost investment 77 KF		Low cost investment 55 KF	
boiler efficiency	0.75	0.7	0.75	0.7
energy saving kTh	42	49	42	49
energy economy KF	8.4	9.8	8.4	9.8
payback time years	9.2	7.8	6.55	5.6

CONCLUSION

The research work made at IFP in the field of absorption heat pump lead to the main following results :

- for collective housing heating : significant improvement of the COP is possible by use of an auxiliary fluids. Absorption heat pump is attractive for the owners and remains the best way to save primary energy. A 100 kW prototype will be soon build.

- for individual housing heating, new working pairs are constinously investigated. Non toxic and easily available CO_2 appears to be an attractive working fluid. A test loop is collecting practical data with varied solvents.

REFERENCES

(1) C. JANSSEN MFG Van der JAGT
VEG - GASINSTITUUT n.v. Appeldoorn (The netherlands)
16 th Internation Congress of refrigeration - Preprints commission E2 Paris 1983.
Experiment-based seasonal performances prediction of water ammonia absorption heat pumps.

(2) R.M. LAZZARIN, L. ROSSETTO, L. VIERO
Universita di Padova, Padova (Italy)
16 th International Congress of refrigeration - Preprints commission E2 Paris 1983.
Heating by absorption heat pumps - Analysis through a simulation model.

(3) H.J. LEIBUNDGUT, J. SCHAMAUN, R. SCHUEPBACH, H. STIERLIN
SIBIR Kühlapparate GmbH Switzerland
16 th International Congress of refrigeration - Preprints commission E2 Paris 1983.
Absorption type heat pump with inert gas for domestic heating.

(4) New working paris of absorption Processes
Proceedings of a workshop in Berlin, April 14-16 1982.

(5) ALTENKIRCH E.
Reversible Absorption smaschinen
Zeitschrift für die gesante Kälte Industrie Vol. 20 (1913)

(6) ALEFELD G.
Double effect, triple effect and quadruple effect absorption machines
16 th International Congress of refrigeration - Preprints commission E2 Paris 1983.

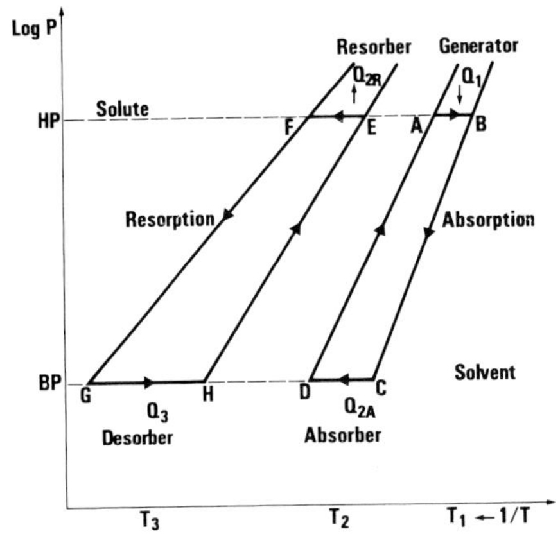

Fig. 1

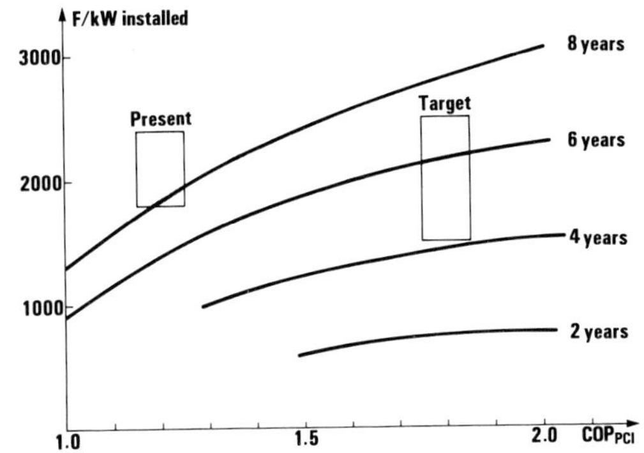

Fig. 2

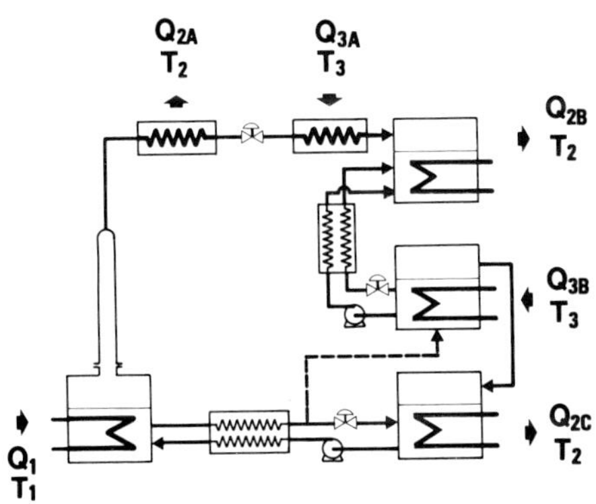

Fig. 3: Two-stage absorption heat pump

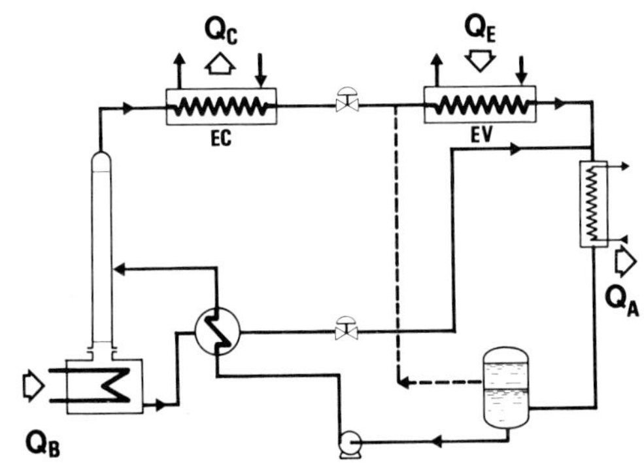

Fig. 4: IFP absorption heat pump

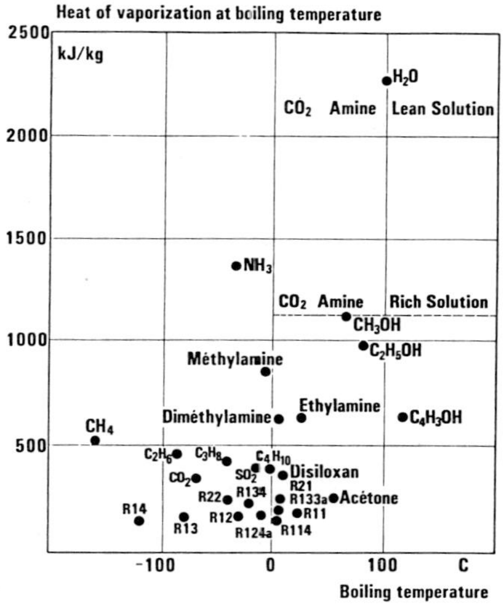

Fig. 5

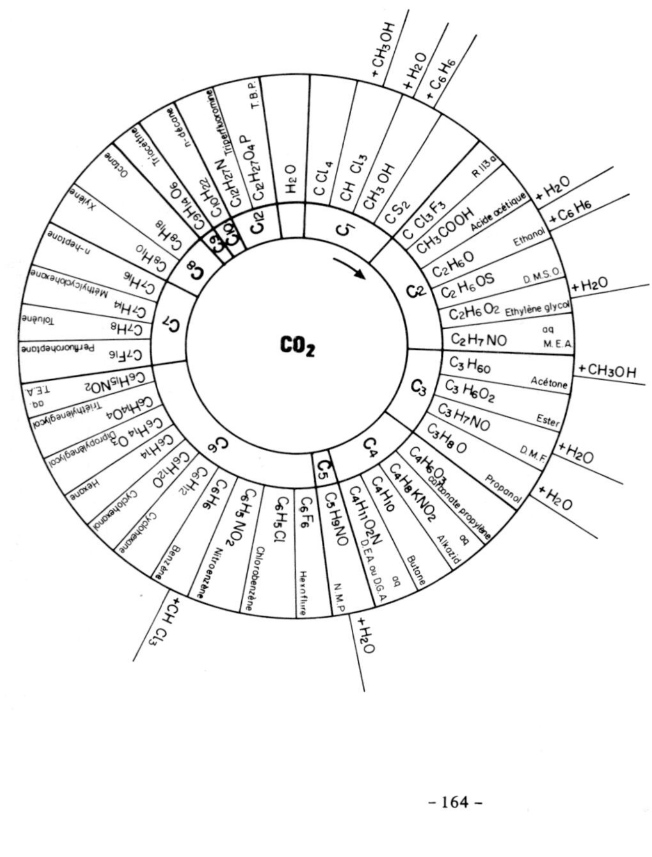

Fig. 6:

Couples CO_2-Solvants

PERIODICALLY OPERATING ABSORPTION HEAT PUMP

K. F. KNOCHE, D. STEHMEIER and
H. B. GRABENHENRICH
Institute for Technical Thermodynamics
RWTH Aachen, Germany

SUMMARY

A periodically operating absorption heat pump has been designed and built for space heating applications. The heat input is accomplished directly by means of a natural gas burner sized at approximately 10 kW and the low temperature heat source is furnished by the ambient air.

Prior to the design extensive investigations have been performed to determine the heat and mass transfer rates in the different components.

In a laboratory size unit the transport phenomena for the ternary mixture Lithiumbromide-Methanol-Water ($LiBr-CH_3OH/H_2O$) have been measured first.

Furtheron boiling tubes with variing geometries have been examined concerning their suitability as a generator with natural circulation.

In this report the results of the preliminary investigations, the possible modes of operating of a prototype unit, the setup and the design of the components as well as the instrumentation of the overall experimental array are described.

By now the solution cycle has been put into operation and at present the necessary operating characteristics with variing boundary conditions are determined. Finally the planned further research and the economical aspects of this special type of heat pump will be discussed.

1. INTRODUCTION

Periodically operating absorption heat pumps are preferable towards conventional heat pump systems especially for small heat requirements. Moving parts within the unit are not necessary because good mass transfer rates are achieved within the solution cycle by natural convection and thus the feed pump is not required any more. Previous absorption heat pump research proved that the feed pump is of critical importance for an overall good operation. The switching mode of operation of the heat load heat exchangers within the condenser, the absorber and the exhaust gas stream is accomplished easily by conventional multiple pass mixing valves.

However, this periodically operating heat pump meets the required heat load continuously. Either the heat is generated by the condenser and the exhaust gas heat exchanger or within the absorber during the period of the evaporator operation.

2. WORK PROGRAM

The objective of the research program "Periodically operating absorption heat pump", which has been funded by the Commission of the European Communities for a period of 16 months,was the design, the construction and the investigation of a directly heated prototype sized at 10 KW for single family dwelling space heating.

One focal point of the overall investigation was the optimization of the transport phenomena within the individual components to minimize the size of the whole unit and to reduce the heat storage in those components whose temperature levels are cycled periodically. Parallel theoretical investigations showed that these heat storage effects result in smaller heat ratios and therefore the mass-heat capacity product ($m_i \cdot c_{pi}$) of these components should be minimized.

The ternary mixture $LiBr-CH_3OH-H_2O$ has been selected because the unit can be operated at low temperatures and thus low pressures as well as any rectification of the refrigerant is superfluous and also no cristallisation of the solution occurs in the projected concentration and temperature ranges.

The ambient air furnishes the heat source and the heat sink is simulated by a water heating system.

The overall experimental setup consists of a prototype unit which is integrated in a controllable air cycle which furnishes the ambient air at defined conditions. Furtheron the experimental setup is completely instrumented with temperature, pressure, flow and level transducers and the obtained analog data are fed into a micro-computer system which accomplishes the data acquisition and the experiment control.

At present the prototype is in operation and the load characteristics are determined for variing operating conditions.

3. METHODS OF INVESTIGATION, RESULTS

3.1 PAWP-Laboratory-Apparatus; Discussion of results

In a simple "laboratory heat pump" consisting of two glass cylinders connected by stainless steel tubing and with inserted heat exchangers of different design, the occuring phases of operation and the influencing system parameters were investigated intensively. The used laboratory apparatus with its pressure and temperature transducers is shown in figure 1.

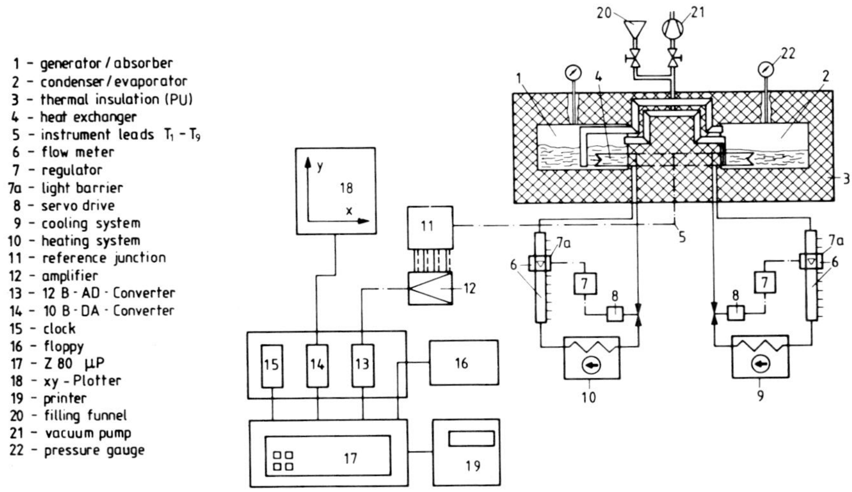

Fig. 1 Test Rig for PAWP-experimental arrangement with computerized data acquisition

With this simple setup the occuring heat and mass transfer coefficients were determined experimentally for variing operating conditions. The data acquisition and the experiment evaluation was performed simultaneously by a Z-80 microcomputer system.

The different absorbing solutions LiBr-H_2O, LiBr-CH_3OH and LiBr-CH_3OH/H_2O were examined in the following temperature ranges:

Generator, heat transfer agent:	110 - 140°C
Condenser:	30 - 60°C
Evaporator:	-10 - 40°C
Absorber:	20 - 60°C

During a period of about 6 month methanol has been used as a refrigerantin conjunction with a lithium-bromide solution. Solution temperatures of up to about 120°C and cycling temperature levels did not affect the operating behaviour of the unit and also a decomposition of the methanol could not be observed.

For the different heat exchanger geometries the following heat transfer coefficients have been measured:

(typical values for LiBr-CH_3OH/H_2O)

Evaporator, finned tubes, flooded:	50 - 150 W/m^2K
Condenser, nest of tubes:	250 - 450 W/m^2K
Bubble absorber:	150 - 250 W/m^2K
Generator, nest of tubes, flooded:	200 - 300 W/m^2K

In figure 2 the results of 11 complete test cycles are presented. The heat ratio COP, defined as the heat load by the condenser and the absorber, divided by the heat input in the generator, the averaged net output $\bar{\dot{Q}}_N$ and the achieved final concentrations ξ_{END} are summarized in this graph. For all these tests the refrigerant mass ratio was $x=0,1=m_{H_2O}/(m_{H_2O} + m_{CH_3OH})$ and the initial concentration was $\xi_o=0,45$ kg LiBr/kg solution.

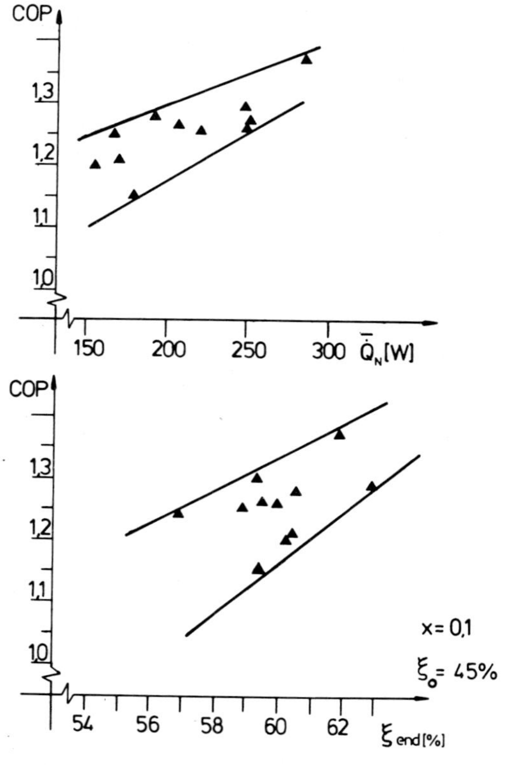

Fig. 2

COP, mean output and final concentration of 11 test cycles with LiBr-CH_3OH/H_2O

3.2 Investigation of the Thermosyphon-Generator-Tubes

The degasification of the working fluid mixture in vertical boiling tubes (thermosyphon-tubes) allows a circulation of the solution without a feed pump (natural convection). Because of a variety of influencing parameters on the transport mechanism, a set of different stainless steel tubing and Duran-glass tubing was fitted in the experimental setup shown in figure 3 and investigated experimentally.

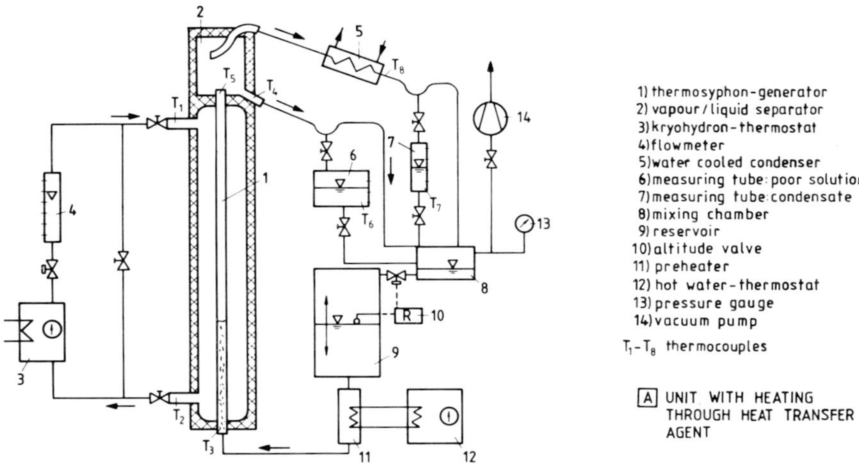

Fig. 3 Test setup for the investigations of a thermosyphon-generator

The operating parameters have been variied as follows:

heat transfer agent, inlet temperature:	97 - 137°C
heat capacity flow:	5o - 1oo W/K
rich solution, inlet temperature:	52 - 1o3°C
temperature of condensation:	9 - 5o°C
material of tubing:	V4A, DURAN
tube diameter:	1o - 15 mm
number of tubes:	1 + 2 parallel
height of delivery:	o,3-o,5 m
angle of inclination:	o°, 5°, 1o°, 15°

As one example of the test series "variation of the angle of inclination" test results are presented in fig. 4. In this graph the mass flow of the refrigerant $\dot{m}_{km}$, the rich solution $\dot{m}_{rL}$ and the width of the degasification $\Delta\xi$ are plotted versus the height of delivery.

During the operation of the boiling tubes with a LiBr-CH_3OH/H_2O solution (mass ratio, 46 : 51 : 3) heat transfer coefficients of 1oo to 28o W/m^2K have been measured. The results were reproducible and thus they could be immediately utilized for a safe design of the directly fired generator of the planned prototype.

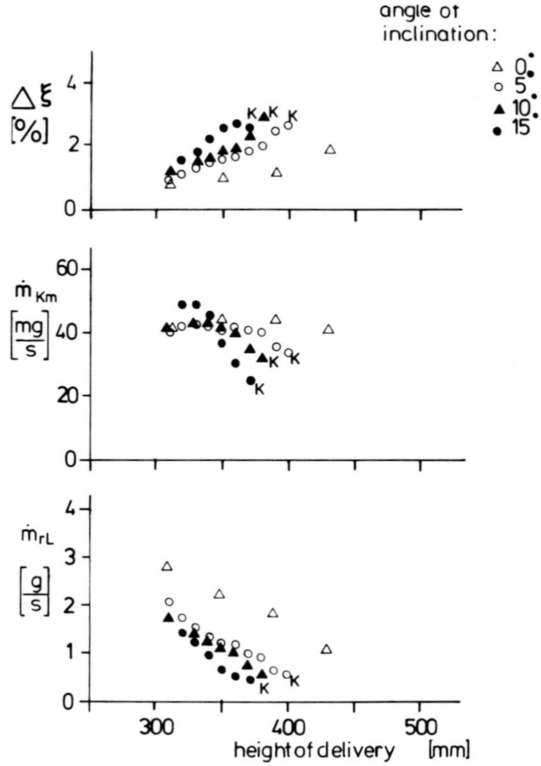

Fig. 4

Influence of angle of inclination on the transport mechanism in a stainless steel generator tube.

Solution:

LiBr-CH_3OH/H_2O

3.3 Modes of operation and design structure of the 10 KW prototype

The results of the preliminary investigations, especially the achieved mean heat output and the heat transfer coefficients, proved that a technically scaled unit can be designed and constructed with an acceptable overall size.

The design of a PAWP-prototype is based on the following concept:

Heat input :	directly fired; natural gas boiler sized at 10 - 12 KW
Heat source :	ambient and exhaust air; app. 0-15°C
Heating system:	low temperature water circuit with impeller pump and conventional mixing valve:30-50°C
Mean heat output:	app. 5 KW during "Heat pump operation"
Modes of Operation:	a) Generator mode of operation as shown in fig. 5 with utilization of the exhaust gas heat flow; effective heat input from condenser and exhaust gas cooler, whereby the solution is cycled between the absorber and the generator.

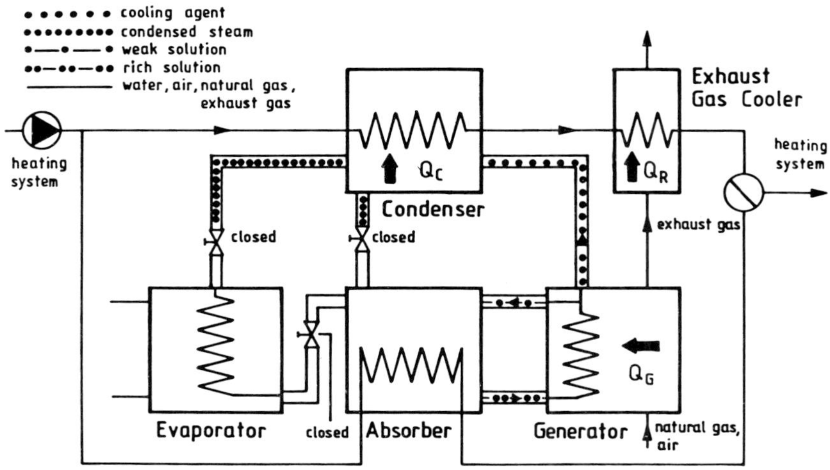

Fig. 5 Periodically operating absorption heat pump
Generator in operation

b) Absorber mode of operation as shown in fig. 6 with effective heat input from the ambient air via the evaporator and effective heat output at the absorber (heat pump mode of operation)

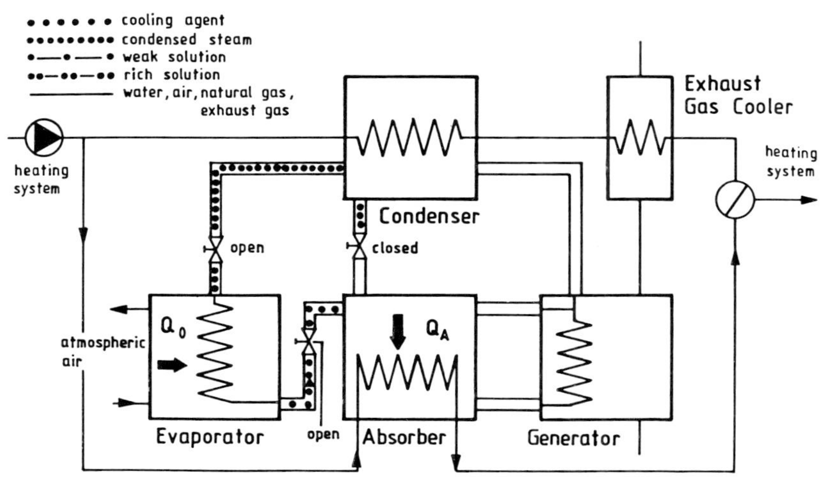

Fig. 6 Absorber in operation

c) Boiler mode of operation with maximum effective heat output at extremely low ambient temperatures. During this mode of operation the solution serves as a heat transfer agent within the apparatus, whereby the produced condensat is directly passed back into the absorber.

The absorber mode of operation is preferably connected to a low temperature heating system. During the generator and boiler modes of operation higher heating system inlet temperatures can be easily accomplished and also utilized for different heat input requirement e.g. process water.

The different modes of operation are regulated and initilized by an on/off-control for the boiler as well as the opening or closing of the specific valves in the refrigerant circuit and heating system.

The individual components are equipped with especially constructed heat exchangers whereby stainless steel materials (1.4571) in all components where the heat exchange is in direct contact to the corrosive solution and refrigerant.

Fig. 7 shows the present experimental setup including the necessary instrumentation.

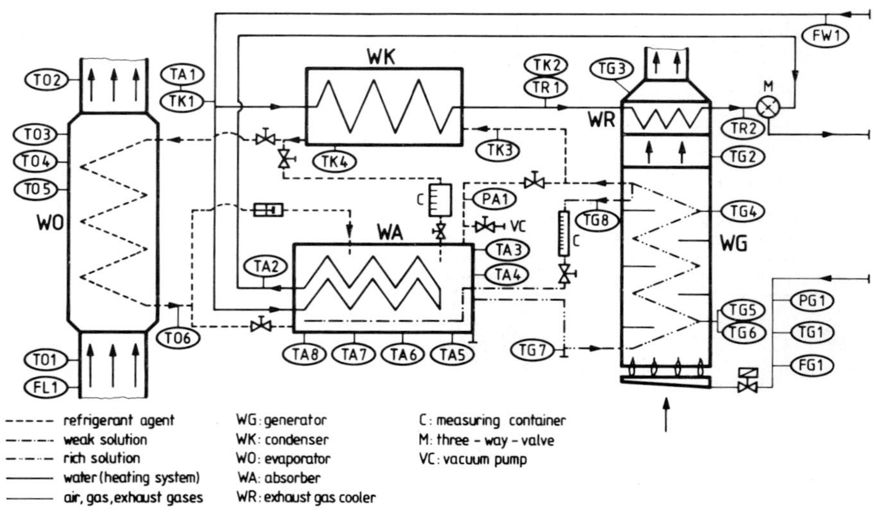

Fig. 7 PAWP-Prototype

Installation and Instrumentation

Meanwhile the prototype is included in a heat pump testing rack which also employs an air channel to stabilize the thermodynamical properties of the ambient air to the evaporator.

Furtheron this testing array allows all functions for a realistic simulation of the heat load and is also furnished with the equipment to control the natural gas flow rate.

Fig. 8 shows the entire test rack with the included prototype and one part of the data acquisition system.

Fig. 8 PAWP-Prototype and Test Rig

4. CONCLUSIONS AND PROSPECTS

The performed investigations showed, that a periodically operating absorption heat pump operates well and can be improved furtheron to become a heating device of high efficiency. However, the mechanical design remains quite simple and the operation is easy to regulate and to control. Thereby the main positive attributes are:

- The apparatus operates in a self contained mode of operation. At low ambient temperatures, where a heat pump mode of operation is not adviseable any more, the included gas boiler meets the required heat load.
- The unit does not need any support energies and no moving parts are employed. Thus noise emission is of no concern.
- Low maintenance costs can be expected.
- The control and regulation can be done in a usual on-off mode and with conventional valves.

5. ECONOMIC ASSESMENT

In respect to the simple construction of the periodically operating absorption heat pump, the simple control, and its self contained operation, the production of such a unit does not require large manufacturing costs to the consumer. Because the device has no movable components, a long lifetime and low maintanence can be expected. If the system is hermetically sealed in the manufacturing process, service and installation only needs to be done from a heating/ventilation technician.

THERMOELECTRIC HEAT PUMPS FOR SPACE HEATING

P. DUBOIS, R. RAVELET, M. BERTHET
Laboratoires de Marcoussis, Centre de Recherches de la CGE
Route de Nozay
91460 Marcoussis - FRANCE
Tél. (6) 449 12 00

Summary

The reported work on thermoelectrical heat pumps has been conducted on extracted air/water pumps for space heating. It includes :

- the analysis of fabrication conditions of large thermoelectric pellets,
- the dimensioning of heat pumps with a computer program,
- design, fabrication and test of a vertical column with 34 pellets,
- design and cost studies of a pump for space heating of a well insulated dwelling and computation of energy savings,
- design and construction of a 3.8 thermal kW exhaust air/water heat pump.

1. Introduction

This study concerns a thermoelectric exhaust air/water heat pump for heating an apartment or a house. The following operations were performed :

- the dimensioning of a thermoelectric exhaust air/water pump for space heating was determined with a computer program. The influence of a number of parameters was analysed during dimensioning of the heat pump.
- the fabrication of suitably-dimensioned thermoelectric pellets was studied. The pellets are produced by sintering ; the influence of doping on the thermoelectric performance of the pellets was analysed.
- a heat pump column was designed and made. This experimental column includes 17 pairs of thermoelectric pellets. The performances of this column were measured on a test bench. It was noted that arrangement in a column did not adversely affect pellet performance. It was also noted that column performances were equal to computer program forecasts.
- the cost of fabrication and the sale price of a thermoelectric heat pump used to heat living space was determined, then a calculation was made of the energy savings allowed by this pump for a house with a living area of 100 m2 in the Paris region.
- design and construction of a 3.8 thermal kW exhaust air/water heat pump.

2. Dimensioning of a thermoelectric exhaust air/water heat pump

The dimensions of a thermoelectric exhaust air/water heat pump were determined for certain operating hypotheses and then efforts were made to get the maximum saving that could be obtained during a one year period by varying a certain number of key parameters.

The thermoelectric exhaust air/water heat pump heats the floor by circulating water at low temperature (outlet 35°C, return 28°C), and takes energy from the exhaust air ; decentralised electrical heaters are used to supply back-up power for heating on the colder days. This product would be used for new habitations, houses or apartments.

3. Analysis of fabrication conditions for large thermoelectric pellets

This study concerned the fabrication conditions for pellets with a 170 mm2 cross section (circular section to ease column assembly) for which we obtained performances at least equal to those of the 5 x 5 mm pellets previously fabricated. The results are as follows :

a) For type N, elements, we reproducibly obtained a figure of merit of $2.6 \times 10^{-3} K^{-1}$, a value which exceeds by 25 % those obtained for pellets with a 25 mm2 cross section. The values of each of the thermoelectric coefficients are entered in table 1 where they are compared with the corresponding values determined for 25 mm2 elements.

For type P elements, we obtained a figure of merit of $2.4\ 10^{-3}\ K^{-1}$ which is fully comparable with the performances measured for the pellets with smaller cross-section (cf table 1).

TABLE 1

Comparative thermoelectric performances of pellets with high and low cross-sections

Thermoelectric characteristics	Type N		Type P	
	170 mm2 cross section pellets	25 mm2 cross section pellets	170 mm2 cross section pellets	25 mm2 cross section pellets
α ($\mu V.K^{-1}$)	223	200	218	200
k (W/cm.K)	7,2	12	7,9	12
ρ (mΩ . cm)	2,6	1,6	2,5	1,4
Figure of merit (K^{-1})	$2.6\ 10^{-3}$	2.10^{-3}	$2.4\ 10^{-3}$	$2.4\ 10^{-3}$

4. Design, construction and testing of an exhaust air/water module

4.1 Description of the exhaust air/water heat pump module

- Overall description

The heat pump module is a vertical column in which the air flux is parallel to the column axis. There is a photograph in figure 2. The column includes 17 pairs of thermoelectric pellets which are cylinders with a cross-section of 170 mm2 and 1.5 mm thickness.

- Assembly of air/water module

The thermoelectric pellets are soldered between the air and water heat exchangers in a single operation. After assembly the electrical resistance of the thermoelectric pellets was measured and it was found to be identical to the value obtained before assembly. No variation was noted after positioning of the polyurethane thermal insulation.

4.2 Comparison of test results with the computer program forecast

The calculations of performances expected for the heat pumps described in chapter 2 were carried out using a computer program devised before the tests.

The results of the computer program were compared with the test results obtained on the test bench ; the data used in the computer program is as follows :

. the characteristics of the thermoelectric pellets,
. the thermal characteristics of the heat exchangers,
. the temperature and flow of the air and water at the module inlet.

The test and the computer results were found to be in close agreement for all tested operating conditions of the air/water module (deviation less than 2 %) (cf figures 3 and 4).

These results show that the design used for the air/water heat pumps does not degrade the elements and that the performances of a heat pump in a house or apartment can provide a 40 % saving over a one year period.

5. Technical and economic advantages of thermoelectric exhaust air/water heat pumps

5.1 Application under analysis

We considered the case of a house with a living space of 100 m2, with a space heating need of 250 W/°K with an air renewal rate of 210 m3/h.

We first determined the cost of fabrication and the sale price of a thermoelectric exhaust air/water heat pump used to heat a floor by circulation of water at low temperature (outlet 35°C, return 28°C) which takes energy from the exhaust air of the habitation ; decentralised electrical heaters are used to supply back-up power for heating during the coldest days. This product would be used in new habitations, houses or apartments.

Using the computer program which had given results very similar to experimental values obtained for the module, we then calculated the savings made by the heat pump installed in the house or apartment.

5.2 Description of the adopted heat pump design

The heat pump design adopted consists of a thermoelectric core, a power supply and control device, a fan, a water circulator, and an expansion vase áll of which are mounted in a steel housing.

We first analyzed the cost of a heat pump incorporating vertical thermoelectric columns in which the air flux is parallel to the columns. Assessment of the cost of fabrication of these columns showed this solution to be too expensive. We then designed air/water horizontal columns with simpler air heat exchangers in which the air flows perpendicular to the columns and the water flows parallel to the columns.

5.3 Cost of air/water heat pump

The cost of the air/water heat pump with vertical columns was determined and then the cost of a similar pump with cheaper horizontal columns. The cost evaluation is based on a production of 1 000 units per year and for the economic conditions of June 1981.

The pump with horizontal columns incorporates 24 columns of 8 thermoelectric pairs each with a cross section of 170 mm2 and a thickness of 1.5 mm. It is equipped with an air extraction fan of 210 m3/h leaving a pressure drop of 8 mm of water available for the rest of the air loop, and a 475 l/h water circulator leaving a 2.3 m water pressure drop available for the rest of the water loop.

a) The fabrication cost breaks down into 7 units :

. 192 air heat exchangers	718,50 F
. 216 water heat exchangers	525,80 F
. 384 thermoelectric pellets	307,20 F
. miscellaneous and thermoelectric core assembly	837,90 F
. accessories (fan, circulator ...)	428,00 F
. housing and mounting	367,30 F
Total	3 737,20 F (Taxe free)

b) The following multiplying coefficients are used to determine the public sale price.

. Factory operational costs	1,3
. Commercial expenses and margin	1,4
. Wholesaler	1,15
. Installation	1,35
. V.A.T.	1,186

c) Costs are therefore, for the economic conditions of June 1981

. Factory sale price	3 737,2 x 1,3 x 1,4	=	6 802 F (T.F.)
. Sale price to installers (T.F.)	6 802 x 1,15	=	7 822 F (T.F.)
. Public sale price (I.T.)	7 822 x 1,35 x 1,186	=	12 520 F (I.T.)

5.4 Comparison of prices with freon heat pumps

The computed price for the thermoelectric heat pump is practically equal to the price of a freon pump providing the same energy saving. A thermoelectric pump has the built-in advantage of bringing no extra noise.

5.5 Performances of the exhaust air/water thermoelectric heat pump

The thermoelectric heat pump is equipped with pellets with a cross section of 170 mm2 and a thickness of 1.5 mm and with a figure of merit of $2.5 \times 10^{3}\ K^{-1}$. Performances over the space heating period in Paris are given in table 2 for a house with a living space of 100 m2, a volume of 250m3 and a space heating need of 250W/°K for an air renewal rate of 210 m3/h.

Performances obtained for a thermodynamic pump are given for comparison.

6. Design and construction of a 3.8 thermal kW exhaust air/water heatpump

We have designed and built an exhaust air/water heat pump for a house with a space heating need of 250 W/°K and an air renewal of 210 m3/h.

The heat pump, which takes "free" energy from the exhaust air, heats the floor by circulating water at low temperature (maximum outlet temperature 35°C, maximum inlet temperature 28°C). Decentralised electrical heaters are used to supply back-up power on the colder days.

The heat pump (cf fig. 5) includes :

- 24 horizontal columns with 8 pairs of thermoelements with a cross section of 170 mm2,
- a water circulator and an exhaust air fan,
- a rectifier able to supply up to 140 Amperes, a control system which gives priority to the heat pump over the back-up power and controls the electrical power supplied to the thermoelements as a function of a temperature of the house.

The heat pump has been built and will be tested at the end of 1983. One can see on figure 6 the view of the 24 thermoelectric columns.

7. Conclusion

We have designed and built an exhaust air/water thermoelectric heat pump capable of supplying 3.8 kW heat power to a floor heating water loop with a maximum outlet temperature of 35°C and extracting energy from 210 m3/h exhaust air. An incorporated control device adjusts water outlet temperature as a function of a house inside temperature and gives priority to the heat pump over back up electrical heaters. This pump will be tested at the end of 1983. A saving of 40 % is expected during the months when heating is required in the Paris area. The installation of the pump in the heated part of the habitation is possible with no risk of discomfort thanks to the absence of compressor noise, and so, one can be sure that no heat leaks outside the heated space will reduce the 40 % computed saving. The sale price of this thermoelectrical pump is comparable to that of a freon heat pumps of the same power.

TABLE 2

Exhaust air/water heat pump with 192 pairs of pellets

Performances over the space heating month in the Paris area

CHARACTERISTICS OF HABITATION				ENERGY BALANCE				
Area	Volume	G	Non-heating temperature	Electrical power consumed during space heating months		Heat requirements	Pump seasonal coef.	Saving
100 m2	250 m3	1 W/m^{3}K	14,5°C	Thermoelectric heat pump	5 875 kWh	10 485 kWh	1,785	44,0 %
				Freon heat pump	5 882 kWh	10 485 kWh	1,783	43,9 %

NB : Heating requirements correspond to 2 500 degrees.days.

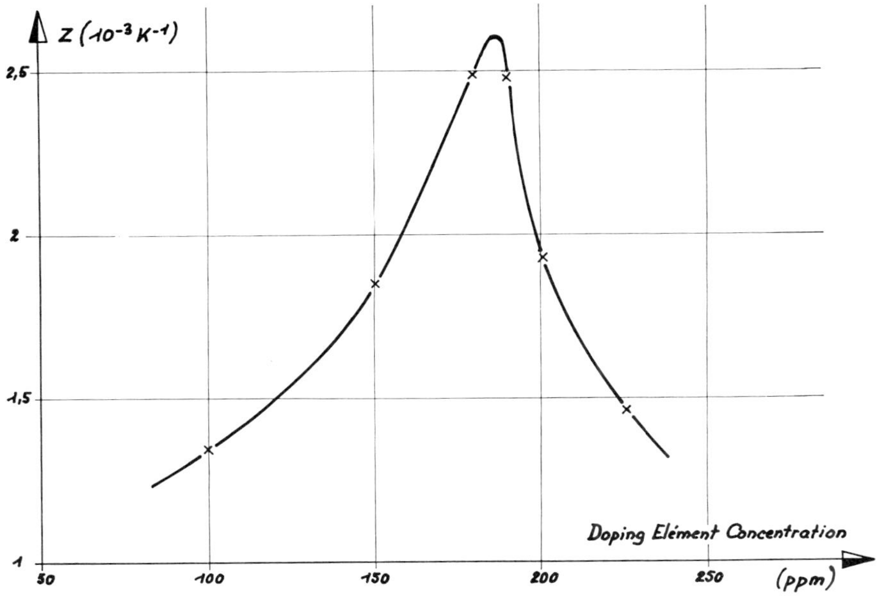

Fig. 1 - Influence of the doping element concentration on the factor of merit of ·N· type pellets

Fig. 2 - Air/water vertical module

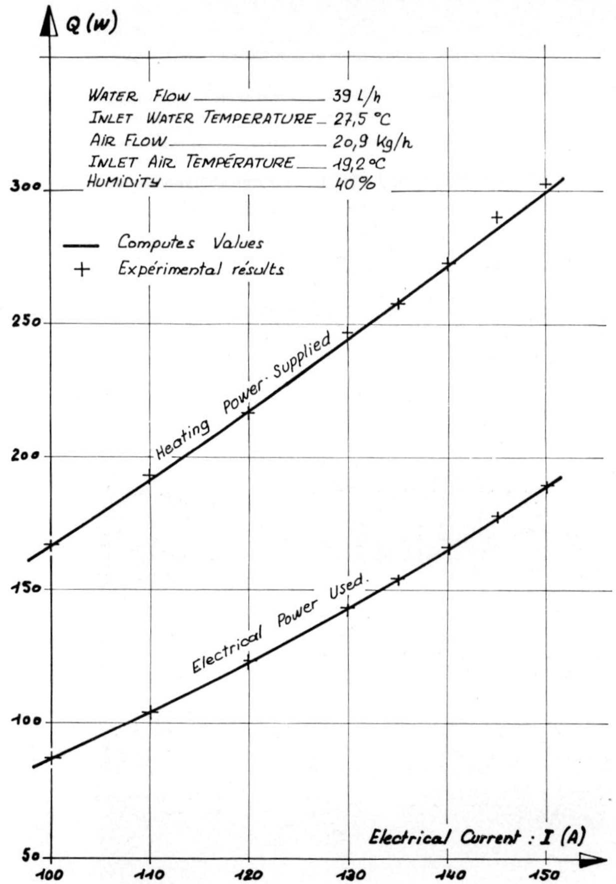

Fig. 3 - Heat power supplied and electrical power used versus electrical current

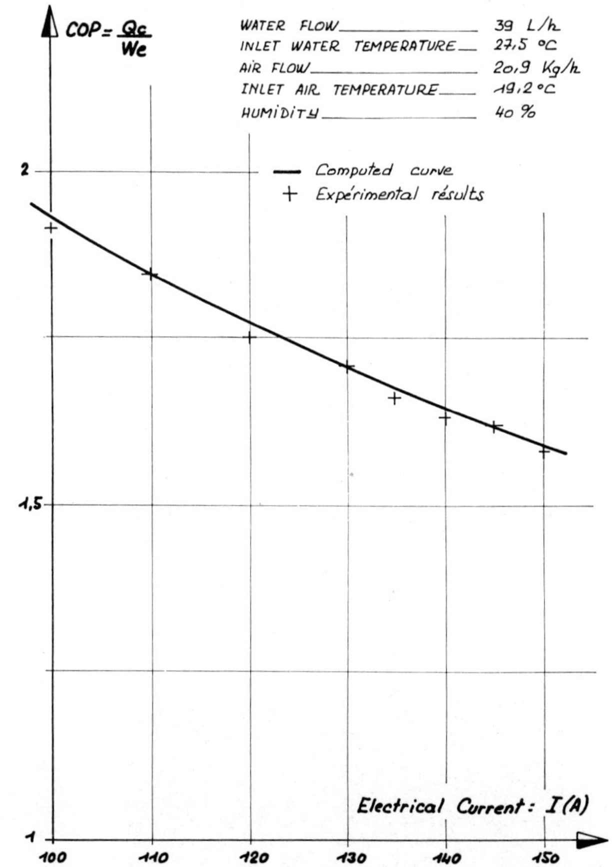

Fig. 4 - Coefficient of performance versus electrical current

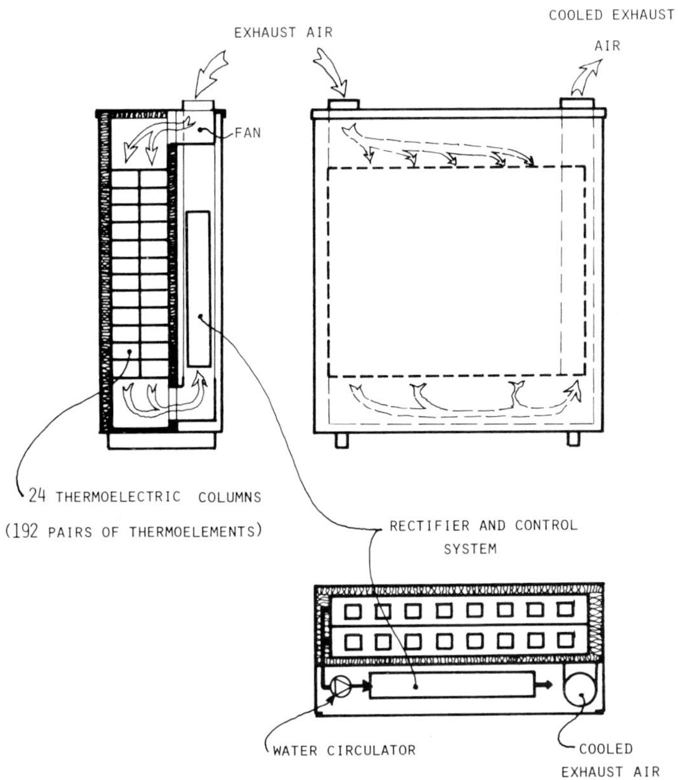

Fig. 5 - Exhaust air/water heat pump

Fig. 6 - 3,8 kW heat pump with 24 horizontal columns

SYSTEMATISCHE KONSTRUKTION UND ENTWICKLUNG EINER KOSTENGÜNSTIGEN, VERLUSTARMEN UND REGELBAREN LÖSUNGSPUMPE FÜR ABSORPTIONSWÄRMEPUMPEN

Karl-Heinz Schrader
Stiebel Eltron GmbH & Co. KG; D-3450 Holzminden 1

1. Zusammenfassung

Markterhältliche geeignete Lösungspumpen für Absorptionswärmepumpen stehen nur als kostenintensive Kolben-Membran-Pumpen zur Verfügung. Die Prüfung ebenfalls markterhältlicher Alternativpumpen warf das Problem der Wellenabdichtungen und der Materialverträglichkeit der gebräuchlichen Elastomere auf. Das beste Förderverhalten wiesen Innenzahnradpumpen auf, wobei die Innenzahnradpumpe einer Hydraulikpumpen-Firma durch Laufruhe, Baugröße und volumetrischen Wirkungsgrad überzeugte. In Zusammenarbeit mit dieser Firma und einer weiteren Firma als Elektro-Motor-Spezialist wurde das Projekt zum Abschluß gebracht.

Mittelfristig ist eine Zwischenlösung, bestehend aus gleitringabgedichteter Innenzahnradpumpe mit angeflanschtem polumschaltbaren Elektromotor, realisiert worden. Das Endprodukt der Entwicklung ist ein vollhermetisches, fördermediengeschmiertes System, bestehend aus dreifach polumschaltbarem Spaltrohrmotor mit angeflanschter Innenzahnradpumpe. Die durchgeführten Funktions- und Verschleißprüfungen erbrachten positive Ergebnisse und bestätigen einen erfolgreichen Projektabschluß.

2. Einleitung

Die Firma Stiebel Eltron ist im Rahmen geförderter Forschungs- und Entwicklungsvorhaben unter anderem auf dem Gebiet Bau- und Versuchsbetrieb von Absorbtionswämrepumpen-Prototypen tätig. Schon bei der Erprobung des ersten Prototyps kristallisierte sich die Lösungspumpe als Hauptproblempunkt heraus. Geeignete markterhältliche Pumpen standen nur als kostenintensive Kolbenmembranpumpen zur Verfügung. Membranmaterialunverträglichkeiten, Lebensdauer, Pulsation, Größe sowie Gewicht und nicht zuletzt die Geräuschentwicklung machten es erforderlich Alternativlösungen zu suchen.

Die Ergebnisse einer Marktanalyse waren unbefriedigend, da die befragten Hersteller für das spezielle Förderproblem nur ähnliche Membranpumpen geeignet sahen, wobei die Kosten dieser Produkte schon jeden Rahmen sprengten, der einen Einsatz in späteren serienmäßig gefertigten Absorptionswärmepumpen zuließ. Eine Eigenentwicklung wurde notwendig, da von der Pumpenindustrie mangels konkreter und auch lukrativer Stückzahlen keine schnelle Entwicklungsreaktion zu erwarten war.

3. Betriebsbedingungen für Lösungspumpen in Absorptionswärmepumpen

Die hier anzuführenden Pflichten sind abhängig vom jeweils verwendeten Zweistoffgemisch. Die unten angegebenen technischen Daten beziehen sich auf das Stoffpaar R 22-E 181 und auf einen Einsatz in einer Luft-Wasser-Absorptionswärmepumpe mit $\dot{Q}_{WS}$ = 15 kW Heizleistung bei ϑ_L = -20 °C Außentemperatur und einer Heizungsvorlauftemperatur von ϑ_{HV} = 60 °C (Spreizung → $\Delta\vartheta_{WS}$ = 20 K) mit kontinuierlicher Output-Leistungsregelung. Bis auf die Volumenstromangaben sind die Daten auch annähernd für das Stoffpaar NH_3 - H_2O zutreffend.

Technische Daten

Fördermedium	:	Tetraäthylenglykoldimethyläther (Ansul E 181) - Difluormonochlormethan (R 22)
Konzentration	:	ξ = 0,15 - 0,5
Dichte	:	ϱ = 1,05 - 1,12 g/cm³
Volumenstrom	:	$\dot{V}_{min}$ = 100 l/h bei Δp_{max} = 12 bar
		$\dot{V}_{max}$ = 500 l/h bei Δp_{max} = 23 bar
Saugstutzend.	:	p_1 = 2 - 7 bar
D.-Differenz	:	Δp_{max} = 23 bar
D.-Stutzend.	:	p_{2max} = 25 bar
Medientemp.	:	ϑ = 30 - 60 °C
Dyn. Viskosit.	:	η = 4 cP (bei 20°)
Leckrate	:	λ = 1 (⟹ vakuumdicht)
V.-Wirkungsg.	:	η_V > 0,9
G.-Wirkungsg.	:	η_G > 0,7

3.1 Zusatzanforderungen unter Berücksichtigung der Anlagenregelung einer Absorptionswärmepumpe

Die Hauptpflicht der Anlagenreglung einer Absorptionswärmepumpe besteht darin, eine außentemperaturgeführte Heizungstemperatur mit zur Außentemperatur umgekehrt proportionaler Leistung zu fahren. Zu dieser Aufgabenstellung durchgeführte Prototypversuche belegten optimale Wärmeverhältnisse einer Absorptionswärmepumpe bei jeweils optimalen Entgasungsbreiten, unter Berücksichtigung der Zusatzforderung möchlichst hoher mittlerer Konzentrationen der Lösung. Die optimale Entgasungsbreite der Anlage vergrößert sich allgemein mit steigender mittlerer Konzentration. Eine Variation der Entgasungsbreite ist dabei nur austreiberseitig durch Variation der Austreiberendtemperaturen möglich, da absorberseitig die erzielbaren Konzentrationen der reichen Lösung - unter Berücksichtigung des Absorbergütegrades - von den außentemperaturabhängigen Verdampfungsdrücken und Wärmesenkentemperaturen festliegen. Bei konstantem Lösungsvolumenstrom der reichen Lösung, bewirkt eine Brennerleistungsveränderung eine proportionale Austreiberendtemperaturänderung und als direkte Folge eine ebenfalls proportionale Veränderung der Entgasungsbreite sowie der Absorptionswärmepumpen-Heizleistung. Die Proportionalitätsfaktoren sind allerdings unterschiedlich und abhängig davon, ob die Veränderung eine Annäherung oder eine Entfernung, bezogen auf die vorhandene optimale Entgasungsbreite, bewirkt.

Wird noch berücksichtigt, daß die max. Heizleistung bei zwangsläufig geringer Entgasungsbreite und demnach hohem erforderlichen Volumenstrom der reichen Lösung zu erbringen ist, führt eine außentemperaturabhängige Leistungsregelung der Anlage mittels einer taktenden oder kontinuierlich arbeitenden Brennerregelung bei sonst konstantem Förderstrom der Lösungspumpe zu einer Entfernung von der optimalen Entgasungsbreite, da diese mit steigender Konzentration der reichen Lösung stärker ansteigt als die durch die notwendige Herabsetzung der Austreiberendtemperaturen erzielbare Entgasungsbreite.

Der Anstieg der Wärmeverhältnisse im Teillastbereich liegt damit unterhalb des möglichen Anstieges bei parallel zur Brennerleistungsverminderung durchgeführter Verringerung des Lösungsstromes. Zur Erlangung stets optimaler Wärmeverhältnisse einer Absorptionswärmepumpe über ein größeres Leistungsspektrum ist daher eine Volumenstromregelung der reichen Lösung zwingend erforderlich (siehe Abbildung 1; bei $\dot{V}$ = 500 l/h). Den Idealfall stellt eine kontinuierliche Volumenstromregelung der Lösungspumpe dar. Dieser Idealfall wurde deshalb auch in der ersten Phase der Projektdurchführung verfolgt.

Um außerdem die absolute Vakuumdichtigkeit über lange Betriebszeiten gewährleisten zu können, wurde als Zielpumpe eine vollhermetische Motor-Pumpen-Einheit-gefordert.

4. Auswahl des Pumpenprinzips

Nach den mit sinkender Gewichtung aufgezählten Kriterien Kosten, Wirkungsgrad, Pulsation, Lebensdauer, Geräusche, Baugröße und Dichtungsmöglichkeit, wurden in der Literatur angegebene Zwangsförderpumpen beurteilt. Diese Betrachtungen pendelten sich infolge der hohen Gewichtung der Kosten auf Zahnradpumpen ein, da hier mit geringem Materialaufwand und wenig Einzelteilen eine sehr stationäre Förderung erreichbar ist. Näher betrachtet weisen Innenzahnradpumpen gegenüber Außenzahnradpumpen den Vorteil eines geringeren Restvolumens zwischen den Zahnflanken auf, und dieser Vorteil wird noch durch die besseren Einströmverhältnisse und der reibungsarmen Berührung der Zahnflanken ergänzt (13/14). Die Prinzipienanalyse führte also zur Auswahl des Innenzahnradpumpenprinzips als Grundprinzip einer Lösungspumpe für Absorptionswärmepumpen.

4.1 Prüfung ausgewählter Pumpenprinzipien

An einem eigens entwickelten lösungsbeschickten Pumpenprüfstand, der es erlaubte mehrere Pumpen gleichzeitig unter praxisnahen Testbedingungen zu prüfen, wurden überarbeitete Pumpen (Installation einer Gleitringdichtung) mit prinzipiell unterschiedlichem Aufbau eingehend untersucht. Die Prüfergebnisse bestätigten die theoretische Auswahl und belegten deutlich die Überlegenheit der Innenzahnradpumpe bezüglich des volumetrischen Wirkungsgrades. Weitere Vorteile dieser Pumpenart lagen in der geringen Kavitationsanfälligkeit, der recht geringen Baugröße und dem äußerst ruhigen Laufverhalten.

Die geforderte kontinuierliche Volumenstromregelung wurde mit elektrischen Regelmotoren unterschiedlicher Prinzipien erreicht. Diese konstenintensive Regelungstechnik war für das Zielsystem noch nicht zufriedenstellend, so daß es zu prüfen galt, ob nicht gestufte Drehzahlveränderungen bzw. Volumenstromveränderungen mit geringer Stufenzahl ausreichen, um einen annähernd optimalen Wärmepumpenbetrieb zu gewährleisten.

5. Regelungsprinzip mit vermindertem Aufwand

Parallelversuche an einer Komponentenprüfanlage für Absorptionswärmepumpen hatten gezeigt, daß ein kostensenkendes Regelungskonzept mit gestaffelter Fördermenge der Lösungspumpe annähernd die Wärmeverhältnisse einer kontinuierlichen Volumenstromregelung erreicht, wenn mit drei Volumenstromsprüngen gearbeitet wird. Die Faktoren zur Verknüpfung der ermittelten Volumenströme entsprachen etwa denen der erzielbaren Nenndrehzahländerungen vom zweipoligen auf vierpoligen zum sechspoligen Betrieb eines Drehstrommotors (siehe Bild 1).

Der Einsatz einer gleitringabgedichteten Innenzahnradpumpe verbunden mit einem Regelmotor in einer kompakt ausgeführten Wasser/Wasser-Absorptionswärempumpe ($\dot{Q}_{WS\,max}$ = 15 kW) bestätigte die ausgewählten Volumenströme auch für diesen Wärmepumpentyp im Dauertest (Bild 2+3). Die Konkretisierung der Förderrandbedingungen ließ also eine Begrenzung auf drei Volumenströme $\hat{=}$ drei Nenndrehzahlen des Antriebmotors zu (1. Stufe: n= 1000 min^{-1} $\hat{=}$ 167 l/h, Δp_{max}=12 bar; 2. Stufe: n= 1500 min^{-1} $\hat{=}$ 250 l/h, Δp_{max}= 18 bar, 3. Stufe: n= 3000 min^{-1} $\hat{=}$ 500 l/h, Δp_{max}= 23 bar). Als Antriebseinheit der vereinfachten vollhermetischen Lösungspumpe bieten sich polumschaltbare Spaltrohr- oder Tauchmotoren an.

6. Gemeinsame Entwicklungsarbeit mit externen Partnern

In einer Dreiersplittung mit Schwerpunktsverteilung in die Aufgabenbereiche Pumpe, Antriebsmotor und Prüfung sowie Koordination, gelangte das Projekt in die Realisierungsphase. Neben dem Pumpenhersteller (Fa. Bucher, D-7895 Klettgau) der bisher in Eigenleistung umgerüsteten Innenzahnradpumpen mit den besten Prüfergebnissen, begründete sich die Zusammenarbeit mit dem ausgewählten Elektromotor-Hersteller (Fa. Hanning, D-4811 Oerlinghausen) in den Erfolgen früherer gemeinsamer Entwicklungen.

Nach Festlegung des Kostenziels eines Serienaggregats auf $K \ll 500$.-DM setzte eine umfangreiche Prüfung modifizierter und teils vereinfachter Pumpengrundtypen ein, um Kompromisse zwischen den gegensätzlichen Forderungen Kosten und Verschleißfestigkeit sowie Wirkungsgrad zu erreichen. Hierbei mußten zur Gewährleistung einwandfreier hydrodynamischer Fördermedienschmierung Wirkungsgradverringerungen toleriert werden (Bild 4), da weitere Spaltminimierungen zum Abriß des Schmierfilms führten und somit permanent abrasiven Verschleiß einleiteten.

Die Entwicklungsarbeiten zum Vollhermetikmotor, daß heißt die Klärung der Prinzipfrage Spaltrohr- oder Tauchmotor, dekkten erhebliche Kostenvorteile des Spaltrohrmotors auf, da fertigungstechnisch bedingte Lackbeschädigungen beim Wickelvorgang zu hohe Ableitströme der geprüften Tauchmotormuster verursachten und Abweichungen von Serienbelackungen, bzw. standardisierter Wickeltechniken zu extremen Kostensteigerungen einer späteren Serienfertigung führen. Die daraus abgeleitete Prinzipentscheidung fiel also zu gunsten des Spaltrohrmotors aus.

Nach Schnittstellendeffinition zur Verbindung eines dreifach polumschaltbaren Spaltrohrmotors mit einer opimierten Innenzahnradpumpe liefen Motor- und Pumpenentwicklung bzw. Optimierung getrennnt voneinander. Das konstruktive Konzept basierte auf einem Antriebsmotor mit zweifach gelagerter Hohlwelle, an derem freien Wellenende das Antriebsritzel der Pumpe befestigt ist. Auftretende resultierende Axialkraftkomponenten werden über die Ritzelstirnseiten und dem hydrodynamischen Schmierfilm vom Gehäuse aufgenommen; die Schmierung der beiden Motorgleitlager übernimmt ebenfalls das Fördermedium.

Im Bild 5 ist das Endprodukt der Entwicklung dargestellt. Die durchgeführten Funktions- und Verschleißprüfungen erbrachten positive Ergebnisse und bestätigten einen erfolgreichen Projektabschluß. Die Übertragung der Prüfergebnisse auf den Praxiseinsatz in Absroptionswärmepumpen wie anfangs geschildert führt zu einer elektrischen Leistungsaufnahme $P_E \leq$ 500 W, die nur in der dritten Stufe, d.h. bei Außentemperaturen $\vartheta_L <$ -12 °C (siehe Bild 1) mit geringer Häufigkeit überschritten wird und im Maximum $P_E \approx$ 900 W beträgt. Zur Bewährung wird eine Vollhermetikpumpe in der vorhandenen Kompaktanlage (Bild 2) einem praxisnahen labormäßigen Dauertest unterzogen.

SUMMARY

The only suitable coolant pumps for absorption heat pumps available on the market are cost-intensive piston-membrane pumps. Tests on currently available alternative pumps revealed the problem of shaft seals and the material compatibility of common elastomers. Internal gear pumps showed the best pumping performance, and the internal gear pump produced by a firm making hydraulic pumps was most convincing owing to its quiet operation,size and volumetric efficiency. The project was completed in collaboration with this firm and with a second firm specializing in electric motors.

In the medium term, a temporary solution has been found consisting of a friction ring-sealed internal gear pump with a flange-connected pole-reversible electric motor. The end product of the development work is a fully hermetic system lubricated by the pumped medium consisting of a triple pole-reversible canned motor with a flange-connected internal gear pump. The operating and wear tests carried out showed positive results and confirm that the project has been successfully completed.

7. Literatur

(1)	Anonym	Tauchpumpen für die Wasserwirtschaft Fachber. Hüttenprax. Metallweiterver. (1978) 12, S. 1058-1059
(2)	Anonym	Wie lange leben Tauchmotorpumpen für Abwasser Betr.-Techn. 19, (1978) 9, S. 53 -56
(3)	Bohl, W. u. a.	Pumpen und Pumpenanlagen (2. Auflage 80 Expert-Verlag G.)
(4)	Krämer, R. Neumaier, R.	Spaltrohrmotorpumpen mit interner Teilstromförderung für Flüssiggase Masch.-Markt, 85 (1979) 44,S. 863-865
(5)	Krämer, R. Neumaier, R.	Spaltrohrpumpe - ein wesentlicher Beitrag zum Umweltschutz Pumpen-Tagung K lsruhe 1978 VDMA Fachgemeinschaft Pumpen 1 (1979) Sep., S. 1 - 48
(6)	Litzenberg, D. P.	A modern solution to chemical pump leakage Pumps (1981) 1/2, S. 31 - 33
(7)	Pohlenz, W.	Pumpen für Flüssigkeiten und Gase, Grundlagen für Pumpen (Verlag Technik,B)
(8)	Pohlenz, W.	Pumpen für Flüssigkeiten und Gase, Pumpen für Flüssigkeiten (Verlag Technik, B)
(9)	Schmotz, D.	Wellentauchpumpen für Salzschmelzen Betr.-Techn., 20 (1979) 11, S. 44-47
(10)	Schulz, H.	Die Pumpen (Springer, B),
(11)	Schulz, H.	Axialschübe und Läuferstabilität in Spaltrohrmotorpumpen Masch.-Markt, 83 (1977) 70, S. 1344-1347
(12)	Trinkel, H.	Axialkräfte und Läuferstabilität von Spaltrohrmotorpumpen Masch.-Markt, 83 (1977) 60, S. 1166-1169
(13)	Truninger, P.	Innenzahnradpumpe für Wasserbetrieb (Fluid, Januar 1977)
(14)	Truninger, P.	Innenzahnradpumpe für schwer entflammbare Flüssigkeiten (Konstruktion & Design, April 1978)

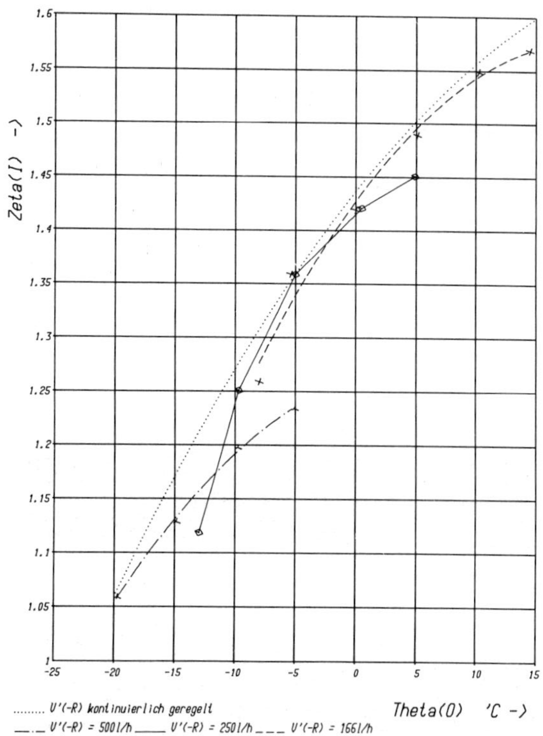

Bild 1: Einfluss der Volumenstromregelung auf das Wärmeverhältnis

Bild 2: Wasser/Wasser-Kompaktabsorptionswärmepumpe

Bild 3: Stirnansicht der Motor-Pumpen-Einheit

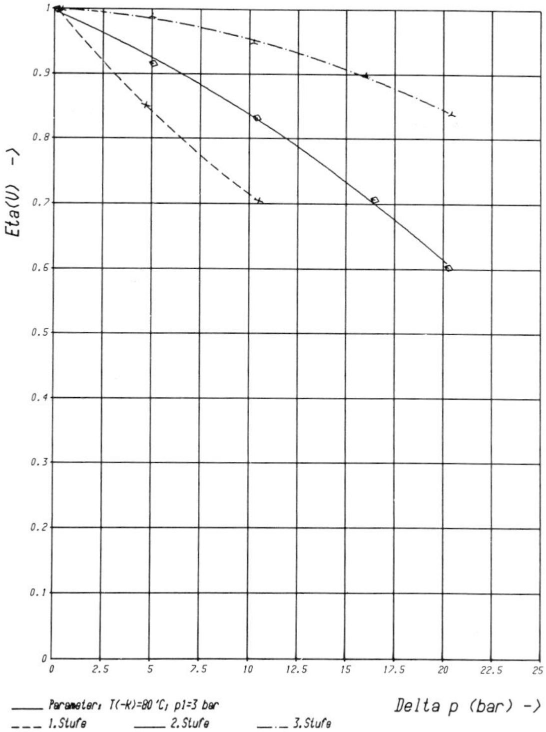

Bild 4: Volumetrischer Wirkungsgrad

Bild 5: Vollhermetikpumpe

WORKING FLUIDS FOR SORPTION HEAT PUMPS

H. BOKELMANN, H.J. EHMKE, M. RENZ, F. STEIMLE
Institut für Angewandte Thermodynamik und Klimatechnik
Universität Essen

Summary

This report presents the results of an extensive search for new working systems for absorption heat pumps in order to overcome the drawbacks of working pairs, which are used nowadays in these plants. Out of a great number of possible combinations of refrigerants and absorbents some favourable systems were selected for detailed measurements which enable a first comparison of the possible improvements by these new systems.

1. INTRODUCTION

The use of absorption heat pumps for domestic space heating offers a potential energy saving of about 33% when compared to conventional gas- or oil-fired boilers. But the up-to-now available absorption heat pumps are very poorly optimised for the use in space heating applications under European conditions especially for hydronic heating systems. This can be explained by the disadvantages of the most widely used working fluids ammonia-water and water-lithiumbromide (1,2). Therefore the search for new working fluids for absorption heat pumps is a matter of vital interest in order to solve the problems of the optimisation of sorption plants for heat pump application.

The presented paper reports about an extensive experimental research program with the aim to find new working fluids and of the measurements of most important properties, which are pressure, solubility, density, viscosity and thermal stability.

2. EXPERIMENTAL PROCEDURE

For the experimental program a great number of refrigerants and absorbents were taken into account. The pure compounds as well as possible combinations of these were selected with regard to their thermophysical properties and their chemical structure (1,2). To estimate the suitability of a refrigerant-absorbent combination for the use in an absorption heat pump the most important criterion is the solubility of the refrigerant in the absorbent. Therefore measurements of solubility at constant pressure were performed with a great number of selected working fluids. The working fluids which show the most favourable results are then selected for detailed investigations. Besides the solubility the next important selection criterion is the thermal stability which was determined, if no literature data were available, for refrigerant-absorbent combinations with good results in solubility. In case of positive results in thermal stability, measurements of thermophysical properties (vapour pressure, density, viscosity) for the whole field of application were performed. The results of these measurements are presented in tables, figures and numerical equations.

In further going experiments the absorbent content in the vapour phase, the enthalpy of the pure compounds and the enthalpy of of mixing were measured in order to calculate the need or the amount of rectification and in order to establish an enthalpy-concentration chart.

3. RESULTS

Results of the measurements on working fluids with methanol and methylamine as refrigerant and additional measurements for selecting working fluids for detailed investigations are published earlier (4,5,6,7,8). This paper presents the results of the measurements of the thermophysical properties of working fluids with the refrigerants trifluoroethanol (TFE), dichlorotrifluoroethane (R 123a), ammonia (NH_3), and monochlorodifluoromethane (R 22). Finally the results will be discussed and compared with some well-known working fluids.

3.1 TFE-N-METHYLPYRROLIDONE (NMP)

Because of its most promising results in solubility and thermal stability (6,8) the system TFE-NMP was selected for detailed investigations. To establish a numerical equation of the p, t, x-properties the vapour pressure of the pure refrigerant (65 points between $-20^{\circ}C$ and $+100^{\circ}C$) and of the system TFE-NMP was measured (139 points between 1 and 1000 mbar) /8/. By using this equation a p, t, x-chart of the system has been constructed (figure 1).

The system offers some favourable properties for the use in heat pumps like an even distribution of the vapour pressure lines in the equilibrium chart and a high weight fraction of refrigerant in the range of application. Therefore even at low evaporation temperatures high absorber temperatures can be reached.

Measurements of density and viscosity were performed for the whole concentration field of interest. The values of viscosity are in a favourable range (8).

The refrigerant TFE offers a sufficient high value in heat of evaporation (449 kJ/kg at $0^{\circ}C$). Under typical conditions the pressure levels in the absorption cycle are below atmospheric pressure and there is also the opportunity to use the system in double-effect absorption cycles.

As shown in table 1 the refrigerant-absorbent-pair TFE-NMP offers compared with ammonia-water excellent values in the specific mass flow rate, pumping energy and heat transfer rate in the solution heat exchanger, whereas the boiling point difference between refrigerant and absorbent is in the same range (need of rectification).

TFE is inflammable and its toxity seems to be in the same range as ammonia. NMP is nontoxic and like TFE less corrosive.

3.2 R 123a - DIMETHYLETHERTETRAETHYLENEGLYCOL (DTG)

R 123a-DTG was selected for detailed investigations because of the safety of the refrigerant and of the relative good solubility (6). With regard to the experience with the systems R 22-DTG and R 123a-ETFE it can be expected, that the system R 123a-DTG also will be stable.

The measurements of the vapour pressure, density and viscosity are performed by now, whereas the computer calculations in order to establish numerical equations are not finished in moment. Therefore the presented p, t, x-chart of the system (fig. 2) is provisional.

With the help of this equilibrium chart the values in the table 1 are calculated, which shows excellent value in pumping energy, whereas when compared to ammonia-water the transfer rate in solution heat exchanger will be remarkably higher.

A great advantage of this working couple is its high boiling point difference, hence there will be no need of rectification, in opposite to ammonia-water and to the discussed system R 123a-ETFE.

3.3 AMMONIA-(LITHIUMNITRATE)/WATER

The investigations using the refrigerant NH_3 were concentrated on the application of aqueous salt solutions as solvent. The advantage of this solvent type towards water as solvent is a lower heat capacity in the solution and also a reduction of the water content in the vapour phase to diminish the amount or even avoid rectification.

As described in (7) several aqueous solutions of LiBr and $LiNO_3$ were investigated by variing the water content. Besides equilibrium measurements there was a determination of limits of crystallization and also of the viscosity. The equilibrium measurements showed that aqueous $LiNO_3$-solutions were superiour to aqueous lithiumbromide (LiBr) solutions as absorbent. Taking viscosity and crystallization into account it was decided to measure a ternary system NH_3-$LiNO_3/H_2O$ with a water content of 25% in the absorbent. The vapour pressure and the density of this ternary solution have been measured and are described by empiric equations. The p, t, x-chart shows that under heat pump operation the working field will be far away from the crystallization line and indicates also that the refrigerant content is higher than with NH_3-H_2O when high temperature differences between evaporator and absorber are desired.

Measurements of the vapour phase lead to the result that for heat pump operation the water content of the vapour will be around 1%.

3.4 R 22-TRIETHYLENEGLYCOLDIMETHYLETHER (DTrG)

Measurements of equilibrium lead to the result that the characteristics of DTrG are slightly better than those of the well known pair R 22-DTG. The difference in boiling points between R 22 and DTrG is 265 K so that there is no need for rectification expected. The viscosity of DTrG is also lower than the viscosity of DTG.

Comparing the results of equlibrium measurements of R 22-DTG with literature data lead to poor agreement especially in the range of high temperatures. Obviously there is a lack of experimental data for R 22-DTG which cover the whole range of heat pump application. This is the reason why a comparison between R 22-DTG and R 22-DTrG is involved in uncertainty and so in table 1 the values for R 22-DTG are in brackets.

4. DISCUSSION

In order to allow a comparison between the working couples which are discussed in the previous chapter, and the systems NH_3-H_2O, R 22-DTG(9), R 123a-ETFE (10) the following factors and properties are defined and presented in table 1.

r_0 - specific heat of evaporation of the refrigerant at $0^{\circ}C$

$p_{50^{\circ}C}$ - the vapour pressure of the refrigerant at $50^{\circ}C$ (condenser conditions)

$$n_p = \frac{f \cdot \Delta p}{\rho_r \cdot r_0} \quad ; \quad f = \frac{1 - x_a}{x_r - x_a}$$

the (minimum) pumping energy which is needed for generating equal cooling effects.

$$n_l = \frac{(f-1) \cdot c \cdot \Delta t}{r_0}$$

the heat to be exchanged between weak and strong solution for generating equal cooling effects.

The values in table 1 are calculated at the evaporation temperature of $0^{\circ}C$, the condenser and the absorber temperature of $50^{\circ}C$ and the generator temperature of $150^{\circ}C$.

In addition to these values, table 1 indicates, whether a rectification has to be expected and gives also an information about the toxity of the system. Table 1 shows that the four systems, which were established in the frame of this research program might be an alternative to the well-known or discussed system NH_3-H_2O, R 22-DTG or R 123a-ETFE.

The system TFE-NMP is not able to avoid the disadvantages of rectification and toxity of the working pair NH_3-H_2O, whereas the pressure level under heat pump working conditions is much lower and the values of pumping energy, specific mass flow rate and heat transfer rate are much better.

The values of the system R 123a-DTG in specific mass flow rate, pumping energy and heat transfer rate are higher then the values of the system R 123a-ETFE, but the great advantage of the working pair R 123a-DTG is its high boiling point difference, so that there will be no need of rectification. Another advantage of this systems is its safety.

The ternary system NH_3-$LiNO_3/H_2O$ has towards the system

NH_3-H_2O the advantage of a significant reduction of the water content of the vapour phase and the heat flux in the solution heat exchanger is also lower, whereas the pumping energy is equal. These advantages become even more significant when a comparison is made at higher absorber temperatures. A disadvantage of the ternary system is the higher value in viscosity which is in the same range as the system H_2O-LiBr.

A discussion of the system R 22-DTrG leads directly to the desire fore more accurate data for R 22-DTG. A general view shows poor results for solutions of R 22 because of the high pressure difference between absorber and generator and the relatively low heat of evaporation of R 22.

CONCLUSION

Within the frame of this research program some new systems were found and discussed, which are able to avoid some of the disadvantages of the up-to-now known systems.

To calculate the overall performance of an absorption heat pump which uses these systems as working fluids it is necessary to perform additional measurements of the vapour-liquid-phase-equlibria and of the calorific properties. These measurements are planed for the next future.

Nomenclature

- c heat capacity
- f specific mass flow rate
- p vapour pressure
- Δp pressure difference between generator and absorber
- r heat of evaporation
- t temperature
- Δt temperature difference between generator and absorber
- x mass fraction of refrigerant in solution
- ρ density

Subscripts

- a weak in refrigerant
- L solution
- O evaporator
- p pump
- r rich in refrigerant

REFERENCES

(1) Renz, M.:"Eignung von Arbeitsstoffpaaren für Absorptionswärmepumpen." in Wärmepumpentechnologie Band II, Vulkan-Verlag, Essen 1978

(2) Bokelmann, H. u. Ehmke, H.-J.:"Stoffsysteme für Absorptionswärmepumpen. Derzeitiger Stand und neue Entwicklungstendenzen." gwf-Gas/Erdgas (erscheint im Dezember 1983)

(3) Renz, M.: "Bestimmung thermodynamischer Eigenschaften wäßriger und methylalkoholischer Salzlösungen." Forschungsberichte des DKV, Nr. 5, Stuttgart 1981

(4) Bokelmann, H., H.J. Ehmke und M. Renz:"Working Fluid for a Sorption Heat pump". Proc. Four Contr. Meet. Heat Pumps, 13-13 Mai 1982, Publ. by Commission of the European Community, Brussels 1982

(5) Bokelmann, H., H.J. Ehmke, M. Renz u. F. Steimle: "Investi-

gations of Working Fluids for Gas-Fired Absorption Plants" International Gas research Conference, London 14-16 Juli 1983
(6) Bokelmann, H., M. Renz u. F. Steimle:"Working Fluids for low Pressure Absorption Systems" IIF-IIR Congress Commission B1, Paris, 31. August bis 7. September 1983
(7) Ehmke, H.J. u. M. Renz:"Ternary Working Fluids for Absorption Systems with Salt-Liquid-Mixtures as Absorber". IIF-IIR Congress Commission B, Paris 31. August bis 7.9.83
(8) Bokelmann, H. u. M. Renz:"Thermophysikalische Eigenschaften von Trifluorethanol-Stoffsystemen für Absorptionswärmepumpen." KI 11 (1983) Nr. 10
(9) Kriebel, M. u. H.J. Löffler:"Thermodynamische Eigenschaften des binären SYstems Difluormonochlormethan (R 22) - Tetraethylenglykoldimethylether (E 181)." Kältetechnik 17 (1965) Nr. 9, S. 266
(10) Murphy, K.P. u. B.A. Phillips:"Development of a Residental Gas Absorption Heat Pump." ASHRAE - Transaction 89 (1983), Pt. 1

system	refrigerant $r_{0°C}$ kJ/kg^{-1}	refrigerant $p_{50°C}$ bar	f	$10^3 \cdot n_p$	n_L	rectification	toxity
TFE-NMP	449	0,384	3,04	0,205	0,86	yes	yes
R123a-DTG	183	2,05	6,14	5,38	5,62	no	no
R22-DTrG	205	19,3	5,59	36,89	5,60	no	no
NH_3-$LiNO_3/H_2O$	1261	20,3	5,25	6,49	0,98	no	yes
NH_3-H_2O	1261	20,3	4,28	6,56	1,15	yes	yes
R22-DTG	205	19,3	(4,94)	(32,57)	(4,88)	no	no
R123a-ETFE	183	2,05	4,29	4,14	3,66	yes	no

evaporation temperature 0°C   absorption end temperature 50°C
condensation temperature 50°C   generation temperature 150°C

Table 1 - Characteristics of working fluids

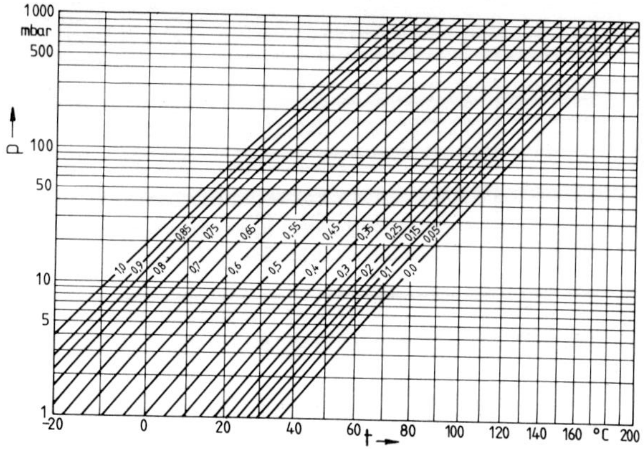

Fig. 1 - Equilibrium chart of TFE-NMP

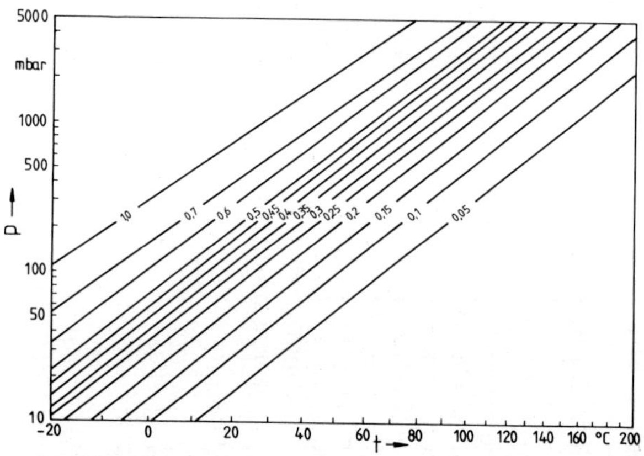

Fig. 2 - Provisional equilibrium chart of R123a-DTG

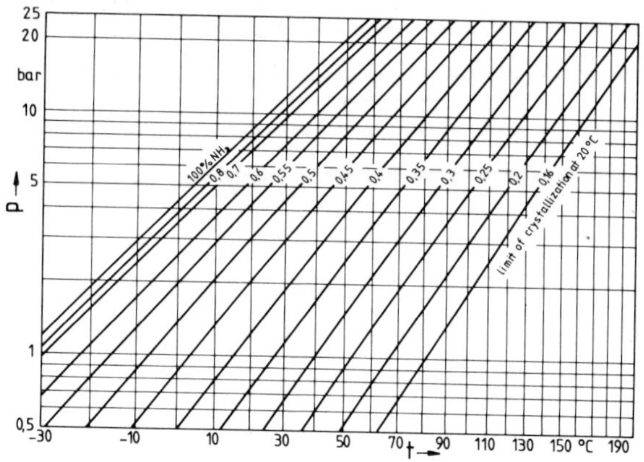

Fig. 3 - Equilibrium chart of NH_3-$LiNO_3$/H_2O(25%)

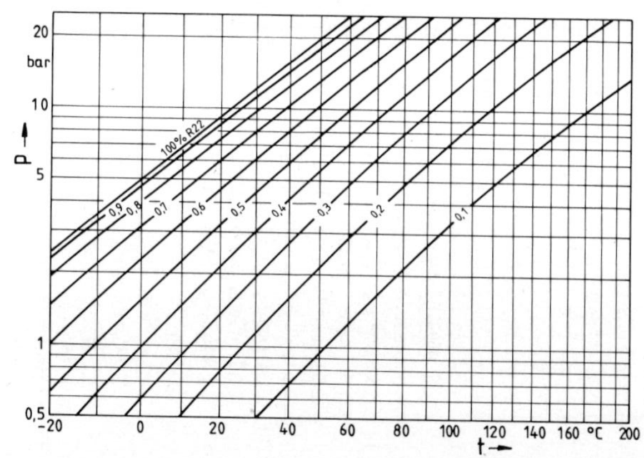

Fig. 4 - Equilibrium chart of R22-DTrG

UNTERSUCHUNG DER AUSWIRKUNGEN DES BETRIEBES VON WÄRMEPUMPEN-SYSTEMEN AUF DIE UMWELT

Dipl.-Ing. P. Müller
Vereinigte Elektrizitätswerke Westfalen AG, Dortmund

Zusammenfassung

Für die Wärmequellen Grundwasser, Erdreich und Luft werden die beim Betrieb von Wärmepumpenheizungsanlagen mit der jeweiligen Wärmequelle auftretenden Auswirkungen auf die Umwelt quantifiziert. Z.B. zeigt sich, dass, abhängig von hydrogeologischen Parametern und der benötigten Wassermenge der Durchmesser der 7°C-Isotherme im Bereich des Schluckbrunnens nach mehrjährigem Wärmepumpenbetrieb am Beispiel eines 1-Familienhauses ca 25 m beträgt. Negative Auswirkungen auf das Grundwasser durch den Betrieb einer Wärmepumpe konnten nicht belegt werden.

Für die Wärmequelle Erdreich hat das Vorhaben Ansatzpunkte für neue Wärmetauscherkonfigurationen geliefert. Geringfügige Expositionsunterschiede wie z.B. Beschattung durch Gebäude u.ä. haben einen grösseren Einfluss auf den Pflanzenwuchs als der Betrieb einer Wärmepumpe.

Bei Verwendung von Luft als Wärmequelle für eine Wärmepumpe sind Massnahmen zur Verminderung der Geräuschemission bei zwangskonvektiv-beaufschlagten Verdampfern unerlässlich. Lufttemperaturunterschiede können bereits bei sehr geringer Windbewegung hinter dem Verdampfer einer Wärmepumpe mit Wärmequelle Luft nicht mehr nachgewiesen werden.

1. Technik der Wärmepumpe mit Wärmequelle Grundwasser Allgemeines

Grundwasser ist durch sein relativ zu anderen Wärmequellen hohes Temperaturniveau hervorragend als Wärmequelle für Wärmepumpen geeignet. Es kann trotz eines hohen theoretischen Potentials dennoch nicht in beliebiger Menge verwendet werden. Die wichtigsten Gründe für Einschränkungen sind:

- Grundwasser ist ein lebenswichtiges Gut. Daher stellt sich die Aufgabe, es in seiner Quantität und Qualität zu erhalten und es vor Verunreinigungen zu schützen. Die Sicherung des Grundwassers zur Trinkwasserversorgung hat Vorrang vor allen anderen Maßnahmen.
- Die Nutzung des Grundwassers ist nur sinnvoll in Gebieten mit reichhaltigen Vorkommen in geringer Tiefe.

Bild 1 zeigt das Schema einer Brunnenanlage für eine Wärmepumpe. Das Grundwasser wird durch eine im Förderbrunnen installierte Unterwasserpumpe gefördert, über eine Rohrleitung zum Verdampfer der Wärmepumpe geleitet, dort abgekühlt und danach zum Schluckbrunnen geleitet und dem Untergrund wieder zugeführt. Damit ist gewährleistet, daß Grundwasserquantität nicht verändert wird.

1.1. Aufgabenstellung

Beim Betrieb von Wärmepumpen bildet sich durch die Wiedereinleitung des in der Wärmepumpe abgekühlten Wassers in den Grundwasserleiter um den Schluckbrunnen ein abgekühlter Bereich aus, eine negative thermische Anomalie. Gleichzeitig bildet sich in der Umgebung des Förderbrunnens ein Absenkungstrichter und um den Schluckbrunnen staut sich ein Grundwasserberg (Bild 2). Daraus ergeben sich folgende Fragen:

- Welche Temperatur- und Druckverteilung ergibt sich im Grundwasserleiter im Umkreis um Förder- und Schluckbrunnen?
- Wie groß ist der Abstand zwischen Förder- und Schluckbrunnen zu bemessen, damit keine gegenseitige Beeinflussung eintritt, die bewirkt, daß abgekühltes Wasser zurückgefördert wird?
- Wie groß ist aus dem gleichen Grund der Abstand der Brunnenanlage zur Nachbaranlage zu wählen?
- Treten physikalische, chemische und biologische Beeinflussungen des Grundwassers durch den Wärmepumpenbetrieb auf?

Im Abschlußbericht der Untersuchung sind Aussagen zu den einzelnen Teilaufgaben getroffen. Da die Temperaturverteilung des Grundwassers im Bereich des Schluckbrunnens viele Fragestellungen von besonderer Bedeutung ist, wird nachfolgend ausschließlich dieser Aspekt der Untersuchung zusammengefaßt behandelt.

1.2. Hydrothermische Grundlagen
Einflußgrößen

Aus den tiefen Bereichen der Erde fließt ständig Wärme zur Erdoberfläche. Es handelt sich dabei im wesentlichen um Wärmeleitung, lokal auch um Wärmekonvektion. Die Temperaturgrenzen des thermischen Feldes sind durch die Temperatur der Erdoberfläche einerseits und die Temperatur der Wärmequellen im Erdinnern andererseits festgelegt. Die Temperaturen an der Erdoberfläche werden weitestgehend durch die Einstrahlung der Sonne bestimmt. Der den Erdboden erreichende Anteil der solaren Einstrahlung beträgt in unseren geographischen Breiten rd. 1.000 kWh/m² oder im Jahresmittel etwa 10^{-2} W/cm².

Die Erdwärmestromdichte ist dagegen sehr gering. Sie liegt in der Bundesrepublik Deutschland zwischen $5{,}4 \cdot 10^{-6}$ W/cm² und $8{,}0 \cdot 10^{-6}$ W/cm². Der Einfluß des Erdwärmestromes auf die Temperatur an der Erdoberfläche ist dementsprechend sehr gering (ca. 0,02 K).

Von der Erdoberfläche nimmt die Temperatur - entsprechend dem örtlichen geothermischen Gradienten - zur Tiefe hin zu. Der Mittelwert des geothermischen Gradienten liegt in Deutschland bei 0,03 K/m. Die Grundwassertemperatur t_0 in der Teufe h kann damit abgeschätzt werden.

Im erdoberflächennahen Bereich werden die Untergrundtemperaturen durch den Jahresgang der Lufttemperatur beeinflußt. In Bild 3 wird dieser Effekt schematisch anhand von vier Temperaturkurven verdeutlicht: die jährliche Lufttemperaturwelle pflanzt sich mit abnehmender Amplitude und einer Phasenverschiebung in der Tiefe fort. In 10 m Tiefe betragen die jährlichen Temperaturschwankungen nur noch etwa + 1 K.

Die erdoberflächennahen Untergrundtemperaturen werden weiterhin von der Art der anstehenden Gesteine, deren Exposition, der Sonneneinstrahlung und der Vegetation beeinflußt. In besiedelten Gebieten sind die Untergrundtemperaturen durch die Abwärme von Gebäuden, Abwasserkanälen, Fernheizungsleitungen, Kühlwasserversickerungen usw. flächenhaft bis zu 4 °C erhöht.

Wirkt im Untergrund eine künstliche, positive oder negative Wärmequelle, so überlagert sich deren thermisches Feld mit dem thermischen Erdfeld. Im Falle der Versickerung von kaltem Wasser kommt es zu Abkühlungen, d.h. es baut sich ein negatives thermisches Feld auf. Diese negative thermische Anomalie breitet sich zunächst rasch aus und strebt dann einem stationären Zustand zu. Dieser ist erreicht, wenn die pro Zeiteinheit aus der Umgebung in die Anomalie einfließende Wärmemenge gleich der bei der Kaltwasserversickerung dem Untergrund zugeführten "negativen Wärmemenge" ist. Bild 4 zeigt die der Anomalie zufließenden, positiven Wärmeströme. Dabei ist im Mittel q_n = 1,5 HFU (Heat Flow Units) = 6,3 · 10^{-6} W/cm² und es ergeben sich zwischen den Wärmestromdichten die Beziehungen $q_n < q_h < q_s$ (q_n = natürlicher Erdwärmestrom, q_s = solar bedingter Wärmestrom von der Erdoberfläche her, q_h = aus der horizontalen Umgebung zufließender Wärmestrom). Die Wärmestromdichten von q_s und q_h sind - außer von der Wärmeleitfähigkeit der Gesteinsschichten - besonders von der Stärke der Anomalie abhängig. Da sich das Grundwasser normalerweise im fließenden Zustand befindet, tritt noch ein konvektiver Term q_w auf, der direkt proportional der Grundwasserfließgeschwindigkeit ist. Das zuströmende Grundwasser vermischt sich mit dem kalten Wasser und vermindert dadurch die Anomalie. Gleichzeitig wird deren Form elliptisch verzerrt und in Grundwasserfließrichtung ausgelängt (Bild 5).

1.2.1. Die thermische Anomalie

Beim Betrieb einer Wärmepumpe mit dem Energieträger Grundwasser wird Grundwasser mittels eines Förderbrunnens aus einem Grundwasserleiter entnommen, in der Wärmepumpe beim Energieentzug abgekühlt und durch einen Schluckbrunnen wieder in den Grundwasserleiter eingeleitet. Dabei wird die Umgebung des Schluckbrunnes abgekühlt, es baut sich eine negative thermische Anomalie auf.

Das Verhalten einer derartigen Anomalie wurde in 2 hydrothermischen Versuchsfeldern meßtechnisch ermittelt. Es wird beispielhaft an der Wärmepumpenanlage zur Beheizung eines freistehenden Einfamilienhauses mit ca. 200 m² Wohnfläche verdeutlicht. Der Grundwasserleiter des Beispielobjektes besteht aus Fein- und Mittelsanden, die Grundwasser-Fließgeschwindigkeit beträgt ca. 0,2 m/d, und der Grundwasserspiegel schwankte im Untersuchungszeitraum zwischen 0,4 und 1,5 m unter Gelände.

Der Betrieb begann Ende Juli 1978. Um den Schluckbrunnen entwickelte sich eine leicht in Grundwasser-Fließrichtung verzerrte thermische Anomalie. In den Bildern 6 - 10 ist das Verhalten der Anomalie anhand von jeweils zwei senkrecht zueinander liegenden und durch den Schluckbrunnen führenden hydrothermischen Profilen aufgezeigt. Im März 1979, nach einer strengen Frostperiode, war die Temperatur im Schluckbrunnenbereich bis auf 4,4 °C abgesunken. Im Frühjahr und Sommer 1979 wurde die Anomalie bei abnehmender Kaltwasserversickerung und zunehmender Luft- und Bodentemperatur in wesentlichem Umfang regeneriert. Die wieder in zunehmendem Umfang betriebene Kaltwasserversickerung dokumentiert sich im November 1979 mit Temperaturwerten unter 6 °C im Schluckbrunnenbereich. Der Vergleich der thermischen Zustände im Herbst dreier aufeinander folgender Jahre zeigt, daß sowohl im September 1980 als auch im September 1981 die Anomalie deutlich kleiner war als im September 1979, obwohl die Wärmepumpenanlage seit Beginn der Untersuchungen ständig in Betrieb war - ausgenommen jeweils 2 bis 3 Monate im Sommer. (Bild 11)

Aus den hydrothermischen Profilen ist zu ersehen, daß eine "Wärmepumpen-Anomalie" nicht stationär wird, sondern ständig im Jahresverlauf pulsiert. Dabei ist sowohl die jahreszeitlich veränderliche Kaltwasserversickerung als auch der Temperatur-Jahresgang von Einfluß.

Die stärksten Abkühlungen traten in der Anomalie im Februar-März 1979 nach einer strengen Frostperiode im Januar 1979 auf.

Der größte, in Grundwasserfließrichtung liegende Durchmesser der thermischen Anomalie - als Maß für die Ausdehnung der Anomalie genommen - erreichte maximal 45 m und stellte sich ab Sommer 1980 auf ca. 33 m ein.

Die Flächengröße der Anomalie stieg bis auf 1.000 m² an, verringerte sich dann aber bis zum Frühjahr 1980 bis auf ca. 550 m². Die pro Jahr versickerte Kaltwassermenge läßt sich in Beziehung setzen zur Maximalfläche der Anomalie. Daraus ergibt sich im vorliegenden Fall ein spezifischer Flächenbedarf von ca. 0,2 m² pro 1 m³ versickerten Kaltwassers. Dieser Wert kann nicht verallgemeinert werden, er gilt nur bei den zugrunde liegenden hydrogeologischen, hydrothermischen, meteorologischen und betriebstechnischen Parametern.

Der Wärmezufluß in die negative thermische Anomalie, d.h. die thermische Regeneration, erfolgt im vorliegenden Fall überwiegend durch Wärmeleitung (Konduktion) von der Erdoberfläche her. In Gebieten mit höheren Grundwasser-Fließgeschwindigkeiten ergibt sich ein zusätzlicher konvektiver Regenerationseffekt.

1.3. Rechnerisches Modell einer Kaltwassereinleitung

Auf die horizontale Ausdehnung der thermischen Anomalie einer Grundwasserwärmepumpe haben verschiedene Parameter einen Einfluß. Qualitativ läßt sich zeigen, daß die Anomalie umso kleiner bleibt

- je geringer die Temperaturabsenkung durch die Wärmepumpe,
- je geringer die wiedereingeleitete Wassermenge,
- je geringer die Einleitungstiefe, d.h. der Abstand zwischen Filteroberkante und Gelände,
- je größer die Filterstrecke des Schluckbrunnens und
- je größer die Grundwasserfließgeschwindigkeit.

An Hand der Meßdaten der hydrothermischen Versuchsfelder wurde ein Rechenprogramm entwickelt, mit dem sich der Einfluß einer Einleitung von Wärmepumpenwasser - bzw. ganz allgemein von Kaltwasser - auf das natürliche Temperaturfeld eines Grundwasserleiters quantitativ bestimmen läßt.

2. Technik der Wärmepumpen mit Wärmequelle Erdreich Allgemeines

Erdreich ist ein natürlicher Wärmespeicher, der von der Sonne gespeist wird und als regenerative Wärmequelle sich hervorragend zur Nutzung durch Wärmepumpen eignet. Eingeschränkt wird die Nutzung dieser Wärmequellen im wesentlichen von folgenden Faktoren:

- Zur Beheizung von 1 m² Wohnfläche werden z.Z. 2-3 m² Erdreichfläche benötigt . Aufgrund der häufig nicht optimierten Wärmetauscherkonzeptionen müssen deshalb Grundstücksflächen zur Verfügung stehen, die im vorhandenen Baustand fast nie und im Neubaubereich selten vorhanden sind.

- Die o.g. pragmatischen Wärmetauscherkonzeptionen führen zu relativ - zu Grundwasser oder Luft - hohen Erschließungskosten dieser Wärmequelle.

Diese Faktoren waren Hauptursache dafür, daß dieser Wärmequelle häufig nicht die Bedeutung zukommt, die sie verdient. Dies gilt insbesondere aufgrund der Tatsache, daß Wärmepumpen mit Wärmequelle Erdreich im Betrieb weitestgehend keine Wartungskosten verursachen.

Bild 12 zeigt das Schema einer Wärmepumpenheizungsanlage mit Wärmequelle Erdreich. Die Umweltwärme wird dem Boden durch einen Wärmetauscher entzogen der gebäudenah in den Erdboden verlegt wird. Dieser Wärmetauscher, auf dessen Konzeption nachfolgend noch eingegangen wird, wird von einer Kühlflüssigkeit (Sole) durchströmt. Die Sole wird in der Wärmepumpe abgekühlt und von der Soleumwälzpumpe wieder zur Erwärmung in den Wärmetauscher gepumpt.

Als Sole wird im üblichen Sprachgebrauch eine frostsicher gemachte Flüssigkeit bezeichnet. Je nach der für den dimensionierenden Fall zu wählenden Grenztemperatur wird bei der Herstellung der Sole vor Ort eine bestimmte Menge Zuschlagstoff beigemischt. Dabei sind die Zunahme der Viskosität und die Abnahme der spezifischen Wärme gegenüber Wasser zu beachten.

Am Anfang der Nutzung der Wärmequelle Erdreich wurden Rohrschlangen mit Unterflurabständen von 1,2 bis 2,0 m und Abständen der einzelnen Rohre zwischen 1,0 und 2,0 m vorsichtig gewählt. In Bild 13 ist eine mögliche Konfiguration schematisch dargestellt. Üblicherweise werden die einzelnen Rohrschlangen bifilar verlegt. Als bifilar wird die Technik bezeichnet, bei der Vor- und Rücklauf im Wechsel mit gleichbleibenden Abstand verlegt ist. Damit nimmt man z.Z. noch wenig Rücksicht auf die Parameter, die Einfluß auf Rohrabstand und Rohrlänge ausüben. Vielmehr orientiert man sich häufig am praktisch Machbaren. Die bifilar verlegten Erdreichwärmetauscher haben den Nachteil, daß aufgrund des thermischen Kurzschlusses zwischen parallelgeführten Vor- und Rücklaufleitungen das Temperaturniveau der Sole am Eintritt in die Wärmepumpe niedriger als möglich ist. Nicht an die Bodenverhältnisse angepaßte, zu lange Einzelrohrschlangen bringen dabei keinen Wärmegewinn, wenn die Soletemperatur vor dem "Umkehrpunkt" die Erdreichtemperatur erreicht hat. Die zusätzlichen Rohrstrecken vergrößern dann nur die aufzubringende Pumpenergie.

Aufgrund von Teilergebnissen der vorliegenden Untersuchungen wurde schon früher auf verbesserte Anordnungen von Wärmetauschern aufmerksam gemacht. Diese Anregungen, die durch Ergebnisse der Untersuchung

"Erdreich-Wärmepumpen mit Umschaltung zur Außenluft und Regeneration des Erdreichs"

bestätigt wurden, wurden vielerorts aufgegriffen. Neben dem Vorschlag der Registerverlegung mit räumlich getrennter Vor-/Rücklaufleitung wurde eine Verlegung des Wärmetauschers in zwei Horizonten angeregt. Obwohl bisher noch keine ausreichend gesicherten Daten aus dem praktischen Betrieb solcher Wärmepumpen vorliegen, scheinen sich die positiven Annahmen, die insgesamt zu einer deutlichen Verringerung des Wärmetauscherflächenbedarfs führen, zu bestätigen. Danach scheint es möglich, ein Flächenverhältnis von 1 m^2 Wohnfläche zu 1 m^2 beheizter Fläche zu erreichen.

Bei einer mittleren maximalen Wärmeentzugsdichte von 40 W je m^2 Wärmetauscherfläche wird über die Wärmepumpe mit Wärmequelle Erdreich rd. 35 % der Globalstrahlung und rd. 7 % der jährlich von der Sonne eingestrahlten Energiemenge nutzbar gemacht. Dabei ist der Wärmegewinn durch fühlbare Wärme insbesondere aus Regen und Wind noch unberücksichtigt. Der Bodenwärmestrom mit ca. 7 W/m^2 bzw. ca. 60 $kWh/m^2 \cdot a$ ist an der Gesamtbilanz nur wenig beteiligt.

2. Aufgabenstellung

Beim Betrieb von Wärmepumpen mit Wärmequelle Erdreich wird dem Erdboden Wärme entzogen. Damit bildet sich im Bereich des Wärmetauschers eine Temperaturanomalie aus. Daraus ergeben sich folgende Fragen:

- Welche Bodentemperaturen sind im Bereich des Wärmetauschers zu erwarten?
- Wie ist der Verlauf der Isothermen im Randbereich des Wärmetauschers, d.h. in welchem seitlichen Abstand werden Nachbargrundstücke noch beeinflußt?
- Ergeben sich Auswirkungen auf den Pflanzenwuchs?

3. Ergebnisse

An acht Orten des Münster- und des Sauerlandes wurden während 2 Heizperioden auf unterschiedlichen Böden Untersuchungen unterschiedlicher Intensität zur Entwärmung des Erdreichs durchgeführt. Die dabei durch Einzelmessungen an vielen Standorten gewonnenen Ergebnisse lassen sich sehr gut mit den gleichfalls an 2 weiteren Orten durchgeführten kontinuierlichen Messungen und den Wärmehaushaltsberechnungen in Einklang bringen.

Ziel des Meßprogrammes war es, einen Einblick in das Verhalten unterschiedlicher Böden unter verschieden klimatischen Bedingungen bei Wärmeentzug zu gewinnen. Hierzu wurden im Abstand von zwei Monaten alle beschriebenen Standorte angefahren und Bodentemperaturen in den Tiefen 2 cm, 10 cm, 20 cm, 30 cm, 50 cm, 75 cm, 100 cm, 125 cm und 150 cm gemessen. Die Messungen wurden auf Flächen, die mit Wärmetauschern versehen sind und auf möglichst gleichartigen Flächen ohne Wärmetauscher durchgeführt. Zu den entsprechenden Terminen wurden für die jeweiligen Tiefen Feuchteproben entnommen, die im Labor auf Gew. % Feuchte analysiert wurden.

Für alle Standorte wurden für verschiedene Tiefen an Hand von Proben die Korngrößenverteilung ermittelt. Weitergehende bodenphysikalische Untersuchungen waren nicht sinnvoll, da abgesehen von einem Standort an allen Meßorten die Lagerungsverhältnisse der Böden einem laufenden Sackungsprozeß unterlagen und durch die umfangreichen Erdarbeiten einen ausgesprochen heterogenen Aufbau aufwiesen.

Die genaue Lage der Entzugsrohre war nicht mehr festzustellen, weshalb alle Messungen im Entzugsbereich als Zufallsstichprobe aufzufassen sind. Um Probenahmefehler möglichst gering zu halten, wurden alle Messungen in demselben Bereich der Entzugsfläche durchgeführt. An mehreren Standorten durchgeführte Parallelmessungen ergaben in diesen Bereichen eine max. Streubreite von 0,4 °C.

Das Fehlen von Meßwerten in den Tiefen 125 cm und 150 cm liegt in der mangelnden Risikobereitschaft begründet, da bei Messungen in diesen Tiefen immer mit der Beschädigung von Entzugsrohren gerechnet werden mußte.

Durch die umfangreichen Erdarbeiten bei der Verlegung der Tauscherflächen wird das Bodengefüge und die Bodenartenschichtung extrem gestört. Für vergleichende Untersuchungen zur Ermittlung des Entzugseinflusses auf den Wasserhaushalt sind die zu fordernden vergleichbaren Bedingungen zwischen Entzugs- und Vergleichsfläche nur unzureichend gegeben. Durch wechselnde Vegetationsschichten bzw. fehlende Vegetation treten weitere Fehler auf.

Zusammenfassend kann für den bei diesen Messungen erfaßten Tiefenbereich von 0 - 100 cm gefolgert werden, daß nach den bisherigen Ergebnissen mit keiner wesentlichen Änderung der Bodenwassergehalte gerechnet werden muß.

Anders sehen die Verhältnisse in der Nähe der Entzugsrohre aus. Dort liegen die Wassergehalte gegenüber der Vergleichsfläche direkt am Rohr bis zu 17 % höher. Die Wassergehalte stellen sich jedoch schon im Abstand von 15 cm auf den normalen Wert ein.

Durch die Temperaturprofilmessungen soll ein Spektrum von Reaktionsmustern erhalten werden, das eine Einordnung der kontinuierlichen Messungen zuläßt und die Ableitung allgemeiner Gesetzmäßigkeiten stützt.

Die Auswertung der Meßergebnisse erfolgt deshalb unter dem Gesichtspunkt des zeitlichen Temperaturverlaufs der einzelnen Profile, der Ausprägung der Temperaturprofile zu den jeweiligen Meßterminen und unter dem Gesichtspunkt der bodenartlichen-klimatischen Beeinflussung, die, wie zuvor schon erwähnt, durch Heizgewohnheiten und Urlaubszeiten der Besitzer stark modifiziert werden.

An den für die einzelnen Stationen dargestellten Jahresgängen der Bodentemperaturen sind Gesetzmäßigkeiten zu erkennen, die in ähnlicher Weise an allen Standorten auftreten. An den Vergleichsflächen findet bekanntlich in den Monaten September Oktober und April/Mai eine Temperaturumkehr im Profil statt. Damit verbunden ist eine Umkehr des Wärmeflusses. In den Wintermonaten fließt Wärme am Temperaturgradienten zur Bodenoberfläche und wird dort an die Atmosphäre abgegeben. Ganz anders sind dagegen die Verhältnisse auf den Entzugsflächen. Eine klare Temperaturumkehr im Profil findet nicht statt. Die Temperaturen im gesamten Profil nähern sich einander an und sinken dicht gebündelt bis zum Minimum ab. Zu Beginn der Heizperiode erfolgt der Temperaturabfall im gesamten Profil der Entzugsfläche sehr schnell, um sich dann mit einer Differenz von $\vartheta_B \approx 4{,}5$ K parallel dem Temperaturverlauf in 125 cm ohne Entzug anzupassen. Mit steigender Lufttemperatur Ende Februar-März erwärmt sich die Bodenoberfläche und die Temperaturen im Entzugsflächenprofil beginnen sich zu spreizen, während die Vergleichsflächentemperaturen im Temperatur-Umkehrbereich zusammenlaufen. Der Temperaturanstieg auf der Entzugsfläche erfolgt schneller als der Anstieg auf der Vergleichsfläche.

Ende August sind mit Abschluß der Messungen zwischen Entzugs- und Vergleichsflächen nur noch geringe Temperaturdifferenzen gegeben und mit Einsetzen der kälteren Jahreszeit stellen sich die anfangs beschriebenen Bedingungen wieder ein.

Nach den vorliegenden Ergebnissen zu urteilen ist der Einfluß bodenbedingter und klimatischer Faktoren von wechselnden Entzugsbedingungen so stark überlagert, daß für die verschiedenen Meßorte keine prinzipiellen Unterschiede abzuleiten sind.

Für Aussagen über die Wärmetransportvorgänge und das Regenerationsverhalten der Böden sind die für alle Meßorte und Meßtermine dargestellten Temperaturprofile aussagekräftig (s. Beispiel Bild 14).

Die Richtung des Wärmetransportes kann aus der Steigung der Kurven abgeleitet werden. Transport aus dem Profil zur Oberfläche und damit Wärmeverlust an die Atmosphäre ist bei negativer Steigung gegeben. Bei positiver Steigung wird Wärme zur Entzugsebene fließen. Verläuft das Temperaturprofil parallel zur Tiefenachse, so ist der Temperaturgradient Null und es erfolgt damit kein Wärmetransport. Die Intensität des Transports ist dem Temperaturgradienten proportional. Bei Ausbildung einer inneren Temperaturscheide (Minimum bzw. Maximum) im mittleren Teil des Profils) kann zwischen Unterboden und Oberfläche kein Austausch erfolgen. Die Profile lassen sich in 4 Gruppen teilen, die entsprechend der Zeit ihres Auftretens als Abkühlungstyp, Gleichgewichtstyp, Übergangstyp und Erwärmungstyp bezeichnet werden sollen.

Der Abkühlungstyp bildet sich zu Beginn der Heizerpiode und bei geringen Entzugsintensitäten aus. Sowohl das Entzugs- als auch das Vergleichsprofil haben einen negativen Gradienten und geben dauernd Wärme an die Atmosphäre ab. Auf der Vergleichsfläche ist der Gradient jedoch jeweils steiler. Die Ausbildung innerer Temperaturscheiden ist häufig und von kurzen Witterungsänderungen abhängig. Der Wärmegewinn beruht auf der Abkühlung des Bodenspeichers.

Beim Gleichgewichtstyp geht der Temperaturgradient des Entzugsprofils gegen Null und damit der Wärmefluß zwischen Oberfläche und Tauscherebene gegen Null. Die Größenordnung des Wärmeentzugs durch die Wärmepumpe entspricht dem Wärmeverlust des Vergleichsprofils an die Atmosphäre sofern keine Latentwärme gewonnen wird.

Bis auf einen geringen Wärmefluß aus tieferen Schichten zur Tauscherebene hin entstammt die gesamt gewonnene Wärme dem Phasenwechsel von Wasser zu Eis. Bei längerfristigen Außentemperaturen unter 0 °C wird sich dieser Typ entwickeln.

Der Übergangstyp bildet sich immer dann aus, wenn die Oberflächentemperatur über der Tauscherebenentemperatur und unter der Vergleichsflächentemperatur des Unterbodens liegt.

Entlang der Temperaturgradienten fließt in den Entzugsflächen Wärme zur Tauscherebene, während die Vergleichsflächen nach wie vor eine negative Energiebilanz aufweisen.

Der Übergangstyp trägt wesentlich zur Schonung und sogar zur zeitweiligen Regeneration des Speichers bei.

Übergangstypen kommen insbesondere bei stark belasteten Anlagen im Dezember und sonst vorwiegend in den Monaten Februar und März vor.

Als letzterer und für die Regeneration ganz entscheidender Profiltyp muß der Erwärmungstyp angesprochen werden. Er tritt nach Temperaturumkehr des Vergleichsflächenprofils auf. Durch intensive Erwärmung der Oberfläche werden sowohl auf Entzugs- als auch auf Vergleichsflächen sehr steile positive Temperaturgradienten aufgebaut. Entlang dieser Gradienten fließen große Wärmemengen in den Boden, die sehr schnell zu einem Abtauen des Eises und zur Erwärmung des gesamten Bodenvolumens führen. An allen Standorten liegt dieser Typ in sehr schöner Ausprägung während der Messung Anfang Juni vor. Ende Juli ist die Erwärmung des Bodens weit fortgeschritten und die Temperaturdifferenzen zwischen Entzugs- und Vergleichsflächen liegen in 125 cm selbst bei Wärmeentzug im Sommer unter 2 K.

Zusammengefaßt sind 4 Phasen des Wärmetransportes bei Erdreich/Solewärmetauschern und die zugehörigen Wärmebilanzen in Bild 15 dargestellt.

Aufgrund der durch Messungen belegten, bis in die obersten Bodenschichten reichenden Temperaturabsenkungen auf Entzugsflächen ist es denkbar, daß dies auch an Schnee, Reif und Pflanzen zu sehen ist. Es wurden deshalb bei Wartungs- und Meßterminen genauere Beobachtungen angestellt.

In der Regel war es nicht möglich, eine Verspätung der Vegetationsentwicklung oder längeres Liegenbleiben von Schnee und Reif festzustellen. Geringfügige Expositionsunterschiede bzw. Beschattung durch Gebäude zeigten häufig einen bedeutend größeren Einfluß. Bei Pflanzen kommt als Unsicherheitsfaktor noch die natürliche Variabilität innerhalb einer Art hinzu.

Aufgrund der Untersuchung kann festgestellt werden, daß nach Wärmeentzug im Winter im Verlauf des Sommers mit schneller und fast vollständiger Regeneration des Systems zu rechnen ist. Langfristige negative Veränderungen sind nicht zu erwarten. Während der Heizperiode stellt sich der Wärmehaushalt des Bodens sehr rasch auf einen um einige °C tiefer liegenden Gleichgewichtszustand ein. Der Temperaturverlauf mit und ohne Entzug in vergleichbaren Tiefen ist annähernd parallel.

Die Verlagerung des Gleichsgewichtes führt zu höheren Wärmeflüssen in den Boden, zu einer Reduzierung der Wärmeverluste an die Atmosphäre in Herbst und Winter und zu einer Verlängerung der Zeiten mit positiven Wärmefluß.

Die Entzugsraten selbst werden in Abhängigkeit von den vorhergegangenen Wärmebilanzen und dem aktuellen Witterungsverlauf zu jeweils unterschiedlichen Teilen aus dem Wärmefluß in den Boden, dem Latentspeicher Wasser-Eis und dem Temperaturspeicher Boden und Wasser gespeist.

Angaben über potentielle Entzugsraten können deshalb nur unter Vorgabe von definierten Randbedingungen gemacht werden. Die Tiefe der Tauscherfläche und die Anordnung der Entzugsrohre sind hierbei mit zu berücksichtigen. Nach den bisher durchgeführten Messungen herrschen entlang der Entzugsrohre äußerst unterschiedliche Bedingungen, die sich nicht linear ändern.

Das durch die Wärmetauscherkonfiguration zu lösende Problem von Wärmepumpen mit Wärmequelle Erdreich ist es, während der Gleichgewichtsphase des Wärmetauschers genügend Wärme zum Betrieb der Heizungsanlage zu gewinnen, um die Zeiten zu überstehen, in denen über die Bodenoberfläche keine Wärme nachkommt.

4. Technik der Wärmepumpe mit Wärmequelle Luft Allgemeines

Luft als Wärmequelle für eine Wärmepumpe ist in weitaus stärkerem Umfang als Erdreich und insbesondere als Grundwasser weiten Temperaturschwankungen unterworfen. Dieser Nachteil war Ursache dafür, daß diese Technik in der Bundesrepublik Deutschland erst vor relativ kurzer Zeit entwickelt wurde. Aufgrund der Gegenläufigkeit von Lufttemperaturniveau und benötigter Heizleistung, - bei sinkender Lufttemperatur steigt der Heizenergiebedarf - ist bei Wärmequelle Luft neben der monovalenten auch die bivalente Betriebsweise sinnvoll. Beim bivalenten Betrieb wird unterhalb einer bestimmten Außentemperatur, häufig bei - 5 °C bis + 3 °C ein zweites Heizsystem mit einem lagerfähigen Brennstoff eingesetzt.

Luft ist als Wärmequelle für eine Wärmepumpe keinen Einschränkungen unterworfen, wie sie für die Wärmequelle Grundwasser und Erdreich gegeben sind. Die universielle Einsatzbarkeit macht diese Wärmequelle besonders interessant. Allerdings sind die erzielbaren mittleren Jahresarbeitszahlen relativ zu den anderen Wärmequellen am niedrigsten. Wärmepumpen mit Wärmequelle Luft werden in mehreren Ausführungsvarianten angeboten. Aufgrund der hohen benötigten Luftmengen werden monovalente Luft-Wasser-Wärmepumpen meist in Splitform angeboten. Bivalente Luft-Wasser-Wärmepumpen, meist in Verbindung mit einer Ölheizungsanlage, werden häufig in Kompaktbauweise angeboten. Das Schema einer Luft-Wasser-Wärmepumpe zeigt Bild 16.

Danach wird die Luft über einen Luftkanal dem Verdampfer der Wärmepumpe zugeführt und abgeleitet. Diese Bauvariante bietet sich für die Aufstellung in Keller- oder Dachräumen an.

In Tafel 1 sind für sechs ausgewählte Temperaturen die theoretisch je kg trockener Luft entziehbaren Wärmemengen zusammengestellt. Sie sind aufgeteilt in die Wärmemenge, die durch die Abkühlung der Luft als fühlbare Wärme und durch das Tauen des in der Luft enthaltenen Wasserdampfes und ggfs. die Vereisung dieses Wassers als latente Wärme gewinnbar sind. Es zeigt sich, daß die insgesamt bei Abkühlung um 10 K je kg Luft gewinnbare Wärmemenge mit sinkender Lufttemperatur kleiner wird. Können bei einem Außenluftzustand von + 20 °C und einer relativen Feuchte von 70 % rd. 4,6 Wh je kg Luft gewonnen werden, beträgt dieser Wärmegewinn bei - 15 °C und 90 % relative Feuchte noch rd. 3 Wh je kg Luft. Die Verringerung des möglichen Wärmegewinns ist nahezu ausschließlich auf die Abnahme des in der Luft enthaltenen Wassers zurückzuführen.

Während je kg Luft bei einer Temperatur von 20 °C und einer relativen Feuchte von 70 % 10,3 g Wasser in Form von Wasserdampf vorhanden sind, enthält 1 kg Luft von - 15 °C und einer relativen Feuchte von 90 % nur noch 0,9 g Wasser.

Kalte Luft enthält eine relativ geringe, warme Luft hingegen eine relativ hohe Wassermenge in Form von Wasserdampf. Bild 17 zeigt die Tagesmittelwerte der auskondensierten und vereisten Wassermenge in Klimazone I im statistischen Normaljahr an einem Verdampfer einer Luft-Wasser-Wärmepumpe im 24-h-Betrieb in Abhängigkeit von den Monatsmitteltemperaturen für die einzelnen Monate. Im Mittel der Wintermonate werden ca. 200 kg Wasser pro Tag ausgeschieden, die zum weitaus größten Teil auch gleichzeitig die Verdampferoberfläche vereisen. Während der Sommermonate beträgt die ausgeschiedene Wassermenge bis ca. 300 kg pro Tag, ebenfalls bei Dauerbetrieb der Wärmepumpe unter den angegebenen Randbedingungen. Dabei fällt auf, daß bis zu Monatsmitteltemperaturen um 14 °C der Verdampfer der Wärmepumpe vereist. Der Wärmegewinn aus der Wasserkondensation und der Vereisung ist erheblich am gesamten Wärmegewinn aus der Luft beteiligt. Bei Lufttemperaturen unter + 5 °C ist die gewonnene Energiemenge wieder zur Abtauung der Verdampfer aufzuwenden. Der absolute Wassergehalt ist vom Druck der Luft und vom Sättigungsdruck des Wasserdampfes abhängig. Der Sättigungsdampfdruck ist um so niedriger, je tiefer die Temperatur der Luft ist. Aus diesem Grunde ist bei höherer relativer Feuchte der Luft während der Wintermonate der absolute Wassergehalt niedriger als in den Sommermonaten. Deswegen ist der mögliche Wärmegewinn aus der Außenluft außer von Temperatur und Wassergehalt auch vom Luftdruck abhängig.

Das Abtauen des Verdampfers bei Luftemperaturen unter + 5 °C ist notwendig, da sich bei zunehmender Vereisung der Wärmeübergang verschlechtert und der Luftdurchsatz bei einem lamellenberippten Verdampfer vermindert. Bei Außentemperaturen über + 5 °C ist kein zusätzlicher Aufwand an Energie zum Abtauen notwendig. Bei dieser relativ hohen Außentemperatur der Luft reichen hier die Stillstandzeiten der Wärmepumpe aus, das Eis abzutauen. Der energetische Aufwand zum Abtauen des Verdampfers ist abhängig vom Abtauverfahren und insbesondere von der Steuerung des Abtauvorganges. Bei Optimierung dieser Komponenten kann bei elektrischen Wärmepumpen die zum Abtauen aufzuwendende Energie ca. 3-5 % des Gesamtenergieverbrauchs betragen. Dies setzt eine optimale Erfassung des Abtauzeitpunktes und eine angepaßte Abtauregelung voraus. Beide für den Betrieb von Luft/Wasser-Wärmepumpen wichtigen Voraussetzungen scheinen derzeit noch nicht optimal gelöst zu sein.

Beim Betrieb von Wärmepumpen mit Wärmequelle Luft wird Luft mit Hilfe eines Ventilators durch den Verdampfer gefördert und abgekühlt. Üblicherweise liegt bei monovalenten Luft/Wasser-Wärmepumpen der Luftvolumenstrom bei m_{Luft} = 3 - 6000 m³/h. Dabei ergibt sich die Frage, wie weit sich die häufig um 10 K abgekühlte Luft hinder dem Verdampfer ausbreitet und damit ggfs. den Transmissionswärmebedarf von Nachbargebäuden vergrößert. Gleichzeitig ergeben sich Luftabreißgeräusche an den Rotoren des Ventilators und an den Verdampferlamellen. Es wurde deshalb in einem Meßprogramm untersucht

- in welcher Entfernung vom Luftaustritt des Verdampfers noch eine meßbare Absenkung der Lufttemperatur nachzuweisen ist und
- welche Geräuschemissionen bei zwangskonvektisch mit Luft beaufschlagten Verdampfern mit und ohne Schalldämmung auftreten können.

Die Abkühlung der Außenluft durch den Betrieb einer Wärmepumpe mit Wärmequelle Luft läßt sich in einer Entfernung von 15 m vom Verdampfer meßtechnisch nicht mehr nachweisen. Wie die Auswertung von 2 Versuchsreihen mit Wärmepumpen von je 15 kW Heizleistung zeigt, sinkt dieser Grenzabstand bereits bei geringen Windgeschwindigkeiten auf ca. 2 m ab. Bei der Schwierigkeit der meßtechnischen Erfassung der Lufttemperaturen muß beachtet werden, daß diese Ergebnisse aufgrund nicht reproduzierbarer Randbedingungen an konkreten Objekten ermittelt wurden. Sonneneinstrahlung u.a. Fremdeinflüsse bewirken wesentlich größere Temperaturunterschiede als durch den Wärmepumpenbetrieb nachgewiesen werden konnte.

Demgegenüber kommt der Geräuschemission zwangskonvektiv mit Luft beaufschlagter Verdampfer Bedeutung zu. Die an einem Wärmepumpenaggregat mit nicht schallgedämpften Verdampfer durchgeführten Messungen zeigen, daß insbesondere in Frequenzbereichen von ca. 160 bis 500 Hz (Terzanalyse) vergleichsweise hohe Schalldruckpegel nachgewiesen wurden. Dabei ist die Höhe des bewerteten Schalldruckpegels in dB (A) von der Drehzahl der Lüftermotore, Lufterkonstruktion und vom Fabrikat abhängig. Zur Einhaltung der in der TA-Luft vorgegebenen Grenzwerte ist es in der Regel erforderlich geeignete Maßnahmen zur Geräuschdämmung vorzunehmen.

SUMMARY

Where groundwater, earth and air are used as heat sources, the effects on the environment produced by thermodynamic heating plants with each of these heat sources are quantified. For example, it is shown that, depending on hydrogeological parameters and the water supply required, the diameter of the 7°C isotherm is approximately 25 m in the area of the drainage well after a heat pump has been operating for several years, taken on the example of a single family house. There was no evidence of negative effects on groundwater resulting from heat pump operation.

In the case of earth used as a heat source, the project has provided basic ideas for new heat exchanger configurations. Minimal differences in exposure, e.g. being overshadowed by buildings and the like, have a greater influence on plant growth than heat pump operation.

When air is used as a source of heat for a heat pump, measures must be taken to lessen noise emission from forced-flow evaporators. Differences in air temperature are no longer detectable behind the evaporator of a heat pump using air as a heat source even when there is very little wind movement.

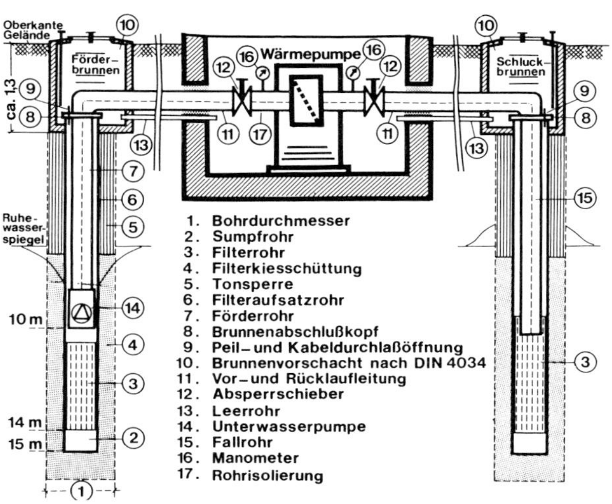

Bild 1:

Schema einer Brunnenanlage für eine Wasser/Wasser-Wärmepumpe

Temperatur d. Luft vor Verdampfer	°C	20	4	7	0	-7	-15
Feuchte der Luft vor Verdampfer	g / kg Luft	10,3	7,5	5	3	1,9	0,9
Wärmegewinn aus fühlbarer Wärme	Wh / kg Luft	2,86	2,83	2,81	2,80	2,78	2,79
Wärmedampf-kondensation	Wh / kg Luft	1,73	1,73	1,18	0,97	0,76	0,20
Vereisung	Wh / kg Luft	-	-	0,16	0,13	0,10	0,03
Gesamter Wärmegewinn	Wh / kg Luft	4,59	4,56	4,15	3,90	3,64	3,02

Tafel 1: Wärmegewinn je kg Luft durch eine Wärmepumpe bei verschiedenen Luftzuständen und einer Abkühlung der Luft um 10 K, Luftdruck 1013 bar

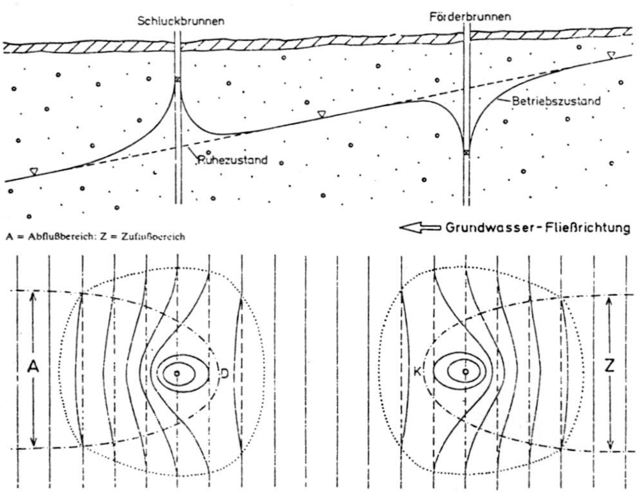

Bild 2: Schematische Darstellung der Grundwasserverhältnisse im Förder- und Schluckbrunnen

Bild 3: Schematische Bodentemperaturprofile im Jahresverlauf, tiefenabhängige Temperaturabweichungen vom Jahresmittel

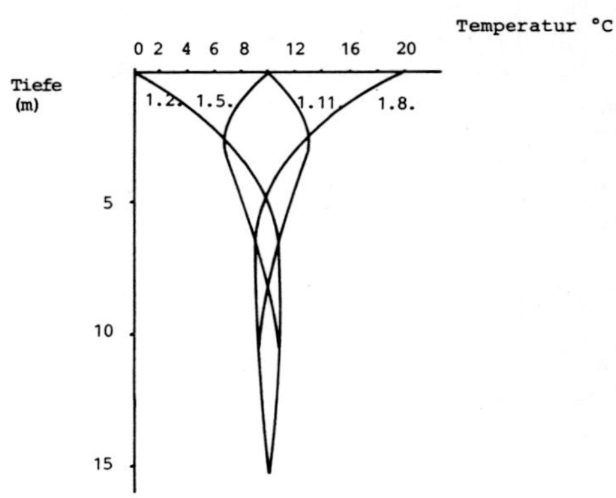

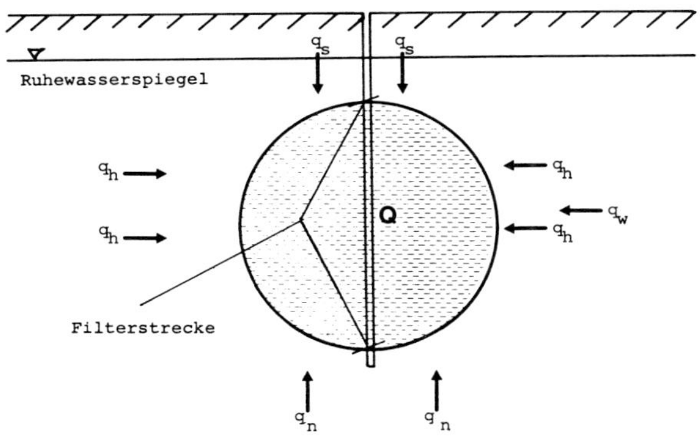

Wärmeströme im stationären Zustand:

$Q = q_s + q_h + q_h + q_w$

q_s = solar bedingter Wärmestrom von der Erdoberfläche her

q_h = aus der seitlichen Umgebung horizontal zufließender Wärmestrom

q_u = von unten zufließender Wärmestrom

q_w = konvektiver Wärmezufluß durch Grundwasserstrom

Bild 4: Schema der Wärmeströme bei einer negativen thermischen Anomalie im Untergrund

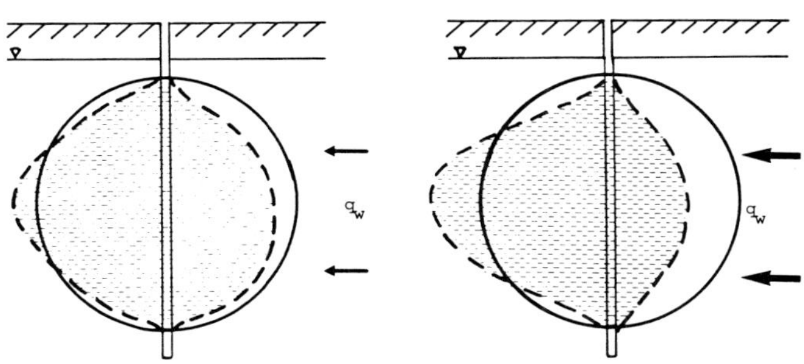

niedrige Grundwassergeschwindigkeit   hohe Grundwassergeschwindigkeit

Bild 5: Einfluss des Grundwasserstroms auf Grösse und Form einer negativen thermischen Anomalie im Untergrund

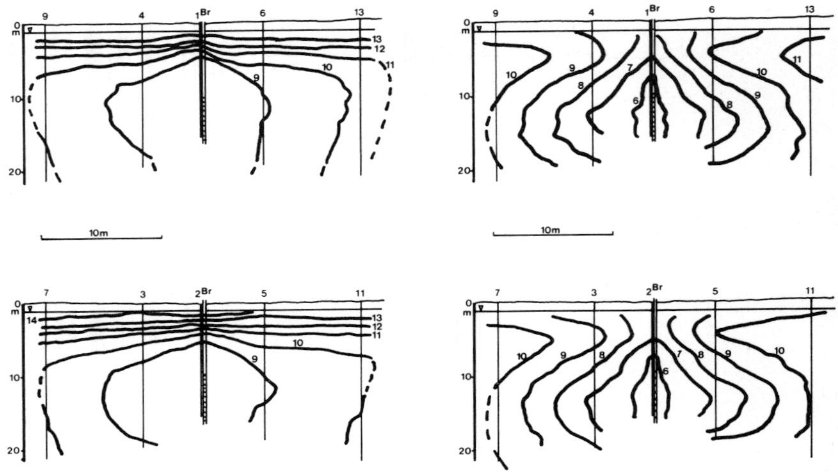

20.8.1978 Hydrothermische Profile (°C) 10.12.78

Bild 6: Hydrothermisches Versuchsfeld II

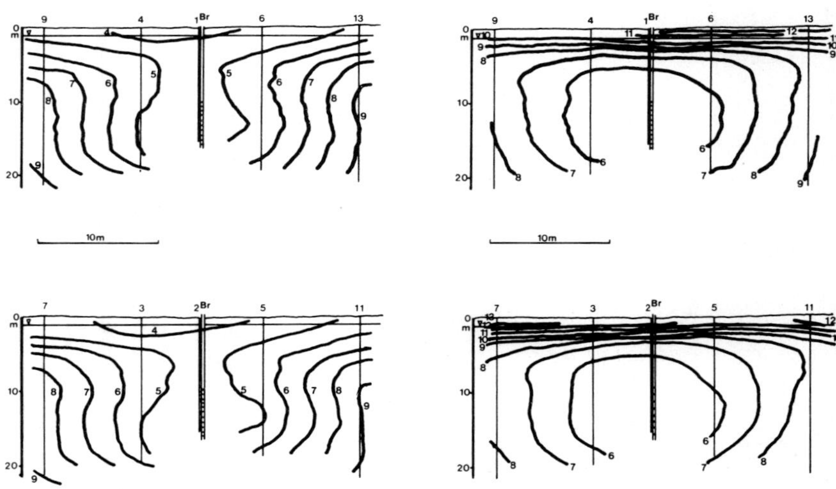

18.3.1978 Hydrothermische Profile (°C) 10.6.78

Bild 7: Hydrothermisches Versuchsfeld II

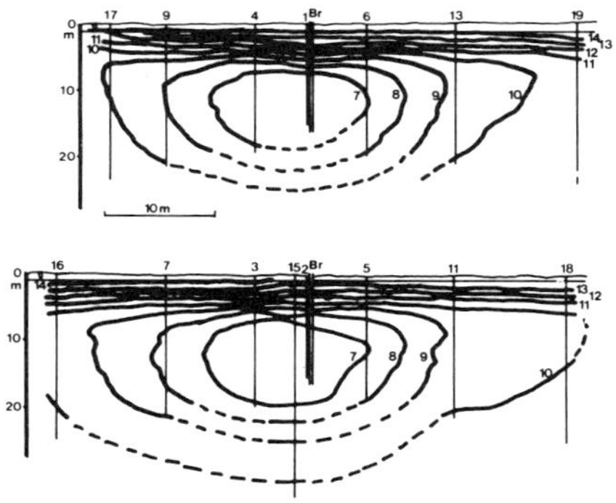

Hydrothermische Profile (°C) 5.8.1979

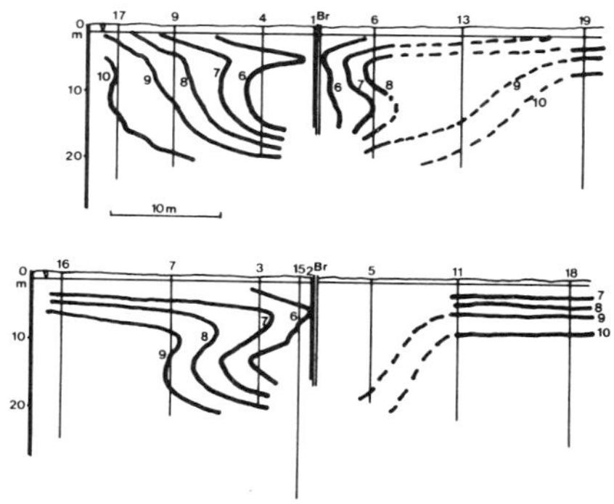

Hydrothermische Profile (°C) 9.3.1981

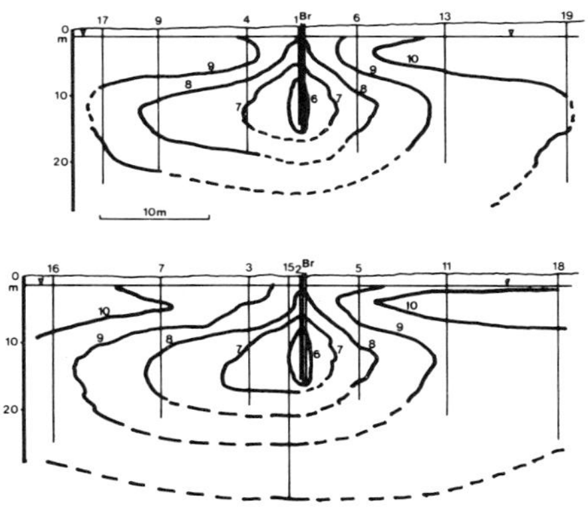

Hydrothermische Profile (°C) 25.11.79

Bild 8

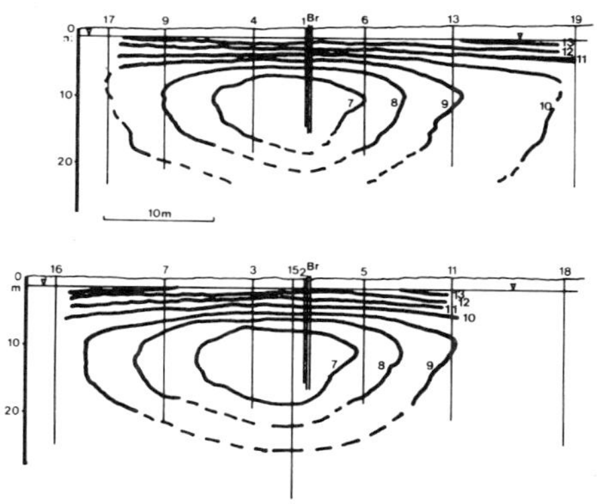

Hydrothermische Profile (°C) 2.9.1979

Bild 9

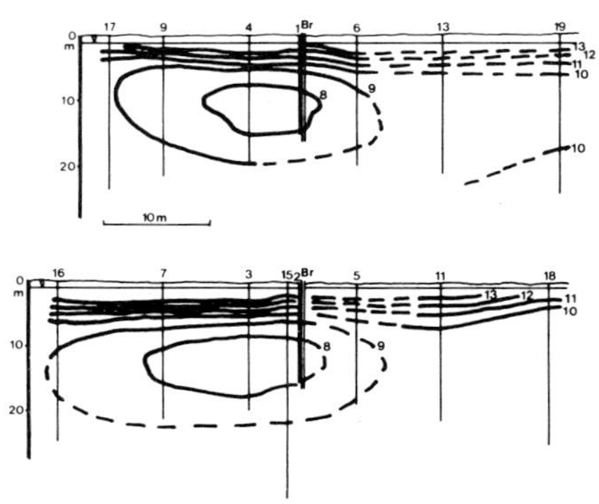

Hydrothermische Profile (°C) 4.8.1980

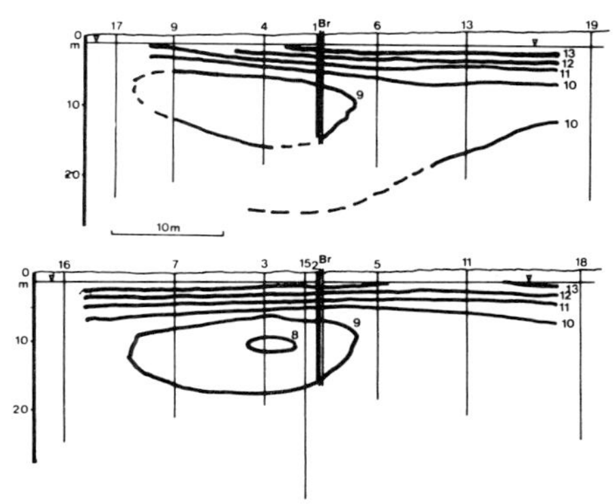

Hydrothermische Profile (°C) 13.9.1981

Bild 10

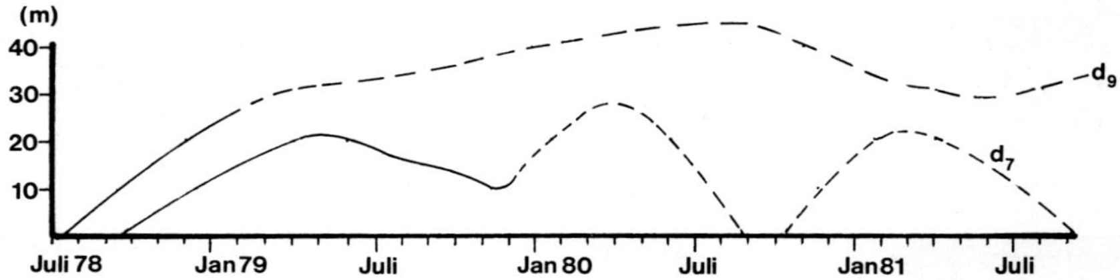

Anomalie-Durchmesser

d_9 = 9 °C-Isotherme

d_7 = 7 °C-Isotherme

Maximale Anomalie-Durchmesser, d_9 = 9 °C Isotherme, d_7 = 7 °C Isotherme

gestrichelte Linie = interpoliert

Bild 11

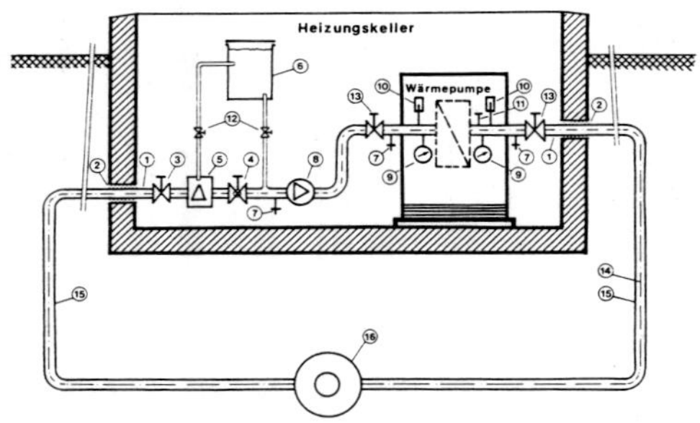

1 Vor – und Rücklaufleitung
2 Mauerdurchführung
3 Absperrschieber
4 Drosselschieber
5 Luftabscheider
6 Sole – Ausgleichsgefäß
7 Revisionsventil
8 Sole – Umlaufpumpe
9 Manometer
10 Temperaturmeßstelle
11 Durchflußwächter
12 Absperrventil
13 Geräteabsperrschieber
14 Sole
15 Vor – und Rücklaufleitung , außen
16 Wärmetauscher

Bild 12:

Schema einer Erdreich-Wärmepumpenanalge

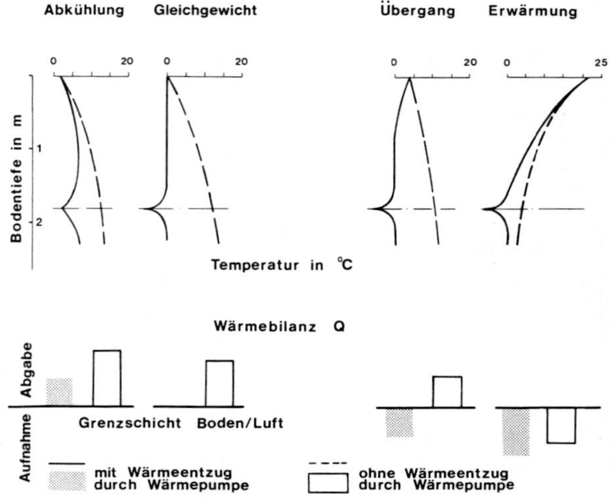

Bild 13:

Phasen des Wärmetransportes bei Erdreich/Sole-Wärmetauschern, Bilanzen.

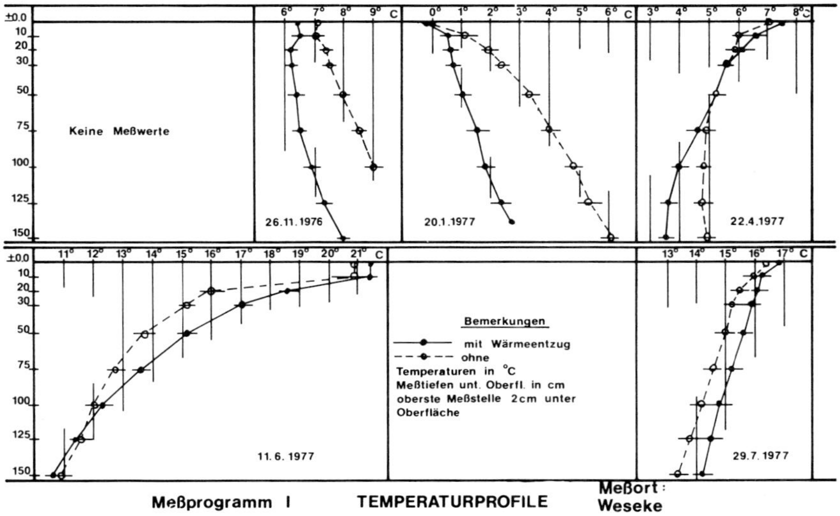

Bild 14

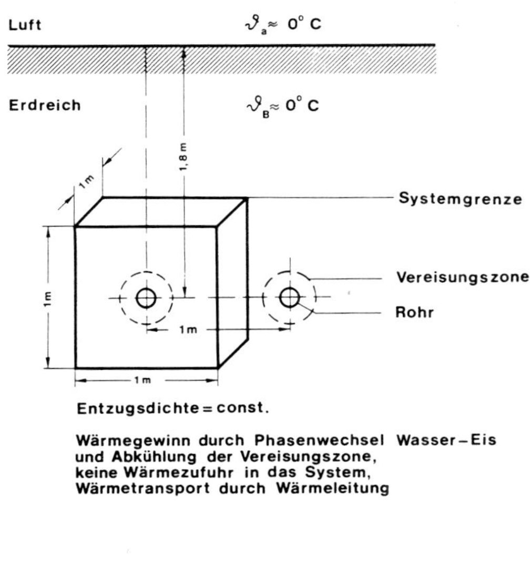

Bild 15: Erdreich/Sole-Wärmetauscher, Geometrie und Ausgangsbedingungen

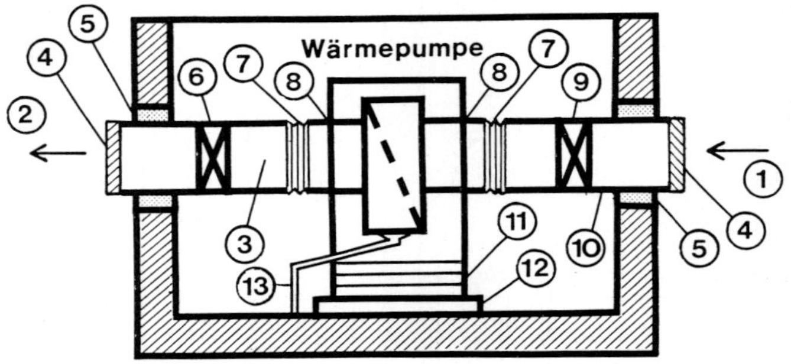

1. Lufteintritt
2. Luftaustritt
3. Luftkanal
4. Schutzgitter
5. Mauerdurchführung
6. Schalldämpfer
7. Flexibler Kanalverbinder
8. Anschlußflansch
9. Filter
10. Luftkanalisolierung
11. Schallschutzmatte
12. Fundament

Bild 16: Schema einer Luft-Wärmepumpenanlage

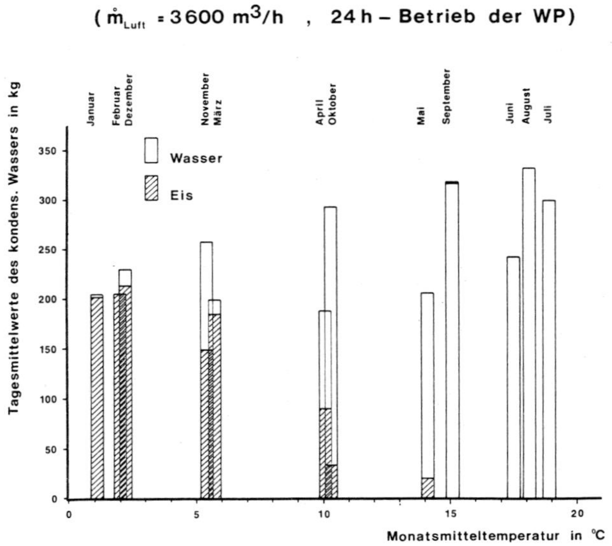

Bild 17: Wasser und vereiste Wassermenge in Klimazone I nach dem statistischen Normaljahr

OVERVIEW OF HEAT PUMPS R AND D FOR DOMESTIC HEATING CARRIED OUT BY THE EUROPEAN COMMUNITIES FROM 1979 TO 1983 IN THE FRAMEWORK OF THE SECOND EC ENERGY R AND D PROGRAMME

P. ZEGERS
Commission of the European Communities
DG XII

INTRODUCTION

The primary energy consumption in the European Community for domestic and commercial heating is 26 %. When we discuss heat pumps we discuss a reduction in energy use of this 26 %. Presently heating is predominantly provided by conventional boilers which have a seasonal efficiency of 60 % to 70 %. As a comparison the technically feasible efficiencies are given for three types of heat pumps (Table 1). If each of these heat pumps would replace the conventional boilers, this would result in energy savings which are given in the second column of the table.

	Efficiency	Energy Saving Potentials
Electrical heat pump	90 - 100 %	± 30 %
Internal combustion engine driven heat pump	150 - 180 %	± 60 %
Absorption heat pump	130 - 150 %	± 50 %

TABLE 1 - Heat Pump Efficiency and Potential Energy Saving

In the long term the energy saving potential of heat pumps as compared to conventional heating may be somewhere between 30 % and 60 %. This is of course a rather theoretical figure, and in practice there will be a large number of barriers which prevent achieving these objectives

Economic feasibility for heat pumps in Europe is yet rarely achieved. The cost (investment, installation, maintenance) of heat pumps should therefore be reduced and the efficiency should be improved.

Reliability is also an indispensable prerequisite.

Maintenance of heat pumps should be assured and a network of installers should be built up.

In order to guarantee a fast introduction of heat pumps on the market, heat pumps should be made suitable for cheap retro- fitting in existing houses with existing radiators.

About 60 - 70 % of the dwellings in Europe are heated individually and heat pumps replacing these conventional systems should presently have a size of 10 - 20 kW. The remaining heating systems are used for central heating of apartment buildings, hospitals, schools and industrial buildings and will require much larger heat pumps or district heating.

PRESENT STATE

In Europe presently about 150.000 heat pumps have been sold. The major part are electrically driven compressor heat pumps which have air as a heat source; they are predominantly used or heating only.

In the USA, 3.000.000 - 4.000.000 heat pumps have been installed and most of them are electrically driven air source heat pumps used for both heating and cooling; air is the heat (or cold) distribution medium.

The success of heat pumps in the US may be attributed to two factors :

- Heat pumps are used for heating in the winter and cooling in the summer. This increases the economic feasibility as compared to heat pumps which are only used for heating.

- Heat pumps in the USA are 30 - 50 % cheaper than in Europe; a large part of this lower cost may be attributed to the economy of scale brought about by the much larger production number in the USA.

In Europe heat pumps are marginally attractive; in particular small electrically driven heat pumps. In Germany heat pump sales in 1983 are only 25 % of the sales in 1981. In France the introduction of heat pumps is strongly backed by the government, which with the Perche programme tries to come down the curve, which gives the cost as a function of the production size, in order to produce heat pumps at lower cost. But even in France heat pump sales have decreased during the last years.

In 1979, in the beginning of the second four-year Energy R and D Programme, it was evident that R and D was to concentrate on three types of heat pumps :

Electrically driven compressor heat pumps although commercially available were only marginally economically feasible and R and D was needed to reduce the cost and improve the energy conversion efficiency.

Internal combustion engine driven heat pumps were available in larger units of typically 100 KW. Here the compressor is driven by an ICE engine of which the waste heat is recovered and added to the heat produced by the heat pump. The efficiency is therefore higher than that of an electrical heat pump and the economical feasibility is somewhat better. Due to the large number of moving parts maintenance costs tends to by high. In this programme the problem was addressed whether small ICE heat pumps (± 20 KW) can be developed which are cheap, reliable and economically feasible.

Absorption heat pumps are more advanced heat pumps and the main R and D topics were : identification and characterization of working fluids and the development of suitable and cheap solution pumps.

The work carried out in the European Community's Second Energy R and D Programme (1979 - 1983) consisted in 28 projects with a total cost of 8 Mio ECU. The EC contributed 50 %. 21 projects dealt with heat pumps for domestic applications and are to be discussed here.

This session has been arranged in such a way that we will first discuss the electric and ICE compressor heat pumps and then the absorption and other advanced heat pumps.

COMPRESSOR HEAT PUMPS

Before discussing the results it is useful to first shortly describe the principles of the comopressor heat pump. A scheme is given in Figure 1 and the basic idea is that a fluid is selected which at a pressure of one bar evaporates at temperatures somewhere between - 10°C and + 10°C. The vapor is then compressed to higher pressures where it has a higher evaporation/condensation temperature. In the condensor the vapor condenses and releases the latent heat at higher temperatures (typically 50 - 70°C) where it can be used for the heating of buildings. The fluid is then expanded in the expansion valve and returns to the evaporator where it again evaporates by extracting low-grade heat from air, soil, waste heat, etc. The compressor is driven by an electric motor or an internal combustion engine (petrol, gas, Dieseloil). In the last case waste heat of the ICE engine can be recovered and added to the heat for domestic heating. It is clear that this heat pump may be used for cooling by means of air from the evaporator; the heat released in the condensor is then waste heat.

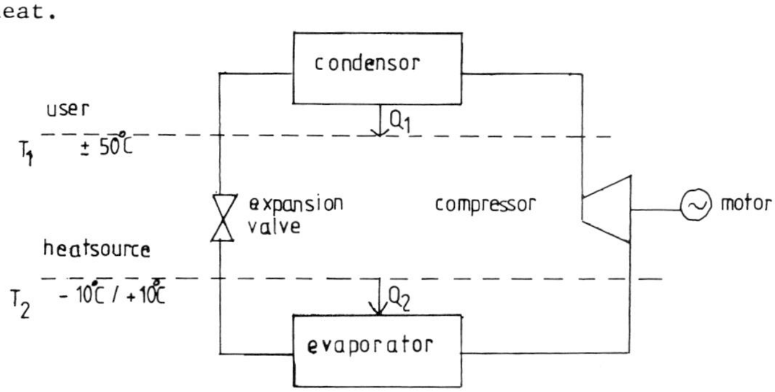

FIGURE 1 - Operation Scheme of a Compressor Heat Pump

The efficiency of heat pumps is expressed in COP (Coefficient of Performance) or PER (Performance Effectiveness Ratio). For an electrical heat pump with electricity as energy input the

$$COP = \frac{\text{heat produced}}{\text{electricity required}}$$

For the ICE driven compressor and absorption heat pumps which have fuel as energy input :

$$PER = \frac{\text{useful heat output from the complete installation}}{\text{energy heat input from fuel}}$$

Below the theoretical COP value of electrical heat pumps is given as a function of the difference in temperature (T_1) of the produced heat and the temperature(T_2) of the heat source (Figure 2). In practice the seasonally averaged COP value is much lower and lies somewhere in the area surrounded by the dotted. It is clear that there is scope for big improvements of the COP.This together with a reduction of the cost were the main objectives in the R and D programme carried out on compressor heat pumps.

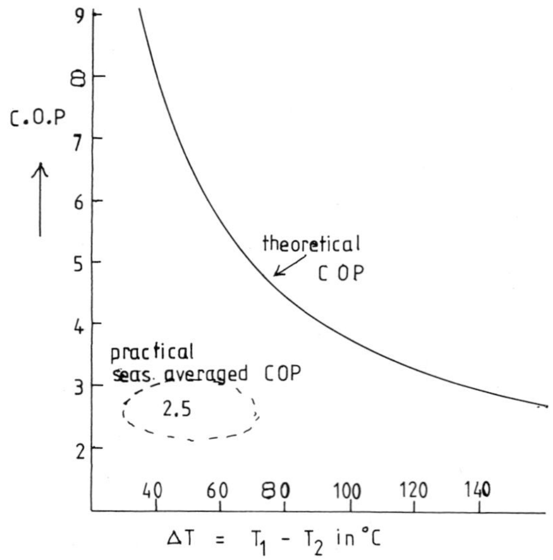

FIGURE 2 - COP of an electrical heat pump as function of $T = T_1 - T_2$

Results of R and D on Compressor Heat Pumps

An important project to identify the possibilities for improvement of compressor heat pumps was carried out by BERTIN, France. Tests were performed with a 12 KW electric heat pump with air as a heat source. It was established :

- that frosting of the evaporator heat exchangers below 4°C and the required defrosting decreased the efficiency

- that fluctuations in the refrigerant flow with a period of three minutes impair efficient operation

- that components such as the evaporator, condensor and the expansion valve do not operate efficiently

- that modifications in the compressor design may lead to an increase of the efficiency from the present 30 - 50 % to 60 - 70 %.

THE UNIVERSITY OF DORTMUND, Germany, studied the optimized operation of a frozen air heat exchanger and the use of a micro-computer controlled expansion valve for a better control of the refrigerant flow. The various air flow rates obtained by varying the speed of the evaporator fan had no impact on the coefficient of performance. Down to 1,5°C it is however possible to defrost the evaporator by the fan alone. This leads to energy savings. The micro-processor controlled expansion valve gave satisfactory results but the cost of this thermo-electric expansion valve and its control has not been discussed.

The fact that oil in the compressor dissolves in the refrigerant results in a less efficient operation of the heat pumps. This problem was studied by the NEW UNIVERSITY OF ULSTER. The fraction of oil dissolved in the refrigerant varies strongly and depends on the type of compressor (e.g. 8 - 15 % in a sliding vane compressor, which is often used in industrial heat pumps, and 1 - 2 % in a reciprocating compressor, often used in domestic heat pumps. It has been shown for a sliding vane compressor that oil can reduce the heat transferred at the evaporator by as much as 30 % and that this can have a corresponding effect on the COP value. A much better understanding of the effects of lubricating oil on the performance of the heat pumps has been acquired and techniques have been developed for improving the performance of the existing designs by up to 25 % through the introduction of a compressor discharge oil separator and through advanced control of evaporator conditions.

Another way to avoid the oil/refrigerant problem is to develop compressors which do not use oil. A lubrication oil free screw compressor of 25 kW was developed by OMPHALE, France (lubrication is brought about by the refrigerant). Efficiencies of 75 % have been obtained at full load and also the partload efficiency was very good.Endurance tests of over 2000 hours and stop-start tests (500 cycles) have been carried out to full satisfaction.

FIGURE 3 - Oil Free Single Screw Compressor

One of the most important developments in the improvement of compressor heat pumps is the use of fluid mixtures instead of a single refrigerant. IFP, France, started working in this field six years ago and fluid mixtures are now being commercialized. With a single refrigerant the condensation temperature is constant, where as the temperature of the fluid, with which heat is exchanged, is increasing from C to D (Figure 4). A similar situation occurs in the evaporator where the evaporator temperature is constant and the heat source temperature (e.g. air) varies from A to B. In this way heat transfer both in the condensor and the evaporator is inefficient. By using a suitable fluid mixture there is a condensation and evaporation temperature range which is parallel to the fluid with which heat exchange takes place. In this way heat exchange is more efficient which results in a higher COP or a higher heat production capacity and thus a lower cost.

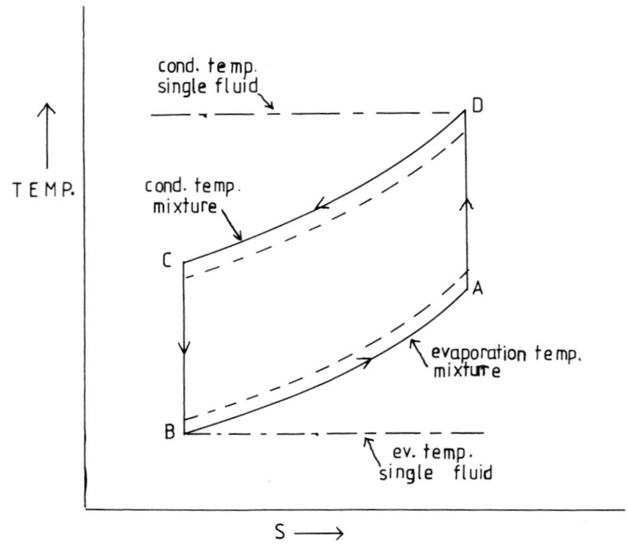

FIGURE 4 - Compressor Heat Pump Operation with Fluid Mixtures

For a heat pump with ideal evaporator and condensor heat exchangers the theoretical improvement of the COP by using a fluid mixture is given as a function of the condensation and evaporation temperature variation $\Delta T' = T_D - T_C = T_A - T_B$ (Figure 4) and of the temperature difference ΔT between the condensor and evaporator (Figure 5). This graph shows a strong improvement for heat pumps with fluid mixtures where the heat source allows a large temperature variation $\Delta T'$ (e.g. waste heat in industry). For domestic heat pumps with air as a heat source, $\Delta T'$ is rather small and improvements limited. In reality improvements are less pronounced due to not ideal heat exchangers. Even so the effect remains very interesting and deserves a thorough investigation.

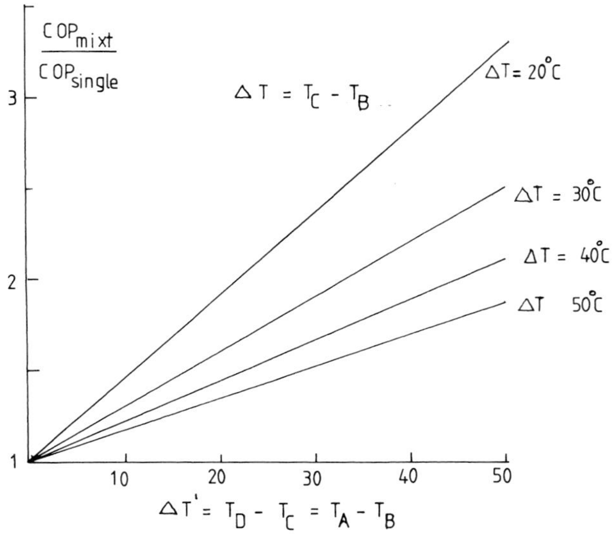

FIGURE 5 - COP mix/COP single as a Function of the Variation of the Condensation and Evaporation Temperature

In this project IFP together with the heat pump manufacturer LEROY SOMER choose for an increase of capacity in order to reduce the cost. With a mixture of R_{23} and R_{22} a capacity increase of 28 % was obtained. In this way smaller and cheaper heat pumps can be used to produce the same amount of heat as larger units. In this way a cost reduction of 10 % was realized. This fluid mixture is now being commercialized.

Contrary to electrically driven heat pumps, internal combustion engine driven heat pumps allow the recovery of waste heat which is connected with the production of power to drive the compressor. This leads to considerably higher efficiencies. Also the heat output can be varied more easily by regulating the speed of the motor and switching off one or possibly more cylinders of the compressor. The heat output can thus be matched to the heat requirements of the house which leads to energy savings. This type of heat pump was up till now limited to large installations above 100 - 200 kW. FICHTEL AND SACHS demonstrated the technical feasibility of a small 20 kW ICE driven heat pump. The heat output can be regulated by varying the speed of the motor by switching off one of the two compressor cylinders. This is done with a microprocessor which also takes into account the outdoor temperature and the heat demand of the house. Monovalent operation was shown to be possible down to - 15°C outside temperatures. The heat is delivered at 80°C which makes the heat pump suitable for retrofitting in existing houses. Eight of these heat pumps have been used in houses and three heat pumps were tested in the laboratory. The maximum operation time is 12 000 hours. From the very limited wear found up till now, a lifetime of 40 000 hours is expected. The cost is estimated to be 20.000 DM and the pay-back time 5 years. The coefficient of performance (COP) was 1,2 for - 10°C outdoor temperatures and 1,5 for 5°C.

Heat sources is one of the subjects left, after having dealt with the evaporator, the condensor, the compressor and the driving motors. It is clear from Figure 2 that the temperature T_2 of the heat source should be as high as possible in order to reach the highest possible COP value. Air is here at a disadvantage as the COP is low in cold days due to the low outdoor air (heat source) temperatures. This at a time when a large heat production is needed. Energy roofs, stacks and fences have the same problem.

In this programme LABORELEC, Belgium, investigated simple unshielded collectors placed against walls as a heat source for heat pumps. Different types were tested such as copper absorbers and galvanized steel plate absorbers clipped on copper tubes. The absorption measured was of the order of 20 W/m 2°K. Also enclosures and stacks made of radiators where tested. They gave a considerably higher absorption: enclosures 35 W/m^2 °K and stacks depending on the wind velocity: 20 - 80 W/m^2 °K. An original concept is a Danish stack made of plastic tubes, with small canals in the tube wall where the water/glycol mixture circulates. The cost of this stack is of the order of 130 ECU per kW low grade heat extracted.

The UNIVERSITY OF PADOVA developed cheap and simple metallic roofing tiles which can be used as a heat source for heat pumps, but which have the disadvantage that a back-up heat source consisting in a vertical soil heat exchanger, is needed for cold days.

Generally heat sources with high and constant temperatures are to be preferred in order to increase the COP. Sewers have been investigated as a possible heat source but their potential is only limited to 5 % of the heating need. Soil below one or two meters has constant temperatures of around 5 - 10°C and constitutes an ideal heat source for heat pumps. Heat extraction from soil can be realized with horizontal and vertical tubes with a circulating fluid (e.g. glycol-water mixture). Results of a series of experimental and computer studies carried in the first programme can be summarized as follows :

- The optimum depth of horizontal tubes is as close to the soil surface as pracically possible (± 0,5 m) in order to profit of the solar heat, the warmth of the rainwater and wind.

- The required soil surface is 2 x heated surface for horizontal tubes and 1/2 x heated surface for vertical tubes (10 m).

In the second programme EUROPEAN HEAT PUMP CONSULTORS, Denmark, and TNO, The Netherlands, both worked with vertical heat exchangers. Six different methods to sink the vertical tubes were studied. The lowest cost for a vertical heat exchanger to a depth of 15 m is around 7 ECU/m. The need for artificial heat recovery in the summer depends very much on local circumstances, but is mostly not necessary. Heat extraction up to 40 - 60 W/m is possible. LABORELEC, Belgium, also investigated horizontal heat exchangers. Their cost is 3 ECU/m for a single layer of tubes of a depth of 60 cm. The construction of two layers of tubes (60 cm and 120 cm) is cumbersome and costly.

Other interesting heat sources for heat pumps are waste heat from industry and power plants, geothermal low grade heat and exhaust air of dwellings with mechanical ventilation.

Figure 2 shows that the temperature of the heat distribution system (T_1) in the house should be as low as possible in order to optimize the COP of a compressor heat pump. Already conventional heating at low temperatures with floor heating, low temperature radiators and warm air, brings energy savings of 10 - 15 % , as compared to conventional high temperature distribution systems of typically 70 - 90°C. If used in combination with heat pumps, low temperature heat distribution systems give an additional bonus in the form of an improved COP value.

In existing houses the installation of a heat pump with a low temperature distribution system will be generally too expensive. Heat pumps should therefore be suitable for operation with existing radiators. In recent years improved insulation in buildings has lowered the heat demand. Operation of heat pumps with existing radiators at 50 - 60°C is therefore often possible instead of 70 - 90 °C for which the radiators have been designed.

Future R and D

About 60 - 70 % of the dwellings in the European Community are individually heated and will require small heat pumps. Due to an increased insulation during recent years the required heat output will be around 7 - 10 kW. In addition, large heat pumps of over 100 kW will be needed for the heating of apartment buildings, hospitals, schools and industrial buildings.

For the small heat pumps an increased COP and a lower cost is indispensible in order to achieve economic feasibility. It is believed that with further research (non azeotropic mixtures, better control and improved compressors) the seasonally averaged COP value can be brought from the present 2,5 to 4 or 5 in the next five years. Further research (e.g. azeotropic mixtures) and larger scale production will allow a reduction of the cost by 30 - 50 %. Finally it is clear that ICE driven heat pumps will never be economically feasible for sizes of 7 to 10 KW.

For large heat pumps further work and possibly the combination of oil free screw compressors and non azeotropic mixtures should give promising results.

ABSORPTION AND ADVANCED HEAT PUMPS

Absorption Heat Pumps

The basic operation of an absorption heat pump is illustrated in Figure 6. Like in a compressor heat pump a fluid B is selected which evaporates at temperatures from - 10 to + 10°C. It is then brought at a high presure where it has a high evaporation/condensation temperature and condenses at typically 50 °C in the condensor where it is used for heating. The absorption heat pump differs with a compressor in the way the fluid B is brought at a high pressure. In the absorption or chemical heat pump this is realized with a set of reversible physical or chemical reactions. In the absorption heat pump the fluid B is absorbed in fluid A in the absorber under discharge of heat at around 50°C which can be used for heating. The mixture is brought at a high pressure with a solution pump and brought in a generator, which is basically a boiler, where the mixture rich in B is heated to 200 - 300°C to release B vapor at a high pressure. The generator may be fired with coal, wood, gas, oil or even waste heat. The solution poor in B is then returned to the absorber. The efficiency of this type of heat pump is 130 - 150 %. As compared to an electrical heat pump much less low grade heat has to be extracted. For an electrical heat pump with a COP of 3, about 2/3 of the heat output is derived from the heat source. This value is for an absorption heat pump with a PER of 1,5 only 1/3. This makes the operation of the absorption heat pump as a cooler much less attractive and less suitable for applications in the USA.

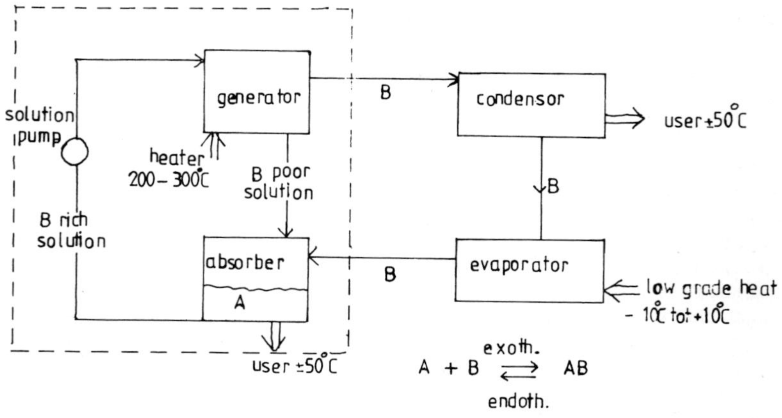

FIGURE 6 - Operation of an Absorption Heat Pump

The advantages of the absorption heat pump are thus :

- High efficiency of 130 - 150 % with the possibility to reach higher values
- Use of a variety of fuels is possible : coal, oil, gas, wood and even waste heat

- No compressor : long life, no noise
- Limited maintenance
- Less expensive heat sources
- May turn out to be cheap
- Suitable for small and large heat pumps.

In the beginning of the second programme very few commercial absorption heat pumps were available. Most of the research work had been carried out on NH_3/H_2O and $LiBr/H_2O$ absorption heat pumps which both have disadvantages.

The NH_3/H_2O absorption heat pump where NH_3 is the refrigerant operates at high pressures which require expensive tubing. NH_3 is toxic and also the use of copper heat exchangers is not possible.

The $LiBr/H_2O$ absorption heat pump with water as the refrigerant has the disadvantage that evaporator operation below 0°C is difficult due to the freezing of the water. Also corrosion is a problem.

For this reason the search for new working fluid pairs was a focal point in the second programme. A systematic search for new fluid pairs was carried out by the UNIVERSITY OF ESSEN. The most important requirements for the fluid pairs are :

- Long term chemical stability well over the generator temperatures
- Large evaporation heat of the refrigerant.
- Non toxic
- Cheap
- The vapour pressure of the solvent should be negligable at generator temperatures
- Large concentration difference between the solutions rich and poor in refrigerant

Several promising working fluid pairs have been identified.

The CRANFIELD INSTITUTE OF TECHNOLOGY, UK had the objective to develop a 10 kW absorption heat pump which is suitable for retrofitting in existing houses and can be used with existing radiators.This requires a heat pump which delivers heat at 60 - 70°C. A first working fluid pair which was investigated (methanol/mixed bromides) did not operate satisfactory due to the decomposition of methanol at temperatures over 190°C. The combination $NaOH/H_2O$ turned out to be promising; an interesting solution was found for the operation below 0°C.

A refrigerant, which has very interesting characteristics is CO_2. It is largely available, neither toxic nor inflammable. It has however a low condensation temperature (- 78°C) and very high pressures at 10 - 40°C. This makes CO_2 not very suitable for an absorption heat pump. If however the condensor is replaced by an absorber and the evaporator by a generator (using waste heat) this problem may be solved. Such an absorption-resorption heat pump with CO_2/amine as working fluids has been calculated to have an efficiency between 160 % and 190 %, with waste heat at 0°C, delivered heat between 35 - 75°C and a generator temperature of 180°C. Work at IFP, France is now in progress to construct this heat pump.

Development of a solution pump which is cheap and efficient is another important aim in absorption heat pump research. About four years ago STIEBEL ELTRON, Germany started R + D on the development of a 15 kW absorption heat pump which uses a fluid pair E181/R22. One of the major problems was to find a suitable and cheap (less than 500 DM) solution pump with a Δ p of 23 bar and a mass flow of 800 kg/h. After systematic testing of different pump types the internal gear pump was found to be most suitable. Two ways were followed to come to an efficient pumping system : In the first solution the pump inside the fluid reservoir is driven by an electric motor outside. Up till now no satisfactory shaft sealing methods were found. In the second solution the electric motor is completely submerged in the working fluid together with the pump (hermetic model). After some time the conductivity in the liquid increased by a factor 100 - 100, possibly due to the formation of ions, to values where hermetic operation is not possible. In the final solution the rotor of the electric motor is separated from the stator with the electric wires, in such a way that the rotor is in the working fluid mixture and the stator is not. This turned out to be a satisfactory and cheap solution. In mass production the cost of this solution pump is estimated to be 500 DM. This pump is expected to be suitable for other working fluid pairs. After 1.400 hr of operation the E181/R22 mixture became chemically unstable. The reasons are now being investigated.

FIGURE 7 - Final Version of the Solution Pump

A way to avoid the use of a solution pump in an absorption heat pump is to use a periodically operating absorption heat pump. This heat pump consists of two reservoirs where liquids are separated, but where vapors can freely move from one reservoir to another. They are used as a generator and condensor for half a period and as an absorber and evaporator for the other half period. During the generator/condensor phase, heat is given to the generator and vapour is condensed in the condensor where heat is extracted. In the absorption/evaporator phase low grade heat is used to evaporate the fluid which is absorbed in the absorber where now heat is extracted. The cycle time is 1-1 1/2 hour. COP values of 1,3 have been measured in an experimental set up. Initially the combination LiBr and pure CH_3OH was used. Major problems were mass and heat transfer. By adding water to LiBr/CH_3OH the operation of the heat pump improved considerably. A 10 kW unit has now been built. The advantages of this periodic heat pump are : no solution pump, continuous heat output, low maintenance.

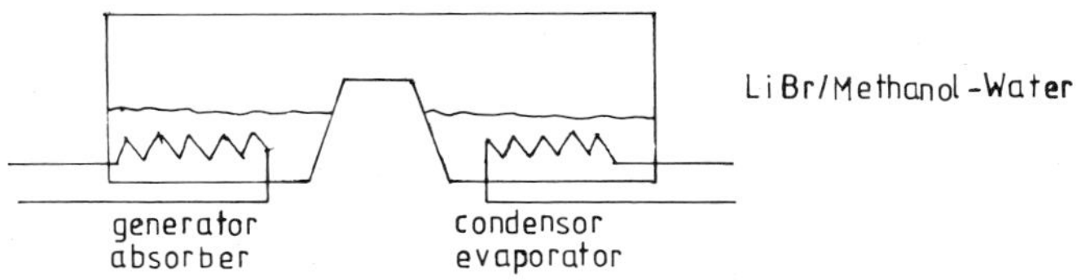

FIGURE 8 - Scheme of a Periodic Absorption Heat Pump

Advanced Heat Pumps

A thermo-electric heat pump has been developed which is based on the Peltier effect. When two different metals from a circuit, a current in that circuit will induce a temperature difference between the two metal interfaces (Peltier effect). Such thermo-elements were produced with two types of pellets where one interface is connected with a heat exchanger which extracts waste heat from exhaust air and the other interface with a heat exchanger by which heat of 35°C is delivered to a water current used for floor heating. Pellets of 170 mm^2, consisting of tellurium, bismuth and antimony (type P) and tellurium, bismuth, selenium (type N), have been developed and were found to operate satisfactorily.

A thermo-electric heat pump of 3.8 kW heat production has been defined where an electric current through a circuit of pellets induces a temperature difference between the heat extraction and heat discharge side. Experiments with a small heat pump module of 17 thermo-electric couples which delivered a maximum heat output of 300 W, showed that the COP value (power $\rightarrow$ heat) varied between 1,5 and 2,0 for 300 W and 150 W heat output respectively. Though this is not very high, this heat pump has advantages such as silent operation, simple maintenance, and a simple and continuous regulation of heat output by varying the current in the thermo-element. The cost is approximately the same as for electrical heat pumps of equal size.

An ORC engine driven heat pump of 10 kW was to be developed by GLYNWED, UK. The ORC turbine and the vapor compression of the heat pump are on the same axis. This heat pump may be fired by oil, gas, wood or waste heat and COP values of 1,2 have been obtained. Problems with rotors and bearings turned out to be insuperable and the project has therefore been stopped.

Future R + D

Future research on absorption heat pumps will be carried out along the same lines as in the second programme. Focal points will be :

- Component R + D : absorber, evaporator, generator

- Identification, characterization and testing of working fluid pairs, with particular emphasis on chemical stability

- Control systems

- Periodic absorption heat pumps.

SESSION I - HEAT PUMP SYSTEMS

B - DEMONSTRATION PROJECTS

Chairman: Prof. Dr.-Ing. K.F. KNOCHE
Rapporteur: Mr. J. DURANDET

REMARKS BY THE CHAIRMAN

Prof. Dr.-Ing. K.F. KNOCHE

The demonstration projects began rather late, partially only around the middle of 1982.

The objective of this programme is to monitor, over a long period and at least a single heating season, operating installations, thereby obtaining data on their functioning and efficiency; this has not yet been accomplished for all the installations.

Electricity- as well as combustionmotor-driven heat pumps, whose thermal production capacities range from less than 20 kW up to more than 1 MW, have been examined.

For some projects, consumption values are already known and the efficiency has been calculated.

Only when the data from all the projects are available, will it be possible to compare all the systems with each other and appreciate, to a certain extent, their effiency.

HEATING PUMP WITH GAS TOPPING UP IN INDIVIDUAL HOUSES

A. DUTTO AND M. GUILLOT
Technical Managers
PHENIX AND ELISE SOCIETIES

SUMMARY

When the heating of new single houses is made by heat pump, topping up heating has up to now been made by electric coils.
This solution requires a lot of primary energy and the topping heating always starts during energy consumption peak hours.
The object of the present experiment is to try a double energy heating solution using as topping-up energy source a stockable combustible, which not only remedies the foregoing drawbacks, but also allows a better overall heat pump efficiency. The described project associates a air-air heat pump and a heating boiler using gas through a hot water heating coils.

1. INTRODUCTION

The experiment has been made in a 30 single houses group, "Le Hameau des Pins" in Valence, France. The housing program includes 19 units equipped with heat pump and 11 units heated by gas.

The process is presented with MAISONS PHENIX RHONE-ALPES, which built the houses. It was put into shape by LEROY SOMER, heat pump manufacturor MAISON PHENIX, and ELISE, engineering office.

A measurement campaign was conducted during 81-82 and 82-83 heating period.

2. DESCRIPTION

The LEROY SOMER OCEANE 60 type heat pump draws heat (by coding outside air) through an outside evaporator (1 400 m3/h).

Heat is injected into the house by condensation. Inside ambient air is recirculated (1 100 m3/h).

The condensing heat and the heat pump efficiency are reduced when outside temperature falls. Therefore to allow a good heating inside the house even on very cold days, the system requires a topping-up energy.

This energy is supplied by a hot water coil inserted in air duct.

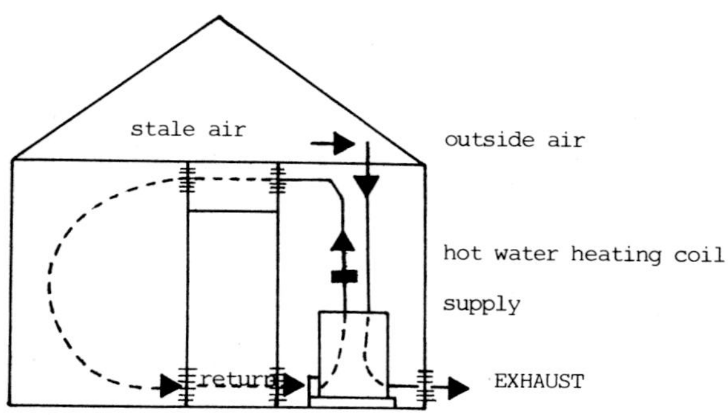

HEAT PUMP AIR SYSTEM

The water is heated by a natural gas boiler.

The boiler also heats sanitary water. Topping up heating only works for outside temperature below 5 °C. The bulb is located in fresh air duct.

A thermostat in the living-room control the start of the heat pump and the natural gas boiler.

When the bulb in outside air heat pump duct allows it, the thermostat controls the start of the boiler.

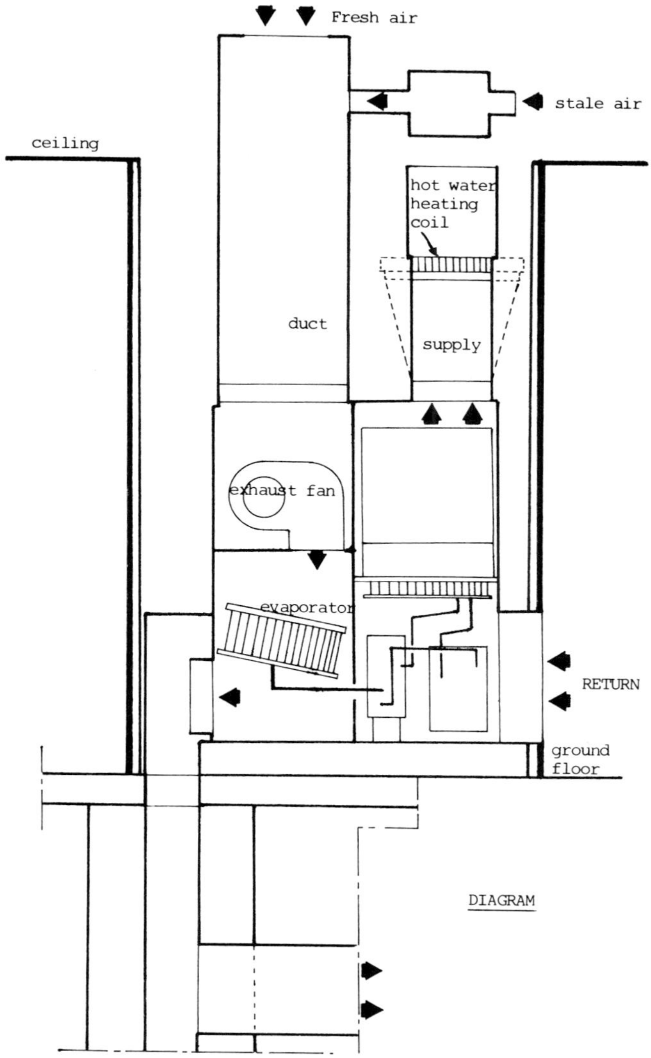

DIAGRAM

3. SUMMARY OF CALCULATIONS OF HEATING LOADS FOR TWO TYPES OF HOUSE TYPE 303 AND TYPE 406 TP :

HOUSE Type 303
- House nb 22
Sh = 74,99 m2
V = 187,50 m3

G = 1,352 W/m3 °C

Zone H2 - Middle inertia -

$$D' = G - Ai = 1{,}352 - \frac{1{,}5}{10} = 1{,}202 \text{ W/m3 °C}$$

Glass area :

Facing ENE = windows = 3,06 m2

French windows = 2,51 m2

Facing WSW = 3,49 m2

Equivalent area south :

	Orientation	A m2	S'	f1 x f2 = f	C1	Ss	Ss/V
EXTERIOR GLASS	WSW	3,49	0,51	0,65 x 0,89 = 0,58	0,71	0,733	0,0039
	ENE	3,06	0,51	0,76 x 1 = 0,76		0,463	0,0025
		2,51	0,54	0,76 x 1 = 0,76	0,39	0,402	0,0021
					TOTAL	1,593	0,0085

Exterior walls = 0,003
TOTAL Ss/V = 0,0115

$$X = \frac{0{,}0115 \times 8{,}5}{1{,}202} = 0{,}081$$

F = 0,065 B = (1,202) x (1 - 0,065) = 1,124 W/m3 °C

HOUSE Type 406 TP
- house nb 5

| G = 1,329 W/m3 °C |

B = G - Ai - As

$$= (G - \frac{1,5}{t})\ (1 - F)$$

Zone H2 - Middle inertia

$\Delta t = ti - te = 10$ °C

$$D' = G - Ai = 1,329 - \left(\frac{1,5}{10}\right) = 1,179 \text{ W/m3 °C}$$

V living = 223,68 m3

3,49 m2 of glass area west

7,10 m2 of glass area east — windows = 4,59 m2, french windows = 2,51 m2

Equivalent area south :

	Orientation	A	S'	f1 x f2 x f			C1	Ss	Ss/V
EXTERIOR GLASS	WEST	3,49	0,51	0,68	1	0,68	0,54	0,654	0,00292
	EAST	4,59	0,51	0,83	0,91	0,76	0,54	0,961	0,00429
		2,51	0,54	0,83	0,91	0,76	0,54	0,556	0,00249
							TOTAL	2,171	0,0097

exterior walls (usual value) = 0,003

TOTAL Ss/V = 0,0127

$$X = \frac{Ss/V}{G - 1,5/t} \times \frac{e}{t} = \frac{0,0127}{1,179} \times 8,5 = 0,0916$$

F = 0,073 B = (1,179) x (1 - 0,073) = | 1,09 W/m3 °C |

4. RESULTS OF MEASURES 1981/1982

I) CALCULATIONS OF CONSUMPTION FOR A STANDARD YEAR

2 pilot houses have been taken : 406 TP

303

Details of calculation in appendise has been made :

a) from G coefficient without solar gain,
b) from B coefficient with solar gain,
c) from B coefficient and we took care of kind of use.
It appeared that, in all houses, heat pump and exhaust of air was stopped from 10 p.m. to 6 a.m.
Calculation has been made with G1 coefficient without heat pump from 10 p.m. to 6 a.m. and with B coefficient and heat pump from 6 a.m. to 10 p.m.
For outside air $>$ + 5 °C the boiler is not used.

Beginning of occupation and measurements were different for each house so far calculations of consumption we touk :
- standard year (Dju 2 314),
- inside temperature : 19 °C,
- heating stop at 16 °C (with heat gain).

II) MEASURED CONSUMPTIONS

The table enclosed herein show :
- winter period,
- heat pump consumption : kWh,
- topping up consumption : kWh gas,
- other consumptions.

"Degree x day" for 1980 - 1981 period : 2 097

Excluding nb 2- inhabit only 2 months
nb 18- not inhabit
nb 20- not inhabit

Average measurements on 7 houses for winter period :

	HEAT PUMP	TOPPING UP
$\sum$	= 18 579 kWh	17 580 kWh gas
$\bar{x}$	= 2 654 kWh	2 511 kWh gas

At the normal "degree x day" : 2 314

HEAT PUMP	TOPPING UP
= 2 928 kWh	2 770 kWh

Middle values are near calculated values with solar gain.
Heat pump and fresh air are stopped from 10 p.m. to 6 a.m.

Operating cost for heating :

We considered that houses nb 5 and nb 22 were typical for theoretical calculation.

Operating cost heat pump :

House nb 5 : electricity = 2603 x 39/100 x 1,246 = 1265 F. TTC

406 TP gas = 2816 x 13,97/100 x 1,186 = 467 F. TTC

TOTAL = 1732 F. TTC

Operating cost for the same house with electical heating.

Consumptions : $\frac{2603 \times 1,9 \times 0,75}{0,69} = 5376$ kWh

0,75 = coefficient of intermittence

Cost : 5376 x 0,39 x 1,246 .... = 2613 F. TTC

Gain/year : 2613 - 1732 .... = 881 F. TTC

House nb 22 : electricity = 1930 x 39/100 x 1,246 = 938 F. TTC

303 gas = 1221 x 13,97/100 x 1,186= 203 F. TTC

1141 F. TTC

Operating cost for the same house with electrical heating

Consumptions : $\frac{1930 \times 2,07 \times 0,75}{0,75} = 3995$ kWh

Cost : 3995 x 0,39 x 1,246 .... = 1941 F. TTC

Gain/year : .... = 800 F. TTC

PAY OFF :

First cost .... = 32 928 F. TTC

If we take a price change of 13 % a year for energy cost the payback is :

for 1 house 406 TP : 13 years - 7 months

for 1 house 303 : 14 years - 3 months

HOUSE TYPE 303

Base of calculations	Without solar heat gain G = 1,31	Wih solar heat gain B = 1,12	With solar heat gain B = 1,2 Heat pump an exhaust air stopped from 10 p.m. and 6 a.m.
Energy needs ΣBXH kWh (1)	9 840	8 406	5 604 6h - 22h 2 255 22h - 6h 7 859
Energy given by heat pump kWh (2)	9 769	8 395	5 597
Energy consumption by heat pump kWh	4 717	4 057	2 705
COP	2,07	2,07	2,07
Topping energy (1) - (2) kWh	71	11	7 6h-22h 1120 22h-6h 1127
Topping up consumption gas $\frac{(3) \times 1,11}{0,8}$ Θ is $>$ 5 °C	98	15	1564

% given by heat pump $\frac{5\ 597 \times 100}{7\ 859}$ = 75 %

% topping up : 25 %

HOUSE TYPE 406 TP

Base of consultation	Without solar heat gain G = 1,33	With solar heat gain B = 1,09	With solar heat gain B = 1,09 Heat pump an exhaust air stopped from 10 p.m. and 6 a.m.
Energy needs ≤ BXH kWh (1) kWh	12 629	9 734	5 858 (1)6h-22h 2 570 22h-6h 8 428
Energy given by heat pump kWh (2)	12 343	9 671	5 816
Energy consumption by heat pump kWh	5 603	4 676	3 118
COP	2,2	2,07	1,9
Topping Energy (1) - (2) = (3) kWh	286	63	42 6h-22h 1276 22h-6h 1378
Topping up consumption gas $\frac{(3) \times 1,11}{0,8}$ θ is > 5 °C	397	88	1 828

% given by heat pump $\frac{5816}{8428}$ x 100 = 69 %

% topping up : 31 %

COMPRESSOR TYPE HEAT PUMP, WITH THE GROUND AS HEAT SOURCE, FOR HEATING A LOCAL AUTHORITY BUILDING

J. GUNIA
Vereinigte Elektrizitätswerke Westfalen
Aktiengesellschaft
Dortmund

Summary

The installation of the gas compression heat pump plant with heat source soil has been completed since March 1982. Different problems being explained in detail in die following description have led to that the plant was not in a position to take up the continuous heating operation. It follows from this that the results of the measuring program up till now do not allow a definitive evaluation of the energetic effects of the plant conception at the moment.

The following report ist therefore limited to the description of the actual state.

1. Plant description

1.1. Dimensioning

The total heat demand of the building amounts to 1,800 kW and is composed of 1,150 kW transmission heat demand -determined with a dynamic heat load calculation - and 650 kW ventilation heat demand for the plants concerning air technique. The optimum being economic with respect to the actual state of the energy price for the design of the heat pump heating power lies at about 60 % of the total heating power. As the ventilation heat demand for the ventilation and de-aeration plants has to be covered for an essentially lower number of hours than is the matter for the transmission heat demand, the heating power of the gas heat pump is designed for the transmission heat demand, the heating power of the gas heat pump is designed for the transmission heat demand of 1,150 kW what corresponds again to the economic optimum of 60 % of the total heating power. Concerning this design, it is possible to cover 97 % of the annual heating heat demand of the entire plant with the gas heat pump plant. This part of the covering has been ascertained by means of the annual line of the outside air temperature for Münster under consideration of the limited operating mode of the ventilation plant. This means that the gas heat pump plant supplies 2,080 MWh/a and the boiler plant 65 MWh/a of useful heat concerning the annual heat demand of 2,145 MWh represented in Table I.

1.2. Technical data of the heat generation plant

1.2.1. Gas heat pump plant

The heat generation plant of the district house in Warendorf consists of the combination of two gas heat pump aggregates with three gas boilers to cover the peak load and for the preparation of useful hot water. A soil heat exchanger is installed as heat source for the heat pumps. Fig. 1 shows the connection diagram of the heat generation plant. The gas heat pump is planned in such a way that both aggregates operate on a common evaporator and condenser. According to the plant conception developped by VEW, the planning has been carried through under the direction of VEW together with the Engineering Offices Zimmermann and Becker, Bochum (soil heat exchanger) and Hüneke and Freitag, Münster (gas heat pump and gas boiler plant). Concerning the installation of the single plant groups, the companies Gebrüder Neunert, Elmshorn (gas heat pump), Autermann, Selm (soil heat exchanger) and Stienemeier, Hamm (gas boiler) have been put in charge.

The basic data are as follows:

design heating power (total)	1,150 kW
number of aggregates	2 pieces
compressor	screw compressor
. product/type	Mycom 160 LUDM
. power regulation	by slide valve regulation
. necessary motive power	121.6 kW
. refrigerant	R 12
. refrigerating capacity	218.5 kW
. number of revolutions	4,032 min^{-1}
driving motor	gas-Otto-engine, 6 cyl. self-absorbing
. product/type	Waukesha F817Gu
. max. motive power at 1,800 min^{-1}	143 kW
. permanent power	130 kW
evaporator	flooded
. evaporative power	440 kW
. volume flow brine	86 m^3/h
. evaporation temperature	- 10 °C
. brine inlet temperature	0 °C
. brine outlet temperature	- 5 °C
. material sheath	steel H 2
tubes	St.35.8
. product Neunert	
condenser	tube bundle
. condenser power	680 kW
. volume flow heating water	98.9 m^3/h
. condensation temperature	55 °C
. material sheath	steel H 2
tubes	St. 35.8
. product Neunert	
heating power motor cooling water heat exchanger	270 kW
. material sheath	steel H 2
tubes	St. 35.8
. product Neunert	
heating power waste gas heat exchanger	200 kW
. material sheath	steel H 2
tubes + bottoms	high-grade steel
. product Neunert	
Oil cooler (screw greasing)	integration into the NT-heating circuit
. transmission power	78 kW

rotating air cooler	segment heat exchanger
. transmission power	40 kW
layer boiler as decoupling and regulating organ	vertical type
. boiler volume	5 m³

1.2.2. Heat source plant

Concerning the gas heat pumps, the soil serves as heat source.The heat absorption power from the soil amounts to about 440 kW. The heat absorption takes place by means of two spatially separated soil heat exchangers. The soil heat exchanger "A' with a basal surface of 6 800 m^2 is situated on the southwestern plot below the paved areas of the parking places planned here. The exchanger "B" with a basal surface of 2 100 m^2 is located in parallel to the nothern boundary of the plot below the green areas which are not paved.

The soil exchangers "A" and "B" consist of PE-tubings lain in parallel one beside the other in one plane. The laying distance amounts to 0.5 m and 0.75 m resp. The length of the tube register amounts to about 105 m. 10 single tubes lain in parallel (circles) are combined to one register by head tubings. Each register contains isolating valves at the inlet and outlet which are arranged in a connecting tunnel. The tube registers are installed in 2 depths of 0.8 and 1.8 m beneath the soil surface. Connecting tubings are lain from the heat exchangers through the outside plants into the building.The circulating pumps are situated in the cellar and the evaporator of the heat pumps is in the central heating in the attic.

The water brought into circulation by the circulating pumps which is cooled in the evaporator of the heat pump absorbs the heat energy supplied by radiation and rain from the soil. The circulating water temperature is about 2 to 5 °K lower than the corresponding soil temperature. The temperature difference of the circulating water on the soil heat exchanger amounts to about 5 °K between inlet and outlet.

With the planned heating powers of the heat pump, the specific heat absorption amounts to maximally 49 W/m^2 of the soil area.

The design data are compiled in the following:

soil heat exchanger	building mode in 2 layers
number of exchanger surfaces	2 pieces (A+B)
. basal surface heat exchanger A	6 800 m^2
. basal surface heat exchanger B	2 100 m^2
heat absorption power (total)	440 kW
heat exchanger tubes	
. material	PE
. diameter	20/25 mm
. laying distance	0.5 - 0.75 m
. laying depth	
- first plane	0.8 m
- second plane	1.8 m
register length	105 m
totally lain tube length	23,000 m
heat carrier	brine out of a water/glycol mixture
design temperature difference of the brine	5 K

specific heat absorption 49 W/m² soil surface

1.3. Functional description of the gas heat pump plant

The heat distribution system is designed for a preflux temperature of 55 °C at a 10 K temperature spreading. The preflux temperature to the preflux distributor is regulated depending on the outside temperature.

Within the heat pump system, the heating water flows through two heating circuits with different temperature level. In the low-temperature-circuit (NT-circuit, ϑ_{max} = 50.9 °C), heat is absorbed from the oil coolers, the condenser and the gear cooling. A partial flow of the heating water is led through the high temperature circuit (HT-circuit, ϑ_{max} = 63.5°C) and absorbs heat from the motor cooling heat exchanger and the waste gas heat exchangers. In the design point, the plant is in a state of equilibrium, that means at an outside temperature of-12 °C the heat absorbed in the NT- and HT-circuit is completely absorbed by the heat distribution system.

The buffer store integrated into the heat pump system stores the HT-heat and supplies the impulses for switching on and off the heat pumps and the boiler. The store is always charged then if the heat generated by the gas heat pump is not completely absorbed by the heat consumers. The store which is formed as a layer store controls the single aggregates by 7 thermostats installed in zones.

In order to prevent that at the starting of operation or after a longer shutdown of the gas heat pumps, the boiler plant automatically switches on without this being necessary due to the actual demand of the building, the boiler standby thermostats are locked depending on the outside temperature. In order to be able to maintain the heat supply during an operational trouble of the heat pump, the locking of the boiler plant is again stopped by a timer after about 30 minutes.

1.4. Function of the heating water temperature regulation

The heating water is led into the NT-circuit by the heating reflux or the store and absorbs the oiler cooler and condenser heat there. According to the heat demand of the building, a partial flow is in each case directly fed into the heating preflux and a partial flow is fed into the HT-circuit to absorb the motor waste heat.

If the heat quantity produced by the gas heat pump is larger than the actual demand, then the store is charged with the excess HT-heat. This takes place by means of a 2-way motor valve controlled depending on the outside temperature in the NT-circuit. If the heating water preflux temperature exceeds the given rated value, the water flow of the NT-circuit is increased by a further opening of the motor valve and more NT-heating water enters into the heat preflux. According to the demand, the HT-heat is partially or totally stored. If the heat demand further decreases, the heat generation power is adjusted to the demand trough the thermostats in the storage zones by switching off the aggregates.

If the heat demand increases, the store is unloaded and the aggregates are switched on in reverse order. If the heating power generated by the gas heat pump during deep outside temperatures is not sufficient, then the thermostats installed in the upper storage zone switch on the gas boiler plant.

2. Starting of operation

The experimental run of the gas heat pump plant took place as of December 1981. After completion of the system connection and regulation works, the plant was given up to the building owner on March 22, 1982 after an inspection by the Technical Control Union Essen.

Operational troubles during the first months have been mainly caused by a failure of the control slide valve regulation of the screw compressor, of the motor speed regulation and the ignition generator. Besides, frequently occurring leaks of the gas motor (ejection of oil) led to operational interruptions.

Operational results will be only available after completion of the first heating period 82/83, at which the long operational interruptions (see point 5.5 and 5.6) will strongly limit the capability of making statements with respect to the measuring results.

3. Execution to schedule

The completion of the district house Warendorf took place according to the construction time-table at the due date in the spring of 1982.

The installation works for the soil heat exchanger started on June 10, 1981 with the measuring of the areas and have been completed with the pressure tests on September 30, 1981. The erection of the gas heat pump plant started on July 6, 1981 with the delivery of the gas motors, compressors and heat exchanger units and was completed with the experimental run in November 1981. The partial acceptance of the units soil heat exchangers and gas heat pump took place in December 1981.

In the time-table of the contract between the European Communities and the District of Warendorf, December 31, 1980 and January 1, 1981 resp. are mentioned as the date of completion.

In the course of the building works and the erection plans of the building cranes and site huts, it resulted that the soil heat exchanger could be only installed as of the middle of 1981. Insofar, one can record a delay due to constructional reasons compared to the original time-table for the erection of the gas heat pump plant of 12 months. Opposed to this, the heating of the building by the gas boiler plant could take place at the due date in December 1980.

4. Cost schedule

The amounts of the investment expenses listed in Annex I of the contract are based on the state of price of 1978/79.

The clearings existing up till now show the following erection and planning costs:

planning and execution	232,714.10 DM
gas heat pump plant	922,063.51 DM
soil heat exchangers	473,538.03 DM
gas boiler plant	217,388.58 DM
total amount heat generation	1.845,704.22 DM
incl. 13 % VAT	===============

5. Problems during erection

During the erection phase, problems essentially occurred within the following areas:

5.1. Soil mechanics

The laying of the soil heat exchanger in the parking place area led to that that the company ordered with the planning of the parking place refused the guarantee. Due to this reason, extensive studies on the change of the properties with respect to the soil mechanics had to be ordered.

A test field was set up. From this test field, untroubled samples were taken in 3 depths, and after that in a depth of 1.2 m, a drainage plough was passed through in the distance typical for the heat exchanger conception and in the following the soil was compressed again in 5 crossings and further samples were taken. The Proctor density of the soil was determined with the samples. The original soil condition (loamy fine sand) resulted in Proctor densities between 96.8 and 102.0 %.

The samples after the compression showed that the required value of at least 100 % of the Proctor density could be reached by the compression. In order to achieve the demanded values of the Proctor density (100 %), it would have been necessary to compress the existing soil also without the laying of the heat exchanger tubes prior to the building of the parking place.

5.2. Sound emissions

Extensive measures were necessary to meet the admissible limits for body and air sound emissions.

The gas heat pumpt plant is installed in a roof heating central in the fifth upper storey of the district house. On the building side, the room for the technical equipment was separated from the other building structure trough steel spring elements. All tube passages have been equipped with rubber compensators. The mounting of the heat pump aggregates took also place on spring elements, at which one had to avoid a resonance with the exciter frequency. In order to calculate the foundations, one had to take as a basis a vertical characteristic frequency of 10-12 Hz according to the demand of the responsible specialist for acoustics.

The ventilation and de-aeration plants of the engine room as well as the exhaust plant have been equipped with sound absorbers which have been dimensioned in such a way that the maximally admissible sound level - 45 dB (A) at night, measured in a distance of one meter at the outlets and intakes resp. - was not exceeded.

5.3. Equipment concerning safety technique

At the acceptance through the Technical Control Union, it was found out that the waste gas heat exchangers did not fully correspond to the regulations concerning safety technique. The Technical Control Union considered the waste gas heat exchangers as self-contained heat generators and demanded an own safety circuit. In order to meet this requirement, an additional expansion vessel was installed and the existing safety valve was changed over to a place at which a steam formation could arise. Additionally, a thermostat was installed which switches off the entire plant at water temperatures of 100 °C.

Up till now, these requirements have not been necessary for gas heat pump plants.

5.4. Adjustment of the heating system to the gas heat pump

The use of a gas heat pump makes special demands on the heating system with respect to the hydraulic switching. Beside the aviodance of too high reflux temperatures, the co-ordination of the water volume flows of the heat distribution system to those of the heat pump plant has to be exactly carried through.

Problems resulted from this field which till the present point of time do not yet allow a continuous heating operation of the gas heat pump. The demand of a constant temperature difference at a variable volume flow on the heat distribution side taken as a basis for heat pump conception is not met. Beside the different volume flows, different temperature differences arise, too, so that the heat dissipation of the gas heat pump to the distribution system is affected and as a result of too large heating water volume flows the regulating store of the heat pump is discharged too quickly and thus it cannot perform a systematic regulating function.

As the described operational behaviour of the plant is influenced by the most different modes of behaviour of the heat distribution regulation, additional operational measurements are carried through at the moment in order to ascertain the cause of the occurred problems and to stop them.

5.5. Damage of the soil heat exchanger

During the ground works for the parking place area, damages of the heat exchanger tubes have occurred for several times. Besides, the temperature measuring pick-ups and lines have been damaged by the ground works.

5.6. Corrosion damages at the damages at the waste gas plant

At the beginning of the year 1983, the gas heat pump plant had to be shut down because of defect waste gas sound absorbers. The supplier needed about 4 weeks to procure the replacement and to make the repair. A short time after operation was started again, a renewed operational interruption took place as a result of defect waste gas heat exchangers.

Waste gas sound absorbers and waste gas heat exchangers showed signs of corrosion which after preliminary material examinations and analyses of the motor oil let conclude that hydrochloric acid is formed by the refrigerant burnt in the gas motor.

5.7. Integration of the building of the heat pump plant into the overall building conception

The erection of the gas pump heating plants requires an intensified readiness for co-operation of all units taking part in the building. In many cases, this is not given. At this, one has to consider that the installation of new, energy saving techniques frequently necessitates additional personal and technical expenses which are not taken into consideration in the offer price of units which are not directly concerned.

5.8. Effects of the problems during erection and starting of operation on the time-table

The 31st of December, 1981 has been fixed as completion date (see explanations point 3.). Up to this date, the time delays influencing the erection could be compensated.

The further equipment with respect to safety technique according to point 5.3 delayed the handing-over of the plant by about 3 months.

The failures described under 5.4 to 5.6 have such an effect on the time-table of the measuring program that the measuring results of the heating period 1982/1983 are only usable in a limited way.

	Maximal heat demand per hour kW	annual heat demand MWh/a
District house	918	1,492
Police wing	186	358
Plants concerning ventilation technique	640	207
Hot water preparation	82	88
Total	1,826	2,145

Table I: Maximal Heating Power and Annual Heat Demand

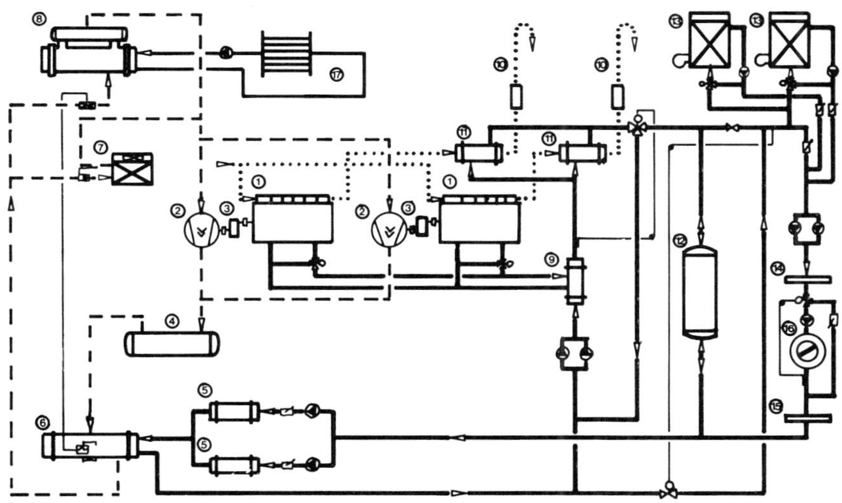

1 Gas-Motor
2 Schrauben-Verdichter
3 Getriebe
4 Oelabscheider/Sammler
5 Oelkühler
6 Kondensator
7 Maschinenraum-Kühler
8 Sole-Verdampfer
9 Kühlwasser-Wärmetauscher
10 Schalldämpfer
11 Abgas-Wärmetauscher
12 Schichtspeicher
13 Kessel
14 Warmwasser-Vorlauf
15 Kaltwasser-Vorlauf
16 Verbraucher
17 Erd-Kollektor

——— Wasser-Kreislauf
- - - Kältemittel-Kreislauf

Figure 1 - Connection Diagram of the Heat Generation Plant

HEAT PUMP DRIVEN BY DIESEL ENGINE

B. Gondro
M.A.N. Maschinenfabrik Augsburg-Nürnberg AG
Advanced Technology Division
D-8000 Munich 50

Summary

Under the European Communities' programme for fuel saving in buildings M.A.N. constructed and tested a heat pump driven by a diesel engine. Test readings were also taken. One of the main specifications to be met was the construction of an installation as compact as possible. This requirement was complied with by installing the heat pump, together with its test and control systems, in a truck container. This container is now situated on the roof of an M.A.N. factory building in Munich. Ambient air is used as energy source. The system has been operating fully automatically since the middle of 1982. Energetic readings have also been taken over this period.

1. Introduction

In the case of heat pumps driven by an internal combustion engine the disadvantage of the extra capital investment required has to be weighed against the advantage offered on the energy side, since in addition to the heat yielded by mechanical energy it is also possible to use the waste heat from the engine.

For a number of years now M.A.N. has been using standard truck engines to drive M.A.N. heat pumps. As fuel not only diesel oil but also natural gas has been used. Oil-injection screw type compressors manufactured by M.A.N. are used in the refrigerant circuit.

Under a research and development programme a heat pump of compact design, driven by diesel engine and using environmental energy from the air is to be tested. This plant is installed on the roof of a building at the M.A.N. factory in Munich. In the building are showers and changing rooms. The heat pump is used to heat water and for space heating.

At every end of a shift cycle there is demand for hot water. The system is connected to M.A.N.'s heating station. Thus the central heating station, as redundant plant, can still replace the entire thermal output of the heat pump.

The heat pump system is of compact design. This has been satisfactorily achieved by installation in a container. This solution offers the following advantages:

- If on subsequent installation of a heat pump there is insufficient space in the existing boiler room, the container construction allows the heat pump to be set up alongside the building, or even on the roof, without difficulty. Noise suppression is achieved by the container itself and with the aid of additional interior insulation.
- If with the container solution the ambient air is to be used as anergy source, as in this particular case, the air compressor can be mounted on the top of the container so as to form one unit.

The heat pump described here has been running fully automatically since the middle of 1982. Test readings have been taken continually since that time.

2. Description of heat pump

The heat pump is installed in a truck container having a floor area of 6 x 2.4 metres. The pump takes up 2/3 of the container's length, the rest being occupied by the cabin containing the test and monitoring instruments. The two sections are separated by a parition. The evaporator, with six blowers, is mounted on the top of the container. To allow induction of ambient air, the walls of the container above the heat pump are removed. The heat pump is covered by a saddle-shaped steel roof. When the evaporator defrosts, the water drips onto the steel roof and runs off.

The system is driven by a diesel engine type D 0224 ME and the screw compressor CF90. The refrigerant used is R12. The control cabin contains the controls and the monitoring system. The latter comprises three blocks:

- for operational monitoring the main data are continously recorded by a 6-channel plotter
- to calculate the heat balance during a year the main heat counters are installed (long-term measurement)
- to check design data the momentary energetic operational status can be called off (short-term measurement)

At design point the heat pump data is as follows:

vaporization temperature	t_o	=	-1^oC
condensation temperature	t_c	=	55^oC
thermal rating of heat pump	$\dot{Q}_H$	=	215 kW
condenser and oil-cooler rating	$\dot{Q}_{cö}$	=	156 kW
engine waste heat output	$\dot{Q}_A$	=	59 kW
engine speed	n	=	2500 rpm

3. Operational mode

The heat pump serves primarily to supply hot water, i.e. it first of all covers the hot water requirements and only then is it available to meet any heating requirements. The hot water passes through 4 boilers (see Fig.)

In order to keep cycling to a minimum the hot water for heating purposes is led off to a buffer store. All 5 stores each have a capacity of 3000 litres. If as a result of low outside temperature t_A the heating system requires a higher flow temperatur t_V than is available from the buffer store, the additional heat required is obtained from the central heating station. The heat pump power is adjustable by speed regulating. The speed regulating range is 1200 - 2500 rpm.

There are the following operational modes:

- Charge hot water boilers.
  If hot water only is needed (summer operation) not only will the heat from the condenser and oil cooler be led off into the hot water boilers, but also the waste heat from the engine. Under this operational mode the regulating range for the boiler temperature is 43° - 46°C (sensor α). Because of the restriction on condensation temperature a higher boiler temperature is not possible.

- Charge hot water boilers and buffer store.
  If both hot water and heating are required (winter operation) only the engine waste heat is led off to the hot water boilers.
  The heat from the condenser and from the oil cooler is transferred to the buffer store.
  In this case the 4 hot water boilers are heated up to 60°C. A higher temperature is not possible because of the need to ensure recooling of the engine coolant.

- Charge buffer store
  If the 4 hot water boilers are charged with water at 60°C, the heat from the condenser and from the oil cooler, as well as waste heat from the engine, is passed to the buffer store. The signal for this is transmitted by sensor γ at a temperature of < 43°C. In this operating mode the store can be charged up to temperature of 46°C (sensor δ). This case rarely occurs.

During winter operation therefore the general rule is engine waste heat to the hot water boilers and heat from condenser and from oil cooler to the buffer store. This leads to the heat pump being in operation for almost 24 hours daily.

4. Description of defrosting process

The evaporator consists of two halves, each having the same rating. This allows one evaporator section to be defrosted while the other continues in normal operation.

If the outside temperature drops below about 3°C the evaporator frosts up, depending on the humidity. It therefore has to be defrosted at certain intervals. This is done by means of a special defrosting circuit.

If the difference between outside temperature and vaporizing temperature exceeds 9 K the defrosting circuit is switched on automatically. This causes the refrigerant vapour produced in the second half of the evaporator to be sucked into the compressor and compressed. After the oil filter and before the condenser the compressed refrigerant vapour is led off to the first evaporator half, which is to be defrosted. Here the refrigerant condenses, giving off heat, and the frost on the tubes and fins melts. The refrigerant condensate is collected and returns to the first evaporator half. During the defrosting process the engine waste heat continues to be fed to the hot water boilers. To avoid too high a concentration in the section operating as evaporator, defrosting is carried out at the minimum speed of 1200 rpm.

The defrosting process takes 10 minutes for each evaporator half.

5. Energetic measurement

After some months of functional testing the system has been running fully automatically since July 1982. Energetic measurements have also been taken since that date.

To enable energetic analysis both long-term and short-term readings were taken during a heating period (July- June). The long-term data serves to demonstrate the energetic effectiveness of the system during a heating period. The short-term data on the other hand are to be used to determine the performance of individual components. Both types of measurement allow continual monitoring of operation and enable evaluation of the quality of energy utilization.

The long-term measurements consist of daily readings of all heat counters, fuel meters and operating hours counters.

The short-term measurements record momentary operating data such as thermal ratings and fuel ratings as functions of mass flows, temperatures and speeds. These readings are of interest because they show how the heat pump's control system performs. Moreover the readings can be compared with the design values so that the extent of deviation of measured values from design values can be seen. These short-term readings were taken on some days during the heating period at varying engine speeds. One of the many readings taken is reproduced below in concise form, the table showing the most interesting data:

n min^{-1}	t_o °C	t_c °C	$\dot{Q}_H$ kW	$\dot{B}$ kW	ζ
1719	-2°	49°	164,9	101,5	1,63
1940	-3°	51°	169,7	109,9	1,54
2180	-4°	53°	187,8	130,8	1,44
2380	-4°	55°	202,0	148,8	1,35

Where

n	=	engine speed
t_o	=	vaporizing temperature
t_c	=	condensation temperature
$\dot{Q}_H$	=	thermal output of heat pump
$\dot{B}$	=	fuel rating
ξ	=	thermal coefficient

In this test series the speed was varied from 1700 to 2400 rpm in steps of 200 rpm. The table clearly shows the falling vaporizing temperature t_o with increasing speed, and consequently the increasing heat output $\dot{Q}_H$.

The increase in condensation temperature t_c is partly due to the increasing return flow temperature t_R. Since as the test progresses the boilers become hotter and hotter, the return flow temperature also increases.

In the course of long-term measurement for the period July '82 - June '83 (heating period 82/83) the following main values were recorded:

operating hours	BS	=	3,861 h
thermal output of heat pump	$\dot{Q}_H$	=	401 MWh
fuel consumption	V_K	=	29,950 litres

Thus for the heating period 82/83 a mean annual thermal coefficient of ξ_m = 1.35 was attained. This thermal coefficient is calculated by the useful thermal output Q_H. The thermal output does not include the heat to defrost the evaporator. 0.9 % of the thermal output Q_H had to be used defrost the evaporator.

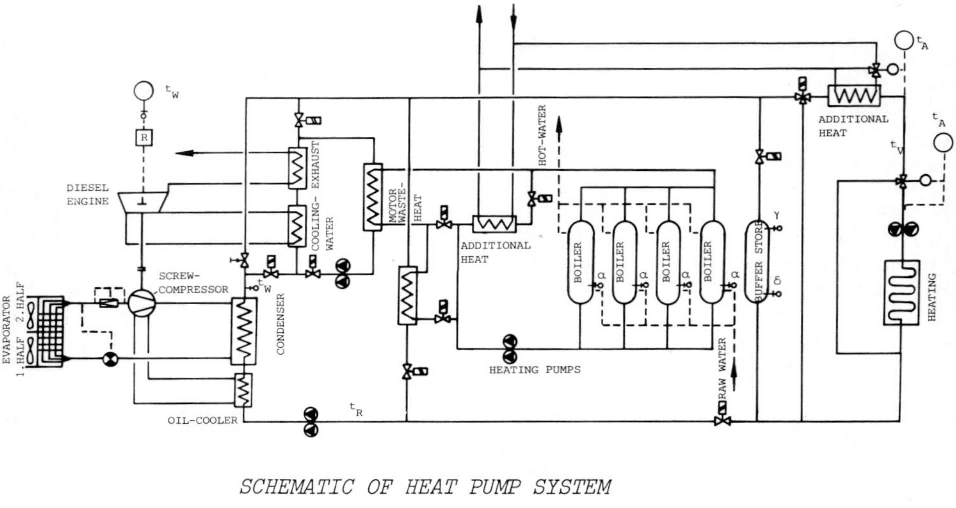

SCHEMATIC OF HEAT PUMP SYSTEM

THE NUFFIELD COLLEGE PHOENIX

P L Martin and D R Oughton

The Oscar Faber Partnership
St Albans, Hertfordshire, UK
AL1 3UT   Tel. 0727 59111

Summary

The installation here described represents the rehabilitation of a heat pump and the associated systems first installed twenty-five years ago. At that time, as now, it was considered that electrical drive from the public supply did no more, in terms of primary energy, than overcome the inefficiencies implicit in power generation. On the presumption - yet to be proven - that manufacturing technology in 1983 has the capability to produce reliable small scale prime movers, economical in operation and easy to maintain, drive to a heat pump compressor by some form of internal combustion engine, with waste heat recovery, must be a preferred solution.

The question of appropriate drive to a heat pump machine and operation of the thermodynamic cycle represent, however, only perhaps one third of the overall problem. Abstraction of low grade energy from a suitable source for use at the evaporator and economic annual utilisation of upgraded potential from the condenser are matters of prime significance. For the installation described, these aspects were considered in the late 1950's and solutions found which experience has shown to be , at least in part, satisfactory. It must never be forgotten, however, that compared with other methods for conversion of primary energy to useful output, the heat pump demands a disproportionate allocation of capital and that methods of both abstraction and utilisation must be such as to counteract this penalty.

1.0 Description of the Work

1.1 Introduction

A heat pump, having an output capacity of approximately 127 kW, was installed at Nuffield College circa 1959. It was arranged to operate in parallel with conventional oil fired boiler plant as a source of supply to the hot water space heating systems within the College buildings. The dual compressors and three water circulating pumps were driven by a commercial diesel engine having a shaft output of 31 kW at 22 rev/s and ancillary equipment was provided for recovery of waste heat. Plant configuration overall, however, required that one large circulating pump be electrically driven and the associated consumption of 7.75 kW from the public supply imposed a relatively high penalty upon the plant performance in terms of both Primary Energy Ratio (PER) and operating cost.

The unique features of the original plant were firstly, the use of raw effluent abstracted from a main public sewer as the base heat source and secondly, the configuration adopted for connection of the water circulation systems within the College to the bivalent heat supply.

In the context of the title to this short introduction, these features are the ashes from which a rehabilitated heat pump has arisen. They have previously been described in some detail (Ref 1) and a brief summary only is included in the present text.

1.2 Experience with the original plant

Over some ten winter seasons, the original heat pump ran for almost 14 000 hours. Those malfunctions which occurred related almost without exception to lack of preventative maintenance to the diesel engine and its ancillaries. The inherent difficulty lay in the fact that Nuffield College is not an engineering establishment and, in consequence, routine maintenance was in the hands of a semi-skilled handyman: any task of significance was dealt with by the several manufacturers of individual major components with the result that the whole was not co-ordinated adequately. Time, obsolescence and the withdrawal of one particular manufacturer from a service contract exacerbated the problem.

The fuel consumed by the diesel engine was to the same specification and unit tariff as that used to fire the contemporary boiler plant and thus a direct comparison of cost-in-use was practicable, operation of the heat pump showing a relatively significant saving in oil consumption when operational. When oil prices escalated in the 1970's however, the original boilers were removed and replaced by units burning North Sea Gas with the result that the economic comparison was distorted. This situation, coupled with the maintenance problems mentioned previously, led to virtual abondonment of the original heat pump as a working unit.

1.3 Rehabilitation

When the prospect of rehabilitation first arose in 1973, the concept of all-electric drive from the public supply was considered. Despite the concomitant advantages of much reduced maintenance demands, the penalties in terms of Primary Energy Ratio were such that this approach was abandoned. A parallel solution using a single gas engine, with heat recovery, driving an alternator to provide a power source to all components was investigated but, although attractive in many technical respects, was not pursued in view of the scale of the installation overall and the requirement that the work undertaken should be confined to rehabilitation rather than development of an alternative concept.

In 1978, attention was drawn to the development by Ford (Ref 2) of a derivative of their production line 1.6 litre spark ignition engine (as used in the Escort car), modified to run on North Sea Gas. The preferred speed was of the order of 45 rev/s matching a shaft power output of approximately 27 kW: a compact waste heat recovery unit was in process of development for mounting to the cylinder block in replacement of the conventional exhaust manifold.

The new heat pump machine in consequence consists of two such engines, works mounted as a packaged unit complete with control equipment, Fig I. One engine powers a Carlyle compressor at 30 rev/s, absorbing 21.5 kW, via a belt drive, and the other powers a 415V, 3 phase, 25 kVA alternator absorbing approximately 14 kW on normal running load, via a similar belt drive. The alternator is oversized to take account of the starting current required by the large sewage circulating pump, powered previously from the public supply. The legendary Phoenix, in its rebirth, was reputed to sing a melodious dirge and this characteristic was echoed by the high frequency sound produced by the two toothed belt drives fitted originally to the new unit.

2.0 Unique Features

2.1 Base Heat Source

The base heat source, raw effluent from the public sewer, is passed through a 100 mm copper-tube heat exchanger at a self-cleansing velocity of 2 m/s. An annulus about the copper tube, arranged in contra-flow, contains water circulated to the shell and tube evaporator of the heat pump. Fig II shows the heat exchanger. The sewage quantity circulated is such that the heat extracted at rated duty does not result in a drop in sewage temperature of more than 5 degrees C, a requirement imposed by the abstraction licence granted by the Oxford City Council at a peppercorn rental.

2.2 Utilisation Arrangements

The heating systems within the College buildings require an energy supply of approximately 400 kW when the external temperature is 0 degrees C and, as has been noted in a previous paragraph, the output of the original heat pump was only about one-third of this total. The systems were designed, at the time of the construction of the College, to ensure that the annual demand imposed upon the heat pump machine remained at a constant level for the longest practicable period. They were thus arranged in three equal and separate circuits which, for design external conditions, are:

: Embedded panels requiring flow water at 49 degrees C.

: "Low temperature" radiators requiring flow water at 60 degrees C.

: "Normal temperature" radiators requiring flow water at 71 degrees C.

The three circuits are connected to the bivalent heat source as shown in Fig. III and it will be noted that each is provided with two motorised mixing valves connected in series. The first valve in each case is arranged to utilise water flow from the heat pump supply in preference to that from boiler supply provided that the former is at a suitable temperature. The second valve in each case operates in a more conventional manner, mixing source water with that returned from the heat emmitting surfaces. Thermostats are fitted in each circuit so set that water at too high a temperature cannot return to the heat pump ring.

The purpose of this arrangement is to provide that the demand imposed upon the heat pump remains relatively constant at design level over the long 40 week heating season required by the British climate for residential occupation. The sequential manner in which the circuits operate produces the load curve illustrated in Fig. IV which shows also frequency data taken from meteorological records.

3.0 Prospects

3.1 Programme

The programme for rehabilitation, supported by the EEC Energy Commission and the British Department of Energy, was due for completion by the spring of 1982. For a variety of reasons, this was delayed and the plant was not commissioned until May 1983 by which time minimum space heating load was available. No significant test results have been analysed in consequence, operation to date being confined to commissioning activities.

3.2 Theoretical Performance

The original diesel engine driven machine, when tested in 1961, had a PER of 1.68 based upon the calorific value of the fuel oil used. This was degraded to 1.25 when the electrical drive, from public supply, to the large circulating pump was taken into account. For the rehabilitated machine, the heat pump proper would have a calculated PER of 1.82 but drive to the ancillary components would as predicted reduce this to 1.42 overall, a value nevertheless almost twice that of the associated boiler plant.

3.3 Monitoring Performance

The performance of the new machine will, as a result of co-operation with British Gas, be monitored by the authors in conjunction with the Department of Engineering Science of Oxford University. British Gas have installed a micro computer on site, complete with a visual display screen and a cassette unit data store. This is interfaced with a 16 channel analogue digital converter which produces frequency counting of pulsed readings of temperature, pressure and flow such that critical input and output data are recorded on magnetic tape for later interpretation.

Over a 30 minute cycle the screen presents alternate 3.5 minute displays of the system diagram, annotated with appropriate readings, as Fig. V, and integrated performance data as Fig. VI. The remaining cycle time provides a record of timing and a warning of cassette run-out. In parallel with this screen display, a facility is provided for digital observation of a single temperature related to the thermodynamic process, as selected from a seven point menu. The parameters available are compressor suction and discharge; evaporator inlet and outlet, heat interchanger outlet and condenser outlet. The seventh point provides for a check reading of the ambient temperature within the data logger cubicle.

4.0 Conclusion

The monitoring facilities available are such that the performance of the new heat pump machine can be fully analysed. This analysis will, initially, be applied to the three system connected load as described previously. It must be emphasised however that a unique facility exists for on-going research into the performance of a heat pump connected either to an embedded panel system, a low temperature radiator system, a medium temperature radiator system, or to any combination of these. Properly exploited and with the co-operation of Nuffield College, an immense amount of data related to heat pump load characteristics and bivalent system operation could be gathered.

5.0 References

1 The Nuffield College Heat Pump. Kell, J R and Martin, P L Journal IHVE Vol 30 p333 1963

2 A Small Gas Powered Engine as a drive for Heat Pumps. Pegley, A C and Rieke, A. Published by Ford, September 1978

Fig. I
New packaged heat pump unit. Engine 1, on right, drives the alternator and Engine 2, on left, drives the compressor. Gas metering equipment is mounted on the rear wall.

Fig. II
Sewage/evaporator-water heat exchanger mounted in insulated space behind heat pump plant room. Air from plant room, warmed by engine radiation, is blown into this space by propeller fan shown top centre in Fig. I.

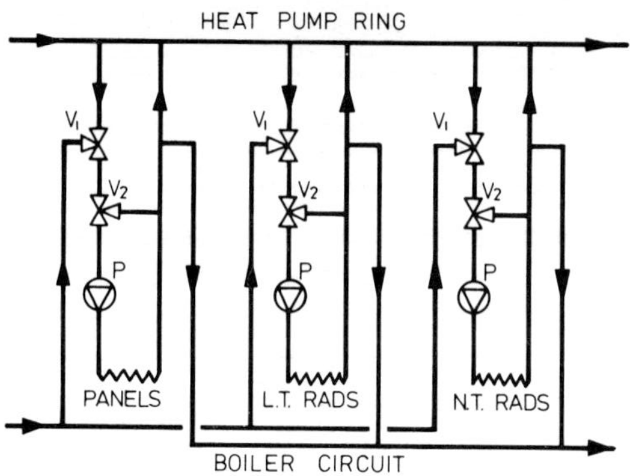

Fig. III - Diagram of arrangement of heating circuits in College

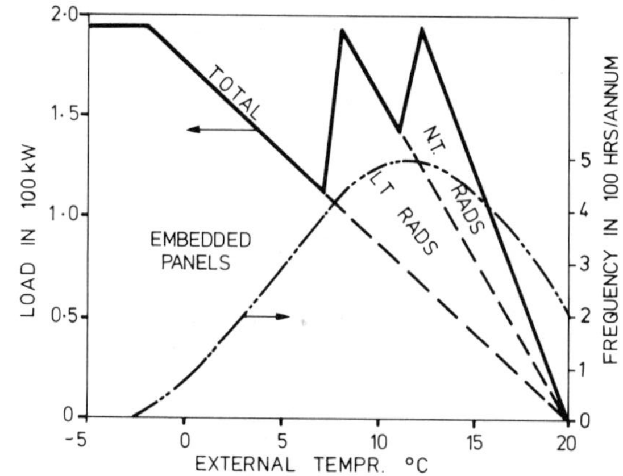

Fig. IV - Load curve and outside air temperature frequency

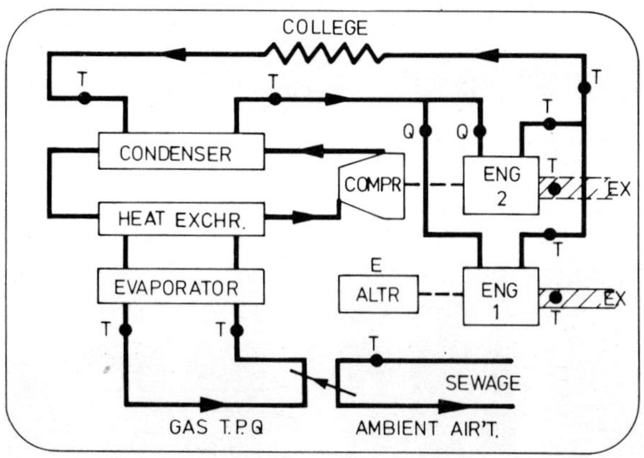

Fig. V - Diagram of heat pump circuit on VDU

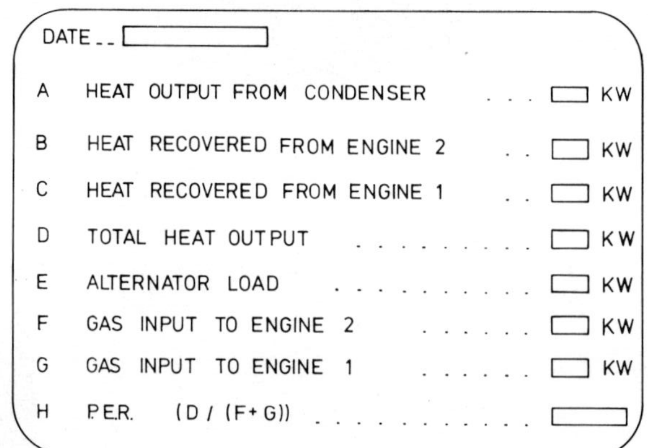

Fig. VI - Integrated performance data on VDU

USE OF A MINI HYDRAULIC POWER STATION TO SUPPLY HEAT TO A HOTEL BY MEANS OF HEAT PUMPS

Ch.G.SPACH
Sté Spach & Fils sa
ROTHAU (F)

Summary

Old textile industry buildings are converted into tourism hotel ; the mini hydraulic power station set since the beginning continues to serve to the electric energy production and yet more, with 3 heat pumps, whose evaporators are immersed in the intake of the turbine, to heat production (space heating and sanitary hot water). We do the balance for 3 utilization seasons, the comparative estimate of the realised savings and the criticism of the equipment.

1. GENERALITIES

11. Objective and description of the operation

As soon as the project was exposed we surrounded the problem:

11.1 the aim to reach : obtaining the energizing autonomy of

a tourism hotel with 16 outstandingly comfortable rooms, a 50 places restaurant, reading and Tv salons, halls and meeting-room, on the whole : 960 habitable m2.

11.2 the available capacities : using mini hydraulic power station set on the next river; but whose power is insufficient to satisfy directly the needs (peak : 60 kVA, mean production : 30 kVA, by 2.4 m fall and 3 m3/sec outflow)

11.3 the solution :

113.1 amplification of the available power setting water-water heat-pumps (two for the space heating, 15 absorbed kW and with 50 and 100% capacities each, to have a best stepped regulation and one, 7,5 absorbed kW, for the sanitary hot water production) and correlatively, optimization of their COP by the immersion of their evaporators in the intake of the power station,

113.2 improvement of the station productivity by adjunction of a mechanical rack-cleaner,

113.3 exploitation of wintry concomitance -in our geo/climatological area - of the best river outflow with maxima heating needs.

12. Exploitation

The difficulties we have met have never called in question again the very principle of the project : they came from the insufficiency or maladjustment of security means used to protect the equipment (for example : many ruptures of compressors'valves..)

The more important improvements have been:

- mounting anti-liquid bottles on compressors' admissions,
- research of the best evporators angle in regard to the water current to facilitate their irrigation,
- mounting compressors' groups in pump-down, avoiding the liquid freon storage in the evaporators and the risk it goes up again on the valves when starting.

13. Production

Generally, we can affirm the previsions have been surpassed

13.1 heating : the equipment has worked during 3 heating seasons and assured permanently a 18/20°C temperature for the whole space until a +1°C outside temperature with continually 15 absorbed kW. As we had foreseen to use the assistance gas-boiler from -5°C outside, it has never been required until -15°C outside (with a 1,2°C primary water temperature) except when the river was so icebound that feeding into the turbine was impossible...

13.2 sanitary hot water : works practically without any problem since july 1980 (38 months) .The assistance gas-boiler also foreseen to reconstitute the 5000 l hot water storage at 48°C in case of an important drawing-off, has never been laid under contribution, even with a 100% prolonged occupation of the hotel.

13.3 subsisting inconvenience consists in the swellings of the river whose current is very torrential : the gravels carried along by high water are drawing towards the intake and, sometimes, cause the locking of the rack-cleaner.

2. GENERAL DATA for ESTIMATE

21. Prices : actualized value at April 83, in french Francs.

- Propane-gas : 3.15 Fr/kg or 0.244 Fr/ absorbed kW
- Electricity : TUP-tariff : power-rate : 246.90 Fr/kVA.year
  HC : 22.90 ,HP: 39.59 Fr/100kWh

  BPS-tariff : power-rate : 109.63 Fr/kVA.year
  HCH: 33.87,HPH: 79.17
  HCE: 9.59,HPE: 19.29 Fr/100kWh
- Electr.sales tariff : HCH: 19.11,HPH: 33.79
  HCE: 9.64,HPE: 12.85 Fr/100kWh

See exploitation's table on next side.

22. Coefficient of Installation

In addition to Coefficient of Performance

$$COP = \frac{\text{restituted power in condenser of heat pump}}{\text{absorbed power in compressor}}$$

was calculed a Coefficient of Installation COI so that:

EXPLOITATIONS TABLE for ENERGY-READINGS in kWh.

PERIODS	TURBINE PRODUCTION	MAINS SALES	MAINS PURCHASE	ABSORBED ENERGY for HEATING electr	for HEATING gas	for HEATING pumps	for HOT WATER PRODUCT. electr	for HOT WATER PRODUCT. gas	for HOT WATER PRODUCT. pumps	OTHER electr
1981 OCT	16 400	8 868	2 039	4 957	-	608	1 606	207	621	860
NOV	19 550	8 147	959	6 831	696	816	1 406	207	601	2 326
DEC	26 300	11 631	594	9 071	1 135	925	1 570	335	621	2 739
1982 JAN	20 540	6 690	1 801	8 140	16 148	1 152	1 984	1 342	621	3 018
FEB	24 150	11 921	422	7 049	5 030	899	1 614	-	561	2 270
MAR	30 420	15 477	53	8 297	-	1 066	1 786	-	621	2 937
Total WINTER	137 360	62 734	5 868	44 345	23 010	5 466	9 966	2 090	3 646	14 150
1982 APR	26 840	12 869	67	5 767	-	666	1 605	334	601	3 134
MAY	17 900	10 252	241	2 945	501	340	1 453	-	621	2 255
JUN	15 420	10 149	370	1 154	167	100	1 213	-	601	2 377
JUL	12 460	7 454	1 539	956	573	162	1 409	-	621	3 153
AUG	9 840	4 269	2 635	1 515	764	252	1 723	-	621	3 845
SEP	505	401	6 578	1 580	287	182	1 324	-	601	2 839
Total SUMMER	80 965	45 394	11 430	13 917	2 292	1 702	8 727	334	3 666	17 603
12 months Oct 81/ Sep 82	218 325	108 128	17 298	58 262	25 301 1960kg	7 168	18 693	2 424 187 kg	7 312	31 753
whose HC	94 750	45 121	5 699	26 569		3 947	7 619		2 436	13 516
HP	123 575	63 007	11 599	31 693		3 221	11 074		4 876	18 237
detail HCH		26 020	1 704	19 169		2 664	4 128		1 218	6 844
HCE		19 101	3 995	7 400		1 283	3 491		1 218	6 672
HPH		36 714	4 167	25 176		2 792	5 238		2 438	7 306
HPE		26 293	7 435	6 517		429	5 236		2 438	10 931

Tariffing periods : HC : night, HP : day, HCH night+winter, HCE night+summer ...etc..

$$COI = \frac{\text{absorbed propane gas power of assistance gas-boiler}}{\text{absorbed power in compressor of heat pump}}$$

for the same work and during same meteorological and hotel occupation conditions.

$COP = COI \times \eta$ $\qquad \eta$ = output coeff.of the gas-boiler.

22.1 Sanitary hot water production (120 h. test)

$$COI_s = \frac{32.78 \text{ kW}}{7.73 \text{ kW}} = 4.24 \qquad \eta = 0.65 \qquad COP_s = 2.76$$

22.2 Space Heating (12 days test or 143 °C.day units)

$$COI_c = \frac{71.48 \text{ kWh/°C.day u.}}{19.60 \text{ kWh/°C.day u.}} = 3.65 \qquad \eta = 0.70 \qquad COP_c = 2.56$$

23. Consequence of Freon-Evaporators Immersion

As river temperature, during greatest heat-pump working time, stands about a mean value of 3,5°C and falls down to 1,2°C his lowering will be limited to 0,1°C (estm.)

23.1 compressors' working

for heating : compr.I 1837 + II 2783 = 4620 hours/year
for hot water production compressor III = 2137 "

23.2 supplied river-heat and necessary outflow

for heating : $\frac{58262 \text{ (absorb.kWh/year)} \times 1.56 (COP_c - 1) \times 860}{0.10 \times 4620 \times 60 \times 60}$

91 000 kWh and 47 l/sec

for hot water : $\frac{18693 \times 1.76 \times 860}{0.10 \times 2137 \times 60 \times 60}$

33 000 kWh and 37 l/sec

23.3 max.primary water pumping power

by total leakage 7 m (3m in height and 4 m in evaporator) and 87% pumping output coefficient.

3,6 kW

23.4 primary pumping energy

3.6 x 6757 hours = 24325 kWh/year

from whose : HC : 10950 and HP : 13375 kWh proportionally and : HCH: 7555, HCE: 3395, HPH: 9495 and HPE:3880 kWh

3. ANNUAL COSTS (in french Francs, Apr.83 value)

31. Standard Installation

Heat production with propane gas-boiler
Electricity from mains.

The correlation between gas and electricity powers is obtained by the aid of our Coefficient of Installation COI.

- Hot Water production :

18693 el.kWh x 4.24(COI_s) = gas kWh		79258		
propane gas purchase	"	2424		
	total " "	81682	x 0.244=	19930.-

- Space heating :

58262 el.kWh x 3.65(COIc) = gas kWh		212656		
propane gas purchase	"	25301		
	total " "	237957	x 0.244=	58070.-
				78000.-

- Electricity (TUP tariff)

HC needs:	heatings pumps		3947			
	hot water "		2436			
	other		13516			
	total	kWh	19899	x 0.2290=	4557.-	
HP needs:	d° total	kWh	26334	x 0.3959=	10426.-	15000.-

Power rate to 18 kVA x 246.90 / year 4450.-

TOTAL of STANDARD INSTALLATION ANNUAL COSTS : 97450.-

32. Savings' Installation

Heat production with water-water heat pumps (primary water pumped out of the river) coupled with propane gas-boiler. Electricity from main.

- Complementary propane gas purchase:

25301 + 2424 = 27724 kWh to 0.244 Fr/kWh 6800.-

- Electricity purchases :
(TUP tariff)

	HC Night kWh	HP day kWh		
sanit.hot water production	7619	11074		
space heating	26569	31693		
primary water pumping	10950	13375		
	45165	56142		
Fr/kWh	x 0.2290	x 0.3959		
	10343 +	22227	=	32570.-

Pumps and other needs (see above §31) 15000.-

Power rate to 48 kVA x 246.90/year 11830.-

TOTAL SAVINGS' INSTALLATION ANNUAL COSTS : 66200.-

and annual savings in regard to standard costs 31250.-

33. Demonstrations' Installation (as described)

- Complementary propane gas purpose : (see above) 6800.-

- **Electricity purchases** :

(BPS tariff)

	Night kWh		day kWh		
	HCH	HCE	HPH	HPE	
total kWh/year	1704	3995	4164	7435	
Fr/kwh	x.3387	x.0959	x.7917	x.1929	
	600	+ 400	+ 3300	+ 1450	= 5750.-

Power rate to 36 kVA x 109.63 Fr/year 3950.-

TOTAL DEMONSTRATIONS' INSTALLATION ANNUAL COSTS 16500.-

and annual savings in regard to standard costs 80950.-

- **Electricity sales** :

	HCH	HCE	HPH	HPE	
total kWh/year	26020	19101	36714	26293	
Fr/kWh	x.1911	x.0964	x.3379	x.1285	
	4950	+ 1850	+12400	+ 3350	= 22550.-

TOTAL DEMONSTRATIONS' EXPLOITATION CREDIT cr. 6050.-

and annual savings' balance in regard to standard costs 103500.-

4. AMORTIZATION OF THE INSTALLATION AND CRITIQUES

The overcost of the savings' installation's investment mounts up 268 000.- Fr , with an annual saving of 103 500.- the return time is of : 2.59 years.

(with heat pumps, supplied by main, this time is longer: 8.58 years)

We must be watchful with electricity purchases tariffing; indead, the application of BPS rate in case of the "savings' installation" §32 would have brought the annual bill to about 75 000.-Fr (+ 13% !) and TUP rate in case of "demonstration" §33 to about 22 000.-Fr (+ 33% !)

We draw your attention on the fact that the compressors had been chosen for the heating on constructors' affirmations who assured an absorbed power at about 60% for a reduced return power at 50% : that's not the case !
No doubt, it would have been more judicious to multiplicate the compressors number instead of having two steps on each of them.

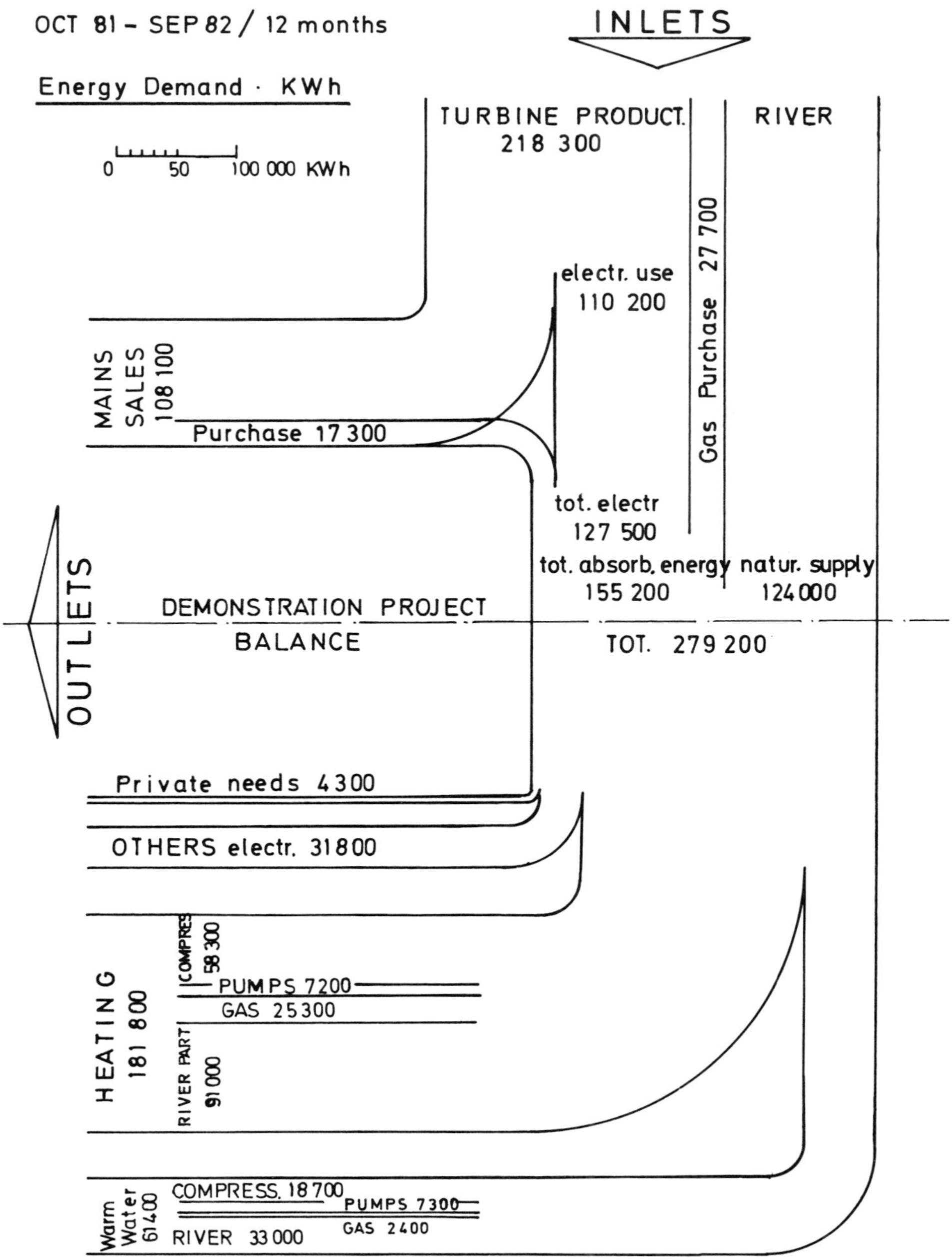
OCT 81 - SEP 82 / 12 months
Energy Demand · KWh
0 50 100 000 KWh
INLETS
TURBINE PRODUCT. 218 300
RIVER
Gas Purchase 27 700
electr. use 110 200
MAINS SALES 108 100
Purchase 17 300
tot. electr 127 500
tot. absorb. energy 155 200
natur. supply 124 000
OUTLETS
DEMONSTRATION PROJECT
BALANCE
TOT. 279 200
Private needs 4300
OTHERS electr. 31800
HEATING 181 800
COMPRES 58 300
PUMPS 7200
GAS 25300
RIVER PART 91 000
Warm Water 61400
COMPRESS. 18 700
PUMPS 7300
GAS 2400
RIVER 33 000

HIGH TEMPERATURE HEAT PUMP

W.F. MALEWSKI
Borsig GmbH, D-1000 Berlin-27
Bundesrepublik Deutschland

Summary

A high-temperature heat pump in compact design using the compression principle with solution cycle is described with respect to its lay-out, operating conditions and energy consumption. In the closed cycle ammonia-water is used as working pair. The installed plant of more than 500 kW heat capacity can be loaded at different temperatures for heat source (20...60°C) and heat sink (60...150°C) to demonstrate its function, operating and partial load behavior. First test-run results and experiences will be reported.

1. INTRODUCTION

Preliminary studies for a high-temperature heat pump in cascade design have shown that technically discharge temperatures above 80 deg.C can be reached, but economically such heat pumps have no future. On the other hand a technology based on the "compression system with solution cycle" offers the possibility to reach discharge temperatures up to 150 deg.C using heat source higher than 30 deg.C with remarkably decreased energy consumption. This process is a closed cycle with the working pair ammonia-water integrating the cycle processes of absorption and compression as well as combining the advantages of both principles. The theory of this process is known for a long time but a technical plant using this process has not been realized so far.

Basing on the experiences of ammonia-water absorption principle for refrigeration and heat pump technology a compact plant for heat capacities of more than 500 kW has to be designed and tested in order to demonstrate its function, operating and capacity ranges.

2. PROCESS

The process is known since the middle of the twenties and was studied theoretically by ALTENKIRCH /1/ for refrigeration. Better operation conditions can be realized in cases of high-temperature heat pumps. A special method of solution injection into the compressor during compression (patent applied /2/) allows component's design according best process conditions.

2.1 PRINCIPLE

Following flow-sheet of Fig. 1, the process proceeds as cycle explained on the base of the working pair ammonia-water hereafter:

Under the supply of heat, the desorber (D) vaporizes rich ammonia-water solution at non-fixed temperatures. The formed ammonia/waste vapour mixture of a high ammonia concentration is compressed oilfree in the compressor (K). During the compression procedure ammonia solution with reduced ammonia concentration coming from the desorber and preheated in the solution heat exchanger (T) is injected by pump force into the compressor. The solution-vapour mixture leaving the compressor in two-phase flow is separated, the solution distributed over the heat exchange surface of the resorber (R) and again brought into contact with the separately supplied vapour so that, under heat rejection, there takes place a resorption. The arising ammonia-enriched solution is cooled down in the solution heat exchanger against cold, poor solution and expanded into the desorber. Expansion vapour is separated from the liquid in the separator and led to the compressor, while the solution is distributed via the heat exchange surface of the desorber.

2.2 PROCESS-BEHAVIOR

The process is running between the pressure of the desorber P_E and of the resorber P_R with a concentration difference between the rich (ξ_r) and poor (ξ_a) solution. The temperature rise $\Delta t = t_R - t_E$ refers to the maximum and minimum solution temperature within the cycle and may be compared with those of single component systems between condensation t_c and evaporation t_o. The coefficient of performance is defined for the system by $\varepsilon = q_R/q_i$ whereby $q_R = q_E + q_i$ is the resorber capacity, q_E the desorber capacity and q_i the power input of the compressor.

The product $\varepsilon \cdot \Delta t$ represents a comparison scale if systems are working with different agents. Figure 2 and 3 show this comparison between single-component (or azeotropic agents) system (KWP) and the binary systems ammonia-water (KML) in dependance on the parameters pressure ratio, concentration difference and poor solution concentration. Figure 4 shows the possible temperature rise $t_R - t_E$ for ammonia-water cycles at different pressure ratios. In dependance on the desorber pressure fig. 5 states the minimum process temperature which would result from a selection of a solution concentration ξ_a.

Figure 2 and 3 show the advantages of the new cycle either in saving energy at same temperature rise or reaching a higher temperature rise at constant energy consumption. The improvement compared with single component systems depends on the working conditions (pressure ratio and mixing concentration of working pair ammonia-water) and exceeds to values between 10...50 %.

The working temperature during compression does not climb over equilibrium values of ammonia-water mixtures due to intensive contact and partial desorption/absorption with cooling effects. Therefore higher pressure ratios ($\leq$ 10) can be operated leading to higher temperature rises ($\leq$ 90 K) (fig. 5.). Two-phase vapour/solution temperature at compressor discharge can rise up to 150°C without higher working temperatures inside the compressor during compression if no other limits for compressor operation reduce the discharge condition.

3. DEMONSTRATION PLANT

Basing on the process calculations a package type unit was designed for a heat capacity of about 750 kW (resorber) at return/discharge temperatures of 60/80°C and a supply temperature of the heat source (desorber) of about 35°C. For test runs this unit is connected with a loading loop which allows via water cycles a variation of the heat source conditions in a range between 20...60°C as well as the heat sink conditions between 60...100°C. It is possible to simulate all conditions with respect to process, operation and different application in house or district heating and LP-steam production.
The individual plant components are designed and selected under following features:

Resorber:	horizontal shell-and-tube exchanger (10 water passes) with incorporated seperator and falling-film-type equipment for outside surface.
Desorber:	same design as resorber
Solution Heat Exch.:	horizontal shell-and-tube exchanger in multi-vessel design (countercurrent flow)
Compressor:	mono-srew type compressor with solution injection under oil-free operation for bearings and shaft seal using cold solution for lubricating and cooling. Displacement stars with plastic coating. Volume flow 850 m³/h Volume ratio 4.9 speed 3000 rpm
Pump:	hermetically operating canned-motor pump (6 m³/h)

Figures 6 and 7 show the installed unit and the loading loop ready for test runs.

4. RESULTS

The test runs started in June 1983 with several start/stop procedures for adjusting control cycles and for becoming acquainted with the new process. Predicted process could be realized with maximum solution temperatures of 120°C. Steady-state operations were only possible for partial load conditions below 50 % due to insufficient suction height of the solution pump during compressor loading and full load conditions. After 20 hours operation with more than 50 starts, there was a damage of the main compressor bearing probably caused by over-loading during start/stop procedure because the compressor was overfilled with solution.

Presently no statements to capacity and consumptions data can be made due to missing of stabilized operation under reasonable load. But it seems possible to maintain a process with discharge temperatures of maximum 130°C enabling a production of steam by flashing pressurized water using heat sources in a temperature range between 40...70°C.

LITERATURE

/1/ E. ALTENKIRCH
Die Kompressionskältemaschine mit Lösungskreislauf
Kältetechnik 2 (1950) 251/259, 279/284, 310/315

/2/ DEUTSCHE PATENTANMELDUNG P 31 00 019.3 (1980/81)
Verfahren zur Durchführung von Kälte- und Wärmepumpenprozessen nach dem Kompressionsprinzip mit Lösungskreislauf

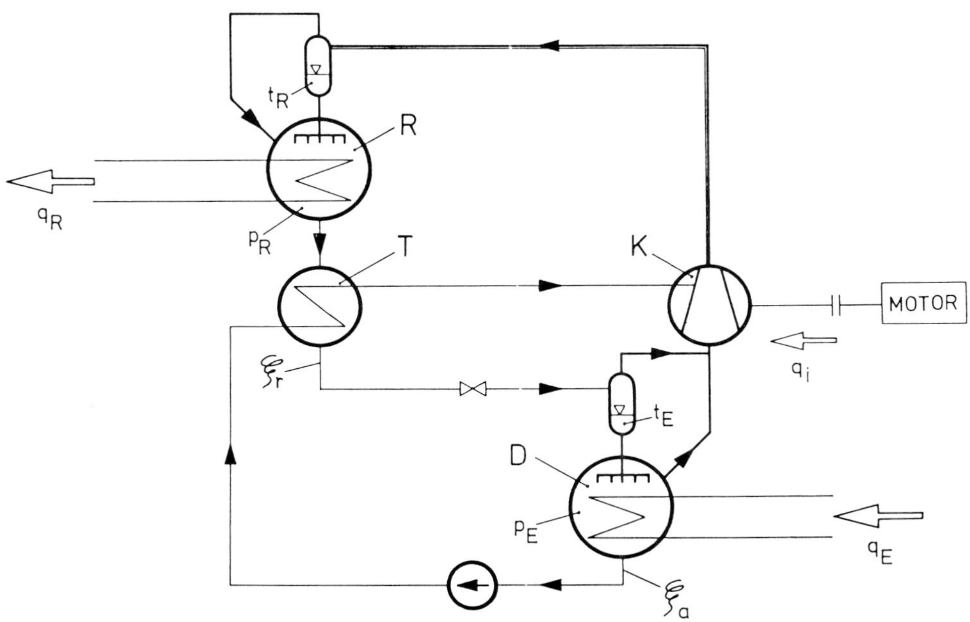

Fig. 1: Flow-sheet of compression system with solution cycle

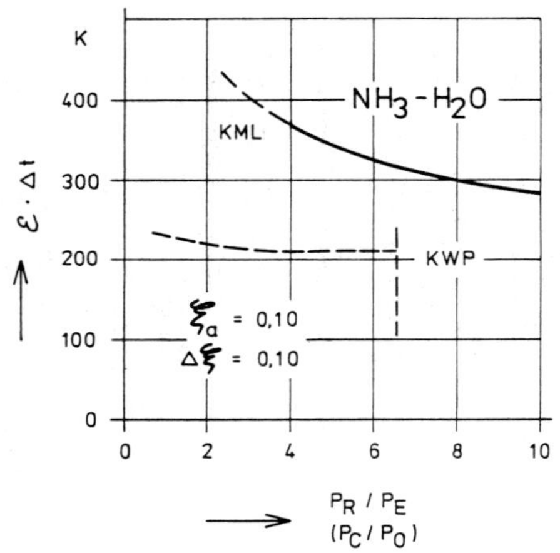

Fig. 2 : Performance comparison between single-component systems (KWP) and binary system ammonia-water (KML) by means of scale ε•Δt in dependance of pressure ratio.

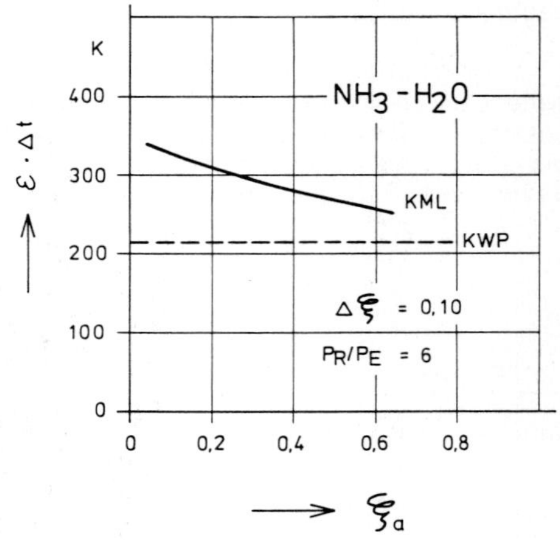

Fig. 3 : Performance comparison between single-component systems (KWP) and binary system ammonia-water (KML) by means of scale ε•Δt in dependance of ammonia concentration.

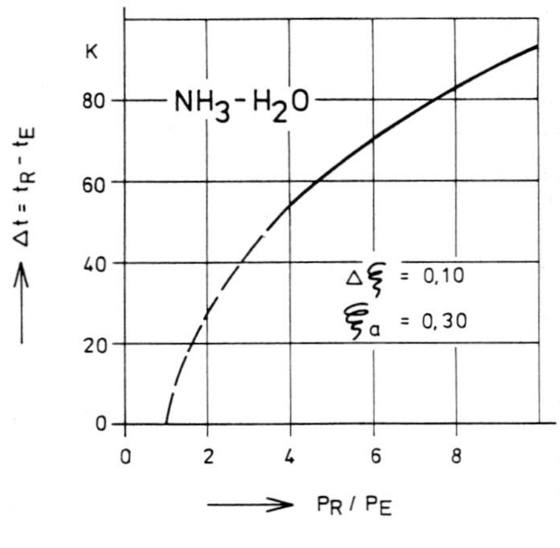

Fig. 4 : Possible temperature rise for compression systems with solution cycles (KML) using ammonia-water.

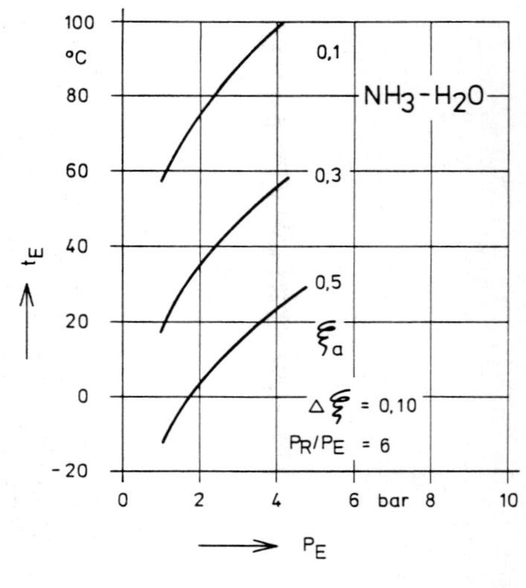

Fig. 5 : Minimum process temperature t_E in dependance of desorbing pressure p_E

Fig. 6/7: Installed unit with loading loop ready for test-run.

ELECTRIC HEAT PUMP WITH ABSORPTION HEAT EXCHANGERS

K. BIASIN, U. JACOBS
Energietechnik GmbH
Federal Republic of Germany

1. SUMMARY

To demonstrate new absorption heat exchanger systems in connection with heat pumps for space heating, to obtain practice-oriented operating results, and to investigate economic conditions, a multiple-family house at Wülfrath has been equipped with such a system. It was planned to show at the same time that it is technically feasible to install subsequently such systems in existing buildings and heating systems. It was the objective to reduce considerably the heating fuel oil consumption and to cut back heating costs. The whole heating system comprising an oil-fired boiler, a brine-water heat pump and an energy stack, is monitored during two heating periods via a measuring instrument. The measured values are evaluated and analyzed by means of a computer. The dwelling house is a good heat-insulated multiple-family house with four storeys and a flat roof. Its construction is of the conventional type. Incorporation of the heat pump and the energy stack pertaining to it, into the available heating system was without difficulties. The estimated capital costs for the total construction measure have not been exceeded. The expected consumption savings have been realized according to the evaluation results hitherto available.

2. PURPOSE TO BE ATTAINED

Four years ago, the multiple-family house comprising eight dwellings with a floor space of 78 m² each required approx. 16,000 l of heating fuel oil per year. With a view to the double-glazed windows and the heat-insulation of essential external components which has been effected subsequently, a considerable reduction in fuel oil consumption seemed not to be possible. Through the installation of an electric heat pump system, a heating fuel oil volume of roughly 10,500 l per year is planned to be replaced. The heating energy gained from the drive and environmental energies exclusively serves to heat the house. Above an outdoor temperature of 0°C the heat pump provides for the required heat for space heating, below this temperature the oil-fired boiler fulfills the same tasks. By means of this bivalent-alternative method the heating power required in one heating period shall be gained to a considerable extent from the environment and the fuel oil consumption be reduced distinctly.

3. DESCRIPTION OF THE SYSTEM

3.1 Dwelling House

The dwelling house is of the conventional construction mode

and has got a basement. Eight identic dwellings with a 78 m² floor space each are available. The external masonry is made of honeycomb bricks, the external wall being 30 cm in thickness. All windows have plastic frames and a double-glazing. The pane spacing is 12 mm. The four storeys are connected via a common stair case. A non-ventilated flat roof forms the upper surface.

3.2 Heating System

The heat producing system consists of an oil boiler and a brine-water heat pump operated according to the bivalent-alternative mode. An energy stack installed on the flat roof of the house belongs to the heat pump. The energy stack comprises horizontally arranged heat exchanger sheets with an exchanger surface of 220 m². The fluid ducts of the heat exchanger sheets are brine-circulated. The energy stack is connected to the heat pump installed in the cellar, by means of a feed and a return line via a collector. The brine cycle has a separate circulating pump. When the heat pump is in operation, the brine is cooled to approx. 6 to 8 K below the outdoor temperature, while heat is extracted from the outdoor air via the absorbing surfaces of the energy stack. Through the heat pump the low-temperature heat gained is raised to the temperature level required for heating purposes. To attain the heat pump COP to be as high as possible, the flow temperature of the heating water is controlled as a function of the outdoor temperature. A heating water buffer storage is installed between heat pump and heating system. It has the task of preventing the heat pump to be switched on and off too frequently. Change-over from heat pump to oil boiler operation and vice versa occurs automatically as a function of the outdoor temperature. The space heating system is adjusted through a weather-dependent flow temperature control. To better adapt the heat pump heating capacity to the specific heat demand of the house, the heat pump is equipped with a three-stage power control. To prevent an excessive formation of ice on the absorber surfaces of the energy stack, the brine temperature is monitored. In the case of the brine temperature falling below -8°C, the heat pump is switched off and the boiler switched on automatically. To prevent air conduction of noise and sounds conducted through solids, the heat pump was mounted onto elastic insulators and the pipes were equipped with bellow expansion joints.

4. MEASUREMENT TESTS

4.1 Method

To acquire the essential operational data, counting measuring devices were installed the counts of which were registered every day at the same time. The measured values are evaluated at appropriate intervals. The characteristic measured variables for rating the overall system are as follows:

a.) Heat pump: heating water volume flow, heating water flow and return temperatures, power consumption, working hours b.) Energy stack: volume flow of the brine, brine flow and return temperatures, working hours of the brine pump

c.) Heating boiler: heating water volume flow, heating water flow and return temperatures, oil consumption

d.) Meteorological data: daily average of outdoor temperature and wind velocity.

4.2 Essential Results

The heat pump system took up operation on 30 October 1981. After initial problems which, however, had no connection with the system design, but were exclusively due to deficiencies in construction and in the internal control of the heat pump, the system has been operating satisfactorily from mid-December 1981. During the period from 1 January 1982 to 31 May 1983, 412 daily values were registered altogether. For evaluating this report, the last heating period from 1 September 1982 to 28 May 1983 totalling 270 daily data records was taken as a basis. In detail, the monthly averages mentioned in the table enclosed, were identified for the months of January to and including May (Table 1). As to the daily average temperatures, the temperature data of the Meteorological Office in Essen were stated. Considered as a whole, the following result proves true for the afore-mentioned period: In the case of a mean air temperature of approx. 7°C and a mean wind velocity of 2.1 m/s, the proportional heating power of the heat pump amounts to some 88 %. During this period the mean refrigerating capacity of the energy stack totalled approx. 21 kW and 95.5 W per square meter heat exchanger surface resp.

4.3 Conclusions

As can be seen from the above-mentioned statements, the originally expected values have been attained for the most part. The applied test methods provided for quite representative values and evident results. In heat pump operation, a heat transition coefficient of approx. 14 - 15 W/m²K is calculated for the energy stack. According to the experience hitherto gained, this value is applicable for dimensioning energy stacks of the same kind. In future, the heat pump components, such as condenser, evaporator and refrigeration cycle control should be adapted more carefully to the energy stack to achieve a better COP of the heat pump. For similar systems, the heat pump COP should not fall below 3.0 for one heating period.

5. OPERATING EXPERIENCE

The heat pump with the energy stack took up operation on 30 October 1981. As a result of the relatively long service life of a compressor between the test run and completion of the system, an unexpected and unusual failure occurred at the oil pressure difference switch of the refrigeration cycle and in the security system of the compressor. When the failure was eliminated and the compressor replaced, the system started operation again on 7 December 1981. Another failure of the heat pump occurred at the beginning of the 1982/1983 heating period, due to the rupture of a defective refrigerant pipe. After repair the heat pump system operated perfectly. The favourable location of the energy stack on the flat roof in a height of more than 10 m and in an exposed position to the wind is the major condition for its good heat capacity. During the operation period from September 1982 to and including May 1983, a mean wind velocity of approx. 2.1 m/s was measured. Attention must be paid to the fact that the static loading of the building part which is loaded further by installing an energy stack, is investigated previously. When doing so, the own weight of the energy stack plus an ice load possibly occurring must be taken into consideration. In the bivalent-alternative mode in connection with a buffer storage, the heat pump operates according to the expectations. The operating time of the heat pump per switching-in sequence could be extended by far through the buffer storage installed between heat generator and consumer so that a too high cycling frequency of the compressor was prevented. This is a assisted by the three-stage outdoor temperature-dependent power control of the compressor. The weather-dependent heating water temperature control and the switch valve (oil boiler/heat pump) which is also controlled by the outdoor temperature, contributed to a troublefree and economic operation of the total system, too. By means of the sound-attenuating lining of the heat pump housing, the mounting of the heat pump on all-metal elastic insulators and by means of fitting bellow expansion joints with sealing material in all pipes, air conduction of noise and sounds conducted through solids to the living rooms could be eliminated nearly completely. Overall noises of the system are below the limits legally prescribed.

6. ECONOMIC CONSIDERATIONS

6.1 Capital Cost

The costs of construction for the heat pump system with energy stack, but without project cost, amount to DM 84,240.00. Capital expenditure relating to a 12 - 15 years service life with a 8 % interest rate, totals DM 10,325.00 per year. The costs of the oil boiler system built according to the conventional method, amount to DM 15,000.00 by comparison. The respective capital expenditure totals DM 1,750.00 per year. Once the series production of the heat pump and energy stack has begun, a considerable reduction of the capital cost to approx. 55,000.00 will be possible, as seen from today's viewpoint. Capital expenditure will be lower by some 35 % as compared with the demonstration plant.
(Table 2 Economy consideration - heating period 1982/83).

6.2 Total Costs per Year

After a first troublefree operation of the heat pump system during the heating period 1982/83, the consumption measurements showed values which were below the original estimations. Annual energy costs of the bivalent heat pumps system are lower by approx. 52 % than those of a pure oil heating system. Accounting for preventive maintenance costs of 10 % of the costs of construction for the conventional oil boiler system and of 2.5 % of the costs of construction for the heat pump system, then the annual operating costs of the heat pump system will be lower by 41 %. Calculating the total costs per year which consist of energy, maintenance costs, and capital expenditure as well, the heat pump system shows costs which are higher by approx. 22 % than the conventional oil boiler system. It has to be taken into consideration, however, that the heat pump system comprises several individual elements and that an extensive test programme was required. As a result the necessary costs of construction are by far higher than those of a normal series production plant. The total costs per year of a heat pump system of a series production are lower by approx. 11 % as compared with the conventional oil boiler system. (Table 2 Economy consideration - heating period 1982/83).

7. CONCLUDING CONSIDERATIONS

The brine-water heat pump with energy stack subsequently installed into an oil heating system already available, clearly shows that heating oil can be saved and operating costs be considerably reduced in this way. For the comparison between the total costs per year which still has an unfavourable balance in this case, it has to be taken into consideration that this individual system has led to higher costs than imaginable series production plants. In addition it has come out that the heating capacity and the heat pump COP were far below the values indicated by the manufacturer. The refrigeration and the control cycles of the heat pump remain to be coordinated with the energy stack as far as possible. Finally the test proved that a large heat volume may be extracted from the outdoor air by means of using the energy stack and the heat pump. The most important result is that approx. 82 % of the annual heating oil volume hitherto required, were saved with this heat pump system.

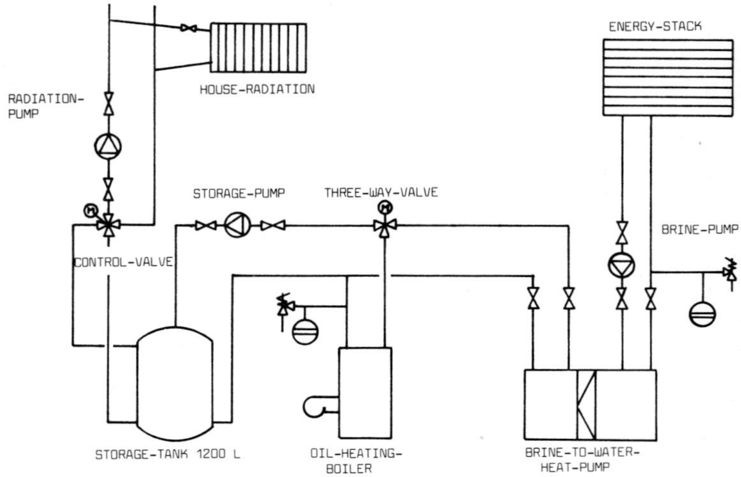

Fig. 1 - LAYOUT OF HEATING SYSTEM

	mean wind velocity m/s	mean day-time temp. °C	heating capacity HP kW	COP	refrig. capac. energy stack kW	spec. refrig. capacity W/m²	environm. energy kWh	proport. heating power HP %	heat oil kWh
Sept.	1.8	14.3	5.08	3.9	3.89	17.70	1244.46	100	0
Oct.	1.7	9.0	42.17	2.6	27,54	125.20	3730.78	100	0
Nov.	2.1	1.3	40.78	2.3	24.58	111.73	4393.85	85.2	1267.5
Dec.	2.3	1.0	38.22	2.1	21.42	97.35	5743.00	77.5	2980.0
Jan.	3.0	2.0	39.70	2.1	22.49	102.23	5941.88	93.8	690.0
Febr.	2.5	5.9	36.81	2.0	19.89	90.40	2991.68	39.93	8330.0
Mar.	2.2	6.8	38.40	2.1	21.79	99.06	5852.9	94,94	550.0
Apr.	2.0	13.8	39.94	2.3	24.00	109.07	4083.4	98.8	82.5
May	1.7	9.1	42.07	2.5	26.70	121.38	2749.75	100	0

Table 1 - Energy balance - heating period 1982/83 - monthly evaluation

Table 2 Economy consideration - heating period 1982/83

	conv. system oil boiler	HP demonstration plant	HP series production plant
1. Capital cost - total	15,000.--	84,240.--	55,000.--
2. Characteristic data			
2.1 Heat req.-space heating			
at -12°C air temp.	49 kW	49 kW	49 kW
at + 3°C air temp.		27 kW	27 kW
2.2 Annual heating req. space heating			
above 0°C air temp.		62,000 kWh	62,000 kWh
below 0°C air temp.	76,000 kWh	14,000 kWh	14,000 kWh
2.3 Load factor	0.7	0.8	0.8
2.4 COP	-	2.5	3.0
2.5 Annual final energy req.			
heating oil	15,430 l	2,190 l	2,190 l
electr. energy		24,800 kWh	20,600 kWh
3. Capital expenditure			
3.1 Service life	15 years	12-15 years	12-15 years
3.2 Annuity at 8% interest			
for 12 years		13.7	
for 15 years		11.68	
3.3 Capital expend.	1,750.--DM/a	10,325.--DM/a	6,830.--DM/a
4. Energy costs			
4.1 Annual energy costs of			
heating oil 0.75 DM/l	11,500.--	1,830.--	1,830.--
electr.power 0.15 DM/kWh		3,720.--	3,090.--
	11,500.--DM/a	5,550.--DM/a	4,920.--DM/a
5. Maintenance costs			
5.1 Share in the capital costs	10 %	2.5 %	2.5 %
5.2 Maintenance costs	1,500.-- DM/a	2,100.--DM/a	1,380.--DM/a
6. Operating costs (4+5)	13,000.--DM/a	7,650.--DM/a	6,300.--DM/a
7. Total costs per year (3+6)	14,750.--DM/a	17,970.--DM/a	13,130.--DM/a
8. Reflux of capital			
8.1 Saving of operating costs	-	5,350.--DM/a	6,700.--DM/a
8.2 Additional capital costs	-	69,240.--DM/a	40,000.--DM/a
8.3 Payback period	-	12.9 years	6.0 years

I.C. Engine Driven Heat Pumps with a heat power up to 20 kW

Diplom-Volkswirt W. Paulick
Fichtel & Sachs AG, D-8720 Schweinfurt

Summary

Fichtel & Sachs AG has developed I.C. engine driven fuel oil and natural gas heat pumps with a heating power of 20 kW. The engine compressor unit is a special development of Fichtel & Sachs, also the microprocessor control equipment. As demonstration constructions 40 heat pumps (27 fuel oil, 13 natural gas) were installed in residential houses in the Schweinfurt region. The houses have a theoretical heat demand between 20 and 60 kW. The installations were realized by handicraft companies - under inspection of employees of Fichtel & Sachs. Each of the heat pumps was supplied with measuring instruments for data processing. The first heat pump was put into operation in December 1981, the last in September 1982. On an average the heat pumps have worked 2 503 h/a; the longest operation hours came to more than 6 000 h until June 1983. All heat pumps are working bivalent-alternatively. The PER (Primary Energy Ratio) of the fuel oil heat pumps amount to 1.32 in the yearly average, the natural gas heat pumps have 1.25 in average; the coefficients of the burner-/boiler system comes to below 70 %. The noise problems were settled satisfactorily; waste gas emissions must be futher improved. The quality of some components gives trouble for they are not always suitable for applications in I.C. engine driven heat pumps. Regarding the present fuel oil and natural gas prices, pay back periods below 5 years are possible, when the saving potential is higher than 4 000 l/a in bivalent-alternative/parallel-installations.

1. THE TARGET OF THE PROJECT

It is the target of the demonstration project to show on a widespread basis under real conditions that regarding the heating of one- and more-family-houses with I.C. engine driven heat pumps, developed by Fichtel & Sachs, about 50 % of primary energy can be saved compared with the heating by conventional systems. Simultaneously it should be demonstrated that I.C. engine driven heat pumps can be integrated into an existing heating system and don't lead to troubles or failures caused by noises or other emissions.

2. SHORT TECHNICAL DESCRIPTION

Fichtel & Sachs has developed a special engine compressor equipment suitable for operation in heat pumps, exclusively. Only by this it is constructively possible to reach the life span target of the I.C. engine of about 40 000 hours (10 - 15 years in operation). Responsible for the development and production of the engine compressor unit is the business group "Engines" of Fichtel & Sachs AG. This group is working on behalf of the business group "Energy Technology" (heat pumps).

The development and production of the heat pump, the air evaporator and the software of the control equipment was performed by our group "Energy Technology"; the hardware of the control equipment was made by an external source by order of Fichtel & Sachs AG.

There is a fuel oil version of the I.C. engine and a natural gas version. The basic configuration of the two is the same. An LPG engine was not installed in the demonstration project; it is still in the development phase.

The I.C. engine driven heat pumps are designed so that they can heat on a monovalent basis a building with a heat demand of up to 20 kW (at an outside temperature of -15 ^{O}C). Houses with a higher heat demand will be heated bivalent-alternatively (houses up to 50 kW).

The heat pump is made of the following essential components:

- I.C. driven engine and compressors (4-cycle-1-cylinder piston engine, 2-cylinder-Boxer piston engine)
- condenser
- induction air silencer
- waste gas silencer
- servomotor
- refrigerant pipes
- water pipes
- fuel oil supply station resp. natural gas supply station
- electric cable assembly incl. sensors
- waste gas pipe
- frames
- noise and vibration absorbers to stabilize the engine and compressor equipment
- sound and heat absorbing hood (hinged for service purposes)
- air evaporator
  The air-evaporator can be installed separately from the heat pump (e.g. garden, garage, adjoining room)
- control equipment
  The total heating system (heat pump, burner/boiler, circulating pumps, mixers, sanitary water boiler, remote control etc.) will be controlled by the SACHS microprocessor control equipment, a special development of Fichtel & Sachs, too. A digital display gives information about the different operating functions (e.g. heating of sanitary water, speed frequencies of the engine etc.); this display shows also the kind of defect in the total bivalent-alternative system. A buffer reservoir was not used because the control equipment allowed an optimal adaptation of the heat power of the heat pump to the heat demand of the building.

3. INSTALLATION AND COMMISSIONING

Starting in May 1981, first contracts have been signed with house owners.
After stocktaking and signing contracts Fichtel & Sachs has prepared for each house an accurate project regarding the installation of the heat pump, of the evaporator (heat source) and of other works which were necessary. Based on these works which required some standard solutions, extensive manuals for the installation instructions, operating instructions, wiring diagrams, electro-installations and other instructions were prepared. These manuals were used to work out special delivery lists. In August 1981 these works were completed for all 40 single projects. Based on the individual measures a standard catalogue was prepared and sent to 34 installation companies to call for tenders. Fixed prices were required for each part of the work. Fichtel & Sachs was the orderer.

Each contracting partner (house owner) was closely involved with the decisions. It was the objective of Fichtel & Sachs to cooperate with only a few companies securing a high level of quality of the installation. Unfortunately it was not possible to concentrate the works on a few companies, because many partners liked the same company to do the installation that had done the heating and sanitary water installation.

Starting mid-December 1981, the first heat pumps were put into operation by F&S people - in the presence of the installation company. Until April 1982, 24 of the total 27 fuel oil heat pumps were installed and in operation; the last 3 heat pumps were put into operation in June and at the beginning of September 1982.

Commissioning of the 13 natural gas heat pumps, originally scheduled for March and April 1982, had to be postponed to May till July 1982 due to special safety regulations of the DVGW which were to consider. The total number of 13 natural gas heat pumps is accepted and released for installation in houses. Every 2 months the gas heat pumps were submitted to a special inspection, at any rate - in the meantime this period has been extended to 6 months by permission of the DVGW.

Within the scope of the EC-demonstration project 40 I.C. engine driven SACHS fuel oil and natural gas heat pumps are in operation.

	fuel oil	natural gas	Total
old buildings	25	11	36
new buildings	2	2	4
Total	27	13	40
of these are with swimming pool	8	1	9

The heat pumps are split versions; the evaporators are installed outside the houses (e.g. garden, yard, roof etc.). All equipment is integrated in the existing heating systems bivalent-alternatively.

4. CONTROL EQUIPMENT

Each of the 40 heat pumps has the following measuring instruments:

- 1 operating recorder for the heat pump (h)
- 1 operating recorder for the hot gas defroster (h)
- 1 operating recorder for the fan defroster (h)
- 1 pulse counter for the heat pump (n)
- 1 pulse counter for the hot gas defroster (n)
- 1 pulse counter for the fan defroster (n)
- 1 recorder measuring the quantity of heat (kWh)
- 1 fuel oil meter (l/h) resp. natural gas meter (m^3/h).

Blanks have been developed for data-recording containing all points of the measuring programm - incl. date and time. In addition, all functions of the installed heat pumps were checked periodically and when problems occurred, using a full range of measuring instruments.

5. MEASURING RESULTS AND EXPERIENCE

Operation hours
The first heat pump was put into operation in December 1981, the last in September 1982; 24 fuel oil heat pumps are in operation since April 1982.

total number of operation hours of all 40 heat pumps (from the day of commissioning until the end of June 1983)	118 766 h
total number of months/years all 40 heat pumps are in operation	569 months 47.5 years
average operating hours per heat pump	2 503 h/y

Operation hours demonstration heat pumps
(from the day of commissioning in December 1981
until the end of June 1983)

	Heat Pumps		
Operation hours h	fuel oil	natural gas	Total
1 000 - 1 500	2	2	4
1 501 - 2 000	2	4	6
2 001 - 2 500	4	4	8
2 501 - 3 000	2	2	4
3 001 - 3 500	5	-	5
3 501 - 4 000	4	-	4
4 001 - 4 500	5	1	6
4 501 - 5 000	-	-	-
5 001 - 5 500	-	-	-
5 501 - 6 000	-	-	-
6 001 - 6 500	3	-	3
Total	27	13	40
thereof: longer than 12 months in operation	25	11	36

Sound

The target of 55 dB(A) was reached. In some houses noise problems occurred due to adverse building substances (wood-concrete construction). Vibrations smaller than 20 Hertz were transmitted into the buildings making changes necessary regarding the flexible suspension of the engine, the concrete substructure and the waste gas pipes. These changes remedied the sound problems.

Waste gas

The exhaust gases will be lead into the house chimney pressureless. Regarding CO and SO_2 the emission values are equivalent to the present standard of the heating technology, but development activities are still necessary concerning NO_x and soot. It is important to note that the total immission load on the environment ("sour rain") will be dramatically reduced, especially for SO_2, because the use of primary energy can be reduced by as much as 50 %.

PER (Primary Energy Ratio)

All selected houses had a DIN-heat demand of 20 kW or higher (DIN 4701/1), but the effective heat demand of some houses was smaller, because these buildings were heated only moderately or partially (long low temperature periods in the night longer than 8 h, to strong temperature reducing in the night longer than 5 h, low temperatures for sanitary water etc.). This additional saving activities caused numerous switch-on-operations and to standstill-losses. The measured burner-/boiler-coefficients of these houses were all lower than 70 %.

The configuration of the heating networks was different (radiator heating: 90/70 to 60/40, underfloor heating: 50/40 or 40/30); private swimming pools were heated from 24 °C to 31 °C water temperature. These different conditions led to different PER's.

Additionally the switch-on/switch-off-power control of the 2-cylinder-Boxer piston engine doesn't work optimal in the total range of all outside temperatures (esp. outside temperature ranges below minus 5 °C); this shows that the full power of the compressors was not used. This led to losses of the PER's of some heat pumps:

	Number of heat pumps		
PER's	fuel oil	natural gas	Total
1.00 - 1.10	-	1	1
1.11 - 1.20	5	3	8
1.21 - 1.30	12	7	19
1.31 - 1.40	8	1	9
1.41 - 1.50	2	1	3
Total	27	13	40
average of PER's	1.32	1.25	-

The results of the demonstration heat pumps caused changes of the compressor concept. Through it the PER's are 10 points higher (instead 1.32 → 1.45) in concrete installed equipment.

Consumption and rentability

The average of all burner/boiler efficiencies amounted to 73 %. Compared with the PER's, there was a reduction of used fuel oil or natural gas of 44 % on an average. But important for the valuation of the rentability of a heat pump is only the absolute saving potential. 50 % saving can mean 1 500 liters fuel oil, but 25 % can be 3 800 liters. The average consumption of an I.C. engine driven heat pump amounts to 0.8 - 1.2 l/h fuel oil resp. 0.9 - 1.5 m^3/h natural gas depending on the speed frequencies, the switch-on-operation times and on the off-cycle-period. Corresponding with the operation hours, a fuel oil heat pump saves 1 liter fuel oil per hour on an average.

Because the total investments for each house are similar to each other (heat pump, installation costs, margins, VAT), the rentability of the heat pumps depends entirely on the following factors:

- the consumption of fuel oil and natural gas in the past
- the PER's of the heat pump
- the energy prices for fuel oil and natural gas
- heating network, temperatures of the heating network
- operation hours of the heat pump per year and consequently the magnitude of the saving potential
- the rate of German income taxation

In houses smaller than 20 kW effective heat demand, the monovalent operation of the heat pumps means too long pay-back-periods. In bivalent-alternative operations and fuel oil savings of 4 000 liters pay-back-periods below 5 years are possible. Esp. of interest is the bivalent-parallel operation of heat pumps, with operation hours up to 6 000 h/a (houses up to 60 kW → heating, sanitary water, swimming pools or commercial buildings like hotels, horticultures etc.). There can be reached pay-back-periods smaller than 4 years.

Herewith the specific economic operational cases and target groups are also defined, corresponding to the present energy prices.

HEAT PUMP INSTALLATION, THE ABBEY, IONA

G.T. Smith
Ian G. Lindsay & Partners, Architects

Summary

In 1975, the Trustees for Iona Cathedral in Scotland commissioned consultants to investigate and determine a suitable method of heating primarily to afford protection to the fabric of the Cathedral which is a building of major historic and religious importance in the country.
After investigation, the optimum solution was determined as being a water to water heat pump which utilises the comparative warmth of the Gulf Stream which flows up the west coast of Scotland past the island if Iona.
The scheme was required to not only provide fabric protection but also domestic comfort conditions for those areas of the Cathedral complex presently occupied all year round, at a more economic rate than is presently possible in view of the building's remote location.

1. INTRODUCTION

Iona is an island measuring 5.5km in length and approximately 2km in width located at the western tip of the Ross of Mull, the peninsula forming the south western part of the island of Mull which is itself a large island off the west coast of Scotland - part of the group known as the Inner Hebrides. While its location makes it extremely vulnerable to the full force of Atlantic storms, the island is also fortunate in that it is situated such that one of the major ocean currents of the northern hemisphere, the Gulf Stream, which crosses the Atlantic from the warm waters of the Gulf of Mexico, passes up the west coast of Scotland directly past its shores, and generates a milder climate than would otherwise be the case in so exposed a location.

The island's importance stems from its being the landing place in 563 A.D. and subsequent base of St. Columba, who it is considered was principal in establishing Christianity as the predominant religion of the adjacent mainland of Scotland.

Situated on the eastern side of the island is a building complex comprising a restored Abbey, Cloister and ancillary buildings, dating from the early 12thCentury, which is the subject of a Trust established in 1899 to ensure the continued well-being of the religious buildings.

The complex, restored from a ruined state to a functioning Abbey and residential complex over the period from 1902 to the present day is now classed by the Scottish Development Department as a Grade A Listed Building being of major architectural and historic importance.

In 1979 after extensive and protracted investigation of the state of the masonry it was determined that major steps required to be undertaken to stabilise the condition of the decorative carved stonework which was decaying rapidly due to the combination of salt contamination and fluctuations in the relative humidity of the air.

The solution to the problem was the stabilisation of the relative humidity by either maintaining a saturated state - unacceptable in a functioning building - or raising the internal temperature by some 6°C over the corresponding ambient external temperature.

At the same time the resident community were experiencing hardship in that not only were the residential areas of the complex underprovided with heating facilities but that the cost of such heating was proving prohibitive to run at the desired level.

Accordingly, it was agreed that some form of heating was the optimum solution if a practical and economic system could be developed.

2. DESIGN PARAMETERS

As Iona is a small island without a ready supply of conventional heating fuels such as oil, coal and gas, a study of alternative energy sources was made including not only timber and peat from the adjoining island of Mull, but also wind and solar power, and heat exchangers.

The use of the majority of these alternatives was precluded on environmental grounds due to the island's classification as a conservation area along with further classification of areas being Scheduled Ancient Monument sites, Buildings of Architectural and Historic importance etc. - no visual intrusion on the landscape being acceptable to the various authorities.

The cost of regular fuels - coal, gas, oil - as well as wood and peat, due to difficulties of transport, along with problems of bulk storage discouraged their use, leaving the only viable source of energy in respect of convenience being electricity supplied from the National Grid through the existing cable from the mainland via Mull.

While it is accepted that the running cost of direct heating by electricity is high the investigation showed that such cost can be substantially reduced when the electricity is used to power a more efficient means of heating such as a heat pump.

The choice of a primary heat source to be used in conjunction with a heat pump is governed generally by temperatures, capital and running costs and availability, and may be air, water, or ground.

In the case of Iona where a network of underground piping was impracticable for various reasons, the choice was either air or seawater.

The problem with using air as a source however is that as the efficiency of a heat pump depends on the temperature of the heat source, the air system has the disadvantage of giving minimum efficiency at the period of maximum heat load.

On the other hand the seawater installation in general gives a better overall efficiency due to the relative temperature stability of the sea as a heat source and such a system proves even more efficient where the influence of the Gulf Stream creates a fluctuation between 7.5°C and 14°C.

In addition to the above considerations in respect of efficiency, the added practical considerations of both the Highlands and Islands Developments Board and the European Energy Commission being interested in the partial funding of a seawater scheme as a demonstration project with commercial potential made the choice of a water source system particularly attractive.

3. SCHEME SOLUTION

Due to architectural restrictions on ductwork within the complex, the system was designed on the basis of a water to water heat pump with seawater circulating through the evaporator and undergoing a 3.5°C drop in temperature before discharging back to the sea.

The heat pump is essentially a heat transfer device which enables heat to be upgraded in temperature and transferred to an area where it can be used. In effect, it uses a refrigeration cycle to extract heat from a low temperaturesource in our case seawater - by using a refrigerant liquid with a boiling point at or below 0°C

A compressor is used to pump the refrigerant around the circuit as shown in the diagram. Heat is removed from the sea-water by refrigerant flowing through the evaporator coil. The absorption of heat by the refrigerant changes its state from a cold liquid/vapour mixture to a cool gas which is then drawn into the electrically driven compressor, where its pressure and temperature are increased. This increase in temperature allows heat to be transferred from a condenser coil direct to water in a radiator circuit. Meanwhile the cooling effect of the condenser converts the gas to a hot liquid still under high pressure and, by means of a thermal expansion valve, into a low pressure mixture of cold liquid and vapour. The cycle is then ready to be repeated.

The efficiency of the system is measured by the term co-efficient of performance (C.O.P.) which is the energy output divided by the energy input to the pumps. The C.O.P. is highest when the temperatures of the heat source and the heat sink are closer, and in this case, with the relatively stable sea-water temperature, works out at 3.2. Thus for every kilowatt of electrical power absorbed, 3.2kw of useful heat are produced.

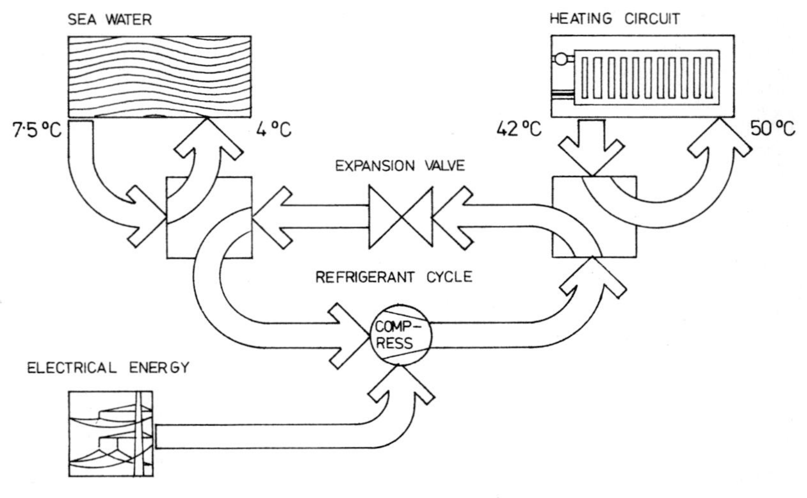

DIAGRAM OF A HEAT PUMP SYSTEM

4. SYSTEM SPECIFICS

The system finally arrived at consists of a 110kw water to water heat pump using a compressor with four cylinders which enables it to operate at a 50% load with only minor reduction in efficiency.

The heating circuit contains 'accelerator' type circulation pumps feeding four separate zones of the complex as follows:-

a) Abbey Church - fabric protection by way of a 6°C temperature increase over ambient using 50mm dia. pipework/radiators around the perimeter of the building.

b) Residences - fabric protection by way of a 6°C temperature increase over ambient using 65 and 80mm pipework.

c) Control Areas - human comfort conditions ranging between 16°C and 25°C using radiators with thermostatic control valves which can be preset as required by the residents.

d) Domestic Hot Water - existing calorifiers to be supplied from a separate pre-heat storage calorifier.

In addition a 50kw multi fuelled back-up boiler is provided to be used during shutdown of the heat pump for whatever reason.

The design data used to develop the system is as follows:

(1) Seawater temperature - 7.5°C flow 4°C return
(2) Low temperature hot water - 50°C flow 42°C return
(3) Heat Pump power - 110kw

comprising:	Abbey Church	35kw
	Residences	20kw
	Control areas	50kw
	Domestic hot water	3kw summer
		15kw winter

(4) Coefficient of Performance (C.O.P.) - 3.2

Originally the system included dehumification plant for use in the Abbey when the heating was not in operation but after a year's monitoring and testing the Department of the Environments recommendation was that dehumidification was not considered necessary at this stage. Nevertheless the possibility still exists that a dehumidification plant will be required for the Abbey at some future date, but this depends on the outcome of further monitoring.

5. SEAWATER INTERFACE

The major element of concern of the design on a technical basis was the design of a suitable seawater intake. Given the limited financial resources of the Cathedral Trust and also the very remote location of the installation it was necessary to achieve a relatively simple means of collecting the seawater in suitable quantity and condition for operation of the heat pump.

After consideration of various alternatives including a shore mounted heat exchanger and a seabed heat exchange unit, it was decided to locate the heat pump adjacent to the Abbey and supply sea-water to it by means of a shore mounted submersible pump. The question of filtration of the seawater was investigated but on assurance from the manufacturer that the heat pump was built to marine specification and was capable of tolerating particles of up to 5mm passing through it, the decision was to omit any sophisticated sub sand filtration system from the point of view of both a capital installation cost and maintenance difficulty and cost.

The solution finally determined was the formation of a wet well just above the line of high water using 900mm dia. concrete sewer pipe sections on end, set with the base approximately 1m below the level of Low Mean Water, Springs.

From the wet well, in an excavated and subsequently backfilled channel, two 225mm diameter marine steel pipes run to approximately 24m below the line of L.M.W.S. to ensure a constant gravity flow of seawater to the wet well.

Within the well a 3.7kw submersible pump lifts the water the approximate 30m to the heat pump at a flow rate of 6 litres/second.

The end of the two seawater pipes are turned up off the seabed and finished with a mesh strainer to prevent unacceptably large items entering the system while it is anticipated that there will be some settling out of particles in suspension in the water once out of the seas turbulence and in the pipe.

Thus the anticipated maintenance to the seawater intake will be reduced to an annual or at the most biannual removal of seaweed and marine growth by scraping off the mesh intake, raking out of sediment, and the dosing of the pipe runs using chlorine to kill any build up of encrustation.

In addition to the annual dosing of the intake pipes the installation includes the continuous chlorination of the seawater flow from immediately after the submersible pump so that the whole seawater section of the system is fully protected against the risk of blockage through internal marine growth.

6. INSTALLATION

No major difficulties have been encountered to date with the scheme from a technical aspect - problems incurred so far have related to the practical aspects of the project such as its remote location, the need for vehicular permits to land on the island and even the weight restrictions on the size of vehicle, weather and tides etc., and also the historic and architectural restrictions - formation of pipe routes through historically important masonry walls, etc., with in fact the installation being delayed at one stage while a further grant aid of Ł25,000. 00 was obtained in order to locate the heat plant room underground where it would not be an unsightly addition to the elevation of the complex - an opinion stated by various interested academic parties throughout the country

7. COSTS

The scheme as originally submitted for consideration by E.E.C. approved by them in July 1980 amounted to Ł134,000. 00. for installation and monitoring, including Consultants fees. Tender figures for the works however came back higher than anticipated and after adjustment of the scope of the works to afford savings the total budget approved by the client amounted to Ł160,606. 00.

Due to the presently incomplete state of the project no overall cost can yet be determined but it would appear that Ł200,000. 00 is a more realistic figure including all builderswork and variations/extras commonly associated with carrying out works to existing buildings where precise pipe routes etc., cannot be accurately determined prior to commencing on site.

Running costs per annum based on electricity costs current at the time of tender acceptance, July 1982, amount to a total of Ł6,322. 00 for a planned 210 day heating year.

CONCLUSIONS

At the time of preparing this report the scheme is not yet operational - the combination of storms and tides having slowed the completion of the seawater intake, and as such no data can be tabled in respect of performance of the system nor can any conclusions be drawn in respect of efficiency.

The installation internally however is substantially complete with the commissioning programme being anticipated as commencing at the end of October,1983, when final connection of the seawater is made.

Thereafter the installation will be monitored for an 18 month period to comply with the requirements of the E.E.C. grant awards.

HEAT PUMP SYSTEMS - DEMONSTRATION PROJECTS

Rapporteur: J. DURANDET

Heat pumps, which are the subject of nine demonstration operations in our session, are of the conventional compression type, except for the compression and resorption device described in text B.1.6. Four papers deal more specifically with the extraction of the heat contained in different environments, i.e., the ground (B.1.2), sewers (B.1.4), the ocean (B.1.9) or the air (B.1.7). Three papers deal mainly with methods of driving heat pumps, i.e. internal combustion engines (diesel in B.1.3, engines powered by diesel oil or gas in B.1.8) or a small hydraulic power plant (B.1.5). Two papers also give the results of campaigns involving the heating of different buildings (B.1.1 and B.1.8).

We will briefly describe each of these papers, with special emphasis on the research that will not be orally presented. We will group the projects by the above approximate classification.

Heat extraction

- Paper B.1.2 presented by J. Gunia concerns the heating of a large public building by a set of two pumps taking heat from the ground. The two heat pumps use the same evaporator and condensor. They are each powered by a 130 kW gas engine. Thermal power amounts to 1150 kW. Heat is extracted from the ground by the two networks of heat exchangers having a total surface area of 8900 m2 and made up of PE tubes buried at depths of 0.8 and 1.8 m. The amount of heat extracted from the ground comes to 49 W per m2 of exchange surface.

This is one of the largest achievements of this type in Europe. The installation cost was DM 1,845,000; a large portion of which, i.e. DM 473,000, corresponds to the buried heat exchangers, i.e. one quarter of the total cost.

The unit has now been installed, following many difficulties during its construction and start-up, due mainly to problems of ground resistance and corrosion. The results of the 1982-1983 test campaign are too limited for an estimate to be made of the real significance of the installation. Annual savings are forecast to be 133 toe.

- Paper B.1.4 by P.L. Martin and D.R. Oughton gives the heating history of Nuffield College, and especially the remodeling of a heat pump and of the hot water distribution network installed there 25 years ago. All that was preserved of the former installation was the exchanger extracting heat from the town sewer system and the heating system for distributing hot water at different temperatures (49, 60 and 71°C).

The new heat pump is powered by a gas engine of about 20 kW, which is a modified version of the model powering the Ford Escort car. A second similar engine supplies the power required for the heat pump circulating sewer water. The heat lost by the engines is used for reheating the hot water coming out of the condensor. The installation has to supply 400 kW for an outdoor temperature of 0°C. A primary energy efficiency of 1.4 is expected.

Work is completed, and a microcomputer will be used to closely monitor the operation of the system. No results are available yet. It can be noted that the use of the surplus electric power for auxiliary heating by electrical resistance is not feasible.

- G.T. Smith (B.1.9) describes the work that has been done to preserve the cathedral in Iona by maintaining its temperature 6°C above the outdoor temperature. The heating system chosen, considering the special geographic location of this monument, makes use of a pump which takes its heat from the ocean. The effect of the Gulf Stream is used to advantage here where the water remains at a stable temperature of between 7.5 and 14°C. The thermal power of the plant is 310 kW. The important element in this system is the seawater offtake which consists of a well 900 mm in diameter and is judiciously situated in relation to the extreme tidal levels, as well as a set of two tubes 225 mm and 24 m long. A 3.7 kW submersible pump feeds seawater (6 liters per second) to the heat pump situated 30 m higher up.

The cost will most probably be £ 200,000. The installation is not yet operational.

- K. Biasin and U. Jacobs (B.1.7) describe the system chosen for heating a building containing 8 appartments (78 m2 apiece). A 40 kW heat pump driven by an electric motor extracts heat from the air by means of two special exchangers located on the roof of the building, at a height of 10 m where it received the full force of the wind. These special exchangers are each made of 20 horizontal aluminum plates providing an exchange surface of 220 m2. The heat extracted is transferred by a brine to the heat pump evaporator. For an outdoor temperature higher than 0°C, the pump heats the building. Below this temperature, an oil-fired heater takes over. So as to prevent excessive ice formation on the plates of the exchanger, the pump stops when the brine temperature approaches -8°C. For an outdoor temperature of 7°C and a wind speed of about 2.1 m/s, the heat captured by the special exchangers amounts to 95.5 W per m2 of exchange surface.

After a few start-up problems, the installation operated correctly and within the time limits planned during the 1982/83 heating season. The results obtained show the effectiveness of the heat-capturing tower accompanied by the heat pump, i.e. 80 % of ordinary fuel consumption was economized.

The economic interest of the system is now limited because the payback time is around 13 years. It has been announced that mass production could greatly reduce costs.

Different way of powering heat pumps

- K. Mötz and B. Grondo (B.1.3) report on their development results with a heat pump driven by a 50 kW diesel engine. The system is to supply 186 kW for heating and 93 kW for sanitary hot water. Construction is compact, and the fact that it is in a caisson reduces annoying vibrations and noise.

The technical goal has been reached. Construction of the unit was completed in October 1981, and it was tested during 1982. The thermal efficiency for primary power is estimated at 1.36, which is very good. Little economic information is available, although the investment is rather high and amounts to about double the amount initially forecast.

- Paper B.1.8 will be presented directly by its authors, G. Brockhoff and V.G. Paulick. It concerns the heating of different buildings by 20 kW heat pumps driven by internal combustion engines. A great many units (about 40) are currently operating, with some having run for more than 2000 hours. The mean efficiency is between 1.2 and 1.3. Problems of noise and polluting emissions must be examined.

- Paper B.1.5 concerning the heating of a hotel by three heat pumps powered by a small hydraulic power plant will be presented by Mr. Spach. This project reuses equipment originally designed for an industrial use. The 105 kW installation has been operating for three years. Allied problems (fouling, corrosion) have created some difficulties. The investment (FF 268,000) was paid off in about 2.5 years.

Heating campaigns : acquiring operating experience

We will not give here results that have already been presented (B.1.8) concerning an important experiment of operating pumps equipped with small thermal engines.

- M. Guillot (B.1.1) describes the results obtained during the 1981/82 heating season for detached houses. Two types of houses were investigated, having respective ground areas of 80 m2 and 100 m2. Heating was provided by commercial heat pumps of the air/air type operating by electricity and extracting their heat from the outdoor air as well as from the air leaving the buildings. Gas wall furnaces take over when the oudoor temperature is below 5°C. The interest of this demonstration lies in the fact that the make-up gas is LPG stored on the site. Therefore, there is no need to make use of the already overloaded public distribution network.

Of the 19 houses originally planned, only 6 were occupied during the heating season. The measurements will thus have to be completed. The economic goal, i.e. investment payback in 10 years, does not seem to have been reached.

- The heat pump described in paper B.1.6 by W.F. Malewski is of a new type. The working fluid is a mixture of ammonia and water. The composition of the mixture varies during its circuit through the pump. Ammonia-rich steam is compressed and then transformed into a liquid by both cooling and dissolution into water. The solution thus produced is then vaporized at a lower pressure. In this compression pump the conventional condensation and vaporization stages have been replaced by resorption and desorption stages which are sometimes used in absorption pumps.

This system enables the temperature of the fluid to be reduced during its compression. Therefore, a compression ratio of less than or equal to 10 is accessible. Under such conditions, the pump can supply heat in a wide temperature range and up to 130°C. The heat capacity of the installation is 500 kW. The COP is higher than that of a conventional pump (by about 1.5). Start-up occurred in June 1983. Few steady-state are available.

* * *

It is difficult to draw general conclusions from these few operations, especially since, for the most part, their schedules have not been respected. Indeed, whereas all the installations have now started operating, their operating times are still quite limited, on the whole. Therefore, we will limit ourselves to the following comments :

- The use of heat sources having moderate and appreciably constant temperature is obviously of great interest. The different environments, however, do not provide the same advantages.

Equipment buried in the ground does not require any maintenance, but its construction seems to be delicate and costly, at least for large-size installations. Recovery of heat entrained by sewers should become quite widespread, although problems of exchanger fouling may arise. The use of sea or river water as a cold source raises several difficulties (proliferation of microorganisms, flooding) but is still an effective solution.

The system proposed for extracting heat from the outside air is all the more attractive as the absence of a fan eliminates noise and vibrations. The development of the device risks being limited by its cost.

- The availability of electricity under economically favourable conditions, either resulting from a large-scale nuclear program such as in France or resulting from suitable local conditions, is in favour of using electric motors to drive heat pumps. The situation is no longer the same when primary fossil energy is used to generate this electricity. Research concerning powering pumps by gas or diesel engines thus becomes of particular importance. Indeed, such systems have good efficiency concerning primary power. Values of between 1.2 and 1.4 have been mentioned by our authors. The technical goal of such operations has thus appreciably been reached, but their economic feasibility has not yet been demonstrated, especially, it would seem, for small-capacity systems (less than 20 kW).

Lastly, we note that the investments originally announced have usually proven to be insufficient because the equipment has required different adaptations in order to complete the projects. Some research, especially concerning the main components of the pumps, thus still seems necessary to improve equipment reliability and to optimize the overall performance of installations. The development of low-capacity pumps (less than 5 KW) also seems desirable because of the market which is opening up for such systems with the evolution of constructions regulations toward high thermal insulation.

Unfortunately, no project concerning absorption heat pumps was submitted to us. Such devices, which are equivalent to pumps driven by thermal engines, effectively have a great many advantages, and in particular they have good efficiency with reduced maintenance.

SESSION II - BUILDING ENVELOPE

A - R & D PROJECTS

Chairman: Prof. C. BOFFA
Rapporteur: Prof. J. LEBRUN

Energy saving by using rolling shutters, with reflective layer and edge seals

Heat reflecting blinds and methods of edge sealing

The measurement of air infiltration rates in large enclosures and buildings

Measurements of intercell airflows in large buildings using multiple tracer gases

Users influence on the air infiltration

Control of ventilation rate in building using H_2O or CO_2 content

Report of the session

REMARKS BY THE CHAIRMAN

Prof. Cesare BOFFA

The envelope of the building plays a major role in the overall energy balance of the building itself.

Suitable interventions of the building envelope coupled with actions on the heating plant and on the user's habits can induce substantial energy savings (up to 30 - 50 %) with a pay-back period of a few years, both in new constructions and in retrofitting of the existing ones.

The results of the EEC R and D programmes, which have been summarized in the first part of this session and which stem from several years of often sophisticated research efforts, carefully stimulated and controlled by DG XII, show that the insight into building envelope is much deeper now than it was when the programme started.

A great deal of information is now available on how to simulate the building thermal behaviour, to design a building according to the climate, how to retrofitt the envelope, both on opaque and transparent surfaces and clear indications have emerged on new methodologies for defining the suitable energy saving interventions in the various climates and for the various building typologies throughout EEC countries.

The results also provide substantial technological know-how related to new types of interventions, both on transparent and opaque surfaces of the building.

More of course has to be investigations particular for offices and industrial buildings, but the level of knowledge already acquired is substantial.

The new proposals which have emerged from research can be directly applied and used by the building industry, architects, civil engineers and planners.

Positive results have been produced by the demonstration projects, sponsored by EEC (DG XVII) ,which have been presented in the second part of the session.

The results of the demonstration projects are based on full scale experiments in actual buildings and provide valuable information related to cost/benefit analysis based on measured data.

The projects cover a wide topical spectrum, and a clear picture emerges of what can be done, at present, in EEC countries and what results can be achieved both from the point of view of energy savings and of cost to benefit ratios.

The results show that several building design and construction techniques and related technologies for energy savings are now mature and applicable on a wide scale to EEC countries.

I would like to summarize this first part of my concluding remarks by stressing that in the sector of building envelope the know-how related to energy savings has shown substantial progress; a large amount of results, data and indications are now available to potential users and particularly to planners, which will enable them to make technically sound decisions and to start bringing into practice, on a wide scale, the know-how provided by EEC R, D and D projects.

However, I feel that there is a wide gap between what we know and what is used and applied practically and I would like to concentrate the second part of my concluding remarks on this topic.

There is a strong need to transfer the vast knowledge already acquired to the users, architects, engineers, building companies, component manufacturers and to the social, political and economical planners.

There is an innovation transfer barrier which poses important problems which of course cannot be solved or even discussed in sufficient detail here, but which we should bear in mind for future actions.

For example I would strongly recommend actions defining and disseminating in the various EEC countries building and retrofitting codes, homogenous and compatible with each other.

User oriented seminars and workshops can be of great help planners and decision makers; they should combine with easy accessible data basis. access.

Another fact should be borne in mind. Generally, the vast majority of potential users are unable to reach and utilize research results. Neither can they identify all the possible fallouts, particularly from R and D projects, unless the users are large organisations with their own team of researchers and innovation specialists.

For small and medium size operating companies and manufacturers results ought be showed to them. It is probably the duty of the research planner to identify potential result users and to convey the relevant results to them.

Finally I would like to conclude with two final comments :

One is related to how R and D results are presented. It would be of great help to define and use a common format in presentation and to adopt identical methodology for the technical and economic evaluations. This indeed would avoid misunderstandings on the reader's side.

Such formats and methodologies can be defined in a more or less stringent manner depending on the subject, but my feeling is that a fruitful effort in homogenization can be performed anyhow on R and D addressing the building envelope.

The second remark can be summed up as follows:

Indeed the building envelope plays a major role in the overall energy balance of the building proper. However, we should also be aware that it is only one part of a system and in no way can it be regarded as a single component.

Here at this seminar the organizers had to split, for technical reasons, envelope and plant (improved conventional heating system), but their performances should always be considered coupled and depending on climate and user's habits as well.

INTEGRATION OF DIFFERENT ENERGY SAVING POSSIBILITIES IN DWELLINGS

M. GUILLAUME
Belgian Building Research Institute
W.T.C.B.-C.S.T.C.
Rue du Lombard 41
1000 BRUXELLES

Summary

Overall efficiency of two heating plantswas measured in two well insulated and identicals houses. The main particularities of these heating installation are
. for the first one : radiant floor heating associated to a high efficiency and low water temperature boiler
. for the second one : direct electrical heating with convectors equiped by electronical thermostat.
It was possible to show the influence of the regulation modes upon the overall efficiencies obtained.
So, in the first house (heated with a radiant floor) we made a reduction of 9 % of the energy consumption of the boiler by using a closed regulation circuit instead of an open one, and a reduction of 11 % of the energy consumption with a night set back from 10 pm to 6 am.
The results also show that the values obtained for the electrical heating system are higher than for the classical system even when this one used low water temperatures.

0.1. Research objectives

The objectives were the following :
- to reduce by all known means which are economically feasible the net energy needs for heating buildings,
- to study and develop various energy emission systems in rooms in order to make an optimal choix of system depending on the energy or heat source,
- to study and develop heating installations with a high seasonal performance and mixed installations with solar collectors ≠ heat pumps ≠ classical heat production. For new systems based on energy collection or transformation, the study begins with the systems requiring the lowest investment
- to study the energy consumption with various combinations, over several climatic periods, and, of course, the investment necessary for each combination.

In order to achieve these objectifs, we constructed two experimental houses, designed to have very low energy needs indeed.

These two low energy houses be with several heating systems (designed as well as possible given our present state of knowledge) in order to make reliable choices between the different systems and in order to increase as far as possible the operating efficiency of the systems or subsystems.

0.2. Symbols

E.C.	: heating energy consumption	(kWh)
dr	: duration of the heating period	(days)

J	:	relation between the overall mean radiation on a horizontal plane and the same radiation under clear sky conditions	(-)
K	:	overall insulation level in accordance with NBN B 62-301	(-)
k_s	:	overall thermal transmission coefficient of the building's heat-loss envelope in accordance with NBN B 62-301	(W/m^2K)
R	:	solar gain factor under clear sky conditions	(°C)
S	:	heat-loss surface area of the building	(m^2)
T_{output}	:	initial temperature of the water leaving the boiler	(°C)
T_{floor}	:	temperature of the upper surface of the floor slab	(°C)
V_B	:	building volume delimited by the heat-loss area A	(m^3)
V	:	habitable volume (including internal partitions)	(m^3)
β	:	air-change rate	(h^{-1})
θ_{ext}	:	external temperature	(°C)
θ_{sc}	:	internal temperature without heating	(°C)
θ_{nc}	:	temperature of no more heating	(°C)
θ_{im}	:	mean internal temperature taking into account the heated volumes and night set-back	(°C)
θ_{iL}	:	internal temperature in the living room	(°C)
η_{gs}	:	seasonal overall efficiency of the heating installation	(-)

1. DESCRIPTION OF THE HOUSE

1.1. General characteristics

These houses which were designed by the architect Professor Dupagne of Liège University, have the following main geometrical characteristics :
- net floor area 107.6 m^2
- cellar volume 212 m^3
- habitable volume (including internal partitions) $V = 227\ m^3$
- glazed area 24 % of ground plan area.

The main characteristics of the house are illustrated in figures 1.1. to 1.2.

1.2. Thermal characteristics

1.2.1. Insulation level - heat losses

Figures 1.1. to 1.3. indicate not only the general characteristics, but also the thickness and location of the insulation.

In addition, the tables in figure 1.4. indicate :
- that the insulation level of the heat-loss envelope (calculated in accordance with NBN B 62-301) is K49 which corresponds to a V_B/S ratio of 0.94 m and a k_s coefficient of 0.49 $W/m^2 °K$,
- that the heat losses (calculated in accordance with DIN 4701) are 6964 W for an external temperature of -10 °C and a mean internal temperature of 19.9 °C.

The house is also characterized by the absence of cold bridges as can be seen from the illustrations of details in figure 1.3.

1.2.2. Real natural ventilation rates

The house, which remained unoccupied during the heating season, is naturally ventilated by the wind and by the stack effect, at the places which are not airtight, due to the internal heating.

The real air-change rates were determined using the tracer gas method and are represented on the graph in figure 1.4. for different wind directions and speeds.

1.2.3. Temperature without heating

The solar gain factor R, i.e. the difference ($\theta_{sc} - \theta_{ext}$) under clear sky conditions, was calculated for different air-change rates and is plotted on the graph in figure 1.5. From among the assumptions on which the calculation were based, we cite below the two main ones :

- insolation under clear sky conditions (*) on the 15th of each month,
- the heat-loss area is defined in terms of the external dimensions of the house in accordance with NBN B62-301.

The temperature without heating θ_{sc} is calculated, under real conditions, from equation (1) (**) :

$$\theta_{sc} = (\theta ext)_m + RJ \qquad (°C) \qquad (1)$$

2. DESCRIPTION OF THE HEATING SYSTEM

We attempted to achieve one of the aims of the research, which was to obtain optimum performances for the installation with the simplest possible heating systems.

2.1. House 1

For the first series of measurements we decided to use the following principal components :

. a boiler fitted with a pump-driven oil-fired burner, capable of producing water at very low temperature (20-50 °C) without the need for a mixing valve; furthermore the boiler had a nominal efficiency of nearly 90 %,
. heat emission by floor radiation (see cross-section in figure 1.2.),
. a regulation system acting directly on the burner and making it possible to operate either :
  + open-circuit mode with an external sensor and a water sensor, or
  + in closed-circuit mode with external and ambient sensors and a water sensor
  each of these two modes can be used with night temperature set-back,
. a pump operating permanently during the heating period and cut off during the night set-back.

2.2. House 2

During the first series of measurments we used an electrical heating installation consisting of direct convectors (without storage) fitted with electronic thermostats and timers making it possible to program the set-back periods.

(*) See bibliography 1
(**) See bibliography 2

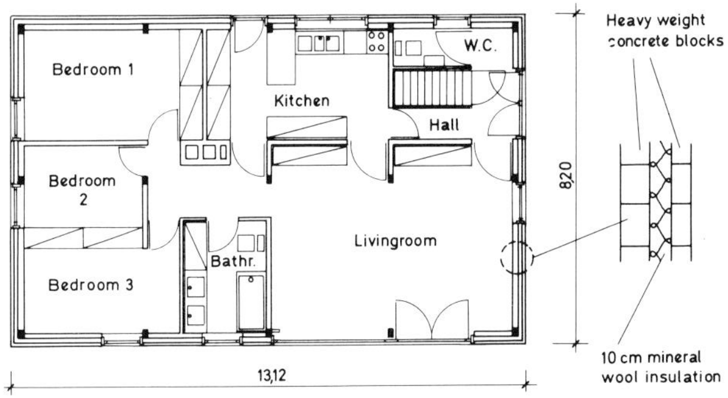

Figure 1.1.- Ground-floor

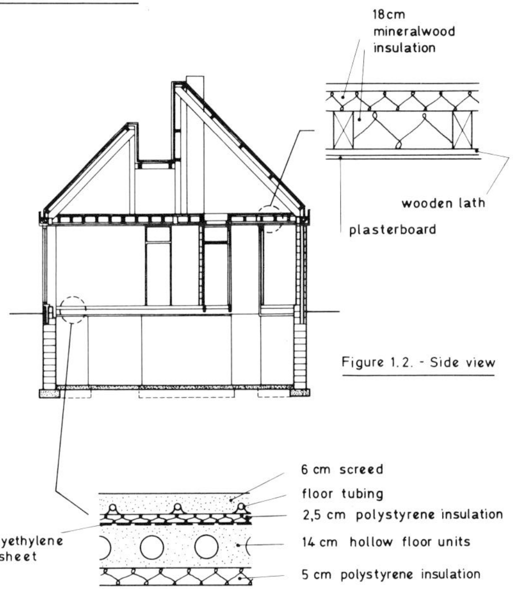

Figure 1.2. - Side view

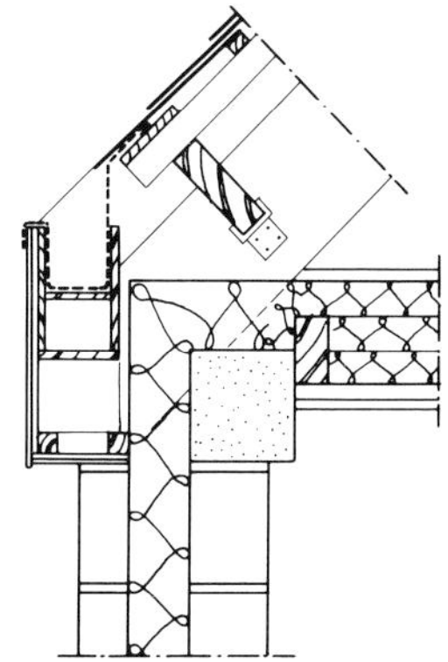

Wall - Attice floor joint

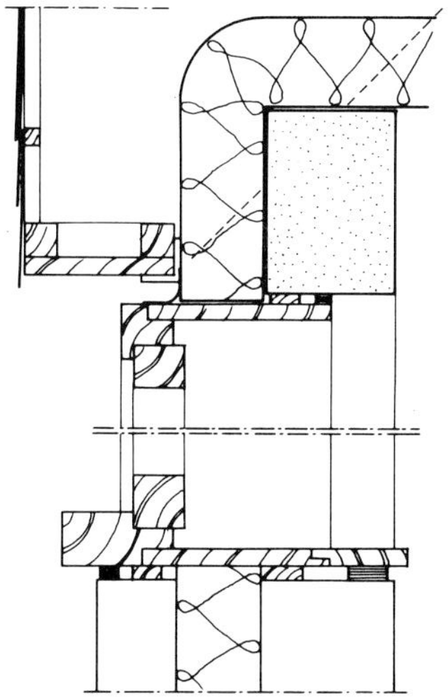

Window frame - masonry joint

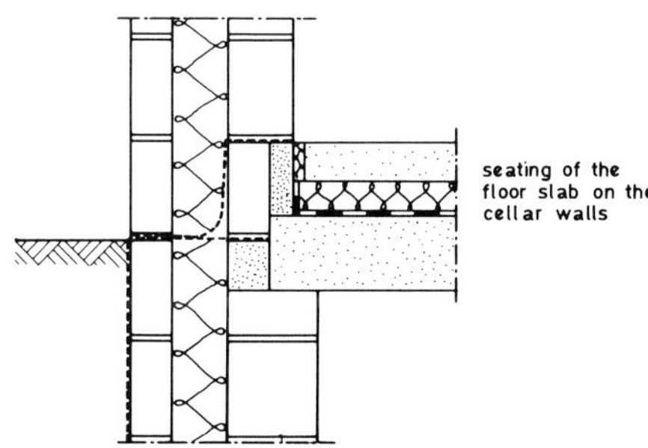

Figure 1.3. - Details

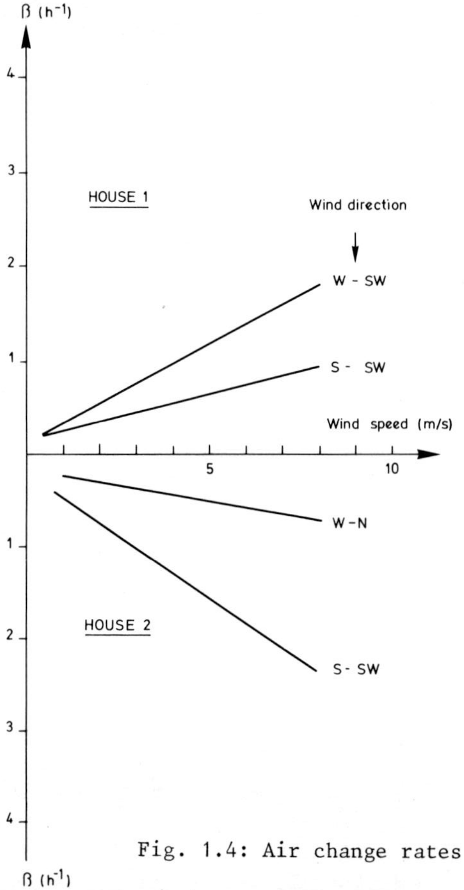

Fig. 1.4: Air change rates

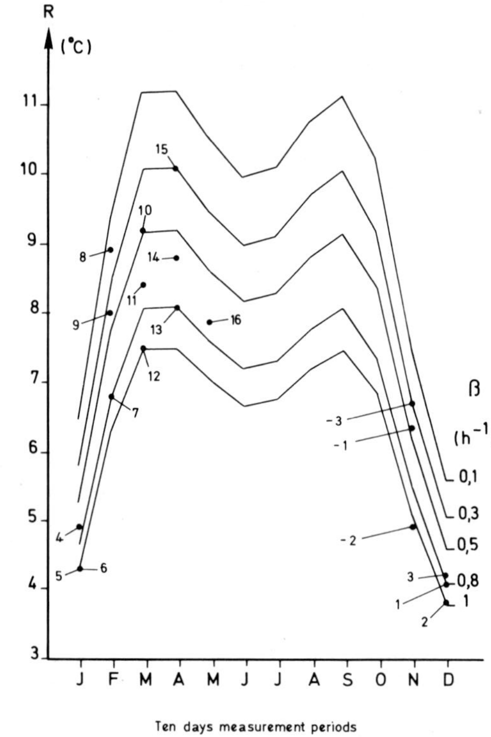

Fig. 1.5: Solar gain factor under clear sky conditions for different months of the year and different air-change rates

3. OVERALL EFFICIENCY OF AN INSTALLATION

The definition of the overall efficiency of an installation is given in equation (2) below (*) :

$$E.C. = \frac{24 \times dr \times (k_sS + 0.36\ \beta V) \times (\theta nc - \theta sc)}{1000\ \eta_{gs}} \quad (kWh/dr\ days) \qquad (2)$$

For the results of the research, the houses being unoccupied and therefore not receiving any free energy of internal origin, the temperature of no heating t_{nc} was taken to be the same as the mean internal temperature θ_{im} which was measured.

Equation (3) therefore becomes :

$$E.C. = \frac{24 \times dr \times (k_sS + 0.36\ \beta V) \times (\theta_{im} - \theta sc)}{1000\ \eta_{gs}} \quad (kW/dr\ days) \qquad (3)$$

Equation (3) can also be expressed in a form which facilitates the plotting of the results on a graph :

$$\frac{E.C.}{dr \times (\theta_{im} - \theta_{sc})} = \frac{24 \times (k_sS + 0.36\ \beta V)}{1000\ \eta_{gs}} \qquad (4)$$

Ten days (**) was taken as the duration of the reference period.

One can therefore plot, the daily consumption CE/dr against the difference $(\theta_{im} - \theta_{sc})$ for each regulation mode, and one obtains a line whose slope (tgα) is a measurement of the overall efficiency corresponding to the regulation mode studied. Thus we obtain :

$$tg\alpha = \frac{24\ (k_sS + 0.36\ \beta V)}{1000\ \eta_{gs}} \quad (kWh/day\ °C) \qquad (5)$$

4. RESULTS OBTAINED

4.1. House 1

4.1.1. Regulation modes

Open-circuit regulation

This mode involves the regulation of the output temperature of the water (leaving the boiler) as a function of the external temperature and in accordance with a load curve displayed on the regulator.

The resultats obtained are given in table 1 and are plotted on the graph in figure 4.1.

Table 1

Ten day period	Mean insolation on a horiz. surface (a)	Clear sky (b)	J (a)/(b)	β	R	θ_{ext}	θ_{sc}	θ_{im}	$\theta_{im}-\theta_{sc}$	CE/dr
n°	W/m²		-	h^{-1}	°C	°C	°C	°C	°K	kWh/j
1	24.1	36.5	0.66	0.8	4.1	6.7	9.4	20.2	10.8	81
2	24.1	41.6	0.58	1	3.8	3.1	5.3	19.9	14.6	121
3	37	40.3	0.67	0.7	4.2	2.5	5.3	20.3	15	119
4	21.7	44.3	0.49	0.7	4.9	6.1	8.5	20.2	11.7	100
5	17.3	50.9	0.34	1.1	4.3	4.8	6.3	19.7	13.4	107
(k_sS + 0.36 βV) average over 5 10-day periods									238.3 W/°K	

(*) See bibliography (2) and (3)

(**) See bibliography (3)

Closed-circuit regulation

This mode involves the determination of the maximum temperature of the water coming out of the boiler as a function of the external temperature (with a load curve displayed on the regulator) and regulation of the internal temperature (of the living room) with an ambient sensor (room-thermostat).

The results obtained are given in table II and plotted on a graph in figure 4.1.

Table II

Ten day period	Mean insolation on a horiz. surface (a)	Clear sky (b)	J (a)/(b)	β	R	θ_{ext}	θ_{sc}	θ_{im}	$\theta_{im}-\theta_{sc}$	CE/dr
n°	W/m²		–	h^{-1}	°C	°C	°C	°C	°K	kWh/j
6	24.8	62	0.40	1	4.3	5.5	7.2	18.9	11.7	83
7	36.1	76.8	0.47	0.8	6.8	1.7	4.9	18.9	14	105
8	69.7	95.5	0.73	0.2	8.9	-2.4	4.1	19.1	15	108
9	63	116.7	0.54	0.4	8	3.6	7.9	19.2	11.3	89
(k_sS + 0,36 βV) average over 4 10-day periods									217 W/°K	

Closed-circuit regulation with night set-back

In order to take into account the thermal inertia of the floor slab, the set-back period finishes at 3.30 a.m. and starts again at 7.30 p.m. the following evening.

The resultats obtained are given in table III and are plotted on the graph in figure 4.1.

Table III

Ten day period	Mean insolation on a horizon. surface (a)	Clear sky (b)	J (a)/(b)	β	R	θ_{ext}	θ_{sc}	θ_{im}	$\theta_{im}-\theta_{sc}$	CE/dr
n°	W/m²			–	h^{-1}	°C	°C	°C	°K	kWh/j
10	75.7	132.8	0.57	0.5	9.2	5.4	10.6	18.7	8.1	54
11	75.4	157.1	0.48	0.7	8.4	7.1	11.1	18.5	7.4	49
12	94.7	182.1	0.52	1	7.5	4.9	8.8	18.3	9.5	63
13	90.9	211.4	0.43	0.8	8.1	6	9.5	18.2	8.7	56
14	133.9	257.5	0.52	0.6	8.8	9.4	14	19	4	35
15	143	255.4	0.56	0.3	10.1	10.7	16.3	19.2	2.9	20
16	144	276.9	0.52	0.7	7.9	10.9	15	19.1	4.1	27
(k_sS + 0.36 βV) average over 7 10-day periods									222 W/°K	

4.1.2. The efficiencies measured

The regression lines obtained and illustrated in figure 4.1. are expressed in equations (6), (7) and (8).

Open circuit

$$\frac{EC}{dr} = 8.1\ (\Theta_{im} - \Theta_{sc}) - 0.4 \text{ kWh/day} \qquad (6)$$

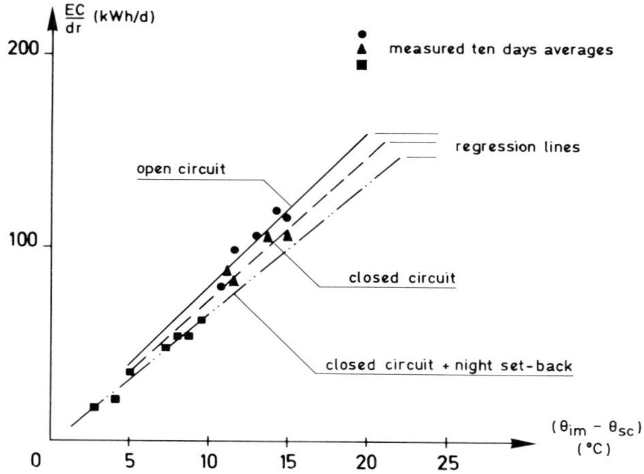

Fig. 4.1: House 1

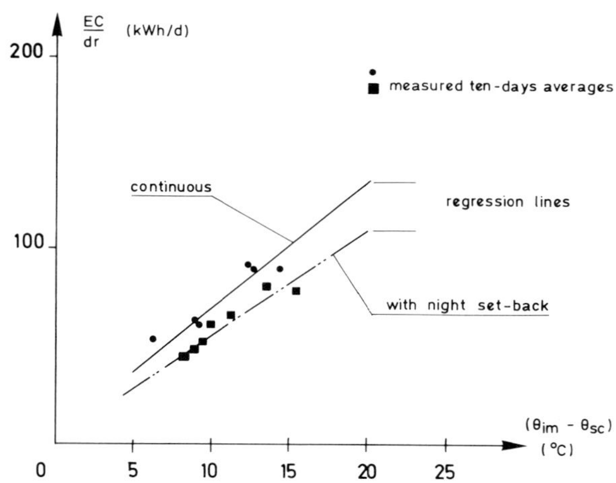

Fig. 4.1: House 2

which gives an overall seasonal efficiency η_{gs} of 70.6 %

Closed circuit

$$\frac{EC}{dr} = 7.35\ (\theta_{im} - \theta_{sc}) + 0.63\ \text{kWh/day} \qquad (7)$$

which gives an overall seasonal efficiency η_{gs} of 70.8 %

Closed circuit with night set-back

$$\frac{EC}{dr} = 6.53\ (\theta_{im} - \theta_{sc}) + 0.64\ \text{kWh/day} \qquad (8)$$

which gives an overall seasonal efficiency η_{gs} of 81.6 %

4.1.3. Comparative efficiency

The efficiencies measured above are only valid for the mean values of the term $K_sS + 0.36\ \beta V$ given in tables I to III.

However, in order to be able to compare all the results one must determine the overall seasonal efficiency of the installation for a value of 209 W/°K for the term $K_sS + 0.36\ \beta V$, which is that obtained with an air change rate, assumed to be constant over the season, of 0.5 h^{-1}.

One then obtains

. with open circuit $\eta_{gs} = 61.9\ \%$ (9)

. with closed circuit $\eta_{gs} = 68.2\ \%$ (10)

. with closed circuit and night setbacks $\eta_{gs} = 76.8\ \%$ (11)

Taking the open circuit regulation system as a reference, one finds by using equations (9), (10) and (11) that :

. the closed circuit regulation mode makes it possible to obtain a 9.2 % energy saving

. the closed circuit regulation mode with night setback enables one to obtain a 19.4 % energy saving.

It can also be noted that the night setback provides an additional 11.2 % energy saving compared to continuous heating with closed-circuit regulation.

4.2. House 2

4.2.1. Regulation modes

The results obtained are plotted in figure 4.2. and given in

. table IV for continuous heating

. table V for heating with night setback

Table IV

Ten day period	Mean insolation on a horiz. surface (a)	Clear sky (b)	(J) (a)/(b)	β	R	θ_{ext}	θ_{sc}	θ_{im}	$\theta_{im}-\theta_{sc}$	CE/dr
n°	W/m2		–	h^{-1}	°C	°C	°C	°C	°K	kWh/j
-3	40.1	20.2	0.5	0.3	6.8	9.8	13.2	19.5	6.3	54.3
-2	40.3	65	0.62	1.3	5	6.9	10	19.2	9.2	62
-1	37.7	53.9	0.7	0.4	6.4	5.8	10.3	19.2	8.9	63.6
1	24.1	36.5	0.66	0.4	4.8	4.3	6.9	19.4	12.5	94
2	24.1	41.6	0.58	0.4	4.8	3.1	6	18.8	12.8	92.2
3	27	40.3	0.67	1	3.8	2.45	5	19.5	14.5	91.8
($k_sS + 0,36\ \beta V$) average over 6 10-days periods									219.6 W/°K	

Table V

Ten day period	Mean insolation on a horiz. surface (a)	Clear sky (b)	(J) (a)/(b)	β	R	θ_{ext}	θ_{sc}	θ_{im}	$\theta_{im}-\theta_{sc}$	CE/dr
n°	W/m²		-	h^{-1}	°C	°C	°C	°C	°K	kWh/j
7	36.1	76.8	0.47	0.4	8	1.7	5.5	19.1	13.6	82.5
8	69.7	95.5	0.73	0.3	8.4	-2.4	3.7	19.2	15.5	81
9	63	116.7	0.54	0.3	8.4	3.6	8.1	19.3	11.2	66.1
10	75.7	132.8	0.57	0.4	9.6	5.4	10.9	19.9	9	48.9
11	75.4	157.1	0.48	0.6	8.8	7.1	11.3	19.6	8.3	44.8
12	94.7	182.1	0.52	1	7.5	4.9	8.8	18.4	10.1	62.5
13	90.9	211.4	0.43	0.7	8.4	6	9.6	19	9.4	52.4
(k_sS + 0,36 βV) average over 7 10-days periods									212,2 W/°K	

4.2.2. The measured efficiencies

The regression lines obtained and illustrated in figure 4.2. are expressed by equations (12) and (13)

. continuous heating

$$\frac{EC}{dr} = 6.7\ (\theta_{im} - \theta_{sc}) - 0.4 \text{ kWh/day} \qquad (12)$$

which gives an overall seasonal efficiency η_{gs} of 78.7 %

. heating with night setback

$$\frac{EC}{dr} = 5.62\ (\theta_{im} - \theta_{sc}) + 0.65 \text{ kW} \qquad (13)$$

which gives an overall seasonal efficiency η_{gs} of 90.9 %

4.2.3. Comparative performances

These are calculated as described in paragraph 4.1.3. above.

One obtains

. continuous heating η_{gs} = 74.9 % (14)

. heating with night setback η_{gs} = 89.2 % (15)

One finds, from equations (14) and (15), that the night setback improves the overall efficiency of the heating installation by 14.3 per cent (from 74.9 % to 89.2 %).

This improved efficiency would result in a 16 % reduction in the energy consumed for heating.

5. CONCLUSION

The results obtained show, when they are expressed in the form of overall seasonal efficiencies, that :

. direct electrical heating has a better efficiency than the traditional type of system (hot water) used even if the latter uses a radiant heat emitter (floor slab) with a high emission efficiency;

. in traditional heating systems installed in buildings with very low energy needs, an adaption of the regulation mode can considerably reduce the energy consumption;
. whatever the energy and installation used night setback produces an energy saving of between 11 and 19 % depending on the reference situation.

6. REFERENCES

(1) Eclairement énergétique par ciel serein des parois orientées et inclinées. Données d'application pour la Belgique.
R. Dogniaux, I.R.M., Publication Miscellanea Serie B - n° R 25 - 1973.
(2) Besoins en énergie pour le chauffage des bâtiments.
P. Caluwaerts, M. Dethier, H. Heikhaus, J. Uyttenbroeck.
Avril 1982 - C.S.T.C.
(3) Equipement - Systèmes de chauffage.
M. Guillaume.
Information des estimateurs n° 12 - Région wallonne
(4) Mesure de la consommation d'énergie dans dix maisons neuves non habitées.
J. Uyttenbroeck, H. Heikhaus.
C.S.T.C. - Revue n° 1, mars 1978.

BASIC CHARACTERISTICS OF HOUSES CONSIDERED WITH A VIEW TO DECREASING ENERGY CONSUMPTION FOR HEATING

J. UYTTENBROECK
Belgian Building Research Institute
WTCB - CSTC
Lombardstraat 41
1000 Brussel

Summary

Since 1976 the Belgian Building Research Institute has undertaken 3 research programms, each of two years. The two latest programms have been realised with the aid of the EEC. The main objective of this important research was to study the structural characteristics of social houses in order to decrease the net energy requirements for heating (passive solar energy recuperation).
Secondly it has been proposed that sufficiently simple calculation methods be developed in order to obtain, at the design stage, a valid evaluation of the net energy demand for heating of houses. The research was completed at the end of 1981 and the final report was published by the Commission of the European Communities in 1983 (1). A summary of the results obtained during the last two year period of the research (1980-1981) was presented during the Contractors Meeting held in Brussels on the 7th of May 1982 (2). The introduction of the present paper summarises the global research over the six years. In the second part special attention is given to some of the more practical results obtained thanks to this important investigation.

1. INTRODUCTION

In a study carried out at the request of the European Commission (DG XVII) and dated 1975 it was already suggested that one of the most important parameters for the energy consumption of the heating plant in a house or a building is the so-called temperature without heating t_{wh}. (This t_{wh} is the equilibrium temperature in an unheated and unoccupied house exposed to the outdoor climate).

In a Belgian Research Programme (1976-1977), called Seneffe I, this hypothesis was confirmed. In this work in situ measurements were carried out in ten new unoccupied identical houses. The first aim of this research was to measure the influence on the energy consumption of the houses due to several improvements in the thermal insulation.

But the most important result of this two year programme was that it was found indeed that the energy needs in the heated houses (temperature t_i) were proportionnal to the temperature difference between the heated houses and the unheated houses (t_{wh}).

In a second two-year programme executed with the help of the Belgian government and of the European Commission, the relationship between the energy consumption and the difference (t_i - t_{wh}) was monitored in a more correct and accurate way.

Twelve models (each of 1 m^3) were exposed in the C.S.T.C.-laboratories to the outdoor climate during two years. These twelve models consisted of 6 pairs of two identical models. In each pair one model was electrically heated at a given constant temperature, in the second model of each pair only the temperature without heating was measured. The six pairs differed from each other in insulation, thermal capacity, airtightness, window orientation. One pair was completely opaque.
A standard model was tested under different configurations in a wind tunnel at the University of Liege.

Thanks to the results obtained it has been clearly shown that the difference (t_i - t_{wh}) is indeed the fundamental parameter for energy consumption.

At the same time we had the opportunity to propose and to verify a simplified calculation method for the temperature without heating of the models. This verification was based on the precisely known characteristics of the models and on the measured climatic parameters (sunshine, outdoor temperature and wind were measured every ten minutes during daytime and every 30 minutes during the night).

The simplified method was tested in the third two year programme (1980-1981) (Seneffe III) on the two real houses of the BRE (United Kingdom) with reasonable accuracy. (The maximal difference between the calculated and measured temperatures without heating in the two houses over two periods of each 10 days was 1.1K).

This third two year programme also gave us the opportunity to study in more detail some specific points (heat loss through the walls in contact with the ground - simplified dynamic mathematical models - and extreme climatic conditions).

2. PRACTICAL RESULTS OBTAINED

2.1. General considerations

2.1.1. Temperature without heating of a building t_{wh}

The temperature without heating of a building during a given period (one month for example) is the mean equilibrium indoor temperature over the same period in the unheated and unoccupied building.

It has been stated (since 1977) that this temperature without heating t_{wh} is probably one of the most important parameters influencing the net energy needs of a building for heating purposes.

The temperature without heating t_{wh} will rise, due to solar radiation incident on the opaque and the glazed elements of the building elements, to a higher level than the mean outdoor temperature t_{em} over the same period

$$t_{wh} > t_{em}$$

2.1.2. Temperature of no more heating t_{NH}

In an occupied building normally heated at a mean indoor temperature t_{im} (over the given period) a fraction of the necessary heat emission will be provided by the internal occupation heat sources (metabolism, cooking, lighting, domestic hot water, T.V. etc.).

The heating plant has, in these conditions, only to assure an indoor temperature which will be situated at a slightly lower level than the desired comfort temperature t_{im}.

This temperature is called the temperature of no more heating t_{NH}.

$$t_{NH} < t_{im}$$

2.1.3. Heat losses and the net energy needs of a building

- The heat losses of a building over a given period of d days are given by

$$\text{H.L.} = \frac{24 \times d\ (U_s \cdot A_{tot} + 0{,}36\beta V)(t_{im} - t_{em})}{1000} \quad \text{kWh} \qquad (a)$$

In this expression :

U_s : is the mean global transmission coefficient of the heat loss envelope W/m^2K

A_{tot} : is the total area (m^2) of the heat loss envelope

β : is the infiltration or ventilation rate (h^{-1})

V : is the volume of the building delimited by the heat loss envelope

0,36 : is the specific heat of the air (W/m^3K) divided by 3.600 s/h

t_{im} : the mean indoor comfort temperature in the building over the period considered (°C)

t_{em} : the mean outdoor air temperature over the same period (°C)

- The Net Energy needs of the building over the given period are obtained from :

$$\text{NEN} = \frac{24 \times d\ (U_s \cdot A_{tot} + 0{,}36\ \beta V)\ (t_{NH} - t_{WH})}{1000} \quad \text{kWh} \qquad (b)$$

where

t_{NH} : the temperature of no more heating

t_{WH} : the temperature without heating

2.1.4. Equivalent degree days

In expression (a) the product $d(t_{im} - t_{em})$ can be considered as the classical "degree days". These degree days only depend on the mean indoor temperature and on the mean outdoor temperature.

In expression (b) the product $d(t_{NH} - t_{WH})$ is defined as the equivalent degree days or as the energy consumption degree days.

These degree days depend on the outdoor climate (temperature + sun) on indoor conditions (desired comfort and internal sources) but also on the building characteristics.

As can be easily understood the equivalent energy "degree days" notion is a more integrated and complete indication for the net energy needs of a building than the temperature without heating (see formula b).

2.1.5. Symposium on temperature without heating

In May 1979 an international symposium was organised by the CSTC and the University of Liege (3).

2.2. The net energy needs of a building

2.2.1. General formulae

The temperature without heating of a house, considered as one volume can, for a given period (10 days or 1 month), be written as :

$$\boxed{t_{wh} = t_{em} + R_c.J} \quad °C \quad (c)$$

where :

t_{em} : the mean outdoor temperature over the period considered (°C)

J : the ratio between the real total solar radiation incident on a horizontal surface during the given period (I'_{Th} W/m^2) and the solar radiation on a horizontal surface for the same period under clear sky conditions (I_{Th} W/m^2)

$$\boxed{J = I'_{Th}/I_{Th}} \quad (d)$$

R_c : is the corrected recuperation coefficient (K)

2.2.2. Corrected Recuperation coefficient R_c

The value of R_c can be computed from :

$$\boxed{R_c = \frac{\Sigma\ A_{ei} \cdot U_{ei}}{(U_s.A_{tot} + 0.36\beta V)} \left[\left(\frac{\alpha_{ei}}{h_e} + \frac{S_i}{U_{wi}} \right) \frac{I'_{Ti}}{I'_{Th}} \cdot I_{Th} - \frac{\varepsilon_i}{h_e} \cdot I_{IRi} \right]} \quad K \quad (e)$$

In this expression :

- the index i indicates the different orientations or slopes of the elements of the heat loss envelope
- A_{ei} : are the areas (m^2) of all the elements of the heat loss envelope which are affected by radiation (from the sun or the sky)
- U_{ei} : are the classical U values of these elements W/m^{2}K
- U_s : is the mean U value of the heat loss envelope (W/m^{2}K)
- A_{tot} : is the total area (m^2) of the heat loss envelope
- $0.36\beta V$: are the infiltration or ventilation losses per K (W/K)
- α_{ei} : are the absorption factors for solar radiation on the outside surfaces of the external elements
- h_e : is the normal heat exchange coefficient on the outside of a building (h_e = 23 W/m^{2}K)
- S_i : is the solar factor of the windows
- U_{wi} : is the surface area of the windows (m^2)
- I'_{Ti}; I'_{Th} are the real solar radiation in (W/m^2) incident on vertical or sloped surfaces (i) and on a horizontal surface (h)
- I_{Th} is the real solar radiation on a horizontal surface under clear sky conditions (W/m^2)

- ε_i : is the emission factor of the outside surfaces of the external elements for infra-red emission to the sky
- I_{IRi} : is the infra-red radiation of these surfaces to the clear sky (W/m^2).

2.2.3. Values of I'_{Ti} and I'_{Th}

The expression (e) contains a number of well-known building characteristics and of well-known solar data (I_{Th} and I_{IR} under clear sky). The ratios I'_{Ti}/I'_{Th} had to be defined).

This research has given us the opportunity to measure (at Limelette), since April 1979, the daily values of I'_{TN}; I'_{TE}; I'_{TS}; I'_{TW} and I'_{Th}.

Thanks to these measurements we are able to calculate, for each month, the mean values of I'_{Ti}/I'_{Th}.

Graph 1 indicates the monthly evolution of I'_{Ti}/I'_{Th}. These values can be used for Belgium.

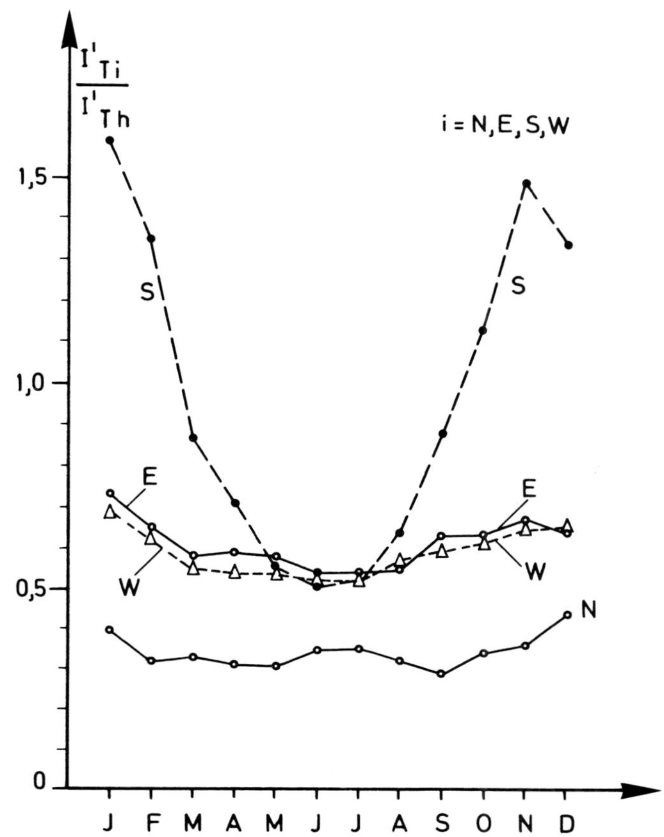

Fig. 1

2.2.4. Net Energy needs - Proposed methodology

The proposed method is a simple steady state one applied to time periods of sufficiant length (10 days or even 1 month). The house or building is considered as a single heated volume at a mean temperature t_{im}.

For each time-period the corrected recuperation coefficient R_c is computed (formula e).

One might consider that it is then very easy to compute, for each period, the temperature without heating t_{wh} (formula c) and further, from formula b, the net energy needs of the building.

However, by doing so one does not obtain realistic results. Indeed given the fact that a steady state method is used over rather long periods, all the available solar energy is distributed over the whole period. One can easily understand that in this case overheating will never appear and thus that all the solar energy is considered as useful. This is not the case; especially on sunny days (spring or autumn) a lot of solar energy will only lead to undesired overheating.

This problem can be solved in several ways. The method used by the research team of Liege University (1) is based on the knowledge of the equivalent degree days.

Thanks to a statistical study of the values of the daily mean outdoor temperature and the daily J value over a period of 18 years (made for the meteorological station of Uccle) it has been possible to calculate the equivalent "degree days" $d(t_{NH} - t_{wh})$ for several values of t_{NH}, for each month and for different values of R_c (R_c = 0; 2; 4; 6; 8; 10K).

Table I indicates the equivalent degree days obtained for Uccle for two values of t_{NH} (18 °C and 17 °C).

Table I : Equivalent degree days $d(t_{NH}-t_{wh})$ Ukkel (1958-1975)

t_{NH}	Month	R_c= 0	R_c=2 K	R_c=4 K	R_c=6 K	R_c=8 K	R_c=10 K
18 °C	sept.	94,7	62,9	42,4	29	19,6	14
	okt.	214,7	177	143	117,8	94,0	75,8
	nov.	350,6	319	289,3	259,8	231,9	204,8
	dec.	451,9	418,8	388,7	357	326,9	296,1
	jan.	461,1	427,4	393,7	360,7	328,5	296,6
	feb.	395,8	362,4	331,2	301,0	270,8	241,2
	march	377,3	339,75	301,6	269,7	238	206,1
	april	265,4	229,3	197,4	168,1	142,3	120,1
	mai	152,1	118	90,6	68,6	52,2	38,6
	june	72,2	49,6	33,4	22,7	16	12
17 °C	sept.	70,2	43,4	28,2	18,3	12,3	9
	okt.	184	148,3	116,7	94,5	73,6	58,4
	nov.	320,6	289,1	259,55	230,3	203,5	177,8
	dec.	420,9	387,8	357,7	326	296,1	265,7
	jan.	430,1	396,4	362,7	329,7	297,8	266,5
	feb.	367,8	334,4	303,2	273,4	243,6	214,9
	march	346,3	308,9	271,2	240,0	209,3	178,8
	april	235,8	200,6	169,9	142,8	119,2	99,6
	mai	124,3	93,5	69,6	51,1	37,9	27,4
	june	52,2	34,1	22	14,5	9,9	7,5

The calculation method can thus be summarised as follows :

1° For each month one calculates the corrected recuperation coefficient R_c (formula e)

2° On the basis of these R_c values and of the t_{NH}-values, table I immediatly provides the equivalent degree days for each month : $d(t_{NH} - t_{wh})$

3° The net energy needs for the season are then

$$\boxed{NEN = \frac{24\ (U_s \cdot A_{tot} + 0.36\ \beta V)}{1000}\ \Sigma\ d(t_{NH} - t_{wh})} \quad \frac{kWh}{seas.} \qquad (f)$$

2.3. Draft Thermal regulations in Belgium

For the French speaking part of Belgium a draft of thermal regulations for new houses and dwellings was prepared at the beginning of 1983 by a working group, chaired by Prof. Meunier (Mons University). The draft is based on the results of the CEE-research work of the CSTC and the University of Liege.

But in order to obtain a reasonable level of simplicity some approximations were adopted :

- It is assumed that solar energy penetrates in the building only via the windows.
- The corrected value of R_c is computed only for the months of March and December. The values of R_c, for the other winter months are obtained by using a sinusoidal law for R_c based on the "R_c-March" and "R_c-December" values.
- t_{wh} and then $d(t_{NH} - t_{wh})$ are calculated for each winter month.
- The problem of overheating on sunny days is solved by using the notion of solar recuperation efficiency (4) which depends among other factors on the thermal inertia of the building.

Thanks to this method the requirements in the draft will be expressed as a maximum for the net energy needs for the heating season.

2.4. Calculation of the I'_{Ti}/I'_{Th} ratio for Belgium

In § 2.2.3. we gave the results of measurements of I'_{Ti} and I'_{Th} during more than 4 years. These measurements (daily values) also provided the opportunity to check a more theoretical proposal made earlier by J. Lebrun (1).

Thanks to this theoretical proposal it will be possible to obtain a valid daily value of I'_{Ti}/I'_{Th} as a function of the daily J value and of the orientation.

In a diagram in which I'_{Ti}/I'_{Th} is plotted against J the relationship obtained is indicated by broken lines passing through the following points

$I'_{Ti}/I'_{Th} = 0.6$ for $J = 0$

$I'_{Ti}/I'_{Th} = 0.6$ for $J = 0.35$ (covered sky value)

$I'_{Ti}/I'_{Th} = \frac{I_{Ti}}{I_{Th}}$ for $J = 1$ (clear sky value)

But for a north orientation it seems that the theoretical law should be :

$I'_{TN}/I'_{Th} \simeq 0.45$ for $J = 0$

$I'_{TN}/I'_{Th} \simeq 0.45$ for $J = 0.35$

$I'_{TN}/I'_{Th} = \frac{I_{TN}}{I_{Th}}$ for $J = 1$

N.B. This study is still going on (ULg)

2.5. Information seminar for Belgium architects

During spring 1982 the C.S.T.C. organised a one-day seminar for the Belgian architects and consultants about the practical evaluation method for net energy needs (in French and in Dutch) (5).

It should be stated that the method proposed at that time was a first draft which has been improved since by using the corrected R_c-coefficient.

2.6. Outdoor temperature for heat loss calculations

Thanks to a statistical study of the outdoor climate in Uccle, between 1958 and 1975 by M. Laret (1) and to a comparison of the results obtained with climatic data from other meteorological stations (6) it was possible to obtain, for Belgium, a more correct evaluation of the basic outdoor temperature for heat loss calculations.

The result of this study indicates that for normal cases it is not necessary to take into account the positive influence of solar radiation on the basic outdoor temperature (the lowest outdoor temperatures are recorded during the months of December or January when the effect of the solar radiation is at a minimum).

Nevertheless it has been possible to increase the basic outdoor temperatures for the heat loss calculations as can be seen in table II.

Table II : Basic outdoor temperatures for the heat loss calculations

	Altitude of the site above sea level					
	0-100 m	100-200 m	200-350 m	350-500 m	500-650 m	> 650 m
Actual values	-10°C	-10	-12	-12	-14	-16
New proposed values	- 8°C	- 9°C	-10	-11	-12	-13

3. CONCLUSIONS

In § 2 we only gave a short review of the most practical results of the research. In the final report (1) one can of course find some other results on a more scientific level on

- the estimation of the heat losses to the ground
- an analysis of the dynamic behaviour of heated and unheated models
- a second-order dynamic approach.

4. BIBLIOGRAPHY

(1) H. Heikhaus; J. Lebrun; L. Laret
Caractéristiques principales des habitations sociales permettant de réduire la consommation d'énergie pour le chauffage

Commission des Communautés Européennes
Energie - Convention n° EEA-5-057 B
Rapport Final EUR 8370 EN/FR - 1983.

(2) Energy Conservation in Buildings - Heating, Ventilation and Insulation.
Proceedings of the Contractor's Meetings held in Brussels on 14-15 December 1981, 6-7 May, 24, 28 and 30 September and 21 October 1982.

Edited by H. Ehringer, G. Hoyaux and P. Zegers.
Directorate-General for Science, Research and Development Commission of the European Communities.
D. Reidel Publishing Company - Dordrecht.

(3) Temperature without heating of buildings.
Proceedings of the Symposium held in Brussels and Liege in May-1979.
Working documents edited by Uyttenbroeck (CSTC
Dupagne (ULg)
Hoyaux (CEC).

(4) P. Claux - J.P. Franca - R. Gilles - A. Pesso - A. Pouget - M. Raoust
Methode 5.000
Règles de calcul du coefficient B : Besoins énergétiques dans l'habitat.
Pyc-Edition - 1982.

(5) P. Caluwaerts - M. Dethier - H. Heikhaus - J. Uyttenbroeck
Besoins en énergie pour le chauffage des bâtiments.
Energiebehoeften voor de verwarming van gebouwen.
CSTC-WTCB avril-april 1982.

(6) J. Uyttenbroeck
Jours et vagues de froid et le calcul des déperditions.
Mars 1982 - Internal CSTC document.

BIOCLIMATIC DESIGN FOR BUILDING SYSTEMS AND COMPONENTS

Contract: AGIP NUCLEARE SPA: G. ZACCARIA
Designer: SYNERGIA PROGETTI Srl: S.LOS, L.AGNOLETTO, N.PULITZER, A.TRESCA
F.NOVI, R.RAITERI, T.ZAMBELLI

Summary

The central thesis of this work is the potential widespread and effective application of Bioclimatic Architecture, which represents an evolution beyond the past energy-bound concept of Passive Solar Systems, restricted to specific devices and components. This approach is applied to different residential building types, influenced by different climatic conditions and adopting many current technological systems, within the scope of the E.E.C countries.
The study, therefore, is an attempt to transfer the new technologies of the Bioclimatic Design Process to current building production. These technological innovations are essentially design rather than device oriented.

1. OBJECTIVES

This work's intent is to demonstrate the applicability of Bioclimatic Architecture within a wide variety of situations, characterized by different multifamily building types, located in various climatic areas and employing a variety of technological systems. In this context, the aim of this work is to transcend the concept of the Passive Solar System as a closed system and to accept instead, the complexity of building production which leads to the development of open systems. The technological systems that have been formulated result, in fact, from a selection and partial adaptation of current building and construction techniques.

2. METHOD

In order to achieve the objectives of the project, it was necessary to devise a design methodology capable of testing a wide survey of building types in relation to a range of climatic and technological situations.
An experimental analysis through the use of test-cells would have brought a limited interpretation of Bioclimatic Architectural Systems for several reasons. Firstly, these systems are not identified with a few specific technological systems, but rather are configurations of building system adapted to different technologies and building types in relation to various climatic contexts. Secondly, Bioclimatic Architecture can not be conceived of in terms of a rigid classification of the solar gain systems (direct, indirect, isolated) used to describe Passive Solar Systems, but, instead, as an integrated ensemble. Finally, the energy behavior of the building must be evaluated by taking into account the occupancy and internal energy sources.
These considerations of Bioclimatic Architecture, together with the great variety of situations to be tested, advises againsts the use of test-cells,

suggesting instead a much more comprehensive method.
The method adopted begins with identifying the correlation between current building types and the various climate areas. The selected building types are examined using a computer simulation program, and revised and corrected through an information feedback process.
The first result consists of a series of projects defined in terms of performance system design, having first verified their energy efficiency.
The next step is to translate the performance system design into technological systems, using selected products from a catalog developed within the scope of this project. The catalog includes a series of products which represent the developing trends in building production, namely the "open system" and "easy construction".

3. PROGRAM

In order to apply the method presented, a set of projects had to be developed.
The program used to define the projects includes the following topics:
- climatic environment;
- building types in respect to the kind of users and their culture;
- building production trends;
- energy performance of the designed building systems, analysed by a computer simulation program.

3.1 Climatic environment

The first operation is to identify the areas in which to locate the building types that will be identified in the next stage. To this aim we have chosen three Italian climate areas, very different in themselves and as such can be generalized and used extensively to describe European climates. The areas identified, in fact, represent conditions of the principal European climatic patterns, that is to say from the combined influence of the Atlantic Ocean and Continental masses and the influence of the Mediterranean Sea at two different latitudes.
The three climatic areas chosen for this work are:
- Suboceanic area with "Continental" characteristics typical of inland and sub-alpine regions (46'N latitude) as can be found in Bolzano and Lyon;
- Northern Mediterranean area (44'N latitude) with a windy climate, cool winters and warm summers such as can be found in Genoa, Nice and Marseille;
- Southern Mediterranean area (38'N latitude) with a mild temperate climate such as can be found in Messina, Athens and Lisbon.

3.2 Building Types

Following the climate analysis, the next step is to identify the building types to be assigned to the different climate areas. The building types selected emerge as a function of the inhabitants and their culture. Later, the mean composition of the nuclear family is used to define the occupancy patterns for the dwellings, needed for the simulation and analysis of the buildings' thermal performances.

This double analysis reveals that the building types most representative of the European building culture are the Compact Tower, the Apartment Block and the Row House.
The Tower building, somewhat penalized in terms of solar collection, is significant for the "conservative" aspect of its form, which approaches the shape of a cube.
The Apartment Block emerges in many forms: from the bi-lateral solution, which aims at an appropriate balance between solar collection and energy conservation; to the uni-lateral solution, which is "specialized" for solar collection.
The split-level feature of the Row House is of particular interest. This solution not only enables sun access for the northern internal living spaces but also provides safety for on-site construction in that the joists are placed by half-floor heights and thus can be used in the place of scaffolds. Each of these building types are resolved specifically in relation to the climate area in which it is located.

3.3 Simulating Model

The building types, previously identified and assigned to climate areas, now undergo an analysis using a validated computer simulation program adopted for this purpose. This allows successive reviews and modifications of the building types to be made, in order to arrive at adequate designs of the performance systems.
The need to evaluate and compare the different solutions to determine the optimum value of the various parameters (including orientation, glazing area, insulation levels) requires a method that can predict the energy consumption and be able to consider the different phenomena operating in the definition of the energy balance.
In the first phase, the NBSLD calculation program is used, followed by the SMECC program, which was derived from the former and developed within the scope of CNR/PFE1. This second model (a method of the utilization factor of free energy contributions) considers the parameters of free energy gains actually used, after having studied, in detail, the energy balance of a wide range of climatic conditions and parameters of building and construction types, using the NBSLD program.

3.4 Technological Context

The last step of the project procedure is to translate the performance system design into a technological system. To carry out this operation, we have purposely avoided creating a special system in order to draw instead from the great variety of current market products.
The global picture of production patterns points out two distinct trends: the "open system" and "esay construction".
The open system consists of building systems characterized by a wide variety of technological solutions, different yet coordinated. The technology can thereby be expressed in the formation of the building systems by "product concepts" rather than by specific products.
The second trend shows a progressive loss of the professionality of the con

struction practice and the consequent need to offer complex and complete products that simplify and reduce the cost of building construction operations.
In response to these trends, we have verified the compatibility between passive solar systems and technological patterns, present and future, using the catalog as a tool. Coherent with the emerging trends, we have in fact assembled a catalog of building products, relevant to the energy problem, to serve in this phase of the Project. For each of these products, both the energy characteristics, which are needed in order to evaluate the correspondence of the technologies to the project performance, and the specific energy content, are noted in the catalog.
Many of the products identified contain a high energy content realized in production that can be off-set by the lower energy costs of on-site construction assembly and operation, thereby justifying their use.

4. RESULTS

The results are contingent on the outcome the development of each project.
The first outcome is the formation of performance systems which are studied and subsequently modified through computer simulation and analysis.
The second outcome attests to the energy efficiency predicted by the computer simulation program.
The final outcome is given by the attributions of current building technologies applied to the performance systems.

4.1 Performance System Projects

The performance design of each selected building type is described in relation to the climate areas considered.
The general characteristics of the bioclimatic strategy adopted is first described for each climatic area, followed by a description of the specific characteristics of the selected building types.
Sub-oceanic Climate Area:
For the Sub-oceanic climate, characterized by low external temperatures and acceptable winter radiation values, the bioclimatic strategy adopted was for energy conservation, combined with a cautious use of direct gain and, above all, a wide use of air collectors, which are characterized by reduced heat loss and high efficiency.

- Tower: Compact in form and highly insulated on the non-collecting surfaces. A light-weight structure which encorporates a mixture of opaque panels, window glass, and air collectors to serve the back rooms, is specified for the South East and South West orientations, having turned the building axis on the diagonal. An enclosed stairway is placed on the tower's North side.
- Bi-lateral Apartment Block: Compact and well insulated. The South façade is light-weight with side glazed apertures to increase the collection area for rooms that experience more heat loss. Channels are contained in the floor slabs to draw warm air from the south-facing collectors and distribute it to the northern zones. The non-collecting surfaces are insulated

to the North. The roof is sloped to allow snow to melt off of the roof surface. The roof-mounted water collectors have a reflective plane beneath to improve their operation. The loggias are all external with an independent structure to avoid thermal bridges.

Northern Mediterranean Climate Area

In the Northern Mediterranean area, characterized by relatively high temperatures and radiation values and by intense and directionally variable winds, a multiplicity of various bioclimatic strategies co-exist, including energy conservation and solar gain, direct (for near-south exposures) and isolated (using sunspaces).

- Tower: Compact and insulated to the north. It is similar to that described for the Sub-Oceanic climate area, using internal loggias that act as sunspaces for direct gain as well horizontal bands of glazing in the South-East and South-West directions.
- Uni-lateral Apartment Block: Less compact but greater solar gain than in the Sub-Oceanic area with an increase in collecting surface. The South-facing exposure is light-weight with air collectors coupled with channels contained in the floor slabs, while the non-collecting surfaces are externally insulated. The geometry of the South-façade is designed to increase the direct gain to the top and end units, which have larger thermal loads due to the increased external surface area. The stairway is completely external to the North and the loggia-sunspaces are internal and South-facing. Solar collectors for domestic water heating can be installed on the roof. North apertures enable cross-ventilation through the dwelling units.
- Row Houses: The Northern zones are served, in virtue of the split-level solution, by direct gain. The roof is sloped against the Northern winds. The sunspaces are internal and coupled with air channels contained in the vertical common wall to transfer and diffuse air to the other living spaces.

Southern Mediterranean Climate Area

In the Southern Mediterranean area, characterized by high temperature and radiation values, the bioclimatic strategy focuses on building thermal mass to absorb daytime heat and dissipate heat at night. The building envelope must therefore be massive, with low surface absorption, reduced aperture and with devices to enhance natural ventilation.

- Uni-lateral Apartment Block: Composed of detached, staggered building bloks to induce natural ventilation by creating differences in air pressure on opposite sides of the building. Ventilation is also fostered by the roof's and base's embrasure. The living rooms are an extension of the "Raumplan" concept, intended as day-time living zones with an augmented height that favors both summer ventilation and winter solar gain. An operable façade to these zones permits variable compactness and serve the ventilation as well. The ventilated end walls provide a place for clothes-lines. The stairway is also ventilated in virtue of its enalarged volume and the windows in the upper side. The mass is exposed and the external surfaces are reflective. The day-time living zones are protected by sun-shades that also provide air flow through the spaces, and can be displaced during the winter-time. The South-facing windows are small with light-shelves to increase interior illumination while the Northern windows are provided with reflective shutters. In both cases the glazing is recessed to the inside wall surface. The ceiling beams and the loggias of the living rooms are light-

SUBOCEANIC AREA-NORTHERN MEDITERRANEAN AREA: TOWER HOUSE

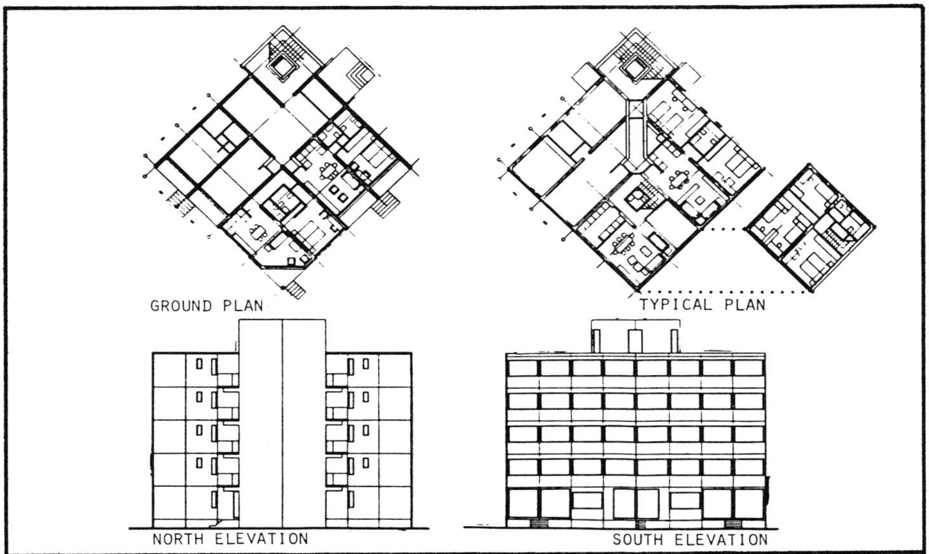

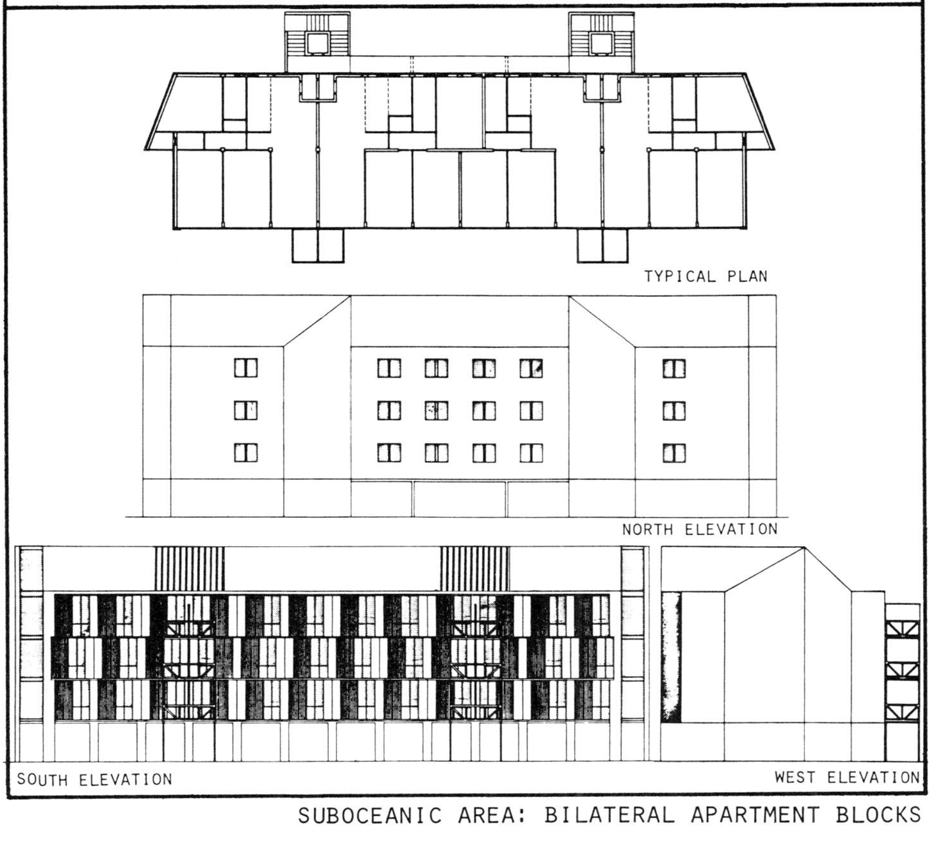

SUBOCEANIC AREA: BILATERAL APARTMENT BLOCKS

NORTHERN MEDITERRANEAN AREA: UNILATERAL APARTMENT BLOCKS

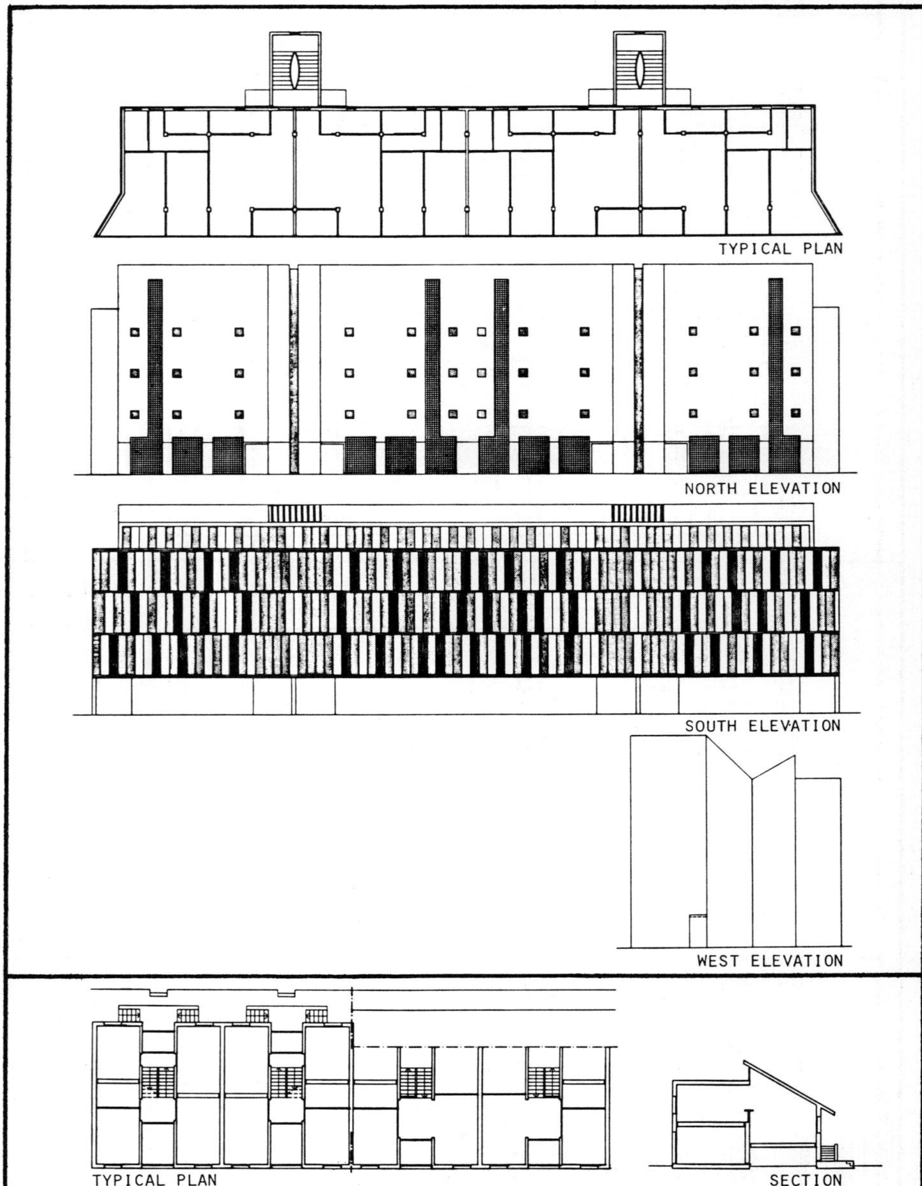

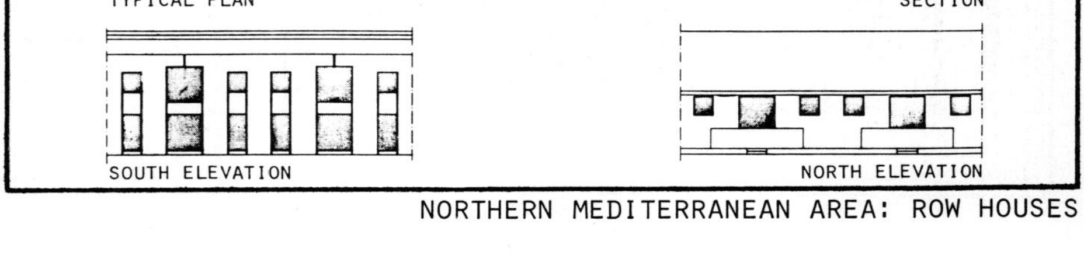

NORTHERN MEDITERRANEAN AREA: ROW HOUSES

SOUTHERN MEDITERRANEAN AREA: UNILATERAL APARTMENT BLOCK

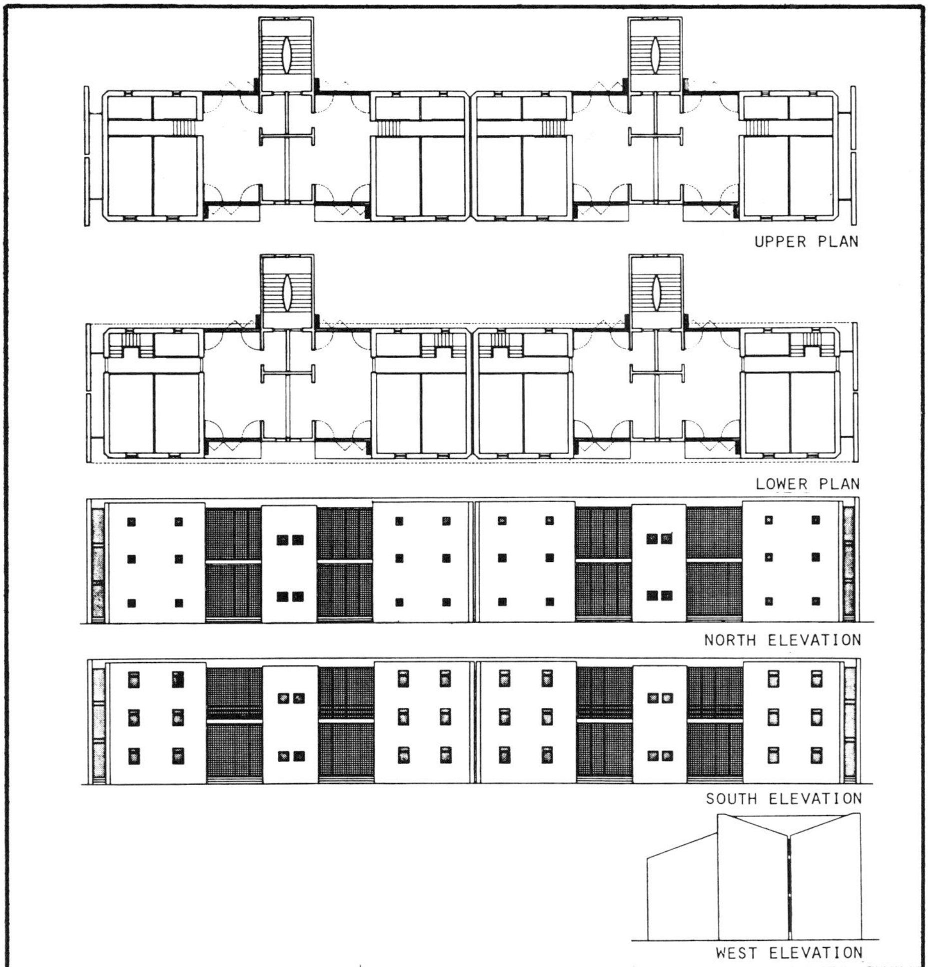

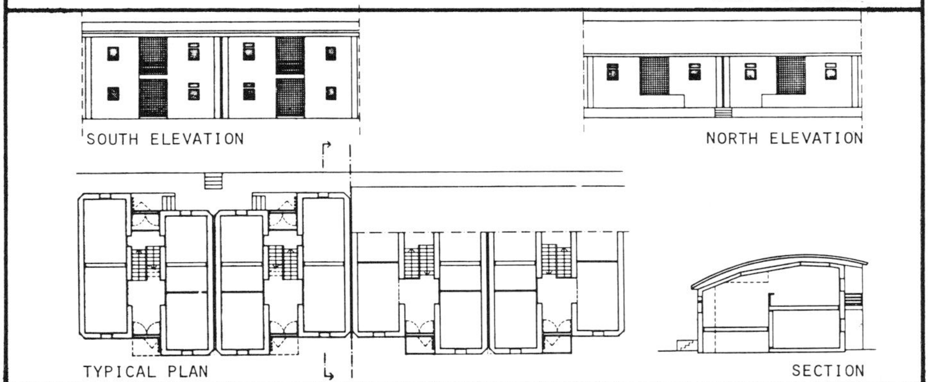

SOUTHERN MEDITERRANEAN AREA: ROW HOUSES

weight and structurally independent to avoid thermal bridges and to reduce the thermal inertia.

- Row Houses: Many of the bioclimatic characteristics are similar to those of Apartment Block: detached, staggered building blocks; variable compactness; reflective outside surfaces; ventilated central zone; reduced fenestration with reflective devices to increase natural daylighting.

4.2 Simulation and Energy Analysis

The simulation of the building types are analyzed for the heat losses and gains and for the overall resulting energy consumption. The building types located in the Southern Mediterranean area are also analysed for summer internal temperature conditions. Gains from internal energy sources for all of the building types in the respective climates are evaluated separately.
The building solutions were optimized to meet the heating demands throughout the year execluding the coldest months, in order to reduce the risks of overheating and to allow further improvements in the building energy performance to be made by the user.
In regards to the passive solar systems, the efficiency of the sunspace increases as the ratio of losses to gains decreases. The performance of the air panels is less linear in respect to this ratio but it is generally more efficient with an increase of the loss-gain ratio.
In the Sun-Oceanic area, the solar gains compensate for 50% of the global remaining months.
The passive systems used in the Northern Mediterranean area are generally more efficient due to the favorable climatic conditions. The contribution of solar gains in this climate area is around 70% during the coldest months.
In these conditions, it is also possible to use sunspace, whose gains can be compared to the direct gain of surface glazing but whose losses are comparably less and which provide comfortable living spaces through out the year.
The conception of the building types for the Southern Mediterranean area were oriented more toward energy conservation and dissipation than for solar collection due to the particular climatic conditions. The ratio of solar gains to global losses is fairly constant during the winter season, about 40%. The solution for this climate zone reduces the problem of over-heating.
The effects of the high building thermal mass and reduced surface absorption, combined with increased ventilation, reduced window area and sun-shading screens, generate a satisfactory summer-time indoor environment.

4.3 Technological System Projects

The selected products used to realize the building systems with very few adjustments, are grouped in three significant product "families":
a) light technology with localized or extended air circulation channels;
b) relatively heavy masonry block technology for the external walls;
c) distributed mass technology combined with operable ventilation devices.
These three product concepts successively form a stronger correlation between warmer climates and unskilled labor.
More specifically, the light technology is adopted for the tower house and bi-lateral apartment block in the Sub-Oceanic area and for the uni-lateral

apartment block in the Northern Mediterranean area.
The relatively heavy external envelope, which lends the possibility to utilize its cavities, is adopted for the tower and row houses in the Northern Mediterranean area.
The distributed mass technology, coupled with the environment and ventilated cavities, typically resolves the uni-lateral apartment block and row house building types in the Southern Mediterranean area.

5. CONCLUSION

The results demonstrate the thesis of possible widespread and efficient adaptability of Bioclimatic Architecture to different multi-family building types, located in various climatic areas, adopting many technologies in the context of current building production.
In particular, it has been demonstrated that neither specific technological innovations not highly qualified enterprising organizations, are needed to realize energy efficient building systems.

REFERENCES

S. Los, E. Foldini, "Progetto guida di casa a torre" from the "Atti del 5° Seminario Informativo CNR/PFE" Feb. 1982, Milano PEG Editrice.
T. Kusuda, "NBSLD The Computer Program for Heating and Cooling Loads in Buildings", NBS Building Science Series 69, 1976.
L. Agnoletto, P. Brunello, R. Zecchin, "Simple methods for Predicting Thermal Behaviour and Energy Consumption in Buildings", ASHRAE-DOE Conference, Florida USA, 1979.

COMPUTER AIDED DESIGN OF LOW-ENERGY SUBSIDIZED HOUSES : PRACTICAL EXPERIENCE OF DESIGNING, BUILDING AND MONITORING 25 HOUSES ON SITE

A. DUPAGNE, J.M. HAUGLUSTAINE, L. LARET and F. SCHMITZ
Laboratoire de Physique du Bâtiment
University of LIEGE, Belgium

Summary

A first research work carried out for the Commission of the European Communities has produced a manual design methodology (1). Mainly orientated to the conception of low-cost houses, it integrates some fundamental thermal concepts into the design process.
The present research work had for target to apply this design method to an actual case, in order to test the practicability of the method, to bring its lacunas to the light and to confront the theoretical estimations of heating energy needs, with actual measurements.
The recommanded design process has been followed along the conception of 25 houses, among which five have been built and monitored. The design guide has been revised by a re-writing of the manual version, and has been computerized on a micro-computer. Any practical application is now considerably speeded up and usable by practicians.

1. DESIGN PROCESS FOLLOWED DURING THE CONCEPTION

1.1. Data collection

The work first concerned the collection of any pertinent information necessary to any author of an architectural project, having the Belgian National Housing Society as a client; contacts had been established with the town authorities where the houses would be built, and with the regional and national housing societies (2). The chosen site was very cheap, but North sloped : cost reasons have thus met the wish to demonstrate that energy saving design is also possible on difficult North orientated sites.

1.2. Site analysis and settlement layout

The existing manual design method gave evaluation methods mainly suited for one house, and barely outlined the site analysis of a group of houses. The badly orientated site offered a good opportunity of devoting some intellectual investment in the site analysis.

The first fundamental options of the settlement layout would dispose the houses in a direction as parallel as possible with the contour lines, in order to reduce the excavation works for the implantation of the houses, clinging to the ground. Furthermore, a good bipolar North-South orientation would be given to the houses, so as to obtain a good solar recuperation in the winter, without any risk of too high summertime temperatures. The plots, usually 8m wide, have been broadened to 10 m, with a resulting increased South and North façade area, for which different design strategies have been used.

The detailed design of the settlement layout has taken microclimate into account, by considering cold wind prevailing directions and cast shadows (3). The settlement layout has reduced the exposition of the houses to the most unfavourable winds to the infiltration losses, by taking into account the Belgian graphs of cold wind prevailing directions during the heating season.

On the other hand, sun-path diagrams have been used for the prediction of insolation and the determination of overshadowings caused by external obstructions (the site or other facing houses). It was important indeed not to priviledge a few houses, better orientated, to the detriment of other ones which were downwards. An energetical point of view has been given to the assessment of relative positions of houses in their own group and to the interrelative position of group : overshadowing periods have been weighted by the corresponding solar radiation intensities. A manual calculation of the overshading risk has assumed that the lowest point of the South façade losts an amount of incident solar energy of maximum 5 %, where 3 % are already caused by the site itself. The resulting settlement layout is given in Figure 1.

The manual estimations consume a long time of calcultation; they have been carried out for only one point representing a whole façade, instead of dividing the façade into a great number of elementary surfaces as close as possible to points. The site analysis itself (and mainly the solar overshadowings estimations) is worth computerizing, in order to become a useful analytical tool. This was the reason of a specific feasability research work to be proposed for development in the future CEC research program.

1.3. Feasibility study of the project and outline proposals

The first stage of the design methodology consists in a feasibility study of the project. No design is studied, no solution is outlined : the whole architect's attention is only focused on the research of acceptable values of main global parameters (shape factor, heated floor area, volume, global heat loss coefficient). These parameters characterize a projected dwelling which must be acceptable from multiple points of view : technic, energy, costs. The feasibility study allows the architect to see if an architectural solution exists or not : he manages the main global parameters in order to obtain satisfactory results. The final product of the feasibility study consists in a set of global parameters representing a lot of solutions which potentially reach the objectives defined by the client at the inception stage. The shape factor, the global heat loss coefficient, the glazing factors, etc., have been silhouetted, in order to become a basis of the outlines to be proposed by the architectural team (3).

On the basis of the conclusions of the feasibility study, the architect roughly sketches several plans of dwellings. At this stage of the design, the architect must give a complete freedom to his natural divergence : any critical mind would bound the search space and throw out possible solutions, which would be later proving full of interest.

The architectural team has designed three different types of designs for the houses of the lowest part of the site. They respect the same general space allocation: : the South portion incorporates both the general entrance and the living-room, with a window area as large as possible, in order to give the house a good solar recuperation ; the cellar level imposed by the ground slope includes two children's bedrooms, unfortunately orientated North-North-West.

The garages are separated from these houses and laid out along the access road, so they neither require additional roads for accessing the house from the street nor take from the house large non-recuperating surfaces (orientated South-South-East). Treated as relatively low volumes, they don't give large masks for the solar recuperation in the house; low volumes located between the public road and the private South-South-East garden place screens which, being usual, increase the garden privacy and, being acoustic, act as obstacles against the noise coming from the street.

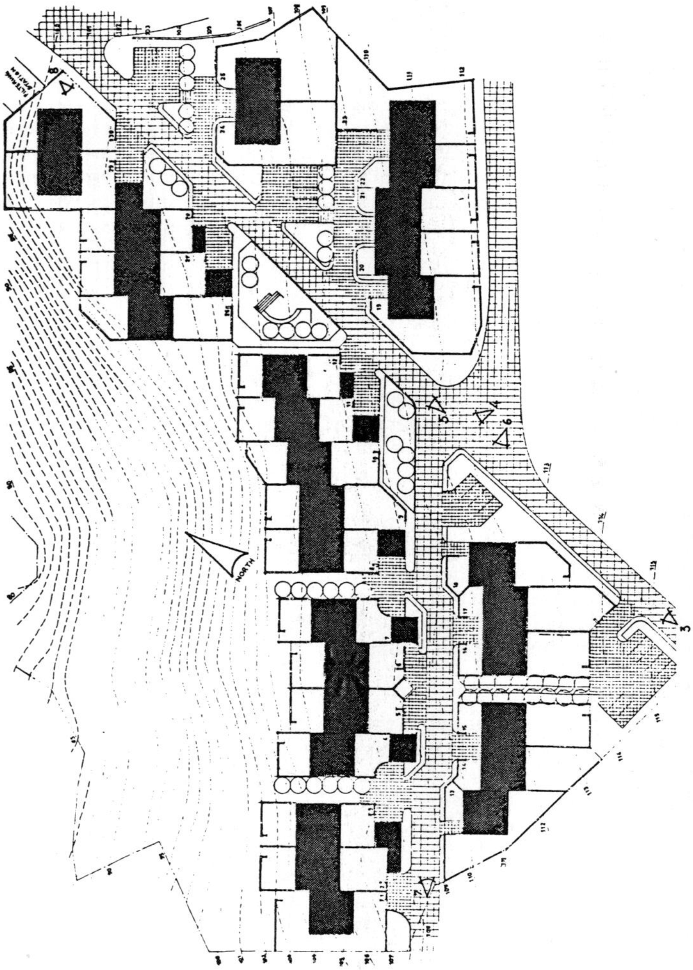

Figure 1. Resulting settlement layout of the 30 houses previously foreseen on the site.

For each of the three designs, the global heat loss coefficient (W/K) including air change losses has been calculated manually. The heating energy needs have also been calculated manually, in a more detailed way than during the feasability stage : the monthly recuperating factors and the equivalent degree days (for heating) have been evaluated month by month. The three designs have so been compared with the same fenestration and insulation hypotheses (4).

On the other hand, in order to see the influence of the site configuration, the same manual method has been used to calculate the annual heating energy needs of the whole group of 12 houses to be built on the lowest part of the site. These 12 houses are distributed in three groups of 3, 4 and 5 houses ; each group has two extreme semi-detached houses and the others are as much in contact as possible, but in respect to the ground contour-lines. The three designs have been first compared from the energy point of view : one of them had a larger volume but needed annually 76 % heating energy more than the others and has so been eliminated. The others had estimated energy needs nearly similar (1,9 % difference). The preference has been given to the design having a larger aperture to the South, giving so the hope of a better daylight comfort, while remaining reasonable in order to avoid glare risks.

1.4. Detailed fenestration.

During the last part of the design, a special effort has been devoted to the fenestration of each room, dealing with daylighting and visual comfort, while avoiding overheating risks, as well as much energy consumption for heating and artificial lighting.

The fenestration of the South windows thus results form a lot of considerations which must be taken into account, such as :

1. conformance with the insulation standards (with the global heat loss coefficient);
2. the appraisal of the acceptable overheating risk;
3. the calculation of the recuperating factor obtained by overcast and cloudless sky conditions;
4. the resulting necessary overshadowing factor (acting only on direct solar radiation);
5. the approach of the maximum accepted temperature by the right dimensioning of the solar protection.

Concerning the North windows, they have a heat balance always negative, from the heating energy point of view. Nevertheless, other considerations are to be taken into account in North fenestration, such as the town planning rules (the respect of which is imperative to receive the authorization to build), the visual comfort (visual contact to outdoors) and the daylighting level, which has a direct impact on the artificial lighting energy needs. In the case of artificial lighting, an optimum fenestration has been searched, in order to minimize the total energy consumed for both the heating of the room itself and the electricity used for artificial lighting. Full calculations of detailed fenestrations for North and South orientations are reported in (4).

1.5. Detailed design.

The detailed design of the houses included the right dimensioning of the insulating materials thickness and the choice of their positioning in the fabric. This last stage of the design has been carried out, by running the LPB4 program (static and multizone), which has given us the sensitivity of the energy consumption to the insulating materials thickness. The right thickness and positioning have been chosen according to the energy consumption, the cost of the insulation and the respect of the Belgian regulations related to the global heat loss coefficient (4).

The type of external walls has been chosen by considering the insulation properties, the access to the inertia, the investment and maintenance costs and other particularities, e.g. related to the risk of thermal bridges.

The external walls are composed of (from inside to outside) :

1. a wall made from heavy concrete blocks (providing the structural support and the thermal inertia) ;
2. insulating panels placed directly against the interior wall and interlaced with a lattice of wood ;
3. an outer "skin" to protect the insulation from wind and water. Between the weather-tight skin and the insulation, a small air space is left by using an additional wood strip to assure for easy circulation of air, which keeps both the skin and the insulation behind it dry (see figure 2).

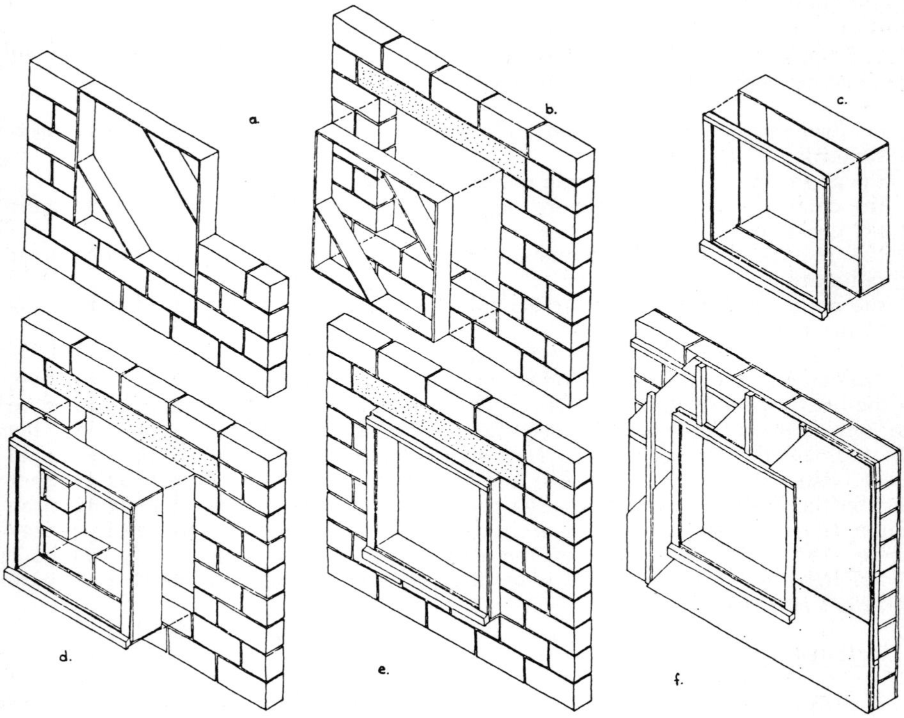

Figure 2. Summary of the Operations of the Window Construction Details.

This type of external wall is relying only on usual construction techniques and materials and so remains cheap. But its installation presents the interest of a continuous control during the construction and gives so the guarantee that the expected properties of the materials could actually be realized (for energy saving reasons).

In spite of this type of wall, the connection of window frame with the external wall would put the window as far as possible from inside, in order to minimize usual thermal bridges, to approach the element recuperating the solar radiations (the glazing) near outside, to give a better visual comfort, etc.

An original, cheap and simple solution has been designed. The window frame system consists in two sorts of problems : the frame supporting the glazing material in the sash and its connection with the surrounding wall. The problems are rather different and must be differently treated. It is easier to separate the connection to the wall from the window frame, in order to affirm the joints, and treat them by the materials which adequately stuff them. This is the reason why the design team has remodelled the conception of the sill and the jambs by replacing them with an intermediate peripheral frame between the window and the wall. This casing presents a lot of advantages (see Fig. 2). Used as a mould by the mason during the building of the masonry wall (a), it is taken by the carpenter having thus the right dimensions of the masoned hole in the wall(b). The carpenter can build the casing and the attached window, in his work-room (c). He comes with the whole in the building site and puts it in the wall (d). The casing provides a peripheral support to the window frame (e) which can be put as far as possible from inside, just in the definitive façade plane (f). Thermal bridges are thus reduced to a minimum, as well as the shading mask. In the same time, it provides the internal finishing of the wall hole, where plaster becomes useless.

2. BUILDING OF THE HOUSES AND SUPERINTENDENCE.

Wallonian budgetary imperatives have reduced the number of houses immediately built (in 1982) from 25 to 5 only , while the other twenty ones will be built later (in 1984). It was important for the research team to have a minimum number of realized houses, in order to be able to make the measurements and so confront the actual results to the theoretically expected ones.

The five houses have been built within 100 days, from August to December 1982. The research team has followed the building works in Marchin, from the energy and the methodology point of view : it has carefully observed the positioning of the insulation layers and -if needed- has given some remarks to the workers, in order to obtain resulting buildings with a thermal quality as close as possible to the expected one, i.e. needing 50% less energy than other common social houses. They indeed must consume 1500 fuel oil liters/yr, in the Belgian climate with an annual amount of 2842 degree-days 18/18. The few problems encountered during the building works concerned especially the difficult rainy conditions during the placing of insulation : additional precautions have thus been taken in order to obtain insulating panels as dry as possible.

The only important problem was the bad care brought by the carpentar during the placing of the frames and casings. Wrong fixings of both the casing in the wall and of the frame in the casing (only nailed, instead of glued and screwed) have emphasized the joints ; furthermore no souple material (although imposed) had been introduced between the casing and the wall. In order to correct these errors and to prevent air infiltrations by these two unfilled joints, the research team has required the placing of silicone between the window frame and the casing, and the injection of polyurethane foam between the casing and the wall. Each window has been checked in details to see if these corrections were actually realized.

3. COMPLETION ENERGY INSPECTION OF THE HOUSES.

The research programme had foreseen a completion inspection of the houses just after the end of the building works, in order to analyse the possibility in the field of completion inspection of newly constructed buildings, especially with regards to the energy consumption. Therefore the Belgian Building Research Institute, charged with this task, has undertaken a number of calculations and measurement actions on the real buildings, in order to verify some possible techniques from a methodological point of view (6).

As the main objective was a completion inspection, and as the buildings were newly constructed, the following actions were undertaken in the first campaign :

1. the classical heat loss calculation for the houses, room by room, according to German Standard DIN 4701 (in common use in Belgium), with different assumptions of indoor temperatures ;
2. the calculation of the solar recuperating factor of the houses, with different assumptions regarding the shadings ;
3. a detailed measurement campaign in 2 houses by using the co-heating technique (electric heaters in each room), in order to relate the energy consumption to the temperatures and the climate ;
4. the check of the assumptions on air infiltration by using pressurization tests and tracer gas measurements and by analyzing their relationship.

In addition to the previously mentioned actions, secundary effects were measured : the relative humidity was particularly measured by direct graph readings, during a period of 16 days. A preliminary analysis of the first results can conclude that the weak correlation obtained with the coheating method is probably due to the problems associated with the newly instructed buildings :

1. the influence on the results of the humidity brought during the construction itself, the drying-out process of which is only starting ;
2. the too big uncertainty factor of the heat balance of really low-energy houses having low ventilation rates, but high losses to uncontrolled spaces (cellars, ground,...).

The pressurization tests, confirmed by the tracer gas measurements, let assume to have an air infiltration rate of 0,25 ac/h. This value is very low in comparison with the rate of 1 to 1,5 ac/h commonly observed in the context of social houses. The special care during the building works seem to have bore the expected fruits.

In a second campaign some of these measurements will be repeated, in order to analyze aging effects on the energy consumption parameters. After the installation of the heating plant, measurements are foreseen on this equipment (nominal efficiency) and consumption measurements with the actual heating equipment. Also a check in situ of the U-value will be tried out.

4. MONITORING OF THE MARCHIN HOUSES.

The aim of the **completion inspection techniques**, as described hereabove, is the direct estimation of some thermal key parameters. It is the reason why this analysis can be achieved on a short time (a few days only) and by the discussion on a few parameters.

The interest of the **monitoring** stands in the evaluation of the significancy of the parameters ; thus any monitoring involves the experimental measurement of their sensitivity and the statistical interpretation of them. During a monitoring, the highest number of data on space and time are to be recorded. All the informations contained in the raw data are to be used as long as possible.

4.1. Main aspects of the Marchin monitoring.

Measurements are made on five attached houses, and could be extended to twenty houses. Three of the five houses are occupied ; the two unoccupied houses are used as references for the analysis of well defined effects. More than 200 points are to be recorded with the logger. It necessitates a network well organized, in spite of the difficulty of interfacing different appliances having a poor information about their hardware details.

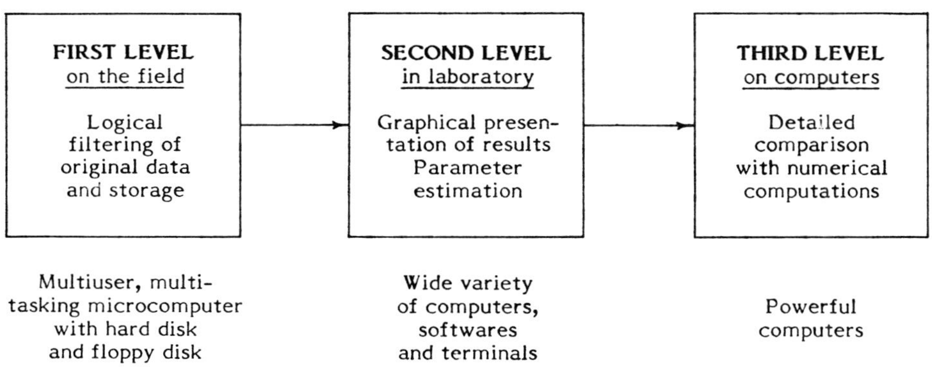

Figure 3. Steps of data analysis.

A three level data process was designed to use the different available computers in an optimal way (Figure 3). Due to the mass of data to be processed, a general software would be strongly desirable, in order to feed the existing standard library (statistics, simulation...) with the raw data provided by the in situ measurements. Such a kind of software does not exist yet ; nevertheless it should be essential at some level of monitoring for a rigourous interpretation.

4.2. Measurement system.

Each room of all the houses has two globe temperature measurements, obtained by thermocouples placed at a 12 cm distance from the wall, in a globe of 3 cm diameter ; the living-room is divided in two zones and, so, considered as two rooms. One globe is positionned 1,80 m above the floor, while the second is placed 10 cm below the ceiling. We obtain so at least 20 points in each house. Furthermore two points are recorded with platinium resistance at 1,80 m height : the first one is in the eating zone of the living-room, the other in the little bedroom (downstairs). Thermocouples record also one cellar temperature, the cold water temperature (and the temperature difference with the hot water), the temperature of the hot water supplied by the boiler and two temperature differences measured in the hydronic circuit. Four electric counters are recorded, giving information on the electric consumption of the small power appliances, the high power ones, the artificial lighting and the total electric consumption of the house. Burner runnings are recorded. We obtain thus 33 measurement points in each house. Two flow counters have been added in each house ; some other temperature probes have been placed in the unoccupied houses.

The weather station allows six temperature measurements : each point is measured with a thermocouple and with a platinum resistance. There are also, at least, two measurements of total solar intensity on horizontal plane and two rain detectors (one horizontal, the other at a slope of 45 degree in West orientation) ; the wind orientation and its horizontal velocity are recorded.

The wiring between each probe and the logger is organized in a network as shown on Figure 4.

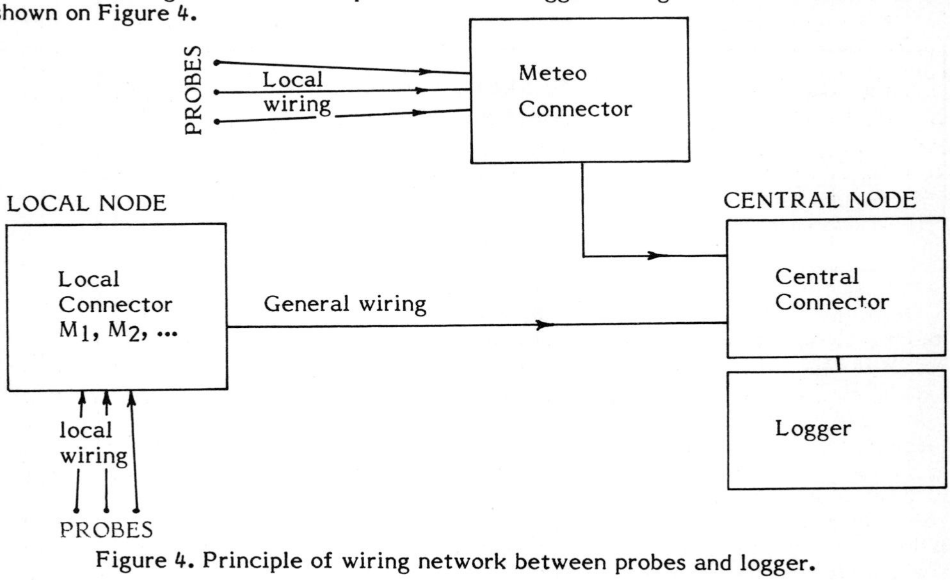

Figure 4. Principle of wiring network between probes and logger.

The data logging system is built around a rather powerfull logger (Solartron ORION), connected with a microcomputer able to work in a multi-tasking and multi-user operating system (MPM II). By the use of a phone and a modem, this system is the source node of a computer network involving the computer of the University (Figure 5).

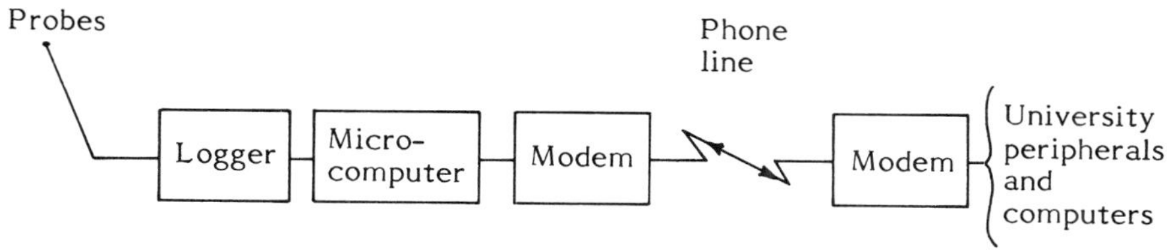

Figure 5. Data logging network.

The multi-tasking possibilities of the microcomputer allows a real time data process on the site, provided that power failure resets on computer are avoided (8).

4.3. Measurements.

Measurements are made simultaneously in the five houses and in the weather station. The sample frequency of data acquisition is 15 seconds for weather data and 10 minutes for data measured in houses. The recorded values are the hourly mean values and the associated standard deviation. With this information, the general behaviour of the group of five houses can be analyzed as well as the individual behaviour of each house and the zoning effects. The reliability of weather data (a critical measurement) is warranted by the multiplied measurements of similar variables. With the phone line and the modems, all operations on the site can be controlled at any time from the Laboratory.

Measurements confirm the good quality of the thermal resistance of the houses. The continuous insulation placed outside and coupled with a thermal mass easily accessible from inside gives a negligeable daily oscillation of t_{WH}. Its measured amplitude of oscillation indeed is varying from 1°C to 2°C maximum in the most sensitive rooms. This thermal behaviour is the guarantee against over-heating during the day or against night discomfort. No overheating has been measured during the dog-days of July 1983 : under clear sky conditions and with a peak outdoor air temperature close to 35°C, the peak indoor globe temperatures were recorded only around 28°C.

5. COMPUTERIZATION OF THE DESIGN METHODOLOGY.

The research program has foreseen the re-writing of the design methodology on a computer, in order to speed up the calculations (up to now only manual), thanks to the computing power brought by the computer. Of course some large programs exist already on big computers ; these programs are quite able to calculate heating (or cooling) loads, heating (or cooling) energy needs, summertime temperatures, etc. But the data required are so complete that these programs become applicable only to buildings which design is finished, i.e. exactly when important modifications are impossible anymore. Any modification on the design must be done as early as possible in the design process, i.e. exactly when the data are not defined yet and the drawings still woolly. This is the reason why the design methodology has brought its major effort on the possibility of simulating the thermal behaviour of a building not designed yet, not drawn yet.

The computer chosen for the support is a microcomputer, more accessible to most architects than mini or big computers. Microcomputers offer a hardware system with very interesting capabilities for a cost remaining accessible to the majority of architects, who can also use it for other software possibilities, such as word processing.

During the feasibility stage, the architect facing the terminal chooses a combination of the main global parameters, i.e. :

1. human parameters (heated floor area, consign temperature and internal gains), deducted directly from the architectural program ;
2. geometrical parameters of the building (shape factor, glazing factor, heated volume) ;
3. thermal parameters of the building (U-values, rate of air change, absorption and transmission values) directly used in the calculation of the global heat loss coefficient.

The building, characterized by the geometrical parameters, will be able to satisfy the client's energy target if its thermal quality allows to obtain satisfactory heating energy needs, for the human parameters previously defined. The whole system (= building + human + climate) giving a good solution will be characterized by a good combination of the different parameters.

In order to guide the architect to a reasonable choice of geometrical and thermal parameters, sensitivity graphs are drawn on the screen and the user of the program is able to manage parameters which are the direct components of more global parameters. For example, during the calculations of the shape factor F_S, the architect is able to visualize (see Figure 6) the influence, on the shape factor, of the geometrical options of his future project (plan aspect factor, grouping factor, number of levels...). He can also decide, with full knowledge of the consequences of his decisions.

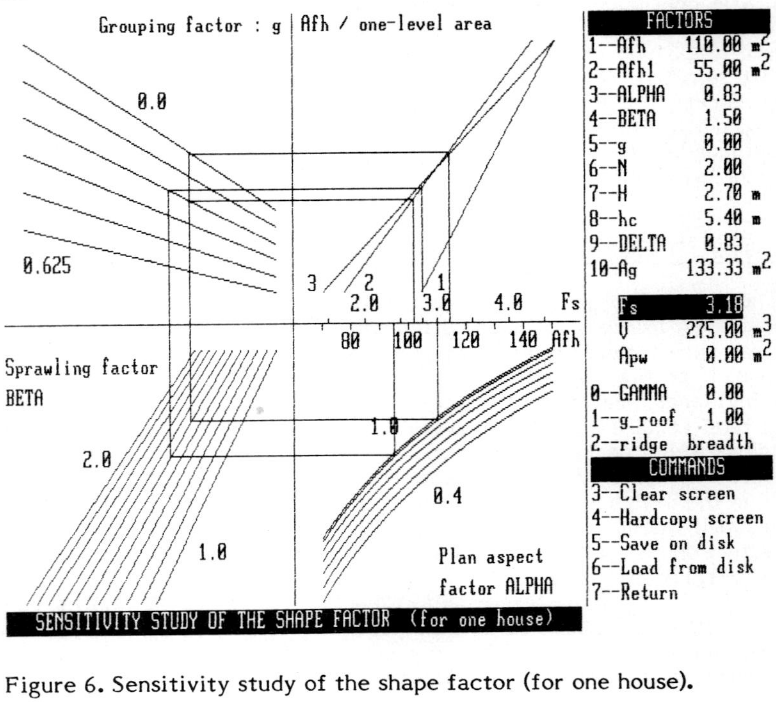

Figure 6. Sensitivity study of the shape factor (for one house).

With the report on feasibility study in hand, the architect must now generate some architectural solutions as divergent as possible, in respect with the good combination of the main global parameters. Each proposal outlined is assessed with the help of the outline proposals study, in order to see if the objectives are actually reached, or not.

The user introduces each outlined project by its geometrical and human data. The geometrical data are introduced in a graphical way. The data are introduced in more details than in the feasibility stage, because the project (now outlined) is more specified :

- the user introduces separately the draft of each level and of each part of the roof;
- the bill of quantities including parameters such as the shape factor F_s and the glazing factor F_{gl} are computed by the program, on the basis of all the surfaces of the walls forming the envelope;
- the thermal data include U-values (for each wall), absorption and transmission values (default values are available);
- some other data concerning the inertia are required, as well as shading masks (if any), for a right calculation of the insulation level according to the Wallonian standard method of calculation.

From all these data, the program itself computes the main global parameters, which are assessed according to the satisfactory values resulting from the feasibility study. If the feasibility targets are not reached, the user can modify the data before the calculation of the heating energy needs, the Wallonian standard insulation level and the mean summertime temperature. The results obtained are gathered in the outline proposals report, which will be used as a basis for the choice of one solution, among the different possibilities outlined by the architect.

6. CONCLUSIONS.

The intentions of the research program were to apply the design methodology to a practical project, in order to test the practicability of the method, to throw light on the lacunas and to confront the theoretical estimations with actual measurements. During the practical application, some developments have already been made, concerning the detailed fenestration, the shading effect on the recuperating factor and the concept of the recuperating factor, now calculated under clear and cloudy sky conditions. Further developments seem to be necessary to take the site into account and to manage groups of houses.

Nevertheless the design methodology, in the manual version, has already brought to low-cost houses needing 50 % less heating energy than commonly ; the study of building details has allowed a cheap, easy and well controlled construction of the houses. Results obtained from the first completion inspection let suppose a good air tightness of the houses, leading to an easier control of the natural ventilation.

Results of the monitoring confirm that the well insulated envelope and the accessible inertia let hope that the heating system will not run during the nights of cold periods in mid-season. The thermal quality of the envelope allows the installation of a small power heating system , without any risk of discomfort.

More generally speaking, energy savings, improved thermal comfort inside buildings, and cheaper -as well as easier- control of the buildings thermal behaviour can be obtained in the residential sector if the new buildings have been designed carefully , i.e. if their design integrates the energy and thermal aspects in the other criteria usually taken into account. The design methodology intends only to give the designers a substantial help, in order to make them energy conscious, without increasing -for this reason- their usual (and already heavy) task.

7. REFERENCES.

|1| A. DUPAGNE, M. CORNET, Basic characteristics for low-cost houses in order to reduce the energy consumption for heating. Link programme - coordination and consequences for the architectural design (synthesis) - CEC Contract N° 615-78-1EEB, Architectural design handbook : design methodology, February 1981.

|2| A. DUPAGNE, J.M. HAUGLUSTAINE, Basic characteristics for low-cost houses in order to reduce the energy consumption for heating. Link programme - coordination and consequences for the architectural design (synthesis) - CEC Contract N° EEA-5-060-B(G), Progress Report N° 1, for the period of 01.01.81 to 30.06.81.

|3| A. DUPAGNE, J.M. HAUGLUSTAINE, L. LARET, Basic characteristics for low-cost houses in order to reduce the energy consumption for heating. Link programme - coordination and consequences for the architectural design (synthesis) - CEC Contract N° EEA-5-060-B(G), Progress Report N° 2, for the period of 01.07.81 to 31.12.81.

|4| A. DUPAGNE, J.M. HAUGLUSTAINE, L. LARET, Basic characteristics for low-cost houses in order to reduce the energy consumption for heating. Link programme - coordination and consequences for the architectural design (synthesis) - CEC Contract N° EEA-5-060-B(G), Progress Report N° 3, for the period of 01.01.82 to 30.06.82.

|5| A. DUPAGNE, J.M. HAUGLUSTAINE, J.B. KNOWLAND, L. LARET, Basic characteristics for low-cost houses in order to reduce the energy consumption for heating. Link programme - coordination and consequences for the architectural design (synthesis) - CEC Contract N° EEA-5-060-B(G) , Progress Report N°4, for the period of 01.07.82 to 31.12.82.

|6| P. CALUWAERTS, D. L'HEUREUX, Completion inspection techniques (Energy aspects), BBRI DEVL N° 4153, July 1983, in |7|.

|7| A. DUPAGNE, J.M. HAUGLUSTAINE, J.B. KNOWLAND, L. LARET, F. SCHMITZ, Basic characteristics for low-cost houses in order to reduce the energy consumption for heating. Link programme - coordination and consequences for the architectural design (synthesis) - CEC Contract N° EEA-5-060-B(G) , Progress Report N°5, for the period of 01.01.83 to 30.06.83.

|8| COMMISSION OF THE EUROPEAN COMMUNITIES, R. FERRARO, R. GODOY, D. TURRENT (Performance Monitoring Group), Monitoring Solar Heating Systems : a practical handbook, Pergamon Press, published for the CEC, 1983.

ENERGY CONSERVATION ON EXPOSED DOMESTIC SITES

F. O' FARRELL, G. LYONS and G. LYNSKEY
An Foras Taluntais,
19 Sandymount Avenue,
DUBLIN 4.

Summary

The internal heat gain from solar radiation in two unoccupied isolated houses were recorded over a two year period at Oakpark Research Centre, Carlow, Ireland. Different recording regimes were undertaken with altered site characteristics mostly windbreaks. Using the monitored data, regression models for each regime were generated which related internal ambient temperature to the external climatic parameters. A model was developed to withstand the difficulties imposed by short monitoring periods and these regression models were used to predict internal temperatures for a standard reference year. Thus resulting energy saving for different site characteristics were estimated. The savings due to the erection of a windbreak were in the region of 5% to 10% and the savings accruing from the use of a standard low energy house was about 70% over a basic house. The possible implications of these savings for Ireland and Europe are outlined.

1. INTRODUCTION

The project is concerned with the conservation of energy in isolated dwellings particularly farm dwellings on exposed sites. In order to do this, it is necessary to examine the improvement in the micro-climate obtainable from different site conditions and thus to try to assess the energy savings that could be expected from the improved siting of single houses in exposed positions.

There are three basic approaches which can be made. First monitor a number of similar houses with different thermal balances. Second, prepare a simulation study involving computer orientated algorithms that balance all heating and cooling terms caused by equipment appliances outdoor weather and other factors. Finally, monitor the external climate and internal climate of an unoccupied cost house for a definite period. From the data collected, develop an analogue formula for predicting internal temperatures from the external methodological factors. Using this mathematical model predict internal temperatures outside of the monitored period.

While the project examines the first approach by monitoring two different test houses with similar siting characteristics, the main thrust is concerned with the third approach. Hence the development of a reliable mathematical model. The project takes the form of selecting two suitable and adjacent sites, examining the climatic conditions of these sites and finally examining two different test houses on these sites. There were five basic monitoring regimes undertaken. After each regime was completed, some aspect of the test house or site was altered and a subsequent regime commenced. From the data generated, an analogue functional relationship, in effect five different basic mathematical models were obtained from which it was possible to predict internal temperatures outside the monitoring period for any given external climatic conditions.

2. BACKGROUND

There has been continued speculation and research on the need to increase the efficiency of utilisation of primary energy resources. It has also been long recognised that the effective siting of buildings will yield improvements in energy conservation. However, it is more difficult to quantify these improvements. The most recent estimates of global primary demand show commercial energy figures in the range 287 EJ to 310 EJ, from OPEC (1) and World Energy Conference (WEC) (2) sources, respectively. We can infer future demand trends from established determinants such as population growth, economic development and lifestyle (3), (4), which would indicate in the year 2000 a global energy demand in the range 560 EJ to 630 EJ.

In Ireland, primary energy demand reached a 1979 peak of 8.80 (MTOE), (5) this has reduced to a 1981 TPE level of 8.16 MTOE. Recent forecasts present primary demand figures of 9.9 MTOE - 12.5 MTOE for 1990, and 13.1 MTOE - for 2020. To assess the energy consumption levels for European housing stock runs into the same type of problems as occur with Irish statistics and the overall annual energy demand for such houses could vary from between 588 - 1476 GJ x 10^6.

3. HOUSE TYPE

Any investigation of houses on exposed sites should concentrate on the most representative type of house, i.e. those designed to accommodate the typical family of each particular country. It is expected that this demand will remain fairly stable in the immediate future. New housing must also take the form of energy efficient low technology passive dwellings. Thus one of the test houses is a passive solar dwelling. Both test houses are similarly sized single storey houses with the similar accommodation of fairly recent construction but different conservation levels, total building fabric conductance of 'A' was 455 watts/^{o}C and 'B' was 217 w/^{o}C.

The test sites are situated in a flat undulating lowland plain near Carlow where the main daily duration of bright sunshine can vary from 2.14 hours in January to 6.62 in May and the annual horizontal global solar radiation is 3.657 Mj/m^2. The average maximum temperature in summer is 19.6 ^{o}C and average minimum in January is 1.3^{o}C. There are an average 2284 degree days below 15°C and the average annual rainfall is 818 mm. The annual mean wind speed is about 4 m/s. The land is virtually flat and there are no hedges or ditches. Both houses were protected on the North and East by stands of trees and exposed on the South and East (see Figures 1 and 2).

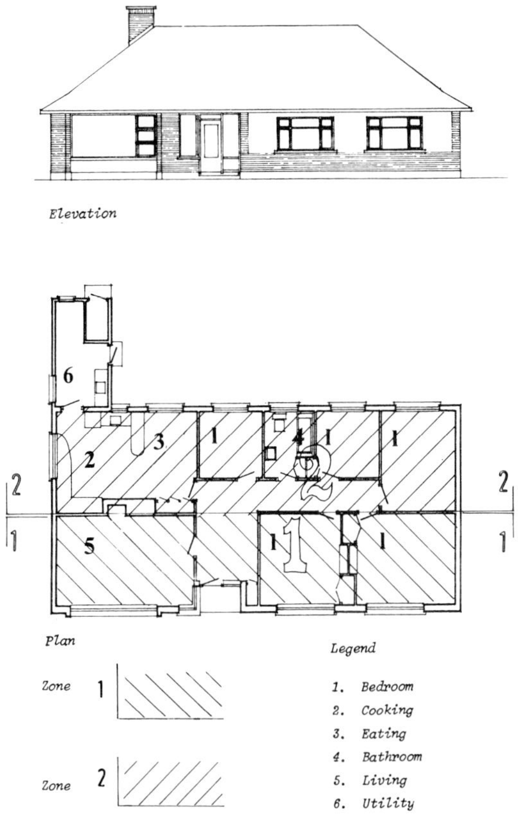

Figure 1: Test House "A"

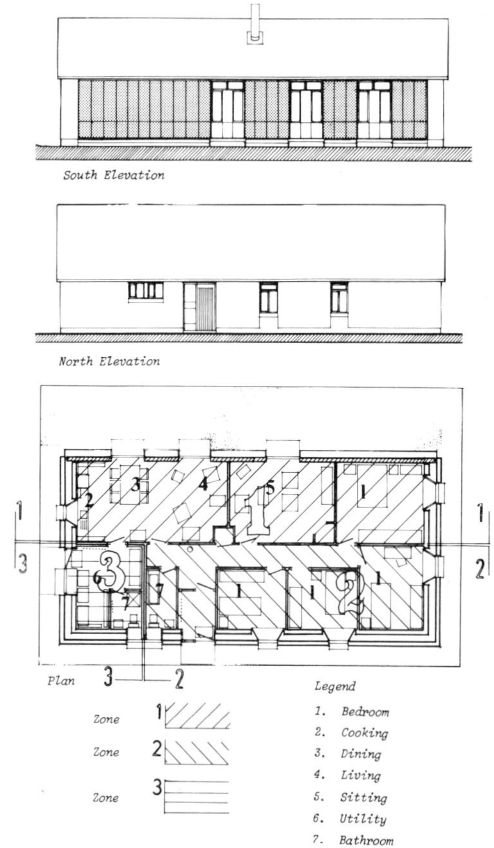

Figure 2: Test House "B"

4. WIND PATTERNS AT THE TEST SITES

Water table tests were undertaken to investigate how these stands of trees and the general topography would affect the wind pattern around the test sites. Further water tests were undertaken to assess detailed wind pattern around test house 'B'.

5. DATA ACQUISITION

The nearest meteorological station was at Kilkenny only 35 km, distant from the test sites. The Oakpark weather station is sited approximately 200 m. from the test sites. A series of statistical comparisons were applied to daily average meteorological records from both the Oakpark and Kilkenny stations, which demonstrated that the Kilkenny data did accurately describe the climatic situation at Oakpark. Not all the equipment was ready at the commencement of the project so for the external parameters particularly Regime 1, the detailed hourly records from Kilkenny were used. The full equipment when installed included aerometers, windvanes and external air temperatures sensors manufactured by Vector Ltd., U.K., sited in four different positions around the test houses. The solarimeters used were manufactured by Li-Cor, Mebraska, U.S.A. The relative humidity probe was manufactured by Vaisala, Finland. The data logger was manufactured by Microdata Ltd., U.K. The tapes from the data logger were read into a DEC/PDP 11 computer, cleaned up as necessary and formatted data written on floppy discs, which were processed on a DEC VAX 750 computer.

6. HOUSE PERFORMANCE TESTS

Because of the particular construction of test house 'B' it was decided to evaluate by experimentation the actual thermal transmittance value for this house. These tests consisted of sealing the house, heating it to a constant temperature and masking it from solar radiation. The overall test period following a thermal stabilisation period of three days was broken into eight hour daily period and sixteen hourly overnight periods for a total of six and a half days. The resultant measured thermal figure of 0.227 was slightly higher than the calculated value of 0.21. Infiltration rate tests were conducted on both test houses to establish the leakage characteristics (6). Having established the air leakage characteristics the infiltration rates may be predicted. The infiltration rate at mean windspeeds in test house 'A' would be .81 ac/hr. The infiltration rate at mean wind speed in test house 'B' would be 0.5 ac/hr. A heat balance test was also undertaken in relation to test house 'B' to assess the accuracy of the predictive formula. The experimental design of the test was similar to the fabric transmittance test with the sealing removed.

7. MONITORING REGIMES

Regime 1 consisted of monitoring test house 'A' for a period of five months. In order to undertake the project with the specified time, monitoring had to commence before the contract itself was signed or some of the equipment ordered. Consequently, the monitoring of Regime 1 was undertaken with limited equipment. This was the basic house.

Regime 2 consisted of monitoring test house 'B' for a four month period. The protection of this house had been increased by the erection of mounds thus altering the micro-climate.

Regime 3 - the fabric conductance of test house 'B' was altered by means of double glazing. This brought this house up to the energy conservation level consistent with the background requirements and was designated the reference house The monitoring period was three months. All the monitoring equipment was fully operable at this stage, but continual trouble with the data logger was experienced during this experimental regime, and the regime had subsequently to be rerun.

Regime 4 - For this regime, a permeable windbreak was erected to the South of house 'B'. The position of this windbreak had been decided from the result of water tests. Monitoring period was three months.

Regime 5 - For this regime the windbreak was left in the same position, but was built up solid. The monitoring was undertaken for a period of three months.

Regime 6 consisted of monitoring test house 'B' without a windbreak and with some of the windows open.

8. MODELLING OF CLIMATIC EFFECTS

Initially the form of model used was that proposed by Siviour (7) and Uttenbroek (8) for daily models. The model related current internal temperature to the internal temperature an hour previously, external temperature, global solar radiation and average wind speed. Hourly regression models were generated for both dwelling zones of Regime 1 using the complete five month data sets for the above variables. These formulae were then used to examine the influence of external climatic parameters on internal temperature without heating during the regime period. It had been realised that the basic regression model would have to be developed and improved further. It was also realised that Regime 1 enjoyed considerable advantages over the subsequent regimes, the main one being the longer period of time involved. An analysis of the heat transfer mechanism operating within test house 'B' was undertaken and the model adjusted accordingly.

To indicate the final form of the model, the results of the regression analysis on the data for Regime 3 are shown in Table 1.

TABLE 1: Y-Variate: INC : Regression Coefficients

Regression Coefficients	Estimate	S.E.	T
Constant	-0.00679	0.00628	-1.08
CDIF	-0.01875	0.00118	-15.87
N	-0.000704	0.000113	-6.25
E	0.000207	0.000236	0.88
S	-0.000718	0.000170	-4.23
W	-0.0003911	0.0000935	-4.18
SREF	0.0026250	0.0000410	64.07
CLOUD	0.004272	0.000745	5.73
TDOT	-0.00438	0.00438	-8.82
SDOT	0.0043703	0.0000804	54.34

TABLE 1 (continued)

--- Analysis of Variance ---

	DF	SS	MS
Regression	9	102.639	11.404335
Residual	1402	9.376	0.006688
Total	1411	112.015	0.079387

Percentage variance accounted for 91.6

The terms of the final model are as follows:

INC	:	Rate of change of internal temperature.
CDIF	:	The composite temperature difference across the building envelope which takes account of both external air and ground temperatures.
N	:	The North component of wind speed multiplied by the difference in internal and external air temperatures.
E	:	The East component of the above factor.
S	:	The South component of the above factor.
W	:	The West component of the above factor
SREF	:	South vertical radiation + (0.3) global solar radiation.
CLOUD	:	Cloud cover (night only)
TDOT	:	Rate of change of CDIF
SDOT	:	Rate of change of SREF

This model was used to examine the shorter Regimes 2 - 6 run on test house 'B'.

9. HEATING LOAD ANALYSIS

Passive solar heat gains and the auxiliary heating requirements were evaluated through the application of an hour-by-hour heating load analysis to the heat houses. The computer program (SOLHET) developed for this analysis uses the general heating demand design approach to estimate the solar gains and auxiliary heating levels.

The models generated for each regime were used to predict hourly internal temperatures for each regime over the same period. The Test Reference Year for Kilkenny prepared by the Irish Meteorological Service was input to the SOLHET program in order to conform with the normal heating season in Ireland. Thus comparable annual results were generated for each regime.

10. RESULTS

The results of running SOLHET on the various regimes are shown in Table 2. Comparing firstly the reference house and the basic house (Regime 1 vs. Regime 3) the percentage solar gain is almost three times greater in the former. In fact this understates the relative advantage of the reference house since the internal temperature of the basic house is often below the external ambient temperature. This is indicated by its extremely low average internal temperature.

The real advantage of SOLHET occurs in the inter regime examination involving the very slight changes of micro-climate. Here the degree of improvement is slight e.g. the predicted average annual internal temperatures for Regimes 3, 4, and 5 only increases from 13.55°C, though 13.83°C to 14.31°C. This would suggest that the reference house with the permeable windbreak is 4.9% better than the reference house on its own. The actual proportional improvement between these regimes is based on the internal temperatures minus the external temperature (the yearly average external temperature for the TRY is 7.8°C). Thus the actual differential improvement is from 5.75°C to 6.03°C. A solid windbreak would yield a 13.2% in energy saving. This seems a greater improvement than there would be expected. The omission of double galzing in this same house would entail a 6.3% increase in energy requirements. All these variations are very slight as would be expected from a consideration of the slight changes to the site surrounds.

TABLE 2: SOLHET Annual Predictions for all Regimes for the Kilkenny Test Reference Year.

Regime	1	2	3	4	5	6
Total yearly heat demand	132917	59809	52806	52692	52885	52348
Total aux. heat demand	109068	29768	24840	23458	22045	35367
Annual useful solar gain (MJ)	23849	30040	27965	29234	30840	16981
Annual excess solar gain (MJ)	0.0	967	1195	1252	2368	91
Percentage aux. heat required	.821	.498	.470	.445	.417	.676
Percentage heat supplied by Solar Gain	.179	.502	.530	.555	.583	.324
Yearly average internal temperature °C	9.2	13.19	13.55	13.83	14.31	11.24
Temperature increase above 7.8°C base	1.4	5.39	5.75	6.03	6.51	3.44

11. ECONOMIC ASSESSMENT

As a first step in assessing any possible saving, the cost of fuel saved must be established. From a consideration of the types of fuel used in domestic housing and their delivered costs an average figure of 0.0514 ECU/kwh was estimated.

The first comparison is between the basic and the reference houses. It is estimated that at current prices the figure of 8240 ECU would represent fairly accurately the difference in cost between such houses. The results of applying economic analysis with these fuel and construction costs is set out below.

Payback period of house = 7 years assuming constant fuel costs.
Internal rate of return house = 10% assuming constant fuel costs and 20 year period.

Present worth value of 34,550 ECU assuming 5% annual increase in fuel price and an annual 15% rate of interest.

The differences between the reference house with or without a windbreak is far less robust. To be economically effective, the cost of the windbreak is of very great relevance. It was decided that the costs adopted for the assessment should be for a natural windbreak formed with reasonably priced plants, planted by local labour. Accordingly the cost of providing additional shelter to conform to Regime 4 was estimated at 410 ECU and at 550 ECU to conform to Regime 5.

When these costs were assessed, the economic returns were as follows:

Reference house with a permeable windbreak added:

Payback period of 21 years assuming constant fuel costs.
Internal rate of return < 1% assuming constant fuel prices and 20 years period.

Present worth value of 550 ECU assuming 5% annual increase in fuel price and 15% rate of interest.

Reference house with a solid windbreak added:

Payback period of 14 years assuming constant fuel costs.
Internal rate of return < 4% assuming constant fuel price and 20 years.
Present worth value of 1145 ECU assuming 5% annual increase in fuel price and 15% rate of interest.

12. CONCLUSIONS

The development of the model was a prime objective. It demonstrates that the use of regression analysis based on short monitoring periods has advantages over direct long-term monitoring and simulation packages especially in studying the effects of microclimate and retrofit changes. However further development of the model building methodology should be undertaken.

A more detailed examination of wind coefficients of exposed building with particulat reference to the use of wind tunnels should also be included in future studies.

Experimental research of this nature should be backed up by surveys and imposed statistics on the national and European housing stock.

REFERENCES

1. OPEC. Facts and Figures. A Comparative Statistical Analysis. OPEC, Vienna, 1981, p. 1.

2. Conservation Commission of WEC. World Energy. Looking Ahead to 2020. Pub. Science, 6 Tech. Press for WEC. London 1978.

3. WAES. Energy : Global Prospects 1985-2000. McGraw-Hill, Hew York 1977. p. 276-280 (App. 1).

4. Hoffman, K.C. Future Patterns of Energy Utilization. N.S.C. Conference. September 1977. Dublin pp. 6 - 7.

5. Energy Policy Project of the Ford Foundation. A Time to Choose. America's Energy Future. Ballinser Pub. Co. Cambridge, Mass. 1974. pp. 133 -134.

6. Warren, P.R. "Ventilation Measurements in Housing". C.I.B.S. Conference National Ventilation by Design. December 1980. pp. 1-34.

7. Siviour, J.B. "Solar Heating in Unoccupied Houses". The Electricity Council Research Centre, Copenhurst, Job No. 461, 1975.

8. Uyttenbroek, J. "Physical Factors Affecting Energy Use in Housing". Symposium Thermal Performance and Energy Use in Housing. An Foras Forbartha, Dublin. April 1977.

9. Clyne, H. "Comparison of Useful Energy Costs". Quarterly Bulletin I.I.R.S., Dublin, April 1983.

EXTENSIONS TO THE ESP BUILDING ENERGY SIMULATION MODEL

J A Clarke
D MacRandal

ABACUS
Department of Architecture
University of Strathclyde
Glasgow, UK G4 ONG

Summary

Over the period 1975-80 ABACUS developed the prototype of a computer based simulation model known as ESP. This system deals by explicit means with the dynamic energy transfers within multi-zone building systems ranging from large commercial developments to housing applications. A grant under the EC Energy Conservation R&D Programme has allowed further developments on two complementary fronts: software enhancements and small machine implementation.

1. Introduction

The foundations for ESP were laid between 1974-77 when ABACUS investigated the various theoretical techniques for modelling energy flow - simple dynamic, response function and finite difference. For a number of reasons the latter technique was favoured: it is easy to preserve the spatial and temporal integrity of real building systems by arranging that whole building energy balance equation-sets (up to 1500 equations say) be solved simultaneously at each computational time-step by direct/iterative matrix techniques based on partitioning methods; the technique can handle non-linearities as well as complex (time varying) boundary conditions; and it allows a building and its plant to be treated simultaneously and by the same theory. By 1977 a working prototype of ESP existed based on implicit finite difference energy balance formulations coupled to custom-built fast matrix inversion techniques capable of achieving stable solutions for any time-stepping scheme.

From 1977-80 three tasks were undertaken:

- Validation of the theoretical basis of ESP, by instrumentation of three test houses to facilitate comparison between measurements and predictions, and by participation in the International Energy Agency's Annex 1, 4 and 10 projects [1,2] concerned with inter-program comparisons, real building comparisons and systems simulation respectively.

- Aided by a UK Government grant ESP was installed in a large multi-disciplinary design practice and, over an 18 month period, passively observed when applied to a number of live projects. This implementation study [3] lead to a number of proposals for user interface enhancements.

- Science and Enginering Research Council grants between 1979-82 allowed a major rewrite of tne package to incorporate tne modifications identified by tne validation and implementation studies. A comprehensive documentation set was also produced [4] to enable external design application.

Since 1981 a grant under tne EC Energy Conservation R&D Programme has allowed further developments on two complementary fronts:

- A range of extensions have been implemented which undoubtedly enhance the model's range of application, ease of use and relevance to design problem solving.

- The system has been reconfigured to operate on the new range of 'super-micros' running the UNIX operating system

ESP has now been acquired by a range of public and private sector practices in the UK, by research organisations in the 8 EEC member states, by a number of central government agencies worldwide and by numerous educational establishments as widespread as the UK and New Zealand. The software is now resident on some 20 computers including the Science and Engineering Research Council's PRIME at Rutherford Appleton Laboratory, England where trial access can be arranged.

2 Overview of the ESP System

ESP dynamically models the energy transfers within, outwith and between a number of zones linked together to form a building. This is achieved by the use of an advanced implicit finite difference technique linked to a proprietory matrix handling scheme which allows the detailed modelling of the zone flowpaths shown in Figure 1. The objective of the ESP system is to make possible a rigorous appraisal of any proposed or existing building either as an aid to an economic selection of installed plant and associated control schedules or, to give an indication of appropriate energy conserving design strategies. The system can be operated either in 'interactive graphics' or 'batch' mode and has sophisticated input management and output facilities to allow the designer to identify potential problem areas, appropriate design modifications, best-fit plant operational strategies, comfort levels, fabric performance, condensation risks and, most importantly, the interrelationship between design and performance parameters which can become the seed solution of future design projects.

Figure 2 shows the connections between the various program and database elements comprising ESP. These include:

ESPIMP	the input management and checking program
ESPCON	construction database organisation of multilayered composites
ESPPRO	casual gain profiles database management
ESPSHD	opaque and transparent surface shading prediction
ESPINS	solar beam tracking
ESPVWF	view factor prediction for complex zone shapes
ESPAIR	air flow simulation
ESPCLM	climatological database management and climatic severity analysis
ESPSIM	dynamic energy simulation by implicit finite difference method
ESPOUT	construction of graphs, tables and statistical relations

3 Contract Funded Software Enhancements

The following enhancements/additions have been made to the ESP system as part of the various contract activities.

User Interface - Input

The move of ESP from single zone simulation capabilities to full multi-zone simultaneous processing (see section 11, appendix 2 of ESP documentation set) introduced complexities in the organisation of the associated computer-based data structure, necessitating development of input management techniques. Stated briefly the developed software communicates interactively with the system user allowing the building description to be entered at a pace which is controlled by the user. Graphical techniques have been incorporated which allow, for example, zone geometry and location checking by full hidden line removal perspective display with automatic photomontage facilities.

Technical Improvements

Much activity has taken place on the technical front. Of this the EEC funded portion has seen the development of independent software systems for the analysis and investigation of exposed surface shading, internal geometry insolation, view factor prediction and the simulation of air movement both internally (zone coupling) and between inside and external environments (infiltration/ventilation). These systems are complete and are represented by the program modules ESPSHD, ESPINS, ESPVWF and ESPAIR respectively (see reference [4] sections 6,7,8 and 9 for operational details and theoretical basis). Recently the task of fully integrating the energy flow module and the airflow module (ESPAIR) was begun. This is necessary to permit the modelling of all temperature driven air movements. A rigorous technique for the modelling of external longwave radiation exchanges with the sky vault, surrounding buildings and ground has also been developed and incorporated within ESPSIM.

User Interface - Output

The graphics output facilities have been extended to allow the display of any predicted parameter in the form of a 3-dimensional surface plot. This move towards an 'experiential' output mode is viewed as essential if model use by the design profession is to be encouraged. Software has also been implemented which allows the instantaneous recall of a detailed energy balance around any point of interest within the building. Connecting software has also been developed between ESPOUT and the ABACUS VISTA package. This allows the display of any ESP multi-zone building image on a colour graphics terminal in colour, fully rendered and textured and lit by any number of user-defined light sources. The same software also allows such images to be coloured on the bases of ESPSIM temperatures or energy flow prediction to again reinforce the move towards 'experiential' thermograph output.

Also undertaken within the project is the implementation within ESPOUT of techniques to allow statistical operations on the simulation results as transferred to the system results database by ESPSIM. Developed statistical operations include linear regression, pattern searching and standard information such as max, min, average, standard deviations, etc.

Climate

The ESP climatological database facility is now complete and allows the establishment of climatic collections and the subsequent management and analysis (statistically or graphically) of the collected data.

Much data has already been entered, for example, the Danish 15 year collection and the UK 40 year collection, and the link with the new UK Science and Engineering Research Council's climatological database has been established.

Plant Component Specifications

Simultaneous plant modelling is a new development field involving, at its most complex, the representation of plant components by discrete nodal schemes (as with the finite difference building model outlined earlier) and the derivation of energy equation-sets which represent inter-node energy flow over time and space dimensions. Such a plant equation system can then be integrated with the corresponding building equation system, subjected to any control function, and solved simultaneously as the model is made to step through finite time-increments. The integration of building and plant at this level is enhanced if both employ the same technical 'engine' and for this reason implicit finite difference methods should be greatly encouraged.

Although the current ABACUS efforts to develop a systems simulation model is funded by the UK SERC, all of the research which lead to the technical specifications of the various plant items was carried out within the EEC contract. Detailed component technical specifications have been drawn up on several plant items including:

- General description of VAV air conditioning system
- VAV Terminal unit
- Ductwork
- Heating and Cooling coils
- Fans
- Humidifiers

- Mixing Box
- Plenum Mixing
- Pumps
- Centrifugal chillers
- Shell and tube heat exchanger
- Cooling tower

These plant component specifications include detailed descriptions of device function, heat transfer laws and miscellaneous technical and economic aspects. They can be viewed as a physical specification suitable for onward processing to establish a mathematical or numerical equivalent. Development of a prototype model, employing the same theoretical basis as ESP, is now underway.

The above technical specifications have been given, as part of the ABACUS contribution, to the ongoing International Energy Agency Annex 10 project.

Simplified Energy Prediction

Emerging dynamic models (such as ESP) are substantially different, in a number of respects, from the previous generation they seek to replace: they are dynamic rather than static; highly predictive in use; and, at the same time, design-oriented and rigorous.

Their increasing flexibility in use has given designers a tool to explore a number of interesting design problems in an attempt to increase understanding of the complex energy subsystem. They also provide the means to develop complementary techniques which, while preserving the dynamic properties of building behaviour, are simple to apply and so overcome many of the problems associated with large system implementation, management and operation.

The ESP system was used as the technical basis of a project to develop an index of Climatic Severity. This Climatic Severity Index (CSI) is building type specific and indicates, by a single number on a dimensionless scale, the stress placed on a building's energy system by an given environment.

The CSI project was a pilot study, now completed, funded by the Scottish Development Department but with substantial input from the EC contract in the form of the resources associated with the long term simulation of a number of distinct house types.

The technique has a number of potential applications in the domestic sector: design location and housing mix decision-making; building control in a way which reflects climatic severity; mechanism for the selection of design climatic data for simulation purposes; and a tool for the determination of some starting design hypothesis.

The major elements comprising the study include:

- the ESP system was used throughout the study to predict the dynamic energy behaviour of selected house types when each were subjected to a range of climates.

- the 10 year climate collections issued by the UK Meteorological Office relating to KEW (Thames Valley), ABERPORTH (Wales) and ESKDALEMUIR (Scottish Borders) were processed by ESPCLM to extract monthly mean values of each of the main climatic parameters for use later in the statistical part of the project.

- a range of representative house types were selected (conventional, well insulated and sealed, passive solar etc) and made ready for computer simulation. These house types were selected to cover the combinatorial possibilities of 'mass and insulation', 'solar potential' and wind permeability.

- at this stage a number of long term simulations were conducted [ESPSIM] to effectively computer process each house type against each 10 year climatic collection.

- then for each zone-month a complete energy breakdown was extracted (ESPOUT) and passed, along with the synoptic climate data, to the statistics section of the project where the data was fed to a linear regression model allowing stepwise regressions in the search for the best fit relationship.

- considering a house of conventional design and a house of passive solar architecture, the regression equations to emerge were:

$$E = -35.8t - 2.9Rd - 1.4Rf + 3.1V + 823.2$$
$$E = -61.9t - 0.4Rd - 2.3Rf + 7.5V + 1116.5$$

  where, E is the monthly energy requirements (KWhrs), t is the monthly mean temperature, Rd is the monthly mean direct radiation, Rf is the monthly mean diffuse radiation and V is the monthly mean wind speed.

- at this stage, the effect on the foregoing equations, of a variety of micro-climatic modifiers, can be determined by directly modifying the climatic data (to simulate shelter belts for example) or by superimposing such factors as shading on the ESP simulations.

- the CSI is now generated by computing the relative contribution of each climatic parameter to the overall energy requirement and then expressing the ratio of these energy contributions (to house totals) in terms of an arbitrary dimensionless scale. The month CSI values for each locality can now be represented in the form of a CSI map.

- for any country for which such a map exists, the CSI for any locality can be read off and used to determine the relative energy consequences of adapting alternative housing mixes or locations.

The process of CSI formulation, its relationship to the commonplace degree day technique and detailed examples of technique application are reported elsewhere [5].

4 Contract Funded Implementation Study

In order to gain any acceptance within the building design profession, computer simulation models will have to be comprehensive, rigorous and, above all, user friendly. These requirements have serious implications for the computer environment in which such models are implemented. In the energy simulation field, to be sufficiently comprehensive and flexible (to cope with a wide range of buildings) and yet maintain sufficient rigour to be useful as a research tool, requires using a large number of nodes to describe the building and small timesteps when examining dynamic behaviour. This results in a large and complex program which is computationally very demanding. On the other hand, the design orientation and user friendliness requirements imply a highly interactive program, using graphics and pre-existing databases to the full. This necessitates very high levels of disk and terminal traffic and a need for a range of peripheral support programs, for example, to maintain the databases, etc.

To illustrate these characteristics consider an existing system, ESP. The main simulation engine, ESPSIM, is a large program containing over 90 subroutines and 13K lines of Fortran. It needs 200K bytes of memory for the program instructions and a further 300K for the data. An average run, for example, 6 zones for 1 week, uses 20 seconds of CPU time on a DEC10, does 4K disk accesses, needs 3K bytes of input data from other modules (for example, ESPIMP) and produces 160K bytes of output data for analysis by another module, ESPOUT. The other two main modules, ESPIMP with 13K lines of code and ESPOUT with 11K lines of code, use approximately the same CPU time, do more disk I/O and also do substantially more I/O to the terminal.

The effect of this size factor is that most advanced energy models have been developed on large mainframe computers and so, due the the cost of time on a mainframe, have been inaccessible to the vast majority of potential users. It is clear that, in order to make such sophisticated models accessible to the bulk of the design profession they have to be implemented on hardware affordable by an average-sized design office. However, it is only over the last year or so that sufficiently powerful hardware has become available at a sufficiently low cost.

'Super-Micro' Hardware

With the advent of the Motorola MC68000 and similar 'super-micro' chips, a new generation of computer hardware has arrived. Based on these 16 bit chips, and using the latest Winchester disk technology, several new computer systems have appeared on the market, giving a performance capable of dealing with thermal simulation of buildings, but at a reasonable cost (£5K-15K).

The systems fall into two categories, depending on the amount of memory available to a user program. Although all the chips have a large address space, for example, 16Mb for the MC68010, and may operate internally on 32 bit data, they all use a 16 bit data word. Thus the largest integer, and hence the highest memory address, that can be held in one word is 64K.

- In some systems, for example, PDP11/24, user programs are only allowed to use 1 word to hold their memory addresses, This restricts the size of a user program to 64Kb, although the system itself uses memory management registers to map a programs 64Kb logical address space onto the much larger physical memory. This allows several independent user programs to reside in memory at the same time.

- However, in the more sophisticated systems, for example, Perq, user programs can use more than one word to hold memory addresses. This removes the 64K size restriction from user programs, although the program must still fit in the available physical memory.

The user environment at which these 'super-micros' are targeted also falls into two categories, single user, highly interactive, graphical 'workstations' and multi user 'office automation' systems. Although it may be necessary, for economic reasons, for energy modellers to develop their software on the second category of system, it should be borne in mind that in the near future each user of their models will have the equivalent of a complete system to themselves. Also, the general environment in which the models will run will be much more graphical and interactive than current systems.

'Super-micro' Software

Given the hardware capable of meeting the computational and I/O demands, the two most important factors affecting energy modelling programs are the operating system and the language used.

The choice of operating system can have very profound implications for the user, since this will dictate the power, flexibility and user friendliness of the computer environment used for development and running of the models. The operating system also has a large influence on the portability of user programs. Even programs written completely within a standard language will have operating system dependencies built into them that can be very difficult and inefficient to remove. On this basis alone, the obvious choice for an operating system is UNIX [6] which is a non-vendor system available in one form or another on nearly every 16 bit computer and on most 32 bit minicomputers. However UNIX is also one of the most powerful and flexible operating systems available today There are over 100 standard utilities supplied with UNIX to perform such tasks as text processing, file and string handling, terminal independent graphics, 4 languages, communication/networking, etc. UNIX completely hides the hardware from the application program, so that all program I/O, whether to disk, terminal, or even to other programs, is treated identically by the program. It also allows several independent programs to run simultaneously and provides a mechanism for a program to communicate with each other. Thus an energy model could be developed as a series of independent, single function modules for example, insolation model, matrix reduction software, database handling, etc, which could be tested in isolation, using data files and terminal input, and then integrated into a complete system with each module getting its data from other modules. This power and flexibility, coupled with the systems user friendliness, is one of the main reasons why UNIX is rapidly becoming the standard operating system for the new generation of small computers.

The second major influence on user programs is the language in which they are written. Two factors operate here, firstly the ease of the initial implementation and secondly the ease of maintaining and extending the program. The most common language used today, and certainly the most widespread in terms of compiler availability and user experience is Fortran. It is the traditional language of non-computer scientists and the latest version, Fortran 77, is rapidly gaining acceptance as a standard, for example, SERC common base policy. In spite of its obvious shortcomings, there is no clear alternative to Fortran. The obvious contender is Pascal, but, although it is a good example of a modern language, it is not sufficiently numerically orientated for large energy simulation programs. Probably the best language, although not widely known, is 'c', especially given the suitability of the UNIX environment. Designed for implementation of large systems, i.e. UNIX itself (95% written in c) it has all the facilities lacking in Fortran and is an extremely efficient language. It also integrates very well with UNIX. The power and flexibility of the c / UNIX combination and the ease of maintenance of a well structured program will encourage more and more people to switch to c, especially as UNIX continues to proliferate as the standard small machine operating system.

ESP Implementation

In the light of the above comments about the new generation of hardware it is clear, perhaps, why it was decided to install ESP on a small 'super-micro' running UNIX. As ESP already existed as a reasonable well structured program written in fairly standard Fortran, it was decided that it was not worthwhile to convert ESP to c unless sufficient resources were available to rewrite the program to take account of new developments in energy modelling and fully exploit the power of c and UNIX. It was therefore decided to produce a UNIX based, Fortran 77 version of ESP and to forgo the efficiency advantages of a tightly coded c version (expected to be a factor 2 faster than the Fortran version).

The first task was to convert the code to standard Fortran 77. This required a very large number of trivial modifications to the source code, for example, parentheses around the parameter list in PARAMETER statements, as well as a few not so minor changes, for example, character handling. Thus there now exists a version of ESP which closely adheres to the Fortran 77 ANSI X3.9-1978 standard. This task, although laborious, has already had some beneficial spin-off in facilitating the implementation on ESP on a Prime 550. It has been found that a number of so-called Fortran 77 compilers do not, in fact, adhere to the standard, implementing only a subset of it and adding extensions which violate the standard.

The second task was to convert ESP, and especially the terminal and graphics libraries, to interface with UNIX. This caused a lot more difficulty as in several cases it required changes to the source code of ESP modules. Some changes were relatively straightforward, for example, changing the date format, but others had more serious implications. An example of this was file handling, in particular random access files. Several changes had to be made to reconcile the different philosophies of UNIX and the Dec10. Another problem was terminal handling, where UNIX provided much simpler and efficient ways to control terminal characteristics than the Dec10 environment in which ESP was initially developed. Due to deficiencies in the Fortran 77 standard several classes of error which could be trapped by ESP on the Dec10 could not be caught and had to be left to the standard UNIX error reporting. This may cause difficulties in future non-UNIX implementations based on the Fortran 77 code.

The third step was the actual implementation. Two separate implementations were considered, corresponding to the two addressing schemes mentioned earlier.

- Firstly consider hardware restricted to 16 bit addressing. This restriction causes severe problems because of the size of ESP. Attention was concentrated on ESPSIM, the simulation engine, as it was felt that, being the biggest and the least susceptible to being split into pieces, the most difficulty would occur with it. A closer look at the sizes of the subroutines and their data requirements revealed several problems. Quite a number of subroutines required 8K of text space (program instructions) and up to 40K of data space. This was further complicated by the need to have the Fortran library available most of the time. The only feasible solution to this problem was to develop a triple overlaying scheme, as shown in Figure 3. In this 8K of physical memory was allocated to the overlay handler and the most commonly used library routines. The next 8K of physical memory was allocated to the ESP subroutines, which were loaded from disk as required. The next 8K of logical address space was allocated to the Fortran library. The library resided on six 8K segments in physical memory and the required segment was mapped into the libraries 8K of logical space by manipulating the memory management registers. Finally each subroutine was made responsible for loading the data it required from disk into the remaining 40K. Fortunately, all the subroutines and data concerned with the matrix reduction could be fitted into 8K of text space and 40K of data space, but this still requires up to 8 context changes (200 disk accesses) at each timestep. This would slow the execution of ESP by a factor 25, approximately. It was felt that a system running as slowly as that would have a very limited usefulness.

- However, when dealing with hardware not restricted to 16 bit addressing, the picture is completely different. The only requirement is that the system has enough memory for the operating system and ESP. Although 512Kb memory should be enough, at least 1Mb is required to avoid bringing the whole system to a halt while a simulation is running.

As it stands now, there is a UNIX based version of ESP which will run in any UNIX based machine with at least 512Kb of memory available to user processes. Since it is written in Fortran77, it should simplify implementing ESP on non-UNIX systems.

Conclusions

The ESP building modelling system is now fully developed, tested and documented and the following are now available:

- A version of ESP which closely adheres to the ANSI standard for Fortran 77

- A magnetic tape containing a version of ESP capable of running on any UNIX based 'super-micro' with an effective user address space of 512Kb or more.

- An overlaying scheme and handler to allow ESPSIM to be implemented on a UNIX based computer with a 16 bit address space.

General conclusions from the study are

- The new generation of large address space 'super-micro' computers are eminently suitable for dynamic energy simulation programs.

- The UNIX operating system provides a versatile, portable and productive environment for development of large software packages and therefore is rapidly becoming the standard operating system for the new generation of hardware.

- Because of the inherent drawbacks of the Fortran language, an attempt should be made to encourage the use of block structured, modern languages. In the UNIX environment, this should be 'c'. One method of achieving this would to have available a selection of simulation 'tools' already written in c.

- It is to be hoped that the mechanism can now be found to allow the use of ESP and similar systems as the technical basis of other research projects aiming to recommend energy saving techniques after careful analysis of the combinatorial design options.

References

1 Oscar Faber and Partners, St. Albans, England, 'IEA Report Annex 1 - Results and Analysis of Avonbank Building Simulation', April 1980

2 Clarke J.A. & McLean D.J., 'IEA Annex 4: Results from the analysis of the Collins building by the ESP system', ABACUS publication, University of Strathclyde, Sept 1983

3 Maver T.W. & Ellis J., 'Implementation of an energy model within a multidiciplinary practice', 'Proceedings of CAD82', Butterworth Technical Press, 1982

4 Clarke J.A., 'ESP Documentation Set' ABACUS publication, University of Strathclyde, Sept 1983

5 Markus T.A. et al, 'Climatic Severity', Ocassional Paper OP82/5, Dept. Architecture, University of Strathclyde, March 1982

6 Complete issue devoted to UNIX Bell System Technical Journal 57(6) 1978

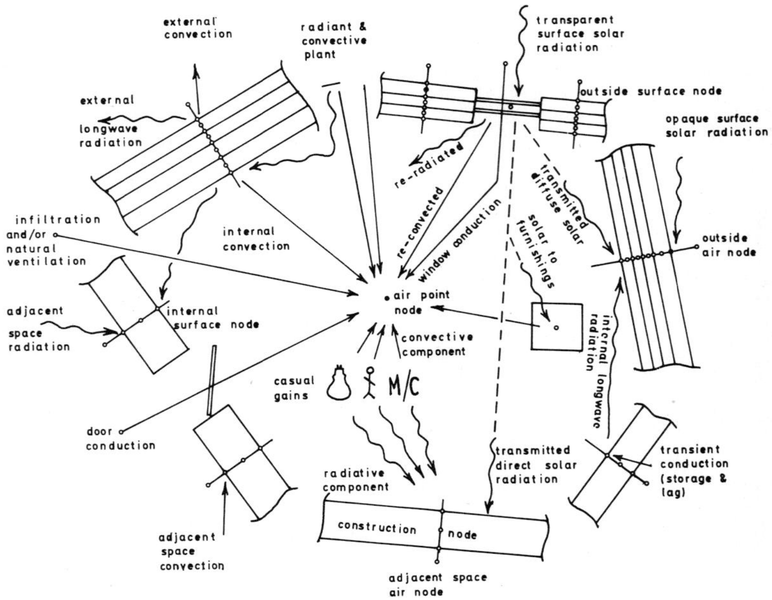

Fig. 1 - Energy Flowpaths

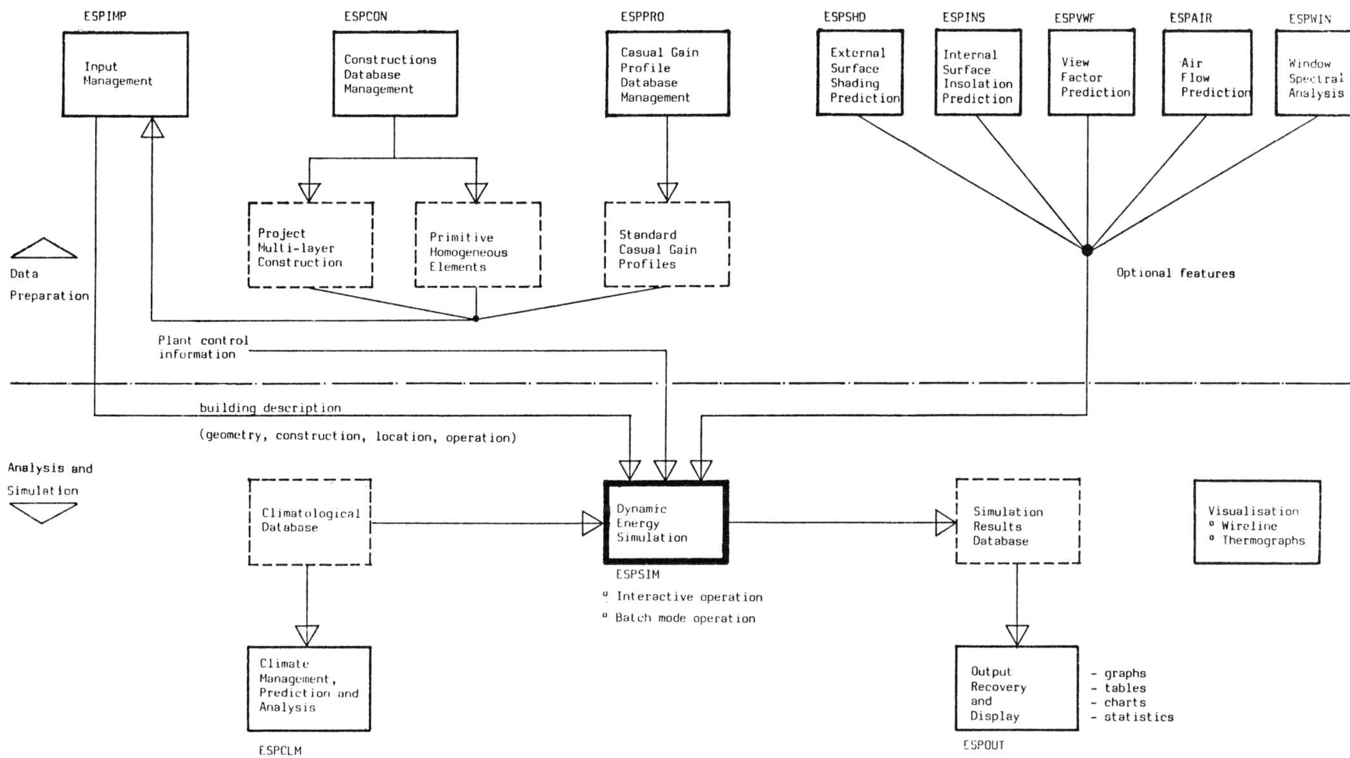

Fig. 2 - ESP System

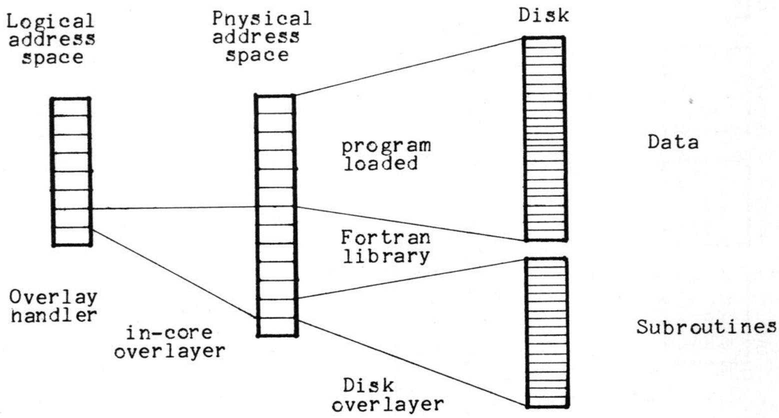

Fig. 3 - Overlay scheme

REINFORCED PHENOLIC FOAMS FOR THERMAL INSULATION

P. DUBOIS et C. REINAUDO
Laboratoires de Marcoussis, Centre de Recherches de la CGE
Route de Nozay
Marcoussis - 91460 FRANCE
Tél. (6) 449 12 00

E. MOREL et C. CHAUVELIER
Institut de Recherche de Chimie Appliquée
BP 1
Vert-le-Petit - 91710 FRANCE
Tél. (6) 493 24 75

Summary

The reported research is related to phenolic resins panels for thermal insulation. This research started from previous results on the fabrication of foams from powdered Novolaque phenolic resins without acid catalyst. We have developped the fabrication of 25 x 25 cm2 panels of reinforced phenolic foam with a specific weight of 80 kg/m3. These panels have a thermal conductivity equivalent to that of the most commonly used polystyrene foams and markedly better mechanical properties, thermal behaviour and fire resistance.

The use of micro-wave or high frequency heating instead of resistor heating has already provided encouraging results on samples : more homogeneous distribution of glass and foam and increased expansion, leading to lower thermal conductivity and lower specific weight which will allow lower prices.

O. Foreword

This research is related to phenolic resins panels for thermal insulation. These panels are made of reinforced foams possessing excellent fire resistance as well as good mechanical characteristics. This research started from previous results on the fabrication of foams from powdered "Novolaque" phenolic resins without acid catalyst.

Phenolic foams possess excellent fire behaviour : inherent non-flammability and absence of toxic or corrosive droplets or fumes ; up to now their development was hampered by :

- the acid nature of the catalyst used with liquid resins, causing residual acidity in the finished product which could lead to corrosion of the substrates,
- mechanical weakness of the foams, which crumble.

We will successively present :

- . the selection of the materials and the selection of the fabrication process for the panels,
- . the characteristics of the reinforced foams thus obtained.

1. Material and fabrication process

Novolaque thermo-setting formo-phenolic resins are based on three essential components : the basic resin, the blowing agent and the fillers.

1.1 The basic resins

Novolaque resins are characterized through three essential parameters :

- the melting temperature, which is a consequence of molecular weight ;
- the hexamethylenetetramine ratio, which characterizes how easily the resin sets ; in theory, 1.2 percent HMTA suffices to obtain a well reticulated lattice but the amounts used in practice vary between 2.5 and 11 percent by weight ;
- the pouring behaviour which depends on the viscosity in the molten state.

We have made a comparison between three resins : PB 10, PB 15 and PB 17, whose characteristics are as follows :

Resin	Melting temp. (°C)	HMTA ratio (percent)	Pour at 125°C (mm)
PB 10	90 - 96	9.7 - 10.3	19 - 25
PB 17	77 - 83	10.3 - 10.9	54 - 66
PB 15	79 - 85	5.4 - 6.0	63 - 77

This comparison has led to the selection of resin PB 15 for the following reasons :

- low melting temperature,
- low viscosity.

Figure 1 shows that the density of the foam obtained varies inversely to the viscosity of the resin used. This is due to the fact that the released gas must exert a lower effort to overcome the viscosity forces within the molten resin when this latter is less viscous.

1.2 The blowing agents

Expansion of the Novolaque resin is obtained through release of a gas within the molten polymer. The expanded structure is preserved through hardening of the thermo-setting polymer. Novolaque being powdery, we have looked toward powdered solid expansion agents. The relative expansion and reticulation kinetics determine the quality of the foam. In addition to the nature of the resin and that of the blowing agent, the following parameters play a part : temperature, viscosity and superficial tension of the molten resin.

As blowing agent, we have selected dipentax, which can release 240 cm3 of gas per gram. Figure 2 shows the influence of the amount of blowing agent used on the density of the foam obtained, at a constant curing temperature of 160°C ; when the blowing agent content increases, the foam density decreases and then stabilizes from a certain value of the blowing agent ratio, this being due to the bursting of pores when the pressure of the released gas becomes too high with respect to the resistance of the walls.

The results of figures 3 to 6 have been obtained through differential enthalpic analysis with an increase of the furnace temperature of 10°C per minute. Figure 3, relative to a mixture of resin and dipentax, shows two exothermal peaks corresponding to two successive reactions : resin reticulation between 120 and 185°C and chemical breakdown of dipentax at 200°C. When comparing this plot to those of figures 4 and 5, relative to resin and dipentax, one can see that plot 3 combines plots 4 and 5. The temperatures of the two peaks are quite different so that the reaction kinetics do not allow to obtain a satisfactory foam. It is thus interesting to try and modify the reaction kinetics so as to obtain better simultaneity between reticulation on the one hand and expansion on the other hand. For this purpose, we have successively incorporated four additives into the blowing agent : zinc oxide, alumina, carbon black and zinc stearate. Alumina and, mainly, zinc stearate (see fig. 6) allow lowering the dipentax breakdown temperature and bringing it closer to the resin reticulation temperature.

1.3 The fillers

Two types of fillers have been investigated :

- reinforcing fillers (glass mats or fibers) ; glass mat has been selected as mineral filler since it strengthens the material mainly by increasing its bending resistance,
- fillers which would lower the cost of the foam ; these charges lead to two opposite effects : they enhance foam expansion by lowering the breakdown temperature of the blowing agent (see fig. 6) and they make the foam heavier through the addition of a denser mineral product as shown by figure 7 in the case of addition of alumina. It must be noted that the addition of alumina also leads to an increase of the foam oxygen index.

1.4 Obtention of low-density foams

a) The expansion of unfilled Novolaque powders can lead to low specific weight foams (up to 20 kg/m3) if the components of the mixture are well homogenized in the molten state in a mixing machine. The expansion process is however slowed when glass, mat or chopped fibres, is incorporated into the mixture. A density of 0.08 has been obtained for the 25x25x3 cm3 panels when the expansion took place in a oven. Observation of the expanded foam plates obtained through oven curing shows a color difference between the skin and the interior of the plates. The foam becomes all the more insulating as the foam expands and thus limits the thermal flow from the outside toward the center. The heat released by the reaction is too low to sustain expansion and hardening. The resin polymerization and blowing agent breakdown kinetics are disturbed. This observation is confirmed by a result : the density of thin plates is always lower than that of thick plates.

b) This has led us to investigate microwave or high frequency heating in order to obtain a higher in-depth temperature so as to obtain reinforced foams with lower density.

We have tried microwave heating at 2.450 MHz with a 1.2 kW generator. Starting from a 5 mm preimpregnated glass mat, we obtained, after a 3 minutes heating, a closed cells foam with a specific weight of 35 kg/m3. With a thinner preimpregnated mat, the results were not as good. It is probably due to the fact that, the heat imput in the mat being lower, the heat losses to the outside were relatively higher and the temperatures in the mat lower.

We have also tried high frequency heating at 27 MHz with the electric field perpendicular to the mat. We have obtained, after a 20 seconds heating, a foam with a specific weight of 50 kg/m3. Better results can be expected with an electric field parallel to the mat.

In both cases, the glass and foam are distributed in a more homogeneous manner and the cross-section of a panel of foam shows a dark-colored material at center which points to good polymerization.

c) We have designed a new equipment for the production of panels of .5meterx1meter. This equipment, which will be operated in December 1983, provides high frequency heating at 27 MHz with the electric field parallel to the preimpregnated mat.

2. Characteristics of the reinforced Novolaque phenolic foam

2.1 Material composition and panel fabrication

The various components are powdery, except for one the tensio-active agents which is liquid. Mixing is realized, in solid state, in a fast mixer whose blades rotate at 1800 rpm ; this powdery mixture is then mixed in a BUSS machine at 75°C. The mixing proportions (by weight) are as follows :

- Phenolic resin PB 15 = 100
- Dipentax blowing agent = 6
- Additives = 6

We then realize a pre-impregnated glass mat as follows : 30 parts glass (b. wt) are incorporated into 100 parts (b. wt) of the above mixture. The glass mat is then pressed under pressure (3 kg/cm2) at 100°C for 4 minutes to obtain a 3 mm thick plate.

Finally, expansion is performed for 30 minutes at 200°C in a oven. The 250x250x30 mm panels thus obtained have a specific weight of 80 kg/m3.

2.2 Mechanical characteristics

We have characterized the bending and compression mechanical behaviour of panel elements subjected to contraints applied perpendicularly to the plate :

. elasticity modulus 23 kg/cm2
. yield point 1.2 kg/cm2
. strength for 10 % crushing 1.8 kg/cm2
. strength for 50 % crushing 5.2 kg/cm2

The values of the bending resistance show the beneficial effect of the skin present on the surface of the panels :

With skin	. σ_{max}	15 kg/cm2
	. $E_{bending}$	1330 kg/cm2
Without skin	. σ_{max}	7.5 kg/cm2
	. $E_{bending}$	55 kg/cm2

2.3 Porosity and thermal characteristics

a) Porosity

The measurements taken under an air overpressure of 400 mm Hg show a 24 % closed porosity.

b) Expansion coefficient

The plot of figure 8 shows the variation of the expansion coefficient according to temperature. This coefficient is equal to 1.6×10^{-5} at 20°C.

c) Thermal conductivity

The measured thermal conductivity is 0.04 $W.m^{-1}.K^{-1}$ at 40°C for panels with a density of 0.08. Our measurements have shown that the thermal conductivity is a linear function of the density. For a density of 0.04 the thermal conductivity decreases to 0.032 $W.m^{-1}.K^{-1}$. So, decreasing the density of the foam will lower both the price and the thermal conductivity.

2.4 Ageing

2.4.1 Water absorption

a) Capillarity

The base of the sample is kept in 2 mm of water, the remainder of the sample being in the atmosphere of the enclosure at 20°C and 90 % relative humidity. After 1 000 hours ageing under these conditions, the ratio of moisture picked-up by the foam tends toward an asymptotic value of 1 % by weight, which corresponds to the water picked-up by the resin. The thermal conductivity of the foam increases but returns to the initial value after one hour at 40°C.

b) Immersion

The specimens are immersed into water at ambient temperature. After 1 000 hours ageing, the ratio of water picked-up by the foam is equal to 24 % by volume, which corresponds to partial filling of the open pores.

2.4.2 Weatherometer behaviour

Phenolic resins are stable even at high temperatures. Specimens have been subjected to ageing tests in a weatherometer under conditions selected as typical of the extreme temperature, humidity and solar radiation conditions to be encountered by thermal insulating material placed externally :

. temperature : 70°C
. relative humidity : 90°C
. radiation : 1 kW/m2

No modification of the foam properties has been noticed after 500 hours of testing.

2.4.3 Temperature ageing

a) Temperature cycling

No modification of the mechanical and thermal properties of the samples occured after 100 cycles between - 10°C and 50°C.

b) High temperature ageing

- No modification of the mechanical and thermal properties of the samples occured after
  . 1 000 hours at 100°C
  . 1 800 hours at 140°C

- No modification of the thermal conductivity but an important decrease of the mechanical properties of the samples occured after
  . 700 hours at 160°C
  . 700 hours at 180°C

2.4.4 Ageing in an alcaline solution

No modification of the mechanical and thermal properties of the samples occured after 2 500 hours ageing in an alcaline solution with a PH of 12.

2.5 Fire resistance

a) Measurement of the limit oxygen index (LOI)

We measured the minimum oxygen ratio in a nitrogen/oxygen mixture required to sustain flame propagation in a vertical specimen.
Measured LOI = 39

b) Measurement of flame propagation (UL 94)

We measured the ignition time and the flame propagation speed on a vertical specimen.
The material is classified V_o.

c) Epiradiator test

Three sets of 20 x 20 cm2 samples have been subjected to a thermal flux of 6 W/cm2 generated by a radiating panel, in the presence of a pilot flame ; this flux is roughly equivalent to that received by the ceiling of a room at the moment of general inflammation.

Set 1 : 30 mm thick reinforced Novolaque foam panels, directly exposed to the thermal flux.

After 30 minutes exposition, photographs 1 show that the sample has been carbonized throughout its thickness ; the glass mat remains in place ; the entire sample never ignited.

Set 2 : composites consisting of a 30 mm thick reinforced phenolic foam panel with a 10 mm thick protective plasterboard plate (Placo plâtre).

This sample exposed to the same 6 W/cm2 flux for one hour, never ignited. The plasterboard has split but the foam carbonized only over a limited thickness. Photographs 2 show the condition of both faces of the foam after this test.

Set 3 : composites consisting of a 30 mm thick expanded polystyrene bonded to a 10 mm thick, protective plasterboard plate (Placoplâtre). After 12 minutes exposition, the plasterboard has split and the polystyrene has sublimated in spite of the protection of the face exposed to the radiation. Photographs 3 show the condition of both faces after this test.

Figure 9 shows the evolution, during these tests, of the temperature of the face not exposed to the radiation for sets 2 and 3, these sets consisting of phenolic resin or polystyrene protected by means plasterboard.

One sees that :

- the temperature is stabilized at 150°C after 1 hour testing in the case of phenolic foams,

- a sharp temperature rise occurs after 12 minutes testing in the case of polystyrene ; this is when the polystyrene sublimates.

3. Conclusion

The cost of the materials used to compound the reinforced phenolic foams studied amounts to 9.40 FF/kg for a composite containing :

. 70 % by weight of phenolic resin costing 9 FF/kg,
. 22 % by weight of glass mat costing 9 FF/kg,
. 4 % expansion agent dipentax costing 13.65 FF/kg,
. 4 % additives costing 15 FF/kg.

This materials cost of 9.40 FF/kg compares with 6.65 FF/kg for polystyrene and 10.5 FF/kg for polyurethane. The ex-factory cost of expanded plates is 12.85 FF/kg for polystyrene and 17.5 FF/kg for polyurethane.

We have investigated a heating mode allowing to obtain more homogeneous temperature within the panel during polymerization so as to obtain foams with a lower density. The use of a micro-wave or high frequency heating instead of resistor heating has already provided encouraging results on samples : more homogeneous distribution of glass and foam and increased expansion of the phenolic foam in the presence of glass mat, leading to a lower density.

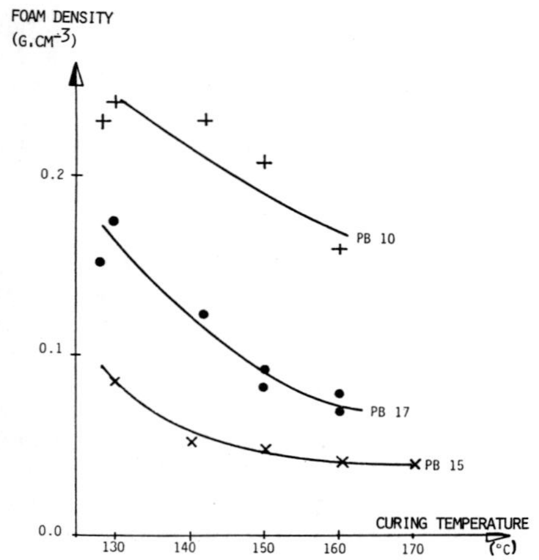

Fig. 1: Foam density versus curing temperature for 3 Novolaque resins (Phenolic resin + 5% dipentax)

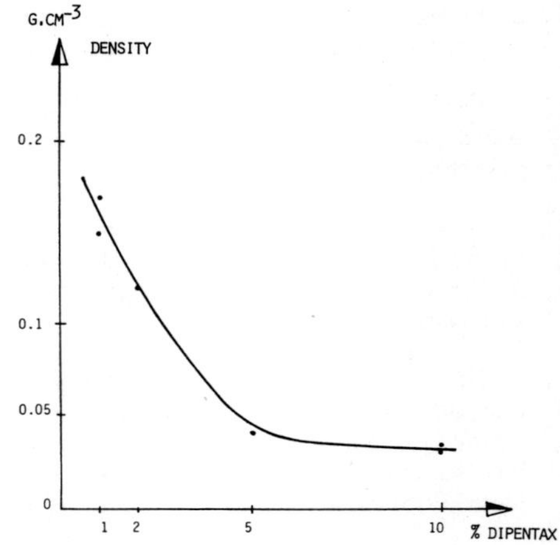

Fig. 2: Influence of the amount of blowing agent on the foam density (PB 15 resin cured at 160°C)

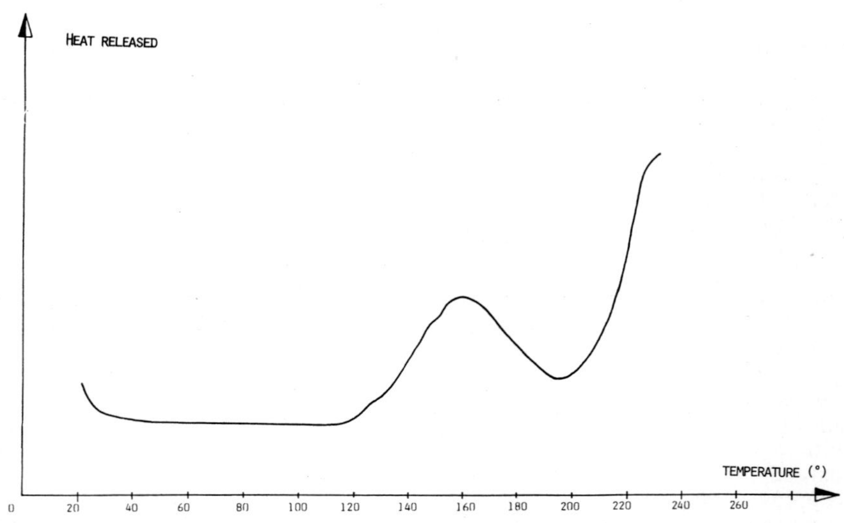

Fig. 3: Heat released by a PB 15 and dipentax mixture versus temperature

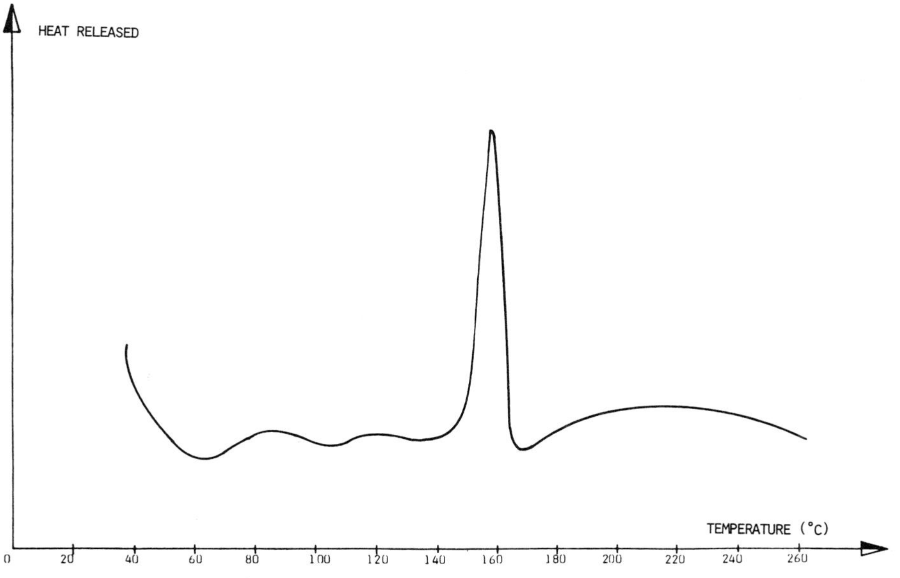

Fig. 4: Heat released by PB 15 resin versus temperature

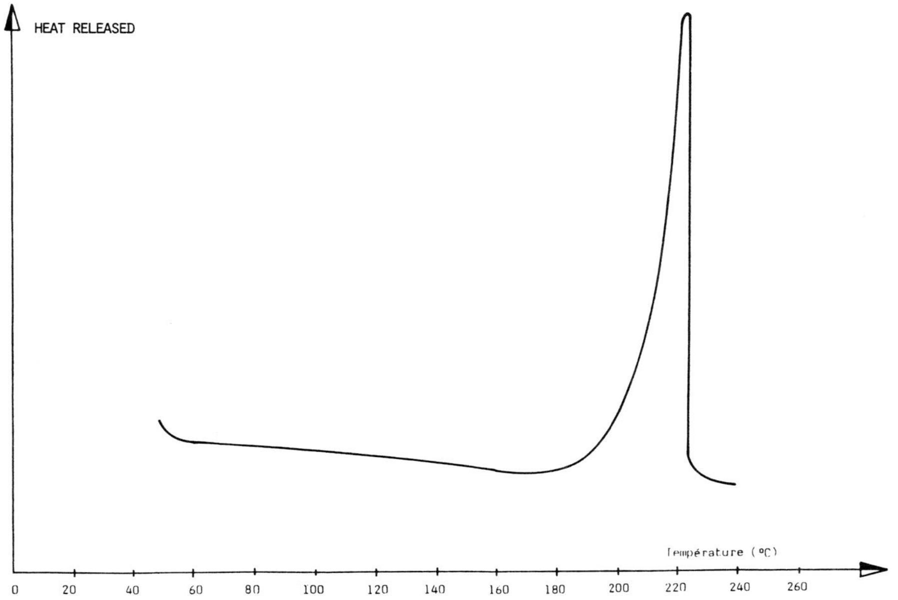

Fig. 5: Heat released by dipentax versus temperature

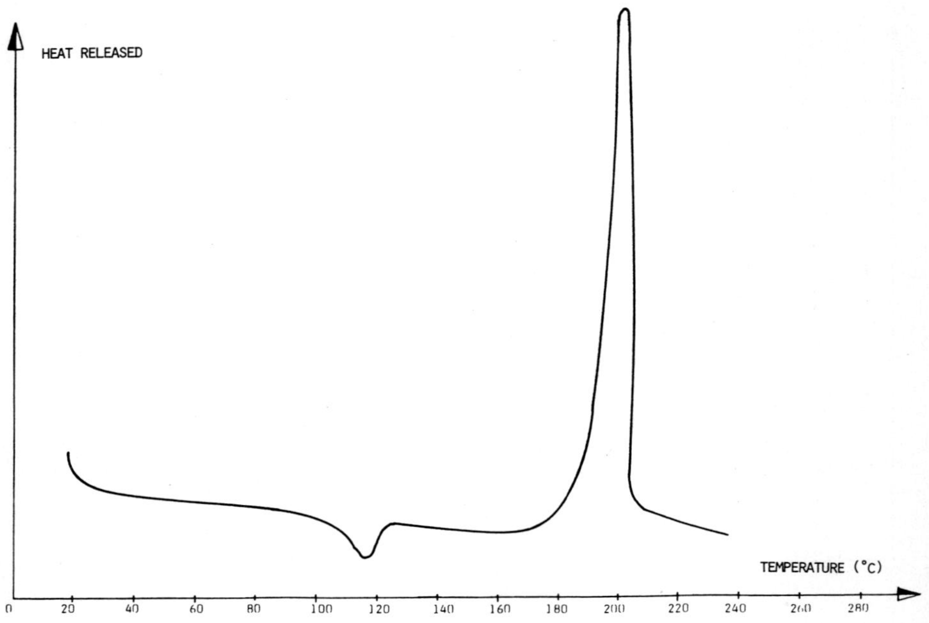

Fig. 6: Heat released by a mixture dipentax - Zinc stearate versus temperature

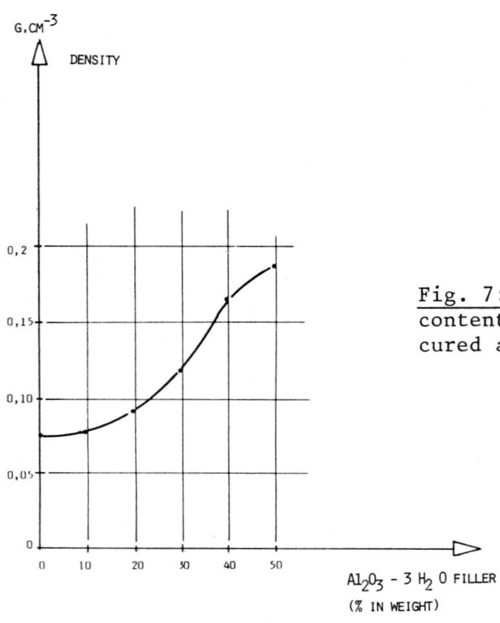

Fig. 7: Influence of the alumina content on the density of foam cured at 160°C

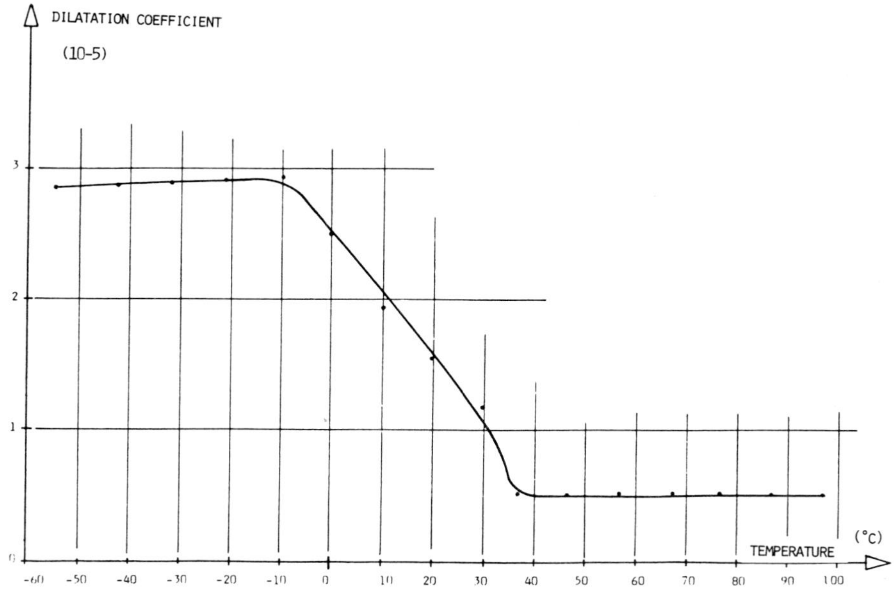

Fig. 8: Dilatation coefficient versus temperature

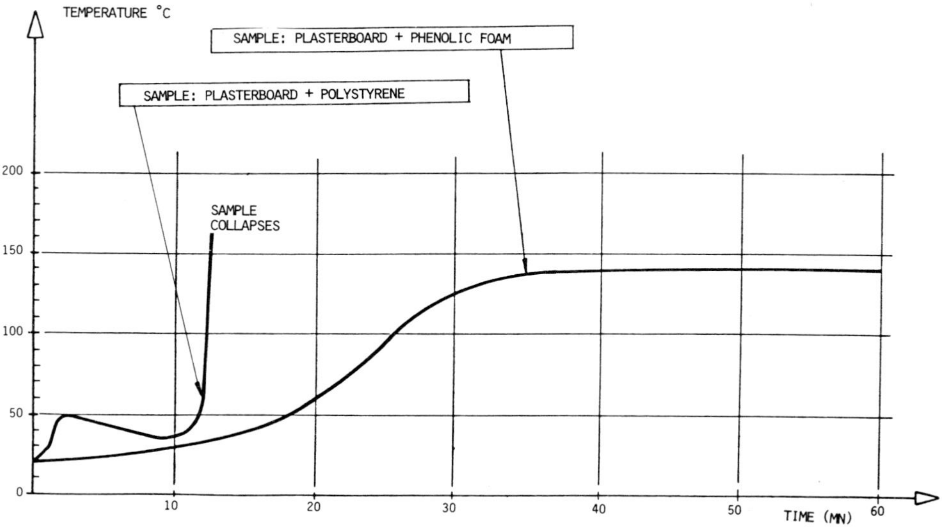

Fig. 9: 20 x 20 cm^2 samples under 6 W cm^{-2} heat flux - Temperature of the side opposite to heat flux

Side opposite to heat flux

30 mm thick phenolic foam reinforced panel after
30 minutes under a heat flux of 6 w cm^{-2}

Side under heat flux

PHOTO 1

Phenolic foam side

Phenolic resin + plasterboard plate
After one hour under a 6 W cm2 heat flux

Plasterboard side

PHOTOS 2

Polystyrene side

Polystyrene + plasterboard

After 12 minutes under a 6 W cm^{-2} heat flux on the plasterboard side

Plasterboard side

PHOTOS 3

BITUMINOUS HEAT INSULATING MATERIALS

V. CASTAGNETTA
Bituminous Products Development and Quality Control,
Industria Italiana Petroli (Italian State Oil Group ENI)

Summary

The project envisages energy savings in the heating of civil and industrial buildings with the use of improved insulating materials, e.g. waterproofed against humidity and water vapour and protected against the attack of polluting agents in the atmosphere, by means of bituminous products. The research aims at the utilization of any kind of granular or fibriform insulating material, where each individual element is completely coated with an impermeable and protective bituminous binder. The latter, as it also binds the elements together, makes possible the manufacture of insulating panels. The coating and manufacturing process can be performed by thoroughly mixing the insulating material with a relatively hot bitumen or with a cold bituminous emulsion and by pressing the mixture so as to form suitable panels. The report covers both the hot and the cold process, as well as the technical characterization of the bituminous heat insulating panels obtained.

1. INTRODUCTION

Our project comes within the general framework for economizing on energy consumption for the heating of civil and industrial buildings. It envisages a better utilization of the properties of the insulating materials through the use of the waterproofing and protective properties of bituminous binders.

In fact, the thermal conductivity coefficient λ values of insulating materials and the normal techniques for their application are known, but some comments have drawn attention to the deterioration of the λ due to the presence of humidity and/or the degradability of certain insulating materials.

As the waterproofing and protective properties of bituminous binders are also known, it is believed they can be exploited with a view to eliminating or at least reducing the λ deterioration referred to above.

Our project covers the use of any insulating material in granules or fibres and its objective is to create - by using bituminous binders - the protection and waterproofing of each individual insulating element against humidity and water vapour, protecting it, moreover, against the attack of any polluting agent in the atmosphere and thus against any degradation capable of reducing its efficiency and life.

Thanks to the impermeableness and indegradability of bituminous insulating materials, energy consumption in the civil and industrial building heating field can drop to and stabilize itself at considerably lower levels.

Our programme considers the two alternatives of the coating process of the insulating granular material (at present expanded polystyrene) either with hot bitumen or with cold bituminous emulsion (the latter with inherent energy saving and no air pollution); in both cases the bitumen, as it also binds the insulating elements together, makes possible the manufacture of insulating panels.

The coating and manufacturing process is performed by thoroughly mixing the insulating granular material with the bituminous binder and by pressing the mixture so as to form suitable panels.

Preliminary studies and experimentations have been carried out with the hot process, e.g.:

- laboratory study of the mix composition and preparation of initial samples of bituminous insulating panels;
- basic characterization of the initial samples (measure of the thermal conductivity λ);
- experimental preparation of a number of panels (prototypes) covered with aluminium paper, bitumen impregnated board or felt, kraft paper, or left uncovered;
- basic characterization of the prototypes (λ measurement), also for acoustic insulation, etc.

Among the main results, the following ones deserve to be specially mentioned:

<u>thermal insulation</u>	thermal conductivity kcal/h°Cm	laboratory
- uncovered bituminous panels	0.040	Padua University
- as above, covered with aluminium paper	0.044	Assoreni
- as above, covered witn bituminous board	0.046	Assoreni
and, for comparison:		
- normal expanded polystyrene panels	0.036	Assoreni
<u>sound insulation</u>	acoustic attenuation dB	
- bituminous panels with aluminium paper	+39.0	Assoreni
- as above with bit. board	+39.8	Assoreni
and, for comparison:		
- normal expanded polystyrene	+22.7	Assoreni
- compressed cork covered with flannel	+31.0	Assoreni
- bitumen and rock wool layers	+39.5	Assoreni

On such bases, the new type of bituminous heat (and sound) insulating materials has been patented in Italy in 1976 and the patent was then extended to other European countries.

The work programme of the present research, carried out with the financial contribution of the Commission of the European Communities, covers the further development of the project with the following items:

hot process:

- careful characterization of the prototypes of bituminous insulating panels
- aging tests on the prototypes in comparison with conventional insulating materials, consisting of exposure to normal (unpolluted), marine and industrial atmosphere
- characterization of the aged prototypes

cold process:

- study and verification of the process in laboratory:
  a) study of the composition of the mixture and preparation of initial samples of bituminous insulating panels
  b) characterization of initial samples (density, thermal conductivity, impermeableness)
  c) preparation of a certain number of optimized prototypes
  d) complete characterization of the optimized prototypes according to UNI 7819 Specification

energy saving and economical assessment.

All the above works have been carefully performed with interesting results, as can be seen in detail hereunder.

2. DESCRIPTION OF RESEARCH ACTIVITIES

2.1 Hot process

Among the prototypes of bituminous insulating panels experimentally prepared with the hot mixing process, two types were chosen for the characterization and aging tests to be performed according to the work programme :

- mixture of polystyrene granules (two sizes), hydrated lime filler and 80/100 pen. bitumen, covered with bitumen impregnated light board
- same mixture , covered with aluminium paper.

For comparison, a normal expanded polystyrene panel was chosen, with a density (volume mass) of 20 kg/m^3.

Characterization and aging tests were carried out as follows:

Characterization of original prototypes:	Test Method	Laboratory
- volume mass (density)	UNI 6349	Istituto sperimentale per l'edilizia ISTEDIL
- compression strength at +18°C	UNI 6350	" "
- compression strength at -20°C	UNI 6350	" "
- response into heat and flame	ASTM D 635	" "
- thermal transmittance(hot box)	ASTM C 236	" "
- thermal conductance (hot box)	ASTM C 236	" "
- thermal conductivity (hot box)	ASTM C 236	" "
- thermal conductivity(hot plate)	ASTM C 177	Assoreni
Aging of prototypes:		
- normal (unpolluted) atmosphere (weather-o-meter)	ASTM G 23	Istituto Sperimentale per l'edilizia ISTEDIL
- marine (sea environment) atmosphere	ASTM B 117	" "
- industrial environment atmosphere (Kesternich)	ISO 3231	" "
- urban atmosphere	conditions as specified hereunder	Industria Italiana Petroli
- hydrocarbon vapours	" "	" "
Characterization of aged prototypes:		
- volume mass (density)	UNI 6349	Istituto Sperimentale per l'edilizia ISTEDIL
- compression strength at +18°C	UNI 6350	" "
- thermal conductivity(hot plate)	ASTM C 177	Assoreni

Test results are indicated in the following tables.

2.1.1 Characterization of original prototypes

Sample A : Normal insulating panel of expanded polystyrene
Sample B : Prototype of bituminous insulating panel covered with bitumen impregnated light board
Sample C : Prototype as above, covered with aluminium paper

See Table 1 : Volume mass (Density)
Compression strength at +18 °C
Compression strength at -20 °C
Response into heat and flame
Table 2 : Thermal transmittance, conductance and conductivity (Hot Box Method)
Table 3 : Thermal conductivity (Hot Plate Method)

TABLE 1 DENSITY - COMPRESSION STRENGTH - RESPONSE INTO HEAT AND FLAME			
Sample	A	B	C
Volume mass (density), kg/m^3	20.4	145.5	115.3
Dimensions, mm	200x150x30	200x150x55	200x150x55
Compression strength at 10% deformation			
at +18°C, N/m^2	0.11	0.04	0.05
at -20°C, N/m^2	0.13	0.10	0.16
Dimensions, mm	60x60x60	55x55x55	55x55x55
Response into heat and flame:	inflammable	inflammable	self exting.
average burning rate, cm/min.	12	5	-
self extinguishing time, min.	-	-	2
Dimensions, mm	25x50x130	25x50x130	25x50x130

TABLE 2 THERMAL TRANSMITTANCE, CONDUCTANCE AND CONDUCTIVITY (HOT BOX)			
Sample	A	B	C
Surface dimensions, cm	100x152	100x152	100x152
Thickness (average), cm	2.9	5.8	5.3
Air temperature - hot side, °C	39.3	39.8	39.5
Air temperature - cold side, °C	-5.2	-5.2	-5.1
Temperature of the hot surface, °C	36.7	38.0	37.1
Temperature of the cold surface, °C	-2.9	-3.5	-3.4
Heat flux, kcal/h	69	44	54
Internal surface coefficient, $kcal/h°Cm^2$	17.45	15.64	14.80
External surface coefficient, "	19.73	17.02	20.89
Thermal transmittance, "	1.02	0.64	0.79
Thermal conductance, "	1.14	0.69	0.87
Thermal conductivity, kcal/h°Cm	0.033	0.040	0.046

TABLE 3 THERMAL CONDUCTIVITY (HOT PLATE)			
Sample	A	B	C
Surface dimensions, cm	15x20	15x20	15x20
Thickness (average), cm	2.26	2.51	2.34
Metering plate diameter, mm	70	70	70
Metering interval, °C	15-38	14.5-38.5	14.5-40
Thermal conductivity, kcal/h°Cm	0.036	0.047	0.040

2.1.2 Aging of prototypes

The prototypes of bituminous insulating panels were aged in order to evaluate the possible variation of their basic characteristics due to aging.

For comparison, aging tests were extended also to a normal insulating panel of expanded polystyrene.

Aging conditions were as follows:

- weather-o-meter : 500 hours (see ASTM G 23 Test Method)
- salt spray (fog) : 500 hours (see ASTM B 117 " ")
- sulfurous anhydride : 5 cycles of 24 hours (see ISO 3231 Test Method - Kesternich)
- urban gasses

urban gasses	time(hours)	concentration*(ppm)	temperature(°C)
CO	160	60	30
NO_2	160	0.1	30
SO_2	160	1	30
Cl_2	160	1	30

* desumed by reports on air quality in a number of towns (FEEMAS mobile laboratory)

- hydrocarbon vapours

hydrocarbon vapours	time(hours)	concentration**(ppm)	temperature(°C)
xylene	160	4,000	30
n-decane	160	4,000	30

** hydrocarbons were added on water surface with a 50/50 ratio, so that - according to their vapour tension curves (in vapour saturated ambient) at 100 °F - a concentration of 4,000 p.p.m. in gass phase might be achieved.

2.1.3 Characterization of aged prototypes

Sample A : Normal insulating panel of expanded polystyrene
Sample B : Prototype of bituminous insulating panel covered with bitumen impregnated light board
Sample C : Prototype as above, covered with aluminium paper

See Table 4 : Valume mass (density)
Table 5 : Compression **strength** at + 18 °C
Table 6 : Thermal conductivity (Hot Plate Method)

TABLE 4	VOLUME MASS (DENSITY)			
Sample	Specimen N°	Dimensions mm	Results kg/m^3	Average kg/m^3
A aged in weather-o-meter	1 - 2 - 3	60x60x30	20.1-20.6-20.4	20.4
A aged in salt spray (fog)	1 - 2 - 3	60x60x30	20.2-20.4-20.6	20.4
A aged in Kesternich (SO_2)	1 - 2 - 3	60x60x30	20.2-20.8-20.1	20.4
B aged in weather-o-meter	1 - 2 - 3	55x55x55	142.6-146.8-145.4	144.9
B aged in salt spray (fog)	1 - 2 - 3	55x55x55	145.0-143.2-146.4	144.9
B aged in Kesternich (SO_2)	1 - 2 - 3	55x55x55	141.2-142.5-144.1	142.6
C aged in weather-o-meter	1 - 2 - 3	55x55x55	111.4-118.8-112.6	114.3
C aged in salt spray (fog)	1 - 2 - 3	55x55x55	113.5-119.3-108.7	113.8
C aged in Kesternich (SO_2)	1 - 2 - 3	55x55x55	110.4-109.6-108.3	109.4

TABLE 5	COMPRESSION STRENGTH AT +18°C, AT 10 % DEFORMATION			
Sample	Specimen N°	Dimensions mm	Results N/mm^2	Average N/mm^2
A aged in weather-o-meter	1 - 2 - 3	60x60x60	0.11-0.11-0.10	0.11
A aged in salt spray (fog)	1 - 2 - 3	60x60x60	0.10-0.11-0.11	0.11
A aged in Kesternich (SO_2)	1 - 2 - 3	60x60x60	0.11-0.10-0.11	0.11
B aged in weather-o-meter	1 - 2 - 3	55x55x55	0.04-0.06-0.04	0.05
B aged in salt spray (fog)	1 - 2 - 3	55x55x55	0.05-0.05-0.06	0.05
B aged in Kesternich (SO_2)	1 - 2 - 3	55x55x55	0.04-0.05-0.05	0.05
C aged in weather-o-meter	1 - 2 - 3	55x55x55	0.05-0.05-0.05	0.05
C aged in salt spray (fog)	1 - 2 - 3	55x55x55	0.04-0.05-0.06	0.05
C aged in Kesternich (SO_2)	1 - 2 - 3	55x55x55	0.05-0.05-0.05	0.05

TABLE 6 THERMAL CONDUCTIVITY (HOT PLATE)*

Sample	Specimen N°	Surface dimensions, cm	Thickness (average), cm	Metering interval, °C	Results (average) kcal/h°Cm
A aged in weather-o-meter	1 2	15x20 15x20	2.400	15-39	0.042
A aged in salt spray (fog)	1 2	15x20 15x20	2.398	15.5-40	0.042
A aged in Kesternich (SO_2)	1 2	15x20 15x20	2.375	15.5-40	0.042
A aged in urban gasses	1 2	15x20 15x20	2.361	16-42	0.041
A aged in hydrocarbon vapours	1 2	15x20 15x20	2.328	15-40	0.041
B aged in weather-o-meter	1 2	15x20 15x20	2.640	15.5-39	0.045
B aged in salt spray (fog)	1 2	15x20 15x20	2.764	14.5-43	0.045
B aged in Kesternich (SO_2)	1 2	15x20 15x20	3.020	14.5-42.5	0.049
C aged in weather-o-meter	1 2	15x20 15x20	2.480	15-44	0.040
C aged in salt spray (fog)	1 2	15x20 15x20	2.526	14.5-39	0.043
C aged in Kesternich (SO_2)	1 2	15x20 15x20	2.374	15-44	0.042
C aged in urban gasses	1 2	15x20 15x20	2.332	14.5-40	0.041
C aged in hydrocarbon vapours	1 2	15x20 15x20	2.335	14.5-39	0.042

* Metering plate diameter 70 mm.

2.2 Cold process

According to the work programme quoted at item 1, the study and verification of the cold manufacturing process in laboratory (items a - b - c - d) were carried out by Tecnocasa and the University of L'Aquila as indicated hereunder.

2.2.1 Study of the composition of the mixture and preparation of initial samples of bituminous insulating panels

The following materials were used for the research:

- spherical granules of expanded polystyrene of 2 to 9 mm diameter, with a volume mass (density) of the loose (uncompacted) material of 7.9 kg/m^3;
- slow breaking bituminous emulsion with a mineral colloidal additive and a 38 % water content.

The mixture of the above materials was carried out at room temperature, with three different bitumen contents corresponding to 80 % - 85 % - 90 % by weight of bituminous emulsion.

The original bituminous emulsion was preliminarily diluted with 12.5 % b.w. of water, in order to achieve the suitable mixing and coating viscosity.

The diluted bitumen emulsion was added to the right amount of insulating material granules and thoroughly mixed by trowel. The insulating granules were completely coated with bitumen and the workability of the mix was good.

Initial samples of bituminous insulating panels were obtained with the above mixtures by using suitable steel moulds with a double bottom, as follows:

a) the mould was filled with the wet mix (Fig.1), after having applied a thin silicon layer on the inside surfaces in order to facilitate the final detaching of the filling material, and it was placed in oven at 50°C for 24 hours for the "breaking" of the emulsion and the complete drying of the mix;

b) a silicon coated steel plate was then put on the mix, which was compressed (1 - 1.5 kg/cm^2) in conditions of confined lateral expansion (Fig.2);

c) the samples of moulded panels were pushed out by means of the mobile bottom plate (Fig.3).

2.2.2 Characterization of initial samples

It was observed that the initial samples, placed on lateral supports and left for three weeks at a temperature up to 35 °C, showed a deformation of their original flat shape.

In order to prevent the excessive tendency to visco-plastic deformation,

FIG. 1

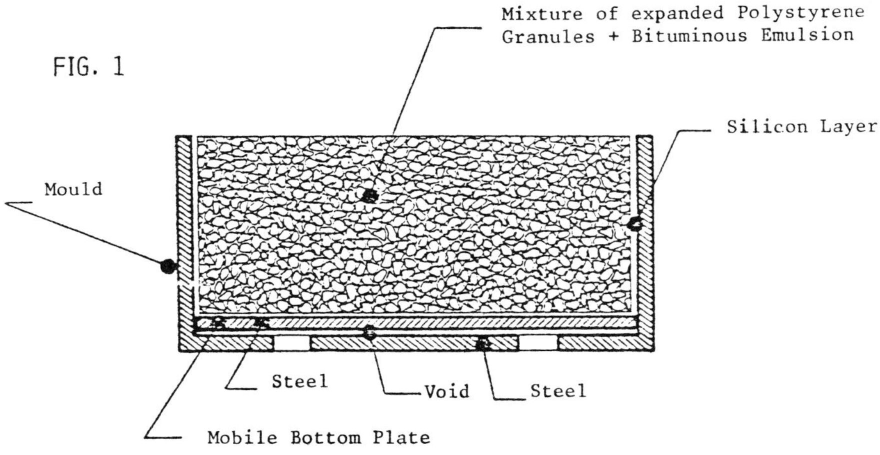

FIG. 2

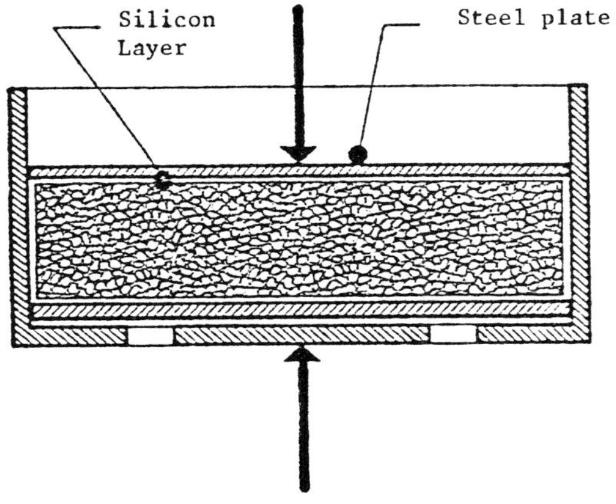

FIG. 3

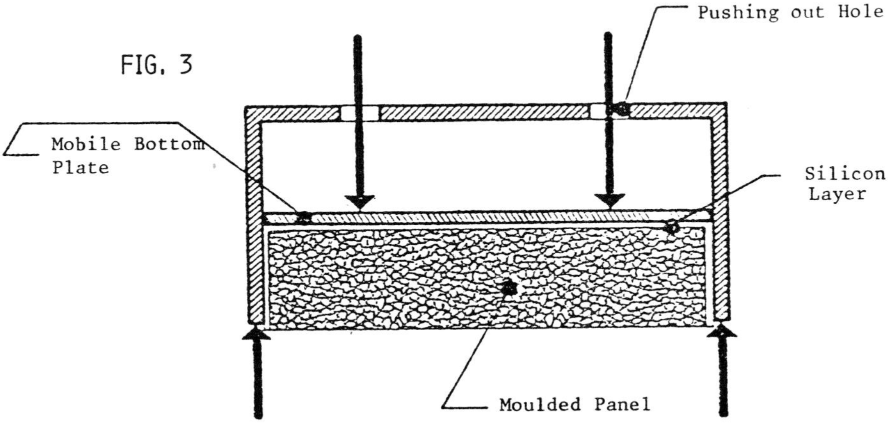

it was decided to increase the viscosity of the bituminous binder by the addition of the mix with 20% b.w. of limestone filler (in practice, a "425" Portland cement was used), with the further result of improving the mixing operations.

Moreover, a Kraft paper was applied on both surface of the panels, with a view of improving their handling and appearance (Kraft paper may be optionally substituted by aluminium paper, bitumen impregnated board,etc.)

New samples of panels (20x30x3 cm) obtained in this way, placed on lateral supports and left in oven at 45 °C for 24 hours, did not show any sign of deformation.

The technical characterization of these samples is included in item 2.2.4

2.2.3 Preparation of a certain number of optimized prototypes

The prototypes of bituminous insulating panels to be tested for the complete characterization, were prepared with the modalities indicated under items 2.2.1 and 2.2.2.

A particular remark is deserved by the fact that a Kraft paper was placed on the bottom of the steel mould before its filling with the mix and a similar paper was inserted between the mix surface and the compression steel plate, avoiding in this way the need of a silicon detaching layer in the same positions.

The adhesion of the Kraft paper to the panel surfaces was achieved by means of the same bitumen emulsion used for the mix.

In order to dispose of all the specimens necessary for the characterization tests, 24 mixtures were prepared in laboratory, differentiated as specified under the following item, and for each mixture and each characteristic to be measured 4 specimens of bituminous insulating panels were obtained.

2.2.4 Complete characterization of the optimized prototypes according to UNI 7819 Specification

The following characteristics were taken into consideration:

- volume mass (density)
- compression strength at room temperature
- tensile strength at room temperature
- resistance to flame propagation (advancement)
- thermal conductivity
- transmission rate (speed) of water vapour
- resistance to water vapour diffusion

In order to evaluate how the above characteristics vary according to the variations of the main parameters ("factors"), a programme of factorial experimentation was performed considering the following factors and levels:

factors:	levels:
- bitumen emulsion content (D), % by weight of the mixture expanded polystyrene + bitumen emulsion	$D_1 = 80$ $D_2 = 85$ $D_3 = 90$
- Portland cement content (C), % by weight of the bitumen emulsion	$C_1 = 0$ $C_2 = 20$
- moulding pressure (P), kg/cm^2	$P_1 = 1.0$ $P_2 = 1.5$
- test repetition (each test was repeated twice)	I_1 = 1st set of tests I_2 = 2nd set of tests

Consequently 24 (3x2x2x2) different mixtures had to be prepared and for each mixture and each characteristic 4 specimens of bituminous insulating panels were obtained.

The test results for each characteristic (average of 4 measures) were collected in tables and graphs. The latter - which show the trend of the principal effects (bitumen emulsion and Portland cement contents, as well as moulding pressure) - were obtained by interpolating (with the minimum squares criterion) the average values of each characteristic corresponding to each level of every factor.

The results of all the laboratory tests carried out are summarized in Table 7 (recapitulating prospectus), which evidences the influence of the main factors on the examined characteristics.

2.3 Energy saving and economical assessment

For the hot manufacturing process, a preliminary study carried out in 1981 has shown that - under the same conditions of initial insulating capacity - the estimated unit proceeds of our panels are practically at the same level of the selling price of the normal expanded polystyrene comparison panel used for the present research, but the latter, when aged (as proved by the laboratory tests), loses 15.5 % of its insulating capacity while our panels maintain their insulating capacity practically unaltered.

As the saving of heating energy in insulated buildings is 40 % (European average), a loss of energy of 15.5%x40% = 6.2 % takes place if a material like the above comparison sample is used for building thermal insulation; in other words, our bituminous panels yield a 6.2 % energy saving compared with the traditional insulating material, in the hypothesis of the same selling price.

For the bituminous insulating panels obtained with the cold manufacturing process, a similar study has not yet been accomplished as the present stage of the research is still at a laboratory level; nevertheless, owing to the analogy between the two processes and the materials produced, we can reasonably expect that the economical/energetic evaluations for the cold process should not differ substantially from those pertinent to the hot process.

TABLE 7 RECAPITULATING PROSPECTUS

Measured characteristic	Minimum value	Maximum value	Factors influencing the characteristics		
			Bituminous emulsion	Portland cement	Moulding pressure
Volume mass (density), kg/m^3	36.4	78.9	+ + +	+ +	0
Compression strength at 50% deformation, kg/cm^2	0.3	0.8	+ + +	–	+
Compression strength at 75% deformation, kg/cm^2	2.2	6.0	+ + +	0	0
Tensile strength: rupture tension, kg/cm^2	1.7	2.3	+	–	+
Tensile strenght: rupture elongation, %	4.0	6.7	– – + +	–	+ +
Resistance to flame propagation: extent of burning, cm	2.7	10.2	– – –	– – –	– – –
Resistance to flame propagation: time of flame extinction, s	8.7	30.5	– – –	– –	– –
Thermal conductivity, kcal/hm°C	0.034	0.044	0	+ +	0
Transmission rate of water vapour, $g/h\ cm^2$	0.42×10^{-4}	2.4×10^{-4}			
Resistance to water vapour diffusion, m/m	4.6	22.8	– – + + +	– –	+ +

\+ + + Marked increase of the characteristic when the factor increases
\+ + Increase of the characteristic when the factor increases
\+ Light increase of the characteristic when the factor increases
0 No influence
– Light decrease of the characteristic when the factor increases
– – Decrease of the characteristic when the factor increases
– – – Marked decrease of the characteristic when the factor increases

The contemporary presence of + and – indicates the existence of a minimum or a maximum inside the measure interval.

3. CONCLUSIONS

3.1 Hot process

The prototypes of bituminous insulating panels obtained with the hot manufacturing process were characterized and aged as indicated at items 2.1.1 - 2.1.2 - 2.1.3: a wide number of data is therefore now available and it allows to evaluate the performance of the new type of insulating material which can be produced with this process.

A particular mention is deserved by the fact that the prototypes covered with aluminium paper result self-extinguishing to flame (according to ASTM D 635 Test Method) with a self-extinction time of 2 minutes, and that prototypes covered both with aluminium paper and bitumen impregnated board clearly improve their compression strength at low temperature (-20 °C).

After various types of aging, the bituminous insulating panels have practically maintained unaltered their thermal conductivity.

The sample of normal expanded polystyrene panel used for comparison has not evidenced particular alterations after aging, but has shown a certain increase of thermal conductivity (15.5%).

3.2 Cold process

The research has given significative conclusions regarding the cold manufacturing process and the bituminous insulating panels which can be obtained.

In the first place it is important to point out the confirmation of the feasibility and convenience of this process, because some doubts existed about the possibility of operating at room temperature (with a cold bituminous emulsion), with the successive elimination of the total amount of water contained in the emulsion.

In this respect, the measure of thermal conductivity - evidencing a value practically equivalent to the one obtained with the hot process - has demonstrated the validity and suitability of the cold process.

Moreover, the prototypes of panels manufactured have shown the following characteristics:

- good workability
- easily cut by blade or saw
- a certain compactness and tenacity
- resistance to stresses comparable to those expected during storing, handling and putting into place.

The flexibility of the new type of bituminous insulating panels could also make them suitable for applications on not flat surfaces.

The technical characterization of the prototypes has been programmed and performed according to the method of factorial experimentation, which has given sufficiently reliable conclusions because the results have been interpreted on the base of statistic methods.

A valuation has been obtained which, besides the values of all the characteristics examined, has provided the possibility of knowing how the various characteristics vary in function of bitumen emulsion content, cement addition and moulding pressure.

3.3 Overall remarks

The research has allowed to obtain a number of prototypes of bituminous insulating panels which, submitted to characterization and aging, have shown a reliable performance, at any rate such as to conform to the objective of the research itself.

The cold process has proved particularly interesting because of its advantages in comparison with the hot process, both for energy saving and for ecological reasons, as well as for its easier technology.

On the basis of the present knowledge, the following further development of the research should be considered:

- investigation of the more suitable fields of application in building and other industries
- design and construction of a pilot plant for the production and industrial optimization of bituminous insulating panels (to be experimented in place).

STUDY OF HEAT STORING BUILDING MATERIALS NUMERICAL AND EXPERIMENTAL RESULTS

G. SOLAINI
Dipartimento di Programmazione Progettazione e Produzione Edilizia
Politecnico di Milano

and

A.M. TONELLI
Luigi Rossi - ARCO S.r.l., Milano

Summary

Energy savings and better conditions for home comfort can be achieved by storing heat in the walls of buildings. Walls can be rapidly charged and discharged by increasing their transversal thermal diffusivity. This goal can be reached by inserting metallic elements perpendicularly to walls' surfaces. Heat storing walls can be charged by sun radiation or by convection. If perimetral walls are to be used external insulation is required. A theoretical and experimental research was conducted on such non-homogeneous structures. Aim of the research was studying their thermal behaviour and calculating their heat storage capacity, as far as different geometrical features and different values of the thermophysical parameters are concerned.
This paper deals with experimental measurements carried out in the second phase of the research, concerning the temperature distribution inside two prototypes (concrete + iron rods, brick+iron networks) externally insulated, charged and discharged by air convection. Such measurements have been performed inside two small test rooms, constructed specifically for this work. Experimental conditions have been simulated by means of a theoretical model previously developed. Numerical and experimental data tally out fairly well. Finally, numerical results are reported, concerning the heat storage capacity of concrete + iron rods walls, as for different values of the geometrical parameters.

1. INTRODUCTION

The proposed heat storing materials have been conceived to be used for the construction of the perimetral walls in buildings and wherever possible for construction of inner walls as well. Briefly, a heat storing wall is composed of a heat storing material (concrete, bricks, etc.) and of metallic elements which are inserted perpendicularly to the wall surfaces (transversal conductors, fig. 1a). Such conductors may be thermally connected to superficial ones (i.e. metallic elements placed in the inner vertical surface of the walls, fig. 1b), so that their heat transfer surfa

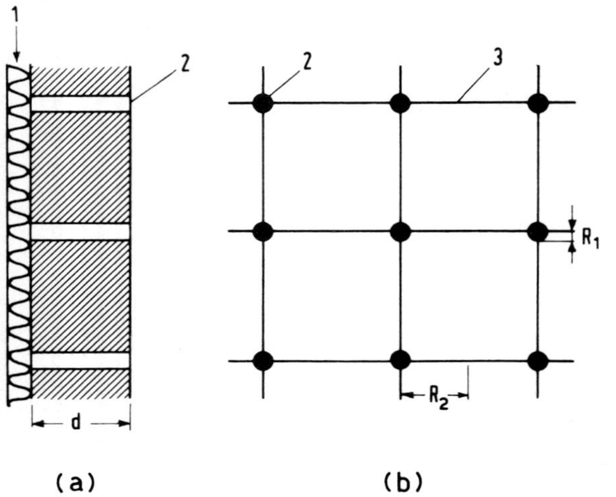

Fig.1 - Scheme of a heat storing wall:
a) section
b) front.
1 = insulating layer,
2 = transversal conductors,
3 = superficial conductors.

ce be enlarged.

During the charge phase, heat is transferred by air convection, by means of heating elements placed all along the wall's base; if solar passive systems are used, heat is supplied mainly by solar radiation. Perimetral walls must be insulated on the outside. During the reverse phase stored heat is released by the wall surface so that the vertical gradient of the internal air temperature remains close to zero. The main advantage of the system is given by the fact that heat can be stored when available (solar systems, heat pump plants working by night, and so on). Secondly, the system allows a reduction in energy need and improves physiological comfort, through a better distribution of room temperature and a better radiative exchange among walls and persons, as well as a lower plant operating temperature.

The aim of this reasearch is to supply elements for evaluation of the appropriate choice to make in materials, their geometrical configuration, keeping both energy performance and technical feasibility in mind.

2. WORK PROGRAMME

The research is concerned with the study of the thermal behaviour of the proposed structures, by means of theoretical and experimental investigations:

a) in the first phase of the research a theoretical model and computer programme have been developed to obtain some results concerning walls' temperature distribution and heat storing capacity for the different sizes and geometrical features of thermal conductors. Furthermore, four large concrete + iron rods, bulkhead walls have been built inside two greenhouses, and an experimental investigation has been performed to test the storing walls' behaviour when passive solar systems are used (1).

b) In the second phase, which is described below, experimental measurements have been carried out in order to investigate the thermal behaviour of two externally insulated, heat storing perimetral walls, charged by air convection, and to validate the theoretical model. Numerical and experimental results have also been used to evaluate the amount of heat stored (and released) by the experimental walls, in given on-off conditions, and to make a comparison with conventional walls.

3. METHODS OF INVESTIGATION

3.1 Theoretical model

The theoretical model has been developed on the basis of the following assumptions (1), (2):

- the wall's volume is divided in several isothermal elementary parts (nodes) considered, from a geometrical point of view, as points;
- each node is characterized by a thermal capacity value, and for each couple of adjoining nodes a thermal conductance value is defined;
- from a mathematical point of view, the system is described, for each time interval Δt, by a set of linear equations, each equation representing the heat balance of a node during Δt ("heat balance method");
- the following initial and boundary conditions are assigned:

  $T=T_o$ at $t=o$; $T=T_e(t)$ on S_e; $\varphi=h(t)\cdot(T-T_i(t))$ on S_i

  where

  S_i, S_e = internal, external surface of the wall

  T_i, T_e = internal, external air temperature

  h = coefficient of surface heat transfer;
- it is assumed that transverse conductors are thermally connected with a metallic network, lying in the wall's inner face. It is also assumed, in a first approximation, that during the surface heat transfer the conducting network be isothermal.

3.2 Computer programme

Based on these assumptions, a computer programme has been developed. The main input for this programme consists of the following data:

- thermophysical parameters of conducting and non conducting materials;
- geometrical size (radius R_1) and half-distance (R_2) of transverse conductors;
- thickness of the wall (d);
- thermal resistance of external insulating layer (R_{ex});
- ratio $A = S_{net}/n\pi R_1^2$, where S_{net} is the surface of conducting network, and n is the total number of transverse thermal conductors;
- values of h, T_o, T_i, T_e.

The main output is:

- temperature distribution inside the wall, heat fluxes through internal and external wall's surfaces, and stored energy, all these parameters being functions of time.

A numerical investigation has been performed, with the following aims:
- simulating the experimental conditions, in order to compare numerical and experimental data;
- evaluating, for different values of the geometrical parameters, the ratio $\Gamma=\Delta E/\Delta E_o$, ΔE, ΔE_o being the amounts of heat exchanged with internal air by the experimental and, respectively, by the corresponding conventional walls, after several identical charge and discharge phases (i.e., in steady cyclic conditions).

3.3 Prototypes

To test the thermal behaviour of walls in building, experimental measurements have been carried out inside four identical cubic rooms (2m x 2m x 2m), constructed in a building yard (fig.2), with the following

Fig. 2- Test rooms (concrete walls).

characteristics:
- of the six walls of each cube, five were built with insulating material (wood and polystyrene) and one wall was built in masonry;
- each stone wall was covered, on the outside, by a layer of 5 cm thick polystyrene;
- of the four stone walls, two were built in reinforced concrete (d=20 cm), so as to be identical, (except for conductor elements if used), and the other two walls were built in solid brick, once again in an identical way;
- the metallic elements used for the concrete prototype (fig. 3) are made

Fig. 3 - Iron elements inserted in the concrete wall.

with iron rod segments (∅ = 10 mm), each bent into a rectangular shape, (20 cm x 10 cm). These elements have been inserted lengthwise in wall thickness and on a parallel with internal surface. They have also been soldered into a rectangular-shaped grid (dimensions: 10 cm x 20 cm), of a double network (∅ = 4 mm);

- the metallic elements used for the brick prototype are constituted by a "Nervometal" iron grid (dimensions: 1 cm x 5 cm; thread section: 1 mm x 0,5 mm), bent into a L shape and placed between each layer of bricks, so as to cover all the horizontal surface and 5/6 ths of the vertical surface on the inside (fig. 4).

Fig. 4 - Iron networks inserted in the brick wall.

3.4 Experimental measurements

Walls have been charged by means of electric fans, placed at their bases. Data concerning temperature distribution inside walls have been taken by means of thermal probes (thermocouples and thermo-resistances) inserted in the walls, at different positions with respect to thickness and height. Internal and external air temperature has been recordered too. Furthermore, in order to find the experimental values of h, the thermal flow through the walls' inner surface has been measured (by means of thermal flowmeters attached to the faces themselves), as well as the temperature at the center C of each flowmeter and the temperature of internal air, at a distance of 2.5 cm from C (fig. 5).

All thermocouples had a common electronic reference junction. Temperatures and heat fluxes have been recorded by means of previously calibrated multichannel recorders.

Finally, in order to evaluate the influence of the superficial conducting elements on air-walls thermal exchange, several infrared thermograms have been taken. Each couple of walls have been cyclically charged (by forced convection) and discharged (by forced or natural convection), with periods ranging from 7 to 12 hours. Identical experimental conditions have been kept for conventional and non conventional walls. All measurements have been carried out in Winter 1983; experimental equipment is shown in fig. 6.

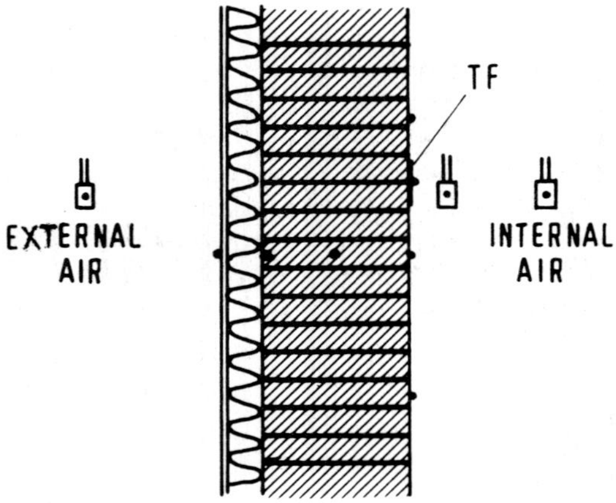

Fig. 5 - Scheme of the disposition of thermal probes (dots) and thermal flowmeter (TF).

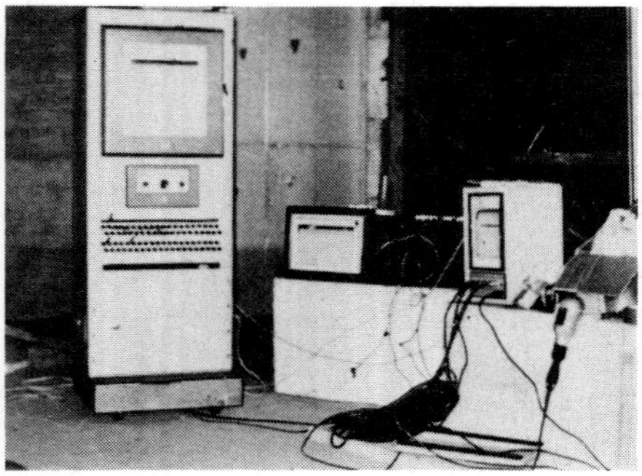

(a)

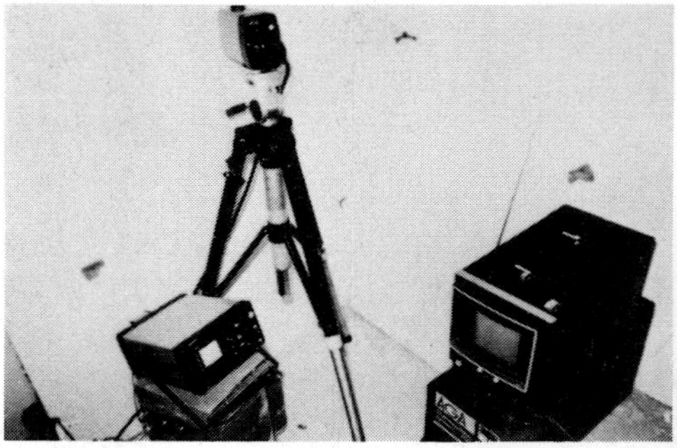

(b)

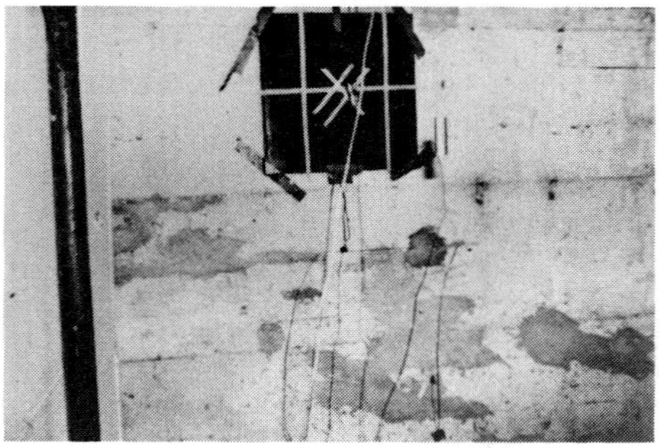

(c)

Fig. 6 - Experimental equipment: a) recorders, b) AGA-THERMOVISION apparatus, c) thermal flowmeter and thermocouples.

4. MAIN RESULTS

Γ values, as deduced by experimental data, appear to be about 1.5 for concrete prototypes, and about 1.3 for brick prototypes.

Furthermore, during discharge phases internal air temperature in rooms with accumulating walls appeared to be about 1°C higher than air temperature in rooms with traditional walls.

Some of the experimental tests have been simulated using the computer programme, in order to validate the theoretical model. Experimental and numerical data agree within an error range which is less than 5%. One example of such a comparison is given in fig. 7.

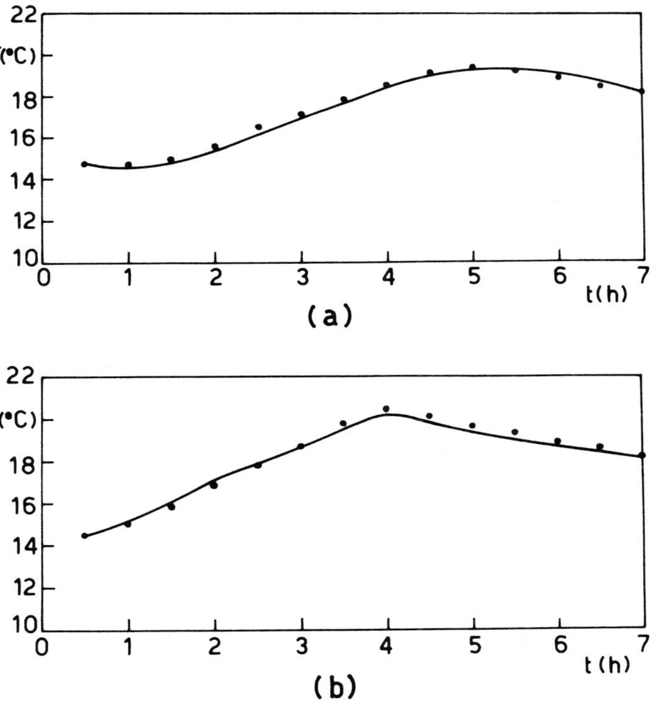

Fig. 7 - Plot of T(t), referred to a charge-discharge test for the concrete-iron wall: a) back, b) center of the wall (——— experimental data; ···· numerical data).

The thermographic analysis confirms that the superficial conductors give a considerable contribution so as to increase the amount of heat transferred by convection, particularly in the case of the L-shaped network (brick wall). Finally, a numerical investigation has been carried out for concrete + iron rods walls, with the on-off conditionsgiven in fig. 8, so as to find Γ behaviour as a function of the geometrical parameters (fig. 9).

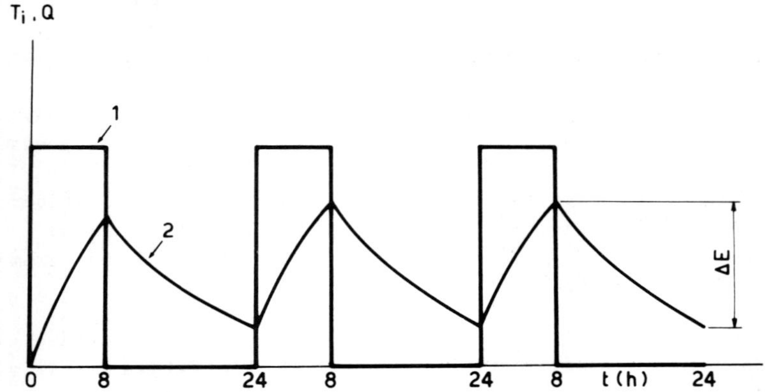

Fig.8 - Assigned theoretical on-off conditions (qualitative plot).
1) $T_i(t)$,
2) transferred heat $Q(t)$ (ref.(2)).

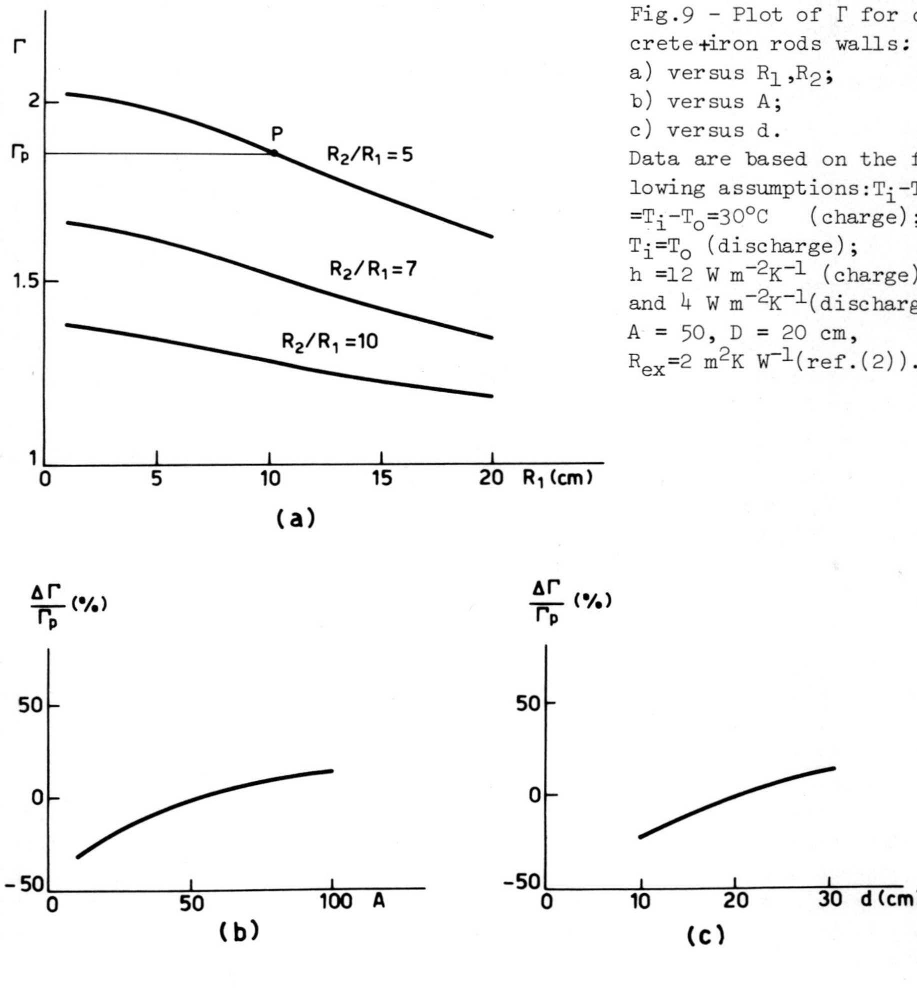

Fig.9 - Plot of Γ for concrete+iron rods walls:
a) versus R_1, R_2;
b) versus A;
c) versus d.
Data are based on the following assumptions: $T_i-T_e=T_i-T_o=30°C$ (charge); $T_i=T_o$ (discharge); $h=12\ W\ m^{-2}K^{-1}$ (charge) and $4\ W\ m^{-2}K^{-1}$ (discharge); A = 50, D = 20 cm, $R_{ex}=2\ m^2K\ W^{-1}$ (ref.(2)).

The contribution of superficial conductors is confirmed by these results, and a practical upper limit-value of about 2 is found for Γ.

5. CONCLUSIONS

As far as the reinforced concrete is concerned, it appears that a proper technical solution consists in a structure formed by two vertical metal networks keeping the conducting elements in horizontal positions, in the direction of thickness. By the way, it is to be noted that such a structure is both able to increase transversal diffusivity and to reinforce the wall.

On the other hand, brick walls seems to require continous conductive plates (close-mesh nets), i.e. low-cost commercial articles, which are very easily packed and transported. These nets work well as heat exchangers on the vertical surface, but the iron mass inserted inside the wall must be increased, in order to obtain a higher transverse diffusivity.

AKCNOWLEDGMENT

The authors wish to thank the Dipartimento di Energetica del Politecnico di Milano for kindly putting part of the experimental equipment at disposal.

REFERENCES

(1) G. Rossi, G. Solaini, A.M. Tonelli: "Heat Accumulating Building Materials", in "Energy Conservation in Buildings - Heating, Ventilation and Insulation" (Proceedings of the Contractors' Meetings), edited by H. Ehringer, G. Hoyaux and P. Zegers, D. Reidel Publishing Company, Dordrecht (1983).

(2) G. Rossi, G. Solaini: "Comportamento termico in regime variabile di una parete non omogenea ad alta diffusività termica trasversale", Atti del 38° Congresso Nazionale dell'Associazione Termotecnica Italiana, Bari (1983).

IMPROVING OF THE THERMAL RESISTANCE OF POLYURETHANE FOAMS

G. ROUX*, A. BALME*, J. RIVERO*, A. SILVENT**, A.M. ANDREANI** and B. FOGG+

* CSTB - 24, rue Joseph Fourier - F 38400 Saint Martin d'Hères -
** CEA - Avenue des Martyrs - BP 85 - F 38041 Grenoble Cédex -
\+ ICI France SA - ZI le Cornage - F 38220 VIZILLE -

Summary

The K value of rigid polyurethane foams increases from about 18 mW/m°K after processing to nearly 30 mW/m°K after a long aging period.
The aim of this study is to reduce this asymptotic value to about 20 mW/m°K by optimizing formulation and process of these foams and following their K value evolution during hot-humid artificial weathering.
This objective is almost reached, indeed results show asymptotics values included between 20,1 to 23,2 mW/m°K. These values correspond to a specific kind of formulations and depend in these limits, on the foaming process used, laboratory or industrial. In fact K value can be reduced by formulation adjustment and by foaming process adaptation as well.
An economic impact of this research is presented and some propositions are made for a farther study.

1. INTRODUCTION

Due to gas diffusion through the cell walls of rigid polyurethane foams, the K value of these products increases with time from the original value. The asymptotic value taken in account for thermal calculation is in France 0,029 W/m°K.

The aim of this research is to achieve a significant improvement in this functional thermal resistance of foams to 0,020 W/m°K by inhibition the process described above without incurring a serious increase in manufacturing costs.

This study is undertaken jointly by the CSTB and the CEN-Grenoble in collaboration with ICI-France SA as raw materials producer and researcher member and SECMER as manufacturer of equipements for processing rigid polyurethane foams.

The programme is divided in two parts. In a first stage and only with commercial raw materials, the decreasing of the initial thermal conductivity, λ_o, measured 24 hours after processing is looked up, as well as the asymptotic value λ_∞.

During this stage the influence of the type and amount of catalyst, surfactant, foaming gas, the amount of air, the polyol type is taken in account.

During the second stage a particular interest is given to the polyols modification to increase their diffusion barrier effect on polyurethane. As a matter of fact gas diffusion kinetic reduction keeps at a lower level the λ_∞ value.

In the same time some test are made on industrial plants with optimised formulations to cheek the actual interest of the results.

2. EXPERIMENTAL PROCEDURE

During this study, about two hundred different formulations were manufactured and tested in our laboratory.

Small blocs (30 x 30 x 30 cm and 60 x 40 x 10 cm) are foamed using a two components SECMER machine with a capacity of 10 kg/minute.

After 24 hours these blocs are cutted is samples of 12 x 12 x 2,5 cm and each sample is marked to know the rise direction.

Just after cutting, dimensions, compressive strength, density are measured as well as the initial K value λ_o using a heat floxw meter ANACON 88 according the ISO 2581 standard.

There samples are weathered into a hot and humid oven at 70°C and 90% HR for one, seven, fourteen and twenty eight days.

After weathering samples are kept for one day at 23°C and 50% HR and then all their caracteristics were measured again.

More over the included cells gaz is studied. For the reference sample and after each aging period the gas composition and the total gas pressure are measured. Then it is possible to follow the evolution of partial pressures of R11, CO_2, N_2, O_2 during the weathering and correlate them with the thermal conductivity evolution.

The total gas pressure is measured with a specific apparatus designed in our laboratory witch pushs the cells gas into a transfert vial for mass spectrometry analysis. These analyses are provited by the CEA-Grenoble, Chemical Analysis Laboratory using a hig resolution VG mass spectrometer.

This volumic analyse can be couverted in partial pressure of each gas using the total pressure measured during the extraction gas operation in addition with the closed cells content obtained in following ISO 4590 standard.

This experimental plan was 30 months long.

3. RESULTS

3.1. Catalyst and surfactant effect

The first part of this programme concerns the classification of 12 surfactants and 6 catalysts. All them are European commercial products.

This classification is made using a lower level of surfactant or catalyst in regards of classical level.

Then the study of the amount of surfactant and catalyst on λ values is developped.

At this stage we have defined an optimum pair surfactant-catalyst respectively 1 and 5, and an optimum reactivity of the foaming process.

One can say that the catalyst level has a great influence on the λ_o value but that the type of catalyst is not very important for a given reactivity. In the same way with 5% of catalyst the λ_∞ value is nearly independent of the catalyst type and the cells size is realy small. This increases the lineic density of cells wall and then their diffusion barrier effect.

It is interesting to note that with a such catalyst level increasing from 1% to 2% the surfactant part gives a better λ_∞ value but beyond 2% there is nearly no more effect. So we get he choice of surfactant n° 1 with 2%.

3.2. Air effect

In the same time the air influence is tested and one can say that, with a classical mixing head, air is a necessity but not at a large amount.

3.3. R_{11} and water effect

Two other factors are then taken in account : the water and R_{11} effect. It is an evidence that the increassing of their rate reduces the density of the foam in free expansion. But one can say that a small amont of water is needful for reducing the λ_{∞} value. With our geographic conditions the optimum is approximatly 40% R_{11} and 1% water.

3.4. Influence of moulding machine

The influence of type of the moulding machine was taking in account too.

It is not an important facteur in regard of the mixing process. With the same moulding conditions, changing the low pressure mixing head have no significant effect but mooving to a high pressure mixing headgives better results. This tendency will be confirmed after the comparison of laboratory and industrial manufactured foams.

3.5. Polyol effect

On the other hand we have checked the influence of the polyol type and of additives mixed with polyols. It was the end of the first part of the study and the connection with the second part.

As it is shown on fig. 1 there is a large influence of the type of the polyol with can be aliphatic or aromatic, amineted or not.

It can be noted that initial and asymptotic λ values depend directly of the polyol type as well as the λ evolution. This shows that it is not possible to evaluate λ_{∞} only from λ_o.

3.6. Gas composition evolution

Looking the table I it can be noted that cells gas composition is also dependent on the polyol type. After 28 days of aging the foam 81 is completely air filled even though with the foam 97 the partial pressure of R_{11} is nearly the same than it was at the beginning.

That is realy very important and shows the great interest of the polyol modification.

3.7. Polyol modification

This second part of the study was provised on the base of the 97's formulation type having a λ_{∞} value near 22 mW/m°K.

A lot of additives, having a specific reaction with MDI and a sufficient compatibility with polyols have been tested.

After resolving these compatibility problems it is possible to get an other decrease of λ_{∞} in laboratory conditions. These results are shown on fig. 2 and, depending on the polyol modification ; the λ_{∞} values are included between 20,1 mW/m°K (foam n° 140) and 21,1 mW/m°K (foam 155).

3.8. Industrial feasability

Finaly the industrial feasability of these foam have been checked. The formulation 155 is then manufactured on an industrial lamination plan in comparison with an industrial formulation.

Results are given on fig. 3. It can be shown up that changing the formulation from the industrial one to a laboratory one have a significant effect on λ_∞. In the same time with a given formulation, changing the process or more exactly the constraint against rise type from the industrial one to our laboratory one have a significant effect on λ_∞.

3. ECONOMICAL IMPACT

Considering that the functional λ value can be seted near 20 mW/m°K instead of 29 mW/m°K one can calculate the annual energy saved for instance in France.

The French annual PUF market is about 6 Mm^2 for roofing and 1,5 Mm^2 for vertical walls. The ΔU value corresponding to the modification of the functionnal value is then about 5.10^3 Tep/an witch correspond to a cumulated value on 10 years of 3.10^5 Tep.

The European market is given as nine time the French one then the Energy saving for the EEC cumulated on 10 years could be nearly 3.10^6 Tep.

This economical impact evaluation is made considering no significant evolution of the PUF market but it is possible to think that this market will increase by reaction with the improvement of the functional thermal resistance of rigid polyurethane foams.

4. CONCLUSION

This study have reached its objective. By adjusting both the formulation and the foaming process in our laboratory it is possible to have PUF with a λ value of 20 mW/m°K.

It seems that it will be difficult to go farther in this study by formulation adjustment when it is easier by foaming process adaptation particulary by the constraint against rise controle. In collaboration with a polyurethane foam manufacturer in EEC, we are looking for an extention of this programme in this way.

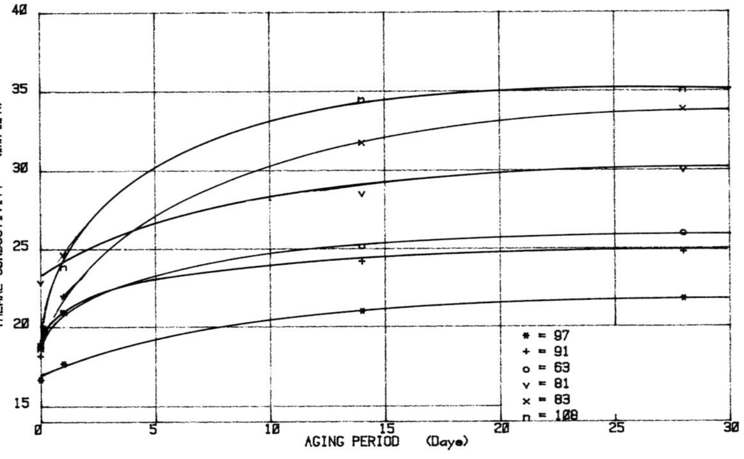

Figure 1

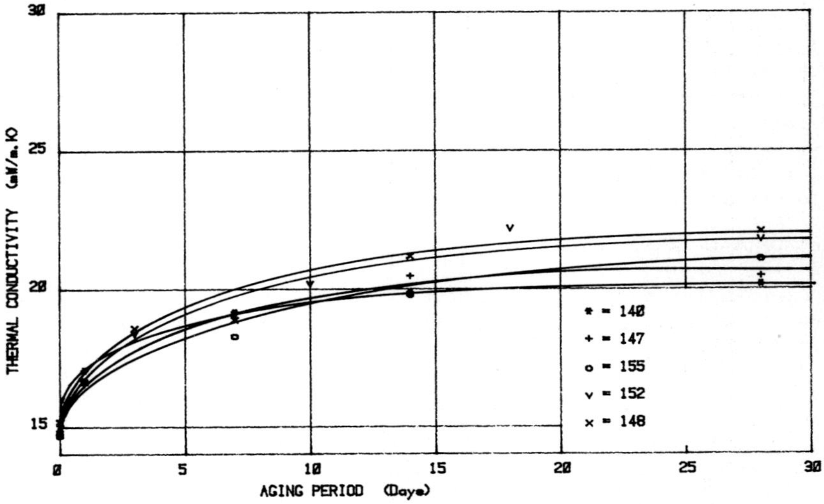

Figure 2

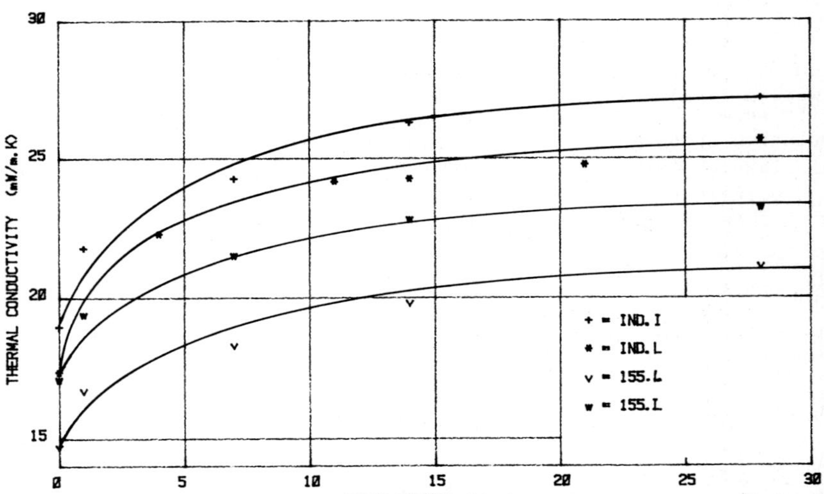

Figure 3

FOAM	Aging period	Total pressure (atm)	$P_{R_{11}}$ (atm)	P_{CO_2} (atm)	P_{N_2} (atm)	P_{O_2} (atm)	$\Delta\lambda$ (mW/m°K)	λ mW/m°K
81	0	0,539	0,33	0,03	0,08	0,09		22,8
	1	0,684	0,27	-	0,27	0,13		24,6
	14	0,868	0,01	-	0,65	0,20		28,4
	28	1,131	0,14	-	0,77	0,22	7	29,9
83	0	0,421	0,18	0,22	0,01	0,01		18,8
	1	0,250	0,13	-	0,07	0,05		20,9
	14	0,750	0,05	-	0,52	0,17		31,7
	28	0,947	0,01	-	0,72	0,20	15	33,8
91	0	0,474	0,18	0,21	0,05	0,02		18,0
	1	0,408	0,22	-	0,09	0,08		21,8
	14	0,829	0,24	-	0,43	0,15		24,0
	28	1,144	0,10	-	0,79	0,24	6,6	24,6
97	0	0,474	0,18	0,27	0,01	0,01		16,5
	1	0,289	0,18	0,05	0,02	0,03		17,5
	14	0,474	0,17	-	0,20	0,10		20,8
	28	0,855	0,15	-	0,49	0,20	5,1	21,6
108	0	0,513	0,22	0,21	0,05	0,02		18,6
	1	0,684	0,25	0,01	0,26	0,16		23,8
	14	0,697	0,02	0,01	0,52	0,14		34,4
	28	1,013	-	-	0,77	0,23	16,4	35,0
63	0	0,553	0,28	0,25	0,01	0,01		18,7
	1	0,539	0,28	-	0,13	0,12		20,9
	14	0,974	0,32	-	0,50	0,15		25,1
	28	1,223	0,29	-	0,71	0,22	7,2	25,9

- TABLE 1 -

RESEARCH ON BUILDINGS GENERAL QUALITY OF INSULATION, ESPECIALLY WITH RESPECT TO DEVELOPING AND IMPROVING METHODS OF POST INSULATION

N.H. BERTELSEN, G.C. LARSEN, T. NIELSEN and A.D. OLSEN
Department for Building Technology
The Technological Institute
Taastrup, Denmark

Summary

In connection with improving methods of post insulation for buildings we have made 3 main investigations.
One of those investigations deals with the construction of a general measuring equipment for buildings. The equipment is able to measure and to calculate the energy balance of a building automatically and continuously. The necessary period of measuring is estimated to be 1 or 2 weeks and during this time the amount of energy, which is passing through the envelope of the building as well as the amount of energy accumulated in the building are measured. The first measurements are planned to start in November 1983.
The efficiency of outer insulation with freely exposed mineral wool on top of roofs and on facades has been measured in the winter 1982/83. On a number of upside-down insulated roof sections with a pitch of 35° east oriented, we have measured the U-values in relation to the influence of the Danish weather. The measured U-values are in the same level as for the U-values for a conventional dry insulated roof. The facades, which are outer insulated with freely exposed mineral wool, have been investigated due to moisture. In the winter 1982/83 the insulation remained dry.
Urea formaldehyde foam for cavity walls insulation has been investigated due to thermal conductivity and due to evaporation of formaldehyde gas. The practical thermal conductivity is about 0.060 - 0.070 W/m°K and the evaporation of formaldehyde gas to the building must not in Denmark be higher than 0.15 mg/m^3. There is doubt about the number of companies and of products in Denmark, that are able to observe this demand. We have made an instruction manual on cavity walls insulation for insulation workers. It describes how the insulation worker himself can control his work during the injection process.

1. CONTINUOUS AUTOMATIC RECORDING OF ENERGY BALANCE

1.1 INTRODUCTION

This partproject deals with the problem of calculating and measuring the energy consumption in any house in Denmark. The main purpose of it is to show the state of the insulation quality of the house and to demonstrate the effect of post insulation.

Basically, the amount of energy flowing from a house depends on the indoor and outdoor temperatures and on the quality of the insulation of the house. In the same way it depends on the airtightness of the house. But the inhabitants of the house influence the energy flow in a very unpredictable way by opening doors and windows etc. and by using hot water in different ways. Furthermore, the sunshine coming through the windows will reduce the need for energy production in the house in a way, which is not a function of the outdoor temperature.

All these circumstances make it impossible to describe the energy consumption of a house as a simple linear function of the indoor-outdoor temperature difference as it is often suggested or done.

1.2 A SIMPLE MODEL FOR THE ENERGY CONSUMPTION

The only way to describe the energy flows into and out from a house is to put a fictive envelope around the house (or part of it) and use it as a reference envelope, through which every component of energy flow is supervised. Thus, the amount of energy that comes in through the envelope or is produced inside it must balance the amount of energy that goes out through the surface PLUS the amount of energy accumulated in the construction in hot water etc.

1.3 IMPLEMENTATION

If all components of energy flow and energy accumulation are measured with sufficient accuracy the balance in the energy account will be zero constantly.

But if the energy calculations are based on incorrect assumptions, f.inst. because of poor insulation quality, low efficiency of oil burner etc., then the balance in the energy account will vary with time, and a graph of this variation will make the detection of the nature of the untrue assumptions possible and thereby it will be possible to find out, what the nature of the problem is, if any.

1.4 TEST EQUIPMENT

An equipment has been developed to measure the energy balance of a house. It is controlled by a microcomputer and is able to measure ......

1) TEMPERATURES on surface and in the air to calculate energy accumulation and flow by conduction.
2) AIR INFILTRATION to calculate energy loss by air flow.
3) HOT WATER TEMPERATURES, to estimate energy loss by waste water.
4) OIL SUPPLY, to calculate energy supply from oil burner.
5) SUPPLY OF ELECTRICAL ENERGY.
6) INSOLATION through windows, to calculate light energy flow.

The BIOLOGICAL ENERGY from human beings in the house is estimated by investigating after the measurement, how many persons were present in the house, and how active they were during the period of measurement.

1.5 NEW ASPECTS OF MEASUREMENT

Three aspects of the test equipment are considered to be new .....

1) The air infiltration rate is measured continuously by means of the tracer gas method (constant concentration of $SF6$). This method has been used regularly for a couple of years at the Technological Institute, Copenhagen.
2) The insolation through windows is measured directly on the inside surface of the glazing unit by means of a number of light energy transducers, responding on the difference between the incoming and

outgoing light energy. Normally 5 transducers are used on each window in the house to give sufficient accuracy.

3) The accumulation of energy in constructions, water etc, is calculated from temperature measurements. In some cases, the temperature of the accumulator can be measured directly, but normally it can not be. In order to calculate temperatures in the inside of building constructions, the temperatures of the surfaces or of the surrounding air are measured and from those, the temperatures inside of the construction are calculated by means of a simple electric analogy, using a R-C network.

1.6 LIMITATIONS

The equipment has been designed to make measurements on buildings where oil or electric power is used for heating. Of course it can be used for any type of power supply, but in buildings with remote heating, it is normally not possible to mount a transducer for the energy supply without dismounting waterpipes and fittings, which should be avoided. So until a handy transducer has been found, remote heating will not be taken into consideration.

1.7 CONCLUSION

At the time of this report, the construction of the test equipment is finished, and the computer programmes that controls the equipment during measurements and make the calculations after the measurements, are being implemented. The aim is to make some on-site few measurements to demonstrate the functionality of the equipment.

In the future, the equipment will be used in the daily working routines at the Institute as a tool for investigation on any energy aspects and functions of buildings. It is regarded to be a powerful means for on-site investigation of any change made on buildings to improve their energy balance, because the equipment is able to make a detailed investigation on all the energy aspects of the building.

2. OUTER POST INSULATION WITH FREELY EXPOSED MINERAL WOOL

2.1 INTRODUCTION

Outer post insulation on roofs and facades with freely exposed mineral wool is interesting, because mineral wool has good qualities to this purpose. Those qualities are illustrated by measurements of moisture and temperature on a number of post insulated constructions, which are affected directly of the natural climate. We have made measurements on 6 roof sections and on 3 facades. The roof sections are built and placed on the area of the Technological Institute. The facades are places on an administration building of Rockwool A/S in Hedehusene. The mineral wool is placed outside unprotected from the Danish weather - temperature - moisture - wind and sunradiation.

2.2 THE UPSIDE-DOWN ROOF

The roof sections are constructed as shown in figure no I. The sections have the following qualities:

Orientation - east

Roof section area - approx. 10 m^2 . Roof pitch - 35°.

Mineral wool density - Rockwool 160 kg/m^3, Glaswool 90 kg/m^3.

On those 6 roof sections we have measured the U-values (W/m^2 ^oK) in the winter 1982/83 as a pilot project.

The upside-down roof no 1 is open for outside air ventilation in the bottum of the sheet at the lower edge. It is airtight at the higher edge of the roof (Figure no I).

The upside-down roof no 2 is airtight at both the higher and the lower edge (Figure no I). Insulating corrugated sheet covering as shown in figure no I can be used for buildings in the industri, agriculture and for house building.

The efficiency of the insulation is shown in figure no II due to rain, sunradiation and wind for a period of 19 days. The measured U-values are calculated on basis of measured values for the heat flow through the actual roof section. The heat flow is measured by means of heat flowmeters, which are constructed in the layer of polystyrene. The flowmeters are calibrated in a guarded hotbox and they are constructed by a 80 mm thick polystyrene plate with thermo-couples on the two parallel sides. The U-values are calculated as follows

$$U = \frac{\emptyset}{\Delta t} \quad \text{, where}$$

$\emptyset$ is the heatflow i W/m^2

Δt is the difference between the indoor temperature and the outdoor temperature against north and in shade.

The indoor temperature is approx. 20° Celcius.

All the data are collected electronically for every half hour and in figure no II the daily meanvalues for roof section no 1 is shown. We notice that the measured U-values for the upside-down roof are at the same level as for the U-value calculated after the Danish Standard for dry materials. On figure no III the measured U-values for the upside-down roof no 2 are shown. We notice that there is only a slightly variation due to the big rainfall in the period. In this period of rain we measured the amount of water coming through the mineral wool to about 2-6% of the fallen rain at all. In figure no IV and V is shown the periods in which the directly sunradiation is zero. Those periods shows that the amount of energy coming from the sunradiation has none or slightly effect on the U-value at the outer insulated roof sections. The U-value for the reference section is very influenced by the sunradiation. The difference in the U-value between the roof sections 1 and 2 is small. The difference is about 0.01 W/m^2 ^oK, which is the same size as our measuring uncertainty.

The improvement of the U-value by the outer insulating with freely exposed mineral wool is

U_{BEFORE} = 0.30 W/m^2 ^oK

U_{AFTER} = 0.18 "

ΔU = 0.12 "

which gives an energy saving per year of approx. 1.2 l fuel oil/m^2 in the Danish climate and with an indoor airtemperature of 20° Celcius.

Figure I

The upside-down roof
Construction
Roof pitch: 35°

Reference roof:

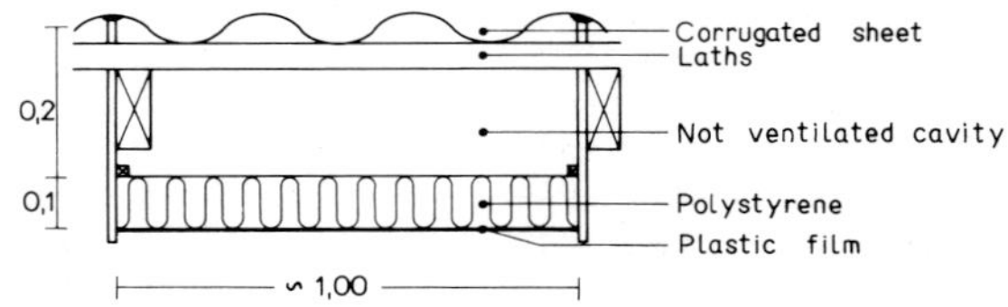

The upside-down roof:

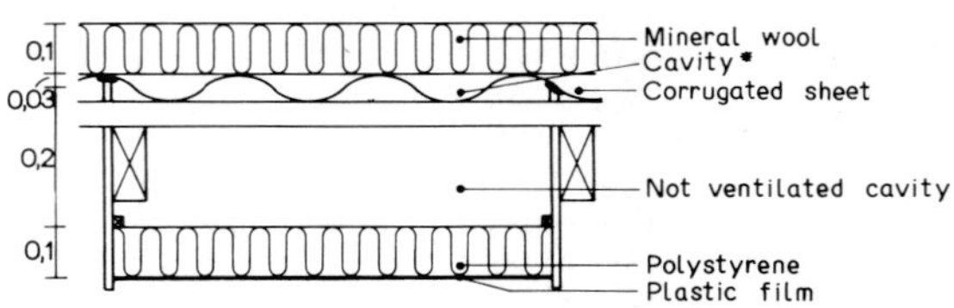

* The upside-down roof no 1. is ventilated in the shown cavity.

* The upside-down roof no 2. is not ventilated in the shown cavity.

Dimension: m

Figure II

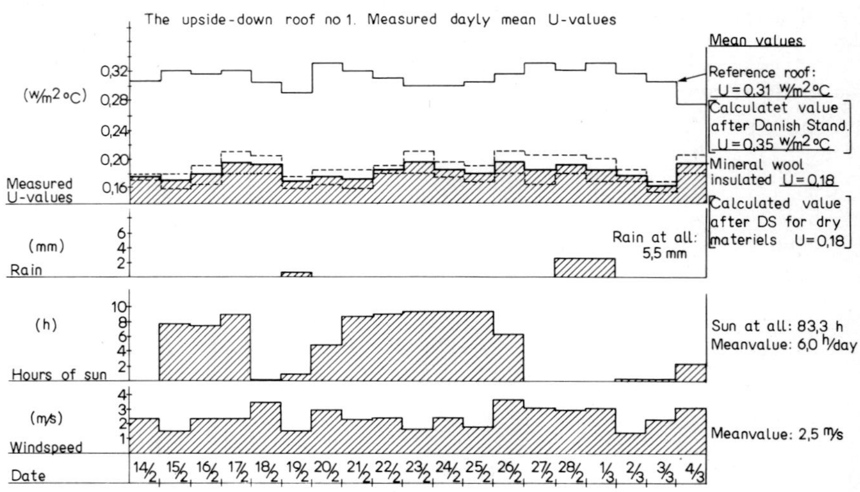

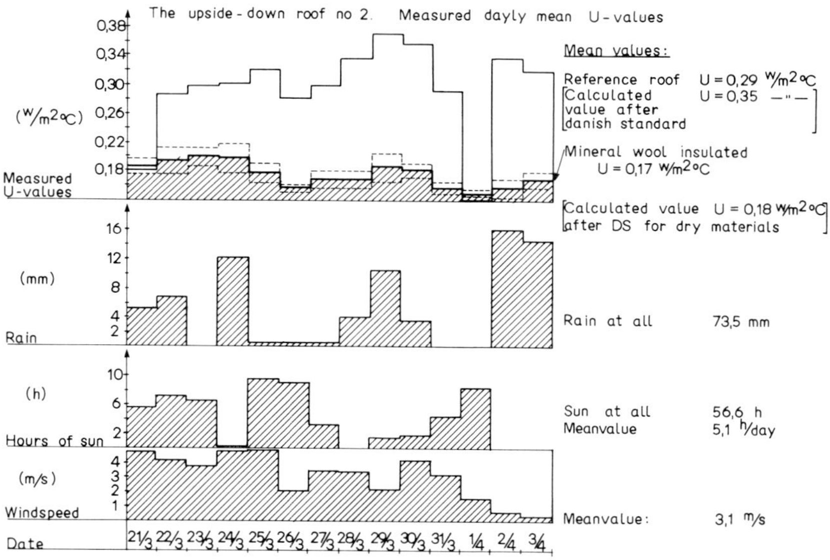

Figure III

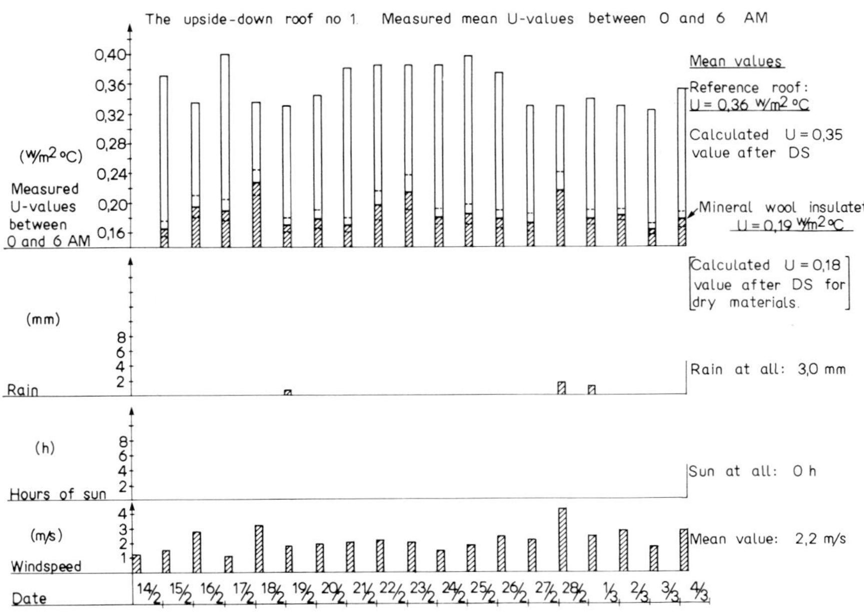

Figure IV

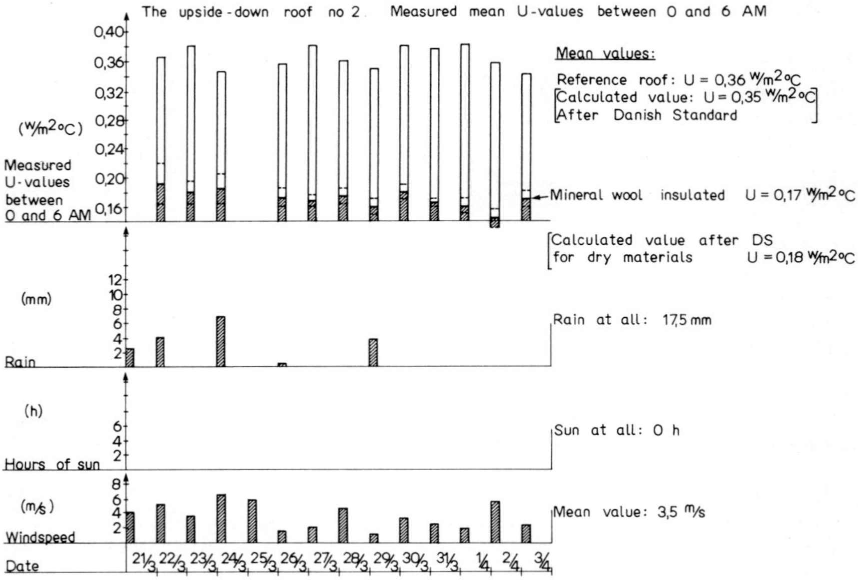

Figure V

2.3 THE OUTER INSULATED FACADES

The facades we have measured are situated at an administrationbuilding on Rockwool A/S in Hedehusene. The construction of the facades is

outside - 100 mm Rockwool (160 kg/m^3)

inside - 200 mm aerated concrete elements.

The Rockwool is placed in contact with the concrete elements with special fittings, and the moisture content in the insulation has been measured in the following depts starting at the outside:

- 10 mm - 50 mm - 95 mm -

The following moisture contents were found in January and February 1983:

Facade orientation	Moisture content in insulation % RH
West	61
West foundation	73
South	64
East	61
East foundation	82

The outdoor climate was in the same period:

Period	Air temperature °Celcius	RH %
January	4.5	81
February	-1.0	83

If we look at the influence on the thermal conductivity due to moisture in the Rockwool material, we find that the increase is as shown belov:

Rockwool used as	Increase in thermal conductivity due to moisture in % of the dry thermal conductivity
Outer facade	approx. 2.5
Outer foundation	approx. 5.0

3. IMPROVEMENT OF CAVITY WALLS INSULATION

3.1 INTRODUCTION

The quality of insulating for cavity walls insulation depends on the performance of the work and on the insulation material that has been used. In an recently finished investigation we have looked at the quality of insulation for 134 single family houses with cavity walls insulation. The result was that the quality almost only depends on the performance of the work. It was shown that approx. 5% of the average cavity wall area per house was bad insulated. Although this area seems to be rather small, a great number of houses had faults in the insulation, that ought to be avoided. An analysis of the faults showed that a great number of the faults was systematic due to

- incorrect method of injection
- bad preliminary investigation of the cavity wall before injection
- wrong placing of the wholes for injection
- missing selfcontrol during and after the work of injection.

There was also a significant difference of the quality of insulation for new and for old injections.

The quality of injection has during the last few years increased in many companies. The quality can be increased more by courses, by education and by an effectiv selfcontrol.

3.2 SELFCONTROL DURING INJECTION CAVITY WALLS INSULATION

The quality control of the work done during cavity walls insulation has in Denmark been devided in 3 levels under the organization - The Danish Quality Control Scheme for the Finished Insulation Work:

- a superior control, which is done by a neutral institute on the finished insulation work
- a sample test control, which is done by a control organization on the finished insulation work
- a selfcontrol, which is done by the insulation worker himself before, during and after the insulation work.

To obtain an usefull tool for selfcontrol by cavity walls insulation, it is necessary to make an instruction manual. This manual shall in pictures and in words explain clearly, what to be controlled, so that the results of the scientific work about cavity walls insulation will be easy accessible for the insulation worker. The following main control items are forming the basis for such an instruction manual.

A cavity wall must observe the following conditions in The Danish Quality Control Scheme for the Finished Insulation Work:

- the cavity between the exterior and the interior wall is at least 30 mm
- there is no organic materials in the cavity
- there is a moisture membrane between the foundation and the wall.

If the actual cavity wall observes those conditions, the selfcontrol can be started.

The selfcontrol is devided in the control
1) Before the beginning of the work
2) During the work
3) After the ending of the work.
No 2) is the most important.
Comments to 1). The main items deal with
- the raw materials, that has to be correct labelled, stored and have the correct conditions
- the injection machinery, that has to be described in an users manual in the national language
- the cavity wall, that has to be investigated, to be frost-proof and to be free of faults in the materials and in the construction

Comments to 2). The control during the work consists of
- the consumption of insulation material, that has to be as calculated
- the injection machine, in which the flow of material shall be steady
- the cavity wall, that shall be well ventilated during the injection
- the insulation material, that must fill up the cavity wall.

Comment to 3). After the ending of the work you have to investigate
- the insulation material, that must fill up the cavity wall
- the injection machine, that has to be cleaned
- the cavity wall, that has to be repaired.

3.3 UREA FORMALDEHYDE FOAM FOR CAVITY WALLS INSULATION

3.3.1 INTRODUCTION

In the summer 1983 Denmark introduced a maximum level of 0.15 mg/m^3 of evaporation of formaldehyde gas from urea formaldehyde foam to houses. The technical investigations of the foam have until 1980 primary been concentrated about the thermal conductivity due to moisture and shrinkage. After 1980 we have concentrated our work on the evaporation of formaldehyde gas from the material.

The work we have done with support from the EEC has gathered the Danish experience at the use of urea formaldehyde foam as cavity walls insulation. The work is divided in 3 main items

- the thermal conductivity due to shrinkage, moisture and different raw products
- evaporation of formaldehyde foam under conditions in the laboratory and in the field
- the basis for a quality control to secure the quality of the foam.

3.3.2 THERMAL CONDUCTIVITY

The thermal conductivity for urea formaldehyde foam, that has to be used for practical purposes is called, the practical thermal conductivity, λ_p. It can be calculated as follows:

$$\lambda_p = \lambda_{10.m} + \Delta\lambda_{unc} + \Delta\lambda_{mois} + \Delta\lambda_{shri} + \Delta\lambda_{age},$$ where

$\lambda_{10.m}$ is the thermal conductivity at 10° Celcius and 50% RH

$\Delta\lambda_{unc}$ is an addition for uncertainty during injection

$\Delta\lambda_{mois}$ is an addition for moisture in the material

$\Delta\lambda_{shri}$ is an addition for shrinkage in the material

$\Delta\lambda_{age}$ is an addition for ageing of the material

From our own investigations and from the literature we have set the values as follows

$\lambda_{10.m}$ = 31-36 mW/m°K (density 10 kg/m^3)

$\Delta\lambda_{unc}$ = 2 mW/m°K

$\Delta\lambda_{mois}$ = 5 mW/m°K

$\Delta\lambda_{shri}$ = 7 · % of shrinkage mW/m°K

$\Delta\lambda_{age}$ = 0 mW/m°K

We hereby get that

$\lambda_p = \lambda_{10.m}$ + 7(1 + % of shrinkage) mW/m°K

3.3.3 EVAPORATION OF FORMALDEHYDE GAS

Evaporation of formaldehyde gas from urea formaldehyde foam placed in cavity walls made of bricks will effect the indoor climate more or less due to the following parameters:

- the air change rate in the house
- the concentration in the house
- the ratio between volume and area of expose
- the resistance of the inside wall against gastransport by diffusion or flow
- the age of the foam
- the concentration in the pore of the material
- the temperatures and the moisture contents

In Denmark we have made a serie of measurements in urea formaldehyde foam insulated houses. We have measured 8 houses in each 3 rooms and with 2 different materials of foam A og B. The results are shown below.

Days after injection		11	25	43	72	148	220	263	310
Material		B	B	B	A	B	B	A	B
Formaldehyde concentration in mg/m^3	Measurement no 1	0.24	0.19	0.29	0.05	0.29	0.23	0.08	0.34
	Measurement no 2	0.19	0.06	0.35	0.03	0.19	0.26	0.03	0.50
	Measurement no 3	0.10	0.08	0.24	0.04	0.06	0.15	0.04	0.35
	Average value	0.18	0.11	0.29	0.04	0.18	0.21	0.05	0.40

The concentration of formaldehyde gas in 8 houses insulated with urea formaldehyde foam.

On the basis of our experience from evaporation of formaldehyde gas from paper pulp sheet and from our measurements of the foam in the laboratory and in the field we have constructed a calculation model for evaporation of gas from the material through the inside wall to the house. The model consider the effect of temperature, moisture, air change rate and the rate of insulation foam and volume of the house. The model has not been calibrated but we have made a plan for the work to be done in the laboratory and in the field to fulfil this aim.

The investigations have lead to a demand in the Danish rules for house-building. The demand is that we in Denmark have to control the quality of urea formaldehyde foam before use. We have to be sure that the concentration of formaldehyde gas is lower than 0.15 mg/m^3 in buildings.

There is doubt about how many companies and products in Denmark which are able to observe this demand.

REFERENCES:

T.Nielsen:
Continuous Automatic Registration of Energy Consumption in houses.
Users manual, Notes 1983.
Calculation and measuring model. Notes 1983.

A.D.Olsen, D.Blanchet, P.F.Collet:
Measurements of the upside-down roof insulated with freely exposed mineral wool.
The Technological Institute. Notes 1983.

P.Andersen, A.D.Olsen:
Moisture measurements in outer insulated facades
The Technological Institute, 1983

H.Samuelsen:
Cavity walls insulation
The Technological Institute, 1983

H.Samuelsen, I.Faldager, N.H.Bertelsen:
Cavity walls insulation by injection
The Technological Institute, 1983

N.H.Bertelsen:
Evaporation of formaldehyde gas and thermal conductivity of urea formaldehyde foam as cavity walls insulation
The Technological Institute, 1983.

THE PRODUCTION OF SELECTIVE OPTICAL COATINGS ON PLASTIC SHEET

R.P. HOWSON, M.I. RIDGE, C.A. BISHOP and G.P. COTTMAN
Department of Physics, University of Technology, Loughborough, Leicestershire, U.K.

Summary

A technique of vacuum deposition is described for producing heat reflecting films onto sheet plastic for application in domestic windows. Thin metal films of copper, silver, gold and titanium nitride sandwiched between dielectric films and electrically conducting oxide coatings of indium, indium doped with tin and cadmium/tin are shown to have the properties required for a heat reflecting coating for visible windows. Durability measurements suggest that conducting oxide films have more promise. Indium oxide coatings on polyester sheet have been made which have been produced onto plastic transferred from roll-to-roll at a rate of 0.25 $m.min.^{-1}$ They are highly transparent and have a low temperature emissivity of about 0.2.
It is shown that their use as a secondary window can improve the performance of the system to give a U-value of less than 2 $Wm^{-2}{}^{o}C^{-1}$ and this can be done with a small weight penalty and, what is most important, can be projected to be able to be manufactured at a cost which makes the unit a very efficient use of energy resources. It should have a payback period of considerably less than one year, both in cost and in energy saved, and have the potential to be made in large areas to make a significant impact on natural energy use.

1. INTRODUCTION

Energy costs have been rising more rapidly than the cost of incorporating insulation into buildings, a trend which is confidently predicted to continue, and a situation has been created where existing buildings are insulated to a level below that which is justified by the cost of heating. This is not only a current problem but one which will be with us for a long time; it is estimated that 80% of the domestic buildings which will be used in the year 2000 have already been built and to inferior insulation standards than those that will be demanded by the price of fuel at that time. Insulation standards of existing buildings have to be raised with methods which are properly effective in that the cost of their installation is justified by the energy that is saved. Solar energy is free and it is important that this be utilised as far as possible. All existing houses have solar collectors installed which can be very effective on occasions, but can also create problems. These are the windows. It is important to consider these elements of a building as they affect energy capture and energy loss.

Domestic windows play a large part in determining the living environment for most of the population of the world. They fill the apertures through which natural light is admitted to a building, which prevents air movement and, hence, the loss of heat. The essential material and design of such windows has changed little from medieval times. This is because of the properties of glass, a strong stable economical material with high resistance to mechanical scratching, which has a good visual transmittance. Unfortunately, it has a high thermal conductance and the thermal insulation it offers is principally obtained from the layers of air at either side and is, as a consequence, quite low and dependent upon air movement. The insulation effect of air, or some other gas, can be obtained by enclosing a layer within the window with two glazing elements and this has been the most obvious thermal development in window design in recent times. Windows, however, have a great advantage as thermal elements in a building in that they allow the energy from the sun to penetrate into the interior. The energy is trapped inside because the glass is absorbing in the infra-red, the region at which the interior re-radiates the heat generated by absorbed solar energy. This is known as the "greenhouse effect". Even for fossil fuel heated housing at Northern European latitudes in winter there is very nearly an energy balance for south facing vertical double glazing elements; they gain almost as much energy from the sun as they lose due to heat loss from the warm interior. There remains the radiation interchange between an individual and the outside environment, which takes place with the window because it is opaque to heat radiation. If the surface of the window can absorb and radiate heat radiation very easily then discomfort will be felt by an individual due to radiation heat transfer with a cold window.

The thermal performance can be improved with a surface coating which gives visual transmittance with a high infra-red reflectance. The high infra-red reflectance has two purposes; to reflect heat radiation and to provide a surface with a very low emissivity which limits radiated heat transfer. For example in a double glazed window two thirds of the heat transfer is by radiation between the two panes. This is so high because of the high emissivity of glass.

Such a surface can be created by the addition of a plastic sheet coated with aluminium to the window surface. The aluminium is vacuum deposited onto polyester film, which is opaque to infra-red radiation, but the protective laminated layer is polypropylene which is transparent and allows the aluminium film to utilise its high infra-red reflectance. Unfortunately, these films may only be used where solar control is also required because good infra-red reflectance is only maintained with considerable loss in visible transparency. In these circumstances the winter solar gain is also lost so that window covered with this film may still be worse than one without in terms of winter energy conservation.

There exists a strong need to create surfaces on windows which have low emissivities and maintain the solar gain, which can be applied to improve the thermal performance and comfort levels and remain cost effective.

The prime requirement is to manufacture, by vacuum coating, a film having the desired optical properties with a process which has a very high rate operating at a high pressure and in a simple way. These requirements are set by the high capital cost of the producing equipment and its influence on the final cost of the product. The process must meet the requirements of the plastic film substrate which will not withstand high temperature operation.

The basis of the methods and materials that need developing is that they provide a technique which can be readily adapted to the equipment used in the vacuum metallising of polyester sheet in the present packaging industry.

2. THE PROPERTIES REQUIRED FOR WINDOW FILMS

The difference between the wavelengths at which energy is

Comfort for a person is dependent both upon air and radiant temperatures. The radiant temperature is determined by the interchange of energy between the individual and his surroundings by infra-red radiation. If the surroundings are at a low temperature and have a high emissivity then considerable heat loss will be experienced and the air temperature will have to be raised to maintain adequate comfort levels. This effect is commonly experienced next to a single pane window on cold days and can be specified by an equivalent perfect absorber at a temperature which represents the energy exchange. The effect can be minimised by raising the temperature of the surroundings or using a low emittance surface to prevent the radiation exchange.

For a temperature difference between the room and the outside of 22°C the equivalent black body temperature seen by an occupant is 14°C below ambient for a normal single pane window. This reduces to 5°C when a coating with an emissivity of 0.3 is put on the inside of the window and 1.7°C for one with a value of 0.1.

3. MATERIALS FOR SIMPLE OPTICAL FILMS

The material has to be non-toxic, cheap, easily available and requires to be used in as thin as possible a layer consistent with durability. As few as possible layers should be used to obtain the desired properties. These requirements exclude conventional dielectric interference coatings and has lead to there being two possible systems.

(a) A metal film, with low intrinsic absorption, used in a sandwich between two high refractive index dielectrics. Suitable materials are: copper, gold, and best of all, silver.

(b) An electrically conducting metal oxide film generally of indium oxide doped with tin.

Both these systems require the production of an oxide film at a high rate with carefully controlled properties, especially in the case of the electrically conducting variety.

4. DEPOSITION TECHNIQUES

Our original hypothesis was that the addition of energy to a growing film by exposure to bombardment by inert heavy ions should lead to properties being attained which were normally seen only at higher temperatures. This was tested with evaporation and sputtering methods. Only reactive methods using pure metal sources were considered in order to achieve the simplest realisable system which could achieve high rates. This gave good results with both systems. An important factor was found to be the control of the oxygen to metal ratio. With the high rate used it was apparent that it was the oxygen gas admission rate that determined the properties obtained.

The precision control of rate, combined with a high sputtering rate capability, and operation at a relatively high pressure to provide the reactive gas, has lead us to the system of d.c. planar magnetron sputtering of the metal or alloy in an atmosphere of argon and oxygen. Ion bombardment of the substrate provided by a radio frequency discharge activates the oxygen reacting on the substrate surface to give higher rates of deposition. In order to achieve optimum properties in the films we have found it necessary to control the power to the sputtering source and the oxygen gas admission rate to within 0.1%. The conditions were adjusted in accordance with the properties of the film continuously monitored in a roll coating apparatus. Planar magnetron sputtering is a

received by a window and those at which it is re-emitted is illustrated in Figure 1. The control of light and energy from the sun is obvious, what is not so clear is the influence of the presence of low emissivity surfaces on the thermal properties of the structure.

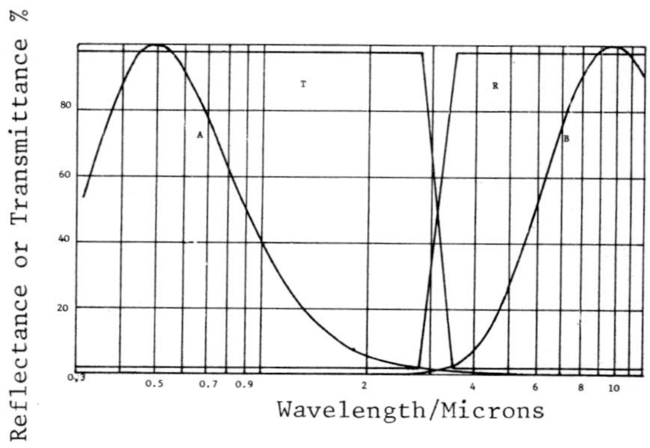

FIGURE 1

An approximate normalised emission spectra is shown for the sun (A) and for room surfaces (B) together with the selective properties required for a selective optical surface for energy conservation. R refers to the reflectance and T to the transmittance for such a filter.

FIGURE 1

The difficulty with adhering a sheet on to a single window pane is that a lower glass temperature will result which is liable to cause condensation on the inner surface. Water is opaque to heat radiation so that the thermal advantage will be lost. In double pane windows, two thirds of the heat transfer is by radiation and considerable improvement can be made to the properties by using a low emissivity surface on the inner pane and facing the outer. This is seen in Figure 2.

FIGURE 2

The U-value for a double pane window with the inside pane having a coating (A) on the side facing the other pane only and (B) on both sides.

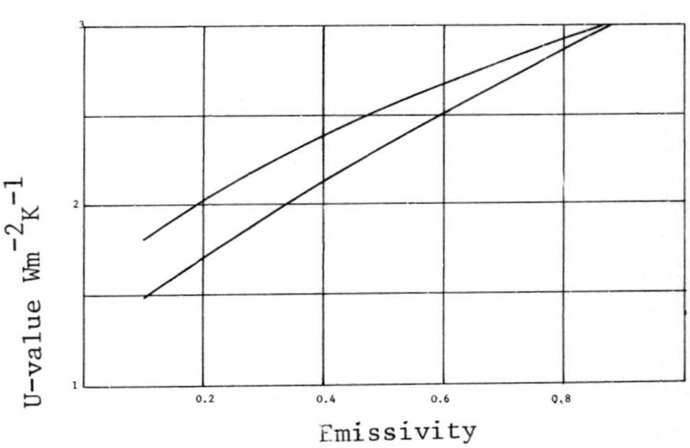

FIGURE 2

The properties required for window films can be specified in terms of the two spectra.

The filter must be designed to allow all the energy from solar radiation to penetrate the building, which requires a transition point where the film goes from completely transmitting to completely reflecting at about 3 μm. This is illustrated in Figure 1.

technique in which a magnetic field is created parallel to the surface of the sputtering cathode which enhances the discharge by trapping secondary electrons.

5. DEPOSITION APPARATUS

Two separate pieces of equipment, including the magnetron source, have been designed and manufactured by ourselves for the deposition of oxide and metal films onto plastic sheet, transferred from roll-to-roll over a drum capable of being taken to high radio frequency potentials.

The design in both cases was conditioned by the initial equipment that was modified. Roll Coater 'A' was constructed to use an external mount magnetron source, 'B' used the motor drive mechanism available in a Balzers 510 Coating Unit, together with the existing electrode feed-throughs for the internal magnetron sources.

The apparatus of 'A' is illustrated in Figure 3. Figure 4 shows a section of a round planar magnetron that was built.

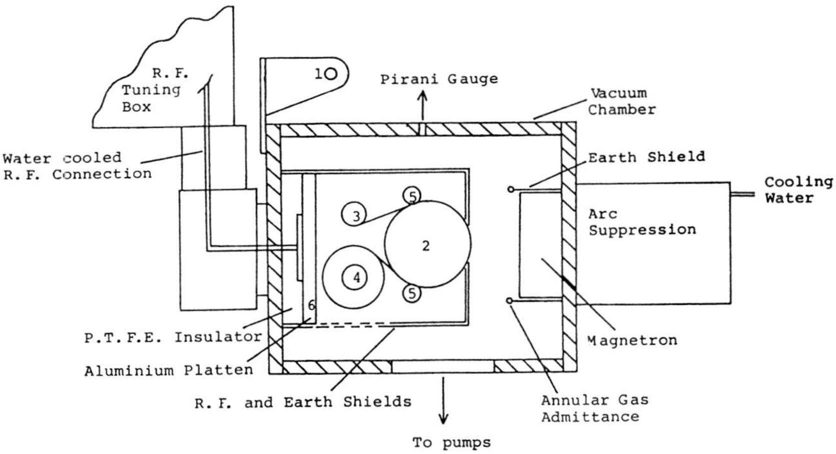

FIGURE 3

The roll-to-roll coating apparatus detail and vacuum system showing the magnetron source in position.

6. WINDOW MEASUREMENTS

In order to assess the effect of changes in design and manufacture on window thermal performance it was necessary to build a U-value measuring facility. The primary requirement of this equipment was that it should be capable of accurate measurement and, hence, be able to detect small changes in overall thermal impedance. The design of the equipment and its method of operation, described below, were chosen to meet the requirements.

The measuring rig comprises an insulated chamber which is divided by a partition capable of accepting a 1 x 1 metre window. The two halves of the chamber each contain a heat exchanger so that a heat flux can be created through the window. The heat exchangers comprise finned tubes through which water is circulated, three d.c. fans and air deflector panels.

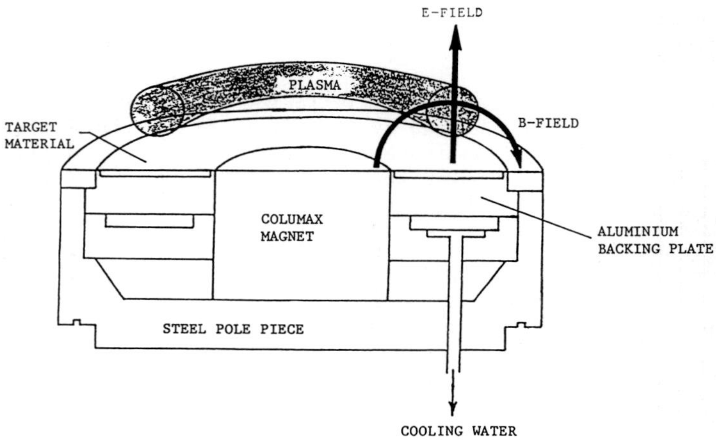

FIGURE 4

Magnetron assembly showing the location of the magnet, the E & B fields and the plasma.

The essential point of the design is that one side of the window is kept at the same temperature as the room in which the chamber is housed, and all measurements of heat flow are conducted on this side of the window. This means that any heat exchange occuring is via the window and not through the walls of this half of the chamber. The equipment is designed to measure the U-value under common environmental conditions and, thus, the other half of the chamber can have its temperature lowered to -5°C if required. This apparatus is shown is Figure 5.

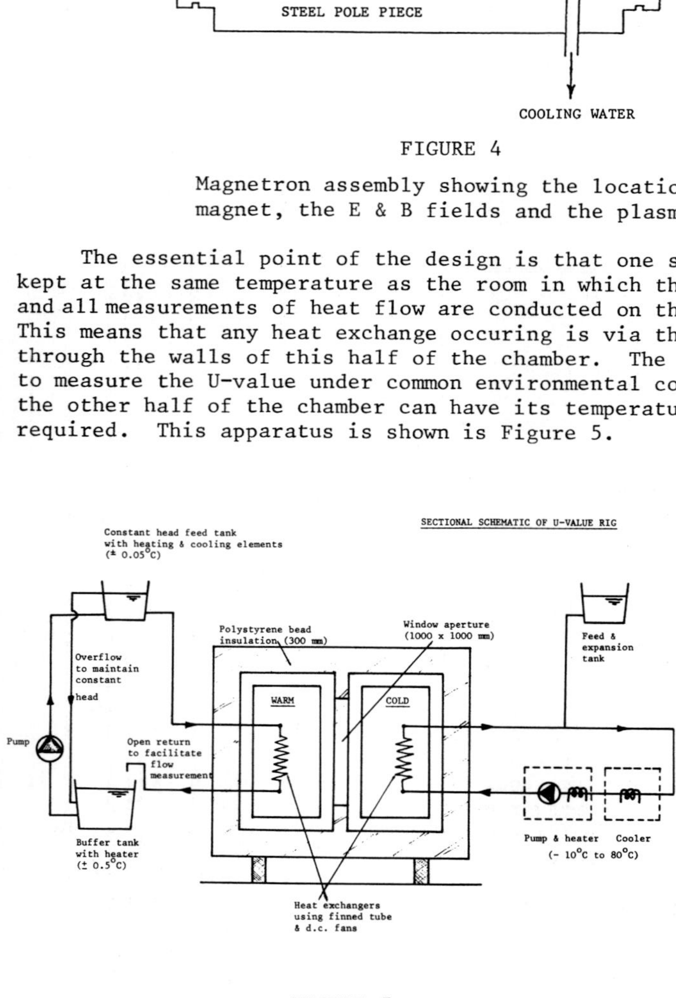

FIGURE 5

Schematic of apparatus to measure the thermal conductance of one x one metre windows.

FIGURE 5

7. RESULTS

The measured spectra of various oxide films are shown in Figure 6 and the related optical and electrical parameters are given in Table 1.

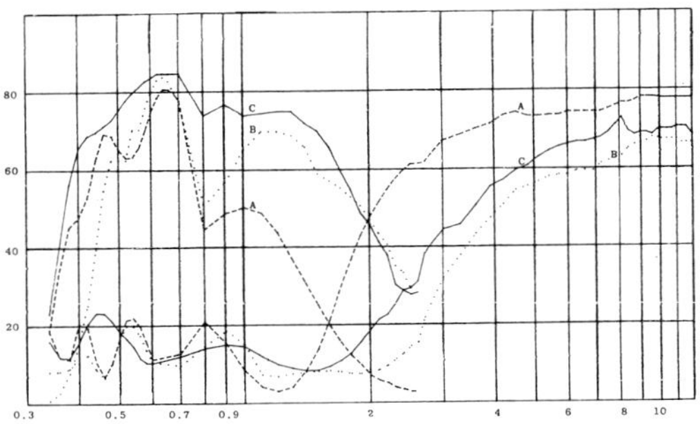

FIGURE 6

FIGURE 6

The selective optical properties obtained with (A) indium 10% tin oxide, (B) indium oxide, and (C) cadmium 2:1 tin oxide.

The sample of indium oxide was taken from a roll of material which was made in a continuous process where 10 metres of the roll of polyester sheet was coated at about 0.25 metres per minute. The rate of sputtering was over 0.5 μm min^{-1} and the power used was about 3.5 kW into the one target. The remaining material was made into a 1 metre square window for detailed testing.

Parameter	ITO	CTO	IO
Relative electric permittivity	4	4	3.5
Plasma wavelength (x 10^{-6}m)	1.3	2.5	2.0
Plasma frequency (x 10^{15})	1.45	0.75	0.94
Sheet resistance (Ω/□)	17	24	30
Film thickness (x 10^{-9}m)	330	280	336
Carrier density (x $10^{26}m^{-3}$)	6.1	3.4	2.1
Electrical mobility (x $10^{-4}m^2V^{-1}s^{-1}$)	17	30	29
Carrier relaxation time (x 10^{-15}s)	2.8	5.8	2.6
Effective mass	0.29	0.34	0.16

TABLE 1

Durability testing has been made using temperature and humidity cycling in the presence of simulated sunlight. These tests were particularly severe because they exposed the film side of the sheet to the 'exterior' environment. Indium tin oxide showed slight signs of crazing after considerable exposure; indium oxide was worse but when examined under a microscope showed that in fact the film was lasting very well except for dimension changes which caused the film to buckle but not to break. Indium oxide has been observed to cause more bending of the plastic sheet and so has greater compressive stress.

The average measured compressive stress was 4 x 10^7 kg.m^{-2} which is compatible to values measured by other workers for other dielectric coatings.

The Hall effect gives the parameters of mobility, μ, and carrier density, η, of a conducting film. These parameters help determine the optical properties of the film through the effect of free carriers, which give the infra-red reflectivity which is required. We have measured these parameters at room temperature and related them to optical properties measured at the same temperature.

Investigation at room temperature of films made at various partial pressures of oxygen indicate that the choice of films with the maximum conductivity to obtain the optical properties that are desired, which has been made hitherto, is not necessarily the best. Figure 7 indicates that films of indium oxide have higher electrical mobilities when made at higher oxygen partial pressures, but show a higher sheet resistance due to a fall in carrier concentration.

The measurement of U-value for a system of a suspended coated plastic sheet added to a single pane glass window gave a value of 1.7 Wm^{2o}C in comparison with a value of 2.7 for 12 mm spacing double glazing. The film was mounted by adhering the 80 mm wide coated polyester in strips onto a one metre wide plastic backing sheet of 75 μm in thickness. The measurement could well be a little low because it was made in summer with an unsealed enclosure which resulted in some condensation being observed on the internal cold face of the "exterior" glass pane. Sealed units incorporating heat mirrors based on thin metal films gave values of 1.25 when a single coated polyester sheet was suspended between two glass panes and 0.9 when two such sheets were suspended. When our coated film was used as secondary glazing in a room of a building which was normally used as an old persons home, and was compared with similar rooms fitted with sealed double glazing and sealed coated triple glazing, very encouraging results were obtained. This was a study performed by Leicester County Council and monitored by the Leicester Polytechnic School of Architecture. In this test our secondary window matched the triple glazing in overall performance which, it is believed, is because of the greater transparency of this coating to solar radiation over that exhibited by the thin metal film used in the triple glazing. An estimate of energy saving in one year provided by this window is 493 kW hr for an English Midland location with a total window aperture of 1.4 m^2. The orientation of the window was 280° from North. A saving on conventional double glazing was given as 361 kW hr and triple glazing as 378 kW hr. The effect of

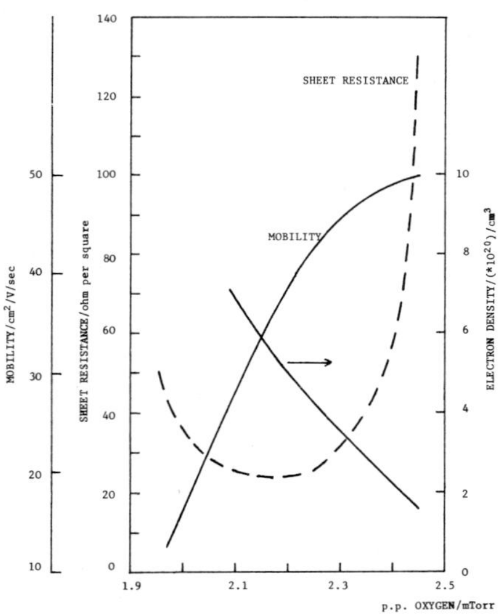

FIGURE 7
The Hall parameters, electron density, η , and mobility, μ , shown as a function of the deposition conditions for an indium oxide film.

having good solar energy transmittance can be clearly seen. This is particularly important at these orientations, which allow considerable solar gain.

We have made a variety of dielectric/metal/dielectric filters whose performance is given in Figure 8. In general these filters can be made with lower emissivities giving lower U-values, but with lower visual transmittance and very much lower solar energy transmittance than oxide films. They proved also to be delicate, in that they are subject to deterioration when exposed to a normal atmosphere, and systems that have been operated have had to be sealed in a controlled, dry atmosphere.

We have investigated one system that has the potential for being much more durable, although its projected performance is not so good. This uses a compound metal, titanium nitride, which can be made with TiO_2 dielectric layers from a single magnetron target of titanium by changing the reactive gas.

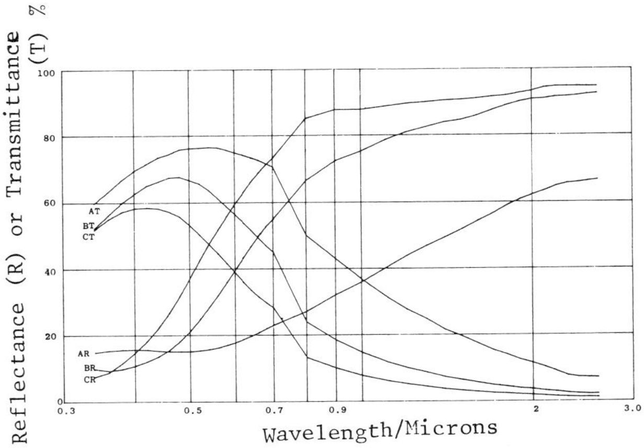

FIGURE 8

FIGURE 8

The experimental spectra of silver films sandwiched between dielectric layers of reactively sputtered tin oxide of an approximate thickness of 22 nm with a refractive index of 1.92 at 633 nm. The low temperature heat emissivity of A is 0.19, of B 0.08 and of C 0.05. The films were deposited on plastic sheet.

The performance of this system is given in Figure 9 which matches theoretical predictions.

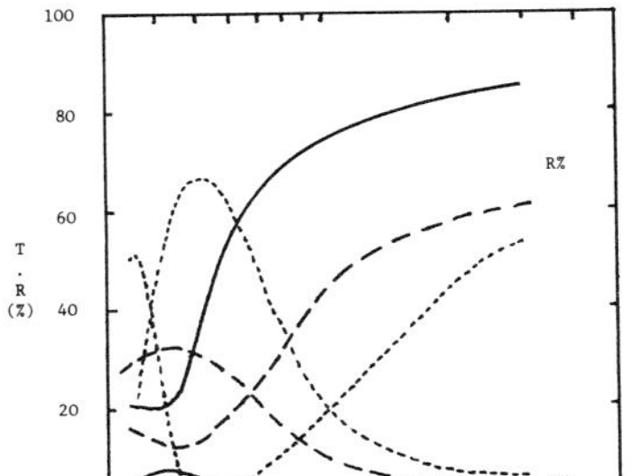

FIGURE 9

FIGURE 9

Transmission and reflection spectra of TiN films with two different thicknesses and TiO_2/TiN/TiO_2 heat reflecting mirror on glass substrate produced by r.f. enhanced d.c. magnetron sputtering. The pressure was 5.4 x 10^{-2} Pa for single TiN film and 3.5 x 10^{-2} Pa for the three layer.

The ability to be able to achieve good TiN onto room temperature substrates lies in operating the deposition system at a very low pressure. This could be achieved by using an additional radio frequency discharge coupled to the plasma through a coil.

The preparation of all oxide films was found to depend critically upon the partial pressure of oxygen combined with a pressure of argon which was used during the sputtering of the metal. This was particularly important in the preparation of conducting films. A typical characteristic is shown in Figure 10.

FIGURE 10

The sheet resistance and magnetron sputtering voltage shown as a function of the partial pressure of oxygen at a constant magnetron current of 3 amps during the preparation of indium oxide films. There was a partial pressure of argon of 3mTorr.

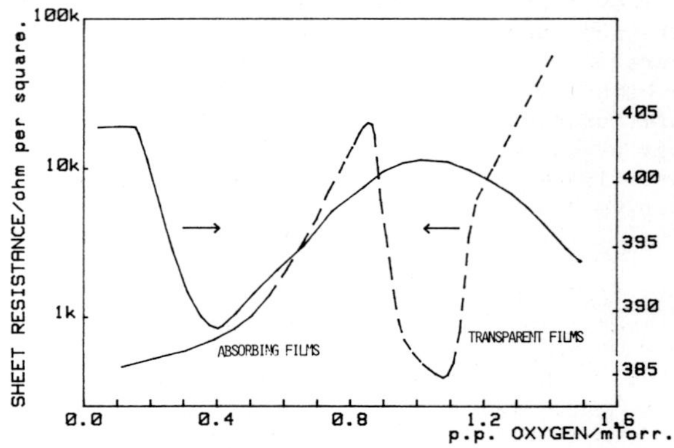

FIGURE 10

10. DISCUSSION

We are convinced that the system of window insulation that makes the most of available resources is that of a coated plastic sheet, stretch-mounted in a frame and attached to an existing window. It would seem at the moment that this film will have to be an oxide because of the durability requirements. Indium oxide has proved to be very attractive for this application; it has good properties which are shown in Figure 11 and can be sputtered very quickly and is not too expensive at its present price. However, if it were taken into large scale manufacture and proved as attractive a proposition to the buying public as its cost effectiveness deserves, then there would be a problem.

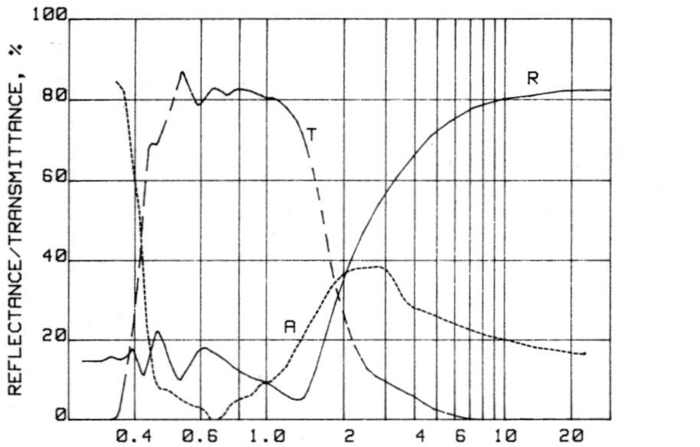

FIGURE 11

FIGURE 11

The transmission, reflection and absorption spectra of an indium oxide film on plastic.

After the installation of the third or fourth machine in Europe, the supply of the indium would be limited, and the price would be expected to rise rapidly. This project would have supplied a technique for providing 30,000 - 40,000 square metres a day at a cost effective price; above this there are requirements for a more easily available metal. For this reason we believe we should continue investigation of other oxide and metal systems.

Commercial systems of window insulation are appearing which use thin metal coatings on plastic other than those of the conventional aluminium. One system uses a thin silver coating and the plastic coating is stretch mounted within the glass panes of a conventional double structure. The film has to be mounted in this way within a sealed dry atmosphere to provide sufficient durability. It does give a very high performance window, but in replacement terms it is not cost effective.

It appears that thin metal coatings on glass have the same problem - they require to be sealed in a controlled atmosphere. What is becoming apparent is that it is being realised throughout the world that window thermal performance can be increased tremendously with the use of coatings, and techniques are becoming available to do this. We believe that the principal contributor to future improved window performance will come from coated plastic film. This is for two reasons - one, the optical base element is inexpensive and allows a cost effective solution, and two, the element is of lightweight and allows the use of secondary windows on existing structures, but perhaps more importantly, it allows the use of multiple coated elements in a window without an undue weight penalty - creating the means where windows can be made which will have a superior thermal performance to that of an insulated wall, but will still allow the light and energy from the sun to penetrate into the building.

It is our belief that the eventual supply of coatings for windows will be of two main types: the on-line coating of hot float glass with a CVD pyrolytic process and the vacuum coating of plastic sheet. Vacuum coating of glass sheet will be used to establish the market for heat reflecting glass and will continue to be used for films where some solar control is required.

The development of an on-line process will be very expensive, but must eventually be taken up by manufacturers. The supply of coated glass for windows will not preclude the use of coated plastic sheet which will allow multiple elements to be built up with little weight penalty as well as providing a low cost secondary glazing facility.

11. CONCLUSION

An approximate cost for an oxide or metal filter can be computed from the data we have acquired.

Calculation soon reveals that the cost of the coated plastic sheet will be determined by the capital cost of the coating equipment and hence, the rate of production. A machine installation could well amount to £1m which will cost about £1,000 per day in charges. With a multiple source machine a rate of 20 metres/min. of 2 metres wide material might be expected. This is ten times slower than the rate obtained with current conventional metallising machines using aluminium. With such a slow rate most of the time will be spent in sputtering so that run times of 10 hours per day should be obtained after development. Then 20,000 m^2 should be produced per day. This may well be halved with machine maintenance but a cost of only 10p per m^2 of coating is then predicted.

The cost of the plastic sheet substrate is about 25p per m^2 and material costs would amount to about 10p per m^2. It is expected that problems will arise in the distribution and use of the volume produced and also the availability of source materials such as indium. One machine will double the U.K. use of indium and the price may rise, but the principal problem will remain that of guaranteeing the supply of the material.

We have investigated how coated film could best be used. We believe that a plastic sheet, coated both sides, and stretched and mounted in an aluminium frame is the most promising system. We have bought a sophisticated frame which is available for approximately £8-00 to create a secondary window of one metre squared. Coated plastic sheet will be durable, resistant to scratching and prevent static build-up, which causes dust to stick to the window, and give the window a performance which will fully justify the use of a good quality frame. The addition of this secondary window will take a single pane window from above 6 to less than 2 $Wm^{-2}\ {}^{o}C^{-1}$. A saving of up to 5 $Wm^{-2}\ {}^{o}C^{-1}$.

For a moderate climate of 2,000 degree days this gives a saving of 250kW hours per square metre of window per year. This is approximately twice the energy content of the coated plastic sheet. With electrical heating at 4.7 pence per unit, a payback period of considerably less than one year will be obtained.

This crude estimate of cost and energy saving does indicate the considerable potential for such a system, without consideration of the other advantages of light-weight, enabling large glazed areas to be treated without any significant extra weight load on the structure. There is the potential of providing windows and lightweight wall structures having a thermal performance considerably in excess of anything that can be achieved today using multiple elements which still allow advantage to be taken of solar input. We believe that this application is a very productive use of our diminishing fossil resources.

ACKNOWLEDGEMENTS

The authors are pleased to acknowledge the other members of the Thin Film Group in the Department of Physics at Loughborough University, without whose work this report would have been impossible.

J.N. Avaritsiotis helped initiate this work, E.M. Stenlake did most of the construction and later measurements on films were made by R. Lewin.

REFERENCES

I have not referred directly to sources of information but have chosen to give an indication of the major sources which allow the subject to be pursued if desired.

1. Fan, J.C.C. and Bachner R.J.
Transparent heat mirrors for solar-energy applications
Applied Optics Vol. 15, pp 1012 - 1017 (1976).

2. Glazer, H.J.
High transparent low emissive coatings on the basis of ZnO/Ag/Zno, ZnO/Au/ZnO, Bi_2O_3/Au/Bi_2O_3, ZnO/Cu/ZnO and Bi_2O_3/Cu/Bi_2O_3
Proc. of the 8th International Vacuum Congress, Cannes 1980 (Suppl. "Le Vide les Couches Minces No. 201). Vol 1, pp 723 - 726.

3. Frank, G. Kaner E and Kostlin.
Transparent heat-reflecting coatings based on highly doped semi-conductors.
Thin Solid Films Vol 77, pp 107 - 117 (1981).

4. Howson R.P. and Ridge M.I.
Deposition of transparent heat-reflecting coatings of metal oxides using reactive planar magnetron sputtering of a metal and/or alloy.
Thin Solid Films Vol 77, pp 119 - 125 (1981).

5. Mahdjuri, B.
High insulated window glazing.
Energy Research Vol 1, pp 135 - 142 (1977).

6. Ridge, M.I., Howson R.P. and Bishop C.A.
New techniques for the roll coating of optical thin films.
Proc. of the SPIE Los Angeles Technical Symposium, 25th - 29th Jan., 1982, Los Angeles, California, U.S.A.

7. Schiller S, Heisig V, Steinfelder K, Strumpfel J.
Using the d.c. magnetron for the reactive deposition of oxides.
Proc. of the International Conference on Ion Plating and Allied Techniques, London, U.K., July, 1979, p. 211.

8. Howson, R.P. and Ridge, M.I.
Heat mirrors onplastic sheet using transparent oxide conducting coatings.
Proc. of the SPIE Los Angeles Technical Symposium, 25-29th January, 1982, Los Angeles, California, U.S.A.

9. Lampert, C.
Heat mirror coatings for energy conserving windows.
Lawrence Berkeley Laboratories, LBL Report No. 11446, August, 1981.

10. Kostlin H. and Frank G.
Optimization of transparent heat mirrors based on a thin silver film between antireflection films.
Thin Solid Films 89 (1982) 287 - 293.

11. Chopra, K.L., Major S. and Pandya D.K.
Transparent conductors - A status review.
Thin Solid Films 102 (1983) 1 - 46.

12. Valkonen E, Karlsson T, Karlsson B and Johannsson B.O.
Selective transmission of thin TiN films.
Proc. SPIE Vol. 401 (Thin Film Technologies" (1983).

13. Howson, R.P., Ridge M.I. and Bishop C.A.
Transparent and heat reflecting coatings
Proc. SPIE Vol. 401-40 "Thin Film Technologies" (1983).

14. Howson, R.P., Ridge, M.I. and Suzuki, K.
Optimized transparent and heat reflecting oxide and nitride films.
To be published - SPIE Vol. 428-32. San Diego (1983).

PYROSOL PROCESS USE FOR DEPOSITING SELECTIVE COATINGS FOR ENERGY CONSERVING GLAZINGS

Gérard BLANDENET, Michel COURT, Yves LAGARDE
Commissariat à l'Energie Atomique
Centre d'Etude Nucléaire de Grenoble
Département de Métallurgie
85 X - F-38040 GRENOBLE

SUMMARY

A thin film deposition technique (PYROSOL Process) designed for the production of selective metallic oxide films is described here. These films give rise to numerous applications, in particular in the field of energy savings, due to their specific optical properties. An assessment of the resulting energy gain illustrates the economic interest derived from the use of these films. The technico-economical aspect of the process will be discussed from the industrialization point of view, and solutions will be put forward.

1. INTRODUCTION

The major energy saving in buildings consists in preventing heat losses. Hence the interest of improving, so as to minimize the coefficient G. However, the thermal resistances of the different parts of a building being in parallel, the coefficient G may rapidly tend towards the one that is imposed by the glazings, especially if the latter represents an important part of the building total surface area. It is therefore necessary to increase the thermal resistance of the glazings. Considerable progress has already been achieved since multiple glazings containing a dry gas between two consecutive sheets were first used. But the presentenergy conjuncture requirements have dictated the further reduction of these heat losses.

The thin metallic oxide films such as indium oxide and tin oxide deposited on a glass surface enable the emissivity of the latter to be decreased considerably, thus reducing the importance of the radiation occurring during the exchange. For example, the heat transmission coefficient K for a double glazing can be reduced from 3 to 1.8 $W.m^{-2}.°K^{-1}$ approximately if one of the inner faces is coated with a .4 μm film of one of these oxides.

In order to deposit these thin films, we propose an original technology PYROSOL process {1} that is now well known by the European Community Commission in the energy field, and that has been the subject of different development contracts† by this body. This process, based on the thermal decomposition of an aerosol obtained by ultrasonic spraying, leads to films having suitable optical and electrical characteristics, partly because of the quality of the aerosol that is obtained (small droplet diameter, homogeneous distribution of the diameters, etc).

† Contracts : 201-77 EEF, 334-78-1-EEF, EEA-2-002F

One of the major difficulties the user of these semi-conducting oxides is confronted with is the increase in the cost of the glazing they lead to; unfortunately, the PYROSOL process does not avoid this dilemma. The problems are of two sorts : on the one hand, the contribution of the basic chemical products to the cost of the glazing, and on the other hand the efficiency of the process compatible with an industrial process ("float glass").

In this paper, we will insist particularly on the properties of the films obtained via this process, their application to energy savings by an assessment of the gain they afford and finally some suggestions concerning the industrialization of the PYROSOL process in this field.

2. SPECTRAL SELECTIVITY THEORY

A thin stoichiometry film material cannot be both transparent in the visible radiation range ($.4\ \mu m < \lambda < .8\ \mu m$) and an electricity conductor.

The only way of obtaining this double property is to create an electronic degeneresoence within an oxide having a wide forbidden band (> 3eV)

We are then confronted with two possibilities for increasing the semi-conducting character of these compounds :

1 - To modify the stoichiometry by increasing the oxygen vacancies.

2 - To introduce in the basic crystal lattice an element having a valence that is greater than that of the atom it replaces.

In both cases, an increase in the number of available free electrons in the conduction band of the material is obtained. Some of the optical properties can be explained by the presence of the latter.

The optical properties of a semi-conductor may be characterized by its refractive index n and its extinction coefficient k, that are related to one another in the form of a complex refractive coefficient : $n^+ = n - ik$.

According to the theory of DRÜDE {2}, the dielectrical constant of a material can be expressed by the relation

$$\varepsilon_r = (n^+)^2 = (n-ik)^2 = n^2 - k^2 - 2ink = f(\lambda)$$

in which :

$$n^2 - k^2 = \varepsilon$$

and $2nk = \frac{\gamma}{\omega} \cdot {}^1 \frac{\omega^2 p}{\omega^2 + \gamma^2}$ — in which ε_1 represents the dielectrical constant of the medium in the absence of free carriers and ε_o the vacuum constant; the frequency ω_N and the relaxation constant γ are correlated to the density N and the mobility μ of the free carriers through two relations :

$\omega_p^2 = \frac{N.e^2}{\varepsilon_o m^+}$ and $\gamma = \frac{e}{m^+\mu}$, in which e and m^+ are the characteristics of a free electron. The plasma wavelength (λp) is defined as the value of the wavelength for which the optical transmission is equal to the reflection. In other terms, when $\lambda = \lambda_p$ the material is no longer dielectric and it becomes a conductor. Mathematically, this means that :

$$n^2 - k^2 = 0$$

that is $\varepsilon_1 - \frac{\omega p^2}{\omega^2+\gamma^2} = 0$

or : $\lambda p = \left[\frac{2\pi.c.(Ne^2)}{\varepsilon_0\varepsilon_1\ m\ +} - \gamma^2\right]^{-1/2}$ in which $\lambda_p = \frac{2\pi c}{\omega_p}$

The position of λ_p that determines the range within which the thin film material considered is transparent to the near infra-red radiation therefore depends on the number of free electrons N.

Besides, if N is fixed, the infra-red reflection level will be greater if the electron mobility is greater. In practice, to obtain a plasma wavelength in the near infra-red range, electron concentrations of between 10^{20} and 10^{21} per cm^3 are required whereas mobilities $\leqslant 30\ cm^2V^{-1}s^{-1}$ are required in order to obtain at least 70 % reflection in the far infra-red ($\lambda \geqslant 2.5\ \mu m$).

3. FILM PREPARATION METHOD

The PYROSOL process requires :

1. A piezo-electrical aerosol generator.
2. A heat treatment furnace to produce the deposition reaction, in general by oxidation.

Figure 1 represents a conveyor-type in-line-machine equipment provided with two consecutive deposition areas, that can be used either separately or simultaneously, for instance for the deposition of composite films.

The solution that is used, of organometallic or mineral origin, is considered in the form of small droplets by ultrasonic spraying, then conveyed by a gas into the deposition zone.

The reaction takes place in two steps :

1. Evaporation of the source compound
2. Reaction of the vapour that has thus been produced with the heated substrate (CVD process).

The optimization of the properties of the films is an operation that is difficult to control because of the existence of numerous experimental parameters, some of which are interdependent, in particular :

- gas phase and substrate temperatures
- dopant nature and concentration
- substrate nature and state of surface
- thermal annealing nature

In spite of the complexity of the operation due to the difficulties that have just been mentioned, it has however been possible to optimize the deposition conditions for these films. The source compounds actually used for depositing respectively indium oxide and tin oxide are indium acetylacetonate dissolved in acetylacetone and tin tetrachloride in solution in methanol. The usual dopants are tin (dibutyl-tin diacetate) and fluorine (ammonium fluoride) or antimony (antimony pentachloride). In both cases, the gas carrier is air. The deposition reaction takes place towards 460° C for tin oxide and towards 480° C for indium oxide.

4. PROPERTIES OF THE SnO_2 and In_2O_3 FILMS

Table 1 summarizes the major electrical properties of different films obtained by the PYROSOL process. We can see on one hand the influence of thermal annealing on the number of free carriers and therefore on the reflecting capacity of the indium oxide films (Figure 2); on the

Film nature	$R_\square$ (Ω)	e (μm)	ρ (Ω.cm)	Hall Effect	
				n(cm^{-3})	μ($cm^2 V^{-1} s^{-1}$)
In_2O_3/Sn	5	0.5	2.5 10^{-4}	9.10^{20}	40
In_2O_3/Sn annealed under vacuum	3	0.5	1.5 10^{-4}	1.1 10^{21}	43
SnO_2/F	12	0.5	6 10^{-4}		
SnO_2/F on SiO_2	10	0.5	5 10^{-4}	4 10^{20}	35
SnO_2/Sb	20	0.5	1 10^{-3}	1.2 10^{20}	35
1/2 In_2O_3/Sn + 1/2 SnO_2/F	6	0.5	3 10^{-4}	8.5 10^{20}	36

TABLE N° 1

other hand the increase in the resistivity of the indium oxide when the latter is coated, immediately after the deposition, with a thin tin oxide film. We must finally mention the beneficial role of the SiO_2 underlayer for the resistivity of the fluorine-doped SnO_2 films. As far as the latter are concerned, the deposition on a soda lime glass raises some difficulties that are inherent to the interaction of the chlorinated products with the glass alcalines during the deposition of this film. These problems can be solved by using protective SiO_2 underlayers, or indium oxide that has the additional advantage of increasing the free carrier density. But these additional operations may reduce or even suppress the economic advantage associated with the use of tin oxide instead of indium oxide.

5. APPLICATIONS

The spectral selectivity of these transparent conductive oxides is very interesting for their use as infra-red radiation reflecting coatings. They are therefore very important from the standpoint of energy savings in a volume that is closed by glazings coated with this type of selective films.

5.1. THEORETICAL ESTIMATION OF COEFFICIENT K

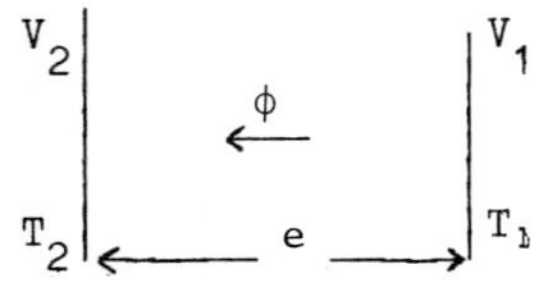

The heat exchanges (ϕ) between both plans V_1 and V_2 are of three types :

- radiation
- conduction within the gas contained between V_1 and V_2
- natural gas convection

If the thickness e is small, the natural convection can be neglected and we will only consider the two kinds of exchange. The conduction losses (k_c) can be expressed in terms of the thermal conductibility of the gas (λ) in the following way :

$$k_c = \frac{\lambda}{e}$$

As far as the radiation is concerned, a simple calculation makes it possible to establish that the heat transmission coefficient due to radiation (k_r) can be expressed as :

$$k_r = 4\sigma\,\varepsilon^2 \cdot \overline{T}^3 \quad \text{if } \varepsilon \text{ is big}$$

and $$k_r = 2\sigma\,\varepsilon \cdot \overline{T}^3 \quad \text{if } \varepsilon \text{ is small}$$

in which σ = constant of Stefan

ε = overall emissivity of $(V_1, V_2) = \dfrac{\varepsilon_1\varepsilon_2}{1-(1-\varepsilon_1)(1-\varepsilon_2)}$

$\overline{T}$ = mean temperature of V_1 and V_2

The total losses can therefore be expressed in both cases mentioned above as follows :

$$k = \frac{\lambda}{e} + 4\,\sigma\,\varepsilon^2\,\overline{T}^3$$

and $$k = \frac{\lambda}{e} + 2\,\sigma\,\varepsilon\,\overline{T}^3$$

The thermal resistance r is $r = \frac{1}{k}$; therefore, if we multiply the number (n) of air coatings, the total resistance will be :

$$R = \frac{1}{K} = \frac{1}{h_i} + \frac{1}{h_e} + nr$$

if n = 0 (single glazing) : $R = \dfrac{1}{h_i} + \dfrac{1}{h_e}$ (h_i and h_e = inner and outer exchange coefficients)

The thermal resistance of a glazing in terms of the number of gas films can thus be calculated. These results are illustrated in figure 4, in the case of an air film.

As soon as a selective film is introduced into one of the faces of a glazing, the principle of the calculation remains the same; the term that is affected is emissivity, the variation of which, in terms of the deposited film thickness, is represented in Figure 5.

Thus, .1 μm of indium oxide deposited on both inner faces of a double glazing result in an emissivity ε = .17, a value that can only be obtained with 0.3 μm approximately deposited on a single face.

5.2. EXPERIMENTAL DETERMINATION OF K (DOUBLE GLAZINGS)

The device that was used for this experiment is represented in Figure 6.

The principle of the measurement is based on the evaluation of the heat flux passing through a sample while maintaining a constant temperature difference T on both sides of the glazing

$$= \frac{A\,\Delta T}{R}$$

$\Delta T = T_1 - T_2 = 20°$ C

A = surface area of the sample = 25 x 25 cm2

R = Thermal resistance

from which $K = \frac{1}{R}$ can be computed

The measurements were carried out with two samples of indium oxide and tin oxide, 0.4 μm thick, having an emissivity of .15 and .2, respectively. The values of K obtained were :

$K_{In_2O_3}$ = 1.8 to 1.9 W m^{-2} °C^{-1}

K_{SnO_2} = 1.9 to 2 W m^{-2} °C^{-1}

5.3. ENERGY GAIN ASSESSMENT

To carry out this assessment, we will take into account the decrease of factor K from 3 to 2Wm^{-2} °C^{-1}

$|\Delta K|$ = 1.W.m^{-2} °C^{-1}

For a building having 30 m^2 of glazings with an inner temperature of 18°C, the heat gain will be :

$$\Delta E = 540 \text{ W}$$

that is ΔE_t = 5400 MW$_{th}$ for 10.10^6 buildings

or ΔE_t = 1800 MW$_e$ if the efficiency is assumed to be 30 %

This represents globally the supply of two 900 MW$_e$ power plants.

From the energy equivalence standpoint

$$\Delta E_t = 5400 \times 2400$$

$$= 13.10^6 \text{ MWh}_{th}$$

if we assume the yearly heating period to be 100-day long, then :

$$\Delta E_t = 13.10^6 \times 0.1$$

$$\Delta E_t = 1.3.10^6 \text{ Tep}$$

or

ΔE_t = 9.5.10^6 barrels

6. ECONOMIC ASPECT

The industrialization of the process is faced with two major problems at present. The technological difficulties mainly concern the obtention of higher deposition rates, in order to make the PYROSOL process compatible with the "float glass" conveying speed (several meters per minute).In the case of indium oxide, there is an additional economic handicap related to the cost price of the chemical products used for its deposition (indium acetylacetonate, solvent).

6.1. ECONOMIC DATA

The reduction in the selective glazing costs studied in detail in the context of a ECC contract { 3 } mainly depends on three factors :
- the price of metal indium : in figure 7 we can see in particular that the latter has come back in 1983 to the price it was in 1974, after the peaks it reached in 1980.
- the optimization of the manufacture of indium acetylacetonate : a judicious selection of indium-bearing compounds in the extraction line and of its minimum admissible purity, reduction of the quantity of acetylacetone.
- the recovery of indium-bearing products that have not reacted and that are eliminated with the gas effluents resulting from the deposition reaction.

The economic evaluation that follows is based on :
- the price of the present and future source compounds and solvents, taking into account the possible reduction of the costs following the optimization of their preparation.
- a coating thickness of .4 µm, corresponding to an approximate emissivity of .15.

The following table summarizes this estimate :

	In_2O_3		SnO_2	
	PRESENT PRICES	ESTIMATED PRICES	PRESENT PRICES	ESTIMATED PRICES
SOURCE COMPOUNDS	2000F/kg	440F/kg**	690F/l	70F/l**
SOLVENT	16F/l	2F/l *	20F/l	2F/l
.4µm COATING :				
SOURCE COMPOUNDS	33 F/m^2	17.6 F/m^2	32 F/m^2	3.2 F/m^2
SOLVENT	25 F/m^2	3.2 F/m^2	6.4 F/m^2	0.6 F/m^2
GLOBAL	58 F/m^2	20.8 F/m^2	38.4 F/m^2	3.8 F/m^2

* This price corresponds to the use of new solvents currently being tested

** The efficiency is assumed to be 20 % for SnO_2 and 60 % for In_2O_3 (because of the recovery of the indium-bearing products).

6.2. TECHNICAL CONSIDERATIONS

The process industrialization requires, as we have seen, an increase in the conveyor speed, so that the selective coating can be deposited at the end of the manufacturing line of the "float glass". This can only be achieved if the film growth rate can be modified either by increasing the kinetics of the reaction or by multiplying the deposition areas.
The increase of the concentrations by increasing the quantity of reacting product supplied does not allow a significant increase of the reaction kinetics. In effect, the latter is controlled by the diffusion of the active species in the boundary gas layer in contact with the substrate, and is therefore very little affected beyond the concentration threshold normally used (.1 mole per liter). On the other hand, an increase in material transportation can be judiciously used to multiply the deposition areas of the installation. We can now deposit .1 μ m of In_2O_3 (R ~ 20 $\Omega_\square$) at the rate of 5 $cm.min^{-1}$ with a single deposition area. By multiplying by 10 the number of nozzles, a figure that is difficult to exceed for technological reasons, it is possible to reach 0.5 $m.min^{-1}$.

The values indicated above (20 $\Omega/\square$ for 0.1 μ m) lead to resistivities of approximately 2.10^{-4} Ω cm, values that can only be obtained by the conventional deposition methods (CVD, pneumatic spraying, deposition under vacuum) by a reducing treatment under vacuum or in an inert atmosphere. The PYROSOL process, however, enables resistivities that are as low to be obtained, through the so-called bi-layer technique {4}. The In_2O_3 coating normally deposited is coated with an SnO film immediately afterwards, to protect the indium oxide coating from oxidation during cooling, thus maintaining the difference in stoichiometry due to the deposition method which is responsible for the high electrical conductivity. This solution, in which the respective thicknesses of In_2O_3 and SnO_2 can be adjusted, has the additional advantage of protecting the glass during the tin oxide deposition, while improving the carrier density of the latter.

As far as glass protection is concerned, another possibility consists of depositing a underlayer of non-conductive SnO_2, approximately .1 μm thick, produced by an organometallic compound of tin, that acts as a diffusion barrier during the fluorine-doped SnO_2 film deposition. This solution, that has been tested at laboratory level, will be experimented in the near future on prototype equipment with two deposition zones.

Finally, the deposition kinetics being more advantageous for tin oxide - 4 fold in relation to indium oxide -, we must also consider this criterion when choosing the final solution that will allow the maximum travel rate to be obtained.

7. CONCLUSION

Taking into account the technical problems : deposition kinetics, glass protection, carrier density, and the economic aspects related to the price of the chemical products - source products, solvents - stated previously, the present optimum solution involves a bi-layer system (~ 0.05 μ m In_2O_3 + ~ 0.25 μm SnO_2) that can be deposited with a travel rate of .1 $m.min^{-1}$ with a single nozzle, yielding an emissivity of .2. The hypothetical increase of the deposition kinetics depends on the use of solvents compatible with a higher deposition temperature.

From the purely economical point of view, the possibility of using a 99.5 % purity indium without affecting the quality of the films obtained and the recovery of the indium-bearing materials, together with a drop in the metal prices, enhances the interest of indium oxide, an interest that the glass manufacturers had laid aside on purpose during these last years, in favour of tin oxide.

BIBLIOGRAPHY

1 - J. SPITZ - J.C. VIGUIE BREVET français n° 7038371 du 23/10/70.

2 - P. DRUDE Z. Phys. 1,161, (1900).

3 - G. BLANDENET, Y. LAGARDE - Compte rendu final, contrat n° EEA-2-OOF Mai 1983.

4 - G. BLANDENET, J.J. CHAZEE, M. COURT - Brevet français n° 83 04565 du 21/3/83.

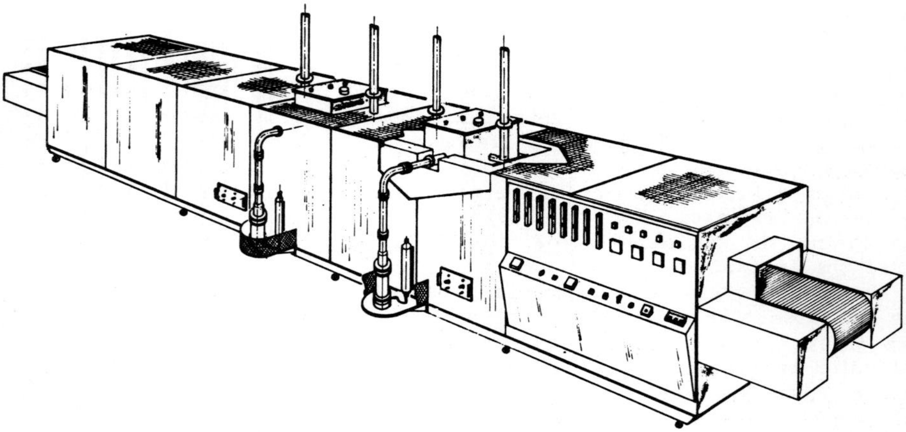

Fig. 1 PYROSOL furnace with two deposition areas

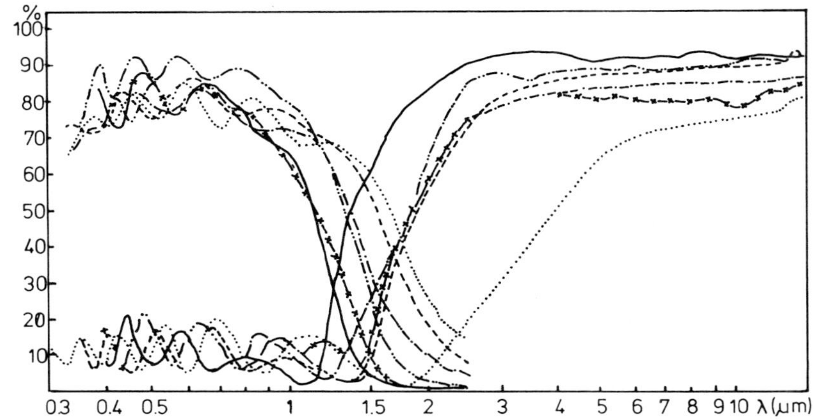

Fig. 2 Transmission and reflection spectra for different films having an approximate thickness of .5 m

—— In_2O_3/Sn annealed, —..— In_2O_3/Sn, ---- SnO_2/F on SiO_2
-.-. SnO_2/F on glass, ... SnO_2/Sb, -x-x 1/2 In_2O_3/Sn + 1/2 SnO_2/F

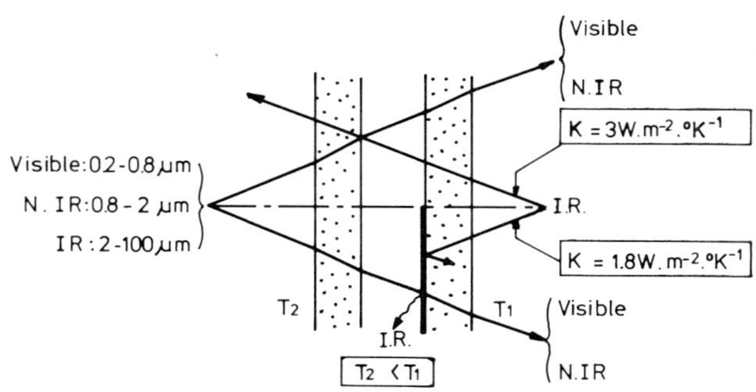

Fig. 3 Principle of a double glazing with and without a selective film

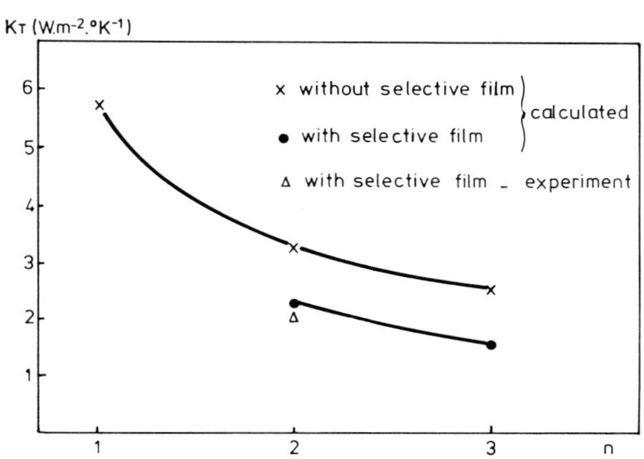

Fig. 4 Variation of coefficient K_T in terms of the number of gas films

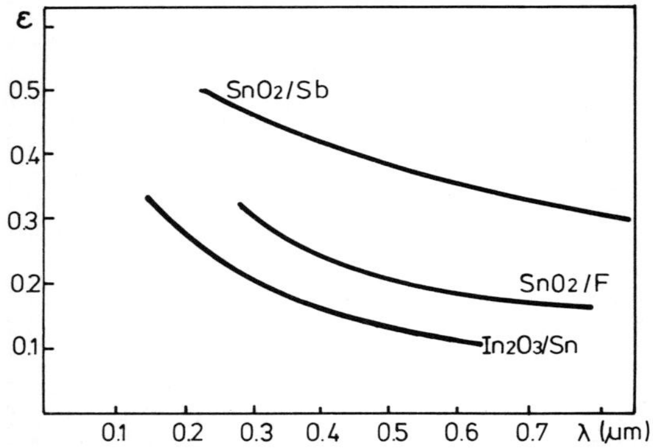

Fig. 5 Variation of emissivity in terms of film thickness

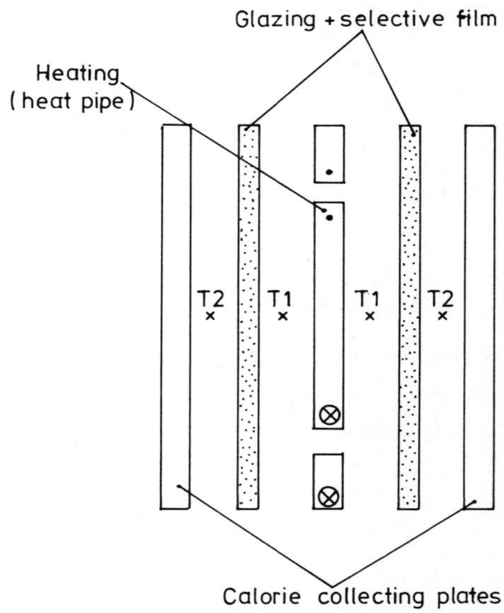

Fig. 6 K measurement principle

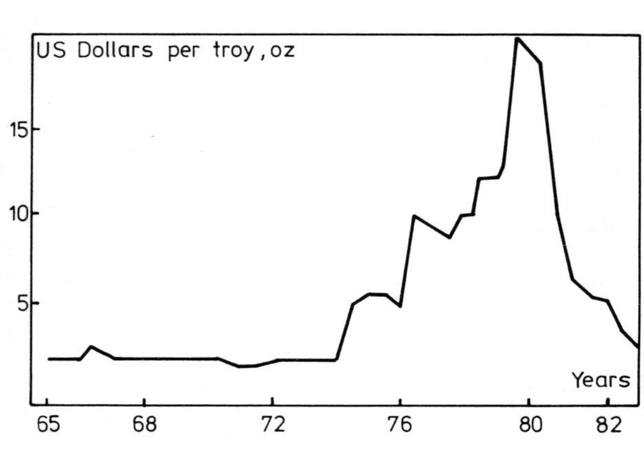

Fig. 7 Evolution of the price of indium

INSULATED SHUTTERS

M.R. Byberg
Thermal Insulation Laboratory
Technical University of Denmark
2800 Lyngby, Denmark

Summary

One of the possibilities for reduction of the thermal transmission coefficient for windows is to use insulated shutters during the night. This paper describes two types of shutters: Internally positioned shutters for the existing building and shutters between two layers of glass for new building, the latter demanding a wall thickness of at least 350 mm. The shutters have been developed and tested in a programme supported by the EEC.
The shutters are extremely lightweighted mainly consisting of PIR-foam. Only 30 mm of PIR has reduced the U-value to less than 40% of the value for a traditional double pane. The design and the jointing material secure that no condensation will occur on the pane when the shutters are used, but when they are removed a steaming up of the cold pane surface is inevitable. Under normal winter conditions the steam will vanish in less than one hour.
Even if the shutters will have a conservation effect of 900 to 1500 kWh/year for a normal Danish single family house with a window area of $20m^2$ the economic analysis indicates that the internal shutter will have to have a lifetime of 10 years to make the investment cost-effective. For the shutters between the panes the lifetime should be 20 years. One of the reasons for this is that it cannot be expected that the shutters are active more than 50% of the hours in the heating season.

1. INTRODUCTION

It is generally accepted that windows thermally spoken is the weekest part of the building envelope. The thermal transmission coefficient for all other non-transparent parts of the envelope has gradually been decreased, mostly as a result of the increasing energy prices but also due to increasing demands for the thermal comfort and the hygiene. The use of insulating materials in heavy thickness is widespread in the Scandinavian countries. The Danish building code thus claims a transmission coefficient of $.2W/m^2K$ or less for roof constructions, corresponding to an insulating thickness of 300 mm of mineral wool.

Contrary to this, the transparent part of the building envelope, the windows, has thermal coefficients more than ten times those of the rest of the building. The use of double panes is claimed in the Danish building code, the use of multy-layer panes or specially treated double panes are not unusual in Denmark and still up to 50% of the energy losses from a single family house will be transmission losses through the windows.What can be done? - In the night time we normally don't use the transparency of the windows, therefore, the windows might be applied with insulating material, and the idea of movable insulation is born.

The project: Mobile Insulation of Windows - Shutters supported by the EEC deals with internal insulating shutters for existing building and with shutters between two layers of glass for new building.

External insulating shutters have earlier been investigated by our laboratory in the Low-Energy House Project where six different low-energy houses were built in 1978. Three of them have insulating shutters resulting in conservations from 800 to 2000 kWh/year. The results have been reported by B. Saxhof: External Insulating Shutters in Conservation Houses [1].

This paper deals with the results of the EEC supported project reported in the Final Report [2].

2. WORK PROGRAMME

As insulation materials are well known and the insulation effect can be calculated with sufficient accuracy for a specific piece of material, if it is installed in the proper way, the investigations have been concentrated on the shutter construction, on the application of the shutter in the window and on the moisture condensation problems in connection with the use of the shutters.

The two types of insulated shutters are both protected against the outdoor climate by at least one layer of glass. Thus the construction can be extremely lightweighted. PIR-foam (polyisocyanurat) has been chosen as this material is manufactured in large pieces with suitable thickness. The weight of a shutter with the dimensions 600x1200x30 mm^3 is about 1 kg including edging. Besides this the PIR-foam has low moisture permeability, sufficient stiffness, good fire resistance and a very low thermal conductivity. With λ_{10}=.022 W/mK it is only 2/3 of the conductivity for mineral wool.

3. INTERNAL INSULATING SHUTTERS FOR EXISTING BUILDING

The internal shutter consists of a 30 mm thick plate PIR-foam with special fibre sheets on the surfaces and an edging with polyether foam protected by nylon cloth.

Much attention has been paid to the jointing material as it shall serve as both airtightening and fastening of the shutter by friction against the window casing. Several traditional weather strips have been tested and have been rejected either due to bad airtightening in the corners or to high friction. Figure 1 shows how two or more shutters can be located beside eachother. At the inside two handles have been countersunk into the surface of the shutter.

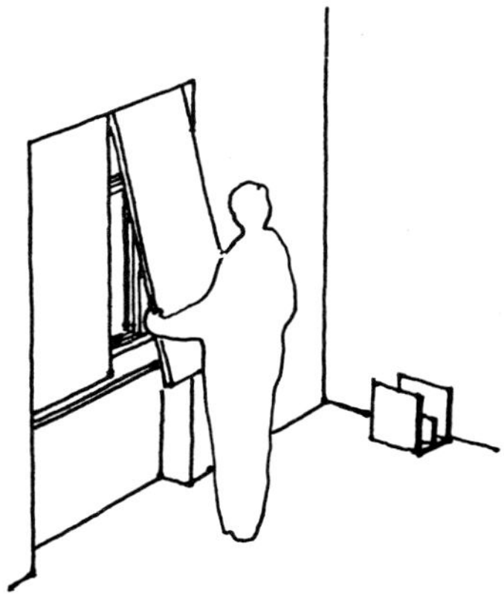

Figure 1 : Shutters for the existing building. Location of the shutters takes place manually - each shutter unit weighs about 2 kg. The shutters are placed in a shutter case during the day. Independence from curtains and the mutual distance between the windows will be achieved by this shutter solution.

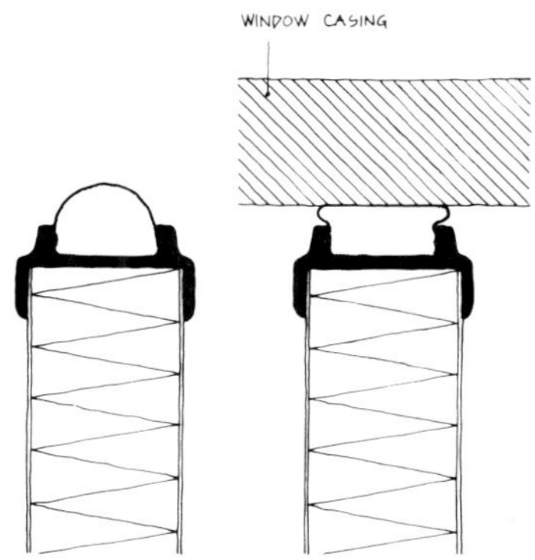

Figure 2 : The edge-profile for the internal shutter.

Used in front of a window with hermetic double panes the 30 mm shutter will reduce the transmission coefficient from 2.5 to 0.67 W/m^2C (measured values).

In the window/shutter construction the shutter will give the main proportion of the thermal resistance of the system. This introduces low surface temperatures on the window pane. If moist air from the room can be transferred to the space between shutter and pane the result is condensation.

Nevertheless, climate chamber tests made at the laboratory have shown that no condensation will occur during the "night time" at any practical inside rel. humidity. On the other hand it is inevitable that the window under normal winter conditions will steam up when the shutter is removed. It lasts from half-an hour to one and a half hour for the steam to evaporate depending on the internal rel. humidity.

4. SHUTTERS BETWEEN THE PANES, FOR NEW BUILDING

For new building the shutter construction is intended to combine the simple and extremely light construction for the internal shutter with the position free of condensation for the external shutter. For this purpose a completely new window has been constructed. It has a distance of 350 mm between the external and the internal pane, thus allowing the shutters to be "parked" perpendicularly to the glasses in the daytime. The construction implies that the outer wall has a thickness not less than 350 mm which will be normal in heavy insulated buildings in the future.

Again the shutter is made of 30 mm PIR-foam with an edging of wood or plastic. As the airtightness of the internal layer of glass with weather strips against the frame prevents moist air to be transferred to the space between the glasses, the airtightness of the shutter is of minor importance. Thus normal brush type weather strips can be used between shutter and window casing.

Due to the unusual high distance btween the panes the U-values of the window without shutters is 2.1 W/m^{2}C (2.5 for traditional double panes). The shutters reduce the U-value to 0.91 W/m$^{2\,0}$C (measured values).

Tests in the climate chamber have shown that there will be no condensation on the panes when the shutters are used at "night". To secure this the lower weather strip of the outside pane has been omitted. In this way the window functions according to the two-stage principle.

However, as the shutters are operated manually in the prototype it has been necessary to open the inside pane during the operation. This causes an immediate steaming up on the outside pane lasting less than one hour. A mechanical device for operating the shutters without opening the window has been sketched out but not yet tested. When using this there will be no steaming up.

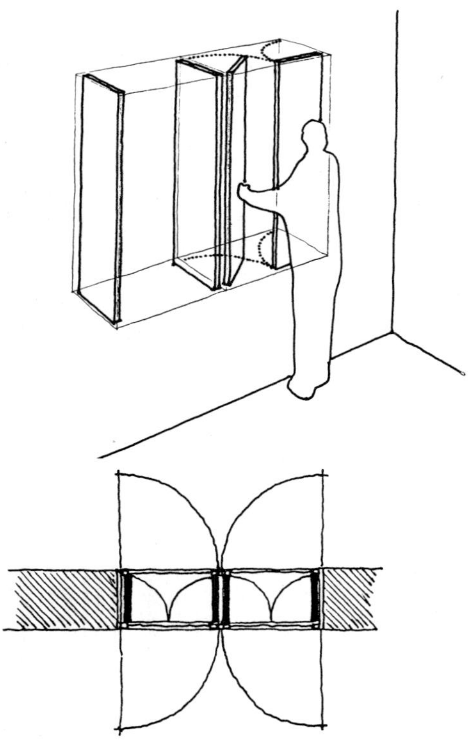

Figure 3: Shutters for new building. The insulating shutters are part of the window construction and are placed between the two layers of glass. In the passive position the shutters will be parked along the casing. The interior windows are opened and the shutters operated manually. The window elements can be located side by side.

5. LABORATORY INVESTIGATIONS

As mentioned prototypes of the two shutter constructions have been tested in the climate chamber facility. The prototypes have the dimensions height x width = 1.6 x 1.2 m allowing for direct correlation to other conservation projects supported by the EEC concerning windows.

Measurements of airtightness of the constructions have shown that the new window can be considerably improved by using an extra hinge and an extra hasp. Even without the improvement the leakage is acceptable, 0.5 m^3/h per m weather strip at 50 Pa pressure difference.

6. ENERGY CONSERVATION

The obtainable energy conservation for a typical Danish single family house with a window area of 20 m^2 is calculated based on the measured U-value reductions. For this purpose it is assumed that the shutters have been used daily from Sep. 1, to May 1, with 14 hours/day in the middle of the winter reduced to 9 hours/day at the beginning and the end of the heating season. The result of this is that the shutters will have to be closed for nearly 52% of the heating season.

The calculations show an annual energy saving of 1400 kWh for the internal shutter and 900 kWh for the shutters between the panes. Compared to a normal double glazing the figure is 1500 kWh.

7. INVESTMENT AND LIFETIME

To calculate the maximum cost-effective investment the present-day value method has been used. In this case no relevant investment information is available as the shutters were handmade prototypes. The calculations are based on traditional assumptions for lifetime, annual rise in energy prices, annual maintenance, annual rise in maintenance and calculation interest after tax. The annual saving of heating expences is calculated by means of the heating degree hours and the measured U-values.

The calculations are further based on the present-day energy prices for oil, 0.45 Dkr/kWh. Based on electricity prices (0.79 Dkr/kWh) the cost-effective investment would be 75% higher.

From a conservative guess on the investments for normally manufactured shutters the necessary lifetime can be calculated.

basis	internal shutters	shutters between panes
oil	8 - 10 years	18 - 20 years
electricity	5 - 6 years	10 - 12 years

Calculated minimum lifetimes for cost-effective investments.

8. CONCLUSION

The project has shown that it is possible to construct leightweighted and well functioning insulated shutters internally positioned for existing building and between two layers of glass for new building. In both cases the shutters are made of polyisocyanurate foam (PIR) with a thickness of 30 mm and with very low thermal conductivity.

Condensation problems are avoided in both cases though the internal shutters demand special edgings to obtain sufficient air (and moisture) tightness. When the shutters are removed temporary condensation is inevitable. U-values for windows with and without insulated shutters have been measured in the climate chamber facilities.

Although the shutter constructions are very simple and the material itself is relatively cheap the economic calculations indicate minimum lifetimes of 10 and 20 years for shutters positioned internally/between the panes, respectively to make the investments cost-effective.

So far only prototypes have been made and tested. Only the future will show whether there is a market for insulating shutters or not, but it is still one of the most effective measures to reduce the transmission coefficient of a window drastically during the night.

9. REFERENCES

1. Bjarne Saxhof: External Insulating Shutters in Energy Conservation Houses. Thermal Insulation Laboratory. Report No. 129, November 82.

2. Insulated Shutters. Final Report of the project: Mobile Insulation of Windows - Shutters. Thermal Insulation Laboratory, October 1983.

ENERGY SAVING BY USING ROLLING SHUTTERS, WITH REFLECTIVE LAYER AND EDGE SEALS

M. Dubbeld

TNO Research Institute for Environmental Hygiene

P.O. Box 214, NL-2600 AE Delft, the Netherlands

Summary

In this research the possible energy saving with rolling shutters, disturbing convective layers and reflective layers on the parapet and back side of radiator is investigated. The acceptance and attitude for the use of rolling shutters by inhabitants is reported. The U-value measurements show a possible reduction of the heat losses till 50 % with two layer shutters. For a boiler efficiency of 70 % the energy consumption is calculated. In the cost benefit calculations is looked to the pay back period and the present worth value of the different situations.

1. Objective and description of work

This investigation is a part of the National Programme which is co-ordinated by the Steering Group for Energy Saving in Buildings and partly founded by the Ministry of Housing and Physical Planning and the Commission of the European Communities. The parts of the investigation are:

1) Inhabitants' acceptation of rolling shutters carried out by Bouwcentrum;
2) Insulation and climate effect measurements, Bronswerk and TNO;
3) Energy saving calculations, TNO;
4) Cost-benefit calculations, Bouwcentrum;
5) Possible integration in the building construction, Bouwcentrum;
6) Summary report investigation E.C., TNO.

The first point is not carried out under contract of the European Commission. The research programme is a follow up of the work for energy saving with curtains and window sill. It can be divided into two parts:

1) rolling shutters carried out in the Bronswerk testroom,
2) window sill and parapet by IMG-TNO.

The aim of the rolling shutter programme is to make cost-benefit calculations of the supplied results of reflective layer, edge seals and second layer and combinations.

In the test room of the TNO Research Institute for Environmental Hygiene the following experiments will be carried out:

- effect of the disturbing of the cold draft along the window surface,
- catch of cold air from the window with the window sill to prevent draft,
- effect of reflective paint on the parapet and the radiator.

2. Methods of investigation

The measurements for the U value of the rolling shutters is carried out by Bronswerk B.V., Research Centre in Amersfoort. The principle of the measurements is the same as used by TNO for these and earlier investigations for the effect of curtains and window sill [1].

The heat loss through the façade is the summation of the heat loss through the parapet and through the window. Because the heat flow through the window cannot be measured directly by heat flow meters it must be calculated as the balancing item of a heat balance. Two heat balances may be set up for the test.

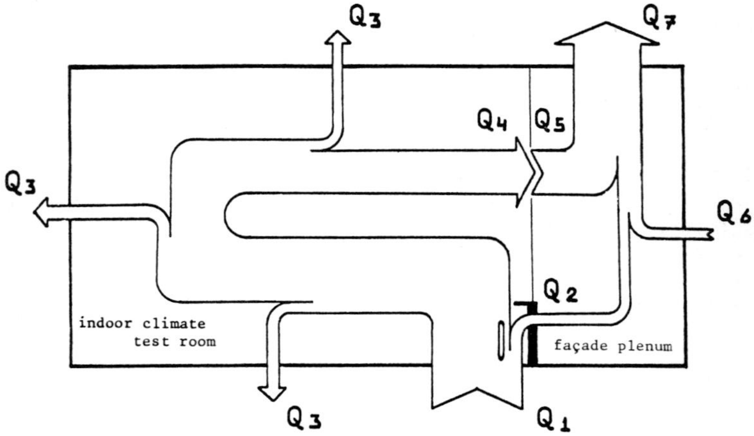

Figure 1 Heat flows.

2.1 Heat balance of the room

The heat flow through the window, Q4, equals the heat flow supplied by the radiator, Q1, diminished by the heat flow through the parapet, Q2, and that through the walls of the room, Q3.

2.2 Heat balance of the façade plenum

The heat flow through the window, Q5, equals the heat flow discharged through the façade plenum, Q7, diminished by the heat flow through the parapet, Q2, and the heat flow from the outside, Q6. The heat loss through the window can be determined by these two ways. The difference between Q4 and Q5 is a measure for the precision of the measurement. The limits are 4 % of Q4 or 25 W. The hall round the test facility can be controlled in temperature.

2.3 Calculation energy consumption and possible use

The energy consumption for the different window and roller shutter systems are calculated for the following conditions. The roller shutter will be closed from 18 till 8 hour. The indoor temperature is 23 °C from 8 till 23 hour and 18 °C in the night. The measured U values are corrected for normal wind velocities around buildings for a mean meteorological velocity of 5 m/s.

The yearly energy consumption is calculated at the mean seasonal outdoor temperature from 8-18, 18-23 and 23-8 hour. For the installation a boiler efficiency of 70 % is used.

In the cost-benefit calculations are given an answer on the economical combination of roller shutter and the possible prices for the used materials. At least the possible integration in existing buildings is evaluated taking in account the differences in the window measures.

3. Main results

The main results are given in table 1 till 3. For each type of shutter the supplied heat resistance and the energy saving is given.
For single glass only a normal shutter can be used. With better shutters it is impossible to avoid condensation. The effect of double glass is greater, 25 till 30 $m^3/m2$, than a roller shutter. For double glass the best results can be reached with double layers, for instance:
- roomside : curtain or normal roller shutter,
- windowside: double reflective layer.

Between double windows the effect of reflective layers is strongly reduced by the higher convective losses. These losses are higher by the stock working of the air on both sides of the shutter system.

When shutters are closed for 24 hours the energy saving is higher. In general, the inhabitants prefer to have the possibility to look outside. Therefore, a half transparant reflective layer can be used. The energy saving will be between the values given in table 3.

With a gas price of Dfl. 0.6 or Dfl. 1.2/m^3 gas and a pay-back time of 5 years the possible costsaving is:

		Dfl 0,6/m^3	Dfl 1,2/m^3 gas
Only used in the evening:			
Single glass	normal shutter, 362 MJ/m^2	Dfl 31,--	- Dfl 62,--/m^2
Double glass	curtain or normal roller shutter	Dfl 22,--	- Dfl 44,--/m^2
	double reflective layer, 250-267 MJ/m^3		
Closed whole day in the bedroom:			
Double glass	normal shutter, 83 MJ/m^2	Dfl 7,--	- Dfl 14,--/m^2
	one side reflective, 195 MJ/m^2	Dfi 16,--	- Dfl 32,--/m^2

The cost-benefit calculations are not available at this moment but the possible cost saving shows that the price of reflective layers must be very low when people will buy it only for energy saving.

For single glass it is measured whether the convective heat flux through the window can be reduced by obstruction. It appeared that no heat loss reduction occurred. Further it is noticed that without window sill and cathes the cold air between the radiator and the parapet no energy savings can be reached.

A reflective paint on the parapet can save energy. For an U-value of 1 W/m^2K and an absorbence of 0,35 the possible energy saving is for single glass 4,8 m^3 gas/m^2 through the parapet and 2,1 m^3 gas/m^2 for the window.

For double glass this is 3,1 and 0,4 m^3 gas/m^2. The energy saving and cost saving in 5 years is for a window : parapet surface proportion of 2:1

		Dfl 0,6/m^3	Dfl 1,2/m^3 gas
- single glass	9 m^3 gas/m^2 parapet, 315 MJ/m^2	Dfl 27,--	-Dfl 54,--/m^2
- double glass	3,9 m^3 gas/m^2 parapet, 136 MJ/m^2	Dfl 19,--	-Dfl 38,--/m^2

Table 1 Inside shutters and others.

Reference	Shutter type	ΔR m^2K/W	Energy saving per year m^3 gas/m^2	MJ/m^2
Single glass	normal shutter	0,085	10,3	362
Double glass	normal shutter	0,050	1,6	59
	with one side reflective	0,142	3,9	138
	.normal sealed edges	0,123	3,5	124
	.double layer reflective between layers	0,177	4,6	162
	.normal roomside double reflective windowside	0,354	7,1	250
	same with sealed edges	0,422	7,8	274
	curtain	0,085	2,6	93
	.curtain roomside double reflective windowside	0,400	7,6	267
Double windows	normal shutter	0,08	2,7	92
	with one side reflective	0,08	2,7	92
	with one side reflective roomside transparant windowside	0,133	4,0	138
	with one side reflective roomside double reflective window-side	0,190	5,1	178
	one side reflective Venetian blinds	0,115	3,5	123

Table 2 Outside shutters.

Reference	Shutter type	ΔR m^2K/W	Energy saving per year m^3 gas/m^2	MJ/m^2
Double glass	normal outside shutter	0,091	2,7	98
	reflective one side outside directed*	0,07	3,8	135

*night radiation is taken in account

Table 3 Inside shutter 24 hours closed.

Reference	Shutter type	Energy saving per year bedroom m^3 gas/m^2	bedroom MJ/m^2	livingroom m^3 gas/m^2	livingroom MJ/m^2
Double glass	normal shutter	2,3	83	2,9	103
	with one side reflective	5,5	195	6,9	242

4. Economic assessment

The cost-benefit analysis is carried out for a window of an average size of 1,62 x 1.20 m^2 (width x heigth).
The cheap operating mechanism model has been chosen.
Installation and maintenance by the occupants (no costs).
Economical/social life: 10 years and 8 % interest.

The system with a pay-back period, PBP, of 10 years and lower are mentioned in Table 4.

Table 4. Cost-benefit rolling shutters.

System	Investments ECU/m^2	(Dfl/m^2)	Saving/ann MJ/m^2	(m^3 gas/m^2)	Cost-benefit price of energy 6.65 ECU/GJ (Dfl 0,6/m^3) PBP	ROR	13.3 ECU GJ (Dfl 1.2/m^3) PBP	ROR
Single glazing (reference)								
Normal roller shutter	9.62	(25)	362	(10.3)	4	21	2	48
Double glazing (reference)								
normal roller shutter and double reflecting roller shutter	21.15	(55)	250	(7.1)	>10	-	6.5	8.9
Double glazing, with curtain (reference)								
Double reflecting foil	11.54	(30)	176	(5)	10	0	5	15

PBP = Pay Back Period
ROR = Rate of Return.

We have to take into account that people will buy roller shutters also for protection, decoration and sunshading.

If a normal PBP of 4 years is used, only a normal roller shutter in combination with single glass can be used economically.

For the different systems the possible investment for a PBP of 4 years is calculated in Table 5.

Table 5. Possible investment for a pay-back period of 4 years.

System	Investment for PBP of 4 years	
	Price of energy	
	6.65 Ecu/GJ (Dfl 0,60/m^3gas)	13.3 Ecu/GJ (Dfl 1,20/m^3gas)
Appliance at the inside		
Single glazing (reference)		
normal roller shutter	10 (26)	20 (52)
Double glazing (reference)		
normal roller shutter	1,5 (4)	3 (8)
normal roller shutter sealed by profiles	3,36 (8.75)	6,72 (17.50)
roller shutter metallized on one side	3,75 (9.75)	7.5 (19.5)
normal roller shutter and double reflecting foil	6,82 (17.75)	13.64 (35.5)
normal roller shutter room-side Double reflecting foil windowside sealed by profiles	7.50 (19.50)	15 (39)
Double glazing, curtain (reference)		
double reflecting foil	4.81 (12.50)	9.6 (25)
Appliances at the outside		
double glazing (reference)		
Vertical roller screen	2.6 (6.75)	5.2 (13.5)

The cost-benefit for reflective layers on the parapet is calculated for a U value of 0,9 W/m^2K. The energy saving through the window is not taken into account. The surface of the parapet is twice that of the radiator.

The calculation is made for existing buildings which have a certain oversize (Table 6) in heat output and for new buildings taking into account the oversize costs of the radiator (Table 7).

Table 6. Cost-benefit of reflective material on the parapet for existing buildings.

Material and surface	Investment		Energy saving		Cost-benefit			
					6.65 ECU-GJ (Dfl 0,6/m³)		13.3 ECU/GJ Dfl 1,2/m³)	
	ECU/m²	(Hfl/m²)	MJ/m²	(m³gas/m²)	PBP	ROR	PBP	ROR
Coloured reflective paint on parapet	3.44	(9)	176	(4.8)	3.1	30	1.6	>40
normal reflective paint on backside of radiator	1.15	(3)	146	(4.1)	1.2	>40	0.6	>40
reflective foil on parapet	3.85	(10)	203	(5.7)	2.9	32	1.5	>40

Table 7. Cost-benefit of reflective material on the parapet for new buildings and enlarged radiators (two panels).

Material and surface	Investment		Energy saving		Cost-benefit			
	ECU/m²	(Hfl/m²)	MJ/m²	(m³gas/m²)	6.65 ECU/GJ (Dfl 0,6/m³)		13.3 ECU/GJ (Dfl 1,2/m³)	
					PBP	ROR	PBP	ROR
Coloured reflective paint on parapet	4.49	(12)	176	(4.8)	4.2	19	2.1	>40
normal reflective paint on backside of radiator	2.2	(6)	146	(4.1)	2.4	40	1.2	>40
Reflective foil on parapet	4.9	(13)	203	(5.7)	3.8	23	1.9	>40

5. Conclusions and prospects.

1) For the Netherlands the following energy savings are possible:

Single glass	normal shutter	362 MJ/m²year
Double glass	one side relfective curtain or normal shutter with second double reflective layer	250-267 MJ/m² year

2) Double glass saves more energy than single with roller shutter.
3) Reflective shutters between double windows have a lower thermal effect than in combination with double glass.
For a high thermal resistance air exchange between the different layers of the system has to be avoided.

4) Only a normal roller shutter in combination with single glass has a PBP of 4 years.
The second solution is a double reflective foil between roller shutter or curtain.
5) The possible investment for energy saving is low.
When with double glass a U value of 1 W/m^2 is attained the possible investment for PBP of 4 years is about:

Investment	Price of energy
9 ECU/m^2 (Dfl 24/m^2)	6.65 ECU/GJ (0,6 Dfl/m^3)
18 ECU/m^2 (Dfl 48/m^2)	13.3 ECU/GJ (1,2 Dfl/m^3)

6) A reflective paint on the parapet with a U value of 0,9 W/m^{2}K saves 297 MJ/m^2 parapet per year for single glass and 129 MJ/m^2 for double glass.
In all cases a pay-back period of about 4 years is realised. Normal reflective paint on the back side of the rdaiator is most cost-effective.
7) There are no problems for using roller shutters in existing buildings.
The width of the window is mostly between .4 and .8 m and the heigth between .8 to 1.5 m.

6. References.

[1] Bossers, P.A. c.s. Energy saving by the use of curtains and window sill. Report C 422, IMG-TNO (Dutch).

[2] Crommelin, R.D. and P.M. Maaskant Determination of possible energy saving in a room by infrared reflective finishing of the parapet behind a radiator. Report C 396, IMG-TNO (English).

[3] Bekker, J. Energy saving by using rolling shutters. Contract meeting, September 1982 (English).

[4] Dubbeld, M. and B. Knoll. Energy saving by using rolling shutters. Contract meeting, September 1982 (English)

[5] Dubbeld, M., c.s. Energy saving by using roller shutters with reflective layer and edge seals. Report C 533, IMG-TNO (English) Including reports of Bouwcentrum and Bronswerk.

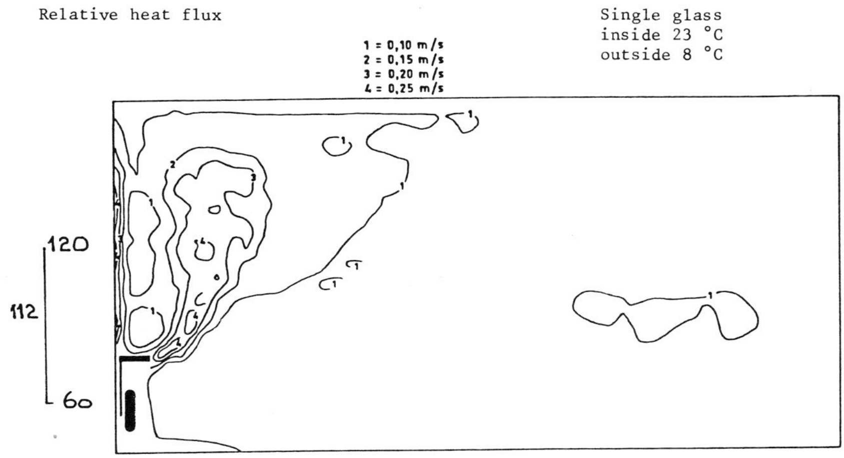

Figure 1 - Results of detailed velocity measurements, window sill 1

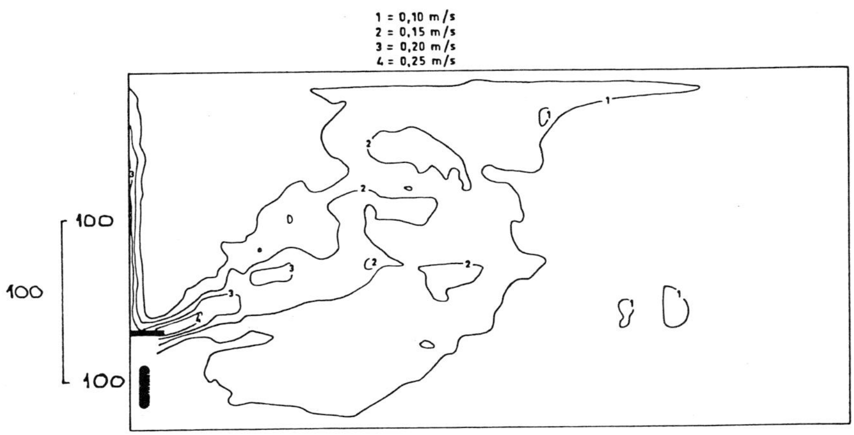

Figure 2 - Results of detailed velocity measurements, reference situation

HEAT REFLECTING BLINDS AND METHODS OF EDGE SEALING

P.A.Ruyssevelt and J.G.F.Littler
Building Unit, Polytechnic of Central London
35 Marylebone Road, London NW1, UK

Summary

Previous work gave rise to a prototype roller blind comprising: two layers of heat reflecting material, brush nylon top seals, magnetic bottom seals, and EPDM soft rubber side seals which may be open and closed to allow the free movement of the blind. The prototype was submitted to thermal transmittance testing in various configurations and the results indicated that a double layer blind is only marginally more effective than a single layer blind within the sealing system.

A patent has been filed on both the single and double layer sealing systems. Production of the single layer blind is expected soon, however the double layer blind needs further refinement.

1.0 Introduction

Previous work has shown that a combination of EPDM (synthetic rubber), magnetic and nylon brush seals applied to an impermeable roller blind can reduce the air leakage rate through a window by 85% (1). Since the closed windows of the average U.K. house account for ~40% of the total air change rate, such a roller blind system applied to all the windows could reduce the total air change rate by ~30%. This would result in a significant reduction in heat loss due to infiltration. Effectively sealing a roller blind serves a second purpose, that of reducing convection currents through the air space between the blind and the window and, in a two layer system, between the separated layers of the blind. Reducing convection currents in turn reduces the likelihood of condensation occurring on the window.

In order to reduce heat loss by radiation, low emissivity aluminised surfaces are incorporated in the fabric. In the single layer blind the aluminised surface faces the window and in a double layer blind the aluminised surface faces onto the airspace between the layers. It is feasable that in the double layer blind the surface facing the window could also be aluminised, this configuration has not yet been tested. As the aluminised surface renders the blind fabric impermeable and the edge seals maintain the layers at an optimum distance apart, the trapped layer(s) of air will reduce heat loss through the window without the need for bulky insulating materials.

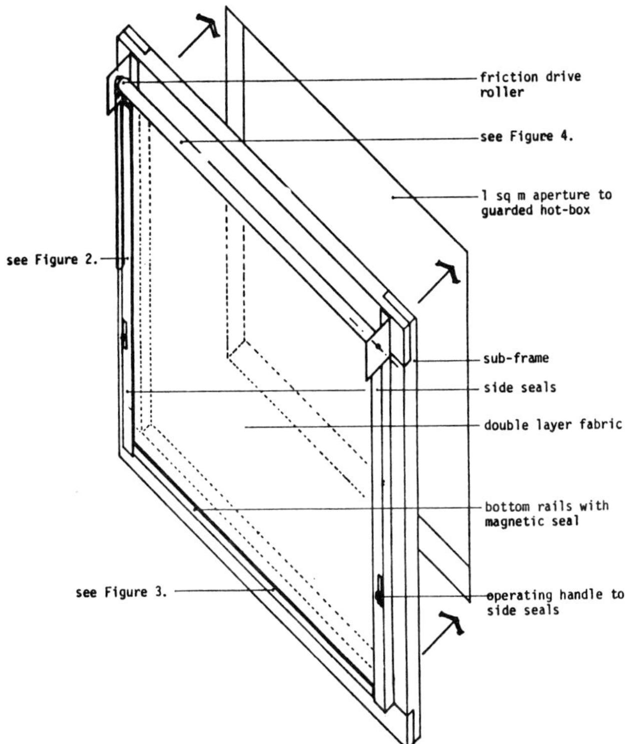

Fig. 1: Prototype double layer blind - 1 sq m

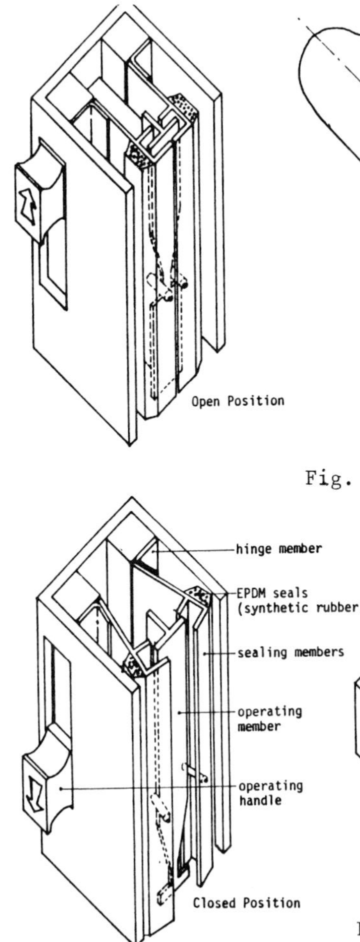

Fig. 2: Side sealing system

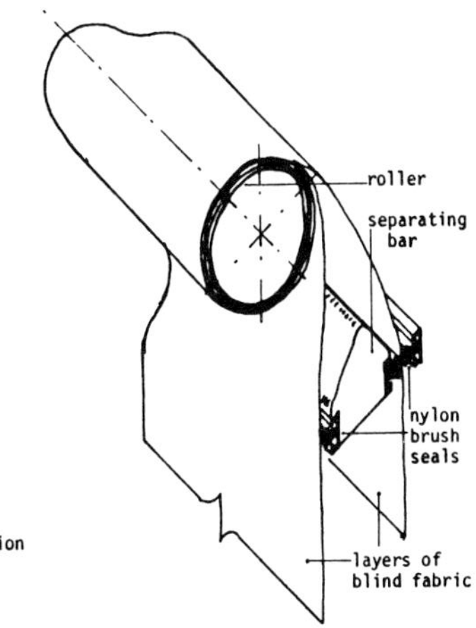

Fig. 3: Top sealing and separating bar

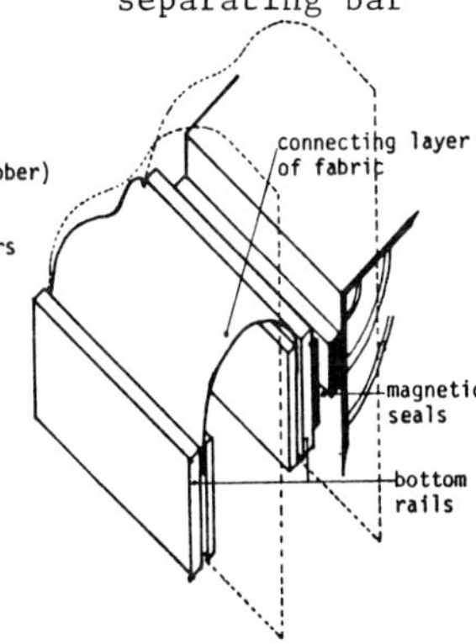

Fig. 4: Bottom sealing system

The configuration of the sealing system is, nylon brush seals to the top, magnetic seals to the bottom, EPDM soft rubber seals to the sides. Previous work has shown (1) that although EPDM seals are by far the most effective, they cannot be continuously applied because they impede the free movement of the blind. Therefore the aim of the side sealing system is to maintain the seals in an open (passive) state whilst the blind is being manoeuvred and cause them to be closed (active) when the blind is in the required position.

2.0 Prototype Development

A 1 sq m prototype double layer roller blind with edge sealing system was built, Figures 1,2,3 and 4 show the design and details of the prototype. The side sealing system underwent several stages of refinement before arriving at the prototype design . Difficulties were encountered in that the ideal material for the system, plastic, was not available in the right shapes and sizes. The prohibitively high initial tooling cost prevented the procurement of tailor made plastic extrusions, and instead existing sections were modified to perform the required tasks.

On completion the prototype required fine adjustment to areas such as the top separating/sealing bar to achieve a smooth operation. The prototype was first fitted with a spring ratchet roller - this gave rise to problems in lowering the two layers of material at an even rate and releasing the ratchet to roll-up the blind. Subsequently a friction drive roller was installed. This allowed full control over the movement of both layers of blind material when aided by heavier bottom rails.

The blind material used in the prototype was obtained from Glamorline International and was developed by the Physics Department of University College, Cardiff (see Figure 5 for detailed description). The material has a range of attractive decorative finishes, is fire resistant, and is no more expensive than ordinary blind material.

The prototype blind was transferred to University College Cardiff for thermal transmittance testing. Tests were performed on the prototype sealing system with double and single layer blinds, using low and high emissivity materials.

Two further prototype blinds were constructed with a reduction in tolerances from the original prototype. The blinds were installed in the PCL Solar Test House at Peterborough. Unfortunately, the tolerances were reduced too far, and as a result the blinds were inoperable. These blinds are presently being modified to single layer systems, and will soon be re-installed so as their performance may be monitored through the heating season. One successful bi-product of these prototypes was an opening/closing device for the side seals, incorporating two solenoids and this will be used in the

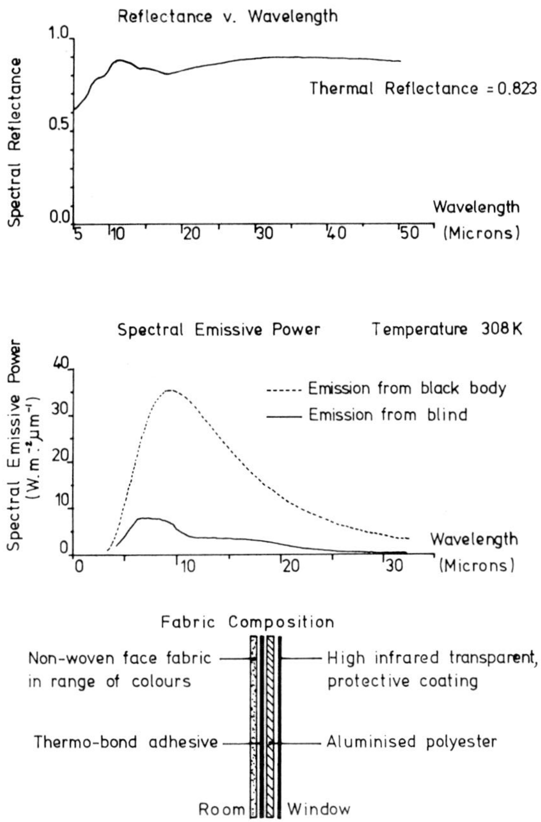

Fig. 5: Details of Energy Saver blind fabric

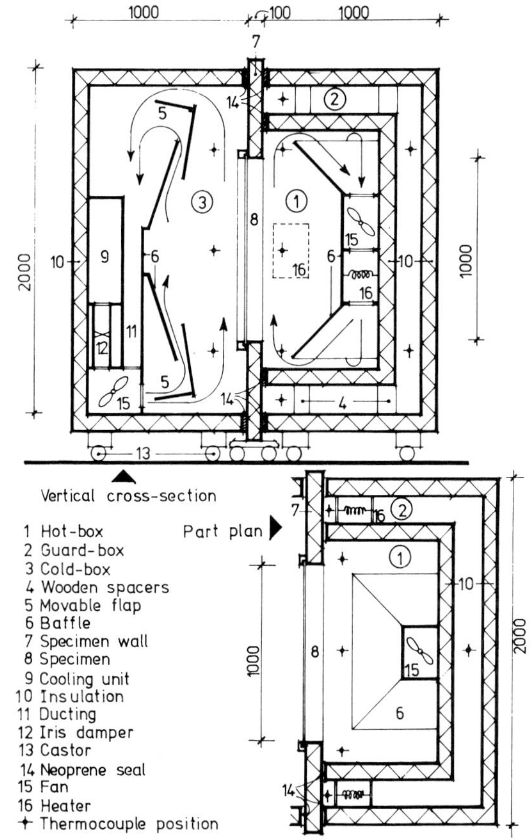

Fig. 6: Thermal transmittance apparatus

automated prototype to be installed in the PCL test cells.

3.0 Thermal Transmittance Apparatus and Procedure

The thermal transmittance apparatus used to measure heat loss is of the guarded hot-box type and is shown in detail in Figure 6. The hot-box whose rate of heat loss is to be measured is placed inside a larger box whose temperature is kept as close as possible to that of the hot-box (~25 deg C). This prevents heat loss through the sides of the hot-box as there is no significant temperature gradient between the two boxes. The open sides of both these boxes face onto the sample whose thermal transmittance is to be measured. A further box of the same dimensions as the guard box (to ensure parallel heat flow through the sample) is then placed on the other side of the sample, this box is maintained at at a temperature of 5 deg C in order to provide a thermal gradient across the sample. The temperature in all three boxes is kept as constant as possible by proportional and integral controllers. Thermocouples are strategically placed throughout the three boxes to examine the temperature distribution. The wind speed in the cold box is variable in order to ascertain the effect of external wind velocities on the thermal transmittance of the specimen. The thermal transmittance of a material is quoted in Watts per square metre per degree Kelvin so that it can be seen that knowing the temperature difference across the sample and the area of the sample, it is only necessary to measure the energy input in Watts to the hot-box to ascertain thermal transmittance or U-value of the sample. The sample only occupies 1m x 1m of the 2m x 2m specimen wall and the method of proportional areas is used to ascertain the U-value.

A test run for a sample may take up to 24 hours as it is necessary to allow the boxes time to equilibrate. Once they have done so, a period of time of 2 hours must occur somewhere in the test where the temperatures in the boxes are constantly to within +- 0.25 deg C. The temperatures in the boxes are continuously monitored along with the energy input to the hot-box and the data stored on a micro-computer.

4.0 Thermal Transmittance Tests

The roller blind sealing system was located on the hot-box side of the specimen window and tests were performed with various combinations of blind material. During the initial testing a period of instability occured where earlier results were not being repeated. However the thermal transmittance apparatus has now stabalised and excellent repeatability is being acheived. The apparatus has been examined by members of the Agrement Board and is almost certain to comply with a new British Standard on the subject, currently being drafted.

University College Cardiff are prepared to quote each set of results relatively, but not as absolute values untill all results been confirmed by re-testing.

Since the Cold-box was maintained at +5 deg C and not at a temperature below 0 deg C very little condensation occured on the glass. When testing the blinds within the PCL sealing system no condensation uccured.

5.0 Results

The results of the thermal transmission tests available to date are shown in Table 1, each result in this table has been repeated to within 0.1 W/m²K.

TABLE 1 RESULTS OF THERMAL TRANSMITTANCE TESTING

Sample Tested	X cm²	Y mm	U-value W/m²K Exposure shelt	norm
1 SG	-	-	4.48	5.52
Blinds Witout Seals				
2 SG + linen blind	350	25	3.61	4.24
3 SG + ES	350	25	2.78	3.20
4 SG + ES (no coating)	350	25	2.57	NT
5 SG + AL POLY	350	25	2.56	NT
Blinds With PCL Seals				
6 SG + linen blind	0	160	2.50	2.79
7 SG + ES	0	160	1.52	1.93
8 SG + AL POLY	0	160	1.23	NT
9 SG + double ES	0	160	1.18	1.33
10 SG + double AL POLY	0	160	0.73	NT

X - area around edges of blind in aperture
Y - distance of blind from glass
SG - single glazing
ES - Energy Saver blind
NT - no test
AL POLY - aluminised ployester

6.0 Economic Assessment

At present a 1 sq m single layer reflective roller blind without seals costs ~£10. The total cost of the single and double blinds will be ~£20 and ~£30 respectively for a 1 sq m blind. However, since the major cost of these two systems is the sealing system which is perimeter dependant rather than area dependant the cost per square metre is reduced as the area of blind increases (see Figure 7).

In order to assess the economic value of the blinds the above costs have been applied to the results of computer simulations (2). From Figure 8 it can be seen that both systems give a pay back period within a technical and economic life of 8 years. Obviously the economical benefits are far higher when electricity is the fuel source; however the figures are those for 1 sq m and do not take account of the fact that larger area blinds are effectively cheaper per sq m. When only selected windows were treated with the blinds a significant increase in the performance and PWV (Present Worth Value) occured. The selection of windows was based on such factors

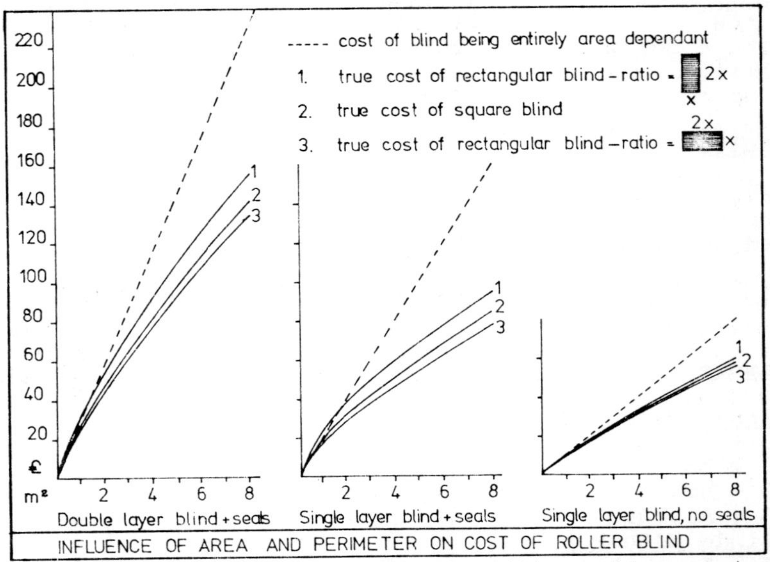

Figure 7. Cost of roller blinds with sealing system

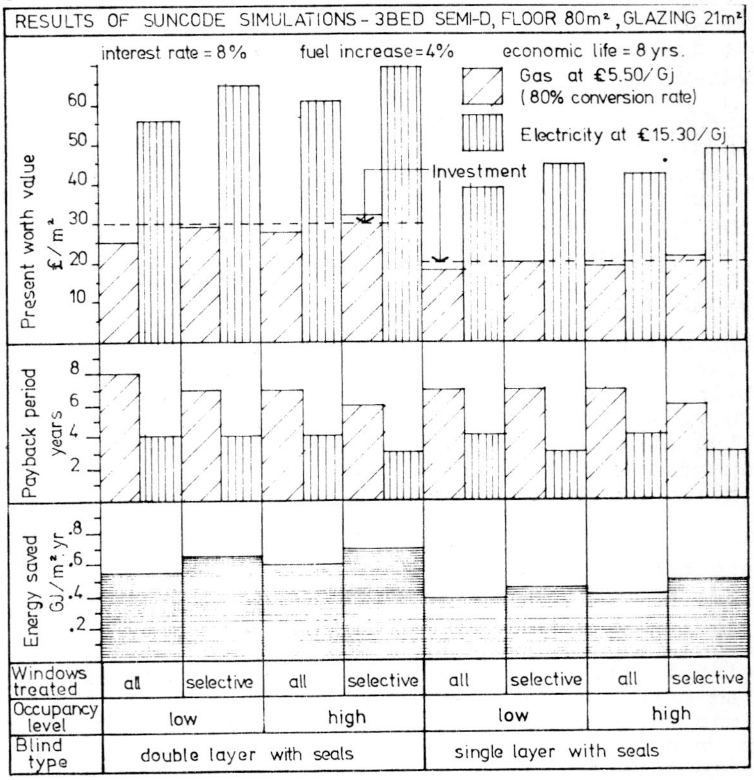

Figure 8. Economic assement of roller blinds

as occupancy pattern, orientation, heating system, risk of condensation etc. This selection process is crucial to the correct application of window improvement systems.

7.0 Commercial Prospects

A U.K. patent (No. 8316503) has been filed on the edge sealing system for single and double layer blinds.

Glamorline International has agreed to work with PCL in the development of a prototype a sealing system for their single layer blind (see Figures 9 & 10). The double layer blind is only marginally more effective than the single layer and will require more intensive development to achieve smooth operation. For these reasons, it is being held in abeyance for the time being.

8.0 Future Activities

When all thermal transmittance tests have been performed at University College Cardiff the prototype will return to PCL for air infiltration testing. The infiltration tests have been performed upon the individual sealing components but not on their configuration within the prototype.

An automated version of the single layer blind will be constructed and installed in the PCL Test Cells. This installation should take place before the winter to allow for a useful monitoring period.

9.0 References

(1) 'Benefits of Insulating Blinds and the Importance of Edge Seals' P.A. Ruyssevelt, J. Littler, Energy Conservation in Buildings, Heating, Ventilation and Insulation, Proceedings of Contractors Meetings held in Brussels, September 1982, pp295-304

(2) 'Passive Solar Design Study 13 - Windows', P.A.Ruyssevelt, P.Clegg, draft Report presented to the U.K. Department of Energy, January 1983.

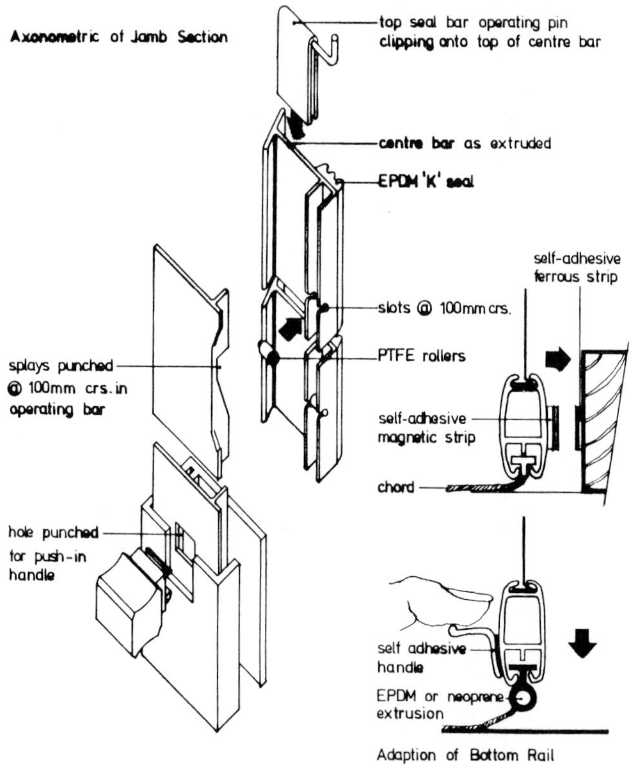

Figure 9. Axonometric of single layer blind

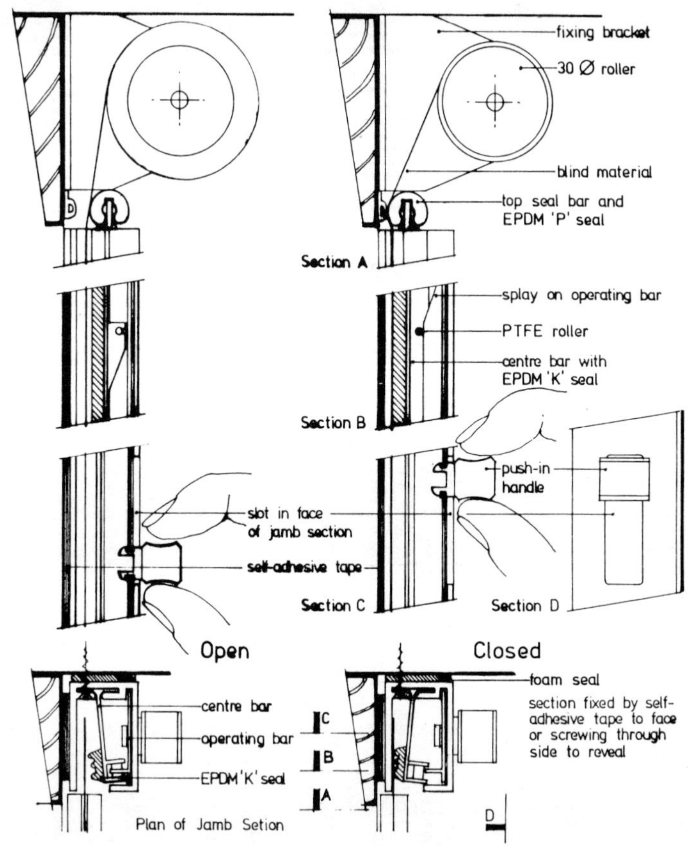

Figure 10. Single layer blind side seals

THE MEASUREMENT OF AIR INFILTRATION RATES IN LARGE ENCLOSURES AND BUILDINGS

I.N. POTTER, J. DEWSBURY and T.J. JONES

THE BUILDING SERVICES RESEARCH AND INFORMATION ASSOCIATION
BRACKNELL, UK

Summary

The report discusses the need for a proven method of measuring air infiltration rates in large enclosures in order to assess the need for and effectiveness of energy saving measures.
The object of the research is to develop such a proven method. Some possible novel instruments are described, together with theoretical predictions of the implications of poor air mixing in the space for methods of measurement and analysis. The measurement system selected and the results of some preliminary measurements are described. The report concludes with an account of work in progress to validate the measurement and analysis methods.

1. INTRODUCTION

The floor area of industrial and warehouse buildings in the UK is about 410,000,000 m^2. The authors have estimated that the heat loss from these buildings due to air infiltration is of the order of 1.6×10^{17} J/year. This represents about 3% of the total UK energy consumption by final users (5.8×10^{18} J/year). Therefore, if the air infiltration rate were to be halved, a saving of 1.5% of the national energy consumption might be made.

However, industrial and warehouse buildings usually consist of a single large space or a small number of large spaces. There is at present no proven method of measuring ventilation rates in these spaces. Therefore, there is no way of knowing whether the energy loss and possible savings estimated above are over or under estimated and no certain way of determining the cost effectiveness of any measures taken to reduce unwanted air infiltration.

2. OBJECTIVES

The objectives of the research are to develop and validate a method of measuring ventilation rates in large enclosures, to use the method to measure ventilation rates in selected buildings and to assemble the results of the measurements into the foundations of a data base.

As well as the ventilation rate, other significant variables will be recorded. These will include temperatures at selected points, wind velocity, factors affecting building air-tightness, etc.

These and future measurements using the validated ventilation rate measurement method will enable the effectiveness of measures for reducing unwanted air infiltration to be assessed, and thus enable energy conservation measures to be directed to where they will have most effect.

3. PROGRAMME OF WORK

The first part of the research consisted of a survey of existing and possible novel techniques for measuring ventilation rates, and laboratory trials of an unconventional gas analyser.

The second part commenced with the selection of suitable instrumentation, and included the construction and commissioning of the measurement system and trials in a medium-sized enclosure in the laboratory.

The third part consists of full scale trials of the measurement system in a large enclosure having a controlled ventilation rate. The measured ventilation rate will thus be able to be compared to the known controlled ventilation rate, and an investigation of factors affecting the accuracy of the method will be made.

The fourth part will consist of using the validated measurement method to measure ventilation rates in selected buildings. The results of the measurements will be used to form the foundations of a data base.

4. INVESTIGATION OF POSSIBLE MEASUREMENT TECHNIQUES

The normal method of measuring an enclosure's ventilation rate is to inject a tracer gas into the air in the enclosure according to some defined strategy, and to measure the concentration of tracer gas in the enclosure over a period of time. Provided the air in the enclosure is well mixed the ventilation rate can easily be calculated from a knowledge of the tracer gas injection strategy and the history of its concentration. This method has been used successfully in many countries, but has been mainly confined to small enclosures. This is partly because of practical difficulties concerning the amount of equipment and labour involved for large scale work, partly because of theoretical difficulties connected with the interpretation of results and control of test equipment when good mixing is difficult or impossible to obtain.

The search for possible novel measurement methods was directed to finding either a low cost substitute for conventional tracer gas concentration measurement instruments, or an instrument with significant technical advantages over conventional equipment.

A method based on analysis of air samples on site by individual low cost fuel cell gas analysers at each sampling point was selected for initial trials. A laser method was also of interest as it would have been able to cover a large volume from a single point, but it was not selected for trials due to the high capital cost of the equipment.

Two fuel cell gas analysers were obtained and tested in the laboratory. The calibration of the analyers were initially subject to significant variation. This variation was reduced when the analyser was maintained at 60°C instead of at room temeprature, with the added benefit of much faster response. Ventilation rates were measured using the fuel cell and a conventional analyser simultaneously, and similar results were obtained.

5. SIMULATION OF IMPERFECT MIXING

Little work on air movement in large buildings has been published, but what there is indicates that poor mixing is likely.

A simple numerical model was developed which simulates a ventilation rate measurement by the concentration decay method in a poorly mixed space. Figure 1 shows the results of a simulation, and Figure 2 shows the air flow pattern assumed in calculating the results shown in Figure 1. The perfect mixing line in Figure 1 is that which would be obtained if the air in the enclosure had been fully mixed. If perfect or near perfect mixing is obtained in an experiment then the ventilation rate can be obtained directly from the slope of the concentration decay line.

It can be seen from Figure 1 that the concentration lines for the individual zones can fall either above or below the perfect mixing line, and that their slopes can be either greater or less than that of the perfect mixing line. The line for the mean concentration has a slope which is different from that of the perfect mixing line.

The results of this and other simulations showed that in a poorly mixed enclosure it would be necessary to know the details of the variation of the concentration of tracer gas through the space in order to estimate the ventilation rate.

The proposed laser method of measuring tracer gas concentrations was essentially an averaging method and was therefore rejected.

6. ANALYSIS METHOD FOR IMPERFECT MIXING

It is assumed that the space can be divided into a number of well defined zones of known shape and size.

The general equation governing the tracer gas concentrations within the space is:

Rate of supply of gas - Rate of removal of gas
= Rate of change of quantity of gas in space

The easiest ventilation rate measurement method to use is the concentration decay method, in which:

Rate of supply of gas = 0

The removal of gas from the space is simply the sum of the air flow rates from each zone to the outside, each multiplied by its appropriate tracer gas concentration:

$$= \sum_{z=1}^{Z} Q_{zo} C_z$$

where Q_{zo} is the air flow rate from zone z to the outside and C_z is the tracer gas concentration in zone z. Z is the number of zones.

The rate of change of the quantity of tracer gas in the space is the sum of the volumes of the zones each multiplied by the appropriate rate of change of tracer gas concentration:

$$= \sum_{z=1}^{Z} V_z \frac{dC_z}{dt}$$

where V_z is the volume of zone z.

$$\text{Thus} \qquad 0 - \sum_{z=1}^{Z} Q_{zo} C_z = \sum_{z=1}^{Z} V_z \frac{dC_z}{dt}$$

During a concentration decay measurement the values of C_z and $\frac{dC_z}{dt}$ will vary with time, so by integrating over T time intervals and inserting estimated value of Q_{zo} the following set of equations can be obtained:

$$\sum_{z=1}^{Z} V_z(C_{z,t1} - C_{z,to}) + Q_{zo} \int_{to}^{t1} C_z dt) = E_1$$

$$\sum_{z=1}^{Z} V_z(C_{z,t2} - C_{z,t1}) + Q_{zo} \int_{t1}^{t2} C_z dt) = E_2$$

etc.

$$\sum_{z=1}^{Z} V_z(C_{z,t1} - C_{z,t(-1)} + Q_{zo} \int_{t(T-1)}^{tT} C_z \, dt) = E_T$$

Here the magnitude of each E represents the magnitude of the discrepancy between the observed values of C_z and the assumed value of Q_{zo}.

Provided that T > Z and that the equations are sufficiently independent, it should be possible to find the values of the Q_{zo} terms which minimise an error function:

$$f(E_1, E_2, \ldots\ldots E_T)$$

subject to the constraint that $Q_{zo} > = 0$ for all z.

The assumption that the space can be divided into a number of well mixed zones of known shape and size is a very great simplification of the true situation. Section 8 outlines the programme of work intended to provide data on the accuracy of the method.

7. SELECTION AND DESCRIPTION OF MEASUREMENT SYSTEM

Of conventional gas analysers, the research team has found infra-red gas analysers to offer the most suitable combination of cost, accuracy and reliability for this application. Preliminary designs and cost estimates were therefore produced for computer controlled multi-point tracer gas concentration measurement systems using infra-red and fuel cell gas analysers.

The capital costs of the two systems were expected to be similar, but a number of significant technical problems were expected with the fuel cell analyser system, mainly concerned with the speed of response of the fuel cell analysers and the extent of the modifications required to them in order to incorporate them into a working system. The infra-red gas analyser system was therefore selected.

The measurement system comprises three infra-red gas analysers with associated sampling valve assemblies, thermocouples for measurement of temperatures and other instruments as required. The sampling valve assemblies permit each gas analyser to be connected to one of a number of sampling points. The instruments are controlled and data recorded by a microcomputer with associated interface equipment. A schematic of the system is shown in Figure 3.

8. VALIDATION OF MEASUREMENT METHOD

Work is in progress aimed at providing data on the accuracy of the multi-point tracer gas method for measuring ventilation rates in large enclosures.

A fan extracts air from or supplies air to the enclosure so as to create a pressure difference between inside and outside. If the pressure difference created by the fan is larger than those caused by wind and temperature differences, then all the air ventilating the space will pass through the fan. The ventilation rate measured using tracer gas can then be compared with the air flow rate through the fan.

Figure 4 shows the results of a test in which good internal mixing was obtained. Measured concentrations from every third zone of thirty are plotted, together with the theoretical line corresponding to perfect mixing and no ventilation except through the fan. Good agreement is obtained between the measured and theoretical concentration decay rates.

Figure 5 shows the results of a test in which poor internal mixing was obtained. Measured concentrations from selected zones of thirty are plotted, together with the theoretical line corresponding to perfect mixing and no ventilation except through the fan. This Figure should be compared with Figure 1. In Figure 5 the concentration decay lines for the various zones have different slopes, generally steeper than the slope of the theoretical line.

Work is proceeding on the calculation of ventilation rates from results such as those shown in Figure 5, by the method outlined in Section 6 of this report.

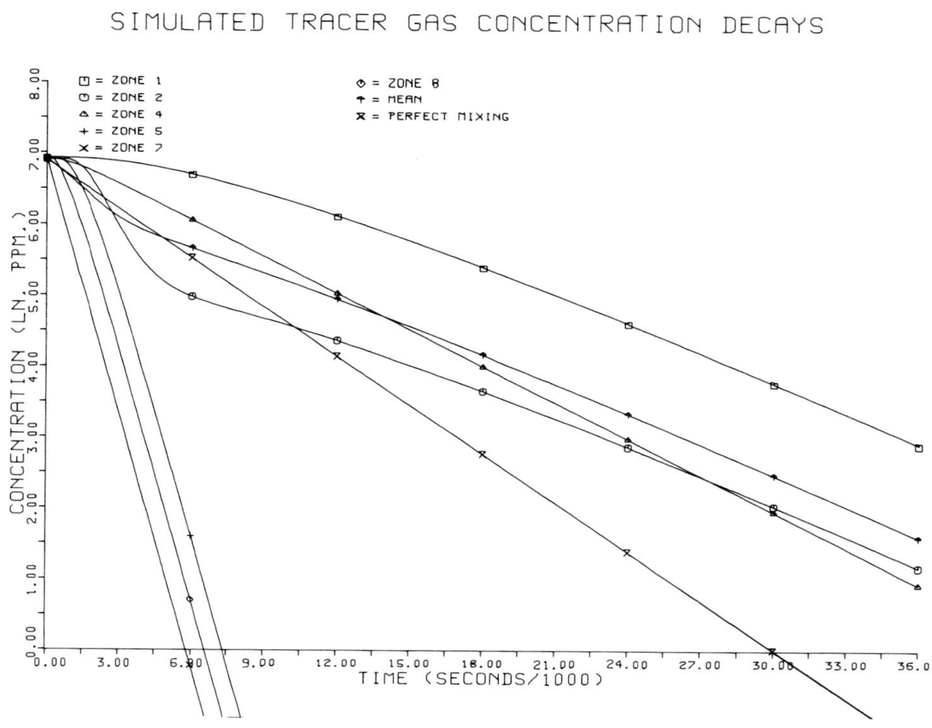

FIGURE 1

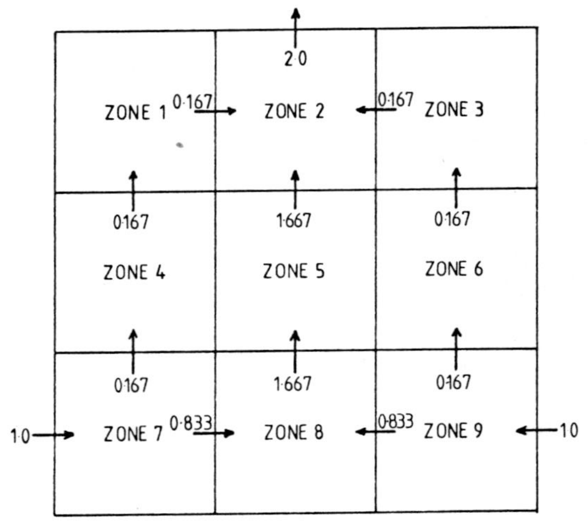

All zones 1000m³
Air flows in m³/s

FIGURE 2

Air flows for figure 1

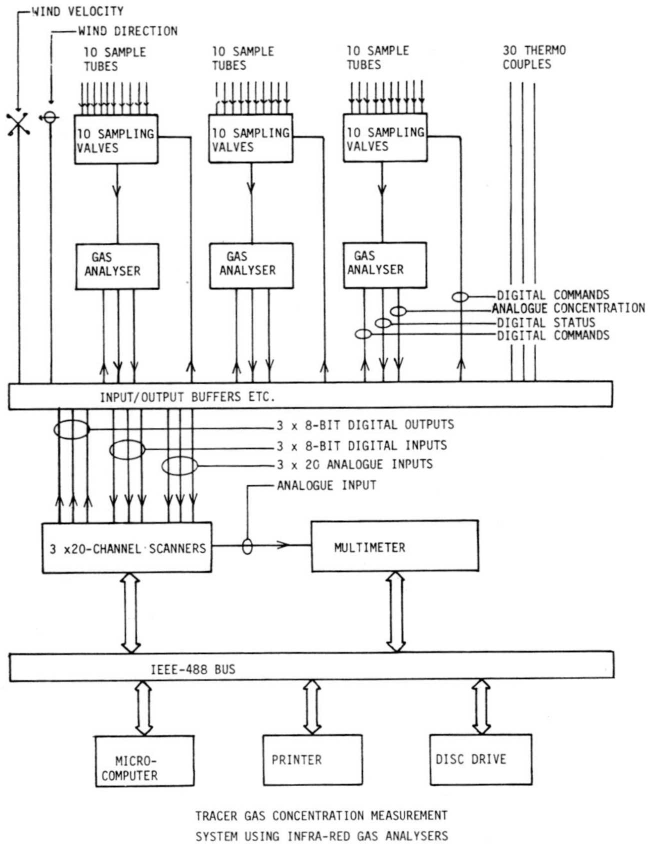

FIGURE 3

TRACER GAS CONCENTRATION MEASUREMENT SYSTEM USING INFRA-RED GAS ANALYSERS

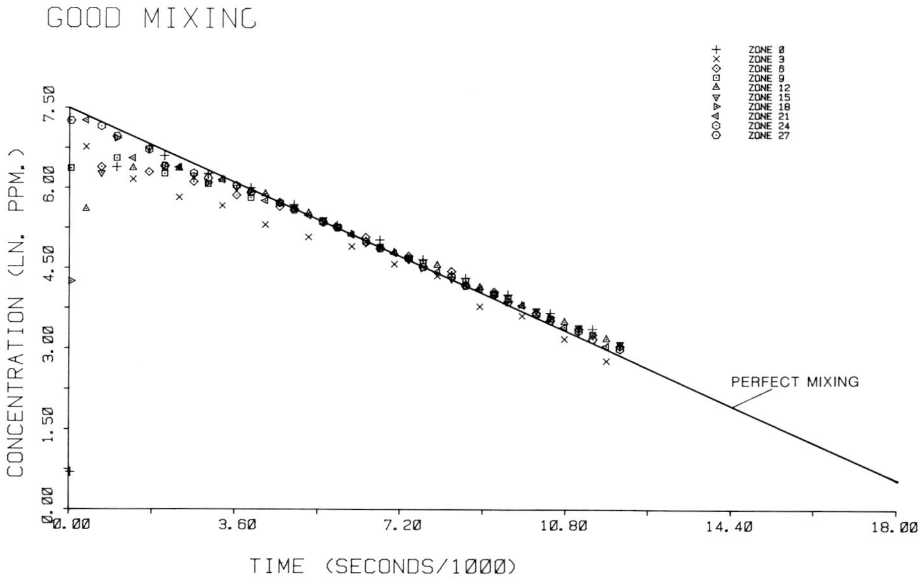

FIGURE 4

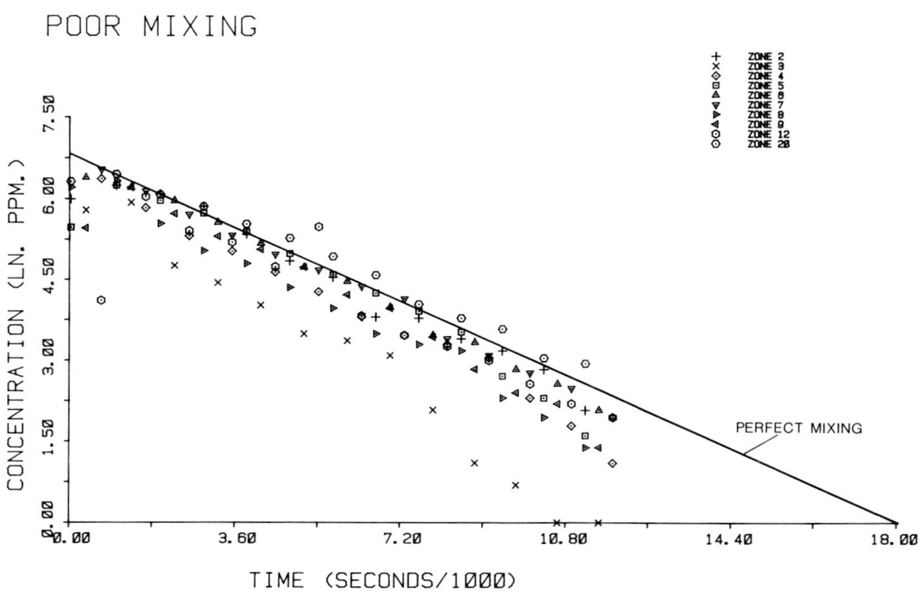

FIGURE 5

MEASUREMENTS OF INTERCELL AIRFLOWS IN LARGE BUILDINGS USING MULTIPLE TRACER GASES

M D A E S Perera, R R Walker and M J B Trim

Building Research Establishment,
Building Research Station,
Garston,
Watford, Herts., UK.
(Tel: GARSTON 74040)

Summary

This paper presents the results to date of the use of the multiple tracer gas technique to determine interzonal airflow and ventilation rates in large, multicelled buildings like offices. This work is part of a wider project designed to extend knowledge of natural ventilation in all types of buildings other than dwellings.

The work presented here includes, (a) an experiment in a naturally ventilated office building using an automated system connected to infrared analysers dedicated to each of three gases, and (b) results from a similar experiment on a mechanically ventilated office building using a more sophisticated version of the system interfaced to a single multicomponent infrared analyser.

1. INTRODUCTION

A previous report (1) has given indications of energy savings that could result from a detailed knowledge of interzone air movements within large multicelled buildings. Such detailed information would also lead to,

(a) identifying the movement of air, and hence contaminants or heat from one zone to another and,

(b) determining the effectiveness of remedial measures or indicating zones where selected remedial measures will be cost-effective.

The same report proposed various techniques, ranging from simple 'grab' sampling to complex multiple tracer gas analysis, to determine such airflows. Extensive details about the tracer gases that could be used in such measurements and the underlying theoretical basis for measuring interzone airflows were also given.

This present paper gives details of two field experiments on office buildings using the multiple tracer technique. The first test was carried out in a naturally ventilated two-storey building and the other in a mechanically ventilated three-storey building. This report describes the automated systems used in these measurements, the method of analysis and the results obtained.

2. TESTS IN A NATURALLY VENTILATED OFFICE BUILDING

The experiment was carried out in a conventional, naturally ventilated two-storey office building at the Building Research Station, Garston. The building is rectangular in plan (40 m x 11 m) and each storey is 2.44 m high.

2.1. Experimental details

For the purposes of the experiment, the building was nominally divided into three zones; two of these zones (each with 783 m^3 volume) incorporated the major portions of the ground floor and of the first floor whilst the third zone encompassed a common stairwell region together with the adjacent offices at one end of the building.

Tracer gases were injected manually to reach target concentration in each zone. The ground floor zone was seeded with nitrous oxide and the first floor zone with sulphur hexafluoride to concentrations of about 200 ppm, and the stairwell zone with carbon dioxide to about 2000 ppm. During injection, small desk-top fans were used to mix the tracer gases within the zones.

Air samples were taken from each zone through equal length 6 mm diameter polyethylene tubes. A sample tube was also placed outside the building in order to provide reference ambient concentrations of the test gases. All tubes were then brought together to a central unit where each sampling line was terminated by a three-way solenoid valve. These valves were controlled by an ITT Director microprocessor. This allowed an air sample from each zone to be analysed by all three infrared Leybold-Heraeus gas analysers where each analyser was dedicated to one of the three tracer gases. The concentrations of the gases in each zone were then recorded on cassettes for off-line analysis. Reference 2 gives fuller details of this automated system.

2.2. Results and discussion

Using the mathematical procedure detailed in Reference 1, the interzone airflows were calculated. It was found that some of the airflows were negative. However, the definitions for these preclude any physical meaning for such values. This is a consequence of experimental error which possibly arises from imperfect mixing of the tracer gases in each zone. Solutions can be constrained in the least-squares sense to have positive values only.

The results (2) show that it is possible to determine interzone airflows as well as the fresh air infiltration to each zone. It also shows that the infiltration rates in each zone are not similar to one another. This kind of information could be obtained from single tracer measurements only with great difficulty.

3. TESTS IN A MECHANICALLY VENTILATED OFFICE BUILDING

A more powerful and sophisticated automated system was built to replace the prototype system. The new system incorporates a multicomponent infrared gas analyser to replace the individual analysers. The new system also includes a microcomputer to control the sampling sequences as well as in abstracting and archiving the data given out by the analyser. A description of tests carried out in a mechanically ventilated office building using this equipment is given below.

3.1. Office building

The three storey building used in these tests was built as a 'low energy' office and is sited at the Building Research Station. The low energy office building is rectangular in plan (60 m x 12 m) and has its major axis aligned east-west. The floor to ceiling height of each floor is 2.6 m. Offices are located on each floor along either side of a central corridor. The construction of the building and the layout of the mechanical ventilation system allows the building to be 'zoned' in many ways. For the experiment reported here, however, each floor was considered as a separate zone with a volume of 1762 m^3 each.

This low energy office incorporates a number of energy saving features. Among these is a mechanical ventilation system which allows a variable amount of fresh air to be taken into the building. This value can be varied from nominally zero (full recirculation) to full (no recirculation) fresh air intake depending upon the setting of mechanical dampers at its air handling unit. It is set to 30% fresh air (about 1.5 ach) during the heating season. Full details of this building and its energy saving features are given in Reference 3.

3.2. Injection and sampling points

The design of the ventilation system enabled each floor to be individually seeded with a tracer gas. These were injected at various strategic points into the supply side of the mechanical ventilation system through 6 mm diameter polyethylene tubing laid out from a central 'control' room.

Using 10 mm inner diameter polyethylene tubes, air samples were taken from various points along the extract side of the ventilation system. These larger tubes were necessary so as to ensure an adequate volume flow rate to the cell of the infrared analyser. In each zone, all sampling tubes from that zone were brought together to a common junction and manifolded together. A single tube from each junction then carried the blended samples into the automated system for analysis.

3.3. Automated system

The automated system (Figure 1) contains three principal components of which the CBM 4032 microcomputer has overall command of (a) the gas line sampler and, (b) the Miran 980 multicomponent infrared gas analyser.

Of these, the gas line sampler consists of two trays of instrumentation. One tray contains sixteen three-way solenoid valves. Figure 2 is a schematic drawing of the pneumatic connections between these valves. The other tray contains the relay control units required to operate them. The relays are in turn controlled by commands from the microcomputer.

Once the operator has fed in the initial operating conditions, the operating sequence is then entirely under the control of the microcomputer. In any one cycle of operation, this involves the microcomputer in (a) selecting the required zone, (b) directing a fresh sample through into the cell of the Miran analyser, (c) initiating an 'analyse' command to the Miran, (d) storing the data relating to the concentrations of the constituent gases put out by the Miran, and (e) repeating this sequence for other zones and other cycles.

3.4. Experimental details

Tests 13, 14 and 15 presented here were designed, amongst other things, to give an initial appreciation of the performance of the mechanical ventilation system. Of these, Test 13 was carried out with the ventilation system on full recirculation, i.e. nominally zero fresh air intake. Test 14 was then made with no recirculation, i.e. full fresh air intake whilst Test 15 used the 30% fresh air intake usually set for the building during the heating season.

These tests were made with all windows and doors, both internal and external, closed. The building was unoccupied during the tests and the heating was off. The internal temperature during the tests was approximately 23°C. All tests were carried out within a space of few hours thereby excluding to a large degree any effects caused by changes in the weather. Table 1 gives the weather data relating to these tests.

Sulphur hexafluoride (SF6), Freon 114 and Freon 13B1 were used as the tracers and were injected into the ground (G), first (F) and second (S) floor zones respectively to reach target concentrations of about 17, 124 and 53 ppm in each corresponding zone. This combination of gases were choosen because of their small values of cross absorption during infrared analysis. The target concentrations were set by the linearity and the strengths of the signals from the Miran analyser for those concentration ranges. The accuracy of the concentration measurement of any one gas in a mixture was better than 1% of the upper value of its linearised range.

3.5. Results

Figure 3 shows the SF6 tracer gas concentration as a function of time in the three zones for the experiment carried out with full recirculation (Test 13). Similar curves were derived for the remaining tracers and tests.

The interzone airflows and fresh air infiltration rates were derived from these graphs using the mathematical procedure given in Reference 1. Figure 4 is a schematic drawing of the results from Test 13. Unlike the results obtained for the naturally ventilated building (Section 2.2), the calculated zonal airflows were always positive. This is believed to be the result of better mixing of the tracer gases which can be expected with mechanical ventilation.

Table 1 summarises the results obtained. In each zone, the fresh air infiltration rate as well as its total ventilation rate is given. The total ventilation rates are calculated by summing either the inflows or outflows into those zones. The whole building infiltration rate is also given. All values are in airchanges per hour (ach).

3.6. Discussion of results

Interzonal airflow information, as supplied by Figure 4, allows the fresh air contibution to any zone to be separated from the total ventilation rate into that zone or to the building as a whole. The differences between these rates can be seen in Table 1.

For this series of measurements, the results tabulated in Table 1 can be compared with 'design' values (given in the same table). It is seen that there are marked discrepancies between them. One striking feature is the amount of fresh air entering the building when the mechanical ventilation system is on full recirculation (Test 13) when there should nominally be no such inflow. It is believed that this inflow of fresh air is partly due to infiltration through the external fabric of the building and partly through leaks in the mechanical dampers and the ductwork housing the mechanical ventilation plant.

Since the tests reported here are of a preliminary nature, it is too early to comment on these differences or in the way they reflect on the energy performance of the building. Further tests, using information as derived from flow rates such as those in Figure 4, would however enable these differences to be clarified.

4. CONCLUSIONS

Details are presented here of the prototype and 'final' automated sytems designed and built to carry out multiple tracer measurements. Results presented from two field experiments, one in a naturally ventilated and the other in a mechanically ventilated office building, have shown that such systems allow interzonal airflows and ventilation rates in complex, multicelled buildings to be determined.

The complexity of these multiple tracer gas systems precludes them from being used widely. Future field work will therefore be confined to in-depth monitoring of a few selected office buildings. Work will however continue on other techniques (1), such as grab and spot sampling as well as the continuous injection of a tracer gas. This will show how the different techniques interrelate and will enable the most appropriate technique to be choosen for any specific circumstance.

AKNOWLEDGEMENTS

This project is funded in part by the EC under Contract No. EE-A-5-050-GB. The work described has been carried out as part of the research programme of the Building Research Establishment of the Department of the Environment and this paper is published by permission of the Director.

REFERENCES

1. M D A E S Perera, Review of techniques for measuring ventilation rates in multi-celled buildings, Paper presented at the Proceedings of the EC Contractor's meeting on 'Natural Ventilation' held in Brussels on 30 September 1982.
2. M D A E S Perera, R R Walker and O D Oglesby, Ventilation rates and intercell airflow rates in a naturally ventilated office building, Paper presented at the 4th AIC Conference on 'Air Infiltration Reductions in Existing Buildings' held in Switzerland on 26-28th September 1983.
3. P M Price, The BRE 'low energy' office building, Building Services Engineering Research and Technology, Vol. 3, No. 3, pp 127-133 (1982).

TEST CONDITIONS

TEST	MECH. VENT. SETTING	TEMPERATURES deg C Int.	Ext.	WIND SPEED m/s	WIND DIRECTION deg N
13	Full recirculation	23.	19.	0.82	47
14	No recirculation	23.	18.	1.12	85
15	30% 'fresh' air	23.	17.	0.31	43

VENTILATION RATES (ACH)

TEST	GROUND FLOOR		FIRST FLOOR		SECOND FLOOR		WHOLE BUILDING	
	Fresh	Total	Fresh	Total	Fresh	Total	Measured	Design
13	1.05	3.10	1.02	3.38	0.55	3.70	0.87	0.0
14	2.33	3.22	2.00	3.22	1.95	2.64	2.09	4.0
15	1.81	4.57	1.35	3.00	1.15	3.19	1.44	1.5

Table 1. Test conditions and results from the mechanically ventilated office building experiment.

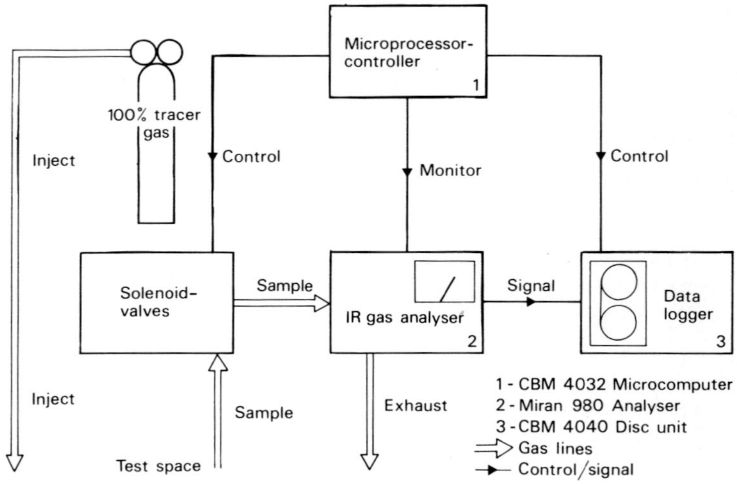

Figure 1 - Schematic of the automated system

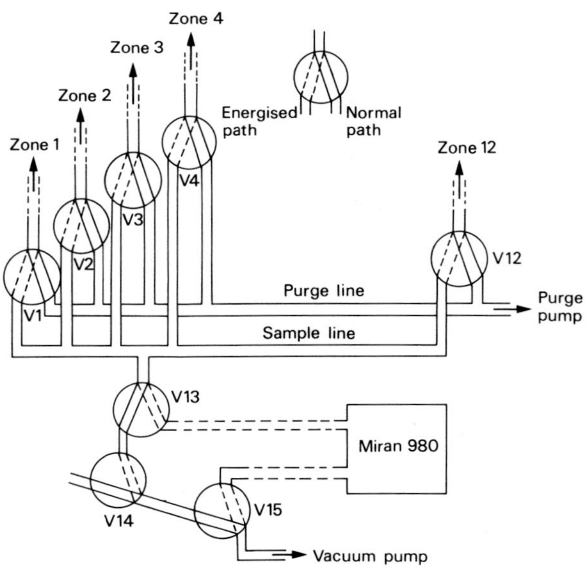

Figure 2 - Pneumatic connections of the solenoid valve unit

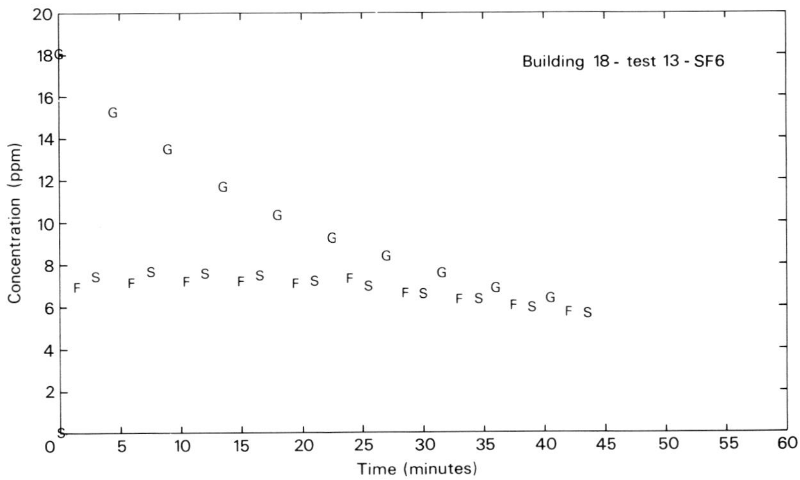

Figure 3 - Sulphur hexafluoride concentrations in all three zones

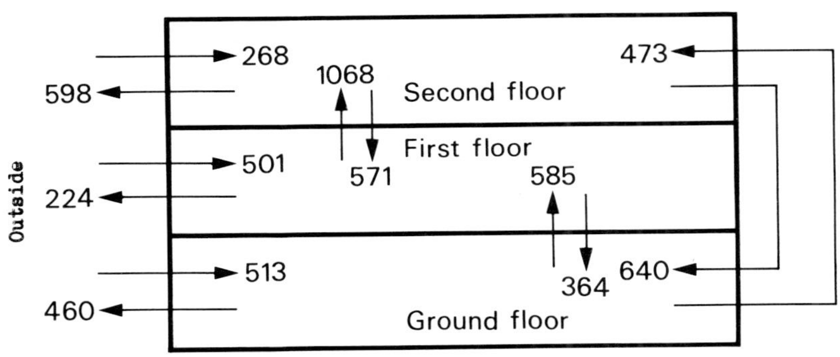

Figure 4 - Interzonal airflows (litres/sec) for Test 13

USERS INFLUENCE ON THE AIR INFILTRATION

P.F. COLLET and B. KVISGAARD

Department for Building Technology
The Technological Institute
DK - 2630 Taastrup

Summary

To measure actual air infiltration in dwellings due to the house and the inhabitants it is necessary to measure continuosly for several days. This can be done by the constant concentration measuring method.
The measurements show that in houses with a reasonably low AIR (0,2-0,5h^{-1}) the inhabitant has the major impact (up to 50-75%) on the total AIR.
This must be taken into consideration when designing mechanical and natural ventilation systems.

Introduction

The aim of this part of the project is to see if it is possible to measure the influence on the AIR (Air Infiltration Rate) from the behaviour of the inhabitants, and to estimate the necessary area of ventilation openings to allocate the fresh air to the rooms wanted.

Measuring Equipment

The measuring equipment has been designed to measure the air infiltration rate continuously in up to 10 rooms. The equipment is controlled by a computer and the measuring - that normally lasts a week - is automatic. The measuring data are gathered on a diskette during the measurement. The measuring principle is tracer gas according to the "constant-concentration-method". During the measurement the equipment will keep up a constant concentration of tracer gas in the room, and the air change rate is then determined on the basis of the quantity of tracer gas that it is necessary to supply to the room to keep up the concentration.

Our analyses of possible measuring errors and our control of the measuring equipment as compared with reliable references show that the measuring method is very accurate. When measuring small to moderate air infiltration rates the inaccurancy is ± 5%.

It appears that the measuring equipment may also be applied to other purposes; it has thus been applied for quantiying the short circuit in case of mechanical ventilation between injection and exhaust.

Measurement of air infiltration rate in dwellings at Galgebakken

Measurements of natural air infiltation rates in dwellings due to the behaviour of the inhabitants, and to ventilation openings.

Summary

The AIR - when the dwellings are used by the inhabitants (called UAIR) - has been measured in 7 houses during a period totalling 61 days from October 1983 - April 1982. The UAIR has been measured for every single room.

The condensed results are shown in table 1.

The variations in the UAIR shown in fig. 1, 2, 3, 4, 5 and 6 are typical for the 7 houses, and the influence of the inhabitant's behaviour on the UAIR.

Fig. 1 and 2 show the difference in the same house during the school holidays (fig. 1) and during school days (fig. 2). In this house the UAIR varies from UAIR - $0{,}7h^{-1}$ in fig. 1 and to UAIR $\simeq 0{,}3h^{-1}$ in fig. 2.

Fig. 3 and 4 show the normal pattern of ventilaton due to excessive use of windows.

The Figures (1, 2, 3 and 4) show that the UAIR exceeds the basic AIR with closed doors and windows by a factor of 2 - 5.

Conclusion

1. The behaviour of the inhabitants has, by far, the major influence on the total air infiltation rate in a house with natural ventilation.
2. The influence of the behaviour will give a UAIR of nearly 1 - 5 time the basic AIR with windows and doors closed.
3. In reasonably thight houses with a basic AIR of the magnitude of $0{,}2h^{-1}$ it is necessary to install ventilation openings to ensure a proper and good ventilation.
4. To ensure a reasonable AIR it is necessary to install ventilation openings of the magnitude of 100 - 200 cm^2 in each room, totalling 500 - 1000 cm^2 for the houses measured.
5. The ventilation openings must not be placed in the same height and in the neutral pressure zone in the wall (due to thermal convections).
6. It is reasonable to assume that properly designed and placed ventilation openings will be used constantly and in agreement with needs.
   The need for using the windows as ventilation openings will then be reduced, thereby reducing the UAIR.

Table 1

House No.	1	2	3	4	5	6	7	3a	Average House 1-7
Terrace house, 2 storeys	x	x			x	x	x		
Atrium house, 1 storey			x	x				x	
Area m^2	75	51	113	113	130	75	51	113	72
Volume m^3	184	130	261	261	299	184	130	261	190
Number of rooms	3	1	5	5	4	3	1	5	3
Number of inhabi-tants	3	2	4	4	4	1	2	4	3
Month (of measuring)	10	12	12	1	2	2-4	2	4	10-4
Outdoor temperature ^oC	6	1	-5	-8	-1	2	0	6	-1
Outdoor humidity % RH	90	87	94	90	91	82	89	73	89
Wind velocity m/sec.	4	6	4	4	5	5	6	4	5
Room temperature ^oC		21	22	22	22	21	23	21	22
Indoor humidity % RH		47	42	47	42	44	45	54	45
Air change n/h	0,52	0,43	0,67	0,43	0,38	0,46	0,54	0,8	0,49
Air change m^3/h	95	56	175	112	114	84	70	208	101
Basic air change n/h	0,12	0,22	0,22	0,27	0,22	0,20	0,25	0,10	0,21
Basic air change m^3/h	22	29	57	70	66	37	33	26	45
m^3 fresh air per person/h	32	28	44	28	29	84	35	52	40

Galgebakken house 1: Oktober 1981

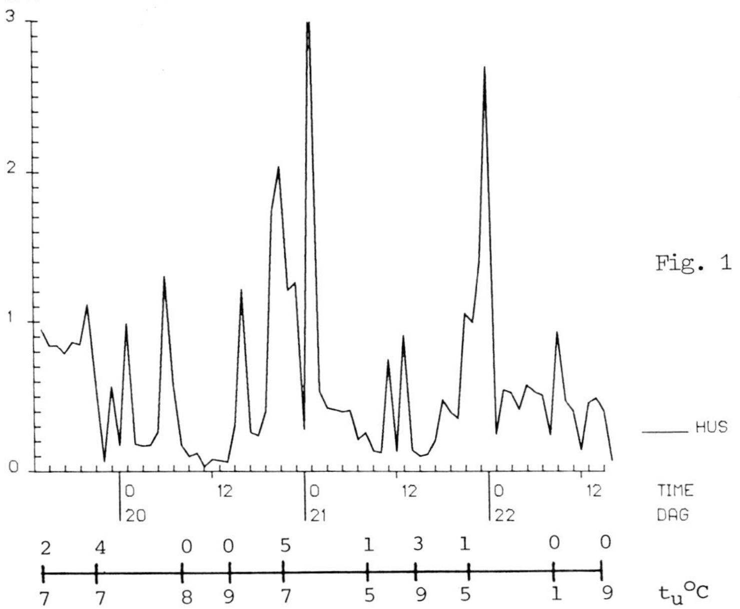

Fig. 1

Note! The autumn holidays of the school are from October 19 - 22.

Galgebakken house 1: Oktober 1981

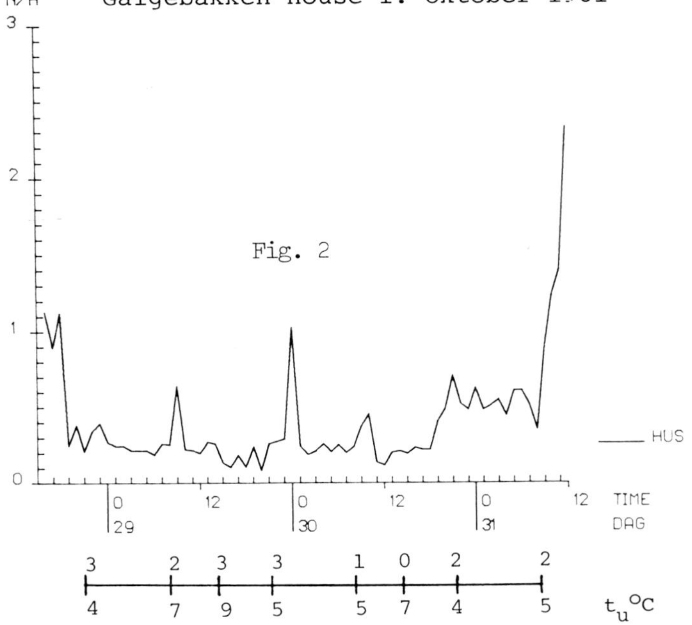

Fig. 2

Galgebakken house 2: November 1981

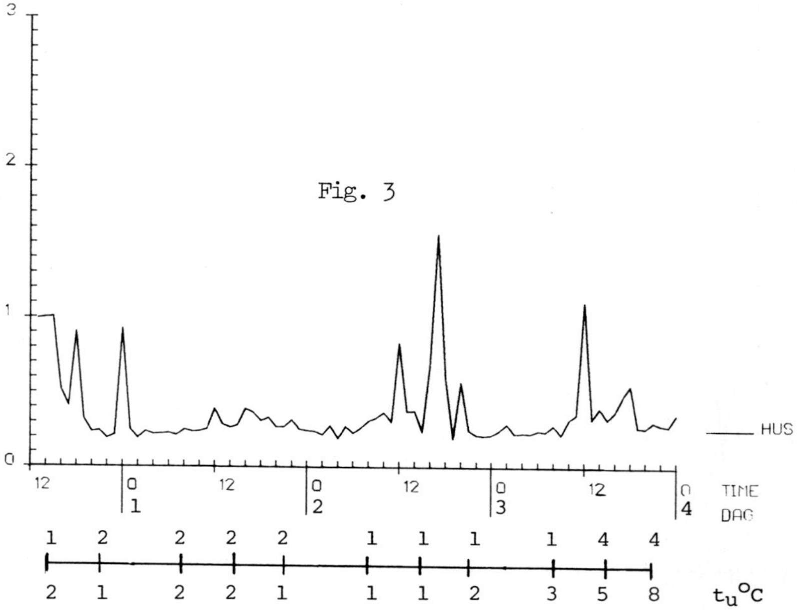

Fig. 3

Galgebakken house 2: November 1981

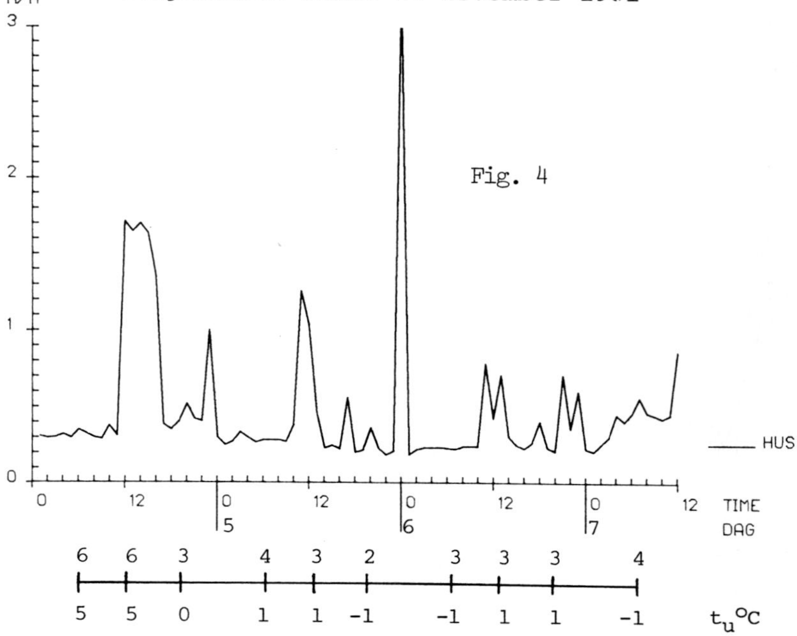

Fig. 4

Galgebakken house 5: Februar 1982

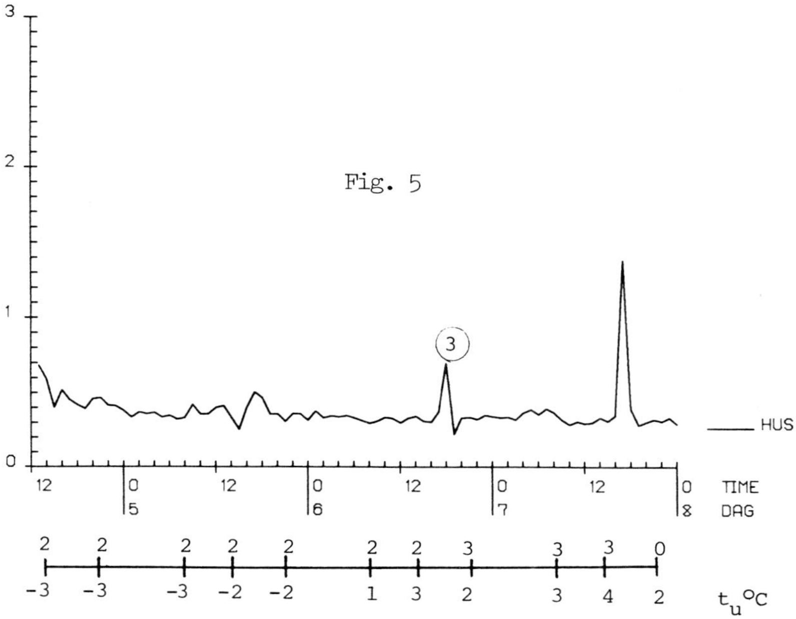

Fig. 5

(3) Ventilation of room, approx. 15 min.

Galgebakken house 5: Februar 1982

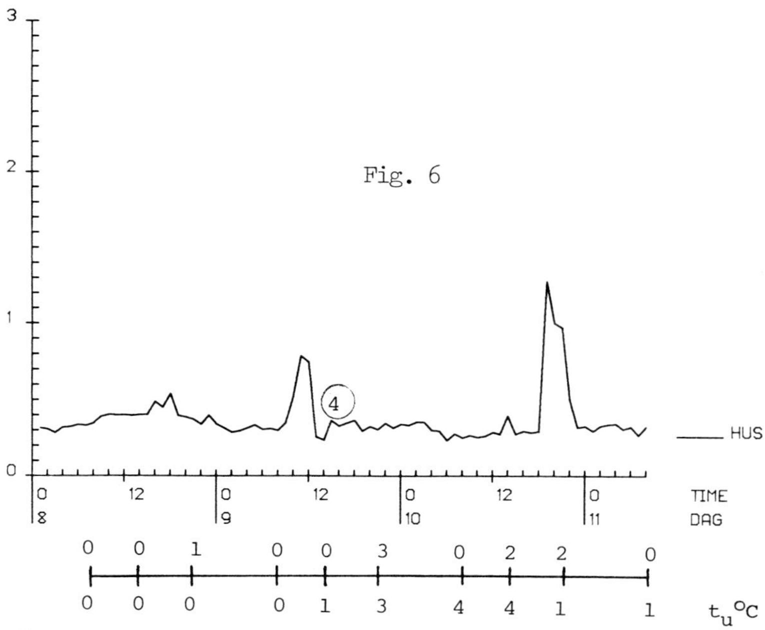

Fig. 6

(4) From 12 - 2 p.m. metal ventilation openings closed, at 2 p.m. ventilation openings open.

CONTROL OF VENTILATION RATE IN BUILDING USING H_2O or CO_2 CONTENT

M. Barthez and O. Soupault

SOCIETE BERTIN & Cie
B.P. 3
78373 PLAISIR Cedex, FRANCE

Contracts n° EEA-S-053 F (D) and EEA-S-061 F (RS)

Summary

A new energy conservation technique based on the control of the fresh air intake through measurements of CO_2 in extracted air was examined for different kinds of buildings. A theoretical study shows that the modulation of the ventilation rate which can be obtained by the control of a two-speed fan or by the modulation of the chimney cross section enables the ventilation rate to be independent of the external conditions (wind, temperature) and to produce annual energy savings of the order of 1500 to 2000 kWhr for a 100 m^2 house. Against this saving, the installation of a mass produced detector should be twice lower than the installation of a double flow system.

The aim of the experimental study is to measure the change in CO_2 and H_2O content in a dwelling and to verify the possibility and the interest of employing one of these two parameters to control a mechanical or natural ventilation system.
This study concludes that the CO_2 content is a most suitable parameter for the ventilation rate modulation according to the occupancy and that the energy saving estimated in the theoretical study is not open to question, since the theoretical changes in CO_2 content have been precisely confirmed by the experiment.

1. AIM OF THE STUDY

The aim of the study is to assess the possibilities of conserving heat energy in a building through control of the fresh air intake. Fresh air flow rate allowed in will be varied according to the carbon dioxide content of the air inside the building which depends on how people are present. Contemplated buildings are large business premises (show rooms, stores, offices) as well as family residences.

2. STANDARDS APPLYING IN VARIOUS COUNTRIES FOR AIR RENEWAL

Standards are extremely various ; the present situation is evolving as the heat insulation scheme is spreading and induces substantial heat losses due to air renewal.

The following ranked instances summarize the situation [data refer to a 100 m^2 area apartment (250 m^3)] :

a) USA, GERMANY, UNITED KINGDOM

Minimum flow rate is stated in m^3/hr per person depending on the activity and the type of premise. For a 250 m^3 apartment, minimum flow rate is 110 m^3/hr.

b) FRANCE, DENMARK

Minimum flow rate per room (Denmark) or per dewelling (France) : for a 250 m^3 apartment, minimum flow rate 170 m^3/hr for France, to be reduced to 120 m^3/hr in 1982. In Denmark, only 80 m^3/hr are required.

c) THE NETHERLANDS

Minimum and maximum flow rate depending upon the type of room (bedroom kitchen, ...). The standard implies 250 m^3/hr for a dwelling of 250 m^3.

ITALY and SPAIN : no standards applicable.

In JAPAN, the standard refers to a maximum CO_2 content not to be exceeded 100 ppm.

3. ANALYSIS OF REAL SITUATIONS

Experiments in foreign countries are scarce.

In JAPAN, Nakahara and his team have experimented such a system in offices. In the USA, experiments have been carried out in a school by a Berkeley team. Data relevant to family dwelling have not been published so far.

On the basis of these few bibliographic elements, the possibility to install flow rate controllers governed by the CO_2 content has been examined for existing buildings (CENTRE GEORGES POMPIDOU, hospital, department store). Such analyses have shown that the prime condition for operating a flow control is to have separate fresh air blowers per volume controlled ; when heating is through warm air, air renewal rate can be controlled with a CO_2 sensor but if the volume to be heated can be considered as having a uniform CO_2 content : such a system can, for example, be installed in a cinema with a single hall · but if several halls are connected to a single heating system, the air renewal common to all halls cannot be controlled to assure comfort in all of them.

4. DAILY NUMERICAL SIMULATION OF A DWELLING EQUIPPED WITH A CO_2 CONTENT-CONTROL

To calculate the reduction in heat loss due to air renewal, brought in by the CO_2 content-governed flow control, a daily simulation allowing to check at every moment the CO_2 content has been worked out.

First, an assumption on rooms occupancy over 24 hours was established; from this assumption, various strategies on start up and stopping of blowers (one or two speeds) were run. The result of the simulation is figured out by an average air flow to be compared with the steady flow rate that would result from applying the standard; the comparison allows to assess the heat energy saved.

Examination of this simulation shows that the mean flow over 24 hours can be reduced by 60 % relatively to the French standard, which means an annual savings of 1,500 kWhr for a 100 m^2 area apartment (250 m^3). Comfort is improved with respect to an air exhauster running at a constant rate of 1 vol/hr : when the rooms are over-occupied for several hours, automatic switch on the high speed of the blower, at 2 vol/hr for instance, allows maintaining the comfort conditions by holding the CO_2 content below 1000 ppm and the odors at a low level.

Moreover, the system compensates automatically for poor airtightness of the house : CO_2 sensing accounts for natural aeration by reducing the forced ventilation so that the CO_2 concentration remains at the desired figure.

5. OBJECTIVE OF THE EXPERIMENTAL STUDY

The objective of the experimental study is to measure the change in the concentration of CO_2 and water vapor in a dwelling.

The measurements have been interpreted as a function of the behaviour of the occupants, to verify the possibility and the merits of employing one of these two concentrations to control completely or partially a mechanical or natural ventilation system.

In fact, the previous theoretical study shows that the modulation of the ventilation flow rate with time would make it possible to obtain a mean flow rate over the heating period of 0.6 vol/hr, with flow rates greater than two vol/hr when the occupation detector requires it, and lower than 0.2 vol/hr (leakage from the house) when the dwelling is not occupied. This modulation, which is obtained by the control of a two-speed fan in the case of a MDV (Mechanical Draught Ventilation), and through the modulation of the chimney cross-section in the case of natural draught, enables the ventilation flow rate to be made independent of the external conditions (wind, temperature) and to produce annual energy savings of the order of 1,500 to 2,000 kWhr, for a 100 m^2 house. Against this savings, the installation cost of a mass-produced detector (50,000/year) should be twice lower than the installation of a double flow MDV.

6. MEASUREMENTS

The tests were carried out in a private house conforming to the electricity standards for an area of 80 m^2, situated at ST MARTIN-en-NIGELLE -28130, France, inhabited by an associate of BERTIN & Cie. This house is fitted with airtight door frames and MDV.

The installation and the recording of the measurements were carried out with the aid of the following equipments :

- a SCHLUMBERGER model ANIR analyser connected to air samplers in the bedroom, the sitting room and the ventilation shaft
- psychrometers installed in the bedroom and in the ventilation shaft
- thermocouples (internal and external temperatures)
- a recorder for the acquisition of measurements

7. ASSESSMENT OF THE VENTILATION

To determine the air change, the ventilation of the house has been assessed by using tracer gas measurements obtained with the assistance of a team of specialists from EDF. These measurements enabled the following two ventilation flow rates to be determined :

- low ventilation flow rate : 0.6 vol/hr = 110 m^3/hr
- high ventilation flow rate : 0.76 vol/hr = 140 m^3/hr

In parallel, we have obtained the following flow rate measurements using a hot wire anemometer placed in the ventilation shaft downstream of the fan :

- low ventilation flow rate : 108 m^3/hr
- high ventilation flow rate : 157 m^3/hr

These two estimates agree to within 10 %.

8. CHANGE IN THE CONCENTRATIONS OF CO_2 AND WATER VAPOUR

During the two months of the experiment (April, May 1983), we investigated the concentrations of CO_2 and water vapour in different rooms and in the ventilation shaft.

These measurements have been interpreted taking into account the behaviour of the occupants.

The outcome of this test campaign is that the variations in the concentration of CO_2 are significant and can be related to the ventilation rate in the house and to its frequency of occupation.

Thus, for an extraction flow rate of 0.6 vol/hr, the measurements reveal the following points of importance :

- when the house is not occupied, a CO_2 concentration below 500 ppm is observed (mean concentration outside the house between 330 and 400 ppm)
- the presence of one person produces a CO_2 concentration of 500 to 600 ppm in the ventilation shaft
- the presence of two peple produces a CO_2 concentration of 600 to 750 ppm in the ventilation shaft
- the response time of the rise in the CO_2 concentration in the ventilation shaft is less than an hour (an increase of 100 to 150 ppm in the CO_2 concentration is observed over the hour which follows the arrival of two people in the house)
- there exists a diffusion phenomenon of the CO_2 from the rooms where it is produced to the ventilation shaft (the concentration of CO_2 which is found in the bedroom, during the night, is 100 to 300 ppm higher than that detected in the ventilation shaft)
- the theoretical and experimental changes in the concentration of CO_2 are quite similar
- the changes in relative humidity or in the concentration of water vapour are difficult to correlate with the frequency of occupation
- a flow rate modulation resulting in an average renewal of 0.6 vol/hr is sufficient to maintain the humidity in the house in the region of 60 %
- the high ventilation flow rate (0.8 vol/hr) is used only for approximately half an hour a day, at the time when the meals are prepared.

9. CONCLUSION

The present measurement campaign has made it possible to demonstrate :

a) that the CO_2 concentration is a suitable parameter for a modulation of the ventilation flow rate. It is, at the same time :
   - sensitive : variation plateau between 400 and 750 ppm
   - independent of the temperature and external conditions
   - capable of being correlated with the frequency of occupation (an increase in the equilibrium of 100 to 150 ppm per person for an air renewal of 0.6 vol/hr)

b) that the modulation of the ventilation flow rate at an average of 0.5 - 0.6 vol/hr over the day is sufficient to maintain a relative humidity within acceptable limits (60 % at 18-20°C), without appearance of local concentrations limiting human comfort and favouring the long-term deterioration of the structure

c) that the energy saving estimated in the previous study using theoretical simulation is not open to question, since the theoretical changes in

the concentration of CO_2 have been precisely confirmed by the experiment without, at the same time, there being local concentrations of CO_2 which can diminish comfort, and without interfering air changes detectable by by-passing the MDV system
d) that the CO_2 detector which is suitable for the modulation of the ventilation flow rate can be defined by :
- a threshold between 600 and 800 ppm
- an accuracy of ± 100 ppm
- a response time of less than an hour

A further stage will consist in defining, constructing and evaluating a complete loop for total or partial control of the ventilation flow rate of an habitation as a function of the CO_2 concentration. This evaluation will make it possible to identify the quality of the comfort provided by this system, and to estimate the saving obtained on the basis of the quantity of air changed daily.

When the test campaign has been completed, the technical data will be collated to make the decision to launch an industrial development of a low-cost CO_2 detector.

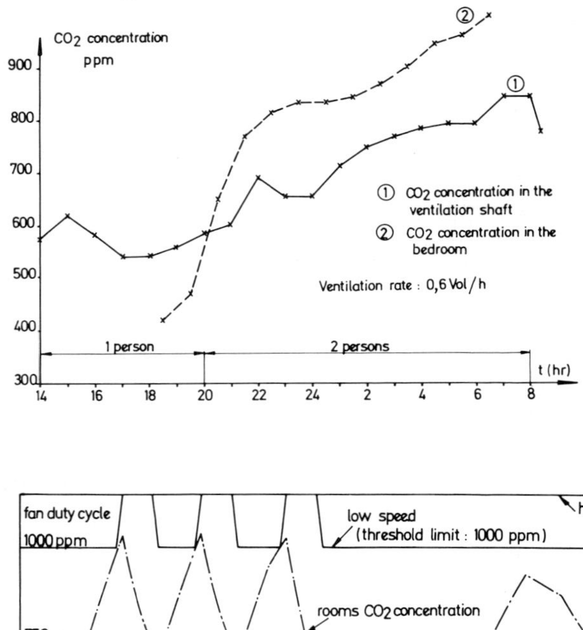

Fig. 1- Variation of the CO_2 concentration as function of time and occupancy (as measured) in an inhabited dwelling with constant ventilation fan speed

Fig. 2- Theoretical operating mode of a CO_2 based ventilation system

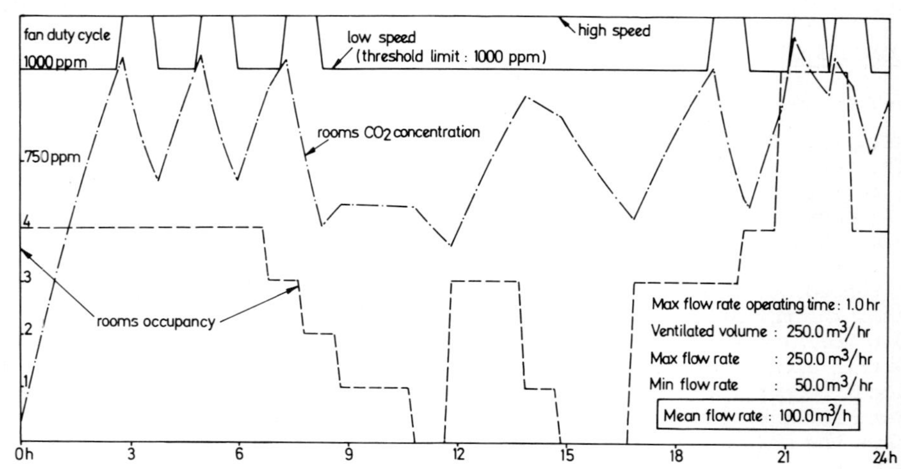

BUILDING ENVELOPE - R & D PROJECTS

Rapporteur: Prof. J. LEBRUN

The object of this report is to give a very short survey of the main results presented by the different research teams and to stress the remaining lacks,discordances and questions still to be answered.

Further information about the research programs can be found in the proceedings of the previous contractors' meetings (Brussels 1981, 1982) and in particular in the excellent synthesis established by P. Zegers (C.E.C.).

The present report is only based on the papers that were made available at the time of the seminar. These papers are hereafter referred to with the same code-number and in the same order as in the program of the seminar.

Under the "Building Envelope" topic of this sessions we are hereafter distinguishing four subtopics :

1. "Global" studies and calculation models ;

2. Insulation (development of products and practical use) ;

3. Improvement of window energy "efficiencies" (coating, movable protections...) ;

4. Ventilation (experimental and theoretical analyses ; effect of the inhabitants, automatic control).

1. "GLOBAL" STUDIES AND CALCULATION METHODS

These two topics are stongly interrelated. Calculation methods are indeed to be used at different stages of any building study :

1) the very first DESIGN makes (or should make) use of some energy calculation ;

2) the experimental assessment should also be prepared with the help of preliminary calculation ;

3) monitoring results can only be rationally interpreted through a detailed simulation of all the system studied.

At that last stage, any significant disagreement between theory and experiment can drive the researcher to several ways :

- some "validation" work (i.e. adjustment of his calculation hypotheses) if the experiment is really accurate enough ;
- a re-design of the experiment on the opposite case ;
- some findings about the imperfections of the building realization which were underestimated at the design level ;
- some change in the design method for further realizations ;
- etc...etc...

In fact these many ways of using (or of "escaping" from) the results of a monitoring still look in themselves like sources of confusion...

Four "global" studies and two energy calculation contributions are reviewed hereafter :

1.1. An experimental study on residential heating systems in two almost identical experimental houses (paper A.2.1.). Two heating systems (electric convectors and floor-fuel oil) are here tested with different control modes. In each case, the author relates the energy consumption to an estimate of the "net space heating demand" (NSHD) through a linear regression law. In principle, this procedure allows a direct comparison among the different cases considered in spite of the changing weather conditions. Due to the influences of varying air infiltrations (not included in the NSHD estimate) and possibly also of transient heat flows, the results have to be handled with great care : they have to be considered as purely "provocative". No statistic test seems have been made to let at last the "confidence limits" of this procedure appear..

1.2. A very simple method for estimating the NSHD : the "equivalent" degree-days (paper A.2.2.).

The "equivalent" degree-days are based on the following "inside" and "outside" references :

- the temperature of "NO HEATING" (t_{NH}) which characterizes all the use of the building (comfort requirements, timing of occupancy and corresponding "free" heat) ;
- the temperature "WITHOUT HEATING" (t_{WH}) which caracterizes the weather as it is "integrated" (radiation and convection) by the building at any time of the year.

The method is particularly efficient to compare different energy saving measures applied or to be applied to a same building.

Having been involved together for ten years in the developments of this approach the author of the communication and the author of the present report could ask themselves why the equivalent degree-days have not been currently used yet in the European Community!

1.3. Theoretical exercises of building design to make some optimal shapes appear in relation with the climate (Paper A.2.3.). It seems that the objective function only contains the NSHD (and not the global energy consumption); it is computed through the NBSLD program. Here no use has been made yet of the simplified method described in paper A.2.2., or of the "reference" proposed in paper A.2.6.

The most crucial questions raised by this paper are:

what about the other objectives and constraints to be considered in the optimization ?

What is the sensitivity of the objective function (if any) to any variation of the design parameters around the optimum proposed ?

How far should we go in the so-called solar passive solutions ? etc... etc...

1.4. A case-study associating design and practice in the field of low cost houses (Paper A.2.4.).

The main purpose of this work is to demonstrate the possibility for an architectural project to be cheap and energy efficient. For evaluating the significance of the optimum reached in the project, here also we would like to have a look at the objective function and at some sensitivity analysis : should we go further in insulation thickness, in tightness and/or in the use of solar gains?...

Here as in many other fields, the demonstration of "feasibility" is much more comfortable at a pure theoretical level than on site. The design energy efficiency can only be achieved through unusual cares during the building of the houses and the final experimental validation still looks as the most hazardous part of the work.

Here, as in other similar studies, (papers A.2.1. and A.2.5.) we strongly feel the lack of good experiment design methods...

1.5. Another experimental study on site is paying special attention to the windbreaks usefulness (Paper A.2.5.).

Some crude regression techniques are used to interprete the experimental results : the inside temperature of an unheated house is hour after hour linearly correlated to its value one hour before and to several meteorological variables.

Such a simple superposition procedure to analyze the effect of strongly interacting variables could be used as a first "cleaning" of the measurements. But it has, of course, very little physical meaning. It must also be emphasized that a higher temperature without heating doesn't necessarily mean a lower heating demand. On the other hand the usefulness of a windbreak is perhaps more to reduce the heating peak demand than in its effect on the seasonal average...

1.6. A very complete reference model for building thermal loads and energy calculation is available (Paper A.2.6.). This model includes a combined representation of the building and its HVAC equipments as an enormous set of differential equations solved by finite differences. This requires an efficient handling of large matrices of structure. As it is presented the model contains almost no simplification : all our present understanding of the phenomenon is included in it.

Here two urgent questions arise :

1) Is it practically possible to extend such a model up to any size of building-and-HVAC system without finding somewhere a "bottle neck" in the computerization?

2) Why has not this reference model been used yet to theoretically support the other CEC researchers ?

Much less convincing is the use of such a "rich" theoretical tool to generate the so-called "climate severity index" This one could be compared to the f-charts generated from solar system simulation, but with a lower physical significance...

2. INSULATION

When discussing building insulation, the first two questions to be answered are : when? and where?

-When? During or after the construction (retrofit): the technical and economical circumstances are obviously much more favorable in the first case than in the second one.

-Where ? In the roof, in the walls or in the floor with at least three possible locations of the insulating layers : inside, outside, or somewhere between the structure components.

The quality analysis of any insulating material should take into account very different aspects : thermal and acoustical properties, fire resistance (and smoke generation!), possible air contamination (gaz, particles), handling and mechanizing easiness etc...

Extreme attention should be paid to the ageing process.

The global impact of any solution would interest us in priority : for example, we have to know if any thermal performance improvement is not nullified by a possible misuse as well as if any further reduction of material cost is not made negligible by the "handling" costs. A better understanding of these various interactions would allow us to orient better certain research efforts...

The first two papers presented hereafter specially concern the development of insulation products ; the third one gives an example of study of the real use of insulating materials :

2.1. Development of bituminous heat insulating materials (Paper A.2.8).

This is an example of research towards cheaper materials. The product proposed has almost the same performances as expanded polystyrene. It could even be slightly better after ageing. But its economical interest from the user's point of view doesn't appear here yet. Moreover the technical comparisons are made with expanded polystyrene only, which is, for some aspects (as for example fire resistance) a relatively poor reference. Neither technical nor economical comparison are possible with the products described in the following papers (A.2.10, A.2.11)...

2.2. Improvement of the thermal resistance of Polyurethane frames (Paper A.2.10).

The objective of this research is to find the way of keeping the thermal conductivity around $0.02\ Wm^{-2}K^{-1}$ after ageing of the foam. The work looks successful, but its economical impact for the user doesn't appear better here than in the previous work (A.2.8). Not even the same technical criteria are considered : for example, the fire resistance should be specified.

The real cost of insulating thickness sould be investigated : what is the cost of each cm to be made available in the walls to receive the insulation? What is the impact of all the construction defects ("thermal bridges")? How are these defects increasing in relative importance when using insulating materials of higher performances?...

2.3. Real use of insulating materials in retrofits (Paper A.2.11).

The first part of this study deals with the early stage of a monitoring whose results are not yet available. The other two parts concern the use of mineral wool in roof and facades and the use of urea formaldehyde foam in walls respectively. The only two practical conclusions proposed by the authors are the following :

1) an excellent behaviour of the mineral wool even directly exposed to the climate aggression (i.e. on the roofs);

2) a much less satisfactory result of foam injections in walls.

The final thermal resistance seems highly uncertain and hardly controlable. From other part, inside air quality rules are hardly respected due to the formaldehyde contamination.

But here again the reader would feel more comfortable if he could make some good "cross-comparisons" among the different energy saving measures : why are all the authors using different criteria and different economical approachs?.

3. WINDOWS

The fashion of almost "transparent" envelopes is probably one of the strongest traumatism developed in building design in the last thirty years!

After having dissimulated the thermal impacts of this fashion for thirty years, more and more building designers are now trying to "sell" us so called "energy efficient" windows...

Having spent some theoretical and experimental research efforts in the field, the author of this report feels that window use should be decided through a careful combination of non-thermal requirements (visual comfort : minimum and maximum of natural lighting, minimum of perception of the outside environment, noise transmission, air quality...) with thermal constraints (heating energy demand and peak power, summertime temperatures and/or cooling demands etc...)

Most of the present research works intend to improve its "selectivity, i.e. its ability to only transmit the climate components which are "welcomed" in the building : some lighting, some fresh air, some heat in wintertime etc... But this developement has also to be associated to a higher "controlability" of the whole system (building and experiments) due to the highly varying weather conditions :

variable natural lighting and corresponding thermal gains, variable air infiltration rate, etc...

For example, the heating cost of a window should not be estimated on the only base of a reasonal NSHD integration: the oversizing of the heating system to face to cold and unsunny periods and its controlability in mid-season are also to be considered...

The first two following papers deal with window coatings (improvement of their radiation selectivity). The other papers present researches of movable outside and inside protections which affect convection and radiation heat transmission flows.

These different approaches are not easier compared with each other (and a fortiori with other energy saving measures) than the previous studies on insulating materials: not only the criteria, the energy calculations (when any) and the economical analyses are still too different ; but even the inside and outside climate conditions are not clearly established : air or globe temperature somewhere in the rooms, solar and sky radiation hypotheses, wind in the vicinity of the window etc...

3.1. Production of selective coating on plastic sheets (Paper A.2.13.)

Selective plastic sheet could be used among others in windows retrofits (addition of a second glazing). As for the product presented in the next paper (A.2.14.), a high radiation selectivity is reached : sharp drop of the transparance of the film when passing from visible to infra-red wavelength. Most of the author's attention is devoted to the production problems. An insight is also given to the economic aspects (crude pay-back time estimates). The thermal comfort allusion made in the paper constitutes another good example of wrong and confusing argumentation...

3.2. Direct coating of glasses (Paper A.2.14.)

Metallic oxydes are here directly deposited on the glass surface. Once again, very promising results are obtained. In their economic analysis the authors make appear a very unstable condition : the Indium international market. The arbitrary reference conditions and the (too simple) energy calculation are still confusing for us...

3.3. Use of outside movable protections (Paper A.2.15.)

The use of efficient rolling shutters is here submitted to a slightly detailed economic assessment. The main conclusion is that these shutters are only cost-effective when associated to single glazing. This makes the reader even more impatient of being able to make some cross-comparison between this technique and other energy saving measures as, first of all, the addition of the second (selection or not) glas pane!

The final accuracies reached in the experiments and in the energy estimates (for example with the assumption of a constant boiler efficiency) also remain questionnable...

3.4. Use of inside movable protections (Paper A.2.17.)

Highly efficient blinds constitute an alternative to the preceeding window improvements. The economical analysis drives us to the same conclusions as before : the technique is cost-effective only when applied to a single glazing. But here again the reader will feel frustrated by the lack of good reference conditions to interprete the experiments as well as to assess the energy balances.

4. VENTILATION

At least three different aspects should be taken into account when analyzing the ventilation of a building : Health, Comfort and Energy. This analysis is made difficult, not only by the contradiction among the preceeding aspects, but also by the enormous uncertainties which subsist in the building aeraulics : pressure distributions, cracks, windows and doors openings, internal air mixings, etc...

In most large buildings the single-zone analysis has little physical meaning : air contamination is varying inside and among the building rooms. In that case, there is also little interest in defining only one global air flow rate (even by mixing artificially the air during the tracor gas measurements) : we have to be able to specifically predict the contamination level in the building occupancy zones.

Contamination modelling in large enclosures is very similar to what has been done for a long time in chemical reactors : the enclosure can be represented by a set of "perfect mixing pots" interconnected by air flows. Mathematically that gives in transient conditions a set of first order differential equations whose coefficients are the air capacities of the different volume and the interconnection air flow rates. It is the purpose of systematic "perturbations" by tracor gas to reveal these interconnections. Such approaches are treated in the following two papers (A.2.18., A.2.19.) with one and several tracor gas respectively.

The ventilation airflow pattern inside a building and the corresponding interconnections are not only varying continuously with the weather ; they are also submitted to the "adjustments" made by the occupants (doors and windows opening). Some indications on this effect are given in paper A.2.20.).

Better understanding of inside air contamination phenomena seems to lead to the development of rational control system using for example CO_2 detectors : this is the subject of the last paper (A.2.21)...

4.1. Multi-zone analysis with only one tracor gaz (Paper A.2.18.).

A multi-zone model can be identified by least square adjustment of its "answer" on decontamination recording in different point of the enclosure. The examples of results given in the paper demonstrate at least the interest of pushing further such an approach.

But how far (into how many zones) we have to go, to get enough accuracy is still an open question.

4.2. Multi-zone analysis with several tracor gas (Paper A.2.19).

The only essential difference between this work and the preceeding one consists in the use of several tracors at the same time which gives of course more instantaneous information. The main advantage of this is a possible identification of the zones "interconnections" on a shorter period of time :
theoretically, indeed, to give a comparable information the single-tracor gas method should require as much consecutive "perturbations" as there are tracors simultaneously injected in the multi-tracor gas method.

A rational choice of the number of tracors involved in the measurements should be related to the importance and the quickness of ventilation variation in time.

Of course, the "zoning" itself is always hypothetical and has to be assessed (or reconsidered) on the basis of the measurements...

4.3. Users' influence (Paper A.2.20.).

This study on site confirms the existence of some air infiltration "control" by the building inhabitants. The efficiency of this process is still nothing but a guess. Air quality measurements should be included in such a study. In relation to the next paper (A.2.21.) , we would like to see here the CO_2 concentration curves in the different zones of the buildings considered...

4.4. Ventilation control (Paper A.2.21.).

The authors of this work propose to use the CO_2 emitted by the occupants as a tracor for controling the ventilation air flow rate. Preliminary calculations and experiments are encouraging : important savings could be obtained by this procedure. Other studies, not mentionned here, would show that the CO_2 concentration in an enclosure is well correlated with the olfactory discomfort of the occupants.

Possible contamination heterogeneity has to be carefully considered to localize the sensor(s) in the building. This point should require more experimental work. The economic analysis is still to be made as well...

5. CONCLUSIONS

As conclusions, here are the main lacks and risks of misunderstanding we have identified :

5.1. Design Methods

The "optimization" examples could still be misunderstood by not making clearly appear their objective function and constraints. Even when the (very partial) "optimization" only consists in minimizing the building heating energy demand, this objective tends to be confounded with a minimization of real energy consumption or even energy costs. These last terms should be differentiated from the average net space heating demand by including effects of peaks, load profiles, different energy prices etc...

5.2. Energy calculations :

Too many different energy calculations methods coexist in the C.E.C. research team. This coexistence should only be preserved if all the methods were complementary and comparable : it should always be possible to "generate" or "validate" the simplest by the most complex methods. An extended "cross-comparison" should be encouraged : the energy calculation specialists could offer their help to the designers and to the experimenters...

5.3. Design of the experiments

The difficulty of experimental studies "on site" has to be emphasized. Too many experiments (and what to say about the so-called "demonstration" projects) on site are without any use by lack of design : as (and even more) in the lab, no measurement should start before the experimenter knows what he can expect from them, i.e. what their significance could be.

Error estimates should be made before and after the experiments. Statistics should be used much more extensively to identify the significance and the confidence limits of each results...

5.4. Reference conditions

Cross comparisons among theoretical and experimental studies are, at present, discouraged by the lack of clear reference conditions. Not only the outside but also the inside climate descriptions should be standarized :

- Peak and seasonal average of outside temperature, solar radiation, sky radiation, meteorological wind, "site" effects...

- Inside comfort requirements or achievements in term of globe temperature and/or other variables etc...

5.5. Economics

Even if a reader is wise enough to achieve some energy comparison among different research outputs, he will surely be lost in the economical interpretation. The only common point in the economic analysis of the research teams is that their project is cost effective! But is it really impossible to use some common tool of economical estimate ? Is not such a tool already available in the C.E.C. ?...

5.6. Sharing the efforts ...

Methods for system design, energy calculations, experimental approaches, economical analyses should not be periodically reinvented by the different C.E.C. research teams. Better profit should be taken from the availability of specialists in the C.E.C. by requiring from them reference tools.

Why not do some more common research work in the future... ?

SESSION II - BUILDING ENVELOPPE

B - DEMONSTRATION PROJECTS

Chairman: Prof. C. BOFFA
Rapporteur: Mr. J.R. BRITTEN

STANDARD BUILDING TECHNIQUES DESIGNED TO SAVE ENERGY IN IRELAND

S.T. McGRATH
Irish Electricity Supply Board

Summary

The Irish Electricity Supply Board undertook a project to design, construct and monitor six energy efficient single family houses. The principal objective was to collect and disseminate information on the costs and benefits of energy efficient measures in house building. The six houses while appearing conventional have three different levels of thermal insulation. One is built to the proposed national standard. Three are built to a good medium term standard and the last two are built to a standard that may only be economic in the long term. Over two hundred different parameters are being monitored including full weather details. The monitoring programme has just commenced but it is intended that the results acquired during an extended period of monitoring will provide useful information on the performance of energy efficient housing and their related systems in Ireland and other north European maritime climatic conditions.

1. INTRODUCTION

Following the oil crisis of 1973, great interest developed in the construction of energy efficient houses, in Ireland as in most of the countries of the western world. There were the usual spate of computer studies, masters' theses and other learned documents produced. This led in time to actual construction projects - not very many in our case but all had one thing in common - they did not look like "ordinary" houses. The other feature of Irish energy efficient houses was that there was little or no attempt made at monitoring - computer studies were again employed to determine what performance "must have been". There was a shortage of factual information about what it cost in practice to introduce energy saving techniques into building houses and what the actual benefits were from such actions.

2. ELECTRICITY SUPPLY BOARD PROJECT

The Electricity Supply Board (ESB) proposed a project to design, construct and monitor six energy efficient single family houses. It is partly funded by the Commission of the European Communities Directorate General for Energy. Its principle objective is to collect and disseminate information on the costs and benefits of energy efficient measures in housing.

Sites were purchased on an estate for which planning permission had already been obtained. This put a constraint of designing energy efficient houses to a given road plan, housing density and basic housing form. It was possible of course to obtain revised planning permission for a lot of detail, but it forced a discipline of designing a house that the general public would see as "ordinary".

A second constraint was that conventional materials and construction techniques only could be used. By involving the builder in the design team at an early stage, this was ensured.

As such capital cost was not seen as the main constraint. Of course there was a budget within which the project had to operate but it was decided that extra measures did not have to be within say 10% of the house price. On any aspect of design for energy efficiency the criteria used were:

(a) Can its cost be separately identified?
(b) Will it be possible to monitor its benefit?
(c) Can it be done without overspending the budget?

If the answer to all three questions was positive then it was included.

3. DESIGN TEAM

The team set up to implement the design process included engineering personnel in the ESB, the project architect, and a builder engaged in private sector house construction. A quantity surveyor also participated to provide cost estimates of various proposals. He will also ultimately ascertain the cost of the various energy-related design decisions which when combined with monitored results will provide the basis for cost benefit analysis.

4. PROJECT AIMS

Within these broad parameters it is intended that the project will yield information on the following:

- The practicability of building highly insulated houses using available materials and techniques.
- The contribution of improved building fabric design and conservatories to savings in space heating costs.
- The energy profiles of space heating and domestic hot water services in energy efficient houses.
- The performance capabilities of a variety of space and water heating systems for low energy housing applications.
- The appropriate ventilation requirements and systems for well sealed low heat loss buildings.

5. SITE

The houses were constructed on the outskirts of the small town of Kilcock in County Kildare, some 38km west of Dublin. The site is less than a kilometre from the centre of Kilcock. It forms part of a housing estate which is in the early stages of development. The type of house already built (two-storey, pitched roofed with 'neo-Georgian' facade) could be described as fairly typical of many contemporary private-sector housing developments in Ireland, and the estate layout is also representative of this type of housing. The site is comparatively flat and without trees or other features. The subsoil is of a firm clay, well drained and with good bearing capacity.

Climatically Ireland enjoys temperate conditions. There is a relatively narrow range between summer and winter values of the main meteorological parameters, largely as a result of the moderating influence of the Atlantic Ocean and the Gulf Stream. Figure 1 shows the mean monthly degree-day distribution (1964-1977) to a base of 15.5 degrees C for the nearest available meteorological station, the mean annual total being 2374.

6. BUILDING FORM

Irish planning legislation regulates physical development through local planning authorities. The estate developer from whom the ESB purchased the site had previously obtained planning permission for six semi-detached houses. Because of the decision to relate the houses to contemporary speculatively-built housing, the starting point in the design process was the existing permitted designs and the earlier developments on neighbouring sites. The aim was to provide similar accommodation including three bedrooms, kitchen/dining and living spaces within the same floor area in a semi-detached house.

The general street arrangement was set by the estate's road layout and minimum set-back distances. The earlier houses bear an angled relationship with the road, and this relationship has been maintained. Individual houses are further stepped to reduce the scale of the facade and create opportunities for southern orientation. The houses have been completely replanned to achieve a more efficient use of space and make better rooms, to provide for increased services and to create lobbies at both entry points. The fenestration has been redesigned to meet architectural and thermal criteria and small conservatories are provided in three houses. Garages have been added to the houses in such a way that ground floor wall heat losses are minimised. The houses are illustrated in Figures 2 and 3.

7. MATERIALS

The construction systems adopted utilise indigenous materials and conventional building techniques as far as possible to maximise relevance to contemporary Irish private housing. In the context of the building fabric, improved energy efficiency is sought by means of better design, integration of thermal insulation and draughtstripping and exploitation of passive solar gains.

The external walls are concrete blocks in cavity construction and rendered externally. The roofs are pitched timber trussed rafters finished with concrete tiles. There are concrete ground floors and timber first floors and timber framed windows. Three different standards of thermal performance are achieved by varying levels of thermal insulation and the introduction of controlled ventilation and heat recovery. One house is designed as a 'Control' house to proposed national standards, three are intended to meet projected medium-term standards, and two are to meet levels of conservation anticipated in the longer term. Table I sets out the outline constructional specification for the three house types.

	House A	Houses B, C, D	Houses E, F
External Walls	300mm Cavity Wall; 100mm Cavity insulated with 40mm expanded polystyrene	As for House A with cavity completely insulated	As for Houses, B, C and D
Party Wall	215mm solid block, plastered both sides	As for House A	As for House A
Floor	Solid concrete, with 25mm polystyrene under final screed + edge insulation	As for House A with additional 25mm of insulation	As for House A with additional 25mm of insulation
Roof	Tiles on battens; Roofing felt and rafters; 10mm plasterboard ceiling on timber joists with vapour check; 100mm mineral fibre between joists	As for House A with additional 50mm insulation in attic	As for House A with additional 100mm
Glazing	Single glazing, hardwood frame	Double glazing with hardwood frame	Triple glazing (or equivalent)

Table I Constructional Specification

8. THERMAL PERFORMANCE

Table II shows the design heat losses for the 6 houses. The heat loss values are based on an average internal temperature of 19.5 degrees C and an external design temperature of -1 degree C. The internal temperature value is a floor area weighted average for the houses based on standard room design temperatures. An average house air-change rate of 1.4 per hour is derived in a similar manner to determine the ventilation losses. In the last two houses where mechanical ventilation with heat recovery is to be used, a ventilation heat loss of 1kW has been assumed.

House	A	B	C	D	E	F
Fabric Loss (kW)	3.1	2.0	1.9	2.0	1.6	1.6
Total Heat Loss (kW)	5.3	4.2	4.1	4.2	2.6	2.6

Table II   Design Heat Losses

9. SERVICES

The houses will be used to evaluate a variety of space and water heating systems in the context of thermally well designed environments. The evaluation will include an analysis of capital and running costs, environmental comfort levels achieved, operating characteristics, and where relevant, demand implications for the Electricity Supply Board. The various systems scheduled for test are outlined below and summarised in Table III.

Space Heating: All houses will have direct electric systems installed for thermal calibration and comparison tests that are taking place during the unoccupied stage of the project.

A water based electrically-heated central storage system is installed in three of the houses - the control house and one of each of the other thermal design types. The system, Centralec, has been designed and patented by the ESB for use in domestic and small commercial premises. It uses mainly off-peak electricity. The water store is accommodated in a services compartment on the first floor of the houses. The Centralec distribution system normally uses steel radiators but in one of the houses an arrangement of polypropylene pipes buried in the floor screed is used.

Two houses have air source heat pumps for space heating. The systems will be tested using a variety of controls and operating regimes. The heat distribution systems are similar to that of Centralec. The ability of the heat pumps to meet domestic hot water requirements is also to be examined. This is particularly relevant in low heat loss houses where heat pump size is not a problem but capital cost may be. In the remaining house a conventional electric storage heating system which incorporates direct electric convectors for bedrooms and top up will be used.

In recent years solid fuel back-boiler type central heating systems have become increasingly popular in Ireland. It is intended that one such enclosed system will be evaluated during the project. All houses will have provision for the later fitting of a solid fuel burning appliance if required by the occupant, including interlinking arrangements with the space heating and domestic hot water systems.

Domestic Hot Water Systems: All houses are fitted with electric immersion elements in well insulated 270 l. storage cylinders. Provision has been made for time switch control to optimize operation for minimum electricity cost. In four of the houses the immersion elements act as back up and auxiliary to other systems. Two of the houses use air source heat pumps to provide the domestic hot water requirements (as already indicated one of these will also be meeting the space heating duty). Two others have solar panels installed - indirect type systems with separate solar storage cylinders.

Ventilation Systems: Construction supervision, improved window joinery, entry point lobbies and appropriate draught sealing ensure that natural ventilation losses are kept to a minimum. Suitable location and sizing of vents and moisture removal at source in bathrooms and kitchens should help to minimise the risk of condensation. In the lowest heat loss houses, mechanical ventilation systems with heat recovery are installed.

	HOUSE					
	A	B	C	D	E	F
SPACE HEATING	CENTRALEC	HEAT PUMP	CENTRALEC	STORAGE & DIRECT ELECTRIC	HEAT PUMP	CENTRALEC
DISTRIBUTION	RADIATOR	RADIATOR	RADIATOR	RADIATOR	UNDERFLOOR	UNDERFLOOR
HOT WATER	IMMERSION	IMMERSION + SOLAR + HEAT PUMP	IMMERSION + HEAT PUMP	IMMERSION	IMMERSION	IMMERSION + SOLAR
VENTILATION	NATURAL KITCHEN EXTRACT	AS HOUSE A	AS HOUSE A	AS HOUSE A	MECHANICAL & HEAT RECOVERY	MECHANICAL & HEAT RECOVERY

Table III Summary of Services Proposals for Energy Efficient Houses

10. MONITORING AND FEEDBACK

One of the major components of the project will be an effective monitoring system to assess the performance of the considerable number of energy related features and systems included in the houses. In excess of 200 parameters will be monitored, ranging from ambient temperatures to heat pump compressor energy consumptions.

Data Acquisition System: An intelligent acquisition system is located on site to collect data from the various transducers. A VDU/keyboard unit is used to set up logging routines and other operational characteristics. Collected data is conditioned in both hardware and software to enable display of data in suitable engineering format. Following appropriate data reduction the required parameter values are stores on a magnetic cartridge for processing on a mainframe computer.

A modem communications link is used to enable remote interrogation of the system. This is particularly important since the houses are located some 38km from the ESB central office and regular inspection of the operating conditions of systems under test in the houses will be necessary.

Feedback: It is intended that the results acquired during an extended period of monitoring will provide useful information on the performance of energy efficiency housing and their related systems in Irish and other north European Maritime climatic conditions.

The project is designed to ensure that these results are disseminated to interested audiences. The ESB's Commercial arm, publicity media and the University School of Architecture are being used to communicate the results of this study to the most relevant sectors - the house buyer, the building trade and future house designers - since it is on their response that the ultimate effectiveness of the project depends.

11. PRESENT STATUS

The present heating season has seen the start of the monitoring programme.

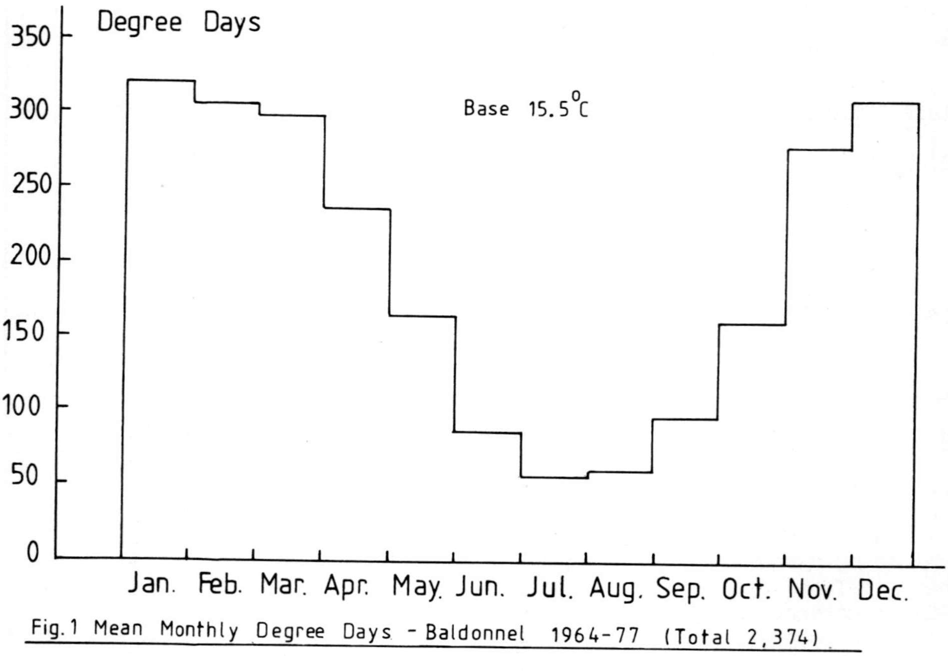

Fig. 1: Mean Monthly Degree Days - Baldonnel 1964-77 (Total 2,374)

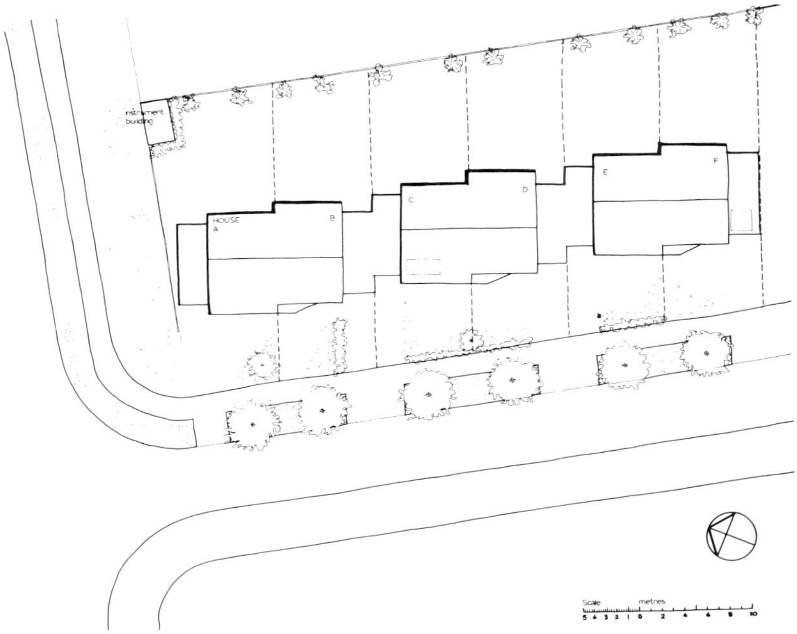

Fig. 2: Energy Efficient Houses - Site Plan

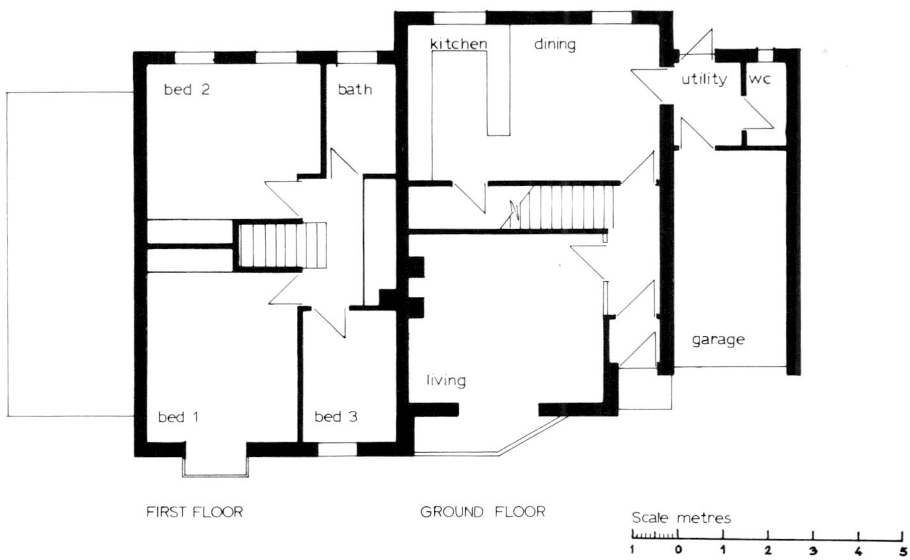

Fig. 3: Energy Efficient Houses - House Plan

LOW ENERGY CONSUMPTION HOUSES MEETING THE COST CRITERIA OF MODERATE RENT HOUSING

SOLPAC project - 12 houses

NOVELERG
12, rue de la Baume, F - 75008 Paris

1. OBJECTIVES

The National Union of H.L.M. (Moderate Rent Housing) Organization Federations (U.N.F.O.H.L.M.) which manages a total of three million units of housing of a social type, and the Compagnie Générale d'Electricité (C.G.E.) intervening through the intermediary of its subsidiary NOVELERG, came together in 1978 to build an experimental progamme of new, energy saving housing.

The level of energy saving chosen as the objective at the time (50%) is close to that which will be made compulsory in France in 1985, within the framework of the H2E 85 programme.

Following a common competition launched by the C.G.E. and the U.N.F.O.H.L.M., a project for building 12 timber framework houses on the TOUR DU PIN site was declared to be the winner. Presented by the SA H.L.M. de VOIRON, a U.N.F.O.H.L.M. organization, and by Mr. GREZES, an architect, this project meets the three following objectives :

a) It harmoniously combines the technique and the architecture without sacrifying the life surroundings to energy savings and without imposing new constraints on the occupiers, moving them away from their traditional way of life. Consequently, a pleasant to live in form of housing has been sought, with plenty of windows despite the high level of insulation adopted, and in which a thermal system with good performances is inserted discretely.
b) It uses in a preferential way the forms of energy available locally and in particular solar energy, to improve the energy balance.
c) It meets the cost criteria for moderate rent housing.

2. TECHNICAL DESCRIPTION OF THE OPERATION

2.1 - Bioclimatic design

The bioclimatic design of the house has led to:

- increased insulation: while preserving easily accomplished construction, the use of a timber framework has enabled installing very good thermal insulation (10 cm of insulant in 2 layers and a blade of ventilated air);
- judiciously inserted inertia in the frame: the light framework in timber has been associated with heavier partitions and floors enabling

smoothing out variations in interior temperature, thereby ensuring optimal use of solar energy and improving summer comfort;

- glazed surfaces directed to the south: by direct recuperation of solar energy, the double glazed windows installed to the south, lead to a positive annual energy balance;
- well located buffer spaces: on the north facade the garage forms a transition space creating a second envelope and ensuring complementary thermal insulation;
- a greenhouse which acts as a transition space and at the same time constitutes a simple means of collecting solar energy. In the SOLPAC system the greenhouse is active, behaving like a better collector than the conventional passive greenhouse.

Alone, the bioclimatic design of the house enables decreasing energy consumptions by around 30%. So as to reach a higher level of savings, this design is completed by active energy recuperation.

2.2 ACTIVE ENERGY RECUPERATION

The active part of the SOLPAC system is based on preheating of the new ventilation air in:

- air solar collectors,
- an active greenhouse,
- a double flux controlled mechanical ventilation (V.M.C.) exchanger.

Air has been chosen instead of water as a thermal vector due to several considerations:

- better thermal efficiency linked with the absence of a threshold in the exchangers, that are necessary in the case of water systems. This enables taking advantage of the low temperature calories contributed by solar energy,
- better reliability of the system both in winter and summer (absence of the risk of freezing and overheating).

2.3 OPERATION DESCRIPTION

In the heating period, depending on the sunshine and the heating requirements, the system's operating phases are the following:

States of the contributions and requirements	Operation phase	Diagrams (1)
Excess solar energy	Stocking	no 1
Solar energy available Need for heating	Direct heating	no 2
Needs for heating Full stock	Taking from stock	no 3

(1) diagrams no 1, 2 and 3 in the annex

The energy thus recuperated is distributed by the renewal air in the main rooms of the house (normal renewal rate).

In certain periods, when the energy available is in excess, it is stocked in HELIOBLOCS to be returned to the house when needs for heating appear, particularly at night. This base heating is completed depending on the needs, by electric convectors installed in each room.

In summer, the air heated by the collectors is used for the production of domestic hot water (diagram no 4 in the annex).

2.4 COMPONENTS DESCRIPTION

The SOLPAC system makes use of only simple and reliable components which are discretely inserted in the house architecture:

- the VITRALERG air collectors (NEUHAUS-NOVELERG patent), forming the cover, have a high temperature vitreous enamel coating granting them long lasting reliability comparable to that of the building;
- the stock, situated at the bottom of the kitchen wall is composed by dry stacking of HELIOBLOC (NOVELERG patent). The HELIOBLOCS can be compared with building blocks; they thus easily constitute self carrying stocking volumes, totally stable throughout time (diagram no 5 in the annex);
- the SOLPAVENT air handler (NOVELERG patent) group together in the same component all the registers: facility of installation and maintenance;
- the SOLPATIC electronic control built with integrated circuits has the operating security specific to these elements. It groups together the control and management of the solar heating system and the complementary heating.

All this equipment has been perfected under NOVELERG's control and tested in the Laboratoires de MARCOUSSIS.

These components are associated with traditional elements: ventilators, registers ... leading to very great operation security, i.e. a low maintenance cost.

3. FORECASTED ENERGY SAVINGS

On the basis of the following characteristics:

- 4 room house with a living surface of .... 94.50 m^2
- orientation .... south
- glazed surface 18.25 m^2 including 16.80 m^2 on the south facade
- partitions .... 70 m^2
- greenhouse .... 23 m^3
- solar collectors .... 15 m^2
- HELIOBLOCK stock .... 5 m^3
- air renewal rate .... 1 Vol/h

The thermal calculations made thanks to the HABITAT computerized simulation model (perfected by the Laboratoires de MARCOUSSIS and approved by the French Public Authorities) lead to the following forecasted results:

- G coefficient of thermal losses .... 0.80 $W/m^3 °C$
  (regulation value in 1978 = 1.25 $W/m^3 °C$)

- B coefficient of thermal needs (after taking into account internal contributions and passive solar contributions) .... 0.53 W/m^3 °C
- heat load .... 15,607 kWh
- internal contributions .... 2,541 kWh
- passive solar contributions .... 2,371 kWh
- active solar contributions .... 1,436 kWh

---

- consumption of complementary heating (electric heating) .... 9,259 kWh

In the region considered, the average consumption of a unit of housing just meeting the regulation in force in 1978 (G = 1.25 W/m^3 °C) is around 18,400 kWh. The consumption reduction expected is therefore around 50% i.e. 9,000 kWh.

This corresponds to an annual saving in FF of around 4,320 F (VAT inc.). To this is to be added the saving made in the production of domestic hot water, estimated as being around 737 kWh i.e. around 30% of the habitual annual consumptions, thanks to the use of solar collectors in summer and the between-season period.

In all, taking into account the new regulations in force in FRANCE since 1983, the VOIRON operation would be worthy of the:

- high energy performance label .... 3 stars
- solar label .... 3 stars

Histogram of consumptions (diagram no 6 in the annex).

4. ECONOMIC INTEREST OF THE OPERATION

Started in 1981, the operation was finished in June 1983, difficulties of a purely local character - resiting of an average voltage electric line - having slightly delayed the site.

The average additional investment for the twelve houses was 60.000 F per unit of housing (V.A.T. included).

This additional investment has been reduced to 40.000 F for a 100 house operation to be launched at a later date, which leads to a return on investments time of around 8 years.

When the houses will be mass built, this extra cost is likely to be reduced to around 25 to 30.000 F.

5. SCIENTIFIC FOLLOW-UP OF THE OPERATION

This experimental character operation is the subject of a detailed scientific follow-up using the TELEDIAG system perfected and run by NOVELERG. Financed in part by the E.E.C., this operation follow-up began in March 1983 and will be spread over two heating seasons. One house has been reserved for analysis of thermal system's operation and it will remain unoccupied. Therefore, the results obtained will be characteristic of the thermal behaviour of the building and the SOLPAC system without undergoing the perturbations necessarily caused by the occupiers.

However, in order to appreciate the effect of the modifications brought along in the internal configuration of the building in the case of occupation, a system has been installed in the house automatically opening and closing shutters and certain interior communication doors.

5.1 - Principle of the TELEDIAG system

TELEDIAG is a computerized electronic system for the collection and processing of the information necessary for use and control of buildings' thermal performances.

TELEDIAG collects data on the site (level 3):

- interior air temperatures
- temperature of the air in the stock
- the sunshine
- consumptions ...

which are transmitted through the telephone network to the Remote Control Follow-up Center where central equipment (level 2 and 1) processes the measurements transmitted making use of preestablished programmes and then edits the results.

5.2 - First results

Curve no 1 - Temperatures and sunshine curve

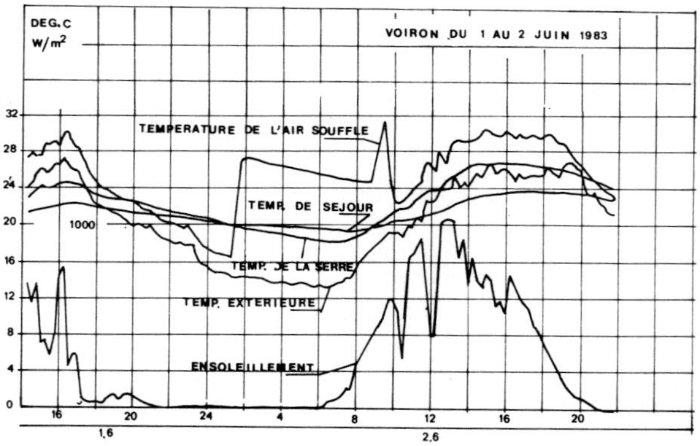

Curve no 2 - Curves of the air temperatures in the stock

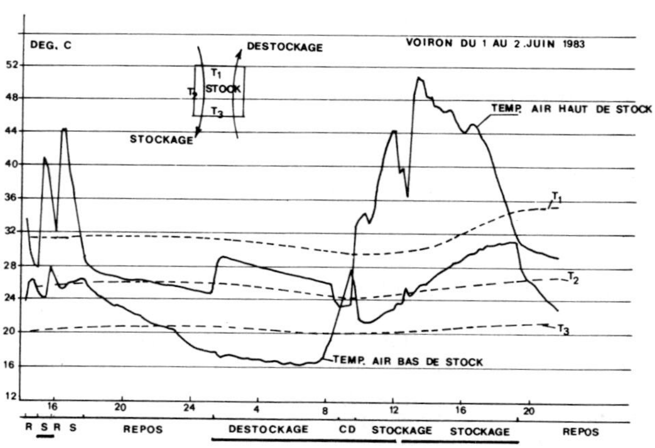

6. CONCLUSION

The VOIRON experimental operation is one of the very first examples of energy saving housing meeting the objectives set by the French Public Authorities for new constructions in 1985.

The level of energy saving expected is around 50% of the level corresponding to the standards that were in force when the operation was launched in 1978.

The operation has also enabled studying the adaptation of energy saving techniques in housing of a social character, by taking into account the specific constraints linked with this housing.

Finally it has been an opportunity for perfecting a new system for detailed remote control follow-up of the energy performances in buildings. For two years it will enable studying in detail the operation of a house.

Operational since May 1983, this system is already showing a thermal behaviour of the buildings in keeping with the forecasts. In addition it has very rapidly enabled detecting certain start-up anomalies in the operation of the active system, and it has provided solutions.

Its use in other demonstration operations is likely to make knowledge of the phenomena involved progress in a significant manner, while at the same time facilitating perfecting installations.

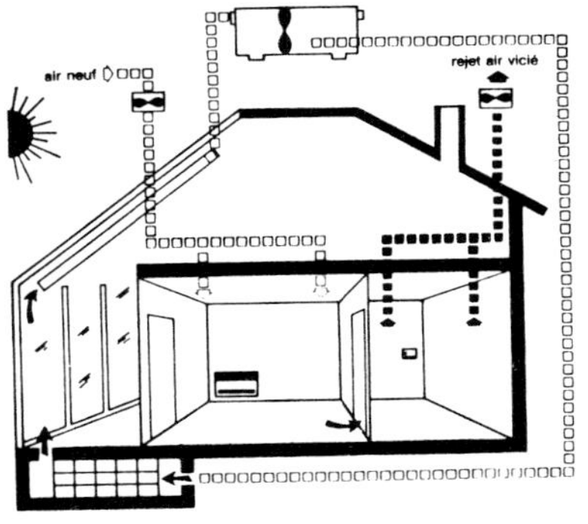

Diagram no 1 - Stocking function

The air in the solar loop (3 vol/h) heated in the greenhouse and the collectors is sent into the stock where it parts with its heat.

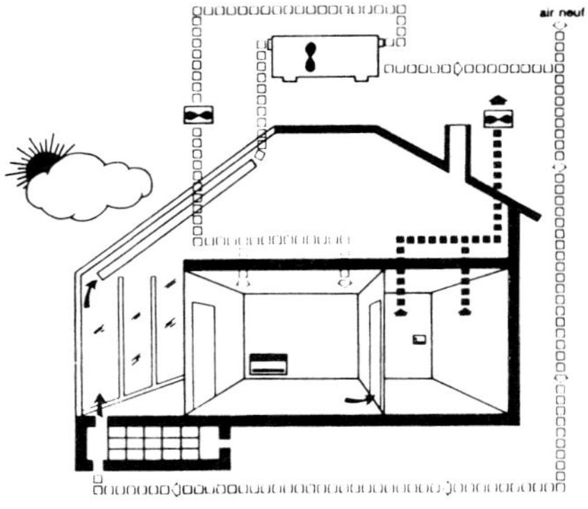

Diagram no 2 - Direct heating function

The air in the solar loop (3 vol/h) at the exit from the collectors is partly introduced into the house (1 vol/h), the 2 vol/h remaining being blown back into the greenhouse.

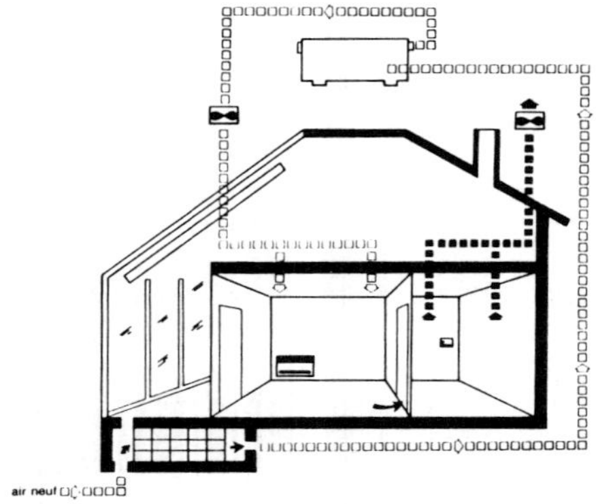

Diagram no 3 - Taking from stock function

The renewal air crosses through the stock in the opposite direction to the stocking direction before being introduced into the house.

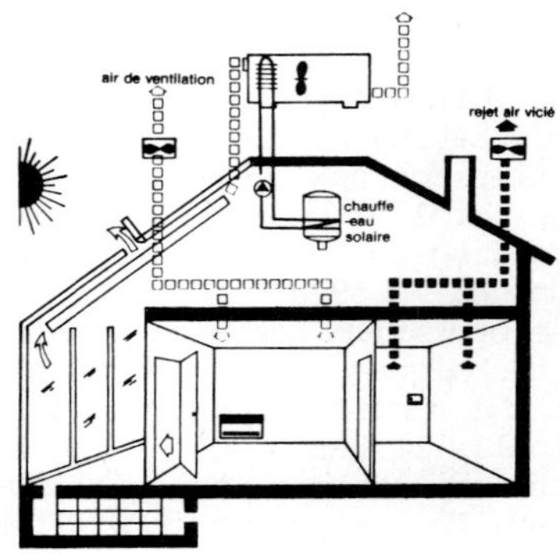

Diagram no 4 - Domestic hot water production function

The air is sucked from the green-house and heats in the collectors. Thanks to an exchanger it exchanges its heat with the primary circuit of the electrostatic tank.

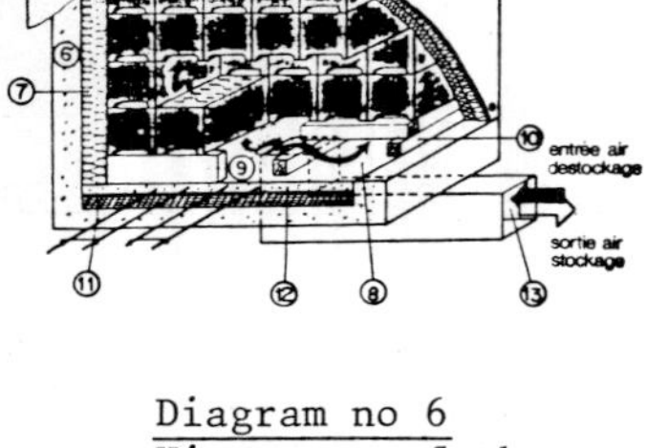

Diagram no 6
Histogram of the consumptions

Diagram no 5 - HELIOBLOCK STOCK

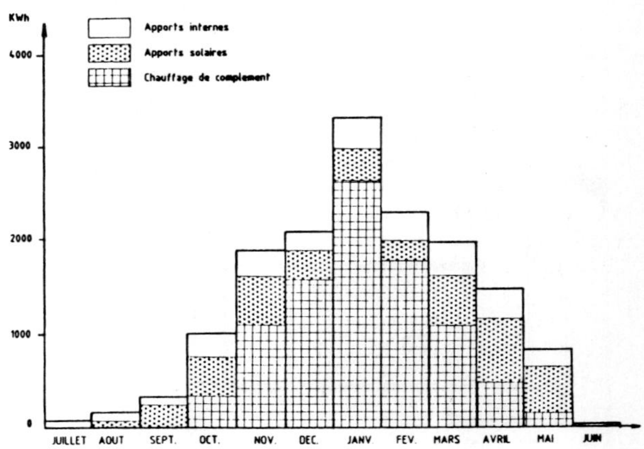

ENERGY CONSUMPTION IN 21 LOW ENERGY HOUSES

L. HALLGREEN
Technological Institute, Denmark

Summary

The 21 low energy houses in Skive-79 were built on the initiative of Skive municipality as a natural continuation of the Skive-77 project which include 9 low energy houses. During a period of nearly two years the Technological Institute has measured the energy consumption of the Skive-79 houses. In this paper the result of measurement is presented which shows that all the houses except one house use less than 60 per cent of the energy consumption stated in the Danish building code BR-77.

1. RESULTS OF MEASUREMENT

The measurements were started on 6 January 1981 and finished on 25 October 1982. However, a start delay occured in several of the houses, partly because of incorrect installation of the measuring system and partly because some houses were not inhabited in the beginning of the measurement period.

The results of measurement have been corrected for measuring errors, which have primarily been caused by bad connections to the temperature sensor or defects in the integrator of the energy meters. Especially, the measurings of the energy consumption for domestic hot water have been defective because of corrosion of the temperature sensors. The corrosion is caused by dezincing of the chromium-plated brass sensor pocket, which should protect the PT-100 sensors. The correction of the consumption of energy to domestic hot water is based on regression of the results of measurement which have been obtained while the temperature sensors were still functioning.

A number of the houses have a woodburning stove or a solid fuel boiler installed. As it has been impossible to make a sufficient control of solid fuel consumption, (mainly wood) the energy supplement from the woodburning stoves is not included in the results of measurement. Where the woodburning stoves with a "heat exchanger" or solid fuel boilers are used, the energy supply, which is carried via the heating system, has been measured.

In houses with oil-fired boilers the operating period has been measured as well as the level of the oil stock. This measurement gives only a coarse indication of the oil consumption. The oil consumption is calculated based on an boiler efficiency of 80%.

The measurement results are stated as weekly sums and shown in three bar chart summaries for each house. Respectively:

- conventional energy consumption (bought energy)
- energy supply to the boiler room
- energy consumption.

It should be mentioned that the domestic electricity consumption is not included in the bar chart summary for the energy consumption. Furthermore, the electricity consumption for heat recovery is included in the bar chart summary for the energy supply to the boiler room. The electricity meter for the heat recovery was not installed at the start of the measurement period. So in the first 10 to 14 weeks the electricity consumption for heat recovery is included in the domestic electricity consumption.

2. TWO- AND THREE-STOREY BUILDINGS (HOUSE NO. 20)

Measuring of the following flows of energy has been done:

- Supply of district heating
- Energy production of solar panels
- Electricity supply to the heat pumps system
- Electricity supply to the circulation pumps and fans
- Energy production of heat pump system
- Heat consumption of radiator system
- Heat consumption for preheating of ventilation air
- District heating for domestic hot water
- Domestic hot water consumption

The energy loss has been calculated as the difference between energy supplied to the boiler room and the energy distributed from the boiler room to the heating system and the domestic hot water. Thus the energy loss includes the heat transfer from the circulation system to the domestic hot water. An unknown quantity of the energy loss will be used for domestic heating. However, it is pointed out that the installations are unverifiable sources of heat and therefore the transfer of heat from here ought to be reduced to a minimum.

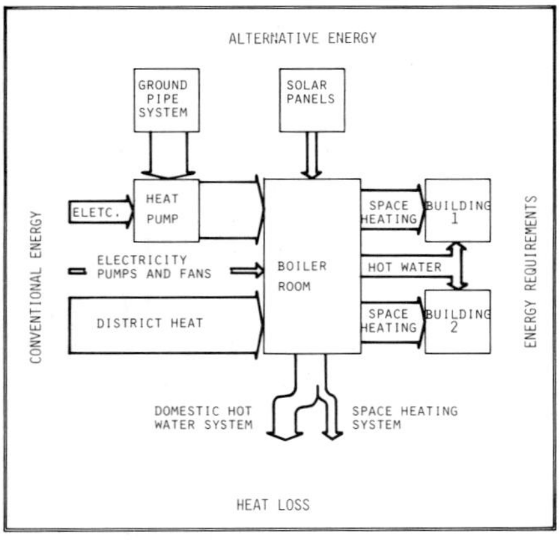

Fig. 1. Energy balance for the two-three-storey buildings.

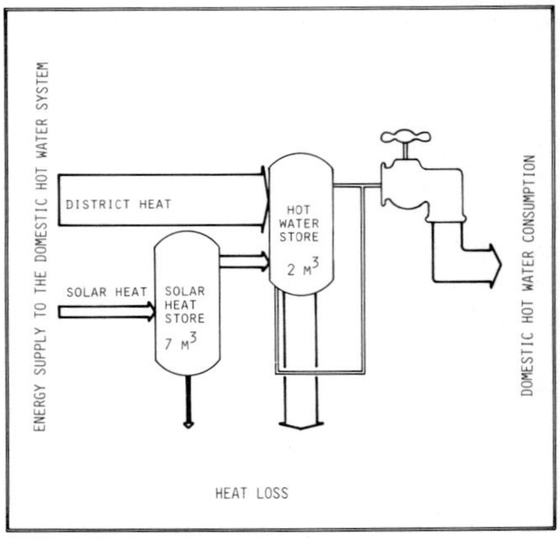

Fig. 2. Energy balance of the domestic hot water system.

Especially, the considerable energy loss from the domestic water system is remarkable, this is primarily due to heat transfer from:

- Circulation pipe
- Hot water tank ($2m^3$ with uninsulated man cover)
- Solar heat store (5 m^3 and 2 m^3 with uninsulated man cover)

The ratio of the energy consumption for heating the consumed hot water and the energy supplied to the domestic hot water system is 0.46 equal to an efficiency of 46%. Thus the energy standard of the domestic water installations corresponds to the experience obtained in other existing flats.

For space heating and preheating of ventilation air the heat pumps have supplied 62% of the energy consumption. The remaining 38% of the energy supply is from district heating. In the measurement period the heat pumps have worked with an average effect factor of 2.58, which must be considered satisfactory. The energy supply for domestic water is divided in 81% for district heating and 19% for solar energy.

Based on the results of measurement it is possible to map the heat losses in the domestic water system as shown in figur 2. It should be noted that the thermal loss from the two solar heat storages - 5 m^3 respectively 3 m^3 - is very moderate caused by the fact that the mean storage temperature does not differ considerably from the room temperature in the boiler room. The temperature level in the hot water tank and the circulation system is considerably higher than the temperature of the surroundings. This is the reason for the big thermal loss from the part system.

The temperature of the heated liquid from the solar panel is normally lower than the domestic hot water temperature. That means that the heat loss from the domestic hot water system primarily is covered by district heating. In relation to the actual hot water consumption the solar energy plants have given a supplement of 39%.

The consulting engineer has estimated a yearly energy production from the solar system in the order of 25-55 MWh. The measurements show a yearly energy production of 18 MWh. The reason for this must be that the efficiency of the solar panels is not as good as assumed in the calculation. Furthermore, the domestic hot water consumption is considerable less than assumed when the solar system was designed.

Figure 3 to 5 shows bar chart summaries of the weekly energy consumption. It is notable that the energy consumption during the summer period is rather high. In Denmark the heating season is from beginning september to end of may. So in june, july and august it should not be necessary to use space heating, especially not in low energy houses. In table 1 the energy consumption in the period 1 june, 1981 to 31 august, 1981 is compared with the energy consumption in the heating season 1981 to 1982.

		HEATING SEASON KWh	SUMMER SEASON KWh
District heat (total)	:	152620	20343
District heat (domestic hot water)	:	70430	17413
Electricity to heat pumps	:	59688	4960
Heat pumps (output)	:	146980	15002
Electricity to pumps and fans	:	11808	2230
Solar heat	:	10322	6888
Solar heat supplement	:	11643	5560
Domestic hot water	:	38159	9907
Space heating	:	218116	15057

Table 1. Energy utilization and requirements of the flats in the period 1 june, 1981 to 31 may, 1982.

CONVENTIONAL ENERGY CONSUMPTION
2 AND 3 STOREY BUILDINGS SKIVE 79

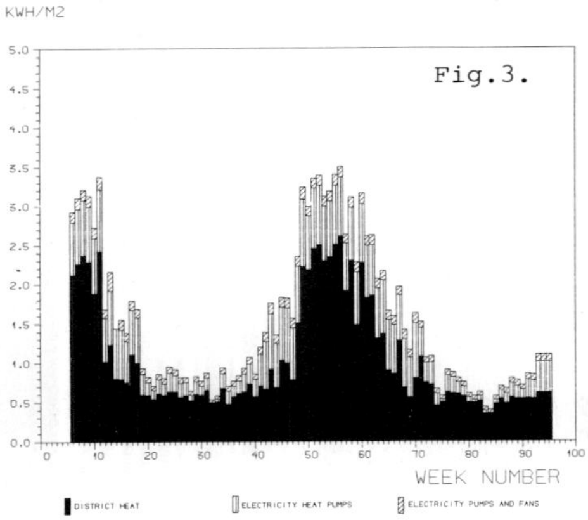

CONVENTIONAL ENERGY CONSUMPTION
LOW/DENSE DWELLINGS

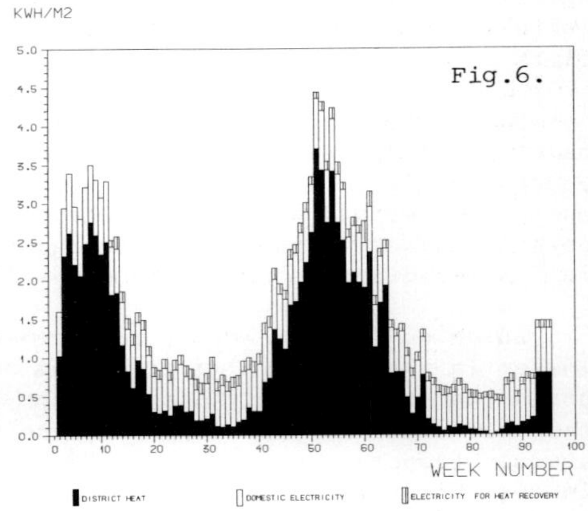

ENERGY SUPPLY TO THE BOILER ROOM
2 AND 3 STOREY BUILDINGS SKIVE 79

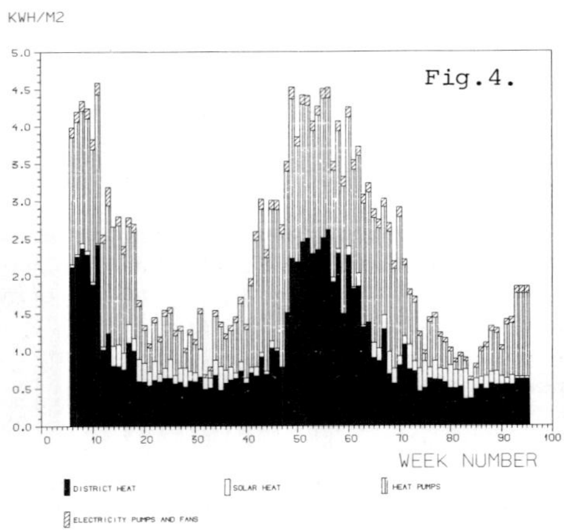

ENERGY SUPPLY TO THE BOILER ROOM
LOW/DENSE DWELLINGS

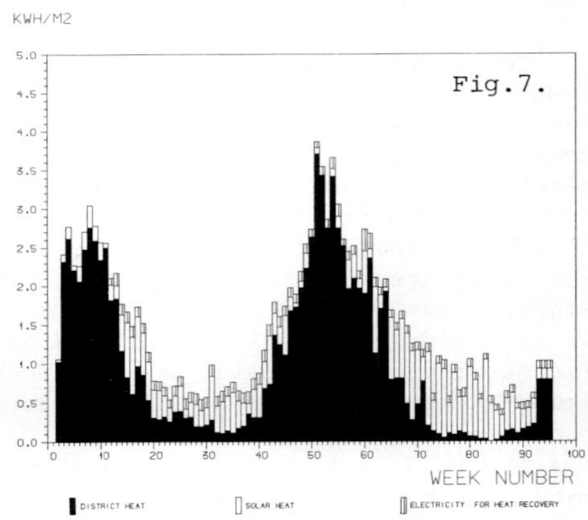

ENERGY CONSUMPTION
2 AND 3 STOREY BUILDINGS SKIVE 79

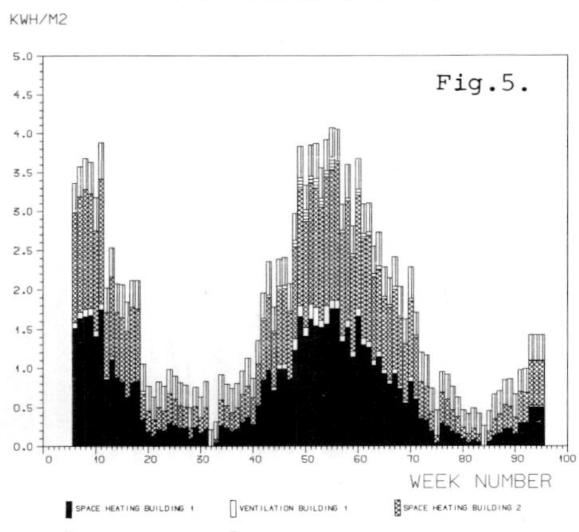

ENERGY CONSUMPTION
LOW/DENSE DWELLINGS

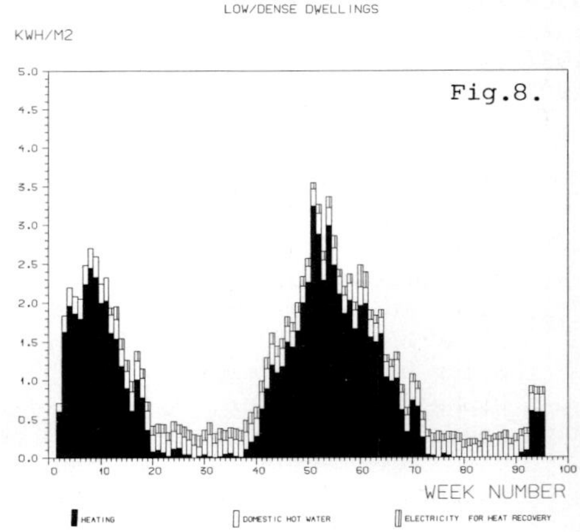

Table 1 indicates that the energy requirements in the summer season is 24964 KWh. During this period the energy supply to the boiler room is 47193 KWh. So in the summer season the energy supply is nearly double as much as the energy requirement. If the building owner will turn off the space heating during the summer season and will change the system so that the heat pumps can be used for heating the domestic hot water, significant energy savings can be obtained.

If we assume that the above mentioned changes are carried out, the energy consumption can be reduced to approximately 8000 KWh. Compared to the measured energy consumption of 27533 KWh, the reduction is 70 percent.

3. LOW/DENSE HOUSES (HOUSE NO. 1 TO 9)

In the low/dense houses measurements of the following flows of energy have been done:

- District heat
- Energy production of solar panels
- Electricity consumption of the heat recovery system (only from week 12, 1981)
- Heat consumption for the radiator system
- Domestic hot water consumption

Furthermore, during the period of 1 march 1982 to 30 November 1982, measurements of the temperature and water flows in the energy system as well as of the space temperature in the living room have been done. For these measurements a microprocessor controlled datalogger has been used. The bar charts in fig. 6, 7 and 8 show the weekly average consumption for eight houses. One house is not included because a woodburning stove is used.

If we compare fig. 5 with fig. 8, it is noticed, that the space heating consumption pr. m^2 in the low/dense houses is less than the space heating consumption pr. m^2 in the flats. Furthermore, it is noticed, that the space heating consumption in the low/dense houses is nearly zero in the summer season. The mean space heating consumption in the heating season 1981 to 1982 is 50 KWh/m^2. It is 36 percent less than the space heating consumption in the flats.

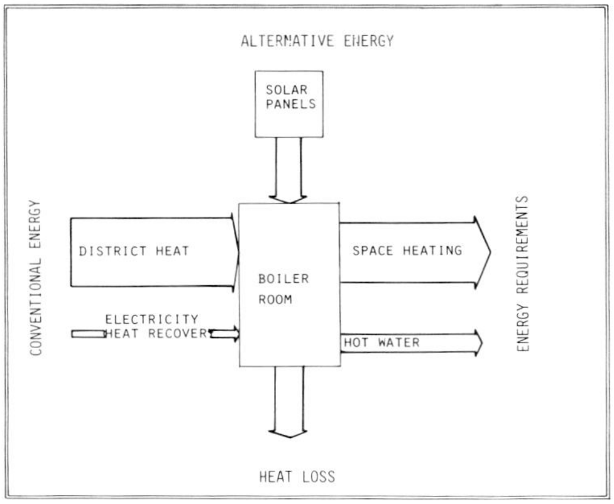

Figure 9. Energy balance for the low/dense houses, based on mean values.

If the energy supplied to the boiler room of the houses is compared to the measured consumption for space heating and domestic hot water a term of the efficiency of the energy system is obtained as illustrated in fig. 9.

The heat transfer from the installation in the boiler room is estimated as the difference between the energy supplied to the boiler room and the energy distributed to the heating system and domestic hot water. The heat transfer from the pipes to the radiators and from the pipes to the taps is not included. Furthermore, 40% of the electricity consumption for the heat recovering system is considered to be utilized for space heating.

In fig. 10 to 14 variations of the different consumptions are shown. The dark bar in the figures indicates the intervals between the lowest registered consumption and the average consumption. The light bar indicates the intervals between the biggest registered consumption and the average consumption.

The conditions that cause the differences in the energy consumption of the houses are only measurable on a limited scale. It is therefore impossible to give a thorough explanation of the big variations in the energy consumption of the houses, based on the existing data information.

However, a closer examination of the results of measurement proves that the nine low/dense houses can be divided into three categories:

1. Houses being significantly below the average consumption (three houses)

2. Houses grouped around the average consumption (four housings)

3. Houses being significantly above the average consumption (two houses)

Generally, the houses in category 1 and 3 have a weekly energy consumption which is below/above the weekly average consumption for all the nine houses. Regarding the houses in category 2 there is however, no unambiguous picture as the weekly energy consumptions in some weeks are bigger than the average consumption and less in other weeks.

In one of the houses in category 1, it is primarily the heat supplement from a woodburning stove which causes the low energy consumption for space heating. Furthermore, this house has the biggest electricity consumption which gives the house a big heat supplement from the electric devices.

The points mentioned below evaluate the conditions which cause some of the houses to consume more energy than others. At the same time we have tried to estimate a priority of the conditions, which have the biggest effect:

1. A low temperature level in the rooms generally
2. Low consumption of domestic hot water
3. Cut off the district heating supply during the summer season
4. High energy supply from the solar heating system
5. Heat supply from current devices, lighting, etc
6. Use of the night setback

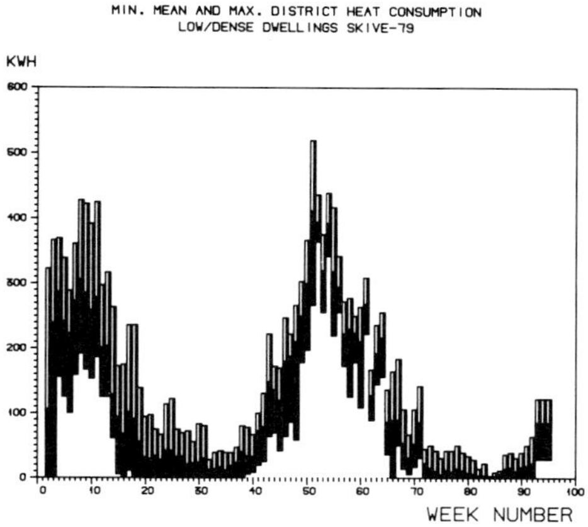

Fig. 10.

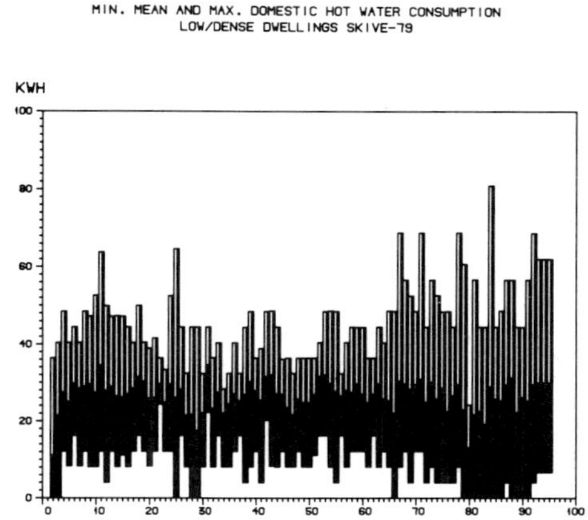

Fig. 13.

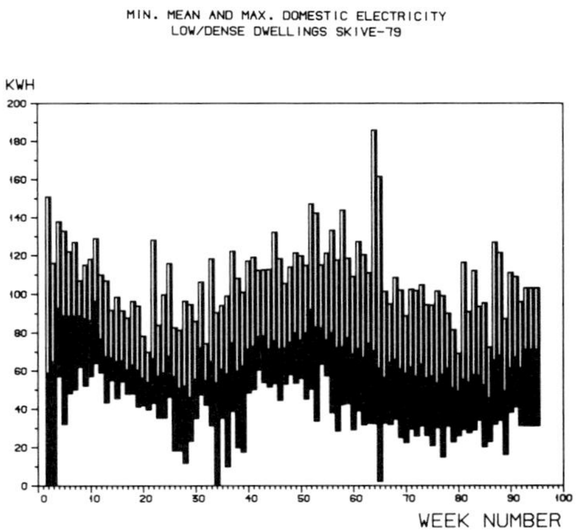

Fig. 11.

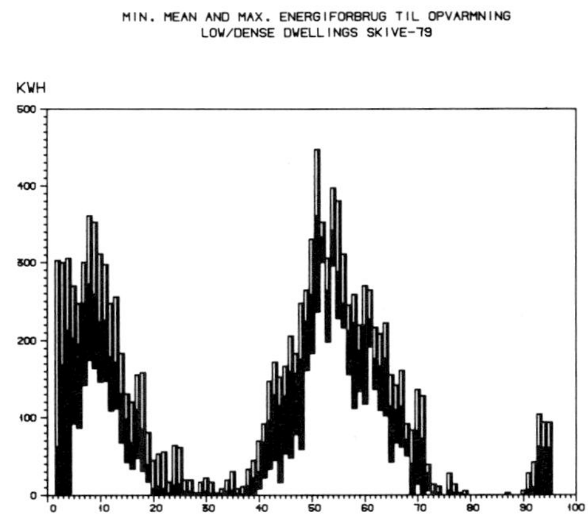

Fig. 14. Space heating.

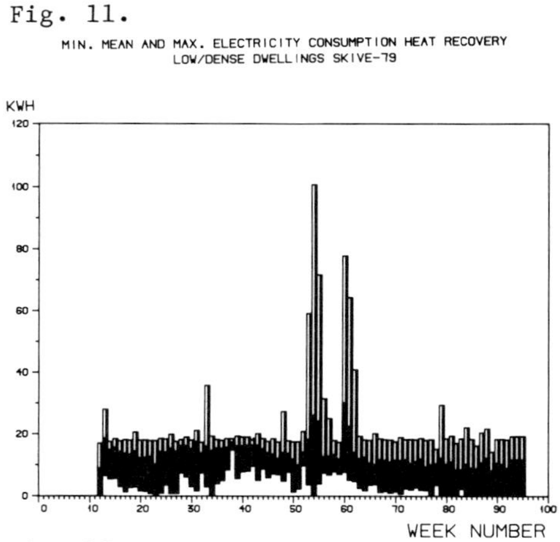

Fig. 12.

It should be noted that point 3 and 4 are connected - as the possibility of cut off the district heating supply is based on the fact that a satisfactory energy supply from the solar heat system can be achieved in the period. In house no 6 the district heating supply is consistently broken off during the summer period. This has been possible because of the solar heating system and because the inhabitants have accepted a lower domestic water temperature for some periods.

In the houses central control equipment for decreasing the space temperature is used, which switch off the circulation pump until the chosen temperature decrease has been reached. The night temperature is hereafter regulated by an on/off control of the circulation pump. By day the control equipment will in the same way switch off the circulation pump if the space temperature exceeds a preadjusted value, which is supposed to be higher than the temperature adjustment of the radiator valves. The effect of the control equipment used is very variable from house, to house which among other things is caused by the following conditions:

1. The control equipment with a built in temperature sensor is placed in the hall, which must be considered as the coldest room in the house. Then the control is based on a space temperature, which is not representative for the most used rooms of the house.

2. The adjustment of the day and night temperature is done by using two dial read cursors, which do not give the consumers sufficient information about the current adjustment values.

No significant connection between the energy consumption of the houses and the use of night setback has been found - besides the above mentioned conditions - which is due to the fact that the effect of periodic decreases of the space temperature in well insulated houses is somewhat limited.

5. SINGLE-FAMILY HOUSES (HOUSE NO. 10 TO 19)

The single-family houses is quite different with regard to the house plan, floor area, insulation standard and the energy system used. Besides six of the ten houses use the woodburning stove more or less. Furthermore, there has been one or more changes of inmates in three of the houses and two of the houses have only been occupied in approximately half of the measurement periods. These circumstances causes great variation in the energy consumption of the houses.

Table 2 shows great variation in the energy consumption for space heating and domestic hot water as well as the conventional energy consumption. The main reason for these variations is the use of woodburning stoves in 6 of the houses. It should be noted that the energy supply from the woodburning stoves is not included in table 2.

It should also be noticed that the use of conventional energy is lowest in houses with heat pumps. That does not mean that the energy bill is less in these houses. Because in Denmark the price of electricity is nearly twice as much as the price of fuel oil for the same amount of energy.

Another reason which could cause variation in space heating consumption is, that some houses use space heating in the summer season and others don't.

HOUSE NO.	HOUSE AREA m^2	SPACE HEATING KWh/m^2/year	DOMESTIC HOT WATER kWh/year	CONVENTIONAL ENERGY kWh/m^2/year
10*+	130	74.1	2165	51
11*+	135	82.3	1987	51
12	150	91.2	2120	116
13	132	50.2	1570	80
14	172	77.2	3828	71
15*	206	38.7	1003	52
16*+	137	31.6	2057	38
17*+	137	84.5	2975	80
18+	150	40.0	2929	60
19*	150	78.5	2764	44

Table 2. Energy consumption (*house with heat pump,+ house with woodburning stove).

The COP for the heat pumps system is here defined as:

$$COP = \frac{\text{Net energy output}}{\text{Electricity consumption}}$$

The COP includes heat loss from the heat pump system, i.e the heat pump, pipes and hot water store and the electricity consumption for direct electricity make-up and for the pumps. In table 3 the mean COP for the heating season and the summer season is calculated.

HOUSE NO.	HEATING SEASON COP	SUMMER SEASON COP
10	1.8	1.6
11	2.0	1.0
19	2.0	0.5

Table 3. The mean COP for the heating - and the summer season.

HOUSE NO.	SPACE HEATING COP	DOMESTIC HOT WATER COP
16	1.2	0.8
17	1.6	0.9

Table 4. The mean COP for domestic hot water - and space heating heat pumps.

In house no. 16 and 17 two heat pumps are used. One for space heating and one for domestic hot water. The mean COP for these two heat pumps is listed in table 4.

The results stated in table 15 and 16 indicate that COP in general is to low. Especially the COP in the summer season and the COP of the domestic hot water heat pump. The reason for these low COP's is mainly short operating periods in the summer period and heat loss from the hot water store. Furthermore, in house 16 and 17 the domestic hot water heat pump is only used when the space heating heat pump is functioning. When the space heating heat pump is not functioning, an immersion heater is heating the domestic hot water store.

The research association of the Danish Electricity supply undertakings (DEFU) has measured the energy consumption in 18 different heat pump systems. The results shows variation between 1.7 and 2.5 of the yearly mean COP. Furthermore, in 60 percent of the heat pump systems the yearly mean COP was less than 2.2.

The COP obtained by the heat pump system in house 10,11 and 19 is almost the same as the COP for similar system of that period. The COP obtain by the space heating heat pump in house 16 and 17 is on the other hand considerably less than expected.

The energy production of the solar systems in all the single family houses, except of the solar system in house 14, is less than estimated by the building contractor. On the other hand if the solar heat production is compared with F-chart calculation based on measured energy loads, a much better agreement is achieved.

HOUSE NO.	MEASURED ENERGY LOAD	SOLAR HEAT SUPPLEMENT MEASURED	CALCULATED
1-9	7151 KWh	1325 KWh	1387 KWH
10	1988	930	1019
12	1961	1018	1146
13	8939	1568	1260
14	10763	4128	4449
18	9972	1481	2027
19	16873	2067	2412

Table 4. Measured- and calculated solar heat supplement in the period 1 sept.,1981 to 31 aug., 1982.

6. DISCUSSION

The main goal of the Skive-79 low energy house project was to demonstrate that it is possible to reduce the energy consumption to less than 60% of the energy consumption stated in the Danish building code BR-77, by the use of today's knowledge with regard to building technique and heat supply. The present discussion is mainly concentrated on the conventional energy used in the Skive-79 houses. To make comparison possible, the measured energy consumptions are adjusted according to the degree days of the normal year (1936-1980).

In fig. 15 the estimated yearly oil consumption pr. m^2 of the Skive-79 houses are compared with the yearly oil consumption pr. m^2 of other Danish houses. The estimation of the oil consumption is based on the measured energy consumption for space heating and domestic hot water. That means that the estimated values in fig. 15 indicate the oil consumption, if the Skive-79 houses have used oil-fired boilers and no energy supply from alternative sources.

The comparison figures derives from a major investigation "Eksisterende bygningers energimßssige standard" made by Building Technical Dpt., Technological Institute, in co-ordination with a big oil company. The results of this investigation are based on oil supplies which have been delivered in the period 1975-1979. Apart from this period the annual energy consumptions are calculated on the basis of an assumed annual energy saving of approximately 8-10%.

It appears from fig. 15 that the Skive-79 houses regarding the energy consumption are divided into three groups. One group comprising the low/dense houses 1 to 9 and house 13. Here the energy consumption is significantly lower than the energy consumption in a typical BR-77 house. In the other group comprising the houses 10,11,12,14,17,19 and 20, the energy consumption is nearly the same as the energy consumption of a BR-77 house.

The last group, comprising the houses 15,16 and 18, has also a significantly lower energy consumption than a BR-77 house. In the last group the low energy consumption is due to a large space heat supplement from the woodburning stove.

Figure 15 indicates the energy saving effects of increased insulation and air-tightness of the Skive-79 houses. Besides of increased insulation and air-tightness, most of the houses use alternative energy sources (heat pumps, solar energy). The total effect of the energy saving efforts is shown in fig. 16. which shows that by the use of the energy saving alternatives - increased insulation and air-tightness, use of alternative energy sources and heat recovery from ventilation air - the energy goal is met in all the houses, except in house 12. If i.e. house 10 has used a oil-fired boiler to provide heat for space heating and domestic hot water, fig. 16 shows that the energy consumption is 114 kWh/m^2. Due to the use of heat pump and solar energy the energy consumption is reduced to 51 kWh/m^2/year.

It should be noticed that only the conventional energy consumption in fig. 11 includes the electricity consumption for heat recovery. In house 13 the electricity consumption for heat recovery is bigger than the energy supply from the solar system.

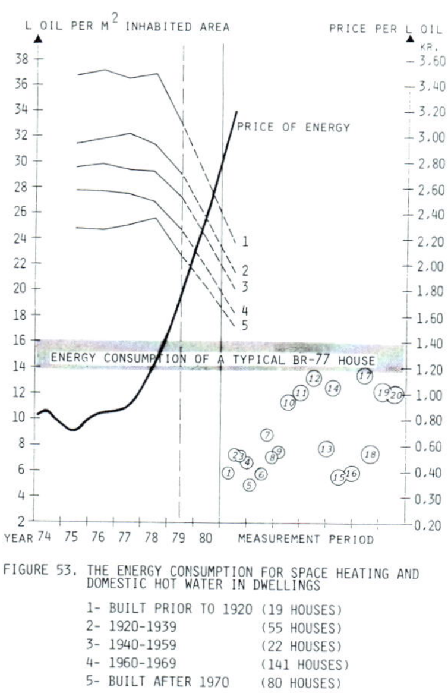

Fig.15. Energy consumption for space heating and domestic hot water.

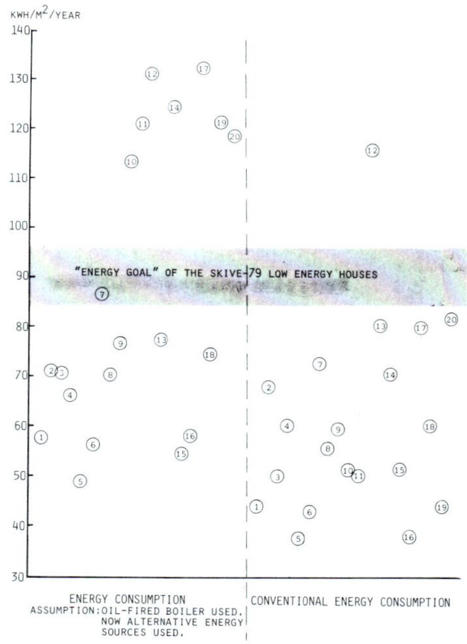

HOUSE NO.	1	2	3	4	5	6	7	8	9	10	11	12	13	14	15	16	17	18	19	20
OIL-FIRED BOILER												Y	Y	Y	Y			Y		
DISTRICT HEAT	Y	Y	Y	Y	Y	Y	Y	Y	Y											Y
HEAT PUMP										Y	Y					Y	Y		Y	Y
SOLAR ENERGY	Y	Y	Y	Y	Y	Y	Y	Y	Y	Y	Y	Y	Y	Y	Y			Y	Y	Y
HEAT RECOVERY	Y	Y	Y	Y	Y	Y	Y	Y	Y	Y	Y		Y	Y				Y	Y	Y
WOODBURNING STOVER					Y					Y	Y				Y	Y	Y	Y	Y	

Fig.16. Energy consumption for the Skive-79 low-energy houses.

DEMONSTRATION OF CONTROLLED SPACE HEATING BY A SOLAR WALL

S. M. Jackson & Dr. W. Houghton-Evans

Stephen George & Partners
Architects & Town Planners
170 London Road
LEICESTER
UK LE2 1ND Tel: 0533 551807

Summary

Stephen George & Partners, Leeds University and Charles Gregory Ltd., building developers, are collaborating in the building of a housing scheme of mixed highly insulated houses and solar heated houses, on a site in Leicestershire, England. The solar heated houses will be heated by pre-heated air from a glazed conservatory, drawn through a hollow concrete first floor slab, by the fan of a gas warm air heating system. The concrete floor will act as a heat store and an automatic valve in the auxiliary heating system will cut over to the gas system when the temperature drops below the required level.

1. Introduction

1.1 Participants

Stephen George & Partners and Leeds University, supported by the EEC under their Demonstration programme, are collaborating with developers, Charles Gregory Ltd of Leicestershire, England, in the building of a group of solar heated houses for sale on the open market.

1.2 Basic Aims

The aim is to provide an energy-efficient house for the average family with a solar conservatory and heat store that is easy to operate and maintain, with the minimal additional capital outlay on conventionally heated housing.

The developer's main criterion is that the house type is comparable in size with the mass private market housing currently being produced by the big developers in this country. For marketing purposes the solar houses have to compare favourably with the mass market housing that is being offered for sale with additional fittings such as carpets, curtains, cooker, fridge etc., which are not to be on offer in this particular development.

It was, therefore, considered that the solar houses would not be aimed at the first time buyer, but should be designed to cater for established families; couples about to start a family and therefore looking for a larger home; and retired couples looking for a smaller economic unit.

1.3 Existing Solar Housing

To date, in the UK, most of the pioneering solar housing has either been 'one-off' houses on green-field sites, or Local Authority housing mainly in the retrofit field. In addition, the concentration has been on solar heated water systems rather than space heating.

In general it has been found that the most successful installations in terms of operation and lasting benefits, have been those that use the simplest principles, are easily understood and operated by the occupants, and use standard building components that can be easily replaced when necessary.

It is gratifying to note that more developers are becoming interested in the idea of energy-efficiency and solar heating as a saleable commodity on offer in their housing packages.

1.4 Proposed Solar Installation

The initial concept was to take a standard small family house type, such as is seen for sale throughout the UK, by the large developers, and simply modify the basic first floor structure to act as a heat store and add a glazed conservatory on the south side. In this way conventional building methods can still be used to produce a cheap pleasant looking house to be sold on the open market.

This has, however, been slightly modified following detailed discussions with the developer involved, on what he considers the best kind of house type to use with the additional solar element.

2. Description of work

2.1 House types & site layout

The house plan must be flexible enough to provide for variations in family size and market demands. It was decided to opt for a standard plan that could variously be a 2 bedroom 4 person; 3 bedroom 4 person; or 3 bedroom 5 person house, thus keeping an overall design concept for the development without limiting the market.

Flexibility in the positioning of the glazed conservatory in relation to the access to the house was considered, but the developer did not consider conservatories anywhere other than on the garden side of the house, as easily sellable. It was, therefore, decided to restrict the conservatory to the rear garden and side elevations of the house, accepting that this restricted the positioning of the house on the site.

Other restrictions, such as vehicle access onto the site, overlooking and plot sizes, produced a final layout of sixteen houses in a mix of half highly insulated but non-solar houses and half solar houses with the glazed conservatory.

2.2 Solar Heating

The main elements of the solar installation are a hollow concrete first floor slab to be used as a heat store and duct, and a single storey glazed conservatory on the south, east or west elevations.

High insulation levels are to be provided with extra insulation in walls and roof and added moveable insulating shutters to the windows.

The concept is that the mass of the concrete floor will act as a heat store both for direct gain solar radiation through the south and west facing windows at first floor level, and for indirect gain from pre-heated air being drawn through the holes in the floor slab from the conservatory and circulated around the house, or stored for later use.

Using the first floor slab as a heat store and duct combined not only cuts the cost of installation, but, because the floor is part of the internal structure of the building, the heat losses are reduced.

Consultations with Johnson & Starley, manufacturers of warm air heating systems, have taken place. The standard size fan used in their system for a family house is sufficient to draw the pre-heated air through the slab and allow heat transfer into the concrete to take place. There is, therefore, no necessity for any additional fan to be installed other than that provided by the warm air heating system.

The only modification required to the warm air heating system, is an automatic valve that can switch over to the gas warm air heater when the pre-heated air coming from the conservatory drops below the required temperature.

Initial computer calculations show a reduction of the energy demand of three weeks off each end of the heating season and an overall reduction of approximately 30% over the whole heating season.

In conjunction with the solar heating element, research is being carried out into the production of a window assembly combining double glazing, venetian blinds and moveable thermal shutters, to increase the thermal efficiency of the building envelope and alleviate overheating in summer.

In addition, research is being carried out on the production of a self-acting cylinder to open ventilators in the conservatory, again, to prevent overheating in the summer.

3. Programme

3.1 Construction

Planning permission for the development has been applied for and it is envisaged that building will commence in the new year. Completion is expected in time for monitoring to commence at the beginning of the 1984/5 heating season.

3.2 Monitoring

The programme of work includes for monitoring the solar installations while in use and this should be done for at least 24 months after the houses are marketed.

Some particularly complex solar heated houses have in the past necessitated long monitoring periods due to the users learning to use the installations to their maximum benefit. In some cases the users' energy consciousness has produced a negative result due to their misunderstanding of how the solar installation operates in relation to the other elements of the house.

In this instance, we do not envisage these sort of problems arising as the solar installation is very basic easy to comprehend and operate. We would, therefore, expect a full 24 month period including two heating seasons to be adequate time for monitoring the installation, to give accurate figures on the effectiveness of the solar energy content of the building.

4. Conclusions & prospects

4.1 Conclusions

The end product aimed for in this development is an attractive, affordable family house that has the added bonus of an additional usable space that provides solar heating in a direct, simple way. In this way, it is hoped that solar heated housing will become more acceptable to the general house buying public, and developers alike, than it is at present.

4.2 Prospects

It is hoped that, in view of the thoughts set out above, that developers will be more amenable to building solar heated housing as part of their general housing stock, especially as the possibilities of future energy savings of upto 30% would be a prime selling point. It is envisaged that solar heated housing could become the norm rather than the exception.

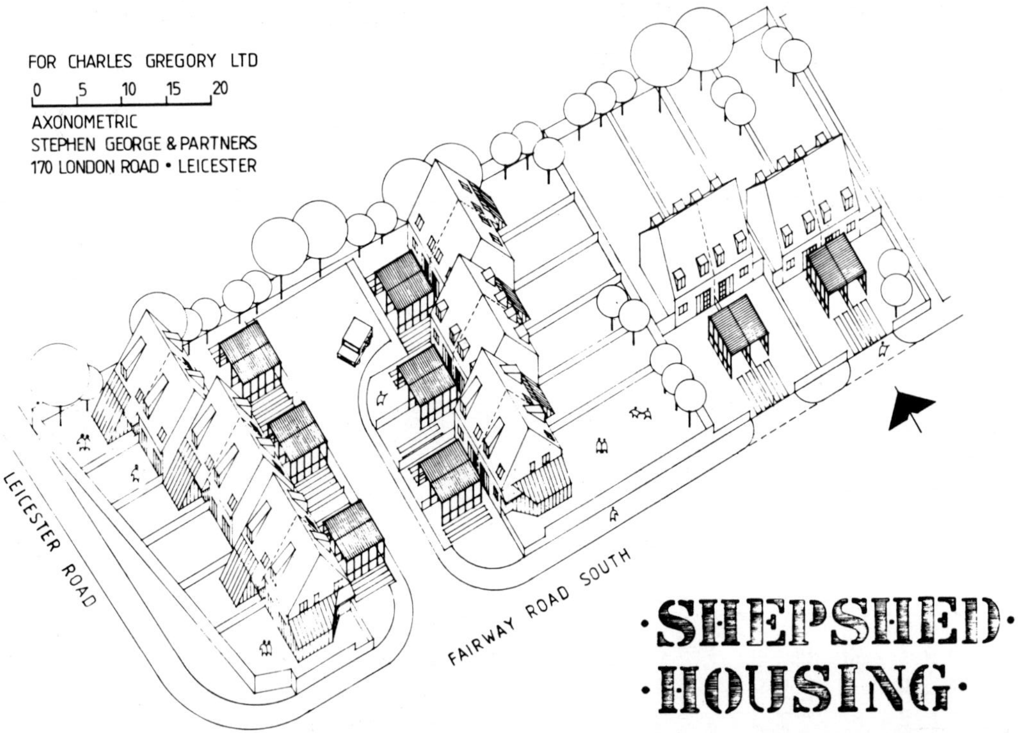
FOR CHARLES GREGORY LTD
0 5 10 15 20
AXONOMETRIC
STEPHEN GEORGE & PARTNERS
170 LONDON ROAD • LEICESTER
LEICESTER ROAD
FAIRWAY ROAD SOUTH
·SHEPSHED·
·HOUSING·

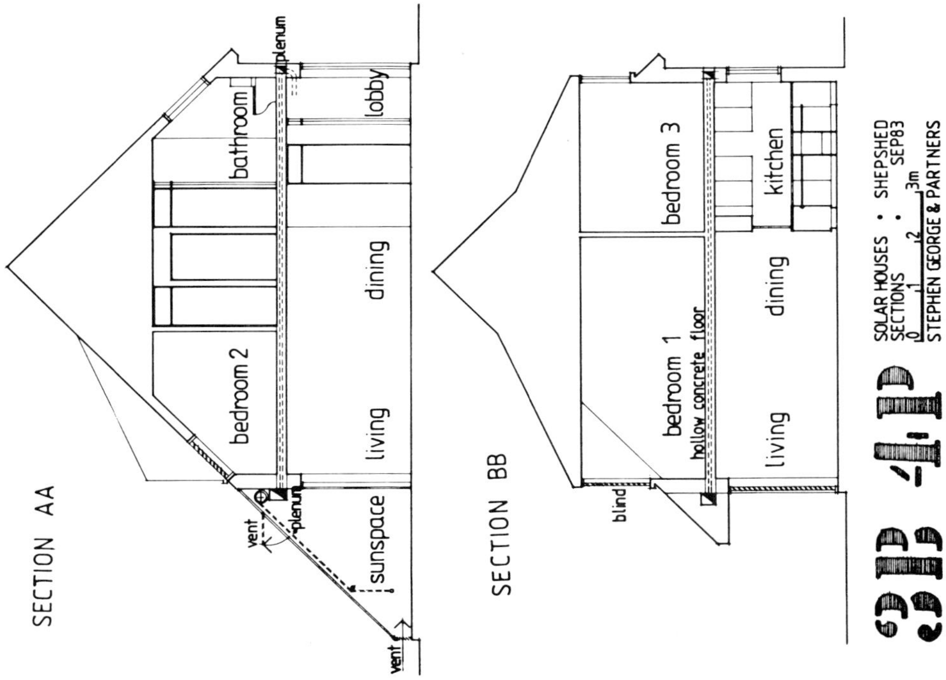
SECTION AA
bathroom
lobby
plenum
bedroom 2
dining
living
sunspace
vent
plenum
vent
SECTION BB
bedroom 3
kitchen
bedroom 1
hollow concrete floor
dining
living
blind
SOLAR HOUSES : SHEPSHED
SECTIONS : SEP83
0 1 2 3m
STEPHEN GEORGE & PARTNERS

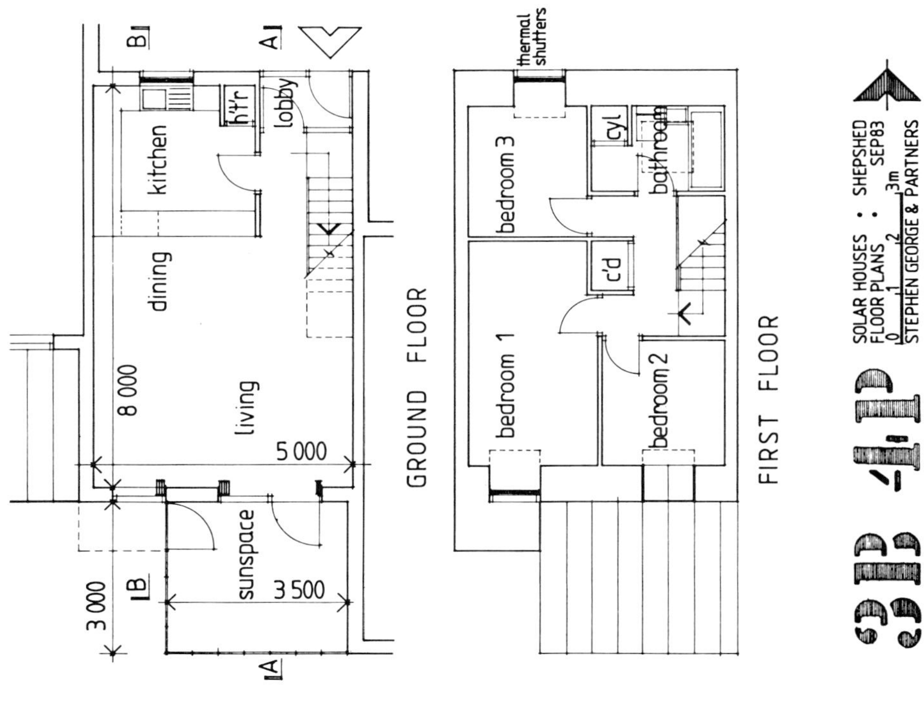
B
A
kitchen
h't'r
lobby
dining
living
8 000
5 000
sunspace
3 500
3 000
B
A
GROUND FLOOR
thermal shutters
bedroom 3
cyl
bathroom
c'd
bedroom 1
bedroom 2
FIRST FLOOR
SOLAR HOUSES : SHEPSHED
FLOOR PLANS SEP83
0 1 2 3m
STEPHEN GEORGE & PARTNERS

INTEGRATED APPROACH TO LOW ENERGY HOUSING

L. Makkar & M. Ince
SLC Energy Group

Summary

The paper deals first with the principles of the Integrated Apprach. Analysis shows that this approach increases the possible savings. Details follow of the new and rehabilitated houses to which these principles were applied and which form the basis of this Demonstration project. Building costs and projected energy savings are given. The necessity for positive Demonstration of low-energy building design and practice was identified in order to bring about widespread change in practice and consequently widespread energy savings. To this end a Demonstration Suite has been set up, incorporating a number of elements designed to bring about this change. The concept of Planned Change and the resultant 'recipe' for the Demonstration Suite forms the third section of the paper. This Demonstration project is funded jointly by the C.E.C., the UK Dept. of Energy and the SLC Energy Group.

1. THE INTEGRATED APPROACH

Integration of the various energy conservation measures for the fabric of the building, incidental and solar gains AND rationalisation of heating and ventilation result in:

- o High percentage of energy saved at low cost
- o Improved thermal comfort for the occupiers
- o Control of condensation

4 Rules have been developed which provide a basis for integrated low energy design:

1. Evenly distributed insulation saves more energy than extra insulation applied only to one element. (see Fig. 1)
2. Draught control saves large amounts of energy and improves comfort. (see Fig 2)
3. Solar gains should be looked at in the context of the total energy conservation package. (see Fig. 3)
4. Ambient and Radiant temperatures should be balanced for energy reduction and condensation control and for thermal comfort. (see Fig 4)

2. THE INTEGRATED APPROACH IN PRACTICE

It is not always possible to integrate all the desirable elements in the design for a specific example and it is generally more difficult in existing buildings than in new ones.

In the London Borough of Lewisham 18 new houses have been built and three large Victorian houses rehabilitated (and converted to 15 flats) using the Integrated Approach. It is expected that the new houses will use 4,213 kWh per year for space heating - a reduction of 61% over normal UK (Building Regulations standard) and that the flats will use 7,755 kWh per year - a reduction of 46%. (The usage in the flats is relatively higher because they have a much greater volume per person than the houses.)

INSULATION applied to all elements achieves more than extra insulation applied to only one element

Assuming:
- 20°C thermostat setting
- 2 air changes per hour
- E-W orientation

Very high level of insulation in the loft only
- 400mm in the loft

Annual Consumption 13,242 kWh

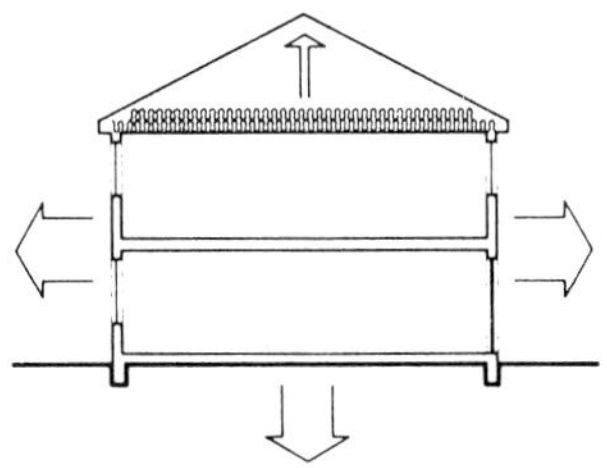

Same amount of insulation evenly distributed round all elements
- 200mm in the loft
- 100mm to the walls
- 50mm under the floor

Annual Consumption 9,259 kWh

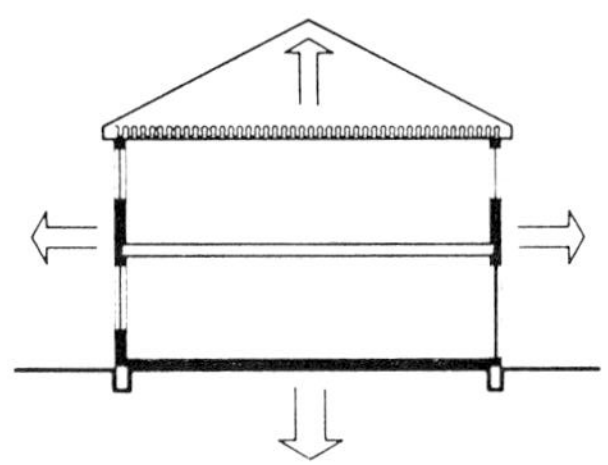

Figure 1

Figure 2

DRAUGHT REDUCTION is a major factor in energy saving

Assuming:
- Evenly distributed insulation
- 20°C thermostat setting
- E-W orientation

Standard structure
- No draught stripping
- 2 air changes per hour

Annual Consumption 9,259 kWh

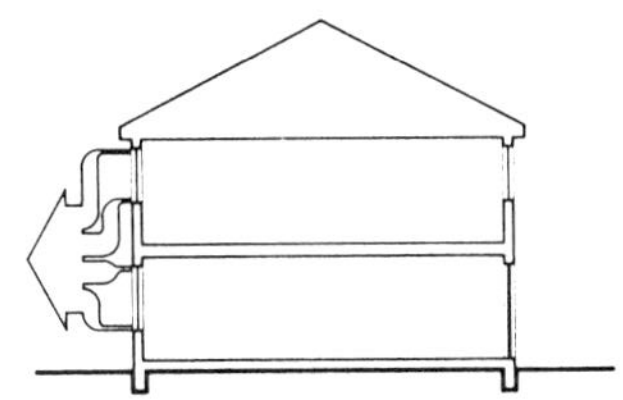

Draught controlled structure
- Windows and doors draught stripped
- Draught lobbies front and back and other measures
- 1 air change per hour

Annual Consumption 4,769 kWh

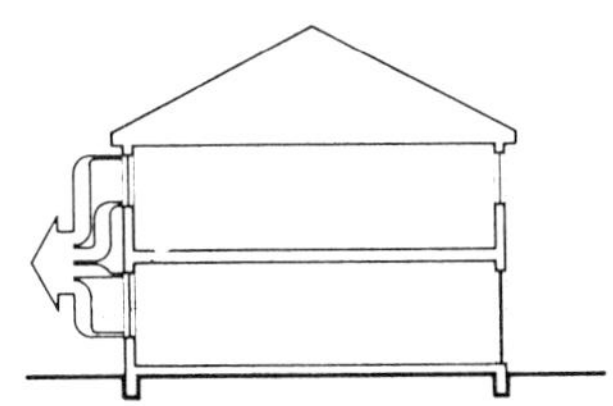

SOLAR GAIN in context

Assuming:

- Evenly distributed insulation
- 20°C thermostat setting
- 1 air change per hour

E-W orientation

- Window sizes remain the same

Annual Consumption 4,769 kWh

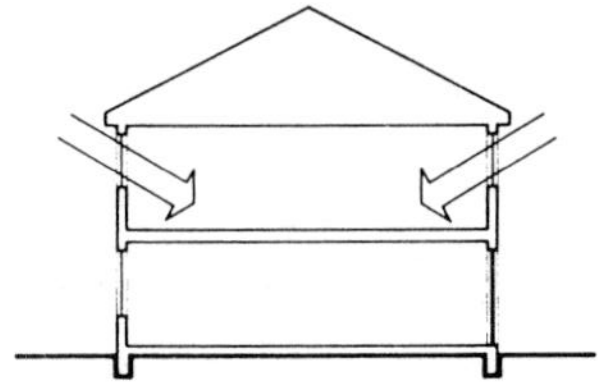

N-S orientation

- Window sizes remain the same

Annual Consumption 4,782 kWh

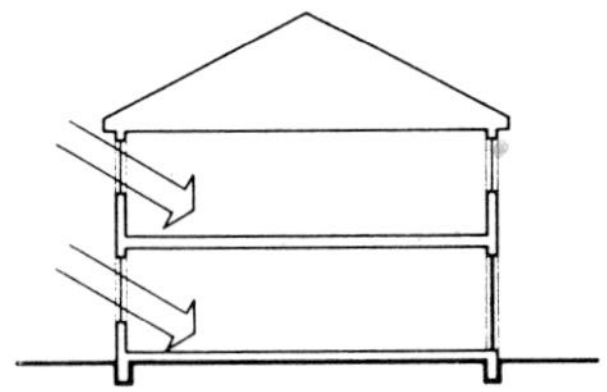

N-S orientation

- Same area of glazing redistributed: Southside increased, Northside decreased

Annual Consumption 4,529 kWh

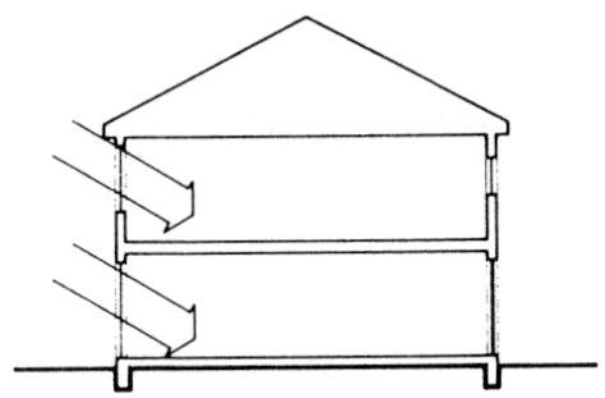

Figure 3

Figure 4

"COMFORT" is improved by balancing ambient and radiant temperatures

Assuming:

- Comfort achieved by resultant temp. of 20°C
- Resultant temp. simply expressed as half the sum of ambient and mean radiant temperatures

Mean radiant temp. 10°C

- Ambient temp. required to give comfort is 30°C

Large net radiant heat loss to "cold" surfaces

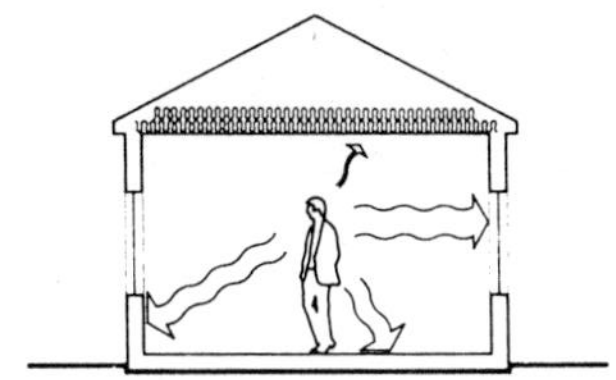

Mean radiant temp. 19°C

- Ambient temp. required to give comfort is 21°C

Minimal net radiant heat loss to "warm" surfaces

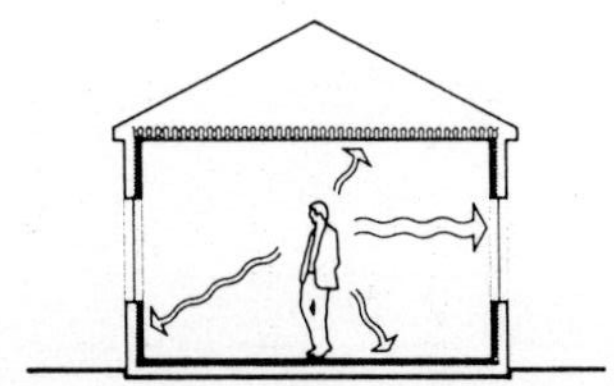

The construction of the old houses was 312mm solid brick walls, timber floors throughout and a pitched slate roof. Figure 5 shows a typical plan. The walls have been insulated internally with, generally, 100mm insulation quilt hung in a 200mm air space created by installing a standard partitioning system. (On small areas 32mm urethane-backed plasterboard has been used). The roof was insulated with 100mm quilt and the sloping ceilings with 80mm quilt. The suspended timber ground floor was insulated with 100mm quilt between the joists, supported by plastic netting. All the windows and external doors are draughtproofed and care has been taken to seal all possible draughts - the design airchange rate is 1 airchange/hour. Space-heating is provided by gas-fired warm-air units which also provide a proportion of fresh air ventilation. The warmed air is ducted to each room and return ducts take stale air back to the heater unit where it is partially recirculated and partially exhausted. Return air is not taken from kitchens and bathrooms, so that positive pressure is created to drive water vapour out of these areas - thus reducing the risks of condensation elsewhere. Figure 6 compares the costs and anticipated energy usage for the 'standard' rehabilitation and the energy saving rehabilitation as built.

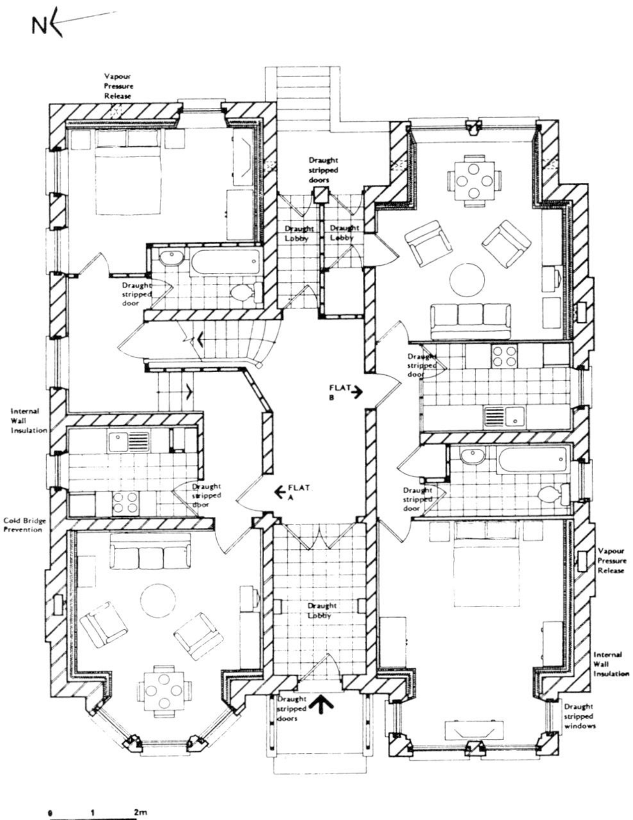

Figure 5 Typical Plan - Rehabilitation - Ground Floor

Energy & Cost Analysis

To London Building Act Thermal Standards

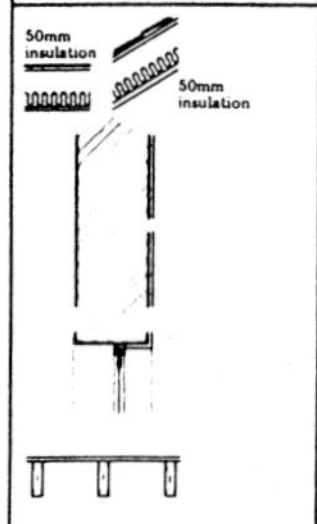

Element	'U' value $W/m^2 {}^oC$	Annual Heat Loss through element- average per flat kWh	Built Cost of element- average per flat £
	0.60	956	836
	1.70	5,354	715
	5.30	3,633	1,372
	0.75	1,147	1,090
Ventilation - 2 airchanges/hour No special measures		8,030	Nil
Space Heating - gas-fired boiler + radiators			1,650
Total cost for these elements			5,663
Annual Heat Loss		19,120	
Annual Space Heating Cost		£287	

To Current Building Regulations Thermal Standards

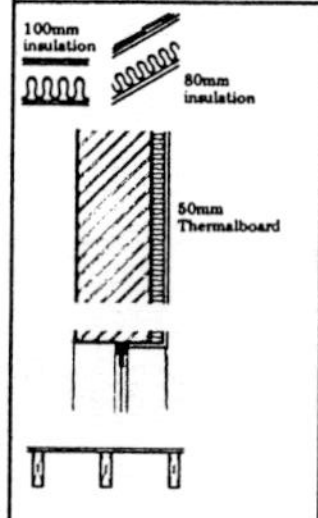

Element	'U' value $W/m^2 {}^oC$	Annual Heat Loss through element- average per flat kWh	Built Cost of element- average per flat £
	0.44 0.34	568	840
	0.60	1,847	1,251
	5.30	3,269	1,372
	0.75	995	1,090
Ventilation - 2 airchanges/hour No special measures		7,531	Nil
Space Heating - gas-fired boiler + radiators			1,650
Total cost for these elements			6,203
Annual Heat Loss		14,210	
Annual Space Heating Cost		£213	

Reduction in space heating cost = **£74** p.a. (26%) at extra capital cost of £540.

As Built

31,33,35, Lawrie Park Road

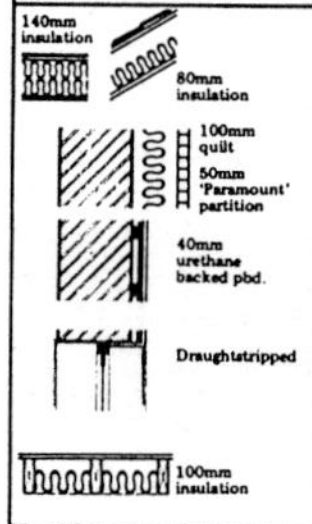

Element	'U' value $W/m^2 {}^oC$	Annual Heat Loss through element- average per flat kWh	Built Cost of element average per flat £
	0.44 0.26	465	856
	0.27 0.47	1,008	1,465
	5.30	2,869	1,464
	0.24	233	1,140
Ventilation - 1 airchange/hour Draught lobbies, draughtstripped external doors (and windows)		3,180	60
Space Heating - gas-fired, warm air system			1,633
Total cost for these elements			6,618
Annual Heat Loss		7,755	
Annual Space Heating Cost		£116	

Reduction in space heating cost of further **£97** p.a. (46%) at extra capital cost of £415.

Figure 6

The eighteen new houses are arranged in three terraces, oriented North-South. (see Fig 7.) The walls, roof and floors are highly insulated, windows are double-glazed on the North side and doors and windows are draughtproofed. The original design included a full width conservatory on the South side of the living room. Unfortunately, cost limits meant the glazed dividing wall could not be built but it is expected that there will still be a net solar gain to this room. The roof of the "conservatory" consists of solar panels for water heating.

The solar water heating is by a thermosyphon system with a "third tap" to allow direct use of the solar heated water. The cost of installing a solar water heating system (approx. £1,500) cannot be justified purely in relation to the fuel cost savings but the cost was incorporated within the overall cost limits. The savings are expected to be in the region of £50 - £70 per year (3,300 - 4,600 kWh).

Figure 8 gives a summary of the specification, costs and space heating fuel savings of these houses as built compared to current (UK) standards.

All the houses and flats are being monitored - this has only just begun as they have only recently been occupied. Temperatures in each room are being measured together with gas and electricity consumption and local climatic data. Data summaries are produced every 24 hours. It is expected that correlation with the estimated consumption figures will be good - based on previous experience on another group of houses.

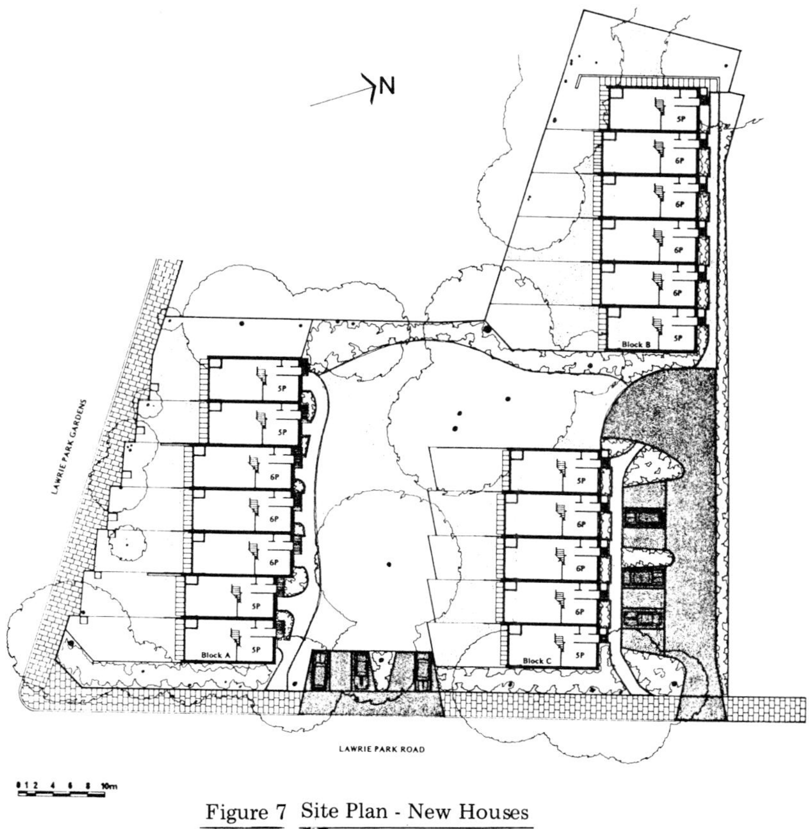

Figure 7 Site Plan - New Houses

Energy & Cost Analysis

To London Building Act Thermal Standards

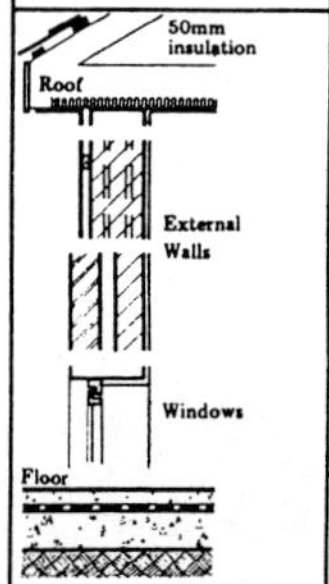

Element	'U' value $W/m^{2o}C$	Heat loss through element per year kWh	Built Cost of element per house £
Roof (50mm insulation)	0.6	1,055	3,190
External Walls	0.94 / 1.0	2,426	2,916
Windows	4.70	3,315	1,931
Floor	0.71	1,216	1,907
Ventilation - 2 air changes/hour No special measures		5,108	Nil
Space Heating - gas-fired boiler + radiators			1,996
Total cost for these elements			11,940
Annual Heat loss		13,120	
Annual Space Heating Cost		£197	

To Current Building Regulations Thermal Standards

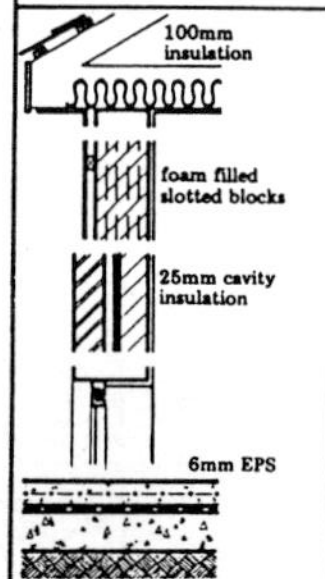

Element	'U' value $W/m^{2o}C$	Heat loss through element per year kWh	Built Cost of element per house £
Roof (100mm insulation)	0.34	552	3,233
Walls (foam filled slotted blocks / 25mm cavity insulation)	0.59 / 0.60	1,412	3,002
Windows	4.70	3,070	1,931
Floor (6mm EPS)	0.60	952	2,092
Ventilation - 2 air changes/hour No special measures		4,730	Nil
Space Heating - gas-fired boiler + radiators			1,996
Total cost for these elements			12,254
Annual Heat loss		10,716	
Annual Space Heating Cost		£161	

Reduction in space heating cost = £36 p.a. (18%) at extra capital cost of £314.

As Built

Lawrie Park Road

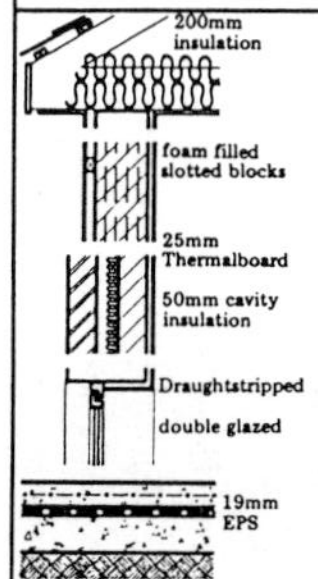

Element	'U' value $W/m^{2o}C$	Heat loss through element per year kWh	Built Cost of element per house £
Roof (200mm insulation)	0.18	209	3,344
Walls (foam filled slotted blocks / 25mm Thermalboard / 50mm cavity insulation)	0.50 / 0.38	650	3,104
Windows (Draughtstripped double glazed)	2.50	1,279	2,477
Floor (19mm EPS)	0.50	522	2,115
Ventilation - 1 air change/hour Draught lobbies, draughtstrip external doors (and windows)		1,553	436
Space Heating - gas- fired, warm air system			1,452
Total cost for these elements			12,928
Annual Heat loss		4,213	
Annual Space Heating Cost		£63	

Reduction in space heating cost of further £98 p.a. (61%) at extra capital cost of £674.

Figure 8

3. THE DEMONSTRATION

The development of the concept of Demonstration resulted from the juxtaposition of two main factors: Firstly, a growing awareness in the minds of the authors that for the widespread adoption of energy saving techniques in buildings, it is not sufficient simply to provide rational proof of the existence and efficacy of the measures and secondly, the introduction (by both the UK Department of Energy and the EEC Directorate - General for Energy) of "Demonstration" programmes afforded the opportunity/funding to explore the problem in depth and propose, implement and evaluate a method of overcoming it.

The designs of the majority of buildings currently under construction, or undergoing major refurbishment, pay scant regard to energy conservation. (At least this is the case in the UK). There are now certain requirements regarding thermal standards incorporated in the Building Regulations and these are, of course, complied with (except they are not always adhered to in London - which has a different type of Building control with no specific thermal requirements). However, the Building Regulations standard is fairly low - lagging a long way behind what can be achieved at a resonable/acceptable cost - in time, effort and money - particularly through use of the Integrated Approach. The materials and techniques are available and cost-effective but are not being utilised, except by a small minority of individuals and groups.

In order for energy conservation to be more widely implemented in buildings it is obviously necessary for change to take place - a change in the behaviour of those responsible for commissioning, designing and constructing the buildings and their services. There are two ways in which change occurs - one is a gradual process whereby new ideas/knowledge/technology are slowly absorbed into society, the other is by "Planned Change". The first, 'natural' process is occuring - society in general is becoming more aware of the need and value of energy conservation but this is an extremely (some would say 'painfully'?) slow process and our aim is to speed it up. We therefore turn to the concept of Planned Change.

First it is necessary to look at which individuals or groups should be the 'target' of the planned change or the potential "change adopters". Ideally, it would be desirable to influence all parts of society at one time so that the change would be wrought throughout society and thus operate more effectively. However, it is generally impractical, as in this case, for the change agent (the authors) to intervene in all groups and at all levels. Therefore, attention must be focused on the individuals/groups/institutions which are: (a) most influential in terms of solving the given problem and (b) most accessible (or can be made so) to the change-agent.

In the area of energy conservation in buildings there are 5 broad categories within society who may have influence on its implementation:

(i) Policy/decision makers.
(ii) Society as a whole including the buildings users and clients.
(iii) The media.
(iv) Industry including the fuel industries.
(v) Designers and constructors of buildings, their services and their environment.

Some of the interrelationships between these groups are indicated in Figure 9.

Of these, the last group - the professionals involved with the building process - was seen to be the one which could most directly affect the energy-efficiency of the buildings and, was also the group in which the change agent could most directly intervene. The behaviour of the other groups with regard to this subject - and indeed, other aspects of societal life - will affect the likehood of the desired change taking place. For instance, should the government and the fuel industries drastically increase or reduce fuel prices, this would affect other's perception of the need and value of energy (fuel) conservation. Therefore, the change agent, whilst concentrating on bringing about change within the building professions, will also attempt to influence these other groups in the same direction, to whatever extent is possible, so as to strengthen the effect of the change strategy.

Demonstration is defined as, "a means whereby, in a non-formal situation, an awareness of the practicality of the subject may be imparted and uncertainties reduced by presenting appropriate information in a suitable format - taking into account the needs, attitudes, previous experience and perception of those to whom it is directed so that they consequently change their existing practices and adopt the desired practices".

In this specific case the desired change is that building practitioners (architects, surveyors, services engineers, quantity surveyors and builders) should incorporate energy conservation measures in their buildings so as to reduce the fuel consumption whilst maintaining or improving comfort conditions.

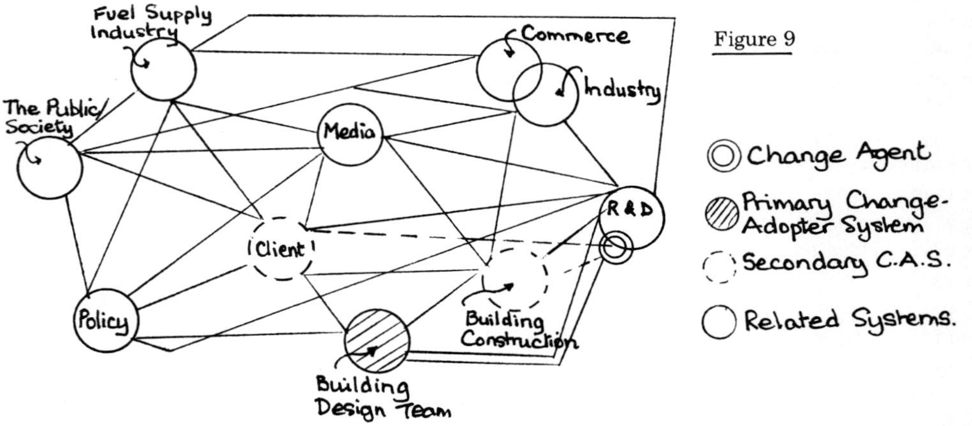

Figure 9

It is, of course, widely accepted that first-hand experience is the most effective strategy for learning/changing, but in the case of energy conservation in buildings there are two barriers to utilising this - firstly, some impetus is required to encourage people to provide themselves with the experience - it cannot easily be given to them and secondly, the time-scale of building (anything from 9 months to 5 years or more) mitigates against the immediacy of experience. The aim then, is to provide a Demonstration which as closely as possible imitates direct experience.

'Demonstration' in this instance is used to describe a particular strategy/mode of presenting knowledge with the intention of bringing about a behavioural change. The desired change is that building professionals will take steps to ensure that their buildings are 'energy efficient' i.e. use the minimum amount of fuel whilst preserving (or improving) the quality of life of the buildings users. There are various strategies which could, theoretically, be employed to bring about this change - such as, for example making more stringent regulations, creating social/moral pressure or doubling the salaries of those who comply! Such strategies, even if seen as desirable, could not be implemented by the particular change-agent in this particular case. The use of Demonstration as the change strategy was indicated both by its characteristics (of being a close substitute for direct experience) and by a particular set of (externally supplied) conditions.

The 'external' conditions which acted in favour of the Demonstration strategy were that the change agent was able to obtain funding for a 'Demonstration project' (from the UK Dept. of Energy and the EEC) and a local authority (London Borough of Lewisham) was prepared to provide the sites for the new building and the rehabilitation of old houses to energy efficient standards. The authority was also prepared to allow the use of part of one old house for the Demonstration for a period of at least 2 years. The Demonstration project is located in a "Demonstration Suite" comprising four large rooms (and ancillary services) being the ground floor of one of the old houses which has been rehabilitated to energy-efficient standards.

Certain restraints were imposed as a result of the location (e.g. from outside, the Suite should appear as an integral part of the house) and by the funding bodies. Within this framework it has been possible to devise a Demonstration made up of a number of elements which together make up this strategy for Planned Change.

The Demonstration includes the following elements, all of which are oriented to take into account the needs, attitudes, previous experience and perception of the potential change-adopters (building professionals):

1. A permanent exhibition presenting:
   (a) The principles of the Integrated Approach to Energy Efficient Building;
   (b) The techniques, materials and equipment used in the new and rehabilitated houses to achieve energy conservation;
   (c) Some of the advantages and disadvantages of various techniques;
   (d) Indications of relative costs and savings of different measures;
   (e) Examples of other energy-efficient buildings;
   (f) Indications of how various measures could be applied in different situations.
2. A film recording the actual implementation of the energy conservation measures during the building process.
3. A full monitoring system - recording internal and external temperatures and fuel consumption - to indicate the effectiveness of the designs in achieving comfort in the dwellings and saving fuel. The monitoring system provides daily summaries of this information so that visitors can relate the previous day's figures with their own recollection of that day's climatic conditions. There is also a computer-generated graphic display where the thermal behaviour of any of the 33 monitored dwellings over any given day can be viewed in a condensed mode - to give a 'feel' for how the buildings respond to climatic conditions over a period of time.
4. A variety of literature for visitors to take away - including reproduction of the information contained in the exhibition; design details ('hints and tips'); lists of relevant trade associations, manufacturers/suppliers of energy saving equipment/materials; bibliography of publications and an indication of individuals and groups with experience in this field who could be approached for design assistance.
5. Facilities for small (up to 25 persons) seminars are available within the Demonstration Suite.

Evaluation of the Demonstration will be carried out during the first 6 months or so of the operation of the project with a view to modifying the elements according to the evaluation results. Further evaluation will be carried out following these modifications. At both these stages it will be possible only to evaluate the immediate effects of exposure to the Demonstration and may not indicate whether any change will endure for a length of time. Therefore, some of the visitors will be contacted some 6-12 months after their visit - to discover whether the change has endured.

COMPARATIVE EXPERIMENTAL STUDY ON HOUSES FITTED WITH A SOLAR ROOF, BURIED PIPES AND A HEAT PUMP

G. ROUSSEAU
ELF Research Centre

Summary :

The heating system studied designated SOL 3 is a system coupling a water/water heat pump, burried pipes, a solar roof, a domestic hot water production and heating floors.
Fives houses sold in the open French market and fitted with this SOL 3 system were the subject of instrumental and experimental analysis during the 82-83 heating season.
The instrumentation is made up of heat meters, electric energy meters, operating time meters, temperature sensors in the soil.
The seasonal performance coefficients for the heat pump were measured at between 2.4 and 2.8 for the houses where the solar roof was not in easy use and came out at between 3 and 3.5 for the house whose solar roof operated well.
In this last case the annual operating coefficient is contained between 2.5 and 2.9.
The consumption taken by auxiliaries represents between 15 % and 20 % of the total electric consumption.
In the immediate vicinity of the burried pipes (less than 10 cm) the threshold for freezing was probably (and certainly for one of the houses) reached.
At 20 cm from the pipes, the temperatures did not fall below 1°C.

1. INTRODUCTION

The SOL 3 System studied at the ELF Research Centre has enjoyed a pre-commercial development.
The aim of this present study undertaken in the framework of the EEC Contract is to follow experimentally the performances of the SOL 3 systems sold in the open market.
The experimental analysis conducted from the 4th quarter of 1982 involved five houses fitted with the SOL 3 System. These houses (except one) were sold in the open market and are, for three of them inhabited normally. This experimentation undergoes the difficulties in the SOL 3 System due to certain poor quality components at the pre-production stage.

2. REMINDERS OF THE SOL 3 SYSTEM

It is a low temperature system. It uses a water-water heat pump which supplies the heat transmitters with the total power necessary. It operates with a high performance coefficient, in drawing the energy from an underground storage, regenerated by the solar roof. We shall briefly describe the elements making up the system.

Solar roof :
The recovery of the solar energy is undertaken by the "SOUTH" slope by means of aluminium heat-exchanging panels with integral channels ("Roll-Bond" technique).
Thesepanels are covered with bitumen shingles traditionally laid (by means of hooks) but in good thermic contact with the heat-exchanging panels. These shingles ensure that the roof is water-tight and that the solar radiation is absorbed. The "SOUTH" slope gives (like the NORTH slope) a traditional appearance making the job of heat capturing invisible.

Underground pipe loops :
The extraction of heat from the ground as well as its regeneration by the solar energy captured in the roof take place by means of an underground heat exchanger made up of small diameter pipes of plastic material. These pipes from the underground pipe loops spread over the soil area.

Water / Water heat pump :
It extracts heat from the ground by means of the underground pipe loops.

Heat distribution by floors :
The heat transmission takes place by the low temperature heating floors. The input temperature into these transmitters never exceeds 38°C (with - 10°C outside) and is regulated as a function of the outside temperature according to a decreasing linear law of heating.
Due to their low operating temperature, heating floors help to improve the performance coefficient of the heat pump and comfort in the living area by removing the hot spots (floor surface temperature is less than 26°C).

Water heater :
The production of domestic hot water is undertaken by an electrosolar water heater coupled to the heat pump in winter and to the solar roof in summer.

3. MAIN FEATURES OF THE HOUSES AND THE EXPERIMENTAL SYSTEMS

	House 1	House 2	House 3
Situation			
.locality	Nandy-Melun (77)	Clermont-Ferrand 63	Morestel (38)
.latitude	48° 50' N	45° 50' N	45° 40' N
.altitude	∿ 40 m	∿ 330 m	∿ 410 m
Features of the house			
. type	T5 (ground+first floor)	T5	T6 (ground+first floor)
. living surface	100 m²	200 m²	150 m²
. living volume	242 m³	520 m³	375 m³
. G coefficient	0.82 W/m³.°C	0.84W/m³.°C(archi) 1.05W/m³.°C(ELF)	0.82W/m³.°C

(foll.)

	House 1	House 2	House 3
Features of the SOL 3 system .Underground loops	. soil : filling with sand . 6 pipes in 3 layers spaced at 40 cm . 2 trenches depth : 2 m width : 60 cm 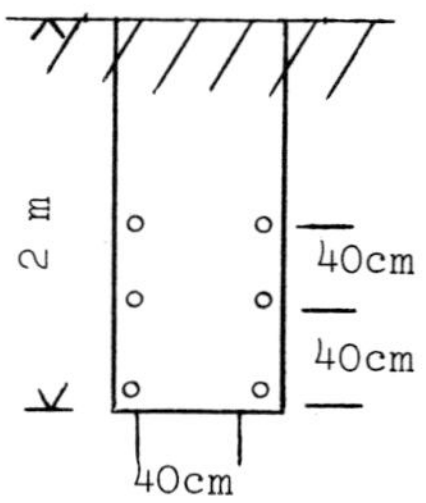	. soil : clay alluvia for 1.50m then clay;filling:sand . 6 pipes in 3 layers spaced at 30 cm . 2 trenches depth : 2 m width : 80 cm 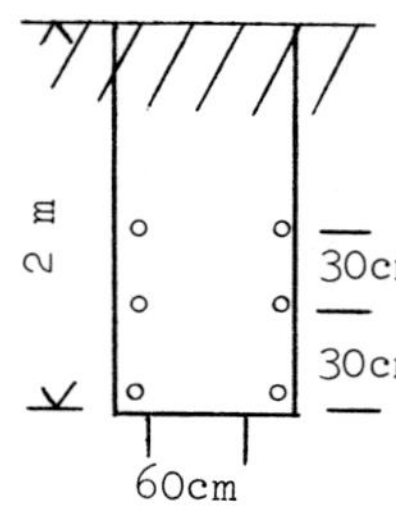	. soil : glacial alluvia . 6 pipes in 3 layers spaced at 40 cm . 1 trench depth : 2.10 m width : 80 to 90cm 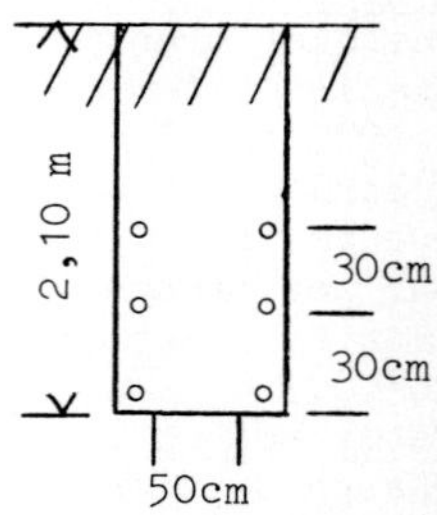
.Solar roof	.overall surface : 48 m² in one slope direction : south inclin. : ∿ 50°	.overall surface : 60 m² in 3 slopes slope A 15.7 m² direct. south-W inclin. 20° slope B 26.2 m² direct. south-W inclin. 20° slope C 17.9 m² direct. south-W inclin. 20°	. overall surface 44 m² in one slope direction :south inclin. : 50°
.Anti-freeze fluid	water+monopropylene glycol (45 %) + additives	same	same
Start of the experimental analysis	Start 4th quarter 1982	Start 4th quarter 1982	Start 4th quarter 1982
Occupation of the house	NO (shutters closed)	NORMAL	NORMAL

	House 4	House 5
Situation .locality .latitude .altitude	Gargenville (78) ∿ 49° N ∿ 70 m	St Nazaire (44) 47° 10' N ∿ 10 m
Features of the house . type . living surface . living volume . G coefficient	F4 81 m² 202 m³ 0.95W/m³.°C	(ground+first floor) 116 m² 381 m³ 0.85W/m³.°C
Features of the SOL 3 system .Underground loops	. soil : filling with sand . 4 pipes in 2 layers spaced at 40 cm . 1 trench depth : 2 m width : 80 cm 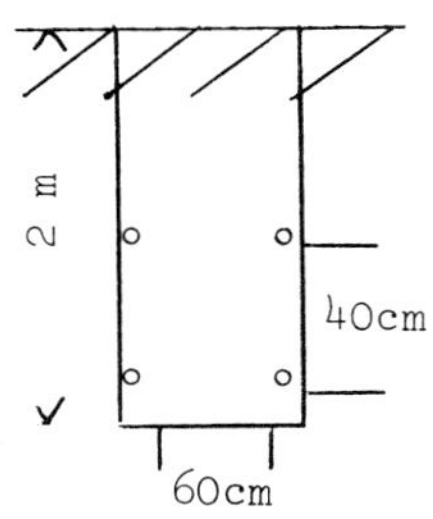	. soil : filling with sand . 8 pipes in 1 layer . 1 layer-trench depth : 1.50m to 3 width : 4 m 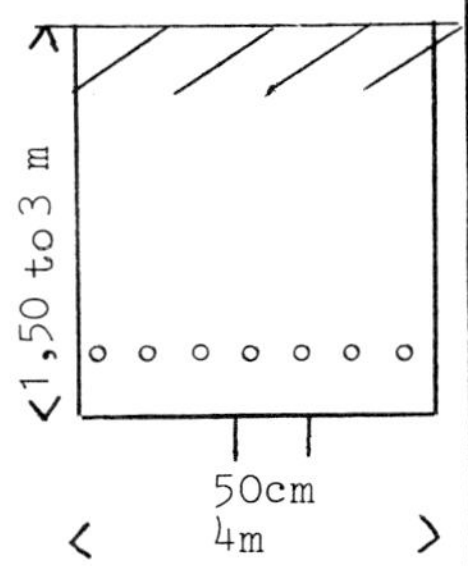
.Solar roof	.overall surface : 44 m² in one slope direction : south inclin. : 45°	.overall surface : 37 m² in one slope direction : south inclin. : ∿ 45°
.Anti-freeze fluid	water+monopropylene glycol (45 %) + additives	same
Start of the experimental analysis	Start 4th quarter 1982	1st quarter 1983
Occupation of the house	NO	NEARLY NORMAL

4. DESCRIPTION OF THE INSTRUMENTATION AND PROBLEMS ENCOUNTERED

4.1. Measuring principle

The instrumentation is made up of 4 types of measuring equipment :
- heat meters at the terminals on the heat pump housing
- electric energy meters
- temperature sensors in the soil
- operating time meters.

4.1.1. Heat metering (see fig I)

The following are metered :
- the heat supplied to each of the 2 heating floors systems :
  metering assemblies : D1, SF1, SC1 and integrator
  D2, SF2, SC2 and integrator
- the heat supplied to the preheating area of the domestic hot water tank
  metering assembly : D3, SF3, SC3 and integrator
- the solar heat captured by the roof absorbers :
  metering assembly : D4, SF4, SC4 and integrator
- the heat stored in the soil by the underground loops :
  metering assembly : D5, SF5, SC5 and integrator
- the heat unstored in the soil by the underground loops :
  metering assembly D5, SF5 twice, SC5 twice and integrator.

4.1.2. Electric energy meters

The following are metered :
- the electric energy consumed by the heat pump by a standard three-phase meter
- the electric energy consumed by the auxiliaries (i.e. the 3 circulators) by a standard single-phase meter.

4.1.3. Temperature sensors in the soil

For certain installations some temperature sensors are implanted in the soil, to be read on a digital display during each visit.

4.2. Problems encountered

The heat metering equipment proved to have very little reliability and suffered numerous breakdowns (in particular on the impulsion transmitters, the temperature sensors). Numerous repair jobs were needed to try and keep the equipment operating.
Nonetheless, this lack of reliability involved an unnegligible loss of data.
On the other hand, the sensitivity of the metering (ΔT minimum of 1°C) led to a sub-evaluation of the heat energies : the results are given by default.

5. WEATHER DATA

	N° 1 Nandy	N° 2 Pont du Chateau	N° 3 Morestel	N° 4 Gargenville	N° 5 St Nazaire
Real data					
. station	Melun Villaroche	Clermont Ferrand	Grenoble St Geoirs	Rouen	Nantes
. period of	27/10to3/5	20/9to29/4	6/10to3/5	7/10to10/9	8/2to1/6
- degrees-days base 18°C	2 152	2 205	2 498	2 430	983
- duration of insolation	524	776	694	632	557
. january					
- average temp °C	5.9	9.7	3.7	5.8	7.5
- duration of insolation	52	110	114	48	77

6. OPERATING CONDITIONS IN RELATION TO THE EXPERIMENTAL RESULTS

House n° 1	. Sometimes insufficient irrigation in the absorbers (various leaks from the Heat Pump, low pressure)
House n° 2	. Continuous operation of the absorbers up to the 26/11 due to faulty regulation . Forced by-pass of the absorbers from the 26/11 to 25/1 caused by too low flow rate . Insufficient irrigation of the absorbers (various leaks from the Heat Pump, insufficient pressure)
House n° 3	. Poor or non-existent irrigation of the absorbers (various leaks from the Heat Pump, insufficient pressure)
House n° 4	. By-pass of the absorbers up to the end of December 82 following a stuck electrovalve . Poor or non-existent irrigation of the absorbers (various leaks from the Heat Pump) . Too low flow rate in the cold circuit (evaporator) leading to a temperature drop . No Domestic Hot Water consumption (house uninhabited)
House N° 5	. Relatively normal operation of the system

7. EXPERIMENTAL RESULTS

7.1. Energy balance-sheets (see table I)

Analysis of the balance-sheets and knowledge of the operating problems show that the systems in houses 2, 3 and 4 worked practically without the solar roof : the heat pumps drew only from the ground without major regeneration by the solar roof.
In this case, the annual performance coefficient of the Heat Pump comes out between 2.4 and 2.8, the annual operating coefficient between 1.9 and 2.3.
House 5 operated with major solar energy capturing. In this case the annual performance coefficient of the heat pump is contained between 3 and 3.5 (*) and the annual operating coefficient between 2.5 and 2.9(*)
The results for the domestic hot water produced for houses 1, 2 and 3 show that the heat pump supplies, by virtue of its gas desuper-heating system, energy for the domestic hot water rising to the order of 7 to 11 % of heat production used for the heating.
The auxiliaries (circulators) consume electric energy representing between 18 and 26 % of the electrical consumption of the compressors of the heat pump, that is between 15 and 20 % of the total electrical consumption (not counting the electric coil of the water heater).

7.2. Evolution of the temperatures in the soil (see figures II, III,IV)

At about 20 cm from the tubes, the soil temperatures fall to the vicinity of 1 to 3°C.
Due to the thermic gradients in the vicinity of the tubes, the sensor placed in contact with the pipe in the case of house n° 2 indicates a temperature 2°C lower than the previous ones.
In the majority of the houses, the soil in the immediate vicinity of the pipes (distance to the pipe < 10 cm) probably reached the threshold for freezing. In contrast at 20 cm from the pipes, the temperatures remained positive.
In January, the differences between temperatures at 20-30 cm from the pipes and at more than 4 m from the tubes (natural soil temperature) settled between 1.5 and 5°C.

8. CONCLUSIONS

The various problems connected on one hand with the measuring equipment and on the other hand with certain components of the system did not enable the balance-sheets to be drawn up in normal operating conditions. Nonetheless the annual performance coefficients were measured at between 2.4 and 3.5 and the annual operating coefficients at between 1.9 and 2.9. The highest coefficients are attributed to the house where the roof operation was normal.

(*) measurement uncertainity

TABLE I : ENERGY BALANCE-SHEETS (kWh)

House	Period	Heat supplied to heating floors	Heat supplied to hot water 2	Total heat supplied 3	Ratio 2/3	Energy restored	Energy stored	Energy captured	Electric Energy for Heat Pump	Performance Coefficient	Electric energy for auxiliaries	Total electric energy	Operating Coefficient	Remarks
1	du 27/10 au 3/5	8140	1099	9239	11.9%	5481		1676	3090	3	630	3728	2.48	Insufficient pressure at roof
2	du 20/9 au 29/4	17330	1901	19633	9.7%				7144	2.75	1356	8500	2.31	Poor roof operation. Low solar energy captured.
3	du 12/1 au 3/5	6838	522	7390	7.1%	4951	1626		2961	2.5 to 2.67	595	3556	2.1 to 2.23	"
3	du 6/10 au 3/5	13741		~14300 (1)					5429	2.71	1159	6588	2.23	"
4	du 7/10 au 10/5	9330				5595	198	392	4209	2.38	1079	5288	1.89	"
5	du 8/2 au 1/6	3855	310 (1)	4165		2934	1709	2250	1379	3 to 3.5	295	1674	2.49 to 2.89	Correct operation

(1) estimation

The missing results correspond to missing or doubtful values.

The heat energy meterings underestimate the real energy : thus, for example, the values obtained for the heat supplied are values by default.

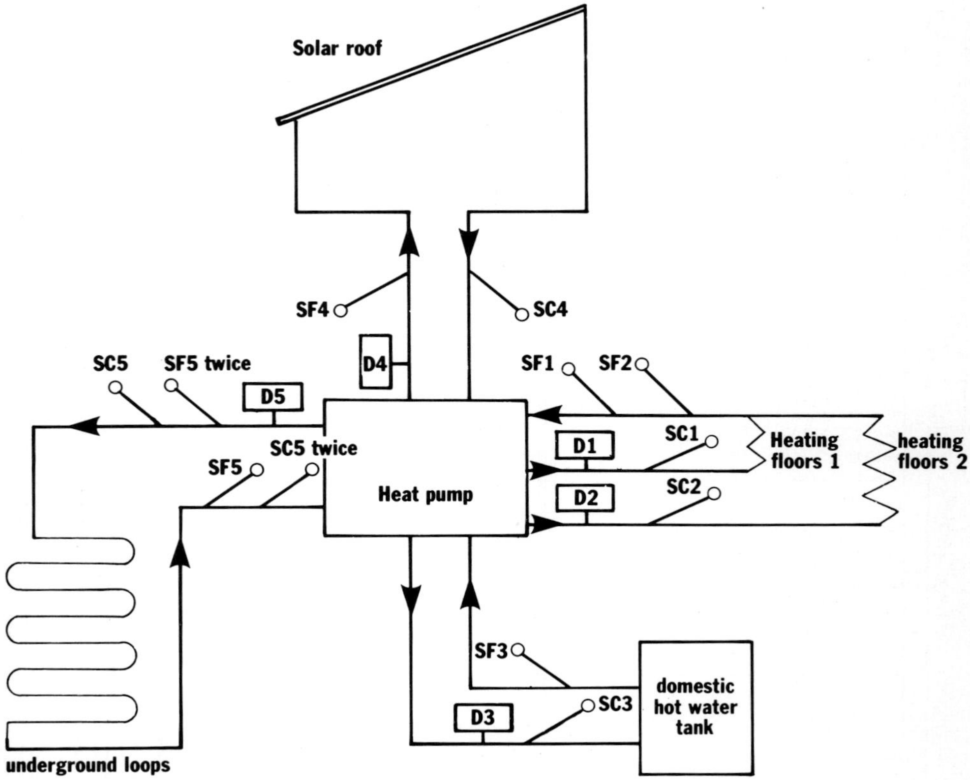

Fig. I: Sensors positions

SFi : Cold temperature sensor for the i metering assembly

SCi : Hot temperature sensor for the i metering assembly

Di : Temperature volume meter for the i metering assembly

Each of these 3 measuring units is connected to an integrator metering energy and volume.

These metering assemblies are made by SCHLUMBERGER, integrator and sensors type CF 21 version 410, volume meters type TG 32 with impulsion transmitters.

In the case of house n° 1 : the instrumentation does not enable the energy stored in the soil to be metered.

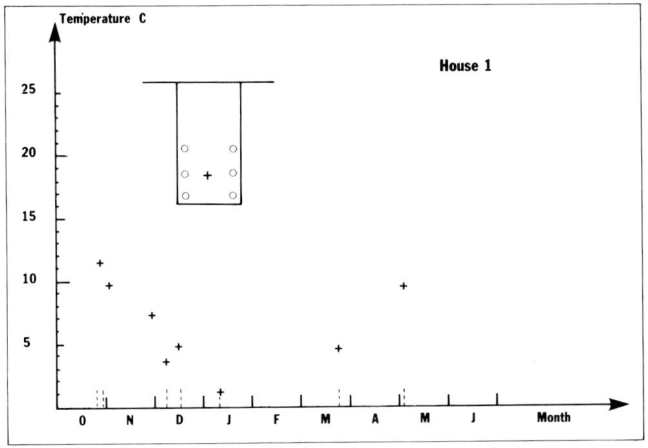

Fig. II: House 1, soil temperatures

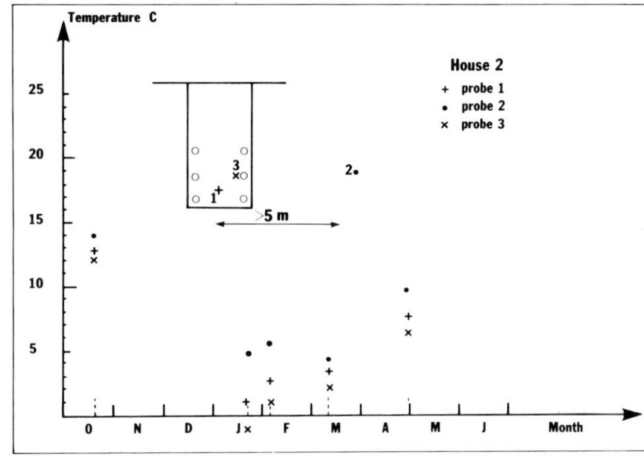

Fig. III: House 2, soil temperatures

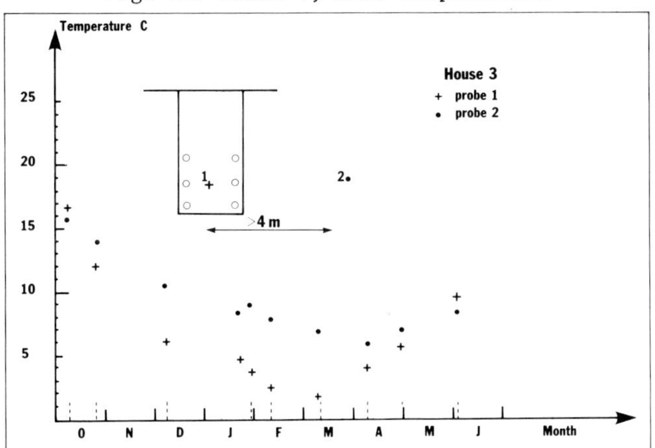

Fig. IV: House 3, soil temperatures

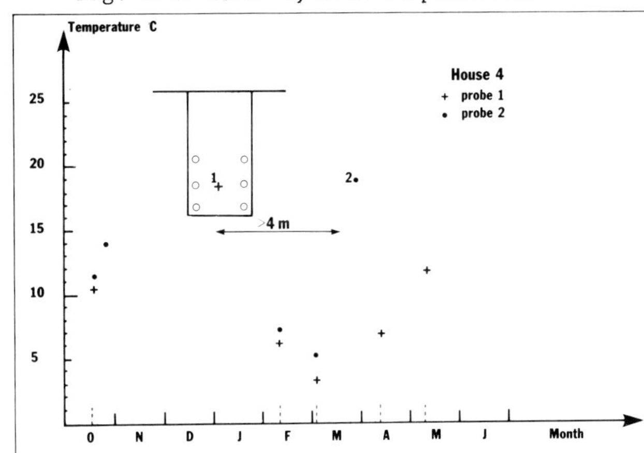

Fig. V: House 4, soil temperatures

THE DESIGN, CONSTRUCTION, MONITORING AND EVALUATION OF A LOW COST, LOW ENERGY DWELLING HOUSE

W.M. Quigley
Project Manager
Cement-Roadstone Holdings Limited

Summary

The aim of this project is to develop a house heating system which will utilise solar energy in a cost-effective way. Points are set out which must be considered if the project aim is to be achieved. The system consists of an air solar collector, integrated warm air heating system, rock bed heat store and electronic controller. These items are described and the present status of the project is outlined. The main conclusion following the construction of the Demonstration Project is that the system looks promising both from the technical and economic viewpoints.

1. GENERAL DESCRIPTION OF PROJECT

1.1 Introduction

The aim of this project is to develop a house heating system which will utilise solar energy in a cost-effective way.

At the outset, two points became clear:

a) Technically it was feasible to provide a reasonably high percentage of the space heating load of a house by utilising solar energy.

b) The cost of doing this with systems that were available, when the project was started in 1981, was absolutely prohibitive.

In order to achieve the project 'aim' the following points were set out as being imperative:

i) The solar collector must be fully integrated into the south facing roof. That is, it must be the structural weatherproof membrane as well as being a well insulated solar collector.

ii) The heat handling system must be integrated with the house space heating system. This includes not only the mechanics of the system but also the heat transfer fluid.

iii) A simple (and cheap) method of heat storage must be provided within the insulation envelope of the house. This includes both short term storage (hour to hour) and medium term storage (day to day). It is felt that long term storage (interseasonal) is not economically viable at this stage.

iv) A sophisticated electrical control device must be designed as it is the core of the total system. Linked to this controller are transducers which not only must have but must hold accurate control points.

v) The system must have a design life of approximately 20 years with as little annual maintenance as possible.

In putting together a system which took account of all the above points two further general points were added. Firstly - the incorporation of the solar heating system into a house should not lead to a radical departure from the traditional form and appearance of an Irish home. Secondly - the net extra capital cost for incorporation of the system should not exceed £2,000 in terms of 1982 money values.

1.2 Background

Interest in solar energy had led to involvement in the design and construction of the low energy passive solar dwelling at Garristown, Co. Dublin. Details of this house are given elsewhere (Ref. 1).

Points which have been demonstrated by that project are principally as follows:

i) The benefits of insulation and the careful detailing of it.
ii) The benefits to be achieved by utilising solar gains; and
iii) The benefits of thermal mass such as concrete floors and walls within the insulation envelope.

The Garristown Project was particularly suited to its rural environment, as the collector was in the form of a solar wall. Following that project a design has evolved which can be used in either an urban or rural environment. Essentially the system consists of a low cost roof air-solar collector which is linked to the house air heating system. Heat storage is by way of a horizontal rock bed store which is placed under the ground floor slab. The total system is controlled by a microprocessor control unit which is linked to the dampers, fan and water pump motors. It is envisaged that the system will be used for space heating and by way of a heat-exchange unit will provide a proportion of the domestic hot water demand.

1.3 Commercial Criteria

Previously this project had been a pure R&D project but when it was realised that it had commercial potential, considerable effort was devoted to developing the components in such a way that they could be mass produced. Not alone that, but also specific attention was paid to using readily available materials and techniques.

Great care must be taken in the selection of materials particularly for the collector. Stagnation temperatures as high as 200°C have been measured on the absorber plate. Durability is of critical importance and hence compatibility of materials must be taken into account.

Flexibility is of considerable importance at this stage. The design is not tied to any particular material or indeed approach. Materials have been selected as being the presently available most suitable for the job. To illustrate this point in the space of one year a number of the materials used were changed and the method of construction has been completely changed. The new method of construction allows production of the units in a factory and hence testing for air tightness prior to delivery - a singularly important point for air collectors.

1.4 Present Status

At present the following major items are being pursued:

- Development of the solar 'box' and methods of commercial manufacture.
- Development of an integrated air heating system including component sourcing and testing.

- Development of the microprocessor control unit.
- Development of the heat store.
- Computer predictive techniques linked particularly to this solar space heating system.
- Formulation of an international patent.

To assist development work in these areas a number of practical projects have been undertaken:

i) A test rig was built in late 1981 at the C-RH Headquarters at Belgard Castle. This test rig has been used to check the characteristics of the air solar collector and the rock-bed heat store.

ii) The total system (including a prototype control system) has been fitted to a single storey dwelling house. This gave experience in making and fitting the solar roof and in fitting out the air handling system together with the rock store. This house is being monitored at present.

iii) In conjunction with the European Commission, C-RH are undertaking a Demonstration Project to assess the solar roof system. This Demonstration Project consists of three houses. The first two houses are of the same design and have an economic level of insulation installed - the only difference being that one house has the solar roof incorporated in its design. The third house has a somewhat lesser level of insulation to meet the minimum requirements of the new Irish Building Regulations. These Demonstration houses are situated in the suburbs of Dublin and construction was completed in August 1983. They have been extensively instrumented and their performance is being monitored over two heating seasons.

2. DETAILS OF SYSTEM AND CONSTRUCTION

2.1 The Solar Collector

The air solar collector consists of an air-tight box which is 5.5m long and 1.0m wide. This 'box' is capable of spanning approximately 3m which means that roofing rafters are not required. Contained in the collector box are two items - the selectively coated stainless steel absorber plate and the insulation system. The insulation level - 150mm of rockwool - is unusually high for a solar collector. The reason for this is that the panel insulation is also the house roof insulation.

2.2 The Air Handling System

The air handling system is integral to the collector, the house heating system and the rock store.

There are six dampers and motors which are set at a particular position to distribute the air in its various modes. The 5 basic modes are:

i) Recirculation of air within the house.
ii) Heating the house from the collector.
iii) Heating the rock store from the collector.
iv) Heating the house from rock store.
v) Heating domestic hot water.

2.3 Description of Test Rig

The test rig consists of $20m^2$ of collector which is linked to a rock bed heat store of $4.2m^3$.

This test rig can be operated in three modes:

i) Heating the house with the collector.
ii) Heating the rock store with the collector.
iii) Heating the house from the rock store.

The IIRS have provided a range of instrumentation which is being used to obtain collector characteristics and system performance. The rig which is very conveniently located has been and is invaluable in giving detailed experience of the system in regard to installation and operation of the system.

A heater has been fitted on the inlet side to the collector for two reasons:

i) To raise the inlet fluid temperature to 20^oC, thereby simulating the return air from a house, and
ii) In order to measure accurately the air mass flow rate. The air mass flow rate can be noted by measuring the electrical energy input to the heater and the temperature rise across it.

2.4 C-RH/EEC Demonstration Project

As mentioned previously, item 1.4, this demonstration project involves building three houses. Details of the houses are as follows:

- House No. 1 is economically insulated and has the solar roof and rock store.
- House No. 2 is of the same design as house No. 1, is economically insulated but does not have a solar roof or rock store. This house acts as the control house for the solar house.
- House No. 3 is of the same design but is insulated to meet the requirements of the new Irish Building Regulations.

The insulation level of the economically insulated solar roof house is as follows:

Item	Insulation	U-Value (W/m^2K)
Floor	150mm EPS + Ins. Rock Store	0 $\longrightarrow$ 0.10
Walls	110mm EPS Bead	0.30
Roof	150mm Rockwool	0.23
Windows	Double Glazed	2.8

The three houses are two-storey detached dwellings with a habitable floor area of $115m^2$. The attic area is within the insulation envelope and this contains the mechanical equipment required i.e. fans, dampers, motors, control unit, pump etc. The front elevation faces due south and the solar roof is directly above this wall. The solar roof is $55m^2$ gross area and is at an angle of 45°.

Energy gain during the heating season is obtained from the following sources:

i) Casual gains.
ii) Direct solar gain (through south facing windows).
iii) Solar roof collector.
iv) Auxiliary heating system.

3. CONCLUSIONS

i) Following the work on this Demonstration Project it is believed that this solar roof system could be installed cost-effectively in a house.

ii) Basic requirements are:
- Roof south facing ($\pm$ 10°).
- Roof slope must be greater than 40°.
- Ideally 0.5m^2 collector to 1m^2 floor area.
- Rock store should contain 0.3m^3 of rock to 1m^2 of collector.

iii) The system can be designed to 'fit-in' reasonably well with architectural requirements.

iv) The questions of durability and strength are being and must be closely watched. Problems in this area can be minimised by using best available materials, good design techniques and by encouraging the development of new materials with improved properties.

ENERGY SAVING LIGHTING SYSTEMS IN EXISTING OFFICE BUILDINGS

A report of a demonstration project carried out in one of the office buildings of the N.V. Nederlandse Gasunie at Groningen, the Netherlands

P. EUSER, H.J. NICOLAAS, H.C. PEITSMAN, TPD-TNO, Delft

H.M. ROOS, W. Ch. MALLON, Nederlandse Gasunie, Groningen

Summary

A demonstration project on energy saving lighting systems has been carried out in one of the office buildings of the N.V. Nederlandse Gasunie at Groningen, the Netherlands. Five different systems were investigated under comparable conditions, each system on a particular floor of the building.

User behaviour, user acceptability to the various lighting systems and operational performance of the systems were also included in the investigation. The experiment demonstrates that by reducing installed wattage and adopting daylight-sensitive controls, it is possible to achieve energy savings of up to 80% on the lighting of existing office buildings. User behaviour can also account for a substantial part of the saving.

Introduction

Not so very long ago is was quite common, particularly in the USA, for offices to be built entirely without individual light switches. All the lighting was controlled from a central point. The saving on wiring and switches more than outweighed the increased energy cost due to the fact that the lights would often be on unnecessarily for a large part of the day.

The sharp rise in energy prices has now changed all that. However, even compared with traditional office lighting systems in which there is individual control over whether lights are on or off, savings of up to 80% on energy consumption can be achieved. This has been demonstrated by an experiment of N.V. Nederlandse Gasunie at Groningen.

The aim of the project was to find an office lighting system that on the one hand would give the maximum energy saving and on the other would still provide a perfectly satisfactory level of illumination. Savings can be achieved in two ways, first by reducing the installed wattage per unit area and second by the automatic control of artificial illumination.

The project was financially supported by the European Commission, Directorate-General of Energy, in Brussels, and a number of Dutch organisations and companies.

The project was carried out by:

-N.V. Nederlandse Gasunie

-TNO-institute for Environmental Hygiene and Health Technology (lighting aspects, survey of user capacibility)

-TNO-TH Technical Physics Dept. (measurement of energy consumption)

Project set-up

The project was carried out in one of the Gasunie office buildings. Five different systems were investigated under comparable conditions, each system on a particular floor of the building. The design of the building is fairly typical, a corridor running the length of each floor with rooms both sides. The size of the rooms is variable in multiples of bays, depending on the position of movable partitions. The original lighting system had been designed with this variability in mind so as to prevent the situation from ever arising where the switch to a given light was to be found next door.

The lighting systems tried provided both general illumination and specific illumination of the working area. The systems differed in installed wattage, design and method of control. The energy consumption of each system was measured over a period of one year. In addition, a survey was carried out among users of the building to ascertain the acceptability of the various systems from their point of view.

The lighting systems

In order to optimise the use of available daylight a number of uniform modifications were implemented on all floors, viz.:

- the wiring was altered - with the original system the lights in each room were wired up in pairs, one pair to the window zone and one pair to the corridor zone; for the experiment, separate push button control was provided for each light in both zones, involving considerable rewiring;
- the lighting circuits to each zone/room were equipped with relays to enable the lights to be switched off automatically by the control system;
- the lights serving the window zone were moved closer to the window.

The various lighting systems are described below. The control systems are discussed in a separate section.

The reference system

One floor initially served as reference. It retained the original lighting with fluorescent tube fittings (with louvred screens) each with two standard 65W tubes and the existing switches enabling the occupant of a room to switch the lights on and off at will. However, during the course of the experiment it proved to be unrepresentative owing to wildly fluctuating levels of occupation and a floor in another indicental Gasunie building (G-block) was later taken as reference. This floor was equipped with ceiling-mounted single-tube trough reflectore units fitted with 58W fluorescent tubes. The wiring was again left untouched.

General illumination

On the second floor of the building the original lighting with the twin-tube ceiling-mounted fittings was basically retained. In addition to the common modifications mentioned above, however, the standard fluorescent tubes were replaced by 58 W high-output tubes (type TLD).

General illumination with dimmer control

On the fourth floor, too, the original ceiling mounted lighting was retained. Here, however, two different dimmer systems were installed, one giving partial dimming and the other full dimming. In order to provide full dimming capacibility, half of the lights had to be adapted by fitting them wiht filament voltage transformers and special fluorescent tubes (type TLM), in addition to the modifications common to all floors.

General background illumination plus lighting of the working area

On the third floor, general background lighting was provided by single-tube low-brightness ceiling mounted units fitted with the 58 W high-efficiency fluorescent tubes. This was supplemented by light from desktop units for individual working areas. The ceiling lights were wired up in the manner already described.

Pendant units

On the fifth floor, pendant units were installed, each fitted with two high-efficiency 36 W tubes. Here, too, the lights were arranged so as to provide separate illumination for the window zone and corridor zone as far as possible. The luminaires of these units are designed to give direct illumination of working areas (desk and tables) and at the same time some indirect illumination of the room as a whole. The users were consulted as to how they would like the lights positioned. The window zone lights were under automatic switch control depending on the level of daylight. The lights for the corridor zone could be switched on and off at will.

Automatic occupancy detection

Sensors were installed in a number of rooms to detect whether anybody was actually present. The detection system automatically switched off the lights if the room was left unoccupied for more than five minutes. This enabled the effectiveness of switching off lights in unoccupied rooms to be accurately determined.

Summary of lighting systems

By way of clarification, table 1 gives a summary of the various lighting and control systems tested during the project.

Table 1

Floor	Lighting system + installed W/m2	Control system
6	original lights 30 W/m2	manual (1st reference system)
G4	trough reflectors 15 W/m2	manual (2nd reference system)
2	original lights 30 W/m2	off: automatic, depending on daylight; seperate control for window and corridor zones on: manual
4	original lights 30 W/m2	on: manual automatic partial dim/off depending on daylight; window and corridor zones controlled seperately

	original lights 30 W/m2	full dim/off depending on daylight; window and corridor zones controlled seperately on: manual N.B.: automatic occupancy detectors installed in two rooms
5	pendant lights 11,5 W/m2	window zone: off: automatic, depending on daylight on: manual corridor zone: manual
3	low-brightness lights 17,5 W/m2 desk lights	off: automatic, depending on daylight; window and corridor zones controlled seperately on: manual manual

The control systems

Automatic "off"

In order to make the greatest possible use of available daylight the above mentioned systems were augmented by automatic daylight-sensitive "off" control whereby the lights were in fact "tripped" off with increasing daylight. The amount of daylight was measured by sensors installed in the stairwell of the building, this being taken as a good indication of the light entering the rooms.

With decreasing daylight the "block" on the lighting was lifted but it required manual intervention on the part of the occupier of a room before the lights were actually switched back on; it was necessary to get up and press a button by the door when more light was wanted.

The division of the systems into window zone and corridor zone was done to take account of the fact that the amount of daylight in a room will tend to be greater closer to the window and it allowed the lighting to be tripped and the block on the lighting to be lifted independently for the two zones. As daylight increased, the lights in the window zone were the first to be switched off, the corridor zone lights not being tripped until a higher daylight level was measured. With decreasing daylight the reverse was the case: the block was first lifted on the corridor side and then lifted on the window side at an even lower level of light.

Oversensitivity of the system was controlled by a built-in delay so that the lights were not switched off with every passing cloud or window cleaner. Independent adjustment of the system delay was provided for.

Automatic dim

On one floor the "off" control system described above was augmented by automatic dimming, which enabled a smoother transition in the overall level of illumination to be achieved since the lights were not switched off at full power. This prevented large variations in luminance. Two different dimmer systems were tried, partial and full dim.

- Partial dim

  The lights connected to the partial dim system were the original units fitted with high-efficiency tubes. With increasing daylight the supply voltage was gradually reduced down to 85% of full power. Further reduction proved impossible with this type of fluorescent tube so that, if the level of daylight continued to increase, the lights were switched off altogether.

When the block was lifted as the amount of daylight decreased, power was restored initially at the 85% level again.

- Full dim
  The full dim system worked along similar lines, the essential difference being that the power supply was reduced to 165 V, which was equivalent to 10% of maximum light output, before the lights were switched off altogether. The daylight levels at which the system switched obviously had to be set accordingly. In addition it was necessary to fit special tubes and install transformers to up the filament voltage, which involved considerable rewiring.

The dimmer systems brought with them a number of specific problems:
- hum and flicker;
- variation of the switching points depending on the weather and cloud cover. The result was that on one occasion there would be too little light falling on the desks while the lighting remained blocked and on another occasion the lights would not be switched off although daylight alone would have been quite enough to work by. The answer here was to set the switching levels such that there was at least adequate illumination of the desks in all circumstances.
  During the course of the project other ways of overcoming the above problems were also tried. One experiment involved using a fully furnished scale model of a two-bay room to measure daylight levels. This did give less scatter in the "on" and "off" levels than was the case with the stairwell sensors.

Masses of data

In order to draw any valid conclusions from the experiment it was necessary to measure and record a great deal of data, the most important factors being listed below:
- The electricity consumption per lighting group (26 groups in all); this was done by kWh-meters fitted with pulse transmitters, a count of these pulses providing a measure of the amount of electricity used.
- The daylight level in window and corridor zones for both east and west sides of the building; this was done, as already described, by daylight sensors installed in the stairwell.
- The intensity of the sunlight incident upon the east and west sides of the building; for this purpose a "solarimeter" was installed on each side of the building.
- The occupancy of the rooms; generally speaking, the presence or absence of the occupants of rooms will greatly influence electricity consumption, especially with systems in which the user has to switch the lights on again manually. The number of persons present at any particular moment was measured as accurately as possible from the clock cards. Everybody clocked in and out and the clock also registered whether a person occupied a room on the east or west side of the building and of course the floor the room was on.

These and other data were recorded as a function of time during normal working hours, i.e. between 8.00 hours and 18.00 hours, Mondays to Fridays. Besides the above data there are other factors affecting electricity consumption such as individual behaviour patterns and the use of window blinds. Individual behaviour will always be a variable that is difficult to quantify but the use of blinds was measured.

The Gasunie office is fitted with venetian blinds at the windows and in order to determine what effect the use of blinds was having the facades of the building were photographed for part of the project, morning and afternoon, involving about 250 individual slides. The numbers of blinds shown to be in use on the slides were counted. It was found that little use was made of the blinds, particularly in the rooms on the west.

Results of the experiment

From the measurement data it was possible to arrive at accurate figures for the energy consumption on lighting. But these figures obviously had to be corrected in order to make comparison possible between the various lighting systems.

Table 2 - Measured energy savings

Floor	installed power W/m2	annual consump. max. possible kWh/m2	h	annual consump. measured kWh/m2	energy savings %
6	30	75	1430	42,9	0
2	30	75	530	15,9	63
3	18	45	429	7,7	82
4part.	30	75	400	12	72
4*full	33	82,5	470	15,5	64
5	12	30	556	6,7	84

Table 2 shows the specific energy consumption of the various systems. The third column represents the maximum amount of energy that would have been consumed by the various systems if the lights had been left on for the whole year during normal office hours (600 h). The substantial differences here are due to the difference in installed wattage per unit area. It will be apparent that the energy consumptions on the third floor (low-brightness ceiling units plus desktop lights) and on the fifth floor (pendant units) were almost equally low. The other floors showed somewhat higher energy consumptions.

Finally as the table shows, the pendant units and the background illumination backed up by desk lights are the best systems from the point of view of energy consumption. However, this does not mean that they must necessarily be preferable since there are numerous other aspects to be taken into account in deciding upon a particular system. For instance, it must be acceptable to the users as well as to the operator of the building. And there is also the matter of the technical side of the installation to consider such as the necessary regrouping of lights and the rewiring involved.

Table 3

Floor	6	2	3	4	4*	5
% of light switched on	57	21	17	15	20	20

Table 3 gives a good indication of the effect of automatic control on lighting energy consumption. This table reveals the length of time for which the lights in the various systems were actually on during normal office hours, expressed as a percentage of the maximum number of hours for which they could have been on (2600 h). It can be seen that the lights were on for 34% of office hours of 1430 h in the case of the reference system.

For all the systems with automatic "off" control, where manual intervention was required to put the lights back on, the equivalent figure was down to 15-20%. This represents a remarkable drop of 35 to 40 percentage points in the length of time that lights were on compared with the maximum of 2600 h.

The differences between the various controlled systems are nog significant. The apparent differences are due in part to the margin of error inevitably involved in correcting the actual consumption.

The effect of occupancy levels and individual behaviour patterns

The lights on the second, third, fourth and fifth floors were automatically switched off when the level of daylight exceeded a set value; to switch the lights back on again required action on the part of the occupants but this was only possible when the block on the lighting had been lifted by the central control system. With this kind of control the lighting energy consumption obviously depends a great deal on whether rooms are occupied since the lights will only be switched on when there is someone in the room to do so, and then only when the need for artificial light is felt.

To compare with the length of time which the reference-floor lights were on, viz. 55% of the possible 2600 h, an analysis was made to establish for how much of the time the block on artificial lighting was lifted in the controlled systems. It was found that the lights were actually free to be switched on for 38% of the 2600 h. In other words, the daylight contribution was not enough to provide the recommended level of illumination for that proportion of the time. The reduction in the artificial lighting time with this system was therefore due to two things, the automatic switch-off (17%) and the fact that the occupants did not switch the ligths back on immediately they were free to do so (18%), since they apparantly did not need the extra light. The further potential saving provided by the full-dim system with automatic switch-off (with respect to the second floor system wiht automatic switch-off alone) is small compared with the saving of 18% that can be described to "individual behaviour").

The most important result of all the measurements must be that the amount of use of artificial light was reduced from 1450 h for the non-controlled reference system to less than 550 h for the controlled systems.

User acceptability

An important part of the project was the survey to establish the reactions of the occupants of the Gasunie building to the various lighting systems. This survey was divided into two phases. The first phase took the form of a general questionnaire put to the occupants before work on the technical side of the project began, in other words, when all floors still had the original lighting arrangements. The questions covered the working environment as a whole and not just the lighting. This was done so as to get as natural a response as possible concerning the lighting. The second phase of the survey was carried out following the changes to the lighting systems.

Results of the user acceptability survey

The number of complaints concerning the lighting during the first phase of the survey was negligible.

The second phase of the survey was carried out after the various lighting systems had been in operation for about six months. The number of complaints about the lighting was found to have increased considerably, particularly on the second, fourth and fifth floors, which were equipped with the standard lighting plus automatic "off" control, ceiling mounted units automatically dimming/switching of and the system with pendant units, respectively. The problems experienced initially while adjustment were still being made to the control systems undoubtedly contributed to the level of dissatisfaction expressed.

On the sixth floor, which initially served as reference and where accordingly no fundamental changes had been made to the lighting system, the occupants were generally satisfied with the lights.

On the third floor, too, with the low-brightness ceiling units and desk-lights, the users were still happy with the lighting when questioned on the second occasion.

The lighting project was made the office users much more aware of the question of light. As regards the level of illumination in the rooms generally and the illimunation of the working area, all the experimental systems basically performed adequately. When it was dark outside and all lights were on, 85-90% of users were quite happy with the amount of light. The majority of users found the level of illumination on an overcast day, when only some of the lights could be switched on, also perfectly adequate. Only in the case of the system with pendant lights was the number of dissatisfied users slightly larger and this had to do with the critical positioning of the lights - not parallel with the long side of the desk and not too low.

Greater dissatisfaction was expressed concerning the lack of control the user had over the ligts, particularly on the second, fourth and fifth floors, where the lights were switched off or dimmed by the central control system, depending on the level of daylight.

On the third floor (desk lights), however, the number of complaints about controllability was less owing to the fact that the users could at least switch the desk light on and off at will. The complaints these users voiced had more to do with the effects of having desktop lighting, such as reflection problems, the space occupied by the light, etc. A quarter of the users said they never had the desk light on and 60% found the design of desk lights not ideal.

Recommendations for reducing or removing user objections to the lighting systems are as follows: to raise the trip points so that the lights are not switched off so soon and the block is lifted earlier; to set a long enough system delay to overcome the problems caused by transient variations in the level of daylight and to provide facilities for switching the lights back on from the desk so that the user does not have to keep getting up to do so. Obviously, these measures lead to higher energy consumption but it must be borne in mind that the search for the optimum lighting system is bound to involve some compromise between the requirements of low energy consumption and satisfactory working conditions.

Of the systems investigated, the low-brightness system with individual desk lights appears to perform best and would be even better if the user objections to the design of the desk light and the space it occupies can be reduced. As already revealed, this system performed particularly well from the point of view of energy consumption.

One form of dissatisfaction not yet discussed stemmed more from the project itself.

Many office users complained about the fact that a great deal of technical work was done on the lighting during normal office hours, which caused considerable disruption. Many also felt confronted with a fait accompli; that there had not been enough prior consultation with those directly affected by the project. It is probable that the number of complaints would have been smaller if it had been explained personally to those who were to be affected by the project what the experiments involved and why.

Technical aspects of the installation

Prior to the project the Gasunie building concerned had an electrical installation comprising ceiling mounted lights, standard light switches and power points. Installing the trial systems consequently involved considerable alteration of the existing installation. In fact, rewiring was called for on most floors so that the lighting in window and corridor zones on both east and west side of the building could be controlled seperately from a central point. This meant quite a lot of work by electricians, which was disturbing to the normal office routine. The work involved included the following:

- moving ceiling mounted lights closer to the windows;
- replacing the normal light switches with suitable switches for the experiment;
- fitting relays to the light units;
- rewiring to obtain the desired grouping of lights;
- replacing light fittings with other types.

This incomplete summary is sufficient to indicate what a major difference there is between dealing with an existing building and a new one, in which the right installation could be build in from the start. In any existing building there is going to be a need for substantial alterations to the electrical system. And in either case, a new or existing building, the installation must take into account the possibility that the office accommodation may at some date have to be rearranged, perhaps involving relocation of partition walls.

Installing dimmer systems demands caution. With normal fluorescent tubes the supply voltage cannot be reduced below about 85% of nominal mains voltage. The dimmer system in fact cuts off altogether at 150% of the set standard light level, which means that in the summer the saving is mainly due to the lights being switched off entirely rather than merely being dimmed. On dark winter days however, the saving due to dimming will be greater. There is also a problem due to the difference in the response of fluorescent lights to dimmer control, depending on whether they have capacitor or choke (inductance) type starter circuits. And the answer is not as simple as using only one type of light since the public utility requires large scale lighting installations to have equal numbers of both types. Briefly then the extra energy saving achievable with partial dimming in addition to automatic switch-off is small compared with a system which simply turns the lights off.

A full-dim system on the other hand does enable a greater saving to be achieved, but not without drawbacks. The fact that full dimming requires a specially developped type of fluorescent light (type TLM) is a serious consideration. These lights necessitate the use of a transformer to up filament voltage and this means extra wiring. The transformer was switched off outside operating hours to avoid wasting electricity. The electronic dimmer system also needed very careful adjustment.

The control curve was too steep, which gave rise to problems. And to overcom problems of getting the lights to strike on lower power, the control range had to be limited which of course limited the possible energy saving.

On the fifth floor there were problems with the pendant light units because there was no simple way of adjusting their position. Suspending the lights from tracks was tried, which did provide a small range of movement longitudinally but it was hardly practicable to enable the lights to be raised or lowered, although some resourceful users tried to position the lights how they wanted them by means of guy-lines and by tying knots in the cords. Installing suitable rise-and-fall units would enable the users to adjust the height of the lights more readily. And to be able to change the horizontal position of the lights to suit internal rearrangements of the office accommodation necessitates special technical provisions such as a system of carefully sited lighting tracks.

Finally as an illustration of the sort of unexpected problem that can arise, the ceiling mounted office lights were basically arranged at right angles to the outside walls except in one room, where they were replaced by ligths placed parallel with the windows. This was found to cause draughts from the air-conditioning outlets, however.

Conclusion

The experiment demonstrates that by reducing installed wattage and adopting daylight-sensitive controls, it is possible to achieve energy savings of up to 80% on the lighting of existing office buildings. User behaviour, however, can account for a substantial part of the saving.

The experience gained during the trial once more emphasises how important it is to ensure that people are fully briefed beforehand about what is involved in the installation of an energy saving lighting system and to maintain good employee relations throughout.

Economic evaluation

Economic evaluation for an existing 6 floor building with 30 rooms per floor.

system (floor no.)	investment Dfl.	annual consumption kWh	annual saving %	annual saving Dfl	pay-back period
6	117,000	154,440	-	-	-
2	143,300	57,140	63	24,300	1.1 (5.9)
3	254,900	27,800	82	31,660	4.4 (8.4)
4 (part.)	173,300	43,240	72	27,800	2.0 (6.2)
4 (full)	181,800	55,600	64	24,710	2.62 (7.4)
5	180,500	24,710	84	32,432	1.96 (5.6)

For the systems on floors number 4 and 5 the calculation is based on modification of the present system. The others are newly installed systems. The replacement cost of the original system was taken as a basis for calculating the pay-back period of the higher cost.
The figures in parentheses concern the total investment.
The price of 1 kWh is Dfl. 0.25.

HEAT RECOVERY SYSTEM AND LOW ENERGY CONSTRUCTION FOR A SPORTS AND LEISURE COMPLEX

DARLINGTON BOROUGH COUNCIL - DOLPHIN CENTRE

G. LOWES
Darlington Borough Architect
C.J. REEVES
Consultant Engineer, Simpson, Coulson and Partners
Professor A.C. HARDY
A.C. Hardy and Associates, Advisers to Darlington B.C.

Summary

The gradual deterioration of existing swimming facilities prompted Darlington Borough Council to determine the most effective design capable of providing swimming and dry sports facilities in a controlled environment for all year comfort. A building of 12900 sq.m. was started in February 1979 on a compact Town Centre site of 6000 sq.m., enclosing, in a high performance thermal envelope heat recovery and management systems. Designed to be cost effective and energy efficient, it was completed in April 1983 with a construction cost expected to be within £9m. The architecture is integrated sympathetically within a historic setting and sensitively houses the extensive equipment installed for the environmental systems. The recovery system is based on a recirculation of air, including that of the ozone treated swimming pools, using run-around coils, plate heat exchangers and heat exchangers and heat pumps. The installations are operated through a Building Automation System. The facility for swimmers, sportsmen and spectators, bars and restaurants have been integrated with a Hall built in 1846 suitable for banquets. Designed to accommodate 2000, the Centre successfully averages 2400 users each day bringing life and income to the centre of the Town.

1. INTRODUCTION

The policy of the Council to rebuild its swimming facilities and add dry sports within the centre of the Town, set many design challenges in all aspects of building and the ambition of realizing a Centre for continuous use in an acceptable environment.

The choice of site with the challenges of being adjacent to an old Market Square and set among buildings of historic importance, had advantages in its compactness and proximity to many local and national bus routes.

Central Hall dated 1846 was successfully integrated with the new building to provide an elegant and cost effective addition to the compact site. Also existing site levels favoured the provision of essential space for large pieces of equipment at ground level.

The structure of the large space elements produced roof voids suitable for ease of location of equipment and, in all, fourteen plant and switch rooms were required. Uses of the various facilities also favoured a low energy design allowing large areas of well insulated wall with few windows.

The building was completed in April 1983 and no monitoring of the completed building has yet been possible, nor will it be able to start until all plant is finally commissioned and operational which should be in the near future. There is, however, general evidence of performance. It is already clear that each item of energy management, recovery or fabric, work better together as a package than individually and as suggested by their predicted performance.

The temperature of the pools falls as little as $1^{o}C$ /24 hours due to the thermal inertia of such a large volume of water and the performance of the building fabric. Records indicate no loss overnight on occasions. Indeed only one of the three 1 Mw gas boilers was needed this summer for 1 hour/day for domestic hot water. Assumptions of use in initial calculations of 300,000 per annum @ 100w/person for an average of 2 hours have changed due to the success of the Centre. Yearly average figures are currently 850,000 and the energetic use of the facilities suggests a metabolic heat input higher than originally envisaged. There has been a clear demonstration of popular use of the building to such an extent that the profitability of local bus service has been significantly improved during the time of day when light traffic has been the norm.

The computer controls installed with grant aid from the EEC will monitor for two years and will start in the near future.

2. THERMAL ENVELOPE

2.1. The compactness of the site and building produces a minimum external surface area. The main structure is a galvanised steel frame with cavity brick external wall giving faced brick finish externally and internally. Insulation of these walls, appropriate to avoid risk of interstitial condensation and minimise maintenance costs, links with insulated pitched and flat roofs.

There are two specific areas which required different design, the dry sports, changing and leisure areas, and the pools and wet changing areas where temperatures and humidity are higher. Although the insulation of the dry area walls is less expensive than that required to meet the different internal humidity levels of the wet sports areas, the performance is close (see tables i and ii). The roofs have high insulation values (see tables iii and iv) and good thermal time lag to reduce, in the summer, the influence of solar radiation on the air conditioning or ventilation load. However, the roofs to the wet areas have a high moisture resistance on the warm side of the roof and a plenum is provided in the roof space to assist moisture resistance.

Most dry sports areas have no glazing and those in the pools areas are designed to a minimum, not only to reduce direct solar radiation, but also to improve competition requirements by excluding solar and sky reflection and glare.

2.2. Actual construction of thermal envelope

Table i

Dry area external walls

90 mm external brick leaf
25 mm ventilated cavity
75 mm glass fibre insulation batts
100 mm inner brick leaf

$U = 0.40\ W/m^2\ ^oC$

Table ii

Wet area external walls

90 mm external brick leaf
60 mm ventilated cavity
50 mm polyisocyanurate plastic foam insulation
90 mm inner brick leaf

$U = 0.37\ W/m^2\ ^oC$

Figure 1

Figure 2

Table iii	Table iv
Pitched roofs (1)	Flat roofs (1)
Metal profile sheet 26 gauge	30 mm concrete paving slabs (1)
80 mm chipboard on timber bearers	25 mm air cavity
Ventilated cavity	Waterproof membrane
40 mm polyisocyanurate plastic foam insulation	12 mm fibre board
	40 mm urethene foam insulation
U = 0.30 W/m^2 ^{o}C	U = 0.40 W/m^2 ^{o}C

Windows are hardwood to avoid internal frame condensation (1). Those in wet sports areas are double glazed.

In all construction, cold bridging has been avoided by isolating the steel frame and insulating the closing of cavities in external walls.

3. HEAT RECOVERY ON AUTOMATED BUILDING SYSTEMS

3.1. Introduction

In coming to the design of environmental services for this complex, there were at work a number of guiding principles:

3.1.1. for the wide range of activities to be undertaken in the complex, people had to be comfortably and safely accommodated so that they would wish to use the building again;

3.1.2. the services were to be flexible and by means of heat recovery and centralised automatic controls, operating costs were to be minimised;

3.1.3. heavy plant was to be housed at ground level and in all cases plant serving a particular area would be located close to the load centre of that area thereby minimising distribution losses and facilitating plant switching to suit varying occupancy patterns;

3.1.4. all plant was to be located in defined equipment rooms thereby encouraging regular maintenance with the minimum inconvenience to members of the public; ease of access was important;

3.1.5. although the building has a useful life of say 60 years, much of the central plant will be replaced in this period. Attention was to be paid to this aspect to avoid excessive building works when such replacement becomes necessary;

3.1.6. services to the Sports Hall and to the older Central Hall were to be capable of dealing with normal day to day use, conferences, seminars, exhibitions, dances, concerts and civic functions.

3.2. Heat Recovery

In realizing these objectives the following features were incorporated:

3.2.1. the use of ozone to control pool water quality. This safely enables the fresh air requirements of the pool hall ventilation system to be considerably reduced and re-circulation to be employed. Control of pool hall humidity is carried out using mechanical re-humidification, so obviating the requirement for large quantities of fresh air and the consequential energy penalty. It is the de-humidification of re-circulated pool hall air which provides the primary heat source within the complex. During winter and intermediate seasons, there is a continual need for such de-humidification which is a cooling load for the heat pumps and consequently provides recovered heat in the form of condenser water

3.2.2. heat recovery from refrigeration equipment serving the squash courts and a beer cellar by means of the heat pump;

3.2.3. heat recovery from exhaust air using one or both heat pumps which also provide cooling to a number of air conditioned areas elsewhere in the complex;

3.2.4. heat recovery from exhaust air using 'run-around' coils to transfer heat to supply air;

3.2.5. recovered heat is used to pre-warm domestic hot water, to warm pool water and to serve the heating coils of air handling units;

3.2.6. any shortfalls in recovered heat is made up by gas fired boiler plant, the latter also topping up the temperature of domestic hot water and part-heating areas of Central Hall;

3.2.7. there is provision to control the minimum fresh air requirement on the basis of Pool Hall carbon dioxide concentration which means, in effect, that the fresh air input is related to the actual number of users rather than to the predicted maximum occupancy;

3.2.8. two-speed fans to reduce power consumption during periods when air flows can be reduced;

3.2.9. an energy management system which, amongst other functions, is used to optimize the control of the various thermal and lighting systems;

3.2.10. during winter the pool hall de-humidification load is relatively large as is the building heating requirement, conversely the building air conditioning load is small. In summer the latter becomes relatively large whilst the building heating load is small. In summer, due to the greater moisture carrying capacity of outside air and the fact that such air is drawn into the building at temperatures close to the desired pool hall temperature, it becomes appropriate to change the pool hall ventilation to a full fresh air system thereby removing the pool hall de-humidification load and allowing the heat pumps to operate as conventional refrigeration machines serving the air conditioned areas of the complex. Any condenser heat which at that time cannot be used is discarded by coils in the exhaust ductwork;

3.2.11. a total of fourteen equipment and switch rooms were provided and in some instances particular areas have a dual use, for instance a climbing wall rising the full height of the building also provides a lifting well for plant and equipment.

3.3. Lighting

Objectives were a good standard of illumination with low energy consumption, long life and accessibility for maintenance;

3.3.1. the pool hall is illuminated by SON high pressure sodium luminaires to achieve 500 to 600 lux with power consumption of 16.3 watts/m^2 and an anticipated lamp life of 8,000 hours;

3.3.2. the Sports Hall is provided with a number of lighting installations to suit the uses made of this space. SON high pressure sodium lamps are employed for day-to-day sports use. When better colour-rendering is necessary, such as for exhibitions, an installation of MBIF mercury blended lamps is used. For dances and similar functions a number of tungsten downlighters and lighting bars, with dimmers, are provided. Details are:

Use	Type	Lux	W/m^2	Anticipated Life
Sports	High Pressure SON	500	14	8,000 hours
Exhibitions etc.	MBIF	500	14	6,000 hours
Dances etc.	HPID	variable	variable	dependant on use

3.3.3. Central Hall has a number of lighting installations to suit differing uses:

Use	Type	Lux	W/m^2	Anticipated Life
General	MBIF	500	24	6,000 hours
Functions	Tungsten	variable	variable	dependant on use
Disco	HPID	variable	variable	dependant on use

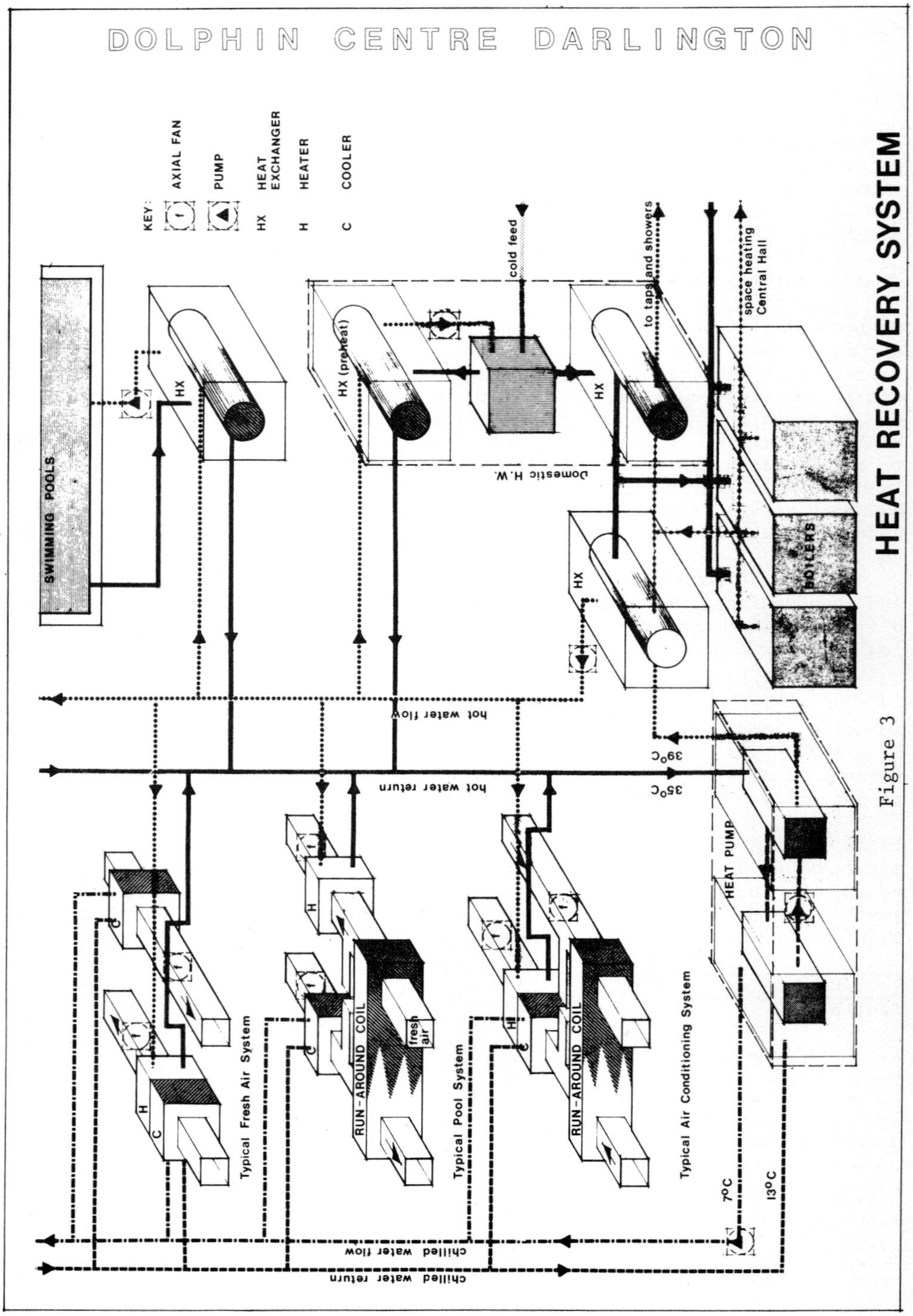
DOLPHIN CENTRE DARLINGTON
HEAT RECOVERY SYSTEM
KEY:
f AXIAL FAN
PUMP
HX HEAT EXCHANGER
H HEATER
C COOLER
SWIMMING POOLS
HX
HX (preheat)
cold feed
Domestic H.W.
HX
to taps and showers
space heating Central Hall
HX
BOILERS
hot water flow
hot water return
39°C
35°C
HEAT PUMP
Typical Fresh Air System
Typical Pool System
RUN-AROUND COIL
fresh air
Typical Air Conditioning System
RUN-AROUND COIL
7°C
13°C
chilled water flow
chilled water return

Figure 3

3.3.4. elsewhere the majority of lighting is by fluorescent luminaires with some tungsten filament fittings in bar and restaurant areas;

3.3.5. except for staff rooms, lighting to all areas is controlled through the BAS computer.

3.4. Controls

In view of the complexity of the building and the flexibility required of its services, an early decision was made to employ a computerised Building Automation System (BAS). The installed controls facilitate:

3.4.5. central switching of plant and of the majority of lighting circuits;

3.4.6. central monitoring of thermal conditions within the building, with print-out and remote adjustment of control set points;

3.4.7. monitoring of alarm conditions on plant, safety and security systems, with visual display;

3.4.8. automatic load-shedding by cycling predetermined plant items;

3.4.9. sequenced control of all plant;

3.4.10. optimum start control of ventilation equipment;

3.4.11. energy management;

3.4.12. ability to add remote monitoring, logging and switching of the heating systems of some 20 local public buildings;

3.4.13. automatic power factor correction which is useful, as due to the flexibility of the services design, the make-up of the electrical load varies with time.

4. DESIGN ASSESSMENT OF EFFECTIVENESS OF ENERGY SAVING MEASURES

Swimming pools, in particular, have in the past been large energy consumers and attention centred on reducing the Pool Hall energy use. The design of the environmental services incorporates a number of features expected to result in considerable savings in annual energy consumption and running costs and which were taken into account in the calculations (2). In order to assess the effectiveness of the measures taken, it was necessary to compare the annual energy consumption both with and without the energy saving features (3). To carry out the assessment, a considerable number of assumptions were necessary regarding the system performance and the operation of the building and during the intervening period these assumptions have changed.

Design temperature differences of 30°C for the Pool Halls and 22°C for the remaining areas were assumed and that during the heating season, heating would be required daily from 0700 to 1100. Lighting is assumed to be on 112 hours per week for 52 weeks per annum.

The energy consumption to make up for fabric and ventilation losses in the Pool Hall was estimated as 4665.6 GJ/annum of which 3502.8 GJ was due to ventilation. Other losses associated with the pools are due to loss of heat from the water in the form of heat transfer through the pool tanks and more importantly by loss of water due to evaporation and by water removed from the pool by bathers and splashing. If we include the heat required for showers and domestic hot water, the total 'water losses' were estimated as 4690.8 GJ/annum (4). The gross annual heating requirement for the Pool Halls was estimated as 9356.4 GJ/annum. The total fabric and ventilation losses for the other areas of the Centre considered were estimated as 11815.2 GJ, giving a grand total for the complex of 21,171.6 GJ. The heat recovery equipment was estimated to be capable of reclaiming 65% of the losses due to ventilation and evaporation of water, giving a total annual recovery of 10958.4 GJ. On the assumption that this heat was recovered using the heat pump and taking a coefficient of performance of 4.5, the net annual heat delivered by the heat pump would be 13392 GJ (5).

The major heat gains to the building are from people, lighting, fans and pumps.

Early estimates of the building use indicated that the annual throughput of people would be about 300,000 per annum (6). If it is assumed that each person occupies the building for about 2 hours then the annual sensible heat gain, at 100 watts per person, would be 216 GJ.

Gains from lighting were estimated on the basis that 25 w/m^2 would be installed over an area of 2320 m^2 and 15 w/m^2 over the remaining 3500 m^2. Lights would be on for 16 hours per day and 7 days per week over the whole year, giving an annual heat input of 2317 GJ (7).

The fans installed are capable of being run at either high or low speed. It was assumed that for ⅔ of the time the fans would operate at low speed and for the remainder at high speed. Total operating time was taken as 16 hours per day, giving an anuual heat input of 2369.9 GJ (7).

The total estimated pump power was 103 kwh, giving an annual gain of 2159.6 GJ (7).

The total estimated gains were therefore approximately 7062.5 GJ. Since the system is able to utilize incidental gains and since the Pools require heating for virtually all the year, it is assumed that 70% of the total gains may be usefully used. The net annual heating required is therefore: 21171.6 - 13392 - 4943.8 = 2835.8 GJ.

The annual electricity consumptions for lighting, pumps, fans and heat pumps were calculated as 2317, 2159.6, 2369.9 and 2976 GJ respectively. In addition, 144 GJ were allowed for comfort cooling in certain areas (7). The total estimated electricity consumption was therefore 9966.5 GJ.

In order to estimate the effectiveness of the energy saving proposals, it was necessary to carry out a similar assessment of energy consumption for the same building without the conservation measures.

The gross heating requirement was increased due to the Pool Hall ventilation required for humidity and odour control to 10476 GJ. Other losses were assumed to be the same. The gross heating energy required would therefore be 21171.6 - 3502.8 + 10476 = 28144.8 GJ per annum.

There would be no heat recovery in this case and heat gains from people, lighting and other equipment were taken as the same as before, except that the fans would run at 'high' speed, providing a gain of 5136.1 GJ. The total gains would therefore be 9828.7 GJ. It is assumed that with a simple system only 50% of these gains could be effectively used. The net annual heating energy would therefore be: 28144.8 - 4914.4 = 23230.4 GJ.

The electrical consumption would be increased by the extra fan power and also by an increased air conditioning load.

The air conditioning load has been approximately assessed by proportioning the figure of 144 GJ previously used by the ratio of incidental gains to give 200.4 GJ.

The total electrical consumption would therefore be:
2317 + 2159.6 + 5136.1 + 200.4 = 9813.1 GJ.

In the following Table No. v, the costs of gas and electricity are taken as £2.89/GJ and £10.14/GJ respectively. (Energy Users Association, (UK) July 1983.) The utilization efficiencies of gas and electricity are respectively taken as 75% and 100%.

No account has been taken of the improved efficiency of operation due to the Energy Management System. If it were assumed to be about 5% on gas and electricity, the annual saving increases to approximately £86,500 (44%) (3).

Table No. v
Analysis of Performance with and without Heat Recovery

Summary	With Recovery GJ/annum	Without Recovery GJ/annum	Saving GJ/annum	%
Gross Heating	21171.6	28144.8	6973.2	25%
Incidental Gains	4943.8	4914.4		
Heat Recovery	13392.0	-		
	18335.8	4914.4		
Net Heating (Gas)	2835.8	23230.4	20394.6	88%
Electricity	9966.5	9813.1		
Total Energy	12802.3	33043.5	20241.2	61%
Gas Costs	£ 10900	£ 89500	£78600	
Electricity	£101100	£ 99500	£- 1600	
Total Cost	£112000	£189000	£77000	41%

5. TECHNICAL DATA

5.1. Heating and heat recovery

Boilers; 3 No., 4 pass gas fired each rated @ 1 MW with efficiency of 82% @ 100% rating.

Heat pumps; open drive screw compressor packaged water chillers. Design of each at: 126 kw input, 686 output (ratio 1:5.4 working ratio 1:4.5/1:5.2 average 1:4.85).

Rub-around coils arranged in pairs with ethylene glycol/water pumped heat transfer medium; operating efficiency up to 60%.

Plate heat exchangers (stainless steel plates) with heat transfer co-efficient in excess of 3000 W/m^2 ^oC serving domestic hot water and medium/low temperature space heating.

5.2. Ventilation

Air speeds; main sections 4 and 6 m/s, sub-sections 2 and 4 m/s.
Temperature range 28^oC Pool hall to 10^oC squash courts.

5.3. Building Automation System

Landis and Gyr Ltd., Visonik 4000 system, Central Computer Digital Equipment Corp., PDP 11/23 Dis Store 5 M byte/disc.

6. FOOTNOTES

(1) The most common feature is quoted. Other types were also used.

(2) Features of environmental services. See 3.2.1 to 3.2.9.

(3) CAVEAT. The consumptions calculated were not intended to predict the performance of the complete building as finally designed and operated, but were limited to the environmental services in the major areas of the sports facilities (Phase 1), i.e. the Pool Hall, Sports Hall, Prac-Hall, Projectile Hall, Multi-purpose Hall and 'Central Hall'.

(4) Domestic hot water use is lower than assumed.

(5) The actual average heat pump co-efficient is 1:4.85. See 5.1.

(6) Actual is 850,000 persons/annum. See introduction pp.2.

(7) Actual.

LOW ENERGY HOSPITAL

H. GOODMAN
Chief Architect, Department of Health and Social Security, United Kingdom

SUMMARY

The intention of this study is to see whether a target of 50% energy reduction could be achieved taking as a datum a number of already energy-efficient standard hospitals. A survey was made of energy usage by examining environmental systems and process loads. This showed about 50% of the total energy was used for environmental purposes the balance being process-related. Analysis was made of measures aimed at reducing demand both of environmental systems and processes, together with a check on alternative building shapes. The team also examined whole hospital strategies, including combined heat and power, thermal storage, heat pumping and the optimisation of these with a view to reducing prime energy input. The results of all these analyses show a theoretical reduction of 59% over datum was technically possible, and in the event a hospital was designed which will produce a 55% energy reduction. A report on all stages of the demonstration project has been prepared.

Until recently, or to be more precise until ten years ago, hospitals were not very energy conscious; the labour element in their running costs was extremely high and when cost of food and drugs was added the energy component represented a very small part of the total running cost. This was despite the fact that high air temperatures are needed for many frail, elderly patients and a majority of the buildings were in use for 24 hours a day and high technology equipment, air conditioning for clinically sensitive areas and high lighting loads, lead to a higher energy use per metre square than in most building types. There was a tendency to disregard energy costs which led to a lack of understanding of energy demands. Traditionally hospitals were coal fired and in many cases engineers responsible for the use of energy were only concerned with the stoking of the boilers and maintaining mechanical and electrical equipment in reasonable condition. Staff, other than the engineering staff, were quite unconscious of the demands they were making in energy terms and overriding considerations of clinical and nursing management meant that energy management was non-existent.

The 1973 oil crisis and the following economic crisis meant that cheap energy would never again be available and the economic basis of hospital design shifted as a result. The Department of Health and Social Security (DHSS), in partnership with the National Health Service (for which it produces guidance on building and engineering services initially turned their attention to reducing the energy consumption in the existing buildings.

When the NHS came into being in 1948 it inherited as the bulk of its estate buildings which were at least 50 years old at that time. Despite a substantial building programme most of the hospitals that provide the NHS today are old and a substantial number still date from the last century. These hospitals are grossly inefficient in energy terms and it seemed that the first priority for the DHSS was to reduce energy consumption in existing hospitals. The scope for improvement in energy consumption was limited due to the age and construction of the buildings but within five years a 20% reduction in energy demand had been realised. Only when this programme was under way did the Department turn its attention to energy efficient design in new buildings; not only in terms of engineering and energy technology but in terms of energy management.

The economic situation had already caused the Department to look again at its building programme and to propose a range of economic hospitals far smaller than those currently being designed. Under the general title of the Nucleus programme (the concept being that the first phase would be the 'nucleus' of a district general hospital) the configuration, building geometry, storey heights, departmental relationships were all considered in an effort to make the building efficient, both in energy terms and in general running costs.

The outcome was a modular concept of a hospital comprising a series of pre-planned cruciform templates which could be arranged in various configurations to respond to particular medical content and site conditions, and at the same time, to allow a piecemeal addition to existing hospitals or whole hospital development. The essential feature of the concept was that whilst a hospital of 600-800 beds would be designed, only the first phase of the development, about 300 beds would be built initially, but this would remain viable until the remaining accommodation was built some years later.

Inherent within the Nucleus concepts was the aim of making the hospital efficient in energy terms compared to previous designs, especially those having high levels of air treatment, and considerable steps were taken to lower consumption. Significant contributions included greater exploitation of natural daylight and ventilation, improved standards of insulation, effective zoning of systems and a move away from air-conditioning, except where medically essential. Notwithstanding these measures, it was when the Department decided to look at energy conservation in new buildings the real challenge seemed to be to improve even further on the energy consumption of these already inherently energy conservative buildings, so the Nucleus hospital was used as the datum for the low energy study. The programme essentially comprised four stages:

i to undertake a short-term pilot study to establish the likely potential energy reduction in Nucleus-type hospitals;

ii to carry out a more intensive long-term study into the design and operation of new hospitals generally and to prepare guidance for designers;

iii to design and construct a proto-type low energy hospital as a demonstration project arranged to incorporate the findings of the study;

iv to complete a programme of controlled monitoring of the proto-type hospital, compare actual and predicted performance, including operational feedback.

The pilot study was carried out by Architects and Engineers within the Department of Health and Social Security. These designers set themselves what seemed to be a hopelessly optimistic target of saving 50% of the energy demands of an already energy-efficient hospital. Within about three months the pilot study had shown that savings of over 50% were theoretically feasible and the project moved into its second stage. This involved commissioning a team of some of the top private sector Architects and Engineers to form a consortium with a target of completing the second stage of the exercise within 12 months.

The initial part of this exercise was to establish the energy usage reference datum by means of looking at the environmental systems and standard processes. This was done in far greater detail than would be normally necessary in design practice. This gave a very detailed estimate of the likely energy consumption of a whole hospital, given good management. Despite the fact that the team were considered experts in the field of hospital planning and engineering the results came as something of a surprise.

About 50% of the total energy demand was required for environmental purposes and it follows that the balance is process related. The bulk of this - nearly 25% of the total energy demand of a hospital - is used in catering; domestic hot water requiring about 20% of the load and the remainder concerned with sterilizing, incineration and other odds and ends.

An analysis was then made of measures aimed at reducing energy demand at the point of use for each of the environmental systems and processes, and for these it was possible to estimate a reduced energy demand for the hospital. A second check was made on a number of different building shapes as to whether they offered any significant energy savings over the Nucleus building form, and likely sources of energy reclaim were looked at to discover the potential of ploughing back energy into the hospitals, which would otherwise be lost. Energy demand use patterns were assessed and once they were established the best ways of meeting these in prime energy terms was considered. The team also looked at whole hospital strategies including consideration of combined heat and power, thermal storage, heat pumping, and these were synthetized with the object of seeing whether prime energy input could be further reduced.

The first stage analysis represents the effect of applying a straightforward energy-conscious design approach, relying on techniques which do not substantially alter the level and standards of technology associated with the current Nucleus engineering practice and at the end of this first stage it seemed likely that an energy saving of about 34% could be made. The energy saving measures adopted for some of the systems and their contribution in terms of reduced energy loads may be summarised as follows:-

a) Lighting
artificial lighting accounted for half the electrical energy required and this was reduced by greater exploitation of natural lighting and the use of higher efficiency light sources. It was found that by increasing the glazing area of outside walls from an average of 26% to 35% the daylight factor was increased from 0.5% to 1% at the bedhead position. In upper storeys rooflights in addition to smaller side windows improved the quality of natural lighting by cutting down glare. The hours of use of electric lighting were thereby reduced; alterations to the pattern of switching and from the use of high efficiency tri-phosphor lamps in suitable luminaires resulted in further reductions in electrical energy use. Over half the original energy demand for lighting was obtained by these measures without appreciably increasing the energy demand for space heating in consequence.

b) Space and Air heating loads:
Over 75% of the rooms in Nucleus design are on the perimeter; space heating to the perimeter areas of the hospital which are provided with low pressure hot water heating systems is about 7% of the total heating load. This was reduced to less than 5% by improving insulation levels of the walls and roofs from about 0.4 watts/m^2 ^{o}k to 0.3 watts/m^2 ^{o}k; by introducing double glazing for the windows and by adopting a higher standard of air tightness for the fabric of the building. Air heating for other areas which are mechanically ventilated was reduced from 20% to 14% of heating energy demand by adopting a single duct zone heating/cooling arrangement and by detailed attention to the control of the system and to their setting:

c) Humidification
The normal arrangement using central humidification by involving steam results in many spaces being unnecessarily treated. Terminal humidification using electrically operated humidifiers resulted in reducing the humidification load to less than 1%.

d) Fan and Pumping Loads
The use of low velocity air distribution and improvements to the water circulation systems has reduced system losses and improved the efficiency of the machinery.

e) Process Loads — The process loads - principally domestic hot water and catering were heavy users of energy. The catering department load was reduced by over 35% by changing from a steam/gas kitchen to one using only gas. Savings were also effected by improving the control and operation of equipment and by careful attention to its sizing. The energy saving in respect of hot water was not so large and is the result of adopting a more stringent approach to distribution losses.

Some energy savings were made by planning changes. Heavy energy user departments, such as theatres, pathology were to be located close to the energy centre to reduce distribution losses. Landscaping was considered in terms of shelter belts and the generation of favourable micro climates within courtyards, both giving a contribution of reducing energy.

The second stage was to consider the energy which would normally be rejected and to discover whether this could be recovered and used. Various sources were identified, such as exhaust air, drainage water, incineration, refrigeration, distribution mains and boiler combustion processes. The whole of the reclaim approach was founded on the facility of pooling resources of energy reclaim and redirecting them at the appropriate time into areas where energy need arises.

The result of this analysis, coupled with the first study showed that a theoretical energy reduction of 59% over the Nucleus hospital was technically possible.

As the results of the study began to emerge they appeared to be so encouraging that the Department made an application to the EEC Energy Commission for European funding towards the project. This application was successful and about £¾ million was allocated towards energy saving measures and monitoring of the project. The Department identified the project within the current building programme and with some modification to the content to make it more typical of a first phase of a district general hospital, commissioned the same design team to produce a low energy solution, using the Nucleus design as the base. This scheme was at St Mary's Hospital, Isle of Wight. The project team started to work in August 1981, and the work on site started almost exactly a year later. This work comprising groundworks, piling, foundation, road works and drainage is complete and the contract for the remainder of the hospital will go to tender this Autumn.

The multi-volume report of the study is now being distributed to health authorities and private sector practices engaged in building new hospitals. Information is technically complex and will undoubtedly take some time to affect the design processes and percolate into all new hospitals. Nevertheless, the results have been quite dramatic and the Department is currently investigating the possibility of building a second low energy hospital with an even wider remit. The first design team were constrained to the use of existing technologies, albeit used in clever combination. The second design team received a remit which allows it to explore the newer technologies with particular emphasis on advanced coal technology. The team has been formed, a suitable project has been identified in the North of England and the feasibility study for this second experimental low energy hospital has now commenced.

TELEPHONIC CONTROL & MONITORING OF ENERGY CONSUMPTION IN SCHOOL BUILDINGS IN ESSEX

E C ENGLISH MA, PhD.
FREEMAN ENERCON LTD
SCIENCE PARK, MILTON ROAD, CAMBRIDGE CB4 4BH
Tel: 0223 315432

1. INTRODUCTION

The object of this demonstration project was to assess the practicability and cost-effectiveness of remote control and energy monitoring of 20 schools based in or near Chelmsford in Essex, U.K. The work involved the design, installation and co-ordination of some 35 separate autonomous building controllers. These controllers - called outstations - are capable of automatic communication over the Public Telephone Network with a supervisory computer based in County Hall, Chelmsford. In essence, the outstations are responsible for all routine day-to-day environmental control of the buildings, whilst the supervisor is available to receive alarms, fuel consumption data and other logged information and to allow remote maintenance surveillance and adjustment.

The novelty of the project lay in the application of relatively low-cost but extensive control to a medium sized pilot scheme. Some problems were caused by the application of tighter control to existing buildings, both in system and human terms; although full comparative fuel consumption data is not available at the time of writing, the indications are that the system is both an energy saver and a useful maintenance tool.

2. PROJECT OUTLINE

Medium sized secondary schools of the type built during the 1950's and 1960's in Essex, are typically heated either by oil or gas, which fuels mixed radiator and fan convector emission systems. Sites will have one or two plant (or boiler) rooms, each containing, perhaps three boilers and associated pumps, mixing valves and domestic hot water storage & distribution equipment. Control is usually of a fairly simple nature incorporating one or two time clocks, weather compensator, and "optimum" start units. These schools are representative of thousands of similar buildings throughout the U.K.

The project had three major components:-

i. the sub-division of the heating pipework to provide separately controllable heating zones.

ii. the design and installation of the intelligent outstations

iii. the central supervisor

Of these three elements, item (i) was to prove the most time-consuming and troublesome to administer because of the large amount of detailed design and site supervision of electrical and mechanical (heating) sub-contractors. Each site was surveyed to establish the heating system layout and building structure, and designs were drawn up for the necessary modifications to pipework and wiring. The main changes incorporated were:-

a. boiler sequencing by load
b. zoning into separately controllable building segments
c. full weather compensation and optimum start/stop on each zone.
d. frost protection
e. Alarm reporting and auto-pump changeover

plus numerous other amendments, some of which were maintenance items conveniently included at the time. The work was put out to tender to approved contractors of the client, Essex County Council, on a school-by-school basis.

The XYNTAX outstations were designed to incorporate in one integrated box all the control, monitoring and communication facilities likely to be needed for each establishment. Early prototype problems arose with electrical interface in the boilerroom environment, but this was overcome with optical isolation of all the vulnerable processor areas. Communication over the British Telecom network has proved fairly reliable, and of great value in practice, in that off-site engineering staff now have ready access to inspect and alter operating parameters from the Area Maintenance Office.

The central supervisor, as well as providing the remote access mentioned above, allows staff to specify, record and inspect data from outstations. Two valuable facilities are temperature graphs and fuel consumption measurements. These latter are generally computed from pulse-emitting rotary oil and gas meters.

3. MODIFICATION OF EXISTING HEATING SYSTEMS

A primary aim of the project was to test our belief that significant savings were possible through division of the buildings into separately controllable areas or zones, and that, just as importantly, these savings could be sustained through the years by remote supervision and monitoring. However, it is difficult to estimate how much re-zoning will cost without extensive pre-contract survey work. Inevitably little record information is available for middle-aged buildings, so pipework runs and existing control details have to be re-established before any modifications can occur.

With the benefit of hindsight, it would have been both cheaper and easier to have used the monitoring facilities of the outstations to carry out much of the pre-contract survey work. Existing patterns of use could have been observed together with more obvious information such as temperature levels and metered fuel useage. The operational zoning for each building (which was established by discussion with the building users) has worked reasonably well in practice and is undoubtedly a prime fuel saving measure.

The simple, linear, nature of many existing heating systems does not lend itself to easy conversion to separate, parallel, zones. As a result, complete separation of heating zones is not usually possible, and the control system must take into account the operational linkage between zones and sub-zones. For example, a zone requiring only frost protection may in fact be related to a sub-zone whose requirements are for boost temperature water or air.

A further "real-life" factor requiring attention is that control and pipework modifications usually suggest and incorporate maintenance and upgrading activities such as re-boilering and pump replacement. These activities, whilst necessary and sensible, tend to obscure the cost of the energy conservation measures alone.

4. THE XYNTAX OUTSTATIONS

The Xyntax outstations designed for the project incorporated

a. a full telephone communications facility, including an integral modem
b. full data-logging and monitoring software and memory

c. inputs, outputs and software to cover all anticipated control functions for H & V purposes.

The outstation uses a Motorola 6802 microprocessor and is designed conservatively from a component performance point-of-view to avoid any problems of specification tolerance. It incorporates a keyboard (which can be disabled remotely) for local operator use and for test and emergency use, a second level of access allows manual override. All electronics are contained in an enclosed secondary chassis which plugs into the control enclosure during commissioning.

So far, the outstations have proved reliable, and apart from the electrical interference problems mentioned above, the dialling relays in the signalling unit are the only components which have needed upgrading.

It is our view that it would have been impossible to have achieved the same sophistication and level of control using conventional analogue components.

The control software includes self-adaptation in the optimum start/stop and weather compensation facilities. As a result, commissioning is greatly eased, although operator intervention is still required to set up engineering related functions such as boiler sequence tables and plant protection thresholds.

5. THE CENTRAL SUPERVISOR

The roles of the central supervisory computer are:

a. to receive alarm and alert calls from outstations
b. to allow engineering and maintenance staff remote access to control parameters and local operating data
c. to store, collate and present energy consumption data as management reports, and to allow budgeting, targeting and tariff analysis using the data-base.

The Essex project has been biased towards functions a. and b., partly because the system is used by maintenance staff, and also because the County Council have their own existing fuel consumption audit procedures. In addition, the system only covers a small proportion of their total of about 1000 buildings.

It is difficult to quantify the manpower savings which accrue from the use of a central supervisory computer. Our "guesstimate" is that the enhanced supervision of the building useage and plant performance which results from close monitoring should produce a sustainable 5-10% reduction in fuel consumption compared with conventional controls, but that manpower savings will depend on the extent to which the system can be incorporated into planned maintenance procedures and also the extent to which local caretaking staff have their work-load reduced as a result of not having to run heating plant.

6. SAVINGS TO DATE AND OPERATIONAL EXPERIENCE

Unfortunately, we do not yet have sufficient data to be able to present fuel savings for the system. This is mainly because contractual delays meant that many of the outstations were only commissioned during 1983, and have not yet had a full seasons operation. Nevertheless, savings are being made even if it is too early to quantify them exactly.

One little appreciated aspect of controls is that they throw into sharp relief any inadequacies of the existing heating systems. For example, local overheating may have been tolerated or even encouraged in the past to compensate for poor circulation or underheating in certain areas. The new controls tend to prevent this happening, and in Essex we found that we received a barrage of complaints from the building occupiers during the early stages. There is also a conflict between statuory temperature limits and custom and practice, when the British "official" 19° C can prove difficult to implement. However, the ability to check temperatures from County Hall and, if necessary, to alter set-points with a simple 'phone call' has enabled staff to prevent too many problems.

After the initial, and understandable, apprehension about the remote surveillance aspects of the system, local caretaking staff have come to appreciate its simplicity of local operation, and the fact that they can now expect plant problems to be diagnosed rapidly and accurately.

Finally, there is no doubt that the operation of such a system places new demands on central staff, which must be anticipated and supplemented by training if problems are not to ensue.

The central operator has intimate access to all the building control parameters, and a clear understanding of their purpose is vital if the controls are to function properly.

7. CONCLUSIONS

1. No firm data is yet available to show whether the expected savings are being met. Current indications are hopeful, but not conclusive.

2. The communications facility is proving to be useful in its own right as a maintenance tool.

3. Administrative and low-level technical supervision is crucial to the success of the project, as is the training and motivation of the personnel who are affected by the new system.

4. Experience shows that the efficiency of conventional control systems generally deteriorates following commissioning due to poor maintenance and interference with the controls by untrained or unskilled personnel. The ability to ensure that controllers are functioning correctly is seen as a distinct advantage of remote communication.

5. The intelligent outstations or controllers are, by cost, a relatively small proportion of the overall project value; decisions on the extent and quality of heating system alterations, therefore, become paramount in the calculations of payback for the project as a whole.

TELEPHONE MONITORING OF ENERGY CONSUMPTION IN 20 SCHOOLS

P.A. PAGE, Assistant County Architect
Technical Services, County Architect's Department,
Essex County Council, U.K.

in association with Dr. E. ENGLISH, Freeman Enercon, Cambridge

I think it very important that the Client recipient of (or participant in) an Energy Demonstration Project should have an opportunity of reporting back as well as the Contractor. (Or perhaps I might say, the courage to support the Contractor in the work that he has undertaken on behalf of the Commission in the Client's buildings).

I represent one of the largest Local Authorities in the United Kingdom having nearly 1500 major premises in its care. About two thirds of these are educational establishments. Building-related expenditure last year amounted to £43 million of which the energy bill accounted for £10 million. Of this sum it is estimated that some 40 percent of our energy expenditure is vested directly in the hands of people who use our buildings and lies outside the reach of any viable technology.

Working on the premise that unless one knows in considerable detail how the energy is being spent one cannot begin to manage it effectively and efficiently. Thus the need for a management system.

Prior to the development of the Xyntax project the most sophisticated control system available to us was the optimiser. Optimisers could only control a single zone, although multiple installations are feasible, the cost is prohibitive and the correct control is not achieved. The development of Xyntax allowed correct control of multiple zone and multiple boiler installations at a cost less than conventional controls. Where the Xyntax controller is installed in existing heating systems, large scale pipework and equipment modifications are necessary. It is our experience that the cost of such modifications can be seriously underestimated but these are essential to make the services viable and efficient as a whole. The Xyntax system has the additional advantage of operating either as stand alone units or linked via the PSTN to a central computer. Where a computer has been installed as a central supervisor it is possible to interrogate outstations and alter parameters remotely.

Client liaison and co-operation is essential to the acceptance and efficient operation of such an installation. The incentive for good energy management does not exist on site since responsibility for paying the bills is not vested in the heads of establishments. Users neither see nor directly pay the bills. Staff, (from the Caretaker to the Headmaster in the case of Schools) need to be fully informed and consulted prior to any work comencing. One enlightened Headmaster actually requested low control settings to start with, in order to test the weaknesses of the system (building + plant + people).

The Xyntax system operation has already been described by Dr. English of Freeman Encercon. Not all sites are under Client control at present. Where heating plant is being controlled from the computer as central supervisor it is possible to fine tune the heating system of each zone according to its response. Considerable variations in responses have been noted confirming the need for a long period of close monitoring. Major fuel savings have been made where evening lettings of buildings are confined to small areas and where, prior to zoning, all the premises would have been heated. Where boilers are oil fired, flow meters have been fitted and it is possible to determine fuel consumption on a monthly and yearly basis and compare this with degree day data. Caretaking staff are now reluctant to switch on heating for unauthorised lettings as they are aware that temperatures are being recorded centrally. Thus the system acts as a policing deterrent.

It is too early to assess what reductions in fuel consumption will be achieved ultimately but it is confidently expected that savings will be in the order of 15 - 20 percent.

As an extension to the E.E.C. Demonstration Project a Xyntax system is presently being installed at St. Johns School, Epping, modified to allow the school's B.B.C. micro computer to access the technical data stored at the outstation. A program has been written by a pupil with the assistance of Freeman Enercon that will display data in a meaningful way, allowing pupils to interrogate and analyse the information presented. This has enormous educational potential. It is envisaged that not only will this support and promote energy project work within the classroom but with further development serve the aims of user education and the legitimate ends of energy management. Appropriate information of course can be made available for on site management directly by the head of establishment and off site in the maintenance contractor's office with automatic alarm signals and first line diagnostic data for the maintenance engineer.

It is my firmly held belief that Energy Management Systems such as Xyntax, as demonstrated in this context by this project, are a prerequisite for good energy management practice now and for the future. Energy expenditure is an essential overhead that cannot be ignored.

On the subject of Demonstration Projects for the future, may I make a plea for a programme of appropriate technology that can readily be demonstrated to be effective and that can in practical terms, simply be applied, furthermore that the scope of these be extended to include human factors - motivation, education and training.

(The author acknowledges information on the installed Xyntax system prepared by Martin Walters of the Energy Conservation Unit, Essex County Council).

BUILDING ENVELOPE - DEMONSTRATION PROJECTS

Rapporteur: J.R. BRITTEN

(Mr Britten is head of the Building Research Energy Conservation Support Unit - BRECSU - which operates the United Kingdom's demonstration scheme in respect of demonstrations in domestic buildings)

Introduction

Delegates will have observed that within all sessions of this seminar there are far more projects of 'Research & Development' than there are of 'Demonstration'. In the present subject of The Building Envelope only eleven demonstration projects have been presented for discussion, compared with twice that number for R&D. And indeed some of these demonstrations might equally have been classed as Research & Development. Only a few of the demonstrations have been sponsored by 'host organisations' - building owners - and yet in order to achieve replication of energy saving measures we need to develop proposals in partnership with both 'hosts' and manufacturers.

Why should there be so few demonstrations? Is it because of greater inherent difficulties in managing demonstrations; does it reflect a lack of interest in the scheme by host organisations, or is that that very few of the schemes which have been put forward actually satisfy the criteria for selection?

Indeed what should we look for in a demonstration - both when selecting proposals and when measuring success? And how should we assess the results of the eleven projects put before us this morning? I shall touch on all these points as I review the various projects.

Some people may well argue that demonstrations are more difficult than R&D. We do not have the luxury of carefully controlled laboratory conditions, nor are we able to dismantle an experiment and start again if things do not go quite as we expected. More importantly we cannot carry out our studies in some back room far from the eyes of those who pay the eventual bills.

Instead we work in the open, with buildings in which people go about their daily life, not as experimental subjects but as individual human beings - Individuals who may behave in an unpredictable and irrational way but whose final approval is essential if our project is to be a success. And the only measure of success in a demonstration is the extent of its replication. We may have shown real energy savings in limited and specific circumstances, but unless we can persuade every-one concerned with buildings to adopt similar energy saving measures on their own premises we have, I regret, failed. Replication and dissemination of results is our responsibility, as contractors and research workers,just as much as it is the responsibility of administrators. When we announce our results we must make sure that the right people are listening.

It is we who must ensure that all architects(*) give as much attention to energy conservation as they do to appearance and form -
We must ensure that builders(*) and developers also consider the energy which will be used in their buildings, and not just the initial cost -
And we must tell manufacturers how to provide efficient water heating systems, so that housewives do not have to economise on the use of bath water like this! (*)

Professor Elias said on the opening day, that innovation can take many years, and that we should not be too impatient for results. By its very nature, replication of the building envelope is an even longer process, and so we should not be too worried that only a limited amount of replication has taken place so far. But we should realise today the difficulties of assuring and confirming replication. We must begin to measure the take-up of the devices that we are demonstrating,so that in later years we can survey current construction and can point with confidence to the benefits we have stimulated.

Current practice

When examining building practice throughout the European Community we immediately find that what is novel in one country is commonplace elsewhere. And measures new to the Community may be well established in neighbouring states. However we should not rush to apply alien solutions without careful thought and trial. In most countries the form of construction that is 'traditional and conventional' is the form which works cheapest and best for that country. A solution which has evolved elsewhere may not be suitable for a different climate, different culture and different building industry, even though the problem may appear to be essentially the same in all cases. Caution is necessary, for buildings have a long life.

TABLE SHOWING MEASURES BEING DEMONSTRATED

solar collectors and heat stores
heat pumps and heat recovery systems
higher levels of insulation and airtightness
more efficient heating appliances and emitters
improved controls and management systems
more efficient lighting and ancillary systems
improved water heating

application to both new and existing buildings
application to both domestic and non domestic buildings

The table (above) shows some of the energy saving measures that are currently being installed throughout the community. Many of them are being demonstrated (either singly or as part of wider packages) in the projects represented here today, and I shall discuss them at the appropriate time.

FOOTNOTE
(*) at these points, the rapporteur showed colour slides to illustrate different aspects of buildings and energy saving projects, but it is not possible to reproduce such slides in this text.

However I am personally disappointed and surprised that only three of the projects address themselves to the problem of existing buildings. The others all involve the construction of new houses, new hospitals, or new leisure centres, and yet we all know that major savings will only come about by reducing the energy consumption of the millions of buildings that were constructed some years ago, in the days of 'cheap energy'.

Fortunately we no longer have as many terraced houses like this (*) heated by open fires burning coal or other solid fuels,but in many member states there are a large number of multi-storey apartment blocks like this(*) , poorly insulated with expensive heating systems which are probably not used sufficiently to prevent condensation of moisture on the fabric.

This is a major problem, and I would have liked to see more demonstrations of the improvements which can be made to existing buildings, I hope that building owners will co-operate with manufacturers and research organisations in proposing further schemes in later years. The United Kingdom Department of Energy runs its own national demonstration scheme and within my own sector of domestic buildings over 60% of the demonstrations taking place are concerned with existing buildings.

I also question whether sufficient attention is yet being given to the energy consumed by building services. The large public sector building projects are certainly attacking this problem, but in many other projects it is mainly space heating consumption which is being reduced, leading to a much greater proportion of energy used for water heating, lighting, and ancillary services. And yet although space heating requirements vary greatly between the climatic extremes of Northern and Southern Europe, there is probably a greater convergence of consumption on other household energy demands, and so a greater opportunity here for demonstration projects which will benefit all members of the Community.

Demonstration projects

Seven of today's demonstrations concern housing ,

- domestic buildings -

B2.1 Energy savving houses in Ireland
B2.2 H.L.M. - SOLPAC project
B2.3 Skive 79 building
B2.4 Space heating by solar wall
B2.5 Integrated approach to low energy housing
B2.6 Houses with solar roof, buried pipes, and heat pump
B2.7 Low cost low energy dwelling

,and the remaining four are predominantly public sector or community buildings

- non domestic buildings -

B2.8 Lighting systems in office buildings
B2.9 Low energy leisure complex
B2.10 Low energy hospital design
B2.11 Remote monitoring of school buildings.

It is notable that only one project takes place in a commercial building.

In selecting the papers for presentation I attempted to strike a balance between housing and non-housing - and between existing buildings and new construction. But there was one further consideration of even greater importance. That is the availability of measured results.

STATUS OF PROJECTS

	Feasibility & Design Complete	Construction Complete	Monitoring Commenced	Results Available
B2.1	*	*	*	
B2.2	*	*	*	*
B2.3	*	*	*	*
B2.4	*			
B2.5	*	*	*	
B2.6	*	*	*	*
B2.7	*	*	*	
B2.8	*	*	*	*
B2.9	*	*	*	
B2.10	*			
B2.11	*	*	*	*

You will see from this table that all eleven projects have been designed, and most of them have now been built. But in several of the projects the monitoring programme only started a few months ago, and so we have no real idea of how successful they have been.Nonetheless, just because there are no measured results yet, this does not mean that the projects are failures. Their promoters have already learned a great deal about the problems of incorporating novel energy saving measures into buildings, and about the construction costs involved.

Happily, in five projects there has been some degree of data collection and analysis - indeed in one project the results have already led to replication of the design - and so I have asked four of their authors to present their papers to you this morning.

I shall say nothing about those four papers - you will see for yourself how they fit into the overall context, but in the remaining time I will attempt to review the essential features of the other seven projects. I apologise in advance to their authors for having to give a very rushed presentation, and I hope that I can highlight the advantages of each one.

South London Consortium

I will begin with the project I know most about - that of the South London Consortium(*).
One of the most valuable features of this scheme is its use of a 'Demonstration Suite', where designers and other building professionals can discuss constructional details with technical staff, where they can examine models and samples of the energy saving measures, and where they can make use of the computer based monitoring system (*) to determine the temperature profiles and the energy consumption of any part of the building. The system is more than just a monitoring device. It also enables visitors to interrogate past records, and to compare the scheme's energy consumption with their own experience.

The scheme includes the construction of 18 new houses, and the conversion of 3 older houses into apartments (*). I believe that this latter aspect is particularly important, as it shows that energy-efficient renovation and conversion can be carried out at little extra cost over the standard rehabilitation. The large rooms permitted the incorporation of 100mm of insulation within walls, together with a further 100mm of air gaps, giving a U-value of 0.27 W/m^2 C for the external walls. When taken with the measures for ventilation control this should give a payback time of about 4 years.

Heating in these flats, and in the new houses, is by a warm air system, with additional contributions in the houses from the solar conservatories. Energy consumption of the houses is expected to be only 40% of that of conventional design and the payback period is approximately 7 years. The houses and flats will all be available at moderate rents to families on the public sector lists.

The SOLPAC project

has similar objectives - that is to provide energy efficient houses at moderate rents - and some aspects of construction and heating system are similar. The houses(*) are built with a lightweight timber frame, but this has been combined with heavier floors and chimney structure, to provide thermal inertia and to smooth out variations in temperature.

There is a separate heat store, built from simple concrete blocks(*) so it can be constructed easily and cheaply on site - and a versatile heating system which can operate in a variety of modes (*) according to demand.

The control and management of these various elements - both the solar heating system and the complementary systems - is handled by a specific electronic microprocessor. This switches between 'storage' and 'recovery' depending on air temperature.

Predicted savings are of 50% of total energy consumption when compared with 1978 standards. This leads to a payback time of approximately 8 years. Measurement and monitoring has only started one month ago, and so today all we know is that the systems work. We don't yet know how good they are

A primary objective of the Solpac scheme has been to provide houses which are pleasant to live in and do not impose new constraints upon the occupants.

Electricity Supply Board, and Cement Roadstone Holdings

Similar objectives were set out by the proposers of the two Irish schemes. An essential criterion of the development by the Electricity Supply Board was to produce a house(*) which 'looked normal' - not one that was in any way strange or unusual in its appearance as a result of its energy saving construction. As a further constraint the houses were fitted into an existing estate, and conformed to previously approved road plans and density. Only conventional materials and methods were to be used. The six houses are well insulated, and the design heat loss through the fabric is as low as 1.6kW in the most highly insulated case. Heating is - understandably - by electricity but a number of alternative systems are being evaluated. Monitoring will establish not only the total energy consumption and the internal conditions produced, but also will provide the Supply Board with information on peak load demand.

The Centralec heating system uses a central water storage cylinder, heated by off peak electricity, with distribution to either radiators or (*) underfloor pipes.

Monitoring of this project also began just a few weeks ago, and so it is too early to present results

Another project from Ireland, by the Cement-Roadstone Company, compares the performance of two similar houses - one fitted with a large area of solar collector on the roof (*), and one with a high degree of insulation - with that of the standard construction. The collector is linked to a warm air heating system, and to an underground rock store, but although the contractors have been very satisfied with the incorporation of the solar roof, they are less happy with the storage and ducting system, which is thought to be expensive and 'non-commercial'(*). As a result of building these houses, they would prefer to use phase change material for heat storage next time, and to mount this in the roof to save on ducting.

Once again, I have to say that monitoring has just begun, but so far the results are very closely in line with predicted values. More importantly - this project has shown already some of the difficulties of trying to accommodate novel measures onto a conventional building site.

This particular demonstration involves the construction of 3 new houses, and you may have noticed how there has been a steady reduction in numbers throughout these projects - 18 houses at SLC, 12 at SOLPAC, 6 at ESB and now only 3 at CRC. Unfortunately the final housing project hasn't yet built any houses at all, although it is hoped that building will take place in the New Year.

Stephen George & Ptnrs are negotiating with British house developers in order to construct a group of solar heated houses on the open market(*). Design work is essentially complete, and should enable the houses to be built simply and cheaply, and also to be maintained simply and cheaply by the owners.

The construction and style will be commercially acceptable. The heating system will use warm air as the distribution medium, heated either by a conventional gas unit or by solar gain thru' the conservatory. An automatic valve will switch between the two heat sources according to the temperature of the pre-heated air. The design also incorporates a simple heat store. The first floor slab is used for this purpose having ducts formed during construction and so reducing the total installation costs.

The designers are aiming for aesthetically acceptable designs, rather than for exotic and sophisticated experimental buildings.I endorse this approach, and can confirm from my own experience with demonstrations in the United Kingdom that some of the most successful low energy houses are the ones which look just like their neighbours.

After all, the people who live in them are no different from their neighbours, and housing should be about people. I find it very pleasing that so many current demonstrations have this point in common.

Summing up on housing demonstrations, we see frequent use of solar gain to supplement a conventional heating system, often the use of a warm air distribution system with microprocessor control, and the incorporation of a heat store, either as a separate item or as part of the construction.

I have not had time today to discuss the importance of careful monitoring in the domestic sector. There are many items that have to be considered such as those in the table below.

MONITORING REQUIREMENTS

Physical conditions	- temperatures, humidities, ventilation
System performance	- efficiencies, C.O.P., solar gain
Energy consumption	- space, water, ancillary use
Constructional aspects	- ease of building, durability & maintenance
Financial aspects	- actual costs in use, value of savings
Occupants views	- behaviour, attitudes, satisfaction
Extent of replication & total community benefits	

Although today's authors have generally given full descriptions of their own monitoring systems, I regret that many demonstration submissions do not provide sufficient information on this topic. Proposals must state just how the energy savings are to be measured, and what other factors (such as durability) will also be assessed by the contractors. We need a careful balance between technical content, and commercial reality.

Non domestic

But now we must temporarily leave the housing sector, and turn to the non-domestic buildings. Within this group of four projects there are two demonstrations in the UK where the energy consumption on services is as important as that on space heating. Both projects pay attention to insulating the fabric to reduce heat losses, but they also make considerable savings in water heating, air conditioning, catering and other ancillary uses.

You have already heard Mr Goodman speak in general terms about the design of the low energy hospitals and about his predictions for the new hospital to be constructed on the Isle of Wight (in the South of England). It is calculated that this hospital will use only half as much electrical energy for lighting as a conventional design, with other electrical savings in fan and pumping loads (*).Space heating costs will be 80% of normal values, and catering loads should be reduced by 35%(*).Overall it is expected that the energy consumption of this hospital will be less than half that of the standard 'Nucleus' design.(*)

It will be some years before these claims can be tested, as the main structure of the hospital has not yet been built; but already plans are well advanced for the monitoring of energy consumption, so that both the overall savings, and the sectoral benefits can be identified.

But in any project, the theoretical energy savings and financial benefits are of little value if the potential occupants of the building don't like it, won't use it or find it uncomfortable. And so I will end by referring to the Dolphin leisure centre at Darlington in the North of England.

This project shows very clearly that energy efficiency is fully compatible with community benefit and popular enjoyment. Within a very compact design(*) the centre provides sporting facilities, swimming pools,(*) conference hall and meeting rooms, together with restaurants and bar. The project also included the refurbishment of an older hall(*) . These different areas serve different groups of the population, and need to be maintained at different levels of temperature and comfort. (*) By careful systems design, the incorporation of heat recovery equipment and heat pumps,(*),and the use of a computer based building automation system(*), these various needs are being met at a lower energy cost than was expected.

Neighbouring municipalities are showing great interest in the performance of the Dolphin centre, (even though detailed monitoring has only just started) and aspects of its design are likely to be replicated in other leisure centres.

It seems that this is a fine example of a demonstration project which will succeed at all levels. It uses less energy than conventional designs of comparable performance, it does so in a cost-effective manner, it is popular with its owners, and with its users, and it is likely to lead to replication of design and construction in other centres.

ACKNOWLEDGEMENT

The views expressed in this paper are those of the author alone, and not necessarily those of the U.K. Department of Environment nore of the U.K. Department of Energy.

Discussion

Only a short time was available for discussion, following the presentations by selected speakers.

Mr English was asked about the data transmission protocols used in communicating between outstations and the central processor; and also about the typical costs of installations. He replied that standard data systems had been developed by his Company, and these worked satisfactorily on the project. The cost of an outstation might vary between £2000 and £5000, but installed cost depends greatly on the extent of mechanical engineering work needed at each location (eg additional pipework, new motorised valves, etc).

Finally, the rapporteur was asked whether he agreed that passive solar design is also a measure being demonstrated throughout the Community. The questioner felt that solar devices had been unfairly criticised by many authors and that their failures resulted from attempts to use them in areas which did not receive sufficient sunshine, whereas elsewhere (eg Italy) solar collectors are more successful.

The rapporteur agreed with this view.

SESSION III - IMPROVED CONVENTIONAL HEATING SYSTEMS

A - R & D PROJECTS

Chairman: Ir. J. UYTTENBROECK
Rapporteur: Dr. S.J. LEACH

REMARKS BY THE CHAIRMAN

Ir. J. UYTTENBROECK

INTRODUCTION

The energy consumption in buildings and thus also the energy savings in buildings depend on thre major factors :

- . Building characteristics
- . Characteristics of the energy consuming equipment
- . Occupants' requirements and behaviour

- It is well known that the last factor, "Human requirements and behaviour", is certainly the one with the biggest influence. Indeed, simply by adapting our living habits, by ensuring that greater intentention is paid to effective use of the building, heating system or control device and by avoiding energy wastage, it is possible to save a lot of energy without investment.

- The first factor, "Building characteristics", certainly was, and still is the factor which has been studied most frequently since the energy crisis in large number of research programmes all over the world. This is not surprising since building characteristics were relatively well known even before the energy crisis (thermal insulation, building heat losses, air infiltration, influence of solar radiation and of the thermal capacity of a building). But the main reason for this preoccupation certainly is that the building characteristics are really the first parameters which must be improved in order to reduce the energy needs of a building.

- Once we are sure that our building are well designed and constructed in order to minimize energy needs, then the following step is to install equipment on these buildings which can, with the highest efficiency, satisfy the reduced energy needs.

Session III is concerned with this problem and an attempt will be made to find ways of improving conventional heating systems.

CONCLUSION

1. The heating plant, ore more generally the building equipment (heating, cooling, ventilation, lighting, domestic hot water), should always be studied as a function of the building and as a function of the real outdoor conditions

   The first statement explains why three research programmes (A.2.1 - A.3.2 - A.3.9) are concerned with studies and measurements in matched pairs of houses. In the case of low-energy and passive- solar houses in particular it will be necessary to be able to examine in a detailed manner the evolution of the very low energy needs and to examine how the equipment can most effectively satisfy these fluctuating energy needs.

2. It is clear that matched pairs of houses in three countries will help to solve a lot of technical problems, but it is also clear that much quicker evaluation methods are needed for practical design of buildings and equipments.

   On this basis of field measurements we should thus be able to produce computer programs to indicate the corrrect boiler plant selection, taking into account all the parameters involved. The A.3.1 (TNO) programme is an attempt to establish such an evaluation method.

3. These prediction or evaluation programmes suppose that all the important parameters of the subelements of the equipment are known and also that ways are known of improving these characteristics in order to obtain an energy efficient plant.

   A lot of programmes in Session III are concerned with these aspects.

   - The A.3.3 (BSRIA) and A.3.4 (MARCOUSSIS) programmes study two heat emission systems, both of which work at low temperatures (warm air heating and plastic radiators).

   - The B.3.1. (OTB) and B.3.4. (SOLAR 77) programmes focuss on the search for better liquid fuel burners.

   - The A.3.7 (HECOS), A.3.8 (CSEA) and A.3.10 (TNO) programmes concentrate their efforts on the development of better adpated automatic controls for heating, cooling and lighting systems in buildings.

All these programmes form a part of the answer to the following question : How can we obtain the most efficient heat production and heat emission, and what is the influence on these efficiencies of the control system.

It can be noted that one should also consider the energy distribution efficiency in a building, but this problem is not examined in teh existing research programmes.

4. Two research programmes are not directly concerned with an understanding of or with an improvement of the efficiency of the system. They are simply interested in the design and the construction of a simple and sufficiently accurate heat meter : A.3.5 (BATTELLE) and A.3.6 (OMI).

   Everybody knows that the correct measurement of the amount of energy which is consumed or which is delivered is a very important parameter for decreasing wastage, because it directly influences human behaviour.

5. The main objective of a heating system however is not to save energy. The main objective of a heating system still remains the realisation of an acceptable comfort level in rooms.

   Given that fact, we should come as soon as possible and at least on an European level to a generally accepted method to define and to measure the obtained comfort level. Indeed such a generally accepted method should be used in order to compare the old, the new or the adapted heating systems (for laboratory as well as for field measurements).

   Personally I think that it is not necessary to make new research work in order to come to such a method. So why should we not do it. This agreement in the field of thermal comfort is necessary just as agreements are necessary in other not so difficult matters as Mr. LEBRUN has explained (vocabularly, references, calculation methods, design of experiments, etc.).

DEVELOPMENT OF A COMPUTERPROGRAM FOR BOILER PLANT SELECTION

A.J. Meijnen
TNO-Division of Technology for Society
P.O. Box 342, 7300 AH Apeldoorn
The Netherlands

Summary

A description is given of the activities concerning the development of a computerprogram for the selection of a heating installation from the viewpoint of a low energy consumption. The program should calculate the energy consumption of a selected heating plant, based on the hourly heat demand of the belonging building. Also an economical evaluation is given, using the present worth value method. Besides the construction of a theoretical model and the computerprogram, also measurements in an existing heating installation have been performed. The measurements partly deal with the analysis of the behaviour of the boilerplant and its components, as a check on the theoretical relations, used in the model. Over and above that, measurements are performed during a longer period, in order to compare actual and theoretical energy consumption, calculated by the computerprogram.

1. INTRODUCTION

From several studies mentioned in literature, it is obvious that composition and control of a heating plant may have an important influence on the utilization efficiency of the plant and therefore on fuel consumption. These studies mainly describe the effect of one or two alterations on the existing configuration (e.g. a flue gas valve or the number of boilers). Heating plants, however, do exist in many designs and compositions of boilers, additional components and control systems. In a former TNO study (1) the effect on fuel consumption of varying boilerplant designs has been calculated. This study, however, dealt with a limited number of alternatives and a fixed heat demand and total installed boiler power. From the calculations it appeared that the fuelconsumption of the most economical design may be about 20% lower than that of the design with the highest fuel consumption. These results stress the necessity of an evaluation of boiler plant design with respect to energy consumption and based on the heat demand of the building belonging to it. This evaluation may be of importance for new plants, but also for existing plants, that need to be improved by energy-saving measures.

In order to make it possible to evaluate a large variety of boiler plants, even with more accuracy, the use of a computer is obvious. The aim of the project is to provide such a computerprogram, that may serve as a tool for boiler plant designers to arrive at a heating installation that covers the heat demand of a given building as efficient as possible (with respect to fuelconsumption and for costs). This computerprogram should be able to operate on the basis of hourly data for the building heat demand. Beside the energetical evaluation, also an economical assesment of the heating plant design should be realised.

2. DESCRIPTION OF THE PROJECT

The project consists of the following four major parts (which will be discussed later on):

a. Development of a theoretical model, that describes the behaviour of various boiler plant components and also contains an economical evaluation procedure for the boilerplant.
b. Effect measurements on an existing boiler plant, the results of which may be used for verification of parts of the theoretical model.
c. Development of a computerprogram that will be able to perform the necessary calculations, based on the theoretical model.
d. Verification measurements, that provide data from an existing boiler plant, on an hourly basis, to be compared with computerprogram calculations for the same heating plant.

2.1. Theoretical model

2.1.1. System boundary and principle

The net heat demand for a boiler plant is the result of uncontrolled (weather, internal heat load and physical properties of the building) and controlled influences (night & weekend reduction, optimizer, heating period). The differences between heat output and fuel input originate from all sorts of losses from the boiler plant, that occur during operation. Once these losses are known, the fuelconsumption may be calculated. This means that the system boundary for the calculations encloses the boiler(s) plus connecting pipes and headers etc. and that the supply and return pipes, the fuel supply and the stack cross this boundary, apart from radiation, transmission and convection losses. (fig. 1)

The basic principle of the theoretical model is, that it calculates the degree of utilization and the "seasonal" or utilization efficiency of the boiler plant per hour, and then the fuel consumption. The interval period of one hour was selected for the following reasons:

a. Existing computerprograms for the calculation of building heat demand usually need hourly weather data as an input and therefore produce the heat demand on an hourly basis.

b. Compared to the simple calculations of utilization efficiency commonly used, based on a year or heating season, a decrease of interval period, will give an increase in accuracy.

The calculation starts from the (given) heat demand and ambient temperature for a certain hour and the operation data of the plant for the previous hour (e.g. boiler water temperature, accumulated heat).

2.1.2. Theoretical relations

To be able to describe the operation of a boiler plant and its components, a large number of relations must be applied. The theoretical relations may be divided roughly into three groups:

- installation dependent
- operation dependent
- economical

Installation dependent relations

These relations are mainly determined by the design and composition of the heating installation and cover the following items:

* Type of ignition - permanent
  - incidental/electric
  - incidental/electric + ignition burner
  - stepwise ignition
* Fan energy consumption - starting phase
  - continuous
* Burner control - on/off
  - high/low
  - modulating
* Standstill-loss - data given by manufacturer, related to burner control, water temperature and boiler design
* Installation constant and watervalue
* Properties of water and fuel

Operation dependent relations

These relations describe the operation of the boiler plant, and cover the following items:

* Degree of utilization - depending on burner control, heatdemand and available boilerpower
* Boilerwater-temperature - constant high
  - weather-dependent controlled
  - varying
* Heating up after period of standstill
* Cooling down of boiler, with or without watercirculations
* Utilization efficiency - based on degree of utilization, corrected stand-still-loss (2) and fullload boiler efficiency
* Fuel consumption - based on maximum boiler power, degree of utilization efficiency.

Economical relations

For the economical evaluation of selected heating installations, the present-worth-value method will be applied. This includes the following quantities:

* Annual energy costs - result from the installation, operation dependent relations, fuelprice level and annual rise in prices
* Annual redemption and interest of the investment
* Annual maintenance costs
* Effect of taxes

2.2. Effect measurements

2.2.1. Heating plant

The effect measurements are performed in the heating installation of a two storey office building. The heating plant consists of two atmospheric boilers (465 kw each), with weather-dependent pre-control of watertemperature, a high/low burner control and a priority selection. Recently one of the boilers has been equipped with a flue gas valve.

2.2.2. Measurements

In order to be able to perform the necessary measurements, the boiler plant has been equipped with instruments, as indicated in figure 2. The signals are collected by means of a Solartron data-acquisition system and recorded on tape after proper conversion to the appropriate units. Data are processed into other relevant quantities, such as:

* mean boilerwatertemperature
* temperature rise of water
* gas consumption under standard conditions
* amount of heat collected by the water
* remaining operating conditions of the boilers

Until now, measurements have been performed on fluegas side efficiency (fig. 3) and standstill-losses; the latter was based on Dittrich's method (3) and the results are shown in table I. Measurement of the ventilation losses are to be performed, to evaluate the effect of the flue-gas valve.

Watertemp. setting (°C)	Ambient temp. (°C)	Burner setting	Standstill-loss	
			measured	normalized
40	11.3	high	0.0135	0.0065
65	15.2	high	0.0381	0.0316
90	15.3	low	0.0955	0.1189

Tabel I: Results of standstill-loss measurements

2.3. Computerprogram

For the calculation of hourly fuel consumption of the boiler plant, the before mentioned theoretical relations (2.1.) have been composed into a calculation scheme, for which a computerprogram is under development now. Beside the hourly heatdemand of the building and outside temperature during the heating period considered, which result from a separate building heatload calculation and may be read from a tape or file, the heating installation lay-out must be fed into the computer. Fig. 4 shows a possible input format for this. The calculation procedure of the program is schematically given in fig. 5. The output of the program basically contains the hourly and/or accumulated energy consumption, on which the economical evaluation is based. Further more the output contains data on boiler plant operation during the heating period.

2.4. Verification measurements

The computerprogram will be verified by measurements on the same boiler plant that has been used for the effect measurements. During a 4 weeks period the boiler plant will be monitored by a data-logging system. From the data thus gathered hourly data will be derived on boilerwater-temperature, degree of utilization of each boiler, fuel consumption, building heat demand and outside temperature. The latter two data-series are input for the computerprogram that already contains all technical data necessary for the description of the boiler plant. The fuel consumption will then be calculated. Adjustments of the computerprogram are possible then.

2.5. Economic assessment

The expected saving of energy consumption obtained by applying the computer model to plants to be modified or new plants to be constructed, is estimated at 5% (average), while savings up to 20% are possible. The energy savings achieved, however, depend also on the choice of a boiler plant made on an economical basis (present worth value of the total investment including energy costs).

References

(1) Ing. W. Adam
"De keuze van een ketelinstallatie uit het oogpunt van energiebesparing"
(The selection of a heating plant from the viewpoint of energy saving)
Verwarming en Ventilatie, april 1980, nr. 4.

(2) Ing. D.J. van der Heeden
"Het gebruiksrendement van C.V.-ketels"
(The seasonal efficiency of Central Heating Boilers)
Verwarming en Ventilatie, februari 1977, nr. 2.

(3) Dipl. Ing. A. Dittrich
"Zum Jahreswirkungsgrad von Ein- und Mehrkesselanlagen"
Heizung-, Lüftung-, Haustechnik, december 1972.

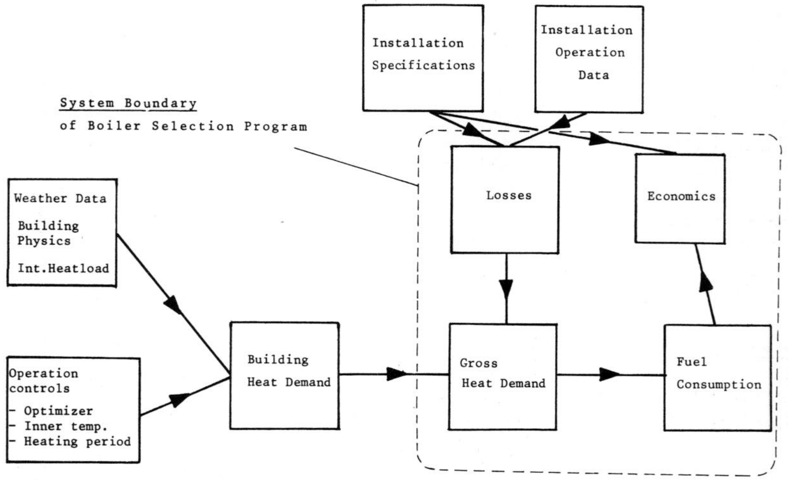

Fig. 1: Principle of Boiler Selection Program

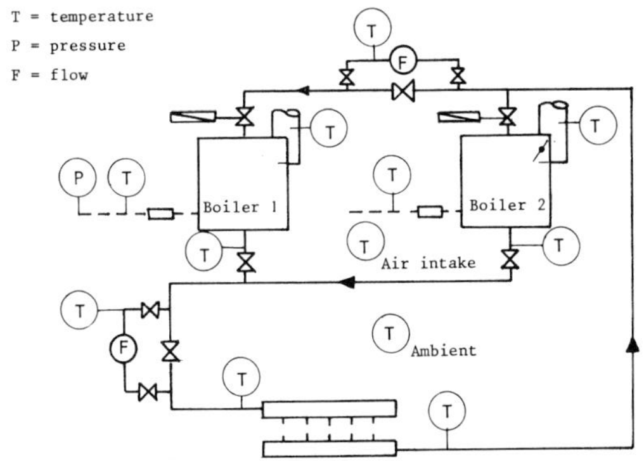

Fig. 2: Heating plant and measuring points; schematic

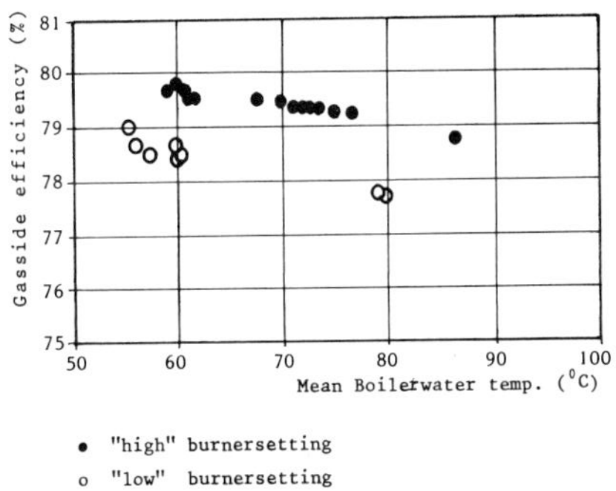

Fig. 3: Efficiency v.s. watertemperature (boiler 1)

Fig. 4 (see next page)

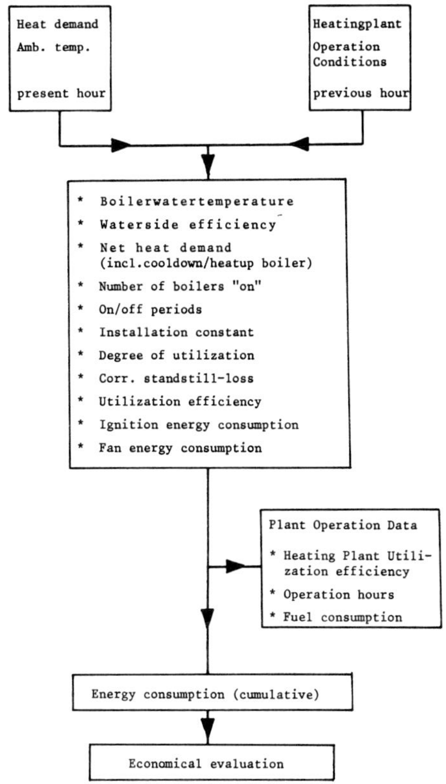

Fig. 5: Program calculation procedure; schematic

```
DATA ON INSTALLATION AND OPERATION
BOILERS       :[_]

WATERCIRCUIT:
      primary   :waterv.(kg):[          ] h.t.coef.(W/m^2.K):[    ] area(m^2):[    ]
      secundary:waterv.(kg):[          ] insidetemp.(C):[    ]

INTERRUPTION OF WATERCIRCULATION:[_]          (ye or no)

MAINTAIN HIGH WATERTEMP. temp.(C):[    ]

BOILERHOUSETEMP.      (C):[    ]    BOILERTHERMOSTATDIFFERENCE   (C):[    ]

BOILERWATERTEMPERATURECONTROL     (select 1 of 3)
      constant :Tbdes(C):[    ]
      weatherd.:Tbdes(C):[    ] Thl(C):[    ] Tambdes(C):[    ] Tbmin(C):[    ]
      variable :Tbdes(C):[    ] Thl(C):[    ] Tambdes(C):[    ]

FUELCODE      :[_]  (1=gas, 2=oil2 , 3=oil6 )
press ENTER to continue
```

```
DATA PER BOILER

BOILERNO:[_]      SAME AS LOWER BOILERNO.  :[_]  (>0 skip rest )
WATERV.(kg):[          ]
H.T.COEF.(W/m^2.K):[    ]
BOIL.AREA(m^2):[    ]
BURNER : Pvent.(kW):[    ]    purge time (s):[    ]
IGNITIONCODE    :[_] (1=permanent, 2=electric, 3=elec.+ign.burner)
      vol.flow perm. burner      (m^3/s):[    ]
BURNERCONTROL       code           (select 1 of 3)
      on/off       :[_](1)
      high/low     :[_](2) Plbel.   (%):[    ]  Th/l(C):[    ]
      modulating:[_](3) Pthrbel.(%):[    ]
BOILERLOAD       : maximum (kW):[      ]  nominal (kW):[      ]
BOILERTYPE                       (select 1 of 3)
  standard        :A0:[      ] A1:[      ] A2:[      ] A3:[      ] (Fst)
  economizer      :A0:[      ] A1:[      ] A2:[      ] A3:[      ] (Feco) Pf(kW):[    ]
  flue gas valve:A0:[      ] A1:[      ] A2:[      ] A3:[      ] (Frgk)
WATERS.EFFECT:C0:[      ] C1:[      ] C2:[      ] C3:[      ] (Fhe)(C0 not equal 0)
WA.EFF. 1/thr:C0:[      ] C1:[      ] C2:[      ] C3:[      ] (Fhe)(C0 not equal 0)
press ENTER to continue
```

```
COST/BENEFIT DATA

FUEL PRICE FIRST YEAR          (Fl/M^3 of kg):[    ]

INVESTMENT                               (Fl):[      ]

NOMINAL INTEREST                         (-):[   ]

PERIOD OF REDEMPTION                     (y):[_]

PERCENTUAL RICE IN ENERGYPRICE           (-):[   ]

MAINTENANCE AND REPAIR COSTS  1e YEAR (Fl):[      ]

INFLATION PERCENTAGE                     (-):[   ]

TECHNICAL LIFETIME                       (y):[_]

TAX REDUCTION FACTOR                     (-):[      ]

press ENTER to continue
```

Fig. 4: Input format of boiler selection program

BRE CALIBRATED HOUSE EXPERIMENTS

Report on secondary windows, electronic room-stats, cylinder-thermostats and new BRE houses

R. RAYMENT
Building Research Establishment, Garston, Watford, Herts., UK.

SUMMARY

A pair of houses with simulated occupancy have been used at the Building Research Establishment (BRE) for a number of years for assessing measures intended to reduce fuel consumption. The technique relies on using one house as an experimental control while conditions in the other house are varied. Results so far have demonstrated that the method gives excellent accuracy in near realistic circumstances. It is particularly suited for the normal UK practice of intermittent heating and for night set-back.

Energy Savings from fitting an additional pane of glass on the inside of existing windows are shown to be around 12% in the houses tested. The savings being brought about by a reduction in U-values and in air-infiltration losses. Generalisation of the result needs to take cognizance of the fact that houses with different glazing areas, heating systems, etc will show different savings. Comparison of actual and calculated consumption in this experiment showed that the calculated prediction was about 20% high.

Electronic room-stats compared with conventional accelerator room-stats showed a small mean saving of about 2%. Further experiments are required to improve the statistical certainty of the result but if the result holds the electronic room-stat would have a pay-back in about 8 years.

The variety of possible fuel savings brought about by fitting hot water cylinder-stat control is demonstrated experimentally. Much depends on the characteristics of the rest of the heating system, but as improved system sizing takes hold this type of control will show up as highly cost-effective.

The new BRE experimental houses now consist of 2 pairs of semi- and a row of 4 terraced houses. These are now fully operational and experiments are underway on comparing single glazing units against sealed double glazing and heat reflecting double glazing units. A programme of work for the next 3 years is being planned.

1 INTRODUCTION

Semi-detached houses with simulated occupancy have been used at the Building Research Establishment (BRE) for a number of years (ref 1 and 2) for assessing the potential of various measures intended to reduce fuel consumption. Whilst not entirely eliminating the need for field trials in occupied houses the technique developed reduces many of the uncertainties caused by variations in the weather and in occupants behaviour. It is arranged that the houses have similar heating systems, controls, insulation characteristics, adventitious heat gains and construction. A high degree of correlation then exists between the energy consumption in the two houses and changes in consumption in one of the houses can be accurately measured. The houses are normally operated with a typical UK intermittent heating schedule and night set back can easily be incorporated.

This paper presents results from experiments on secondary windows, electronic room-stats and cylinder-thermostats. Also a new set of 8 houses, now equipped and ready for further investigations is reported on.

2 MATCHED HOUSE METHOD

The course of an experiment follows a distinct pattern. Two or three weeks are usually needed at the start of a heating season to adjust the heating systems and temperature controls for equivalence in a pair of houses. Where temperature controls are under test, a further period is required to ensure similar temperature levels in relevant areas of the houses. For other measures, such as improved insulation, this additional period can be omitted. If only one measure is under test, a two week cycle has proved sufficient: one week being devoted to a calibration run, with both houses in their base condition, and the second week used to examine the measure when installed in the nominated test house. Whether or not the experiment has to be carried out through the whole heating season depends in part on the sensitivity of the measure to heating load. Also if occupancy behaviour is thought to be important the two week cycle can be extended to simulate different patterns of behaviour.

The basic analysis of the results uses the energy consumption in the nominated reference house to predict that in the test house, and compares this with the actual consumption when the measure under test is installed. This method, compared with a degree-day technique, improves the accuracy of prediction by up to a factor of 7 times.

Analyses reported here have been carried out by comparing regression lines and by examining the histogram of differences in consumption in a paired experiment. More detailed studies can be made with the data which currently include globe temperatures in every room, gas consumptions, solar, wind and external temperature, as well as switching times and temperatures of the space and hot water heating systems and controls. A full account of the heating systems and controls is given in ref 1. Briefly a gas-fired, wall mounted heating appliance drives ten radiators and heats a 160L hot water cylinder. The hot water is controlled by a cylinder-thermostat and the space heating can be controlled by thermostatic radiator valves and/or by a room-stat mounted in the lounge. Heating periods can be specified using a programmable time clock. Both hot water and space heating circuits are fully pumped.

3 ENERGY SAVINGS FROM SECONDARY WINDOWS

3.1 Introduction

A very common form of double glazing used in the UK is to add an additional pane of glass on the inside of an existing window. Usually this is considerably cheaper than installing a complete sealed unit plus frame. It can be arranged that the secondary window reduces any unnecessary air-infiltration through the windows.

The construction chosen for investigation (see Fig 1) consisted of a wooden frame, mounted and sealed to the wall-cill-lintel surround with a gap between the panes of about 57 mm: turnbuckles enabled the windows to be easily demounted for calibration. The houses used were those described in ref 1 and all the windows in one house were so fitted.

3.2 Calibration

As mentioned in Sections 1 and 2 conditions were set up in the houses to be as nearly identical as possible. Calibration runs were carried out periodically throughout the heating season with single glazing in both houses. This takes account of asymmetries between the two houses, due to air-infiltration and solar gains. The results are shown by the black circles on Fig 2 and the regression equation is:-

$$Gt = - 77.4 + 1.320\ Gr \qquad \text{MJ/day}$$

where Gt is the gas consumption in the test house and Gr that in the reference house.

3.3 Results with secondary windows

The gas consumption, with the secondary windows in the test house, are shown by the open circles on Fig 2. As indicated by the dashed line they correspond to an approximate 12% saving with higher savings at high loads and hardly any saving at low loads.

The mean difference in consumption between the single and secondary glazing, across the heating season, is 38 MJ/day and the 95% confidence limits are +- 11 MJ/day. Thus, as expected from inspection of Fig 2, the result is statistically highly significant.

3.4 Conclusions

Reference 3 gives a full account of this glazing experiment and in that, a comparison is made between actual energy consumption and that predicted from steady state calculations. It turned out that the prediction over-estimates the absolute consumption by about 20% but predicts, more or less correctly, the percentage reduction in consumption when the secondary windows are in place. From the analysis it is estimated that the standard U values, particularly for the walls in these houses, are of the order of 10% too high.

Before the main result ie the 12% saving, is extended to other houses the following factors need to be noted.

(a) Though extensive air-infiltration measurements were carried out for single glazing it was not possible to arrange this with the secondary glazing.

(b) The heat loss through the single glazing was about 25% of the total loss. Also the houses had 100 mm of loft insulation but no cavity fill. A different mix of insulation measures and window sizes will affect the consequent savings.

(c) For security reasons it was not possible to simulate window and door opening in the occupancy profiles. A method for doing this, however, is being tried in the latest test houses (see later) and further double glazing experiments are in hand.

4 ELECTRONIC ROOM-THERMOSTATS

4.1 Introduction

Experiments carried out in the BRE houses in previous heating seasons gave indications that conventional room-stats suffered a change in set-point according to the heating load. This change was such that the house temperature tended to be 2-3°C higher in mild compared with cold weather conditions. Claims had been made that directly sensing thermostats using for example thermistors, as opposed to a bi-metallic strip or vapour capsule with an accelerator heater, did not suffer from this effect. Some computer model results reported in ref 1 also seemed to bear this out.

In experiments reported here (see ref 4 for a full account) the electronic and conventional room-stats were set up to give the same living room temperatures in cold conditions, it being implied that this temperature would ideally also be that required in milder weather conditions.

4.2 Calibration

Calibration runs were carried out using conventional room-stats in both houses. The resulting regression line was:-

$$Gt = -4.2 + 1.050\ Gr \qquad \text{MJ/day}$$

where Gt is the gas consumption in the test house and Gr that in the nominated reference house.

4.3 Electronic room-stat installed

A histogram of the difference between the predicted energy consumption in the test house for accelerator room-stat control and the actual consumption with electronic room-stat control shows, for the full set of results, a mean saving of about 16 MJ with a standard deviation of about 4.6 MJ. However, a period of doubtful operation of a 3-way valve meant that some of the data may not have been representative. A more reliable sub-set gave a mean difference of 6.5 MJ with a standard deviation of 4.2 MJ. This result has only weak statistical significance.

4.4 Cost effectiveness

The electronic room-stat costs about £25 (ie 2 to 3 times that of a conventional room-stat) and can be directly fitted in place of an existing room-stat. The saving of 6.5 MJ is about 2% of the annual consumption and for current marginal gas costs would, if the savings could be proved, give a pay-back period of 8 years ie in about half the lifetime of the device.

4.5 Conclusions

As a thermostatic device the electronic room-stat performed entirely satisfactory. It turned out that the sub-set of the results contained a preponderance of higher than average amounts of solar radiation. This would have prevented a clear demonstration of improved control stability as claimed. Thus, further evidence is required to prove the superiority of the device.

5 HOT WATER CYLINDER-STAT CONTROL

5.1 Introduction

In the UK there is still some way to go before fitting cylinder-stat control is more common than not fitting it. Though most leading authorities recommend cylinder-stat control of hot water heating, doubts have been expressed concerning the necessity of this when, due partly to oversizing, many gas-fired central heating systems are able to perform satisfactorily with their boiler-stats turned well down from the upper flow temperature settings.

There are many design alternatives associated with the central heating systems currently being installed which include the thermal capacity of the boiler, the use of gravity or pumped circulation for transferring heat to the cylinder, the efficiency of the water cylinder's heating coil, the use or otherwise of time-clocks and the arrangements for switching off the space heating supply when domestic hot water alone is required.

To try to clarify some of the issues experiments were mounted in the BRE houses for a range of situations where the energy consumption for simulated water run-offs was measured with and without cylinder-stat control at two extreme boiler-stat settings.

5.2 Cylinder-stat systems

The effectiveness of cylinder-stats depends greatly on what is already available or intended for the particular central heating system. In its simplest form it is arranged that a strap-on cylinder-stat switches the boiler on or off according to the sensed temperature of the hot water. This method can only be used when there is no requirement for space heating. Such a case was studied by Whittle and Warren (ref 5) and, in terms of energy savings, compares broadly with that in which the boiler-stat is turned down in the summer and back up again in the winter. This means that in winter the hot water temperatures can quite easily reach levels where care has to be taken to avoid scalding. Pro rata adjustments to energy saving figures given in ref 6 suggest, however, that the cylinder-stat should be more effective than the boiler-stat in controlling hot water temperatures and the measures reported here support this. Nevertheless the implications are that the case for fitting cylinder-stats rests mainly on arranging that they operate in the winter-time as well.

This in turn will depend on the sizing of the heating system and its boiler.

A further level of improvement is to arrange for the cylinder-stat to switch the boiler off when space heating is not required; so called partial cylinder-stat control. There will still be many occasions, however, when the hot water temperatures are uncontrolled.

Only by fitting a motorised valve can the full benefit of the cylinder-stat be realised. A comparable system is a non-electric thermostatic valve fitted into the line to the hot water heating coil with a temperature sensor strapped to the hot water cylinder. Such a device cannot normally switch the boiler off and, since it requires plumbing in, is more expensive to fit.

5.3 Experimental arrangements

Simulated water run-offs to bath, basin and kitchen sink are described in ref 1 and will not be repeated here in detail. Briefly a strap-on cylinder-stat was fitted, one third of the way up, on a factory insulated hot water cylinder; the insulation being equivalent to a 2-3 inch loosely fitting sectional jacket. A schedule of 180 L of hot water over 16.5 hours was chosen and, using both experimental houses, some 34 days were simulated. Space heating was turned off and the cylinder-stat control either took the whole of the pumped output from the boiler or switched it off. Due to the fact that the experiment was carried out in the winter there were large house temperature variations. However, a good correlation was found between the measured daily gas consumptions and the mean temperature in the vicinity of the hot water cylinder's cupboards, enabling a clear separation of the results to be shown.

5.4 Results

In one house (SET 1, see triangles on Fig 3) a comparison was made between with and without cylinder-stat (set to 60°C) for a boiler-stat setting of 60°C. In the other house (SET 2, see circles on Fig 3) a comparison was made between with and without cylinder-stat (set to 60°C) for a boiler-stat setting of 80°C.

A bracketed notation is used to show boiler- cylinder-stat settings in °C ie (boiler-stat, cylinder-stat) settings, where N indicates no cylinder-stat control.

The savings associated with SET 1 (ie (60,N) to (60,60)) range from 4 to 9%. Essentially these savings come about by the cylinder-stat switching the boiler off, leading to a lowered mean temperture for the boiler and hence more efficient operation. In addition the stored water temperature cycles directly through the cylinder-stat's differential bringing about a lowered mean storage temperature than when under boiler-stat control only. Taking 18°C as a representative 24 hour mean temperature outside the hot water cylinder's cupboard shows that a saving of 5% is achieved for the (60,N) to (60,60) comparison.

SET 2 (ie (80,N to (80,60)) relates to some part of the winter-time operation when space heating is not required so that the cylinder-stat can switch the boiler off. This implies, of course, that the system allows the boiler to be switched off automatically. The whole winter season would be covered by SET 2 for a system with a motorised valve fitted in the primary feed to the hot water cylinder and a boiler that operates at its highest thermostat setting. If such a system works with a lower setting the savings will lie between the two cases investigated.

For a surround temperature of 18°C SET 2 shows a saving of 29%. This is brought about by a reduction in storage and boiler losses. Losses from the boiler-cylinder connecting pipework and the draw-off pipes will be proportional to the boiler flow and stored hot water temperatures respectively. There is evidence for an approximate halving of recovery times, after a draw-off, with the higher boiler-stat setting.

5.5 Other measurements

Drawing on published data and other work done at BRE enables a table to be produced for the daily percentage savings for the following changes:-

(80,N) to (60,N) 25%, (80,N) to (80,60) 29%,
(60,N) to (60,60) 5%, (60,N) to (80,60) 0%

As will also be seen in the next section this highlights the large range of savings possible.

5.6 Cost effectiveness

Data from another BRE experiment has shown that in the winter, heat is only required for about 50% of the time that the heating is programmed on. Thus partial cylinder-stat control could, on average, be expected to operate for 50% of the time in the winter.

The foregoing leads to three primary types of cylinder-stat control; summer use only, summer plus half winter (ie partial cylinder-stat control) and summer plus the whole winter (ie full control using a motorised valve). In each of these categories three pairs of boiler-stat settings are sufficient to delineate the range of savings and pay-back periods; these are summer and winter respectively 60-80°C, 60-60°C and 80-80°C, with the cylinder-stat set to 60°C in all cases. Space does not permit showing of the detailed analysis but for an average family and using standard costs the following table of pay-back periods in years results:-

Boiler-stat °C Sum.	Boiler-stat °C Win.	Type: Summer only	Summer + half winter	Summer + winter	System type
60	80	12	3	4	Average
60	60	12	9	16	Oversized
80	80	2	1	2	Fast response

The table reveals the very large range of possible pay-back periods and in part explains why it can be misleading to quote a single figure for the saving from cylinder-stat control in general, unless a knowledge of the mix of existing and new systems is to hand. In conclusion it can be said that as the pressure for better system sizing takes hold it will become more important that cylinder-stat control is fitted.

6 NEW BRE TEST HOUSES

All the foregoing experiments were carried out in a pair of houses that have now been vacated. They have been replaced by 2 pairs of semi-detached and a row of 4 terraced houses, built in the late 1950's and greatly expanding the capacity for this type of research (see photograph).

Apart from one house, reserved for computers and simulation control equipment, all have been fitted out with gas-fired wet central heating systems with facilities for the selection between thermostatic radiator valve, with or without flow-switch and room-stat control, upstairs-downstairs zone control, gravity or pumped hot water services and the option of weather compensator control.

Twin globe thermometers have been fitted in every room plus room-stat and water temperature sensors. Gas consumptions, external temperature, wind speed and direction and solar radiation on the 4 vertical walls of the houses are also automatically recorded. Parallel logging is also made of the operation and switching times of all the relevant control elements enabling a very comprehensive dynamic picture of the systems to be built up. Simulated occupancy is as described in ref 1 plus pairs of fans located in the front and back doors of the houses to simulate window and door opening.

Currently experiments are underway on a comparison of single glazing versus double glazing and heat reflecting double glazing; no loft insulation versus 50, 100 and 200 mm of loft insulation and room-stat control versus thermostatic radiator valves plus flow switch and weather compensator.

7 CONCLUSIONS

By arranging that pairs of similarly constructed houses have identical heating systems, controls, insulation characteristics and occupancy simulation a high degree of correlation can be brought about between the energy consumption in the two houses. Changes in consumption can then be accurately measured when conditions are deliberately altered in one of the houses. The improvement in accuracy compared with using a single occupied house is the result of a marked reduction in the uncertainty caused by the weather and in occupants behaviour. Occupancy effects are present but fixed and as knowledge on this becomes more comprehensive it can be incorporated. The typical UK pattern of intermittent heating is usually simulated but the effects of night set back and other heating periods can easily be included. BRE has developed these techniques over a number of years and the experiments described in this paper and elsewhere have shown the method to be a powerful tool for assessing the potential of various measures intended to save energy.

Initial plans include a large variety of experiments through to 1986. It is expected that important and reliable data will be obtained which will be used to promulgate advice on fuel saving measures, aid questions on legislation an provide good data for computer modelling verification programmes. This work takes direct cognizance of other work at BRE not only on the last mentioned item but also on air infiltration, infra-red and heat flux measurements, energy economics and occupants use of controls.

8 ACKNOWLEDGMENTS

This research is jointly funded by the Commission of the European Communities and by the Department of the Environment and is published by permission of the Director of the Building Research Establishment.

REFERENCES

1 Rayment R et al. 'Basic characteristics of low-cost houses in order to reduce the energy consumption for heating'. EEC Contract (No 363-78-1 EE UK), EEC Ref. EUR 8506, EN, Final Report, 1981

2 Rayment R and Morgan K. 'The performance of heating control systems in experimental houses'. Proc. 'New Ways to Save Energy' Conf., D Reidel Pub. Co., 1980

3 Rayment R and Morgan K. 'Energy savings from secondary windows in experimental houses'. Proc. Contractors Meeting 'Heating, Ventilation and Insulation', D Reidel Pub. Co., 1982

4 Rayment R and Morgan K. 'Comparing conventional and electronic room-stat performance in experimental houses'. Proc. CIB S-17 Conf. 'Heating and Climatisation', Delft, 1981

5 Whittle G and Warren P R. 'The efficiency of domestic hot water production out of the heating season'. Building Research Current Paper 44/78, 1978

6 'Controls for domestic central heating systems'. British Gas Bulletin Supplement, Watson House, 1978

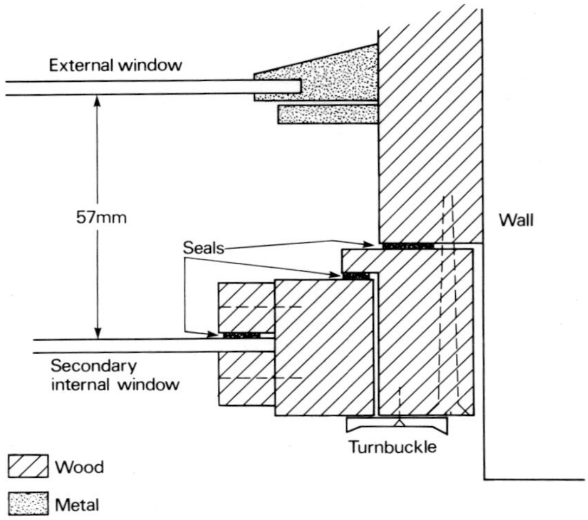

Fig. 1: Schematic diagram of the construction and fixing of the secondary double window

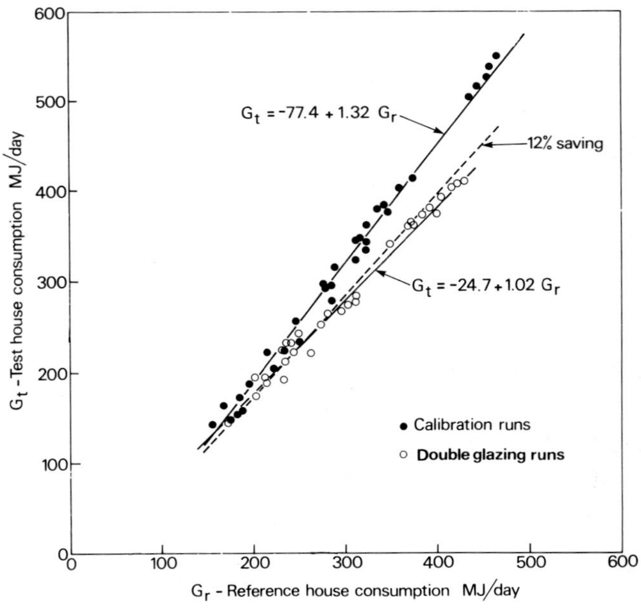

Fig. 2: Daily consumptions during calibration and secondary double glazing experiments

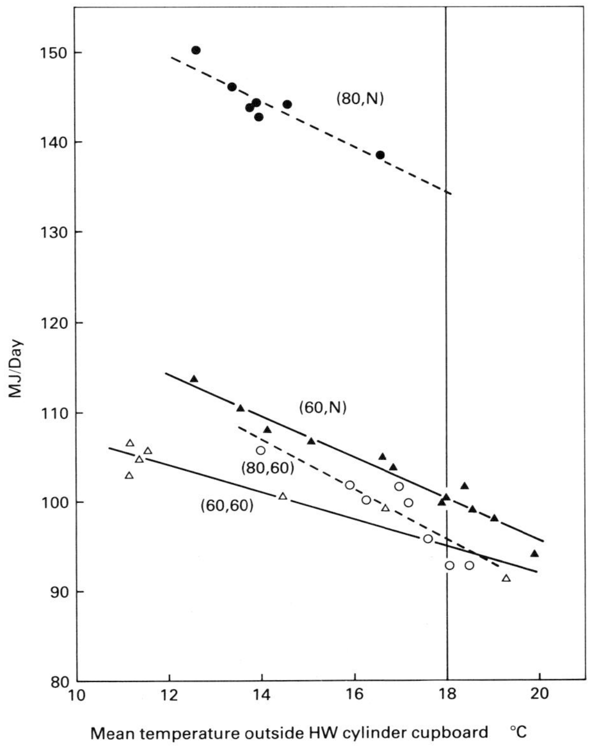

Fig. 3:
Comparison of daily energy consumption for the equivalent production of 180 L of hot water raised through 45°C and delivered at 55°C by simulated run-off

Key: (Boiler-stat, cylinder-stat) settings, N = no cylinder-stat.
In text triangles = SET 1 and circles = SET 2

Plate 1: Building Research Establishment's new field test houses

DOMESTIC WARM AIR DISTRIBUTION WITH LOW-GRADE HEAT SYSTEMS

H.B. AWBI*

THE BUILDING SERVICES RESEARCH AND INFORMATION ASSOCIATION
BRACKNELL, UK

Summary

The performance of a domestic warm air heating system that utilises a low-grade energy source was investigated in a simulated living room of a low-energy house. Because of the high air change rates associated with low-grade energy sources, the main concern was the diffusion of large air flow rates in the heated space without creating unacceptably high air velocities in the occupied zone.
The limits of air change rate and supply temperature within which thermal comfort was maintained were determined for direct discharge into the occupied zone at low level, discharge across a wall at low level, and high level discharge outside the occupied zone. These limits were defined by applying Fanger's comfort criterion to the thermal environment in the occupied zone for the UK winter-design heating load condition.

1. INTRODUCTION

A warm air heating system is probably most suitable for heating low-energy houses using low-grade energy sources. However, for optimum utilisation of the available low-grade energy and to maximise the effectiveness of the warm air heating system, it is necessary to determine the limiting conditions of supply air flow rate and temperature beyond which comfortable conditions within the heated space cannot be maintained. These limits have been determined for a simulated living room of a low-energy house for winter-design conditions.

Three methods of supplying the warm air to the room were investigated:

(a) Low-level direct discharge into the occupied zone emulating conventional practice of domestic warm air distribution.

(b) High-level discharge outside the occupied zone emulating air conditioning practice.

(c) Discharging the air at low level across the wall surfaces.

For each of these three discharge modes, surface mounted air terminal devices of different types were used to investigate their influence on the room environment.

The conditions in the occupied zone were assessed using the Predicted Percentage of Dissatisfied (PPD) method of Fanger's thermal comfort criterion. For this purpose, measurements of air velocity, air temperature and dry resultant temperature were carried out throughout the occupied zone. In addition to predetermined comfort limitations on the air temperature gradient and the air velocity in the occupied zone, this criterion formed the basis for defining the limits of air change rate and supply temperature for each supply device.

*Present address: Mechanical Engineering Department, Napier College, Colinton Road, Edinburgh, UK.

2. THERMAL COMFORT CRITERION

The thermal quality of a controlled environment is assessed by the degree of comfort of the occupants. To achieve comfortable environment in a mechanically ventilated space, it is necessary to provide a condition of thermal neutrality to the human body(1) as well as spatial uniformity of the environment(2). The condition of thermal neutrality is determined by the interaction of the four physical variables: air temperature, mean radiant temperature, relative air velocity and water vapour pressure, and two other factors related to the occupants, namely the activity (metabolic rate) and the thermal resistance of clothing. Fanger included these six parameters in a thermal equilibrium relation (comfort equation) between the body and the environment. If thermal equilibrium is not fulfilled the expected degree of discomfort is expressed by the Predicted Mean Vote (PMV) index which is a seven point scale ranging between -3 (cold) to +3 (hot) with 0 being neutral. Using environmental test chamber results, Fanger found a correlation between the ratio of people dissatisfied with the thermal environment (PPD) and the PMV. The relationship gives a minimum PPD value of 5% for PMV = 0. The recommended maximum PPD is 10% (1,3).

Although the body may be thermally neutral in a particular environment (i.e. the thermal comfort equation is satisfied), there may not be thermal comfort if one part of the body is warm and another is cold (local discomfort). This could be caused by a vertical air temperature gradient, an asymmetric radiant temperature field, local convective cooling of the body (draught), contact with warm or cold floor or temperature swing caused by the room temperature controller.

The thermal comfort criterion adopted for domestic warm air heating was based on Fanger's approach and the recommendations of the ISO and BSI draft standards(3) for asymmetric thermal limits. The criterion summarised below was applied within the occupied zone envelope which was taken to be 0.15m from the room surfaces.

- (i) Operative temperature (defined in reference (3)) between 20 and 24°C.
- (ii) Vertical air temperature difference between 0.15m and 1.8m (ankle and head level) <3°C.
- (iii) Mean air velocity <0.35 m/s.
- (iv) Relative humidity in the range 40 to 70% (although this was not investigated).
- (v) Radiant temperature asymmetry from windows or cold walls <10°C.
- (vi) Surface temperature of the floor between 19 and 26°C.

3. EXPERIMENTAL FACILITY AND TEST PROCEDURE

3.1 Test Room

To evaluate the thermal conditions in a room provided by an air distribution system, it is necessary to obtain measurements of air temperatures andvelocities throughout the occupied zone. Due to the effect on air movement and temperature distribution of both the magnitude and distribution of the thermal load it is usually required to simulate this load in the mock-up. A test room of internal dimensions 5.5m x 3.5m x 2.5m ceiling height was constructed of timber and slotted steel sections with slightly better insulation properties for the external walls than the current design values for UK dwellings. Two adjacent walls and the suspended floor of the test room were enclosed by a compartment in which the air could be cooled to the external design temperature for the UK of

-1^oC. The thermal transmittance coefficients (U-values) for the room are given below with the current UK design values in brackets:

U for exposed walls = 0.47 (0.60) W/m^2K
U for floor = 0.60 (0.60) W/m^2K
U for window = 3.06 (undefined) W/m^2K

The design heat load for the test room was 632 W (33 W/m^2).

Air to the test facility was provided by a cold air system supplying the external compartment and a warm air system supplying warm air to the room. The facility was capable of providing a wide range of internal and external conditions as well as a flexibility to incorporate various configurations of air terminal devices. Figure 1 shows a schematic diagram of the test room and the heating and cooling systems.

Although the test facility was designed for UK design conditions, the results would also apply to European countries with a lower external design temperature and a higher standard of thermal insulation.

3.2 Instrumentation and Data Acquisition

A two-directional traversing system was constructed to transport the velocity and temperature sensors to the required position in the room. It consisted of a 300mm wide x 150mm high gantry spanning the width of the room, within which a small carriage was supported. The instrument carrying pole was attached to, and moved with, the carriage. Stepper motors, controlled by a microcomputer, were used to move the carriage and gantry. Plate 1 shows a general view of the traverse system in the room.

Type T copper-constantan thermocouples were used for the measurement of air temperatures at heights 0.15, 0.5, 0.85, 1.2 and 1.8m in the room, internal and external wall surface temperatures and air temperatures in the air distribution systems. The dry resultant temperature was measured at a height of 1.2m using a thermocouple junction at the centre of a black hollow copper sphere of 111mm external diameter.

The time mean air velocity was measured at the same points where the air temperature was measured using a TSI model 1620 omnidirectional anemometer.

The outputs from the thermocouples and anemometers were fed into the microcomputer through two analogue/digital converters. The signals were processed by the computer and the results printed out, plotted and stored on magnetic disks.

The air supply rate to the room was measured using a convergent nozzle manufactured to AMCA standard.

3.3 Test Procedure

With the supply air device positioned in the required location in the test room, the warm and cold air supply systems were operated to produce the required test conditions. The instruments were moved to the first position and the air temperature and velocity at five heights and the resultant temperature were recorded. These measurements were automatically repeated at 77 positions in the room using 0.5m x 0.5m grid. The time mean air velocity at each position was obtained from the average of 120 samples using 1s sampling period.

At the end of each test, the data were automatically retrieved from the magnetic disk and analysed for comfort using Fanger's criterion. For this purpose, the activity parameters used were for a seated person (metabolism = 58 W/m^2) and the thermal insulation value for clothing was for average winter clothing combination (1 clo or 0.155 m^2K/W). Using these values and the values of air temperature and velocity measured at each point, a PMV value for the point was calculated for a relative humidity of 50%. An average PMV for each position in the room was obtained by averaging the five PMV's calculated at each height and using this value, the PPD for each position was obtained. The mean PMV for the room was found by averaging the PMV's for 77 positions and using this value, the PPD for the room was calculated.

4. RESULTS AND DISCUSSION

4.1 Room Air Temperatures

The difference in the average room air temperature between head height and ankle height ($t_{1.8}$ and $t_{0.15}$) is shown in Figures 2(a) and 2(b) plotted versus the air change rate. The results show that the vertical temperature gradient was dependent on the method of discharging the air into the room but almost independent of the number and type of supply devices used in each case. For each discharge mode, the vertical temperature gradient decreased as the air change rate was increased. For a given air change rate the high-level supply mode produced a larger temperature gradient than the low-level direct discharge mode. The temperature gradient for the low-level wall diffusion mode was between these two profiles. Extrapolation of the vertical temperature gradient curve for high-level supplies to less than about 9 air changes per hour produced a temperature difference in excess of the 3°C limit specified in the comfort criterion.

4.2 Room Air Velocities

The average air velocity for the occupied zone, V_R, is shown in Figures 3(a) and 3 (b) plotted versus the air change rate for the low-level supplies (direct discharge and wall discharge) and the high-level supplies respectively. Allowing for experimental scatter, the results indicate that the mean room air velocity was independent of the type of supply device. The low-level supplies produced higher mean velocities than the high-level supplies.

The effect of the momentum, m, of the supply jet on the mean room air velocity is depicted in Figure 4. As also found from previous work at BSRIA(4), a linear relationship existed between V_R and m. However, for a given jet momentum, a low-level supply produced a higher mean room air velocity than a high-level supply.

4.3 Comfort Assessment

The average PPD for the occupied zone was for some tests greater than the recommended limit of 10%. This was partly due to the thermal non-uniformity, and partly due to the air temperatures in the occupied zone being generally lower than the values required to satisfy the comfort criterion. To examine the thermal uniformity more closely, the room air temperature for each test was elevated at each measuring position to obtain a zero value for the average PMV in the occupied zone. A new 'corrected' PPD value (termed LPPD by Fanger) for each position was then calculated using the new PMV value at that position. This procedure produced more uniform distribution of PPD in the occupied zone for all the tests with maximum values less than 10%.

The supply and extract air temperatures were also increased (normalised) by the same amount as the room air temperatures and the results are shown in Figure 5 plotted versus the air change rate. For each of the three supply modes, the range of air change rate, supply temperature and extract temperature within which comfortable conditions were achieved in the occupied zone are defined in the Figure.

To cover the full range of air change rates shown in Figure 5, it would be necessary to use the low-level supply mode for low air change rates (less than about 10) and the high-level supply mode for high air change rates. The low-level wall-diffusion mode could provide comfortable environment for a wider air change rate than is possible with the direct discharge mode.

These results show that if the supply air temperature were below about 28°C (but not less than 26°C) at the winter outdoor design temperature then to provide comfortable conditions in the occupied zone, a high-level supply would be required. This method of supply produces unacceptable temperature stratification in the room for a supply temperature greater than 28°C and therefore a low-level supply mode would be required in this case. The low-level wall-diffusion mode should provide better thermal environment than the direct discharge mode because the high velocity jet regions remain close to the walls.

The room air temperature normalised using Fanger's comfort criterion is shown in Figure 6 plotted versus the air change rate. As expected, this temperature increased with increasing the air change rate with an average value over the range of about 24°C. This is about 2 or 3°C higher than the indoor temperature recommended by some building services guides(5, 6).

The results reported here were obtained with acceptable pressure losses across the diffusers ($\Delta P < 20$ Pa) and aerodynamic noise levels (based on catalogue data) less then 30 NR which is the maximum recommended rating for living rooms(5).

5. ECONOMIC ASSESSMENT

The most viable low-grade energy source for a low temperature warm air heating system is that from a heat pump. A system using this source of energy will be required to operate at varying supply temperature and air flow rate. For the same level of thermal comfort, the acceptable room air temperatures are somewhat lower at low room air velocities than at high air velocities so it is potentially more economic to supply the warm air at low flow rates. Not only would the heat losses to outside be less but also the fan power consumption would be lower. Furthermore, the aerodynamic noise level of the supply devices would also be lower. In this case a low-level wall-diffusion supply mode would be most suitable.

If the supply air temperature that can be provided were low (such as during peak heat load demand), it would be more economic to supply the air at high-level because lower room air velocities could be achieved with this mode than with the low-level supply mode, and consequently a lower room temperature would be required to achieve comfortable conditions.

To maintain thermally comfortable environment with optimum utilisation of the available low-grade energy, it would therefore be necessary to equip the room with two-level supply terminals, i.e. a low-level and a high-level supply. The air supply to each terminal would then be activated or otherwise according to the value of supply temperature. This may well be regarded as an impracticable solution and if so, a design assessment will need to be made to determine the most advantageous supply terminal arrangement for each specific application.

6. CONCLUSIONS

The results indicate that low-grade energy such as that supplied by a heat pump could be used with a warm air heating system to provide acceptable thermal environment in a low-energy dwelling. Depending on the air change rate and the supply temperature, thermal comfort could be achieved with the air supplied to the heated space either at low level or high level. For UK winter design condition, high-level air discharge can be used for a supply temperature only 2 or 3°C above room temperature. With supply temperature and room temperature difference higher than about 5°C, a high-level supply causes temperature stratification in the room and, in this case, a low-level discharge across a wall would provide better thermal environment.

The 'comfortable' room air temperature for sedentary activity and indoor winter clothing varies from about 23.5°C for an air change rate of about 3 per hour to about 24.5°C for an air change of about 22 per hour. The supply to room temperature differences for these air change limits are approximately 13 and 2°C respectively.

ACKNOWLEDGEMENT

This work was jointly supported by the Commission of the European Communities and the Department of Enviroment, UK.

REFERENCES

1. Fanger, P.O. Thermal Comfort, McGraw-Hill, 1972.

2. McIntyre, D.A. Indoor Climate, Applied Science Publishers, 1980.

3. Moderate thermal environment, determination of the PMV and PPD indices and specification of the conditions for thermal comfort, ISO draft TC 159 N48E, 1981.

4. Jackman, P.J. Air movement in rooms with ceiling mounted diffuser, HVRA Report No. 81, 1973 (BSRIA).

5. CIBS Guide, Section A1, 1978.

6. DIN 1946 - Part 2, Room ventilation technique, technical health principles (VDI Ventilation Rules), January, 1983.

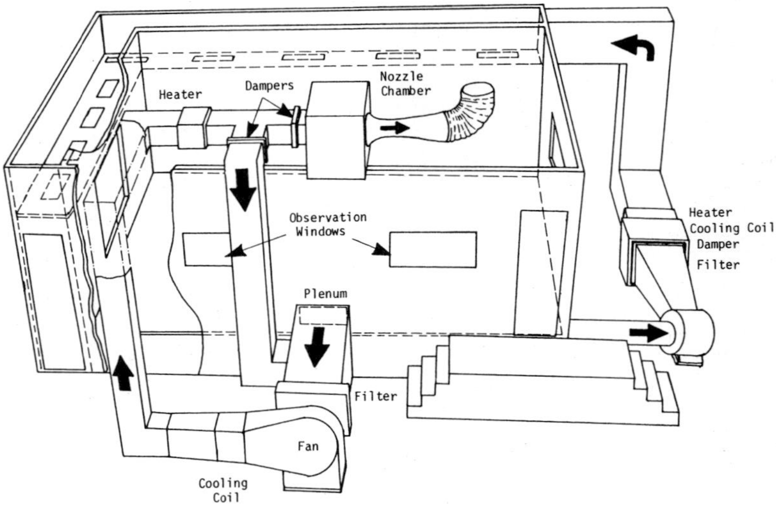

Fig. 1: Schematic Diagram of Test Room and Air Plants

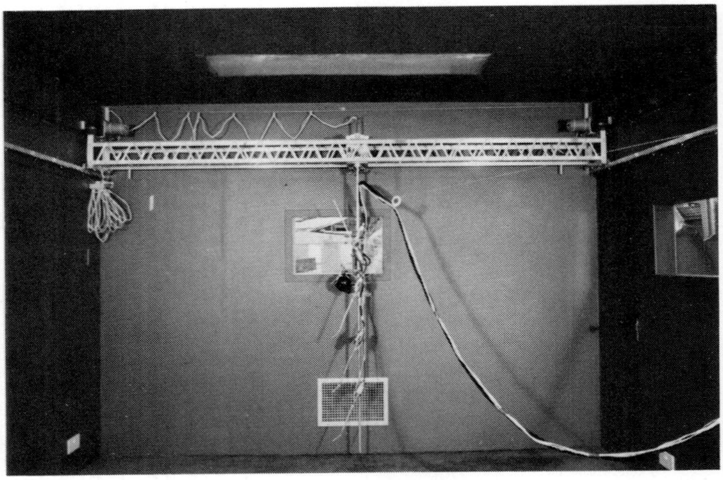

Plate 1: A View inside the Test Room showing the Instrument Traverse Arrangement

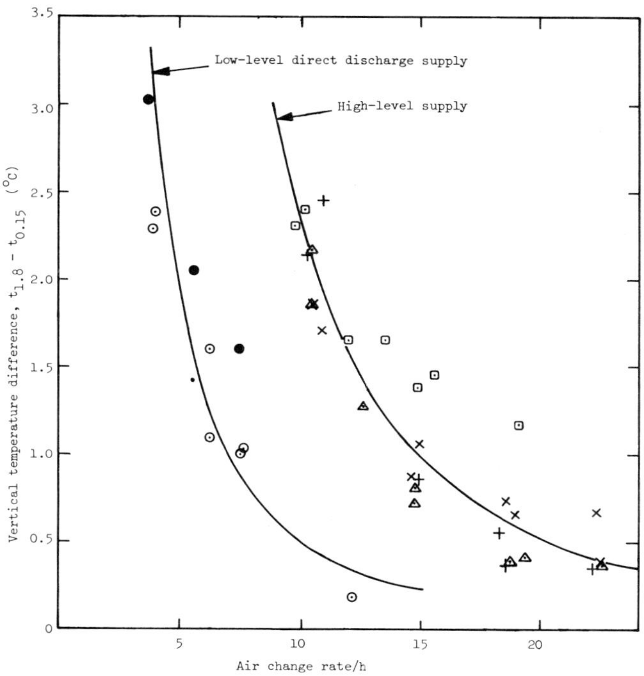

Fig. 2a:

Mean vertical Air Temperature Gradient in the occupied Zone

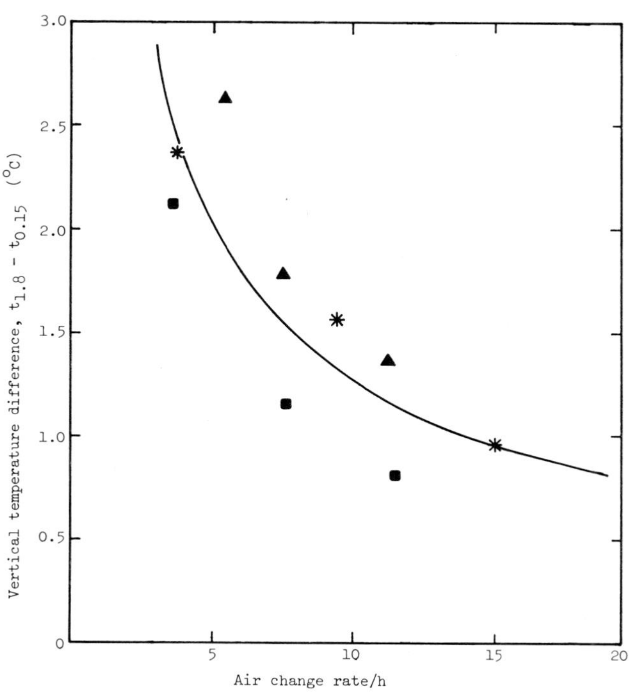

Fig. 2b:

Mean vertical Temperature Gradient in the occupied Zone (low-level Baffle Plate Diffusers)

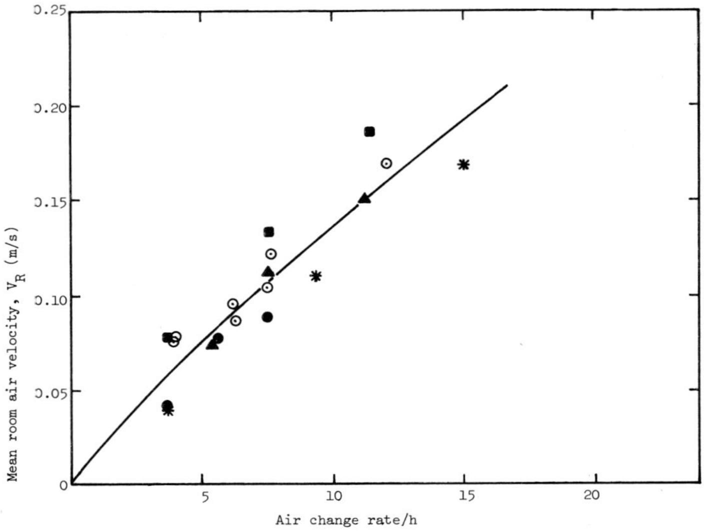

Fig. 3a : Effect of Air Change Rate on the mean Air Velocity in the occupied Zone (low-level Supply)

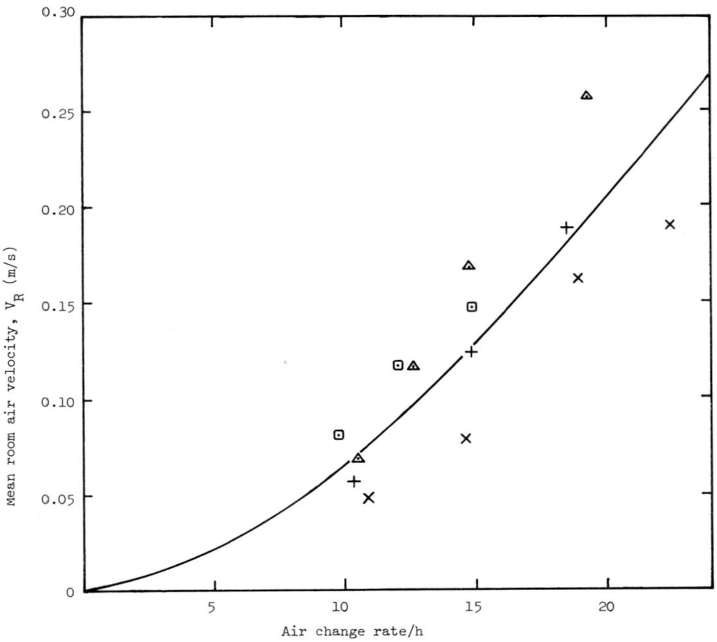

Fig. 3b : Effect of Air Change Rate on the mean Air Velocity in the occupied Zone (high-level Supply)

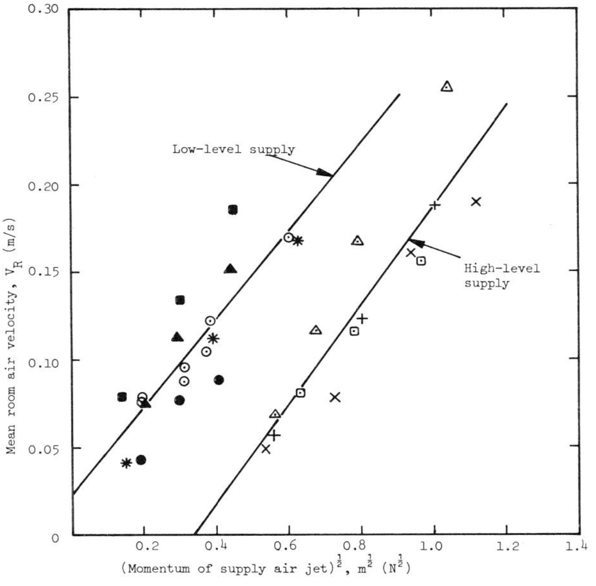

Fig. 4:

Variation of mean Velocity in the occupied Zone with Momentum of Supply Air Jet

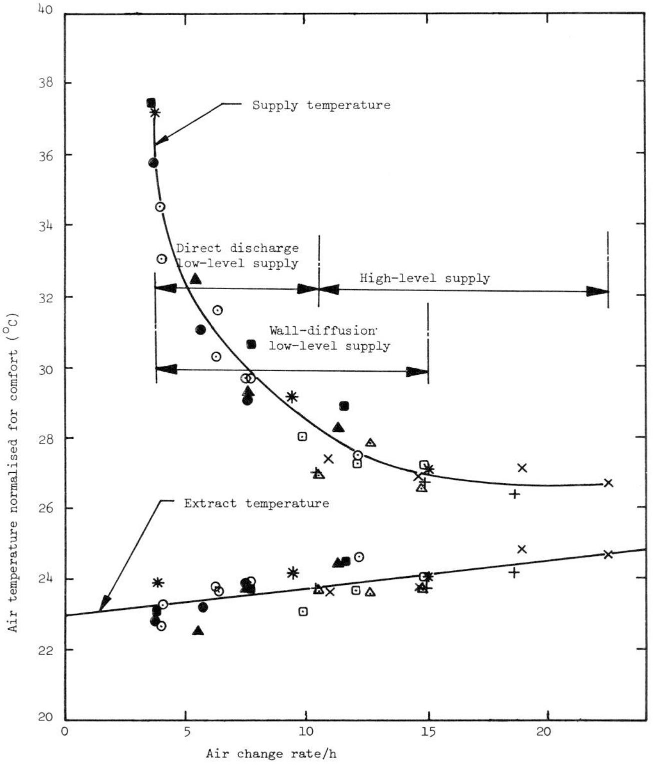

Fig. 5:

Supply and Extract Air Temperatures required for different Air Change Rates

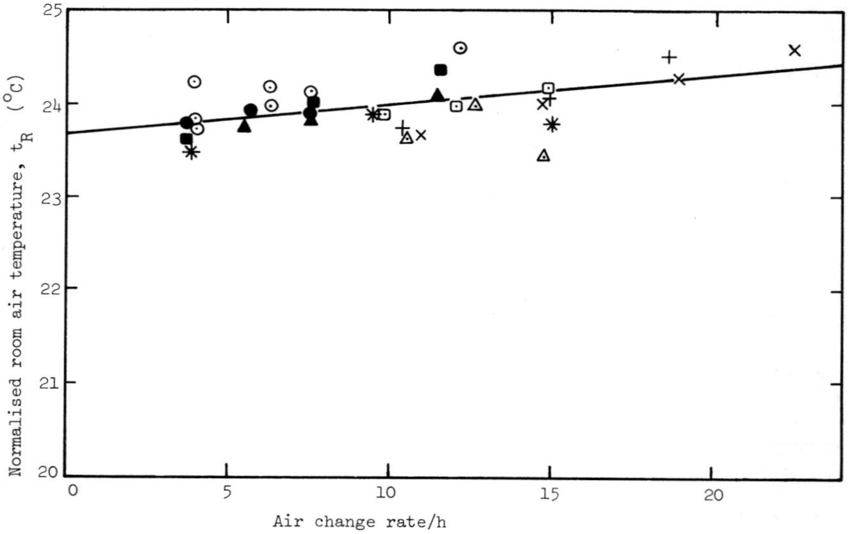

Fig. 6: Effect of Air Change Rate on the "comfortable" Room Air Temperature

WATER/AIR CONVECTOR MADE WITH PLASTICS OF IMPROVED THERMAL CONDUCTIVITY

P. DUBOIS, M. BERTHET and C. REINAUDO

Laboratoires de Marcoussis
Centre de Recherches de la Compagnie Générale d'Electricité
Route de Nosay
F - 91460 Marcoussis (France)

Tel : (6) 449 12 00

Summary

The reported work has been conducted on water/air convectors for space heating by lower temperature water (< 50 °C) made with plastics of improved thermal conductivity ; it includes :

- studies of technological problems which have to be solved for the fabrication of tubes or plates with water channels,
- computer model and design studies,
- test of a 300 watts convector,
- design of 1 kW convectors and economic assessment.

0. Introduction

The program relates to water/air convectors designed for space heating by water at a temperature not exceeding 50 °C and made of plastics with improved thermal conductivity, incorporating a polyethylene matrix with an ethylene propylene elastomer and a carbon black load.

In this water/air convector, the naturel convection of the air is actuated by a chimney effect induced in the casing which contains, in the lower part, the water/air exchanger and, higher up, the back up electrical resistor (cf. diagram (a) of figure 1). The convector is designed to extract 1 kW thermal power from the water entering at 50 °C.

The reported work includes :

- studies of the technological problems presented by the construction and assembly of tubes or of plates with water channels,

- the elaboration of a computer program which could be used as a design and dimensioning tool and the analysis of several preliminary 1 kW convector designs,

- the construction and testing of a 300 W convector model which demonstrated the validity of the computer program,

- the design of a 1 kW convector satisfactory from the technical and economical standpoints.

1. Studies of the technological problems presented by the construction and assembly of tubes and plates with water channels

11. Fabrication of tubes and plates by extrusion

We have studied the technological problems presented by the fabrication by extrusion of tubes or plates :

- by mixing the constituants of the compound in an industrial mixer,

- by extruding these mixtures to fabricate flat plates and tubes whose mechanical properties at 50 °C are comparable to those obtained from mixtures made in a laboratory apparatus :

. strain at breaking point 110 kg/cm^2
. elongation 75 %
. modules of elasticity 3300 kg/cm^2

12. Tubes and plate assembly

Assembly problems were studied :

- hot forming of plates with flowpaths from extruded flat strips,

- measure of the mechanical strength and ageing of tube or plate bonding : standard test pieces were checked to ensure that the breaking lead under traction of bonded parts stuck together with ECOBOND 45 LV glue was not modified by several hundreds of thermal cycles between 20 and 70 °C,

- developing heat welding of hot-formed plates :

. the "heating strip" method consists of bringing the two parts to a temperature sufficient to soften the material (250 to 270 °C) then removing the heating strips while applying a controlled pressure to the two opposite parts. Tensile and shear tests showed that unit assemblies of this kind could withstand without change in their mechanical strength 2000 hours of ageing tests at a pressure of 10 bars in water at 50 °C,

. in order to weld the hot-formed plates which, once assembled, form the plates with water channels for the 300 W model, two prints were made fitting the shapes of the hot-formed plates. They make it possible to keep the initial geometry of the plates during the softening part of process. Development of the welding process involved adjustment of parameters : duration of heating of the material, cooldown during welding and pressure on the material. Since the material is highly loaded and contains a large quantity of elastomer, it is only slightly deformed during welding, which makes it difficult to adjust this parameter.

2. Computer program and premiminary designs of a 1 kW convector

A computer program was devised for the thermal and aeraulic calculation of a "chimney effect" convector. By means of this program, a parametric analysis of the performances of the preliminary 1 kW convector design was carried out.

21. Setting up a computer program

The calculation allows for the following two related physical phenomena :

- air heating, which is a function of the air flow between the plates,

- the driving effect generated by air heating, which dissipates into pressure drops which are a function of air flow. These pressure drops were calculated by totalling :

  . the friction pressure drop between the flat parts of the plates (cf. diagram (b) of figure 1),

  . the extra pressure drop caused by the obstructions represented by the plates water channels,

  . the inlet and outlet pressure drops.

22. Preliminary 1 kW convector design

A premiminary design was made for a 1 kW convector of following dimensions :

Height : 0,75 m
Length : 1,75 m
Thickness : 0,20 m

The water/air exchanger consists of plates with water channels, vertically arranged by zigzagging the water flowpaths to obtain more regular air channels (cf. figure 2a). Each 15 cm high plate, obtained by heat welding of hot-formed plates, includes two water channels zones in a bend configuration (cf. figure 2b).

The exchanger incorporates fifteen 15 cm high plates with water channels of 10 mm outside diameter and 9 mm spaces : spacing between plates is 0,8 mm.

At a water flow of 100 liter/hour entering at 50 °C and with air at a temperature of 18 °C, the power extracted from the water is 965 watts without back up electrical heating and 1035 watts with 500 watts electrical heating.

23. Preliminary parametric study

The 1 kW convector design was subjected to an analysis of the influence of key parameters on the power extracted from the water :

- height of plates and distance between plates : the increase of extracted power with plate height is restricted by the reduction of air flow resulting from the increased pressure drop accompanying the increase of exchange surface (cf. figure 3),

- water flow : the increase in water flow increases extracted power but there is no point in exceeding about one hundred l/h (cf. figure 4),

- back up electrical heater : the use of back up electrical heating allows an increase in the power extracted from the water : + 7 % for an electrical heating of 500 W and a plate height of 2 x 76 mm (cf. figure 5).

3. Construction and testing of a 300 W model

31. Model construction

A 300 W convector, which is a 1/3 model of the 1 kW convector previously designed (cf. 22) was constructed : the cross section of the convector is the same but its length is reduced to about 500 m. Each plate consists in fact of two half-plates of 80 mm height, hydraulicaly in series, to ensure water circulation in a bend configuration. A photo of the model is shown in figure 6.

Nominal operating conditions are :

- water inlet temperature : 50 °C
- air inlet temperature : 18 °C
- water flowrate : 30 l/h
- distance of air inlet nozzle from ground : 90 mm.

32. Model tests

The results of the model tests demonstrate that :

- the power extracted from the water under nominal conditions is 325 W, i.e. slightly higher than the calculated value of 305 W ; it goes up to 350 W when the water flowrate increases from 30 l/h to 45 l/h (cf. figure 7),

- a back up electrical power of 170 W increases the power extracted from the water loop,

- raising the distance from the ground from 90 to 150 mm does not improve performances but reduction to 50 mm causes a 7.5 % drop in the power extracted from water.

4. Design of a 1 kW convector and economic assessment

The acquired knowledge has been used to design a 1 kW convector satisfactory from the technical and economic standpoints. Two designs were made, differing from each other by the geometry of the heating unit, the mode of fabrication and the material thicknesses : one design has hot-formed ane hot-welded plates and the other has finned plates or tubes produced by extrusion. Each design is for a convector whose outside dimensions are as follows :

Height   : 0,75 m
Thickness : 0,20 m
Length   : 1,60 m

41. Convector incorporating water/air exchanger with hot-formed and hot-welded plates

Given the technological constraints of heat-welding, an extruded plate thichness of 1 mm was chosen, leading to flat sections between water channels of 2 mm thickness. With a water channel internal diameter of 4 mm, the outer water channel diameter is 6 mm. With these data, the design leading to the minimum plastic weight for a 1 kW convector is a water/air exchanger consisting of 21 plates of 2 mm thickness, 50 mm height, each incorporating 4 channels with an outside diameter of 6 mm and an inside diameter of 4 mm separated by a flat section of 6.6 mm. There is a 7 mm spacing between the plates.

The plastic exchanger weighs 4.5 kg

42. Convector incorporating water/air heat exchanger with extruded plates or tubes

Fabrication by extrusion of the water/air exchanger components allows lower thicknesses than those used for the other design. In the case of extruded plates with water channels, the thickness of the flat sections between channels can be reduced to 1.6 mm : the tubes have an inside diameter of 4 mm and an outside diameter of 5.6 mm. Study of the plate height and spacing leading to the minimum plastic weight for a 1 kW convector led to the following dimensions : water/air exchanger incorporating 18 plates with a thickness of 1.6 mm and a height of 65 mm, each with three water channels with diameters of 4 mm - 5.6 mm separated by a flat section of 16 mm. There is a 7 mm spacing between the plates.

Heating unit weight is slightly under 4 kg.

43. Economic assessment

Economic assessments were made for both the above-described convector designs :

- the factory fabrication price is 66 F (T.F) for the extruded water/air exchanger and 89 F (T.F) for the heat-welded water/air exchanger ; the weight of the materials is nearly the same in both cases : the price difference is due to the difference in the estimated cost of fabrication ;

- the sale price to the public, obtained by multiplying by 2.5 the fabrication price, is 165 F (T.F) for one of the designs and 223 F (T.F) for the other. These prices should be compared to an estimated price of 310 F (T.F) for a steel heating unit.

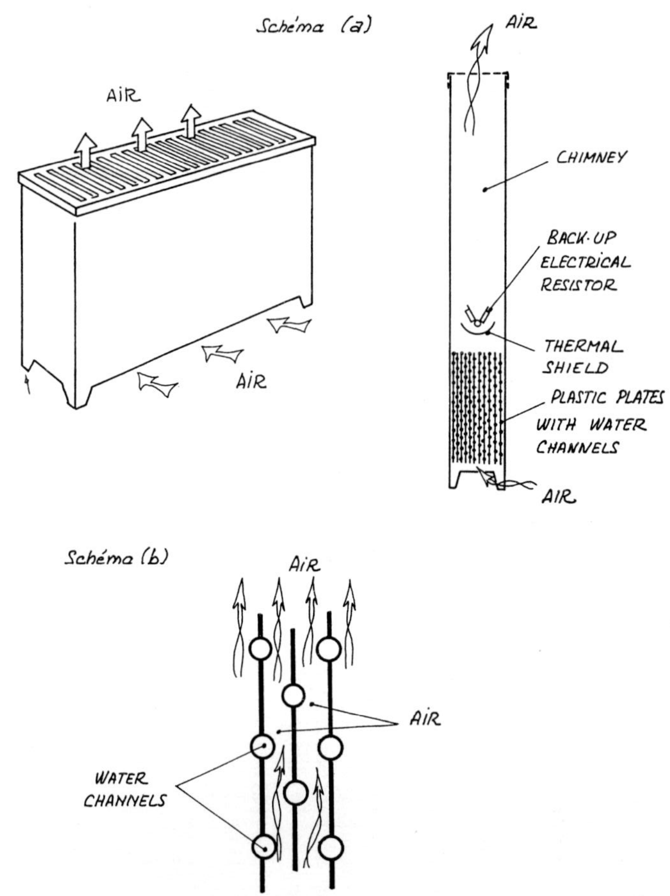

Fig. 1: Water/air convector for space heating

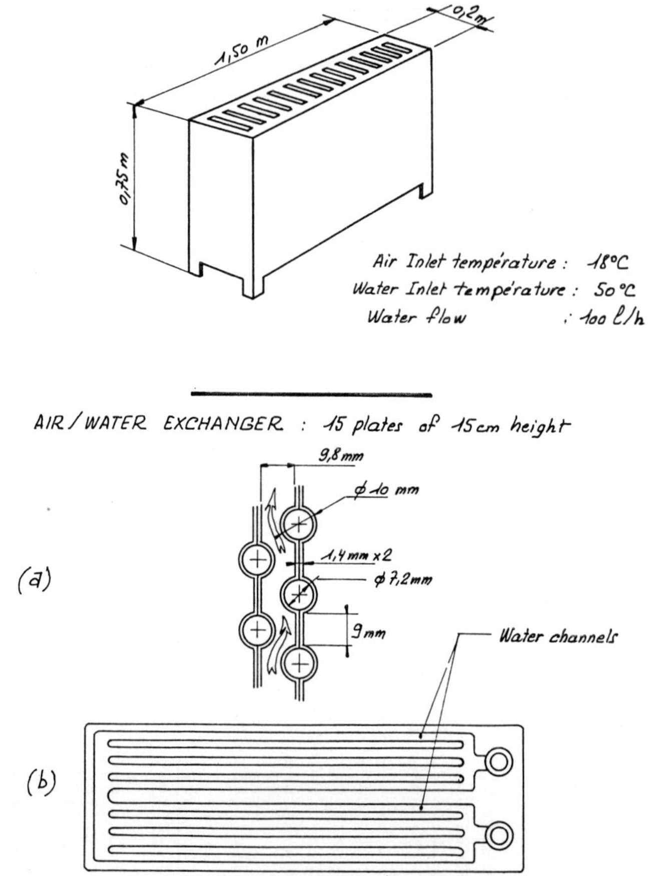

Fig. 2: 1kW convector - preliminage design

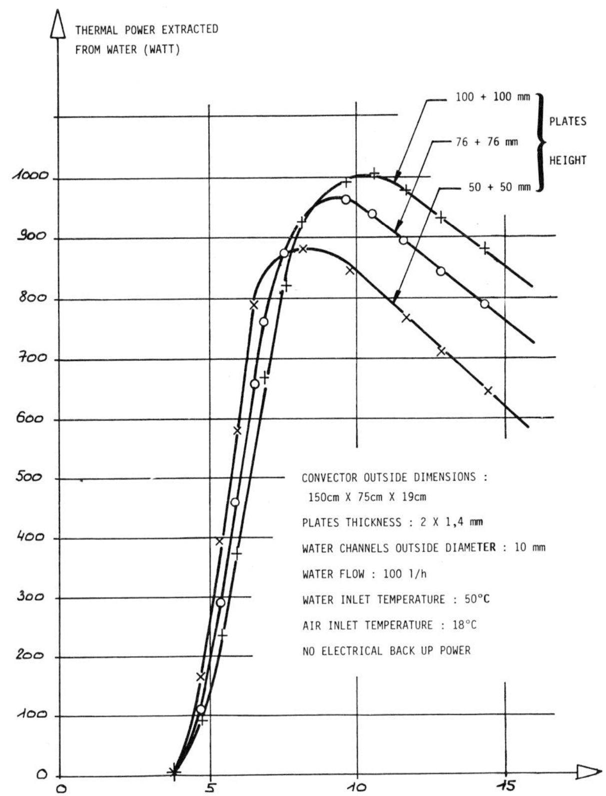

Fig. 3: Influence of plates height and spacing on thermal power removed from water

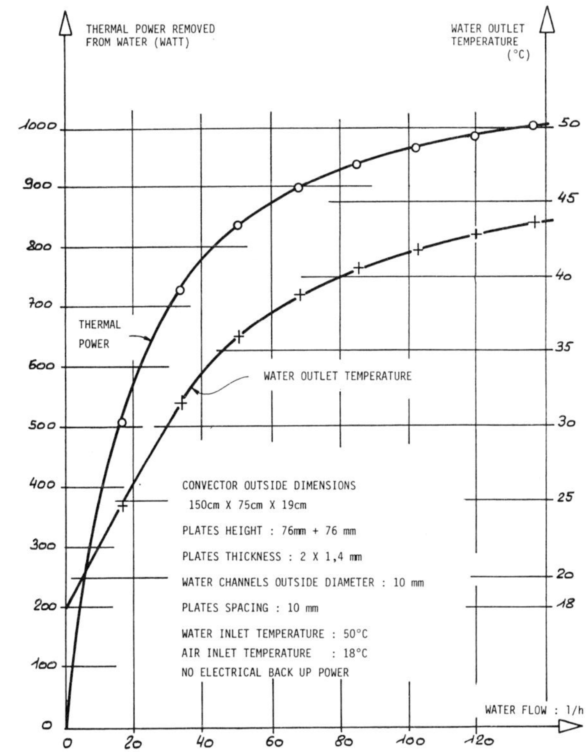

Fig. 4: Influence of the water flow on thermal power removed from water

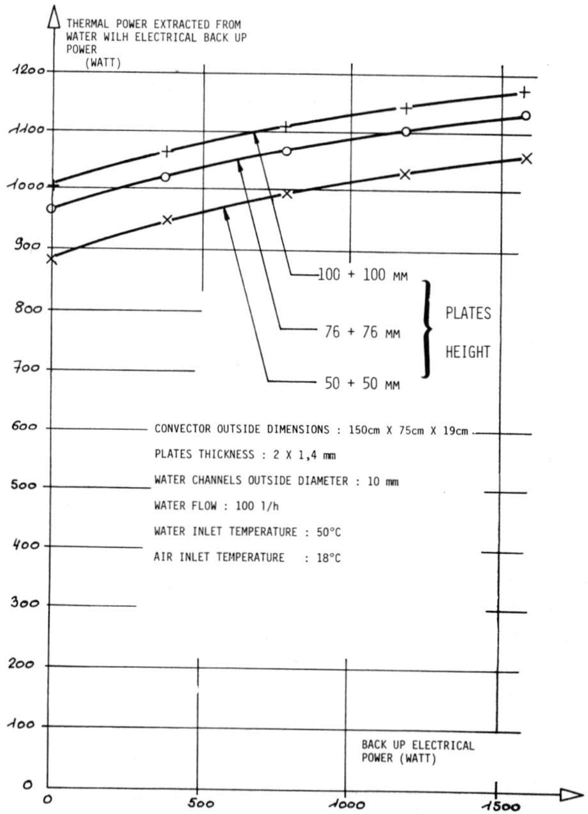

Fig. 5: Influence of the electrical back up power on thermal power removed from water

a) Plate with water channels

b) Water/air

Fig. 6: Heat exchanger of the 300 Watt model

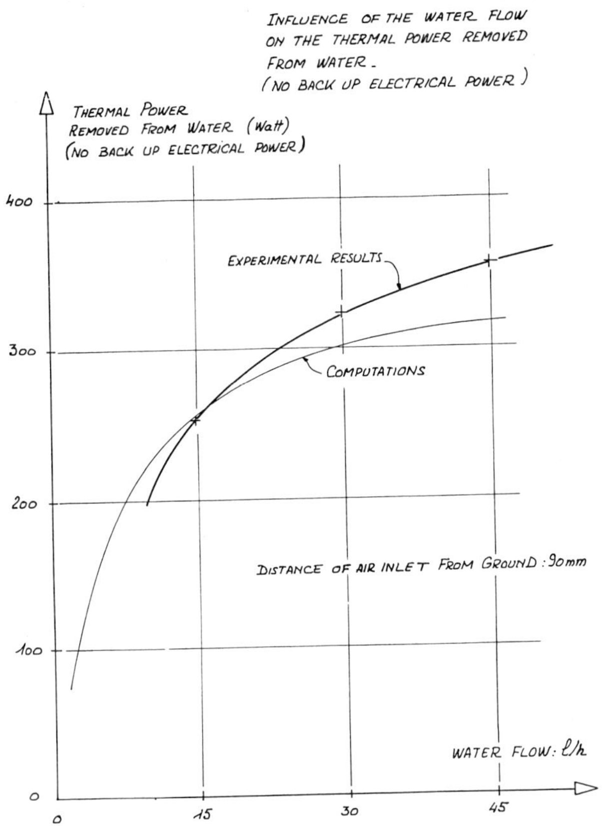

Fig. 7: "300 Watt" water/air heat exchanger

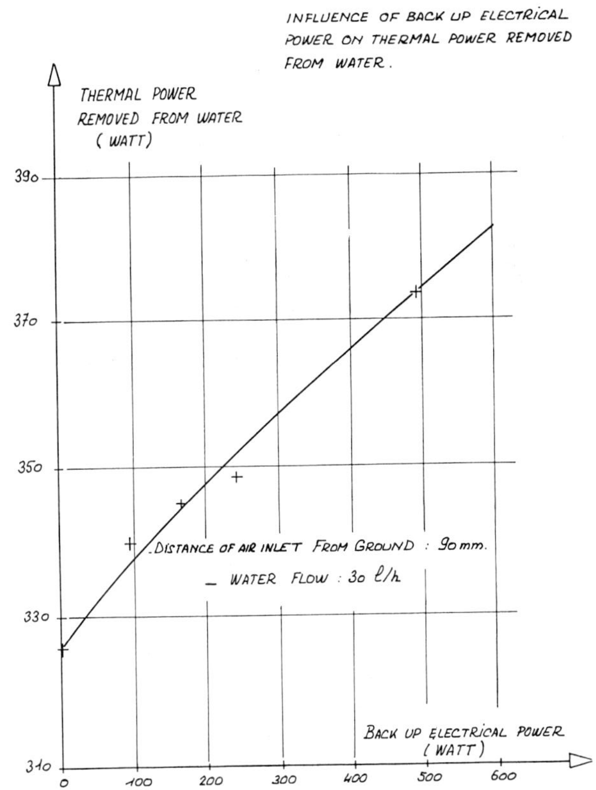

Fig. 8: "300 Watt" water/air heat exchanger

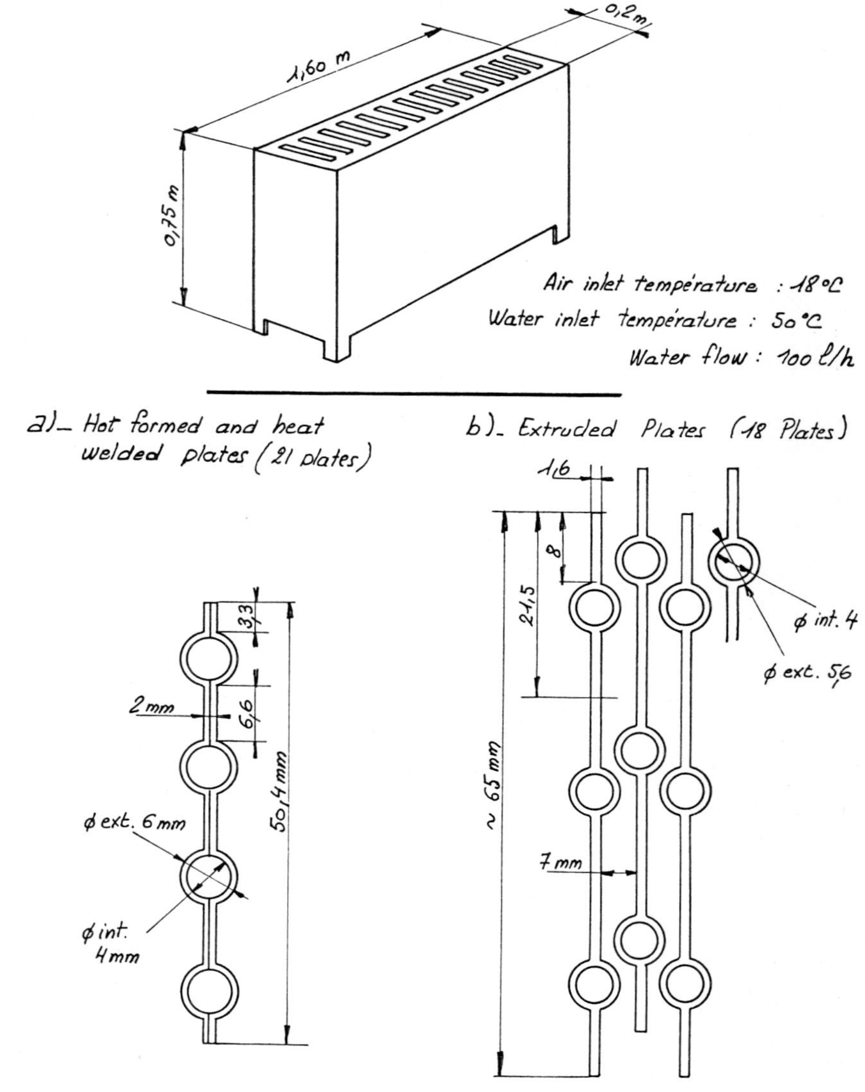

Fig. 9: 1 kW convector

RESEARCH AND DEVELOPMENT CONCERNING A DEVICE FOR MEASURING THE ENERGY CONSUMPTION OF HEATING INSTALLATIONS WITH FLUID HEAT TRANSFER MEDIUM USING A NEW KIND OF CALORIMETRIC SYSTEM

W. GRAMATTE
Battelle-Institut e.V.
D-6000 Frankfurt am Main
Am Römerhof 35
Postfach 90 01 60
Telefon (0611) 7908-0
Telex 04 11 966

Summary

The objective of this project is the conceptual design and construction of a new calorimeter with the necessary laboratory experiments. The operating principle of this calorimeter is based solely on temperature measurements in a heat bridge without moving parts. The new calorimeter is expected to require less servicing and have a longer useful life than currently available devices; also, it is expected to feature a higher measuring accuracy. In the course of the project work it has been found that realising the in fact rather simple physical principle leads to the technical limits of present temperature measurement technology. Experiments with selected temperature sensors Pt-500 and Pt-1000 have shown that temperature differences of 0.1 K can be determined with an accuracy of 1 %.
At the present stage, the work is concentrated on the development of the electronic measuring and evaluating unit. A laboratory trial in which the interaction of the transducers and the electronic components is to be tested is under preparation. Because of the difficulty of translating the theory into technical practice, in-depth investigations into the various factors that influence the temperature measurements had to be included in the proposed research programme.
Based on an estimate of the future provision of buildings with heat measuring devices, the new calorimeter can be expected to have a market of 40,000 pieces per year. Considering the energy saving effect of 10 to 20 percent which will result from the possibility of monitoring energy consumption with the new device the calorimeter will have an amortisation period of 4 years at most.

1 OBJECTIVES

In view of present regulations (1), (2), absolute heat consumption measurements on heating systems are very expensive at present. The currently available calorimeters make separate measurements of the flow rate and the temperature difference between cold leg and hot leg, and in most cases separate electronic units are used for evaluating the measured data. Flow meters with mechanically moved parts, such as impellerwheel or Woltmann meters, are subject to considerable strain by deposits and wear, which results in high error rates and entails significant maintenance and re-calibration costs. Others, which do not include movable parts (e.g. magnetoinductive systems or ultrasonic systems) are too expensive and so far still need to be mains-operated.

Only few systems are fit to determine the actual heat consumption accurately (3).

The new calorimeter to be developed will determine simultaneously the mass flow and the heat consumption of heating systems. It will be battery-driven, and is considered particularly advantageous for the following applications:

- In central warm water system (e.g. connected to district heating systems or in high-rise buildings with central heat supply systems) the actual heat consumption can be read outside the living units (as with electricity meters). Since the consumption data are generated and stored by a microprocessor, automatic reading is possible on-line or off-line.
- Permanent boiler efficiency control enables the heating boilers in single-family houses to be monitored so that periodic expert inspections become superfluous. The calorimeter here also serves to monitor fuel supplies.

Analogous applications are in industrial heating or cooling systems.

2 OPERATING PRINCIPLE OF THE NEW CALORIMETER

A partial heat flow is transported from the hot leg to the cold leg of the heating system by means of a special heatbridge. The resultant temperature variations are measured, and the total heat flow is calculated on the basis of these measurements (Fig. 1). Energy losses do not occur because the heat bridge is insulated towards the outside. The measured thermal values are converted into electrical signals, which are processed by a microprocessor with integrated time clock. The mass flow and the quantity of heat are shown on a display.

3 WORK PROGRAMME

The project work includes six individual tasks, three of which are complete or in the course of being completed:

a) Development of the theoretical basis (including computer programs) for optimising the design of the heat bridge and for constructing the heat bridge.

Using existing calculations and documents all the currently available heat exchanger concepts have been compared with a view to the given boundary conditions. The sensitivity of a heat exchanger system can be easily determined by varying the design parameters. In this context it is important that a heat exchanger system be defined as a heat

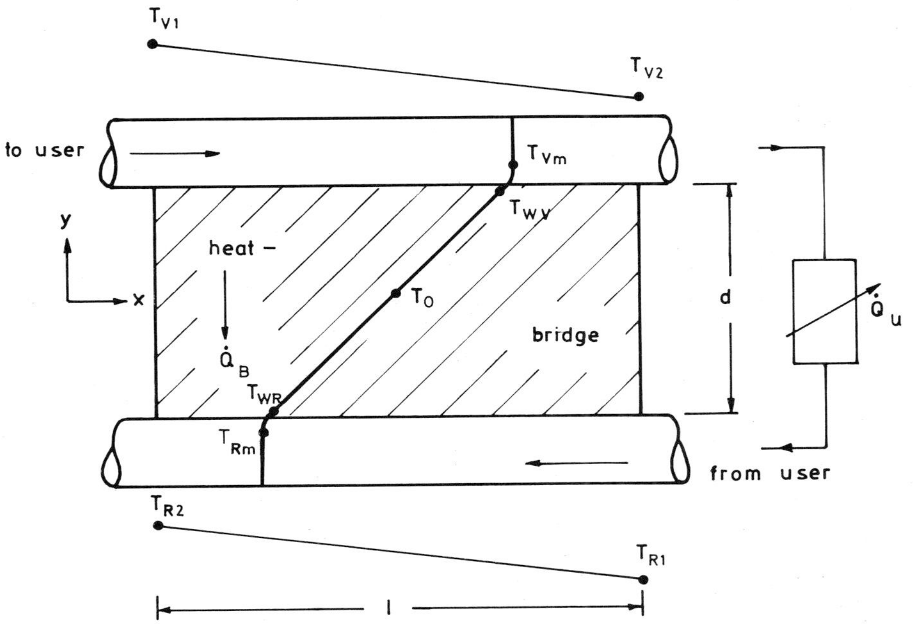

$$\dot{Q}_{user} = \dot{m} \cdot c_w \cdot (T_{V2} - T_{R1}) = \dot{m} \cdot c_w \cdot \Delta T_G$$

$$\dot{Q}_{bridge,\ to\ user} = \dot{m} \cdot c_w \cdot (T_{V1} - T_{V2}) = \dot{m} \cdot c_w \cdot \Delta T_K$$

$$\dot{Q}_{bridge,\ from\ user} = \dot{m} \cdot c_w \cdot (T_{R2} - T_{R1}) = \dot{m} \cdot c_w \cdot \Delta T_K$$

$$T_{V1} - T_{V2} = T_{R2} - T_{R1}$$

$$\dot{Q}_{bridge\ heat\ transfer} = \frac{\lambda}{d} \cdot A \cdot (T_{WV} - T_{WR}) = \frac{\lambda}{d} \cdot F \cdot \Delta T_W$$

$$m = \frac{\frac{\lambda}{d} \cdot A \cdot \Delta T_W}{c_W \cdot \Delta T_K}$$

$$\boxed{\dot{Q}_{user} = \frac{\lambda}{d} \cdot A \cdot \Delta T_W \cdot \frac{\Delta T_G}{\Delta T_K}}$$

Q heat flow
m mass flow
T temperature
c_W specific heat of fluid
A cross sectional area of heat bridge
d thickness of heat bridge

λ thermal conductivity
G large
K small
W wall
V hot leg
R cold leg

Fig. I - BASIC PRINCIPLE OF CALORIMETER

bridge which describes the physical events as accurately as possibly, and which thus provides for a physically valid evaluation algorithm.

b) Construction of a first laboratory model (including the selection of temperature sensors) and performance of first laboratory trials.

The heat bridge which the theoretical calculations have shown to be optimal has been used as the starting point for making the first laboratory model.

Since the operating principle of the new calorimeter centres on temperature measurements, the selection of suitable temperature sensors is of paramount importance. It must be demonstrated that the requirements which are made with respect to the measuring accuracy and long-time stability of the calorimeter are fully satisfied by the sensors.

The first laboratory trials are to prove that the envisaged operating principle is practicable in principle.

c) Development of a computing algorithm for evaluating the measured signals, and conceptual design of the electronic circuitry for the measuring and evaluating unit.

Taking the results of the first laboratory trial as a basis, the design of the heat bridge can then be optimised. The measured values will also be used for possibly required corrections of the evaluating algorithm.

The information gathered in this step is necessary to be able to establish a concept for the electronic measuring and evaluating unit. The requirement that the given error limits be not exceeded is of crucial importance for the arrangement and quality of the electronic components. This part of the work programme is to result in the assembly (in the laboratory) of the evaluating electronics for the new calorimeter. The function of the electronic unit will be tested with simulated temperature values.

d) Construction of a second laboratory model with heat bridge and electronic measuring and evaluating unit, and performance of additional laboratory trials.

A second laboratory model will be constructed after the findings of the previous steps of the work programme have been evaluated; this model will consist of a heat bridge and an electronic unit mounted on a bottom plate. The function of the laboratory calorimeter will be tested in a series of experiments.

e) Optimisation of the device and first field tests.
Two optimised prototype calorimeters will be constructed, taking the measurement results of the previous work steps as a basis. The prototype will be operated under normal conditions in a limited-time field test. This field test is intended to prove the functional reliability of the new calorimetric system and provide basic information for later series production.

f) Documentation and preparation of the official approval procedure at the Board of Weights and Measures (PTB). To be prepared for later series production, the steps to be taken to obtain official approval of the calorimeter are to be identified in this step.

Finally, a documentation of the project work and its results will be prepared with the objective to provide information material for the later industrial utilization of the new calorimeter.

4 CURRENT PROJECT STATUS AND RESULTS

At the present time, tasks b) and c) are being carried out simultaneously. In addition, certain theoretical preparations have been made for task d). The laboratory model is designed for the following boundary conditions:

- maximum hot leg temperature 140 °C
- minimum cold leg temperature 30 °C
- maximum temperature difference between hot leg and cold leg 80 °C
- minimum temperature difference between hot leg and cold leg 5 °C
- nominal flow rate 0.5 m^3/h

The smallest temperature difference to be measured is 0.1 K. If it is assumed that the given calibration error limits of ± 5 % for the whole calorimetric system are not exceeded this minimum temperature difference must be measured with an accuracy of approx. 1 %.

This requirement directly leads to the primary problems to be solved:

- Suitable sensors for differential temperature measurements, i.e. sensors with sufficient accuracy and long-time stability, have to be selected.
- The consequences which deposits in the flow channels of the calorimeter may have on its error band-width have to be identified. (Such deposits will influence the thermodynamic processes involved).
- The effect which these two factors, i.e. temperature sensors and deposits, will have on the evaluation algorithm and on the electronic circuitry will also have to be investigated.

Following is a brief outline of current findings in these three respects.

4.1 TEMPERATURE SENSORS

To be able to determine the usefulness of platinum thin-film temperature sensors (types Pt-500 and Pt-1000), selected sensor pairs were subjected to an accelerated wear test, following the relevant PTB guidelines. The sensor pairs were heated to 140 °C and left to stand at this temperature for 22 hours. After this they had to be cooled down to room temperature within 1 to 2 hours. Overall, the sensors had to withstand approximately 40 such cycles, which corresponds to a total test period of about 1,000 hours.

The results of the experiments with the Pt-500 sensors showed that long-term drift occurs for up to about 220 hours, whereupon stabilisation sets in. Before using these sensors in practice they should therefore be pre-aged at 140 to 150 °C for at least 220 hours. Of the 10 sensor pairs tested, two achieved a constancy of $1.1 \cdot 10^{-3}$ K (4).

The Pt-1000 sensors showed a less pronounced ageing behaviour. However, it was found that some of the sensors behaved absolutely atypically and were characterised by considerable drift. If it is planned to use these sensors in practice, a test method would therefore have to be developed - which would allow the temperature sensors to be checked for freedom from drift.

The results of these investigations confirm that the platinum thin-film temperature sensors Pt-500 and Pt-1000 satisfy the given requirements. However, additional discussions with manufacturers are considered necessary before a final selection can be made.

4.2 DEPOSITS

The relevant literature does not contain any reports of investigations concerning deposits, and effects of deposits, in heating systems. However, ample information is available about deposits and corrosion in industrial water systems and about corrosion problems in heating systems.

In (5) it is said that no damage is likely to be caused by "water in a state of equilibrium and practically salt-free", such as is normally used in heating systems, provided the usually existing protective layer is not damaged (6) and (7). The nature of the protective layer has been described, but no information is given about the usual thickness of this layer, nor about its effects on the flow and heat transmission conditions inside the piping.

Consequently, valid statements about the layer thickness, the growth rate of layers, and the variations in heat transmission which result from deposits in heating systems can only be made on the basis of specific investigations.

The following investigations were therefore carried out to gain insight into the problem:

- Micrographs were taken from a calorimeter which had been in use for five years, and which was made available by an electric supply company. The micrographs that were taken at the brass housing of the device revealed deposits with layer thicknesses between 0 and 60 μm. 80 % of all deposits detected had a thickness between 0 and 5 μm. The deposits on the glass-fibre-reinforced plastic turbine wheel of the volumenometer were less than 2 μm thick.
- To be able to make valid statements about deposits on the copper parts of heating circuits, a long-time test has been carried out with the calorimeter test stand available at Battelle. For about 1,000 hours copper pipe sections were exposed to the same flow conditions as occurred in the laboratory model of the novel calorimeter. Conditioned heating water was used as thermal medium. Microscopic examination of the micrographs revealed that the layer thickness of the deposits is below 1 μm, and that the deposits occur in the area where the natural protective layers form.

In the light of these investigations the following can be said:

- Deposits do form on surfaces of heating systems which are in contact with water, especially in calorimeters.

- The thickness of such deposits depends on the material of which the device or part is made, the surface properties, the nature of the water, and the flow velocity.
- When the flow velocity is relatively low (< 0.1 m/s) - as at the brass housing of the volumenometer which was used for the investigations - and when the materials surface is relatively rough, the deposits are considerably thicker than with high flow rates (> 1 m/s) and smooth surfaces such as occur at the plastic impeller and the copper test tube.
- It is not to be expected that deposits will have negative impacts on the novel calorimeter because deposit layers below 2 μm do not have significant thermodynamic effects. On copper, only the natural protective layer forms which builds up in a very short time and therefore has no long-term consequences. This can be taken into consideration when calibrating the devices, i.e. it is possible to prevent or greatly reduce deposit formation by appropriate design measures and appropriate adjustment of the flow conditions.

4.3 EVALUATING ALGORITHM

A numerical model has been developed with which it is possible to determine the temperature field in the heat bridge and identify the aptness of the system for calibration (see Fig. I). It has been found that when taking the gradients ΔT_x and ΔT_y in the block and the medium temperature T_o as starting points, calibration is definitely possible as a function of the mass flow $\dot{m}$, the linear gradient ΔT_W, and $\Delta T_G/\Delta T_K$. The boundary value problem $T(x,y) = 0$ has been solved purely empirically, and it has been demonstrated that the temperature profiles in the block are linear when the heat conductivity λ in x-direction is zero. The temperature function of the heat bridge is therefore

$$T(x,y) = T_o + \Delta T_y . y + \Delta T_x . x$$

When introducing heat conductivity λ in x-direction into this equation, the boundary value problem in the block becomes anisotropic, anisotropy being caused by ΔT_K and ΔT_G.

It suggests itself to use the following equation:

$$T(x,y) = T_o + \Delta T_K \, . \, f(y) + \Delta T_K \, . \, f(x,y) + \Delta T_g^a \, g(x,y)$$

Since the derivative of the temperature function must be zero at the insulated edges of the heat bridge, the following simple approach to solution results:

$$T(x,y) = T_o + \Delta T_y . y + \Delta T_K . a_1 \sin(ax) \cosh(by) + \Delta T_G^d . a_2 \sin(ax) \sin h\ (by)$$

The calorimeter can definitely be calibrated with the aid of four check points in the heat bridge. However, a very high measuring accuracy is required of the temperature sensors because the first three terms of the above equation already describe the temperature field as T = 0.1 K, and because the overall function necessitates an accuracy of T = 0.001 K.

5 CONCLUSIONS

In the course of the project work it became obvious that the limits of the present state-of-the-art of temperature measurement technology are frequently reached when trying to realize the very simple physical principle of the envisaged calorimeter with heat bridge. This means that considerable extra work is required to gain insight into various influencing parameters. For example, the following aspects had to be investigated:

- the temperature field in the heat bridge,
- the flow processes in the flow channels of the heat bridge,
- the effects of possibly occurring deposits,
- the integration of temperature sensors into the heat bridge (and possibly resulting measuring errors).

Also, a considerable amount of theoretical work is necessary for calibrating the temperature sensors which have to measure temperature differences below 0.1 K. Consequently, the proposed experimental programme has to be reduced to a minimum.

Nevertheless, it can be concluded from the results achieved so far that in the end a prototype calorimeter will be available which will satisfy all the given requirements.

6 ECONOMIC ASSESSMENT

The production cost of the novel calorimeter are estimated at DM 250 to DM 300 per unit; this estimate is based on the assumption that devices in the 20 mm to 50 mm range will be manufactured in series. Interviews of potential users of the new device have shown that approximately 85 % of all currently installed calorimetric systems are within this nominal width range. In fact, the percentage is likely to increase in the near future because the quantities of thermal energy which will have to be paid for will be lower in future.

The market potential for the next 15 years is currently estimated at 600,000 units at least. This estimate has been made only for the Federal Republic of Germany, assuming that as little as 10 % of all central heating systems in existing buildings will be equipped with the new calorimeter. With the new device it will be possible to reduce the fuel consumption per heating system by 10 to 20 % annually. When installed in a single-family or two-family house (with an annual energy consumption of about DM 3,500) the calorimeter will pay off within 4 years at most. This calculation is based on a purchase price per unit of DM 1,300.

7 REFERENCES

(1) Anforderungen der Physikalisch-Technischen Bundesanstalt an Wärmemengenzähler für die Zulassung zur Eichung. PTB-Mitteilungen 1/70, pp. 46-49 and 2/72, pp. 98-99

(2) Zähler für thermische Energie. Erster Entwurf einer internationalen Empfehlung. PTB-Mitteilungen 91, 1/81, p. 45-52

(3) Sauer, E.: Wärmemengenzähler im Testbetrieb. HLH 33 (1982) No. 2, February 1982, pp. 61-68

(4) Bolk, Will T.: Messung von Temperaturen und Temperaturdifferenzen mit Widerstandsthermometern und Gegenstromschaltung.
Messen und Prüfen/Automatik, January: February 1983, pp. 36-39

(5) Ulrich, E.A., edit.: Werkstoffzerstörung und Schutzschichtbildung in Wasserversorgungs- und Heizungsanlagen. Verlag des TÜV Bayern e.V., 3rd revised edition, 1974

(6) Fachnormausschuß Materialprüfung im DW Deutsches Institut für Normung e.V., DIN 50930, part 5, draft: Korrosionsverhalten von metallischen Werkstoffen gegenüber Wasser, Kupfer und Kupferlegierungen.
Beuth Verlag, Berlin, Köln, 1976

(7) VDI-Gesellschaft Technische Gebäudeausrüstung - Heizung -, Klima -, Haustechnik - Ausschuß Korrosionsschutz - Wasseraufbereitung, VDI-Richtlinie 2035; Verhütung von Schäden durch Korrosion und Steinbildung in Warmwasserheizungsanlagen.
Beuth Verlag, Berlin, Köln, 1979

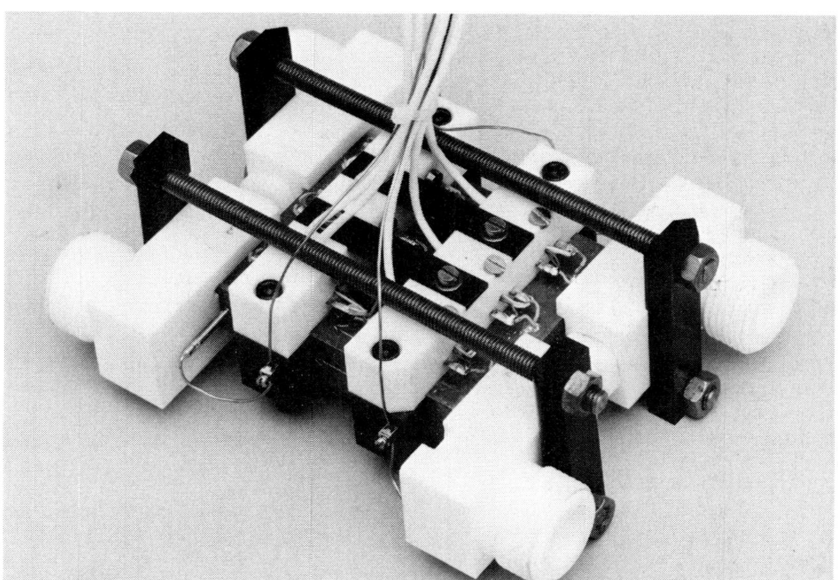

PHOTOGRAPH OF THE FUNCTIONAL MODEL

HEAT METER WITH MICROPROCESSOR: FINAL RESULTS OF THE STUDY

R. LEOPARDO, V. CENA

Ottico Meccanica Italiana OMI S.p.A., Roma

Summary

The aim of the study carried out under the Contract EEA-1-012-I(S) was to demonstrate the possibility of saving energy by means of a low-cost, accurate heat meter which, with the use of a microcomputer would have also the capability of controlling autonomously, the complete heating system to which it is devoted.

An examination of the existing flow meter and of the existing electronic is made, and the contractor choices are explained with the conclusion that the device as it has been developped will allow substantial energy savings at the individual, nation-wide, and Community level.

1. Scope

The research activities carried out in the framework of Contract EEA-1-012-I(S) were aimed at the development of a heat meter suitable for application in residential water-heating systems. The stringent requirements to be met by the heat meter were as follows:

- precise, reliable, maintenance-free flow metering subunit
- electronic integration of the product temperature difference times flowrate, for a precise accounting of the delivered heat
- microprocessor-based control unit capable of implementing complex, self-adapting control strategies.

2. Description of work

2.1 Flow metering section:

The research started from the flow metering subunit. It was realized since the early stages, that the available instrumentation could be divided in two groups:

- high class flow meters for industrial applications
- cheaper flow meters for general purpose applications.

The first group met all the requirements in terms of precision and reliability, but entailed costs well above the level compatible with the residential heating. On the other hand, the second group was not satisfactory, either being unreliable (turbines) or having high pressure drops (orifices); requiring complex indirect reading systems (rotameters), etc.

The results of a detailed market survey led to the conclusion that the best operating principle was the electromagnetic one.

Several manufactures offered instruments based on such a principle, all of them falling in the first group. Therefore it was decided to investigate the possibility of designing and building our own prototypes.

The latter are based on the Lorentz law, which states that the force $\vec{F}$ acting on an electric charge q travelling with velocity $\vec{v}$ through a magnetic field of induction B is:

$$\vec{F} = q\vec{v} \times \vec{B}$$

and it is directed perpendicularly to the plane formed by the velocity and the magnetic induction. In the metering section two electrodes are placed in the direction of the force by opposite sides with respect to the charge flow at distance D. One of them will collect the positive, the other the negative charges. The electrodes therefore behave as a capacitor, in which the electric field is given by

$$E = \frac{\Delta V}{D}$$

where ΔV is the voltage between the two electrodes. The electric field acts on the same charge q with a force

$$\vec{F} = q\,\vec{E}$$

In steady state conditions it will be qE=qvB, and therefore

$$\Delta V = v\ B\ D$$

This relationship allows the calculation of the velocity from the measurement of the voltage. Each prototype was composed by a short flanged pipe (metering section)carrying the electrodes, the coils for the generation of the Magnetic field, a Magnetic circuit outside the pipe, a power supply for the magnetic coils.

The metering section required the usage of a non-ferromagnetic material, not to disturb the lines of flux of the Magnetic field. The first prototypes were built with polymethylmetacrylate; eventually polypropylene was adopted.

As to the material for the electrodes, tests in the typical residential water-heating systems conditions led to the choice of AISI 316 stainless steel. The surface of the electrodes in contact with water is machined so that the curvature is the same of the metering section, to minimize fouling of the electrodes.

The magnetic circuit was built by coiling outside the metering section some transformer-core sheet winding.

The final version of the power supply is composed by a subunit which transforms the 50 Hz sine wave into a 50 Hz square wave, a second subunit which divides the frequency by a factor of 16 yielding a 3 1/8 Hz square wave, and an "instrumentation amplifier" with gain 40 dB and input impendance of the order of thousand megaohms.

Extensive test were carried out to evaluate the response linearity of the flow meter as a function of the water velocity. The results were satisfactory, especially in the range 0-0.5 m/s in which the majority of the flow meters show marked deviations from linearity.

2.2 Control unit:

A situation similar to that of the flow meters existed for the control unit. The market offered traditional, inexpensive controllers based on discrete logics, which could not be programmed with complex functions (self-adjusting set points or telephone calls with plain text messages in case of alarm). On the other hand, pre-programmed microcomputers having all the necessary peripherals (display, keyboard, non-volatile memory, input and output lines) were available, but at a Cost much higher than the target, even in large quantities. The reason for the latter was that existing microcomputers have capabilities going in some cases well beyond the requirements of a control unit, because they are general-purpose devices.

Therefore our choice was to design a custom microcomputer, with its own carefully adapted peripherals.

The functions implemented in the control unit are as follows:

- acquisition of the signals from:
  - two analog sensors which monitor the temperature of inlet and outlet flow from the boiler
  - one analog sensor which monitors the temperature inside the building
  - the flow meter (analog sensor)
  - two digital sensors which trigger the boiler temperature alarms
  - one digital sensor which triggers the fuel low-level alarm
- autodial of a set of programmable telephone numbers in case that one or more of the above signals (and in particular the alarms) are outside the allowable range
- after the autodialling, transmission over the telephone line of a phrase in plain English, stating the number of the calling station, identifying the alarms and other information
- accounting of the delivered heat, by means of integration of the product flowrate times difference between the two water temperature sensors
- switching on and off the burner, the pumps and the valves, according to a programmable schedule (daily, weekly, yearly) and to the actual temperature conditions inside the building
- adjusting the internal set points as a programmable function of the building temperature

  The above functions are implemented by means of a microcomputer having the following architecture:
- Z8 microprocessor, i.e. the Zilog Z8671 Single Chip Microcomputer. Its language is the "Tiny BASIC" interpreter, which allows the implementation of the mathematical and logic activities of the Central Processing Unit (CPU) in a high level language (even if somewhat simplified). Therefore the assembly language programming is limited to the interfacing of the peripherals.
- 4 kbytes of RAM memory, pin-to-pin compatible with EPROM memories. This feature allows to store the management program on non-volatile EPROM'S

and to substitute the latter on the printed circuit board without any hardware modification. The EPROM program is always available to the CPU and therefore allows the autostarting of the control unit even after prolonged power failures

- clock/calendar board, which enables to formulate and keep the heating schedule
- analog-to-digital conversion board with 8 channels. The conversion is performed with a 12 bit + sign bit precision, equivalent to 0.025%
- constant current source, which allows to use the standard PT100 platinum resistance temperature detectors as temperature sensors
- 7.4 MHz clock, which allows, if necessary, a hich acquisition rate and in general a more than adequate elaboration speed
- terminal port, used with a video screen during the phase of program development and with a single-line, 16-character alphanumeric LCD display when the control unit is installed on field
- 20-key alphanumeric keyboard, whereby the installing technician and eventually the user can communicate with the CPU to read the instrumentation, change the schedule, modify the set-points, etc.
- speech synthesizer which outputs on the telephone line phrases, long as necessary in which the CPU talks in plain, perfectly understandable English and tells to the listener (e.g. centrally located technician) the current alarm situation and other essential information
- general power supply (interruptible) and a special power supply (non-interruptible) for the clock/calendar board and a particular memory board. Both these board use the CMOS technique which, being extremely low-power, enables to keep alive the timing and the set-points even in absence of power from the mains.

3. Conclusion:

The activities carried out in the research project led to the following conclusions:

- the flow meter performed satisfactorily as to linearity and precision
- the microprocessor-based control unit is capable of handling the input/output signals and implementing highly complex, self-adjusting control policies on a yearly basis.

Once installed in real world heating stations, extensive field tests will be needed to assess the long-term reliability of the whole system.

Yet we feel that the tests will prove the ruggedness of our design and implementation.

It is therefore expected that in the near future this type of heat meter will allow substantial fuel savings, at the individual, nation-wide, and Community level.

ELECTRONIC HEATING CONTROL SYSTEM

F. FONZI, Pl. SEGALINI
ITALENERGIE S.P.A., Via della Repubblica, 39, Sulmona (Italia)

Summary

The research studies the possible energy savings using decives which at different level of complexity execute the control and maintain the heating system in condition of maximum efficiency. Furthermore the decives must allow a rational use of heating system having thermal energy generation only when there is a real request.

1. Object and descriptionof the research

The research in the field of the system to control the heating has brought us to developsome products, which could cover several possible needs of small and big users. These products have been named: HECOS 1, HECOS 2 and hecos I HECOS 1 control the burner HECOS 2 allows to program the centrlized heating though the measure of the outside temperature, the flow-back, the flow-up and the setting of the wished temperature HECOS I allows to program the wished heating in the house in different time spaces The possible application are shown in the following schemes. The devices HECOS 1 and HECOS 2 have been presented in the meeting of the 13th dec.1982 We refer shortly the main features of HECOS 1

2.1 Main features of HECOS I

Digital clock Indipendent setting of hours and minutes Lighted indication for day week Digital themometer Weekly setting of the chosen program Board control for maintaining a constant temperature at 20°C Possibility to insert batteries to behave the setted program Possibility to pilot the start of the burner and the position of the motor driven mixing valves

2.2 Electrical features

Nominal voltage..................220 V (+/-10%) 50 Hz
Output control
        voltage..................220 V
        Current................. 2 A
Temperature regulation field.....0° - 25°C
Room operating temperature.......0° - 50°C
Room memorized temperature.......-25°, + 65°C

2.3 Structure of the device

The device can be realized in a very compact wai with reduce dimensions, which has as result low costs. The whole consist in one micro-computer, four integrated circuits to pilot the 6-digit of the seven segments display and the 8 leds, 1 integrated circuit which in union wit the sensor gives the signal. The microcomputer uses this signalto measure the temperature of the realis to control the burner. To have a better idea we report a blok scheme of the device.
The keyboard panel and the visual control is indicated in Fig.2

2.4 Operating stages of the set

The set operates fundamentally in three following steps:

SET CLOCK: setting of the clock
HEAT PROGRAM: memorizing the weekly program of the heating
MONITOR: visual indication of the selected heating program
These three stages operate as follows:

2.4.1 SET - CLOCK

This stage is inserted to have the operation of the internal clock of the set, which is activated by simultaneous pressing the keys + and - on the keyboard panel. Thereafter by pressing singulary on the keys + and one obtains the wished value of the clock. It is obvious that by pressing the key + one obtains once and then the upper mean value and by - one obtains the immediately foregoing value. Once set the day one sets the hours by pressing the key → once set the hours, always using the →, one sets the minutes.

2.4.2 HEAT - PROGRAM

Memorizing the weelly program This function is obtained by pressing simultaneously the keys+and → at this point the user starting from the 1st. time interval (monday,hour 00.00) can introduce the wished data according to the following sequence:wished temperature by pressing the keys + or - according to the wished increasing or decreasing value. Once the wished value is reached one presses the key → and introduces the function hours and minutes, that is the upper value of the time interval, in which one wishes the chosentemperature in the room. Thereafter one sets the temperature for the following time interval. It has to be noted that the minutes are set in this step by selecting the fully quarter of an hour. The passage from a day to the next one is obtained automatically ny setting a time interval of a lower value than the foregoing one. Furthermore it is important to put in evidence the possibility for the user to set quickly the days with the same heating program. Once the heating program is set for a fixed day by pressing the key ➞ one has the memorizing of an identical program for the following day.This can be repeated obviously even for all the days the week following monday. The described procedure can be particularly convenient for anyone having a need of a uniform heating for the whole week or for just the first five days. In this last case once set the program for the monday one can press after the key ➞ and the user has the repetition of the program till the friday (included), thereafter can be programmed the heating for saturday and sunday. In the direct procedure the user has the immediate feeling of the passage from a day to the next one because the visual indication lights the chosen day.

2.4.3 Monitor: visual indication of. the set heating program

By pressing simultaneously the keys - and ➞ at this point the visual indicator shows the corresponding values to the first time interval. Thereafter by pressing the key ➞ one controls the situation of the heating so as it has been set in the previous stage described under 2.4.2.

The coming out from this stage as from the others happens in two ways. One way consist by pressing simultaneously the two keys of the individuation of the new stage in which one will enter. The second one is automatic. If the set will be left for about 36 seconds without pressing any key, the thermoregulation will automatically start according to the inserted values.

2.5 Operation and use of the set.

The set " HECOS I " has thee ways of operation, which can be chosen by means of a 3 position switch indicated as " SW operating selection " in fig.1. The three operating ways are the following: Automatic operation with the awich in the position " auto " the set controls continuously the room temperature (TAMB) trough the sensor, compares it with the value of the wished temperature (TDES) set by the user for the running time interval, thereafter through ah "on" and"off" signal after the following relation:if (TAMB - TDES) > or = 0 then the burner is switched off if (TAMB - TDEE) < 0 the burner is switched on. Operation with fixed temperature at 20°C. With the switch in position "20°C" the memorized values during the program are not considered. Only the value 20°C will be considered. This way of operating is particularly interesting in the of suddenly and temporarly use of an increase of the room temperature. Indeed when the necessity is over, one can replace the switch in the "AUTO" position and the set will operate again on the basis of the wished temperature inserted by user.

Manual operation with the switch in "MAN" position the set will be completely excluded. The burner will be controlled directly by its own sensors.

2.6 Application

The programmable electronic thermoregulator HECOS 1 can be applied in several uses, in someones in the basic execution, in some others must be used as well as software as outputs. The main applications are:

- programmable electronic thermoregulator applied to independent heating system.
- programmable electronic thermoregulator applied to branch valve in centralized heating system.
- Programmable electronic thermoregulator in water heater.
- Interface module to the user in a heating calculation system in centralized plant.

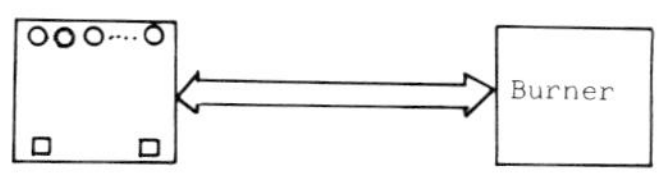

HECOS1:burner control

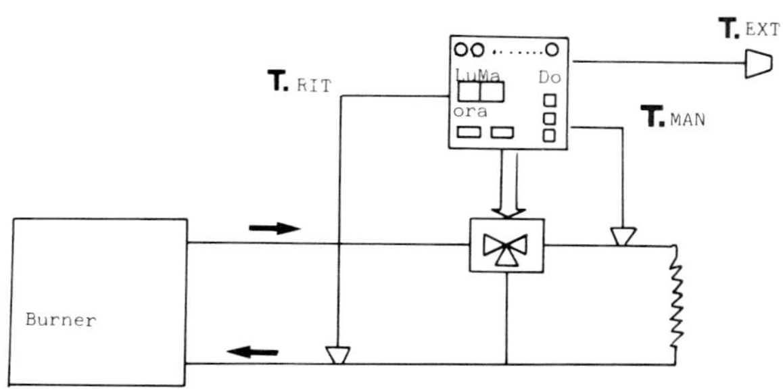

HECOS2:Thermoregulation

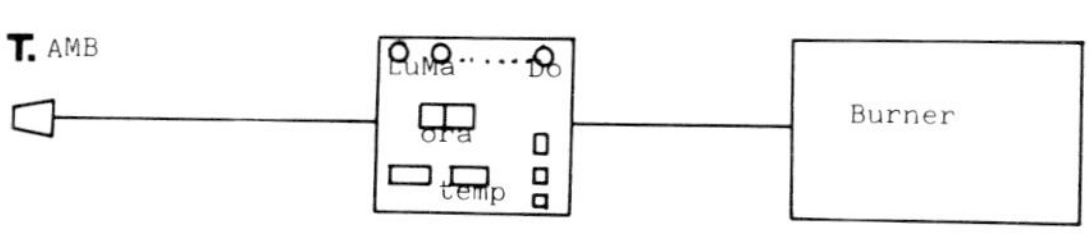

HECOS I:monofamiliar heating system

FIGURE 1

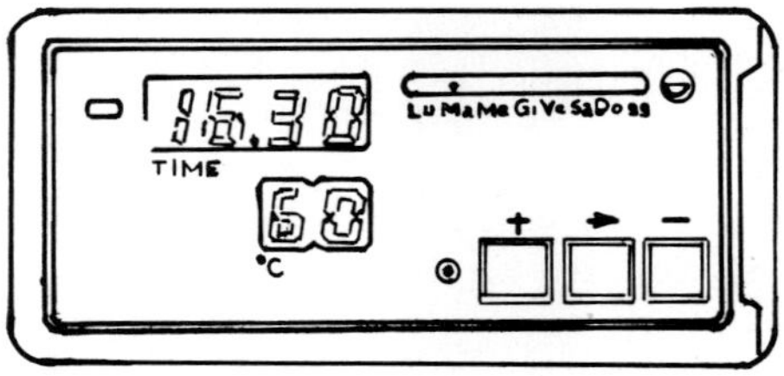

FIGURE 2

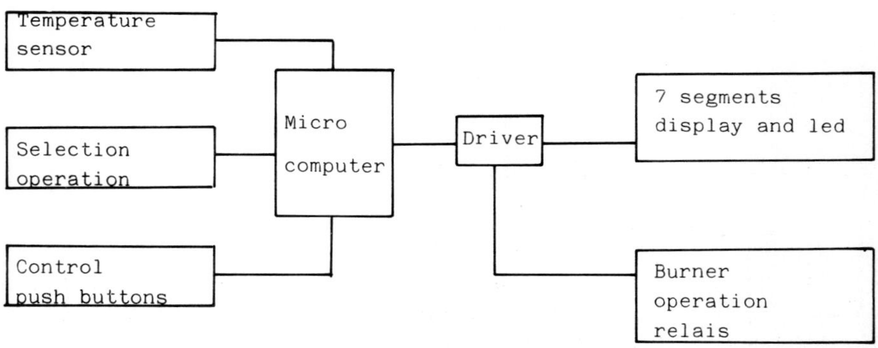

FIGURE 3

MICROCOMPUTER NETWORK FOR HEATING SYSTEM

Prof. Ing. Vito Mauro

CSEA - Consorzio per lo sviluppo dell'elettronica e l'automazione
Via Ventimiglia 201
10127 TORINO (I) tel. : 011 - 637058

Summary

This paper presents the results of a research on a distributed system for energy saving in building heating. The system is composed of a network of microprocessors, one for each apartment, governed by a central microcomputer. The system regulates the temperature in the apartments, taking into account the outside temperature, the users requirements and the buildings parameters and, finally, it provides for the heating bills. It learns both the true habits of the users and the main buildings characteristics.
The results of the experiments, carried out in many buildings with users of different type are reported.

1. Objective and description of work

1.1 Introduction

The system here described is made to save energy in centralized heating system (where heating is supplied by radiators through which hot water circulates) in buildings with many individual users. Typical application are blocks of flats for residential use, schools, office buildings, etc; it is well known that a good plant regulation, together whit a distributed control of ambient temperature can give a substantial energy saving. Moreover, a really efficient system should allow:

- room by room (or area by area) programming
- individual accounting
- hydraulic equilibrium keeping
- user's attitude driving towards energy saving.

1.2 Description of the system

1.2.1. System action

a) Diversification

The system allows programming of differents temperatures for each individual room and for each hour of the day. In such a way unused (or little used) rooms are not heated unnecessarily, while occupied rooms are assured a comfortable temperatures for the periods of use. The user can then graduate the comfort in the various room, by merely asking for higher or lower temperatures.

b) Accounting

The system estimates the quantity of heat actually consumed by the user and therefore allows a carefull division of expenses. The consumption data are always accessible to the user, besides condensed information which indicates if he is consuming more or less than the tenants average.This is done in order to stimulate the user to adopt his own policies of energy saving. The system holds in due consideration the so called "heat thefts" in calculating the amount of energy consumed by each user.

c) Signalling of absence

The user can signal to the system with simple commands, that the apartment (or even a single room) will not be used for a certain periods. This function is particularly useful to save energy further when the house is empty for periods of several hours or one ore more days.

d) Learning

The system is able to define by itself the heating program appropriate to each individual user : this distinctive feature is very useful because few user will continuosly change and adapt their heating programs. To accomplish this, the system analises all the user communications and makes a time map for the various rooms. This map is continuosly upgraded to define "minimun-risk-minimun-energy" policies.

e) Adaptability

The system is "intelligent": it doesn't only execute an imposed heating programme, but it can adapt itself and it learns informations autonomously. For example, the system "learns" how the heating plant is made, the orientation of the rooms and other data in order to correct possible errors in the plants itself.

f) Optimisation

In order to control the central boiler all the request of heat coming from all the served apartments are kept in consideration. Plant temperature is continously regulated, and the boiler temperature is set to the lower limiting value, compatible with safety requirements. The results is a high level of efficience of the boiler.

1.2.2. The hardware

The system is made up of a network of microprocessor (one for each apartment) connected to a central microcomputer. The apartment microprocessor measures the internal temperature and commands, directly, the special electrotermic valves placed on the radiators. The central microcomputer drives the plants acting on the burner and on the mixer valve. All the function of the processing and memorising are distribuited in a suitable way betwen the various units of the apartments and the central unit. See fig. 1 and (1) for more detail.

2. Methods of investigation

The research on the above outlined system has been realised trough the following steps:

2 - a Feasibility study (1980)

A simulation analysis has been made on the basis of average italian data. It gave a forecasting of the possible energy saving as high as 30% and an average time of investiment return for the user of few (3-6) years.

2 - b Prototype production and installation (1980)

A prototype system has been designed. A series of 100 units (where the word "unit" means the hardware for an apartment) has been produced and installed in three buildings (two blocks of flats and one school in the Northern Italy).

2 - c Prototype testing (winter 80-81)

The installation have been fully tested during the winter 80-81, with reference to

- energy saving
- users' acceptance
- effectiveness of the learning algorithms
- effectiveness of the control algorithms
- effectiveness of the accounting algoritms.

2 - d Preseries production and installation (81-82-83)

The system design has been modified and a pre-series of 600 units has been produced and installed, in different italian regions.

2 - e Extended tests (81-82-83)

All the installation have been tested to strenghthen the previous results and to further investigate:

- effectiveness of the accounting methods
- system reliability
- system maintainability
- methods of installation.

3. Results

In this chapter the main results of the research are reported: since the main features of the system are

- the learning algorithms
- the ability to make correct bills for individual expenses
- the hierarchical regulation
- the level of energy saving,

they will be discussed separately.

3.1. Learning algorithms

The learning algorithms, implemented in the system, are devoted to two different aspects:

- to learn users habits and needs
- to calculate the buildings parameters

The later belongs to the more general theory of "linear system identification". The model of the building thermal behaviour, including the water distribution plant, in the range of applicable conditions, is a linear one, in which some of the parameters are unknown. So, the microprocessors of the system, by making use of the informations collected by sensors, calculate the "best fit value" of the unknown parameters. The detail of the implemented algorithms will not be reported here: the interested reader is referred to (2)

The former problem is, by far, more relevant to this research: indeed, the user attitude is one of the most relevant points of the energy saving problem. Moreover, for different reason, correct and continous programming may not occur. Experience in real buildingsshows that user tends, in a high percentage, to estimate the occupation of the apartment, in space and time, higher than the real one and to ask for a higher than needed temperature. This fact, joined to the need of a continuos attention to the changing habits of the family, makes impraticable the correct programming by the totality of users. So, the system here described has autonomous capability of learningthe probability of occupation of ambients in the differents hours of week days and comfort temperature for individual users.

The model for the behaviour of average users, on which the learning algoritms are based, can be roughly defined by the following sentences:

- "the average user, during different week days, dwells, probably, the same rooms in the same hours"
- "the average user is conservative and lazy in programming: the system can trust him, only when he programmes for an absence"
- "the average user reacts, possibly with a delay, to a too low temperature, but he doesn't always react to a too high one"

On the basis of these assumptions, the system can use all the informations collected from the user to update the user's model. The parameters of the model are: a map of A x H probabilities of occupation (where A is the number of hours in a day and H is the number of hours in a day), and a temperature (the comfort temperature). The probabilities are first settled to values 1 (for the couples room-hour for which the user has programmed the presence) and 0 (for the other couples). The temperature is settled to an initial value (normally, 20° C or 21°

c). Then, as time goes, the system makes use of keyboard touch of any type to update this data; but, more importantly, also the lack of information is used, decaing the probabilities (on a day by day basis) and the comfort temperature (on a week by week basis). When the presence probabilities become low, the system regulates to a lower level, so as to solicit the user. The process then goes on, until a reaction is obtained (learning in this way both presence and temperature), or the probability goes to zero and a minimal temperature is reached. The process can be clarified by fig. 2 and 3, which give a visual outline of one decision rule (the sign + means a user touch of the keyboard, asking for "more heat"). The algorithms above outlined have been shown as highly efficient during the experiments. In fig. 3 results are outlined with reference to a sample of 75 apartments.

3.2. Accounting

The user must be directly involved in the energy saving process, since no saving can be made without, or against him. Closing of windows, better programming, closing of unused room door are examples of actions that can be performed only by the user himself. The involvement of users can be obtained through a direct relation between individual energy consumption and bills. Moreover every user must have direct and easy access to his consumption data, both integral and over short period. The system here described computes, every day, the amounts of energy which can be ascribed to each individual user and shows the result through the local unit.

3.2.1. Methods for accounting

In principle, there are three different classes of methods for individual sharing of heating expenses: no one of them can, obviously, be considered as exact since they are, more or less, conventional.

a - Sharing on a fixed "per cent" basis. This method is the most frequently used: as such, it is the most undestood but it does not make the user involved in energy saving.

b - Sharing based on the amounts of energy given from the plants (radiators) to each apartment. This method is used when flow-meters, or similar devices, can be installed in apartments.Application of this method may stronghly involve the user. On the other hand, results are affected from such factors as differences in thermal insulation, differences in the amount of free energy sources, "heat thefts" etc.

c - Sharingbased on the temperature obtained in rooms (or energy lost from apartments). Results obtained from the application of this method are easily accepted from users. On the other hand, this method does not drive the user to a really good attitude toward energy saving (opening of windows and bad insulation do not affect the bill....)

Some correction can obviously be made, to make the methods more applicable in real environment. Method b can be modified by taking into account the differences in thermal insulation, as described in the building design; "heat thefts" can be estimated, etc.. But finally, one can see that in different practical cases different methods could be preferred. Moreover, results obtained by the outlined methods may be, in some cases, definitely diverging: see fig. 4, where the results of the application to an experimental building are shown. After the experiments, it has been decided that the system should estimates, the energy sharing based on all the given methods (with the due corrections) and should compute the bill on a linear combination of the results. The coefficients used in the combination are independently defined for each building.

3.2.2. Technical notes

The system does not make direct measurements of the individual amounts of energy: it estimates continously the "state" of building and plant and then computes the other data. A unique energy flow meter is installed, just after the boiler, which measures the energy released to the plant. Then, the estimated sharing for each period (one hour for methods b, to take care of peak periods, and one day for method c, to minimize the effect of differences in the amount of energy stored in walls) is applied to the measured energy, to obtain the individual amounts. Finally, the latter are summed up over periods to obtain the integral value; they are then filtered to give to the user an information on short term variations.

3.2.3. Experimental results

Experience has shown that

- the algorithms based on method b give a reasonable estimate of the energy that could be measured by flux meters installed on single radiators
- the algorithms for room temperature estimation, which are relevant on method c, give a good estimate of temperatures.Average apartment temperatures where estimated with error less than 0,5° C in all experimental tests.
- the bills obtained by a correct combination of the given algorithms are easily understood and accepted from users.
- the continuos access to consumption data, as provided by the system, is an important factor for energy saving.

3.3. Hierarchical regulation

The structure of the system is relevant also to the methods for regulation. To this point, the building is described, as in fig. 5, as a hierarchical system and a decentralised control is applied, at every level. The "local unit" is at the "apartment" level. It is autonomous in regulating the apartment itself, but it receives from the central unit the limiting temperature values (max. and min.).It can "call for help" when the right apartment temperature cannot be reached: in that case, the "central unit" will rise the water temperature or, if necessary, the minimal temperatures to contiguous local units. Similar rules hold for all the system levels. The experiments show that the hierarchical regulation is efficient:

- it can give a very strict temperature regulation.
- it leads to a decentralised, fault-tolerant, self-degrading system.

3.4. Energy saving

The system has been installed and verified, in many buildings of different use, in Northern and Central Italy. Up to the winter 1982/83, 12 experimental building have been tested, with a total of 273 units. The resultscan be summarized as follows:

- block of apartments: observed saving varied from 17% to 32%. The lower limit is touched when apartments are small, involvement is scarse, and the plant is not well suited to the building. The higher limit is reached in case of big apartments, or when the involvement is high. Other elements affecting saving are the thermal capacity of the building (a well insulated, small capacity building is better suited), the attitude of users toward the system (the reliability of the apparatus and a good service are very relevant), and the time from installation (learning requires weeks, and in some cases months to be effective). Time of investiment return varied from 4 to 8 years.
- schools, offices, etc. : observed saving varied from 20% to 40%. Saving is strongly affected, in this case, from the type of plant, fromthe real building use;the use of the local unit, in most cases, must be left to authorised personnel only (specially in schools). The time of investment return can be very short: the prototype system installed in a school led to a pay back period of 2.2 years. In fig. 6 the results obtained in this school are reported.

4. Conclusions

The research has shown that an efficient system for energy saving in buildings heating can be produced; moreover, the system which has been studied helps to a better comfort in existing buildings. Finally, the system is convenient from an economic point of view. CSEA has then launched production and marketing of the system, with the (registered) trademark "Caldorobot".

References

(1) - Vito Mauro - Computer controlled network for heating system. Energy conservation in Buildings. Reidel Publishing Company.

(2) - Eykhoff P.E. : System identification. Parameter and state estimation (New York, J. Wiley, 1974)

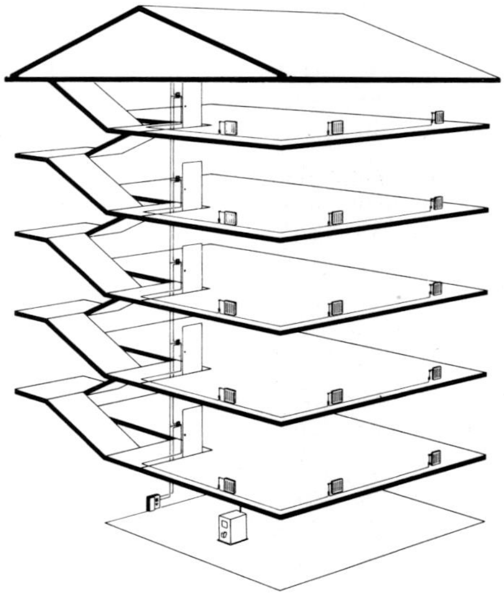

Fig. 1: Structure of the system

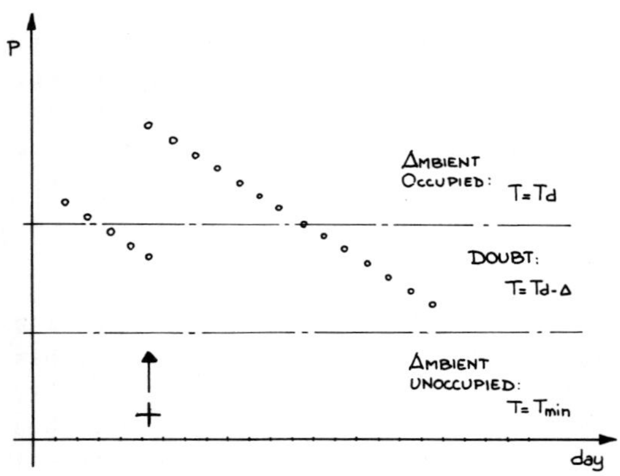

Fig. 2: Learning of probabilities

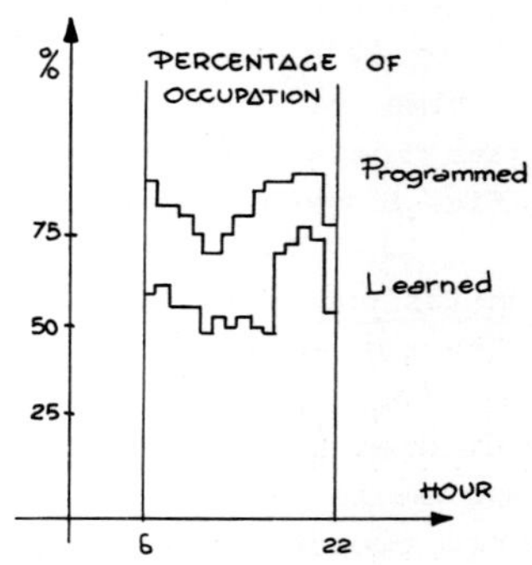

Fig. 3: Effectiveness of learning

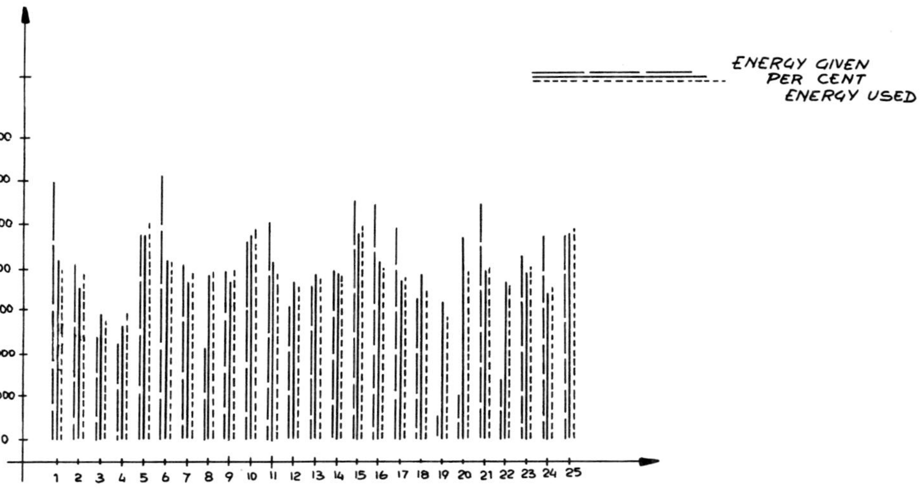

Fig. 4: Application of three accounting algorithms to a real building

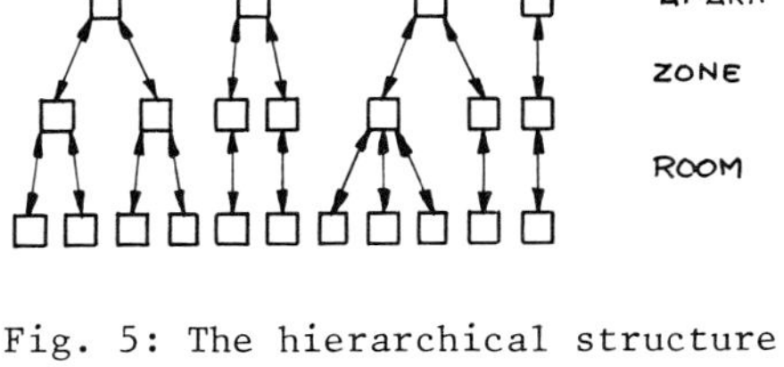

Fig. 5: The hierarchical structure of the building

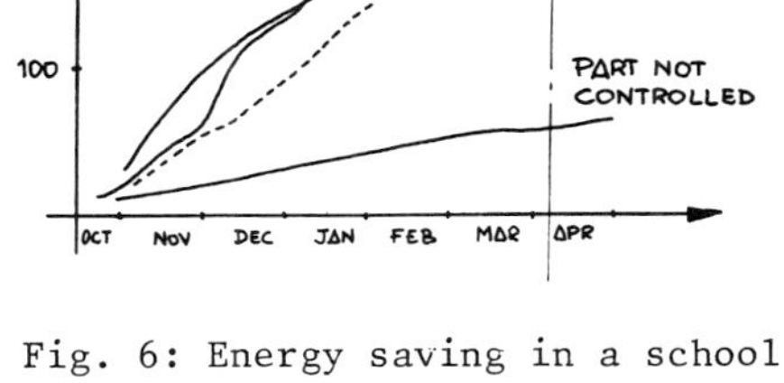

Fig. 6: Energy saving in a school

AN INVESTIGATION OF THE PERFORMANCE OF VARIOUS HEATING SYSTEMS IN A MATCHED PAIR OF HOUSES - INITIAL RESULTS

M. FINN, T. FULLER, P. MINOGUE, D. CARROLL

An Foras Forbartha, St. Martin's House, Waterloo Road, Dublin 4.

Summary

The characteristics of a range of heating systems were studied by using each system to maintain a particular temperature regime in the test house of a pair of matched houses in which occupancy was simulated. Temperatures in the control house were caused to track those in the test house using electric heating, thereby using the control house to measure the total heat input to the test house. The work has illustrated a number of deficiencies in the operation of appliances and control systems and pointed to areas in which improvements may be made in the performance of conventional heating systems. Due to time limitations imposed by the length of the heating season, the number of options for improvement which could be evaluated was restricted. Hence the work was concentrated on two systems, the solid fuel back boiler and the natural gas systems. With the solid fuel back boiler, which is basically a very simple, low cost, heating appliance, the effect on efficiency of improved control of the combustion air supply using a retrofitted firefront will be examined. Also the effect of incorporating a heat-store on the system's response to variation in demand will be studied. The natural gas boiler will be used in conjunction with control systems comprising thermostatic radiator valves and individual water heating controls, in order to examine their effect on energy consumption and on the ability of the system to maintain a desired heating pattern.

1. INTRODUCTION

This paper describes the work carried out to date on a research project which was aimed at investigating and improving the performance of a range of domestic heating systems typical of those currently used in Ireland. The majority of new private houses being built in this country have central heating with the most common systems being solid fuel or dual oil/solid fuel based. Natural gas became available for domestic heating in Ireland late in 1982 and it is likely that it will be increasingly used in the future as the gas distribution network is developed. These factors contributed to the selection of the heating appliances to be studied which were; a high output back boiler on the livingroom open fire, an independent solid fuel boiler, an oil boiler, and a low thermal capacity natural gas boiler.

2. OBJECTIVES AND EXPERIMENTAL METHOD

2.1 Aims

The particular objectives of the project were to examine the ability of each heating system, i.e. the combination of appliance, distribution system, and controls, to maintain various temperature regimes and to quantify appliance and heating system performance in terms of efficiency figures. Based on the results of the foregoing, it was proposed to implement and test modifications to the heating systems which would have the effect of conserving energy while satisfying the demands placed on the system by the user.

2.2 Experimental Method

The experimental approach used was to set up a pair of houses, a control house (house 0) and a test house (house 1), and establish relationships between their energy consumptions for different patterns of heating. The test house was then heated using each heating system to be studied, while the control house was heated using oil filled electric radiators and used as a reference against which the performance of the systems under test could be judged. The process of establishing a relationship between the energy consumptions of the houses, called calibration, was carried out by selecting a number of heating patterns according to which both houses were heated electrically while simultaneously monitoring both houses extensively. Computer control of the electric radiators was used in order to maintain the same temperature in corresponding rooms of the houses. Regression analysis was used on the recorded values for hourly energy consumption in order to establish relationships between the energy consumptions of both houses.

The pair of houses could then be used in two different ways to gain information on the heating systems being studied. In the first method a system of temperature tracking was developed in which temperatures in each room of the control house were made to follow precisely those of the corresponding rooms of the test house using the computer controlled electric heating system. Consequently, during a test run, the effect of the heat input to the test house from, for example, the solid fuel back boiler system was mirrored in the control house. Therefore, because the relationship between the energy consumptions of the pair of houses was known, the magnitude of the heat input to the test house could be calculated from the measured electrical energy input to the control house. Such a measurement method was very useful in assessing the net heat input to the test house from the solid fuel back boiler system whose output comprised direct radiative and convective components in addition to its output to water.

In the second experimental technique the control house is heated electrically according to one of the heating patterns used during calibration with the test house heated by one of the systems under study and the heating controls set to give a similar heating pattern. Again using regression analysis the results of such a test of a number of weeks duration are used to determine a relationship between the energy consumptions of the houses. The effect on energy consumption of modifications to the test house heating system can be judged by repeating the test for a further period of weeks using the modified heating system, establishing a second regression relationship between the energy consumptions of the control and test houses and comparing this to the first relationship.

To summarise, therefore, the experimental work comprised three separate steps:-

(a) Calibration of the pair of houses.

(b) Test work, carried out over a full heating season, aimed at identifying the operating characteristics and deficiencies of the heating systems.

(c) Testing of improvements to the heating systems which would be aimed at reducing energy consumption and improving their ability to meet user demands.

The experimental houses, which were acquired late in 1981, comprised a semi-detached pair typical of modern private estate houses in Ireland in terms of construction, materials used, size and layout. The floor area was 80 m^2 with three bedrooms and two main rooms downstairs. External wall construction was of hollow concrete block, insulated on the inside so as to achieve a U-value of 0.6 $W/m^2\,{}^{o}C$, while the average U-value for all the enclosing elements was 0.85 $W/m^2\,{}^{o}C$.

The pair of houses were fitted with instrumentation to monitor internal air and globe temperatures, relative humidity, and electricity consumption for space and water heating. A weather station was set up which recorded external temperature, global and diffuse solar radiation and windspeed. For each heating appliance, flow and return temperatures, water flow rates, fuel consumption, thermostat and burner operation and on-times, were monitored. A more detailed description of the instrumentation is to be found in reference (1).

In order that the heat demands placed on the heating system should resemble more closely those of an occupied house certain aspects of occupancy were simulated. These were heat gains due to occupants, cooking, lighting, curtain drawing and hot water draw off. The daily energy inputs due to occupants, cooking, and lighting were 6.05, 4.0 and 1.7 kWh respectively.

At the time of writing work has been done on stages (a) and (b) while the results of stage (a) will be extended when analysis of the most recent calibration work is completed. The third stage of the project, in which improvements to conventional heating systems are being evaluated, is underway.

3. CALIBRATION

3.1 Method

In order to use the pair of houses to study the characteristics of a range of heating systems and to quantify the benefit of proposed modifications it was first necessary to establish a consistent set of relationships between the energy consumptions of the houses. This was done by maintaining similar internal temperatures while measuring the energy inputs required to maintain these conditions.

During calibration, the houses were heated by oil filled electric radiators. Control of air temperature was achieved by having the on-site computer monitor the temperature in each room every minute and switch on an electric radiator if the temperature had fallen below the desired temperature. All internal doors were kept closed during calibration, as was the case during heating system tests and opening of external doors was kept to a minimum.

Three patterns of temperature were maintained in the houses during calibration: whole house heating to 18°C for 24 hours each day; a high heating regime which is illustrated in table 3.1: and a low heating regime which is shown in table 3.2. The constant heating to 18°C was used to provide a reference data set using which any significant structural changes, for example crack development, which would effect the relative energy consumptions of the houses could be identified by repeated periods of constant heating throughout the project.

Desired Room Temperature (°C)				
Time Period	23-7	7-9	9-16	16-23
Room				
Living-room	none	18	18	20
Kitchen/dining-room	none	18	18	18
Hall	none	16	16	16
Bathroom	none	16	none	16
Bedrooms	none	16	none	16

Table 3.1 High heating regime.

Desired Room Temperature (°C)				
Time Period	23-7	7-16	16-20	20-23
Room				
Living-room	none	none	20	20
Kitchen/dining-room	none	18	18	18
Hall	none	none	16	16
Bathroom	none	none	none	16
Bedrooms	none	none	none	16

Table 3.2 Low heating regime.

Pressurisation tests were carried out, in order to determine the relative air tightness of the houses, by the Building Research Establishment, U.K. The method was to subject the houses to a cycle of internal/external pressure difference which varied from approximately +50 to -50 Pascals while measuring the rate of air flow required to maintain the pressure difference. The contribution to air leakage of doors, windows, ventilators, etc., was identified by repeating the pressurisation cycle with each set of elements sealed in turn.

3.2 Results

During calibration the control system successfully maintained very similar conditions in both houses. Hourly periods during which the average whole house temperatures were different by more than 0.2°C, due to equipment faults, etc., were excluded from the analysis. The energy consumptions of the houses were found to be quite similar, but with the average hourly energy consumption of the test house being consistently higher for each of the three heating patterns. For the constant and high heating regimes the difference was 4% and for the low regime it was 2%.

Because of the similarity in house construction a fairly simple relationship was expected between energy consumptions. Consequently a simple linear regression model of the form

$$E1 = m.EO + c$$

where
E1 = energy consumption of the test house
EO = energy consumption of the control house.
m = slope of the regression line
c = intercept of the regression line

was fitted to the hourly energy data and the resulting functions are shown in Table 3.3. All the regression lines have a positive intercept and a slope of less than one which means that for low values of hourly energy consumption, that of the test house is greater with the opposite being the case for higher values.

TABLE 3.3 CALIBRATION RESULTS

Heating Regime	No. of Hours	Regression Line E1 = mEO + c	Correlation Coefficient
Constant Temperature	177	EI=0.926EO+0.250	0.919
High heating regime	180	EI=0.921EO+0.306	0.978
Low heating regime	86	EI=0.912EO+0.178	0.974

Using analysis of covariance techniques it was found that statistically the three regression lines were not equivalent. Taking the lines in pairs it was found that the lines corresponding to the constant and high heating regimes were equivalent and both were different from the lines for the low heating regime. Using these simple models the accuracy of prediction of the test house energy consumption from that of the control house when both are heated to similar temperatures was 5%.

Pressure testing of the houses yielded average values for the standard 50 Pa air flow rates of 3616 and 3870 m^3/h for the control and test houses respectively without any sealing of likely leakage paths except conventional draught stripping of external doors and windows. The relative importance of the major leakage paths was identified as follows:- 25 - 30% for doors, windows, flue and fixed ventilators, 20% cross-leakage between houses with the remaining 50 - 55% being leakage through the building fabric, though it should be noted that the importance of the "remainder" is over-emphasised by testing at 50 Pa. From the results of work by the B.R.E. aimed at relating measurements of air leakage and tracer gas measurements of air infiltration (2), estimates for air infiltration rates of 0.91 and 0.98 air changes/hour at an average wind speed of 3.5 m/s were made for the control and test houses respectively.

4. HEATING APPLIANCES AND CONTROLS

A summary of the characteristics of the three types of heating appliance, solid fuel, oil and natural gas, for which test results are presented here, is given in Table 4.1. As one of the main aims of the project was to examine the influence of controls on heating system performance it was decided to start by selecting, as a baseline system, a very simple, commonly used, set of controls for each appliance. For the natural gas and oil systems the baseline controls were a timeswitch and wall mounted room thermostat, of the bimetallic strip with accelerator heater type, which controlled the circulating pump with the space heating and hot water loads connected in parallel. The gas and oil boiler thermostats were set to 88°C and 70°C respectively. With the solid fuel system the circulating pump was switched on manually after the fire had become established, and remained on until the following morning. Additionally the solid fuel back boiler itself had two manually adjustable dampers, one to regulate the supply of combustion air to the fire and the second to control the flow of hot combustion gasses through its two flueways. This latter damper, therefore, permitted a crude level of control over the proportion of the boiler output which went to water heating. For all the heating systems,partial and intermittent heating, as indicated in Tables 3.1 and 3.2, was achieved by switching motorised valves connected to the radiators at the appropriate times thus simulating the action of the householder in turning on and off radiators as heating was required.

TABLE 4.1. CHARACTERISTICS OF HEATING APPLIANCES

BOILER	TYPE	RATING	FUEL USED	CALORIFIC VALUE
High output back boiler	twin flue	7.6 kW	household coal	8.49 kWh/kg
Oil boiler	pressure jet, HTC	12.0 kW	gas oil	10.59 kWh/ℓ
Natural gas boiler	balanced flue, LTC	8.0 kW	natural gas	11.10 kWh/m^3

NOTE: HTC = high thermal capacity
LTC = low thermal capacity

5. GENERAL TEST PROCEDURE

For the natural gas and oil systems the test procedure was to operate each system so as to heat the test house according to the high and low heating regimes (see Tables 3.1 and 3.2) from 7.00 to 23.00 each day for an extended period of days. The controls were as described above with the room thermostat set to 16°C. The time during which the houses were heated was divided into four periods; 7.00 to 9.00, 9.00 to 16.00, 16.00 to 20.00, and 20.00 to 23.00 during which individual radiators were on or off as determined by the heating regimes. Curtain drawing and hot water draw-off were done automatically, the latter according to two draw-off profiles, a high and a low, totalling 180 and 90 litres/day respectively.

For the solid fuel back boiler work the main fueling period was shortened to the period 9.00 to 20.00 for convenience. Therefore the periods during which rooms were heated were adjusted as follows:- the kitchen/diningroom was heated from 9.00, the hall from 15.00, and the bedrooms from 17.00 onwards. Following the lighting up period, which typically took 1.5 to 2 hours, the fire was manually fueled at approximately 1 hour intervals. The fueling rate varied from day to day between 1.0 and 2.1 kg/hour depending on the physical size of fire which was maintained and on how the cumbustion air damper was adjusted.

Throughout the tests temperature tracking of the test house temperatures by the control house, as described in section 2, was used so that, by using the calibration relationship already derived, a measurement of the total heat input to the test house from the heating system could be made. This technique was also used to measure the output of the solid fuel back boiler to the hot water storage cylinder via natural circulation, by electrically heating an identical water storage cylinder in the control house to the same temperatures and recording the electrical energy used.

6. RESULTS: SOLID FUEL BACK BOILER

6.1 Initial Findings

This section deals with the results of 12 all-day tests which were carried out on the solid fuel back boiler system using the procedure outlined in section 5. The testing was done during a period of fairly mild weather, having an average external temperature over the test period of 11°C. Other tests, to a format described in BS 4834 (3), which were also carried out will not be dealt with in detail here.

The average rate of appliance output during the period 11.00 to 20.00, i.e. when the fire was well established, ranged from 2.2 to 5.0 kW and was mainly a function of the fueling and burning rates. It should be noted that these outputs were net heat outputs to the test house as determined using the temperature tracking procedure outlined earlier. The maximum design output rate of the appliance was 7.6 kW, a level which was most closely approached during a six hour BS 4834 type test when an output rate of 7.3 kW was achieved at a burning rate of 2.2 kg/hour.

Of particular interest, however, was the pattern of output from this type of appliance. The pattern was characterised by an initial period of 1.5 to 2 hours during which the output was low, typically totalling 2 to 3 kWh, or 10% of the total daily output. This was followed by the period 11.00 to 20.00 during which typically 70% of the total daily output was delivered, with the output varying slowly in response to changes in fueling and burning rates. From 20.00 to the following morning the appliance output comprised typically 20% of the total daily output, as the fire burned down and the heat stored in the system, firesurround, etc., was given up to the house.

The heating system efficiency, as defined by the total daily output to the house over the gross calorific value of the daily fuel consumption less the energy content of any partially burned fuel, ranged from a maximum of 40% to a minimum of 30%. The latter value was achieved after five test days when the back boiler flueways were found to be partially blocked with soot. Thereafter it was found that daily cleaning of the back boiler was necessary in order to maintain an efficiency in the region of 40%. In addition tests carried out while maintaining a very small fire throughout the day resulted in efficiencies averaging 32%, indicating that, in order to maximise output, the fire must be in good physical contact with the back boiler.

6.2 Discussion

The solid fuel back boiler output was capable of meeting the test house space heating load under steady state conditions during mild weather as the design space heating load was 7.5 kW for an outside temperature of -1°C. However, the pattern of output was not well matched to the typical load pattern, and the heating system could not respond quickly to changes in load. In particular the period of maximum load, the morning period when the house was being brought up to temperature, coincided with the period of minimum system output and the system continued to deliver a significant proportion of the daily output to the house after the end of the fueling period, when in fact heat was not required.

Improved matching of system output to load could be achieved by introducing a heat storage element between the heating system and the load which could supply the high early morning heat demand and furthermore take up the heating system output at times when the load was satisfied. Such a heat-store was designed and installed, and will be tested in the final stage of the project. It comprised a well insulated steel water tank to which the solid fuel back boiler output was directly connected with the radiator circuit similarly connected via a second pumped circuit. In sizing the heat-store a basic design requirement was that its output should be capable of heating the test house from a whole house average temperature of 10°C at 7.00 to 16°C at 9.00 in average winter weather. This typically required an output of 15 kWh of which 25% went to heat up the distribution system. Furthermore a minimum output rate of 4 kW after two hours operation was adopted. Given that the maximum rate of output that could be expected from the system throughout the day would be 5 kW and the requirement of having a heat-store temperature in the region of 70°C at 7.00 it was important that the heat-store was not oversized. These requirements were most closely met by a heat-store of 750 litres capacity.

A fundamental limitation on the efficiency of this simple type of appliance, when using household coal, was imposed by the fact that there was very limited control over the amount of combustion air available to the fire. Because of the high flow rate of combustion air, a large part of the energy content of the fuel was lost in the form of unburned volatiles. Two possible improvements to appliance performance could be implemented. The simplest option would involve fitting an airtight door or firefront, incorporating a combustion air damper, over the fireplace opening which would give improved control over combustion air flow. This measure will be evaluated in the test houses during the coming heating season. The second option would involve modifications to the shape of the back boiler flueways in order to achieve burning of the volatiles before they left the boiler flueways thereby making heat, which otherwise would have been lost, available to the boiler. This second option would effectively mean a re-design of the solid fuel back boiler and is therefore not well suited as a retrofit measure.

7. RESULTS OF TESTS ON OIL SYSTEM

7.1 Initial Findings

The test results which are summarised here are based on an extended period of operation of the oil system to heat the test house using gas oil as the fuel. The boiler output referred to was the output to water, calculated every 10 seconds, and subsequently summed, by using a turbine flowmeter to measure the water flow rate and a pair of platinum resistance thermometers to measure the temperature difference between the boiler flow and return pipes. Boiler efficiency was calculated as the heat output to water over the gross calorific value of the oil used.

The resulting boiler efficiencies are shown in figure 1 which shows the variation of part load efficiency with the percentage of time that the burner was on. Because daily percentage on-times were always below 30% during the test period, efficiencies for higher levels of load were derived using the results of boiler operation over periods of shorter length.

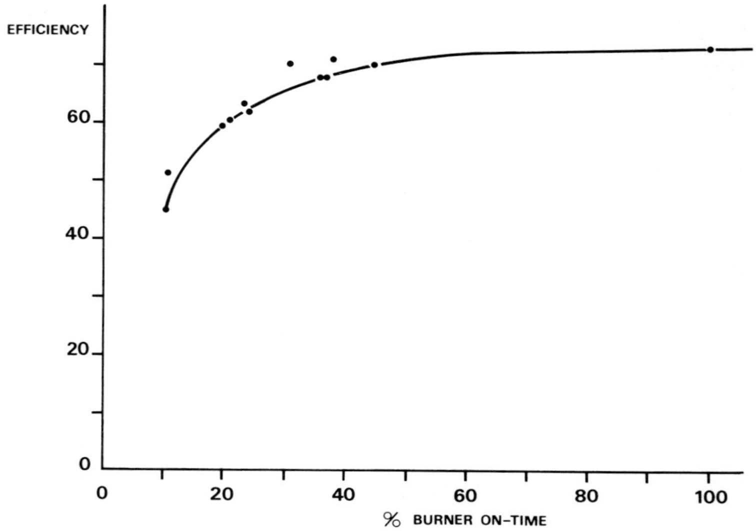

Fig. 1. Oil Boiler Efficiency

The boiler stack efficiency was typically 77% with full load efficiency on the water side of 73%. The boiler standing losses which comprise transmission and ventilation components, were estimated to be 3% of rated output at an average boiler water temperature of 77°C by operating the boiler for a number of days with no load. However, an increased ventilation loss could be expected under operating conditions due to the increased on/off cycle frequency. Because of problems with soot formation and consequent deterioration in full load efficiency while using the burner as supplied and commissioned according to the manufacturer's recommendations, it was found necessary to use a smaller size nozzle with increased atomizing pressure in order to successfully burn gas oil.

The total boiler output over the test period averaged 27.7 kWh/day. The daily output varied mainly due to external temperature, from a maximum of 46.6 kWh for a daily average outside temperature of 4.2°C to a minimum of 11.3 kWh when the outside temperature averaged 9.4°C. The burner percentage on-time varied from a maximum of 23.6% to a minimum of 8.4% with an average for the test period of 15.8%.

7.2 Conclusions

(a) In order to achieve part load efficiencies which drop below the efficiency at full load by less than five percentage points the boiler rating must be closely matched to the load such that percentage on-times do not drop below 35-40%. Where oversizing is such that percentage on-times are significantly less than 20%, part load efficiencies of 10% or more below full load efficiencies can be expected. The test house, being a small fairly well insulated dwelling, is typical of modern estate type houses in Ireland and the boiler rating of 12.0 kW was the lowest commercially available for a pressure jet type using gas oil, the limitation being the oil nozzle size. Hence it is clear that there is a fundamental limitation on the size and type of house to which this type of boiler can be correctly matched, with this type of installation.

(b) For efficient burning of gas oil without soot formation effective atomising must be achieved by careful selection of oil nozzle size and adjustment of atomizing pressure. When adjusting the burner combustion air control for minimum CO_2 content in the flue gasses, smoke tests must be made to ensure that no soot is being formed, particularly during the early part of the burner firing cycle, which would quickly degrade boiler efficiency.

(c) Given the variation that occurs in gas oil viscosity with temperature and that the viscosity of delivered gas oil can vary considerably within current standards there may be an argument for preheating the oil to obtain consistently good atomizing, constant nozzle capacity, and soot free combustion, though this has not been experimentally evaluated. An additional advantage of preheating would be better matching of gas oil boilers with well insulated, low demand, houses in that, with preheating, effective atomizing can be achieved at lower oil pressures and hence lower oil throughput rates.

8. RESULTS: NATURAL GAS HEATING SYSTEM

8.1 Initial Findings

The results of tests with the gas heating system, over ten days, using the high heating regime are shown in Table 8.1. The average space heating load during the period 7.00 to 23.00 ranged from 0.7 kW for a daily average outside temperature of 9.6°C to 3.3 kW on a day when the average outside temperature was 2.8°C. The typical load pattern for this heating regime was a high level of load in the morning period 7.00 to 9.00 followed by a very low level of load for the remainder of the day. In the morning period when the whole house was being heated the average load varied from 3.2 kW to 7.5 kW over the ten days with an average of 5.3 kW, while the average load for the remainder of the day was 1.25 kW. On a number of days the heat demand for a large part of the period 9.00 to 23.00 was satisfied by solar gains or gains due to occupancy simulation.

The boiler was rated by the manufacturer to have an output of 8.2 kW at a gas consumption of 10.4 kW which implied a full load efficiency of 78.8%. In practice when the boiler operated continuously for long periods, efficiencies slightly in excess of this figure were achieved. The part load efficiencies achieved for operation over a sixteen hour period which are shown in Table 1, illustrate that, for percentage on-times as low as 11%, efficiencies in excess of 70% were achieved. Because, in this installation, the room thermostat controlled the circulating pump, and hence the boiler via a built-in flow sensor, the only gas consumption during periods of no load was that due to the pilot light. Hence the pilot gas consumption exerted a significant

TABLE 1 NATURAL GAS BOILER TEST RESULTS

Day	Boiler Output (kWh)	Burner % on-time	Part Load Efficiency (%)	Estimated Full Load Efficiency(%)
1	19.7	11	71.5	77.5
2	20.7	12	71.4	76.8
3	13.7	8	68.1	76.6
4	18.5	11	70.1	76.0
5	11.1	7	66.7	76.6
6	10.4	6	66.1	77.8
7	20.7	12	70.6	76.0
8	26.2	15	71.0	75.1
9	52.1	29	72.3	74.0
10	39.4	22	71.5	74.0

influence on part load efficiency. A theoretical relationship between full and part load efficiency, which assumes constant gas combustion rates on pilot and main flame is given by:-

$$E_f = E_p \frac{1}{1 + \frac{e}{q}(1-\frac{1}{n})} \qquad (1)$$

where

E_f = Full load efficiency

E_p = Part load efficiency

q = constant gas input on main flame (kW)

e = constant gas input on pilot (kW)

n = proportion of time that the burner is on main flame.

The estimated full load efficiencies achieved using this expression, given in Table 8.1, show that, for the first seven test days, the full load efficiency averaged 77%, dropping off for the final three test days. This latter effect was identified as being due to a boiler fault. The difference between the full load efficiencies calculated using (1) and those measured for periods of long continuous operation may be due to the assumptions on which the expression is based not being strictly correct. To summarise, therefore, the boiler operated with high efficiency at very low load, and it can be concluded that a low capacity, rapid response, boiler is well suited to meeting the widely varying demand pattern imposed by the test house.

8.2 Features of the Control System

The baseline control system used, comprising a single room thermostat located in the hall which controlled the circulating pump, was not capable of maintaining the temperature regime of figure 3.1 in the test house. The level of control of house temperature could be characterised by the following features:-

(a) As the temperature sensing element was in the hall, the heat input, and hence the temperatures maintained, in the rooms depended on the heat load imposed on the system by the hall and on the matching of radiator output to room load. Furthermore, as the load imposed on the hall by outside weather relative to the whole house load, varied from day to day, so the temperatures achieved in other parts of the house varied. As the load on the hall increased, causing the percentage of time that the thermostat was calling for heat to increase, so the average whole house temperature increased, while the hall temperature remained close to the desired temperature.

Similarly as the water heating load was connected in parallel with the radiator circuit, the extent of water heating was a function of the heat load on the hall. In order to provide hot water to meet a particular demand profile, an independent hot water control system would be required which could respond to that demand quite separately from the space heating demand.

(b) The inability of such a control system to respond to heat inputs other than the heating system was particularly noted. Overheating occurred in the south facing rooms during period of high solar radiation.

(c) A feature of the control action of the room thermostat was a gradual downward trend throughout the day in the temperature being maintained in the hall. This drop in hall temperature typically began after the house had been brought up to temperature, continuing to 23.00, and generally was of the order of $1^{\circ}C$. It was thought likely that this was caused by warming of the internal wall on which the thermostat was located due to the higher temperatures maintained in the adjacent livingroom and by afternoon solar radiation.

REFERENCES

1. Finn, M. Instrumentation for Monitoring the Performances of Alternative Heating Systems in a Matched Pair of Houses, Proceedings of CIB W67 Symposium "Energy Conservation in the Built Environment", An Foras Forbartha, 1982.

2. Warren, P.R., A Simple Method for Predicting Infiltration Rates in Housing, Proceedings of CIB W67 Symposium "Energy Conservation in the Built Environment", An Foras Forbartha, 1982.

3. BS 4834: Inset open fires without convection, 1972.

TWO APPLICATIONS OF MICROPROCESSORS IN CLIMATE AND LIGHTING IN OFFICE BUILDINGS

H.J. Nicolaas
Technisch Physische Dienst TNO-TH

Summary

In this paper two applications of microprocessors in the climate and lighting control of buildings are given. These applications are:

- Lighting control depending on daylight level and use of the sun-shading. The method is based on measuring the solar radiation on a vertical plane and predictions of the horizontal illuminance in several places in a room.
- The possibility of mechanical ventilation during the night to cool down the building mass and to reduce the maximum inside air temperatures during daytime. Therefore, predictions has to be made for the inside air temperature of the next day. A multiple regression method gives sufficient results. The influence of the ventilation during night was smaller than expected.

1. INTRODUCTION

Recent developments in the field of micro-electronics have provided possibilities to program regulation algorithms on relatively cheap and very reliable microcomputers. The purpose of these regulation algorithms will be the processing of the various stimulants for outside and inside the building e.g.: inside and outside temperatures, solar radiation, humidity, the buildings heat capacity, occupancy and weather forecast. The system conduct, its momentary "condition" and the requirements of an actual situation can be determined by mathematical description and by measuring. Computer simulation and regulation are the obvious aids to achieve an integrated regulation of air-conditioning and lighting plants.

2. TWO MICROPROCESSOR APPLICATIONS

In this paper two microprocessor applications are worked out:

- Lighting control depending on daylight level and use of sunshading. In relation to a demonstration project on lighting control in an office building, measurements are carried out to get better prediction of day-lightlevel on a certain place in a room. This method is based on measuring the solar radiation on a vertical plane.
- The possibility of mechanical ventilation during the night to cool down the building mass and to reduce the maximum cooling load and the maximum inside air temperature during day time. Therefore it is necessary to predict the inside air temperatures in relation to the outside conditions.

3. LIGHTING CONTROL

3.1 Control strategy

The control has been designed for a rectangler office building and as such looking at four orientations. The lighting in the individual offices of the building are controlled depending on the daylight level, the office hours and the wishes of the inhabitants. For cleaning purposes, the lighting of the building can be switched on half its capacity, out of the office hours.

The objective is to optimize the use of the lighting with regard to energy savings, while the lighting level on the working area must be sufficient in all circumstances. This goal is met by controlling the artificial lighting in three fases (on, half on and off) in two areas of the office; on the window side and on the passage side. The artificial lighting is controlled, depending on the daylight contribution to the lighting level on the working areas. The fenominum of a very bright sky has been known to necessitate an extra degree of artificial light on the working area. The control will switch on half the lighting when this occurs. The window blinds (if installed in the building) can be controlled with the system.

3.2 Electrical installation

The lighting of the individual offices are divided in four groups; two groups in the area on the window side and two in the area facing the passage. Two switches, one for the window area and the other for the passages area, make it possible for the inhabitants to swith the lighting off or in the controlled mode. The four groups of the individual offices are controlled separately, while an extra realy in the manual switches can de-activate the lighting in both the window and the passage area.

3.3 Manual control

The inhabitants of the offices have the possibility to manually switch the light on or off. The on position will ofcourse mean that the control will evaluate the necessary artificial lighting level. The control starts at the beginning of the office hours, but the lighting in each individual office will have to be initiated by the inhabitants. After office hours the lighting is switched off centrally, but can also be switched off by the inhabitants. During cleaning hours, the lighting is switched on at half its full capacity. However, the lighting in the individual offices will have to be manually activated by the cleaning staff. The cleaning hours can be preset or started and stopped manually. During times out of the office hours and the cleaning hours, the lighting can be switched on manually in each office on half its capcity for safety purposes.

3.4 Measurements

The daylight level on a vertical plane is measured on each of the four orientations, with calibrated light dependent resistors (LDR). From investigations it is found that there is for practical purposes a lineair relation between the vertical solar radiation level and the daylight level on a horizontal plane. With this relation the average daylight level in the two areas (window and passage) of an office room on working height can be calculated. This daylight contribution to the lighting on the working areas determines the artificial lighting level; on, off or half on.

The window blinds are controlled on the basis of the vertical solar radiation level in those orientations of the building where a window blind is installed.

When the window blinds are lowered, the relation between the vertical radiation and the lighting level on the horizontal working areas are adjusted to this new situation.

The brightness of the sky is evaluated on each orientation with a so called brightness sensor. The brightness in this case can be defined as the sky radiation in a part of an area of the sky which can be seen by an observer from a representative position in a room at eye height for a sitting person. The brightness is measured with a light dependent resistor positioned in a tube in which it can only "see" the part of the sky which is relevant. Care should be taken in the interpretation of the reading of the sensor, since direct solar radiation can also be measured with this brightness sensor.

3.5 Switching frequency

In order to avoid frequent switching of the artificial light caused by an instabel daylight level, delay times are introduced. The respond time of the lighting on-switching must be chosen small, while the off switching responds can be delayed several minutes. The delay times only prevent switching frequencies of once in several minutes, but it is still possible that during a day with changing weather conditions the lighting is switched repeatedly. This is prevented by introducing an extra delay time after an off switching. The next off switching can only occur after this delay time is elapsed.

3.6 Monitoring program

In order to judge the right working order of the lighting control a monitoring program has been developed. The lighting control is installed in a test office with two inhabitants. In this test office two main points of interest are investigated:

- The evaluation of the objective quality of the control.

  The quality of the control can be judge upon, by measuring the lighting level on the working areas in an office. In each of the two working areas in an office (window and passage) two sensors are installed. The lighting level on these sensors must in all circumstances be higher than the minimum level acceptable. The results of the measurements are presented in barcharts, showing the time that a certain lighting level occurs.

  The measurements are only performed when the lighting control is in operation; during office hours, with the manual lighting swiches on. In order to judge the individual controls of the system in this manner, the barcharts are made for several control phases:

  A. Over the total of the operation time

  B. Normal control
     * no window blinds down
     * no brightness correction

  C. Control with window blinds down
     * no brightness correction.

  In each of these cases (A, B en C) the barcharts are made for each sensor and for an average lighting level on the window area and the passage area, separated for the artificial lighting on, half on and off. The length of time in which each part of the control is active, is registrated on a daily basis.

- The energy savings of the system.

  The burning hours of the artificial lighting will be registered. The energy savings can be derived from the ratio of the actual burning times and the maximum possible burning time.

The results are of course only valid over the testing period and under the specific testing condtions.

An important aspect of the quality of the control is of course the judgement of the inhabitants. In order to get an impression of this judgement, a switch is installed in the test room, which when pressed, activates an extra degree of artificial lighting during half an hour. The times when this occurs are registered. With these results it is possible to get an impression about the quality of the control.

4.7 Results

A daily output of the monitoring program is given on page 5 (window side only). The profits of the microprocessor controlled lightingsystem are the easy way of changing setpoints, control of daylight level and changing time delays for on/off switching in the program and the possibility of registration of burning hours.

4. CONTROL OF MECHANICAL VENTILATION

4.1 By computer simulation the influence of ventilation during the night on the inside air temperature was investigated.

For these simulation a building was chosen with a volume of 24 000 m^3, double glazing, 40% glass area and inside blinds. Each floor has 30 rooms (15 on each side) with dimensions of 3.5 x 5.8 m. We will consider now a system with night ventilation in summer to cool down the building mass with outside air.

The study was carried out for two different types of installations:

a. A four pipe induction system.
   With computer programs it is possible to minimize the cooling demand and the number of hours the fans are on during the night period by varying:
   - the starttime of night ventilation hours depending on the temperature,
   - the number of night ventilation hours depending on the temperature difference between inside and outside air temperature.

   With such an optimized control system the energy consumption was calculated for a four pipe induction system.

   A result of this study was that the total energy consumption with or without night ventilation was the same: the increase of fan energy was as big as the decrease of the cooling energy consumption.

b. Heating system with mechanical ventilation (no cooling).
   When in the same building a heating system is placed with mechanical ventilation during office hours (same air quantity as for the four pipe induction system) it is possible to calculate the frequency of the appearing inside air temperatures. When there is nog night ventilation the inside air temperatures are intolerably high during a number of hours (400 hours higher than 26 °C). With night ventilation the number of hours during which the temperature is higher than 26 °C decreases considerably (80 hours).

4.2 Measurements

A result of this study was that a remarkable reducing of the inside air temperatures can be achieved when there is a good control. To check these computer calculation, measurements are done in a building under actual weather conditions to find a simplified mathematical model of the building. With such a model it is possible then to predict the influence of another control strategy on the inside air temperature in the building.

In general also other options are possible such as optimum start of a cooling plant depending on the predicted temperatures during the day.

*********** OUTPUT OF OPERATION TIMES ***************************

FIRST DAYNUMBER: 83297 LAST DAY: 83297

****

		WINDOW ZONE:	PASSAGE ZONE:
LIGHTING :	OFF	314 min.	0 min.
	HALF	148 min.	5 min.
	ON	41 min.	499 min.
OPERATION TIME:		923 min.	923 min.
OFFICE HOURS:		504 min.	504 min.
CLEANING HOURS:		120 min.	120 min.
CONTROL TYPE:	TOTAL	504 min.	504 min.
	NORMAL	504 min.	504 min.
	SUNSHADING	0 min.	0 min.
	BRIGHTNESS	0 min.	0 min.
HOURS OF MALFUNCTIONING		0 min.	0 min.

*************** SWITCHING FREQUENCIES **********************

FIRST DAYNUMBER: 83297 LAST DAY: 83297

****

		WINDOW ZONE	PASSAGE ZONE
LIGHTING	TO OFF:		
	ON to HALF	2 Times	2 Times
	HALF to ON	3 Times	1 Times
LIGHTING	TO ON:		
	OFF to HALF	3 Times	1 Times
	HALF to ON	1 Times	1 Times

******************* LIGHTING CONTROL ***********

OPERATION: Office hours BAR CHARTS

CONTROL : Total of control SENSOR:G

ZONE:Window

FIRST DAYNUMBER: 83297 LAST DAY: 83297

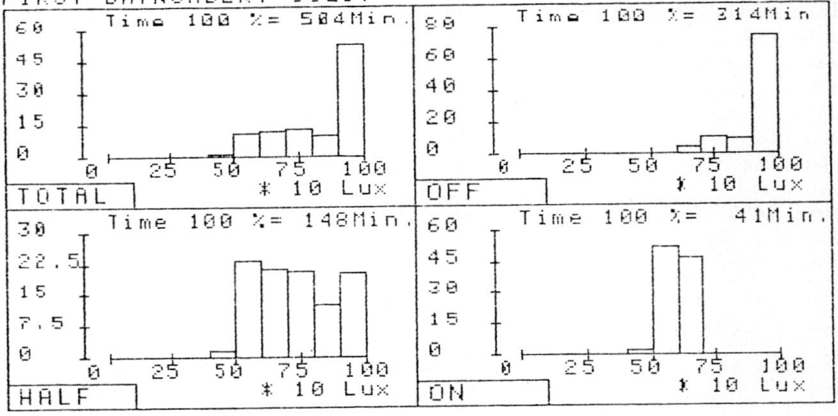

<u>Figure 1</u>: Daily output of the measuring and monitoring program for lighting control.

The measurements are carried out in a building which consists of two parts, a 7 storey part and a 5 storey part (figure 2). The higher part has a mechanical ventilation system. The ventilation rate of the system is 3. The air is supplied in the rooms and exhausted through the corridors and a vertical air channel.

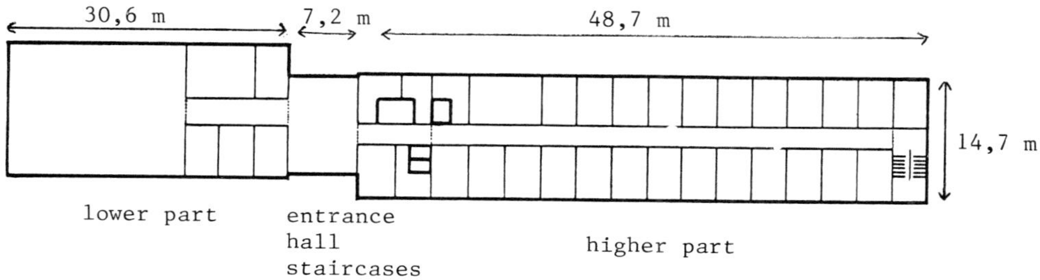

Figure 2: Floor plan.

On 6 places in the building air temperatures are measured.

- On one floor in the corridor and in two rooms on both sides of the building.
- The outside air, the supply air and the return air temperature.
- The solar radiation is measured on both sides of the building.

4.3 Control

During working hours between 06.00 a.m. and 17.00 p.m. the ventilation system was on. Outside working hours (during the night and the weekend) the system was off when the air temperature in the corridor was lower than 19 °C. The ventilation system was switched off too when the temperature difference between supply and inside air temperature was smaller than 6 °C.

4.4 Computermodel

With the measured data several calculations are carried out with simplified mathematical models. The best predictions of the inside air temperatures gives the multiple regression method [1]. An adequate regression equation appeared to be:

$$\dot{T} = a_1 (T_i - T_o) + a_2 (T_i - T_o) + c\ S + d\ I \qquad (1)$$

with:

$\dot{T}$ = Temperature change of the inside air temperature.
T_i = Inside air temperature.
T_o = Outside air temperature.
S = Solar radiation.
I = Internal load.
a_1 = Unknown thermal parameter related to the U-value of the building.
a_2 = Unknown thermal parameter which is o when the mechanical ventilation system is off.
c = Unknown thermal parameter related to the incoming solar radiation.
d = Unknown thermal parameter related to the internal load.

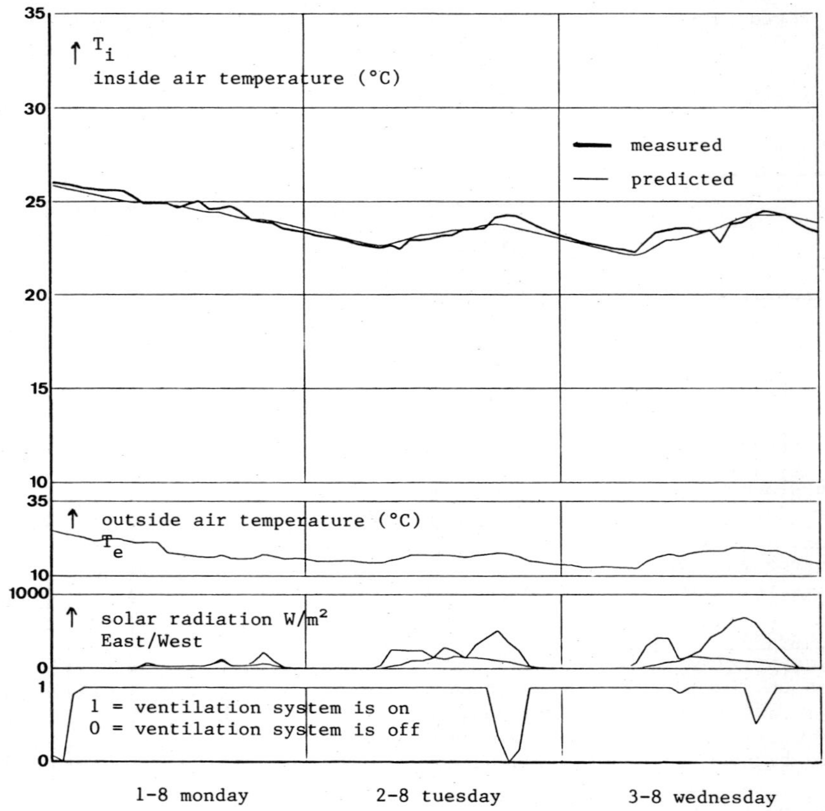

Figure 3: Predicted and measured air temperature in the building.

4.5 Results

With formulae (1) predictions of the inside air temperatures with knowledge of the actual values of the outside air temperatures and the solar radiations give good results. In figure 3 the calculated values and the measured of the inside air temperatures are given.

4.6 Future work

Instead of making use of actual weather data, weather forecasts are bringing in the model. Therefore the model had to know for one or two days in advance the predicted mean outside temperature and the solar radiation for instance cloudy, half cloudy, sunny. With these information predictions of the inside air temperature have to be made.

The influence of the switching of the ventilation system on the inside air temperature is still under investigation. Because of the heavy construction of the building the temperature difference between ventilation system on during one night and off is relatively small.

[1] J. Oldengarm, P. Euser, Test method determining the thermal performance of houses. Proceedings of the international colloquium, Liège Belgium 6-8 May 1983.

FEASIBILITY OF UNDERGROUND HEAT STORAGE IN CAPTIVE AQUIFERS

J.Y.L. DESPOIS* and B. CONSEIL**

* Atomic Energy Commissariat

** The National Elf-Aquitaine Company

Abstract

The Atomic Energy Commissariat (CEA) and the National Elf-Aquitaine Company (SNEA) have undertake studies to establish the feasibility of a system of underground heat storage in deep aquifers, allowing initial temperatures of 180 to 200 °C and providing good efficiency of heat restitution to conventional district heating systems, with return temperatures of 60 to 70 °C. Heat will be supplied by domestic garbage incineration plants or nuclear power stations able to supply excess heat during the summer. In order to demonstrate the feasibility of this type of storage, CEA and SNEA have selected a site in the outer suburbs of PARIS. An exploration well has been drilled near to a garbage incinerator down to a sandy aquifer at a depth of 500 meters. The hydrogeological and geochemical data gathered have provided means of digitally simulating the behavior of a complete system consisting of source, storage system and distribution network. Construction of this system is now to start. Should it be successful, the inter-seasonal transfer of about 15 GWh/year, added to direct delivery, will allow the heating of about 4,000 homes in the neighboring town of PLAISIR.

1. INTRODUCTION

Systems of direct heat storage can make use of sensible heat or of latent heat. The specific heat by volume of substances that can be considered are always very low when compared with the chemical energy contained in a traditional fuel. Volumes are considerable when storage systems are of great size. The cost of the heat-carrying material and of the reservoir needed is all the more prohibitive because the system is basically inter-seasonal thus hampering pay-back of the capital invested. Water appears to be a relatively propitious medium - its cost is low even when treated and its specific heat by volume is advantageous. On the other hand, water requires reservoirs of high cost that can only be depreciated by a considerable number of operating cycles each year (so-called "daily storage systems").

When temperatures considerably exceed 100 °C, pressures increase rapidly and technical obstacles are encountered when storage systems are of large size. This difficulty can be overcome by storing water under pressure in an underground cavity but, here again, the scale of investment is such that the method can only be considered for "daily" systems.

For inter-seasonal storage, we can only consider a system in which the heat-carrying material and the reservoir are supplied by nature - an underground aquifer. If it is under pressure and contained in a porous and permeable medium, there is nothing to prevent consideration of storage temperatures higher than 100 °C.

A preliminary study has shown the advantages of a high-temperature storage system when a suitable heat source is available. The Atomic Energy Commissariat and the National Elf-Aquitaine Company, associated for the circumstances, have undertaken studies on the feasibility of a high-temperature heat storage system (at 180-200 °C) in a deep aquifer.

2. ADVANTAGES OF HIGH TEMPERATURES

It might be thought, intuitively, that thermal losses increase with temperature. This is not certain even in the case of a surface reservoir. Let us consider, for example, an insulated reservoir containing water at 50 °C when the ambient temperature is 0 °C. If we place water at 100 °C in the same reservoir, the losses are multiplied by 2 for any given storage time. But the reservoir might be smaller if we wanted to store a given quantity of energy. For a constant thickness of insulation, losses only increase by 26%. If we accept to use the same weight of insulation in both cases, the insulation of the smaller 100 °C reservoir can be thicker. Calculation then shows that losses are reduced by 21% when compared with those of the 50 °C reservoir.

The theoretical efficiency, the relation between heat remaining in the reservoir at any given moment with heat introduced, may be determined from the calculated losses. This efficiency is equal to 1 at the start and increases with time. It is of little interest to the user. The practical efficiency, when heat is being recovered, may be determined as a function of time - the ratio between heat recovered and heat stored. This efficiency is nil at the start and remains so during the whole system charging phase. It then increases, when heat is being recovered, though it can never exceed the theoretical efficiency because it is never possible to recover the heat lost due to thermal losses. With this reserve, the practical efficiency only depends on the operating conditions and, particularly, on the minimum recovery temperature. We see, for example, that if the latter is higher than the charging temperature, no heat can be recovered - the practical efficiency is nil whatever the theoretical efficiency at any moment. This double notion of efficiency is the reason why two types of storage systems, and only two, are economically advantageous:

1) when operating at very low temperature that requires a heat pump, there is every advantage in storing at a temperature as low as possible; the fact that the practical efficiency imposes no limit leads to research for the maximum theoretical efficiency that is attained at low temperature;

2) when operating at a temperature noticeably higher than ambient temperature, there is advantage in storing at high temperature in order to raise the practical efficiency.

There are three limits on such use of high temperatures:

1) an appropriate thermal source is needed;

2) certain technical thresholds may not be exceeded;

3) there is no need to go further when the practical efficiency

reaches the theoretical efficiency at the end of the cycle - the theoretical efficiency would be lowered and, because it cannot be exceeded, would lead to a correlative drop of the practical efficiency.

These considerations have guided the choice of temperatures at about 180 °C which does not exclude slightly lower figures if imposed by the available source.

In the precise case of storage in aquifers, the problem becomes complicated if we consider several successive cycles because the losses of the previous years that are only evacuated very slowly constitute a "guard ring" around subsequent stocks if the system is correctly designed. We see in every case that there is a convergence of efficiencies to a fairly high value that is practically independent of temperature, after a sufficient number of cycles. But such convergence occurs very rapidly when temperature is high because of the poor performance during the first cycle whereas, at low temperature, a certain degree of equilibrium is only attained after a time that may exceed the system lifetime.

The use of high temperatures also provides a means of reducing the size of the system by a factor of about 10 for the same energy flow while the permeability of the formation is increased by about 80% when passing from 45 to 180 °C in spite of a certain limited clogging. These are factors that cannot be disregarded when the reservoir is provided free of charge. On the other hand, drillings must be sufficiently deep and sophisticated equipment is needed. On account of the effect of size, these drawbacks may appear to be minor when the systems are sufficiently large (several tens of GWh).

3. OVERVIEW OF SYSTEM

The system being considered is quite different from the traditional geothermal pair of wells because of the need to control the position of the stock and of its "guard ring".

It consists of a central well reserved for high-temperature transit (admission or recovery) and at least three peripheral wells operating at temperatures near to return temperature. During the summer, water is pumped from the periphery, heated on the surface and reinjected into the center. Heat is mainly stored in sand and, accessorily, in the interstitial water. During the winter, the water circulates in the reverse direction and heat can be recovered on the surface.

The insertion of heat-exchangers provides a means of totally insulating the industrial water circuits from the water in the aquifer. Various devices placed on the surface or in the wells prevent vaporization of the water and unwanted chemical precipitation.

A prototype system is being built. The site of PLAISER in the western suburbs of PARIS near to VERSAILLES, was selected after rigorous prospection because it has several superimposed water tables than are quite distinct and at reasonable depths allowing selection of the one expected to be the most informative. This is a scientific test necessarily on a large scale with which can be combined a true demonstration while remaining within true operating conditions and which can be made to produce profits if successful. There is, on the site, an incineration plant that can be equipped with heat-exchangers and, in the immediate vicinity, several housing units with collective heating systems that can be combined together. This represents 4,000 to 5,000 homes in all.

4. EXPLORATION BOREHOLE

Thanks to the assistance of the European Community Commission, an exploration drilling has been carried out in order to confirm the suitability of the site for a storage test, to check the assumptions relating to general feasibility by comparing them with a true case, to dimension the demonstration facility and to attempt to forecast its behavior. The horehole can be used as a peripheral well for the demonstration facility. It has allowed samples of cores and of water to be taken, pumping tests to be carried out and logs to be prepared. Simulation processes have been finalized in order to achieve the targets based on the data supplied.

The borehole encountered the expected aquifers at depths between 356 and 530 meters under an overburden of chalk and of shale. It ended at 545 meters in a layer of limestone. After comparative studies, the upper Wealdian water table was selected and prepared for production. It is about 20 meters thick with cold transmissivity of 65 Darcy-meters. The traditional studies and analyses have been carried out on the cores. An apparatus has moreover been specially designed to replace the non-consolidated sandy cores under conditions of stress near to natural conditions and to provide percolation of pressurized water at temperatures between 20 and 180 °C in order to measure variations of permeability. The logs have allowed direct computation, or comparison, of porosity, specific weight and specific heat at different temperatures. Several conductivities were estimated from the core composition.

5. SIMULATION OF PHYSICAL-CHEMICAL OPERATION

A complete chemical and bacteriological study has been made with great care on the water samples by two specialized laboratories working separately. The physical-chemical behavior of the system has been simulated on the basis of these results and on the core analyses. The study was carried out with the assistance of the Louis Pasteur University at Strasbourg and a digital model has been developed from existing models.

Results have been encouraging with regard to behavior of the formation. During the storage phase the quartz is nourished and calcite dissolves in the zone with a high heat gradient. The reverse occurs during recovery. Because the critical zone sweeps over the whole formation during the time of one cycle, the end result is a very slight non-cumulative modification of porosity and, indirectly, of permeability. According to the calculations, this factor should be negligeable.

A few precautions must nevertheless be taken for the immediate neighborhood of the wells. During the storage phase, dissolution of silica may lead to degradation of the gravel pack of the central well that might reduce the life of the system; on the other hand, any incipient transformation of clay into chlorites should have no harmful effect.

During the recovery phase, the reinjection of silica-saturated water might disturb the surroundings of the peripheral wells - there is little to fear from nourishing of quartz at low temperature but the formation of kaolinite in the presence of free aluminium ions existing in the formation, might lead to irreversible clogging and failure of the operation. Precautions must be taken against this possibility.

The situation is not critical on the surface during the recovery phase, contrary to what happens in the formation near to the wells. Silica oversaturation cannot lead to formation of quartz because the speed of reaction is practically nil at the operating temperature. The temperature would have to be lowered to less than 50 °C in the exchanger before a deposit of amorphous silica would suddenly appear and damage the exchanger but the conditions of operation exclude this possibility. On the other hand, during the storage phase, heating of the water leads to over-saturation of calcite with rapid scaling of the exchanger and a risk of clogging when the precipitate is drawn into the strainer of the central well.

It would therefore be unrealistic to underestimate or to neglect the physical-chemical problems which may be resolved into three risks, in diminishing order of importance:

1) precipitation of calcite in the exchanger during storage;

2) formation of kaolinite near to the peripheral wells during recovery;

3) degradation of the central well by dissolution of silica during storage.

The possibility of clogging by sand or fine particles drawn in mechanically must be added to these risks. These problems are not specific to high-temperature storage systems but have a certain degree of importance in this case. Fortunately, high temperature amplifies the problems but also facilitates their resolution - the limitation on water flowrates for a given thermal flux, the relatively high economic value of each cubic meter of water and, above all, the increase of the speed at which chemical equilibrium is attained permits the consideration of simple methods of chemical balancing which would be unrealistic at lower temperatures. The system projected makes use of a simplified model of traditional water-treatment installations based on precipitation of lime and of magnesium-oxide. It only re-establishes the balances disturbed by temperature variations and is therefore totally inoffensive with regard to the quality of the aquifer water. It is completed by filtration and, possibly, by passing over a sand bed.

6. SIMULATION

The project has been simulated by the STOICA model which is semi-analytic and semi-convective and was specially developed for the project. Based on the formation characteristics supplied by the exploration drilling, on the incineration plant heat production curve and on the thermal requirement curve, the model determines what contribution can be made by the storage system. The model computes recovery temperatures and water flowrates allowing for the various internal or external constraints and electricity consumed for pumping. The temperatures in the water table and in the surrounding formation are determined approximately which simplifies the computations. A highly efficient aquifer model was considered to be useless even though knowledge of a number of parameters relating to the formation is incomplete. The adoption of a simple and semi-analytic type of simulation reduces computation time and a number of simulations could thus be carried out at reasonable cost to study the sensitivity of the results to variations of the more doubtful parameters. The time of simulation could thus be extended. It was in fact evident that the results obtained during the first cycle, generally mediocre or bad, were not representative of the later behavior of the system. All simulations covered four cycles per year which would have been practically impossible with an efficient finite-element model.

Within the scope of the study of which we are speaking, 40 simulations have been carried out - an indispensable minimum. Some 20 other have been performed since the end of the study without modifying its conclusions.This complex model that does not take account only of the aquifer but of the whole source/storage/utilization system has led to interesting and original conclusions.

The results are practically insensitive to variations of the formation parameters around the values deduced from the borehole. This conclusion conforms with the idea expressed above on the function of the "guard ring" due to losses during previous cycles. The conclusion is not unexpected, is reassuring and is striking by its clearness. In most cases, recovery exceeds 50% of heat stocked as from the second year. At the fourth cycle, all the results converge towards a recovery rate between 60 and 70%.

On the other hand, the results are very sensitive to certain operating parameters. The minimum recovery temperature which is controlled by the return temperature is, in particular, a very critical parameter - a drop of 12.5 °C leads to an increase of 10% of the recovery rate.

Power used for pumping is very acceptable and does not exceed a few percent of the thermal energy recovered when taking account both of storage and of recovery, on condition that the pump is sufficiently efficient.

The model has also provided a means of determining the optimum system size. In the case under consideration, the peripheral wells should pass through the aquifer at a distance of 85 meters from the central well.

On parallel lines to simulations with the STOICA model, the impact of the storage system on the environment has been studied and it can be considered as negligeable. A natural gas underground storage reservoir is located at five kilometers from the planned site, in the same aquifer. A simulation performed by Gaz de France, who are operating the storage system, has shown that there is no problem in cohabitation of the two systems.

7. CONCLUSIONS

The various problems likely to be encountered during the demonstration operation have been resolved - thermal, hydraulic and chemical behavior of the formation and of the water; stock stability; constructional and operational technical parameters; operational forecasts and impact on the environment.

The simulations carried out have perhaps not allowed for certain unknown perturbative factors but has shown that only a true-scale test can dispel these uncertainties.

Because the system is still only experimental, the economic balance sheet has no particular significance. An economic conclusion can only be made after a prolonged demonstration operation and costs, at the present moment, do not appear to be absurdly high if success is achieved.

It is therefore possible and legitimate to undertake the projected demonstration operation and the boreholes are to be drilled this year.

References

DESPOIS J. and NOUGAREDE F. (1976). Stockages souterrains de chaleur. Société Française Thermiciens. Colloque Grenoble "Aspect thermique des problèmes énergétiques actuels".

CHARROPPIN P., DESPOIS J., FAUCONNIER J.C., NOUGAREDE F. (1976). Underground heat storage in captive aquifers : technology, economical advantages in relation with nuclear heating plants. First Total Energy Congress - Copenhague.

DESPOIS J. and NOUGAREDE F. (1977). Stockages souterrains de chaleur - Revue Générale Thermique N° 184 - Avril 1977.

DESPOIS J. (1977). Transport et stockage de la chaleur. Revue Générale Nucléaire - Mars/Avril 1977 - N°2.

DESPOIS J. and NOUGAREDE F. (1977). Technologie du stockage souterrain d'eau chaude. Société Electriciens, Electroniciens et Radioélectroniciens. Journées d'étude Gif sur Yvette.

DESPOIS J. and NOUGAREDE F. (1978). Stockage souterrain de la chaleur. Techniques de l'Ingénieur. "Premier Jet". B 269-1/4.

DESPOIS J. and NOUGAREDE F. (1978). Etude de la faisabilité du stockage d'eau chaude dans les nappes souterraines ou dans les cavités creusées dans le roc. Colloque DGRST - Ecole des Mines de Paris - Sophia-Antipolis (12/13-10-78)

DESPOIS J. and NOUGAREDE F. (1979). Faisabilité du stockage souterrain de chaleur : compte rendu de fin d'étude (2 tomes) - CEA/DEMT/Saclay.

DESPOIS J. (1981). Les stockages souterrains de chaleur. Conférence Marly-le-Roi : Chaleur et électricité - situation française - CEA-EDF.

DESPOIS J. (1982). Etude d'un stockage en aquifère captif utilisé comme réservoir géothermal. Rapport final CEE - Contrat EE/E1/403/F.

HOUSSE, B.A. (1983). Prototype of an underground heat storage systeme at Thiverval-Grignon(France). Int. Conf. on subsurface heat storage in theory and practice - Stockholm.

Fig. 1 - Drilling of the exploration well

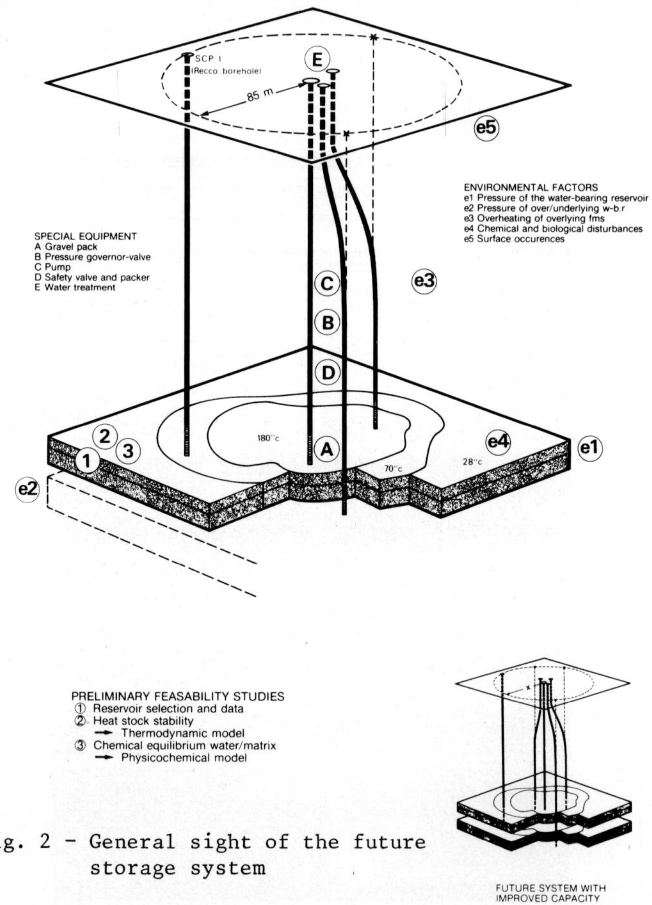

Fig. 2 - General sight of the future storage system

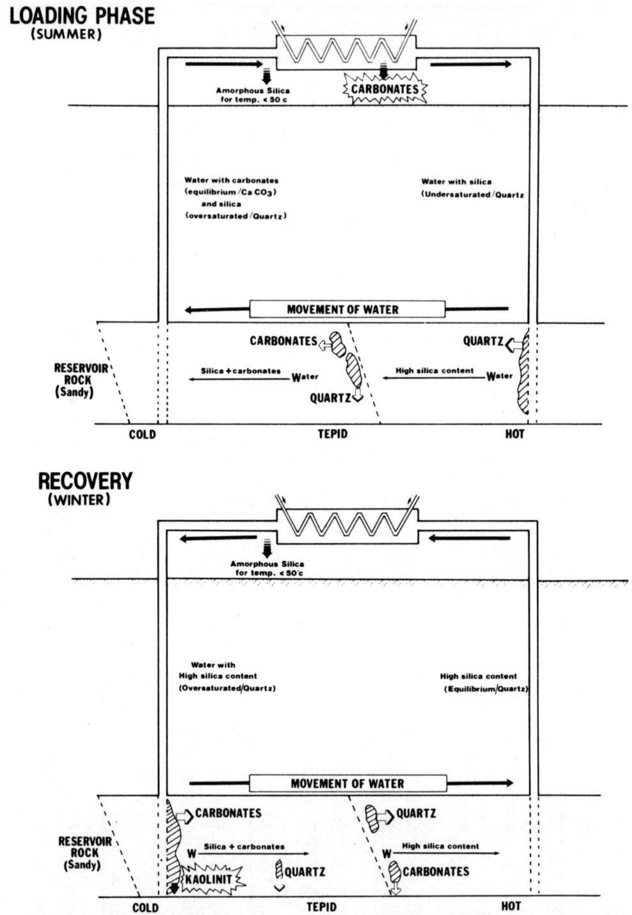

Fig. 3 - Physical-chemical behaviour of the storage system

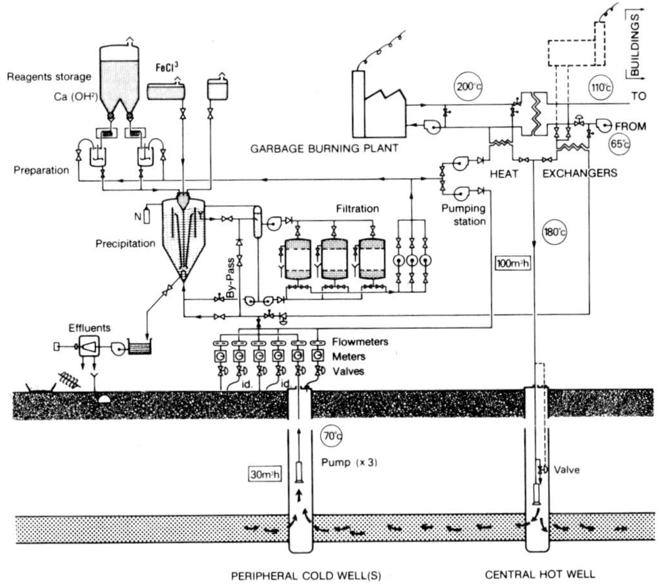

SUMMER

Fig. 4 - Flow-sheet of the future storage system (storage phase)

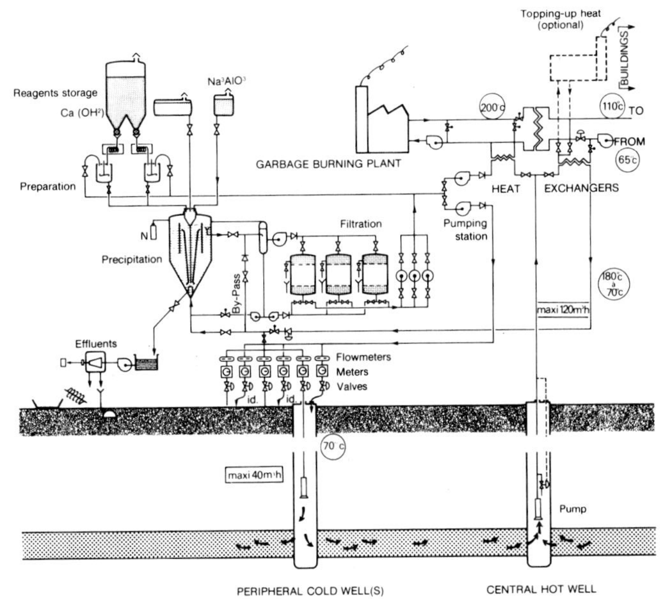

WINTER

Fig. 5 - Flow-sheet of the future storage system (recovery phase)

HEAT STORAGE BY PARAFFIN WAX FUSION FOR SPACE HEATING

P. DUBOIS and M. BERTHET

Laboratoires de Marcoussis
Centre de Recherches de la Compagnie Générale d'Electricité
Route de Nozay
F - 91460 Marcoussis (France)

Tel : (6) 449 12 00

Summary

The reported work relates to heat storage by paraffin wax fusion for water space heating, the heat being supplied by a heat pump. Such a heat store should improve the average performance coefficient of the heat pump due to a smaller number of stops and goes and it should reduce the cost of space heating by shifting electricity consumption from higher to lower tariff hours. The following tasks have been completed :

- research on cheap and thermally efficient ways of paraffin wax packaging,

- theoretical and experimental study of the kinetics of heat transfer between paraffin and water during heat storage and heat discharge,

- design of a heat store for a lodging of 120 m^2,

- test on a water loop of a heat pump with a 1 m^3 heat store containing 0.9 m^3 of paraffin ; the conclusion is that a paraffin heat store operates properly but is unfortunately too expensive for an individual house compared to the electricity savings it provides with today's french electricity tariffs. Heat storage by paraffin fusion would be more interesting, from an economic point of view, in the case of air space heating systems including solar collectors.

0. Introduction

The object of the research was to study a heat store for space heating, in which the storage material is paraffin wax and the heat-transfer fluid is water. A store of this type, combined with a heat pump, should permit :

- an improvement in the APC of the heat pump through a smaller number of stops and goes,

- operation of the heat pump exclusively or mainly during the night, enabling the user to take advantage of more favourable electricity prices.

In the course of a program conducted with the help of the Laboratoire de Thermocinétique of the CNRS at Nantes on modeling and heat transfer kinetics, we have :

- studied and developed high-performance and inexpensive technologies for the encapsulation of paraffin wax,

- modelled the behaviour of stores of different designs and studied the storage and discharge kinetics experimentally,

- designed a store of the appropriate size for a space of 300 m^3 based on the results of stimulations over one heating season, and we have built and tested, on a water test-bench, a 1 m^3 store containing 0.9 m^3 of paraffin associated to a heat pump,

- assessed the economic value of paraffin stores combined with heat pumps for space heating of individual houses.

1. Study of paraffin wax encapsulation technologies

We have fully investigated the encapsulation of paraffin by plastics. We studied encapsulation by rigid plastics and encapsulation by flexible plastics, which permits variations in specific volume of the paraffin without inducing significant mechanical stresses.

The encapsulation materials which could be used were selected following a variety of tests for compatibility and ageing in the presence of paraffin by performing, on the one hand, heat cycles between room temperature and 70 °C and, on the other hand, prolonged periods of residence at 90 °C.

11. Encapsulation with rigid plastics

Our work enabled us to select two rigid plastic encapsulation materials : polypropylene and chlorinated polyvinyl chloride (chlorinated PVC).

a) We showed that the mechanical properties under traction of test-pieces of these two materials did not change significantly following a period of residence of 3,000 hours in paraffin at 90 °C, the values being as follows :

Characteristics at 20 °C	Polypropylene	Chlorinated PVC
Stress at the flow threshold	3 kgf/mm^2	6.8 kgf/mm^2
Elongation at the flow threshold	22 %	7 %
Modulus of elasticity	40 kgf/mm^2	250 kgf/mm^2

Attention is drawn to the greater mechanical strength of chlorinated PVC compared with polypropylene, and to the fact that these two plastics have a good heat stability up to temperatures of the order of 100 °C.

b) From the point of view of the permeability to paraffin, our tests showed that :

- polypropylene, which has a good impermeability up to 70 °C, allows the diffusion, above this temperature, of some light fractions contained in paraffins. This phenomenon, which is also encountered at lower temperatures with polyethylene, caused us to eliminate polyethylene for paraffin encapsulation,

- chlorinated PVC remains perfectly impermeable at 90 °C.

c) We studied techniques for closing tubes at the ends and we showed that this could be done :

- by hot-pinching and welding between jaws in the case of polypropylene,

- by gluing or rotation welding of moulded plugs in the case of chlorinated PCV.

d) Because of the conjugated effect of the large variation in volume of paraffin during melting (15 %) and of the external pressure exerted by the water in the heating water loop (a few bars), the wall thickness of the tubes must be sufficient to prevent the tube from being crushed by buckling. In the case of chlorinated PVC, which has the higher mechanical properties, this makes it necessary to use wall thicknesses of 3 or 5 mm, according to the diameter of the tubes, 50 or 100 mm.

12. Encapsulation with flexible plastics

Following impermeability and ageing tests carried out on several types of monolayer or multilayer films, we showed that the best results were obtained with a triple-layer complex film comprising :

- a 120 µm thick film of high-density polyethylene, which ensures weldability, on the inside,

- a 12 to 24 µm thick film of polyester, which imparts impermeability and improves the mechanical strength,

- a bonding adhesive chosen for its good heat stability (up to 120 °C).

The impermeability, the cohesion, the mechanical properties and the chemical characteristics of a film of this type did not change after a period of residence of 5,000 hours at 90 °C in paraffin. Flow tests under traction, which were carried out on dumb-bell test-pieces of the H1 type, show that, after 1,700 hours in paraffin at 55 °C, a multilayer film including 12 µm of polyester flows 2.5 % under a load of 1.15 kgf/mm^2, and that the recovery after the load is removed is 40 %. As this load of 0.15 kgf/mm^2 corresponds to the stress exerted by the hydrostatic pressure of the liquid paraffin contained in a flexible tube having a diameter of 50 mm and a height of 1 meter, placed vertically in the air, we see that the use of a multilayer film containing a 24 µm polyester film would be desirable in the case of a paraffin heat store with air as the heat-transfer fluid. In the case of a paraffin heat store with water as the heat-transfer fluid associated to a heat pump, the external overpressure exerted by the water eliminates this problem.

These complex 3 layers thin films can be used in a simple manner with the aid of a heat-sealing machine, which makes it possible to produce elements having a very wide variety of shapes : tubular "sausages", mattresses consisting of adjacent tubes, and the like. These advantages, together with the moderate cost of the film, namely 3 to 4 F per m^2, make using a flexible film an attractive method of paraffin encapsulation.

2. Studies of the storage and discharge kinetics

21. Experimental studies on the storage and discharge kinetics

In the chain of heat transfers between the water and the core of the storage material, the heat exchanges with the water are high, namely of the order of 100 to 200 $W/m^2°C$, by comparison with the other thermal resistances. This led us to direct the heat transfer study in the structures to the heat transfer within the paraffin and its encapsulation and at the interface of paraffin and encapsulation.

We studied experimentally the thermal behaviour of :

- structures comprising cylindrical tubes containing the paraffin, the heat-transfer fluid circulating outside these elements,

- structures comprising parallelepipedal elements containing the paraffin, the heat transfer fluid circulating outside these elements.

211. Experiments on tubes

a) The experiments carried out on single tubes placed in a water duct and cycled between 30 °C and 60 °C showed that, for a tube of Ø 116 mm, the insertion of internal heat drains, in the form of aluminium fins occupying 8.35 % of the volume, improves the exchange capacity with the heat-transfer fluid in the first hours (cf. figure 1) : the average storage rate over 2 hours is thus increased by 17 %.

b) Experiments were carried out on a nest of 52 polypropylène tubes of 46/50 mm diameter and 1 meter length : they showed that the vertical or horizontal position of the tubes considerably changed the transitory distribution of the temperatures within the paraffin, because of the convection currents, but did not substantially modify the overall storage and discharge rates and capacities.

212. Experiments on parallelepipedal structures

We studied the fusion kinetics of paraffin contained in a tank of 500 x 500 x 60 mm, receiving a constant energy flux over one of its large face. It demonstrated the benefits of horizontal internal partitions, which permit a better distribution of the fusion of the paraffin between the top and the bottom of the element. From the energy point of view, the curves of figure 2 show that this effect results in a 20 % increase in the energy stored over 6 hours by changing the shape ratio height/thickness of the paraffin cavities of 8 (initial tank) to 2 (tank within 3 partitions). A further decrease of this shape ratio to unity does not improve heat transfers. These experiments were carried out with ARCOWAX paraffin, the characteristics of which are given in figure 3.

22. Modeling of the thermal behaviour of storage elements

Two models were developed : one related to encapsulating tubes and the other related to parallelepipedal elements filled with paraffin.

221 Model of tubes

This model, in pure conduction, is of "enthalpic" type, in which the heat accompanying a change of state is taken into account in the form of an apparent heat capacity varying with the temperature, such as is determined by differential enthalpic analysis measurements. The numerical determination is carried out by the "Crank Nicholson" method and by using a ratio of the intervals of axial space and of time which is close to the speed of the fluid.

This model was compared with the experiments carried out on single tubes and on a nest of 52 tubes. In both cases, we found a very good agreement as regards the change in the temperatures of the fluid passing through the store and in the paraffin temperatures in the central region of the tubes. Differences between the calculated temperatures and the measured temperatures were found at the ends of the tubes, because of the contraction on solidification of the paraffin in these regions. The overall energy behaviour of the nest of tubes is well simulated by the model.

222 Model of parallelepipedal elements

The Laboratoire de Thermocinétique de Nantes (CNRS) developed a model of a parallelepipedal paraffin element heated on one of its large faces. The model uses two units dealing successively with the different heat transfer conditions in the solid or liquid paraffin :

- one unit for pure conduction in the solid and in the liquid,

- one unit taking into account convection in paraffin. For boundary layers, a one-dimensional solution is based on an equivalent rectangular liquid cavity, the thickness of which is equal to the abscissa of the fusion frontin the median part. The change in the fusion front is represented as the removal of the solid zone under the effect of a convection flux in the liquid.

The experiment showed that this model represented better than a purely conducting model the thermal behaviour in the presence of convection movements during storage.

3. Design studies of a paraffin-water store

31. Choice of the heat exchange area between water and paraffin

Using the simulation models developed, we compared the values of heat stored and the values of electrical energy consumed as a function of the exchange area between water and the paraffin, both in the case of tubes and in the case of parallelepipedal elements containing paraffin. This comparison was made when charging the store during 6 hours by means of a 2.5 kW heat pump operating at full power.

The following conclusions were made :

- in the case of the tubular structures, it is not advantageous to use tubes having a diameter of less than 50 mm, and the internal heat drains can only be justified for tube diameters greater than or equal to 100 mm. Figure 4 shows the effect of the exchange area and of the heat drains on the energy stored and on the performance coefficient of the heat pump,

- in the case of parallelepipedal elements, the thickness of paraffin between the exchange walls must be between 60 and 100 mm and that leads to 30 to 50 m^2 of exchange area for a store of 1.5 m^3. The gain in terms of heat stored is only 10 % when changing from 50 to 300 m^3 (cf. figure 5).

32. Choice of the overall size of a store associated with a heat pump for a house space heating

We used a model of house space heating by a system including a heat pump and a heat store in parallel with the heat pump, the heat being supplied to floor heating pipes or to water radiators. This model was applied to the case of a 300 m^3 house requiring 300 W/°C and located in the Paris region. We considered an external air/water heat pump which operated above -5 °C outside temperature, with a water output temperature not exceeding 62 °C.

The model takes into account the solar contributions and the outside temperatures on an hourly basis as well as the regulation criteria on the heating system, the thermal inertia of the house and the dual electricity tariff :

- during 16 higher tariff hours : 0.378 FF/kWh

- during 8 lower tariff hours : 0.220 FF/kWh.

We computed operating costs over a full heating season for several sizes of store and of heat pump and for several heat system regulation criteria. The conclusions are as follows :

- there is no significant difference in the gains in terms of operating costs between the heating systems with heated floors and the systems with radiators,

- with a 2.5 kW compressor heat pump, the gain in terms of operating costs provides by the installation of a store shows virtually no further increase beyond 1 m^3 of store (cf. figure 6),

- an enlargement of the heat pump to 4 kW would lead to greater financial gains, but the additional annual savings made by enlarging the pump and the associated store does not appear to justify the corresponding extra investments,

- when considering a store of 1 m^3 of paraffin melting at 47 °C, associated with a 2.5 kW electrical heat pump, the annual gain in terms of operating costs provided by the store amounts to 436 FF under these conditions, that is to say 19 % of the initial expenditure on electricity (cf. figure 6).

4. Design of a paraffin store and construction of a 1 m^3 unit

41. Choice of the store structure

a) Analysis of various structures for the store led us to the following conclusions :

- the structures in which the storage material is encapsulated in elements which are immersed in a water tank imply constraints of a mechanical nature, which are reflected in the cost :
  . the tank must withstand the pressure of the water circuit, which is a few bars,

  . a rigid encapsulation must be sufficiently thick to withstand mechanically the differential pressure between the outside and the inside of the tubes, which leads to a substantial weight of encapsulating material,

  . a flexible encapsulation requires supporting structures, because of the buoyancy of paraffin in water,

  . the structures in which the heating water flows in exchangers immersed in the paraffin are better suited from a machanical point of view : the use of a thin walled parallelepipedal tank filled with paraffin is thus possible.

b) We compared from an economic point of view several possible designs of exchangers for stores with a paraffin tank. This analysis les us to choose elements made of roll-bond aluminium, the purchase price of which, when used on a rather large scale, would be less than 100 F/m^2.

42. Design of a store and construction of a 1 m3 unit

a) We designed a store for a house of 300 m^3 requiring 300 W/°C for space heating. This store, which contains about 0.9 m^3 of paraffin, is shown on figure 7. Its main characteristics are as follow :

- volume of paraffin : 0.9 m^3,

- exchanger consisting of eight 1,900 x 750 mm sheets of roll-bound aluminium, 61.5 mm apart, giving a wet area of 22.8 m^2,

- parallelepipedal tank : width 860 mm, depth 540 mm, height 2,050 mm,

- thermal insulation by means of 50 mm molystyrene slabs.

b) A 1 m^3 unit was constructed using this technology and filled with 690 kg of TOTAL Inter B paraffin. The photograph on picture 8 shows a view of this store, which was fitted and tested on a bench in parallel with an external air/water heat pump.

5. Tests of the paraffin/water 1 m^3 store

The paraffin/water sotre, described above, which contains about 0.9 m^3 of paraffin, was installed and tested on a bench comprising an external air/water heat pump with a 2.5 kW compressor.

51. Description of the test bench

The test bench makes it possible to reproduce charge/discharge cycles of the store under conditions representing those of an actuel house. The heating water loop, of which the simplified diagrams corresponding to the heat storage and heat discharge phases are given respectively on figures 9 and 10, includes an exchanger with a cooling water loop, which makes it possible to simulate the heating needs of the house by removing a regulated amount of heat from the water loop to which heat is supplied by the heat pump and by the store. The modes of operation representing the different periods in a day are as follows :

- in the lower tariff period (8 hours), the pump covers the heat needs of the house and at the same time ensures the charging of the store with the excess heat power available.

The motorised "three-way" valve VI, controlled by a temperature regulator distributes the flow created by the circulator C1 between the exchanger and the store, so as to keep the return temperature TR from the exchanger at a set value which varies with the external temperature according to the heating curve of the radiators for the house. The heat pump operates continuously up to the end of the lower tariff period or until the temperature of water leaving the heat pump reaches the limit of 62 °C, when the store is full. In the latter case the heat pump covers the heat needs of the house, with control by the value of TR ;

- in the higher tariff period (16 hours), the heat needs are covered by store discharge. The motorised valve V2 distributes the flow created by the circulator C2 between the store and a by-pass, so that the return temperature TR is maintained at the set value which varies with the outside temperature according to the heating curve of the radiators for the house. This heating by store discharge only is maintained as long as the discharge power can supply the needs. Beyond this, in order to utilise the store more completely the needs are covered with the aid of alternate sequances of heat pump operation and heat discharge from the store. As heat discharge is then insufficient to cover all the needs, fluctuations in the return temperature TR are allowed between two extreme values T_{lower} and T_{upper} on either side of the set temperature. The values T_{lower} and T_{upper} were chosen so as correspond to variations in the ambient temperature of the accomodation which were compatible with a good degree of comfort : $\pm$ 1 °C relative to ambient temperature, which corresponds to $\pm$ 2 to 3 °C relative to the heating water return temperature when the external temperature is 1 °C.

52. Conditions of store testing

a) We carried out tests for three successive periods of 24 hours, 48 hours and 24 hours. By dividing the annual curve of the cumulative frequencies of the daily average of the external temperature into three regions, the following values of the external temperature were computed as representing one heating season in the Paris region :

- 13 °C for 24 hours,
- 8 °C for 48 hours,
- 1 °C for 24 hours.

B) During each of these three periods :

- the temperature of the climatic chamber simulating outside conditions was kept constant, as well as the thermal power removed from the water, which represents the needs of the accomodation, this being a known function of the external temperature for a given house,

- that part of the heat produced by the pump which does not have to be used immediately for heating the accomodation was stored during the lower tariff hours.

c) The values of the heat power representing the space heating needs and the values of the temperature of the water returning from the radiators are given for space heating needs of 300 W/°C for a house with an internal temperature of 19 °C and a non-heating temperature of 15 °C :

External temperature	Space heating needs	Temperature of water from the radiators
1 °C	4.2 kW	48 °C
8 °C	2.1 kW	33,5 °C
13 °C	0.6 kW	23 °C

53. Results of tests

The tests carried out showed that :

- for an external temperature of + 13 °C, the transfer of electricity consumption by the heat pump from the higher tariff hours to the lower tariff hours was 100 %,

- for an external temperature of + 8 °C, the store alone provided 2600 W during seven hours, 2100 W needed for space heating and 500 W to cover water loop heat losses,

- for an external temperature of + 1 °C, the store covered the needs during higher tariff hours to the extent of 22 %,

- the average performance coefficient of the heat pump over the whole of these four days was 2.42,

- the distribution of the electricity consumption of 118 kWh was :

. 73.7 % in lower tariff hours (87 kWh),

. 26.3 % in higher tariff hours (31 kWh).

The electricity consumptionof 118 kWh would cost 30.8 FF for the four days in question, taking the higher/lower tariffs into account.

54. Comparison with heating by means of a heat pump without a heat store

We calculated the electricity consumption by the heat pump which, for the four typical days in question, would provide the same service in the absence of storage. To do this, we used the characteristics of the heat pump which we had measured on the bench for various conditions of external temperature, water return temperature and stops and goes of the heat pump. These calculations show that :

- the average performance coefficient of the heat pump would then be 2.35 ; the electricity consumption for the four days in question would increase from 118 to 121.5 kWh,

- the cost of the electricity consumed would increase to 81 kWh x 0.378 + 40.5 kWh x 0.22 = 39.5 FF.

The installation of the heat store therefore makes it possible, under the conditions of the experiment, to make a 3 % gain in the performance coefficient of the heat pump and to reduce the operating cost from 39.5 to 30.8 FF. The gain extrapolated over a heating season of 210 days would be of the order of $210 \times \frac{8.7}{4} =$ 456 FF, very close to the computed value of 436 FF (cf. 32).

6. Economics of paraffin heat stores

61. Paraffin store and water heat pump

We calculated that the factory cost price of a paraffin store of $0.9 m^3$ constructed with the technology described in § 51, is about 5,000 FF excluding the regulating system.

This cost is broken down in the following way :

- 700 kg of paraffin at 3.7 FF/kg	2,590 FF
- 8 exchange plates made of roll-bond aluminium	1,100 FF
- tank	1,300 FF
- heat insulation	80 FF
- water collectors	90 FF
- fitting	122 FF
Factory cost price	5,282 FF

The sale price of a store of this type would therefore be between 10,000 FF and 15,000 FF, to which the cost of the regulating system would be added.

In view of the annual electricity savings, which in France are only about 500 FF, with the current peak/off tariff, this evaluation leads to a negative judgement as regards the economic advantage of a paraffin heat store in the case of a heating systems with water heat pump for a 300 m^3 house.

62. Paraffin store for space air heating with solar collectors

Based on the results obtained on paraffin encapsulation technologies and on modelling the thermal performance of a paraffin store, we have showed that a paraffin store on a air loop for solar heat collection and space heating should constitute and advantageous alternative solution to the usual pebbles bed heat store. We showed that a store with 0.75 m^3 of paraffin weighing 600 kg would provide the same service as a 1.5 m^3 store of pebbles weighing 2,700 kg. Such weight and volume reductions make it possible, without mechanical reinforcement of the structures of the house, to install the store within the heated space of the house, the location being chosen to reduce the length of air ducts connecting the heat store to the solar collectors.

7. Conclusion

We have designed and tested a 1 m^3 paraffin/water store associated with a heat pump for space heating. These experiments have led to positive technical results. Taking into account the performances of such a system over one heating season, such as they have been determined by computations and proved by experiments, we evaluated, with the electricity tariffs existing in France today, the economic advantage of such a heat store associated with a heat pump. It leads to a negative judgement as far as space heating of an individual house is concerned.

Based on the results obtained on paraffin encapsulation technologies and on modelling of the thermal performances of a paraffin store, we have shown that a paraffin store combined with an air loop for solar heat collection and space heating offers advantages over conventional pebbles bed heat stores.

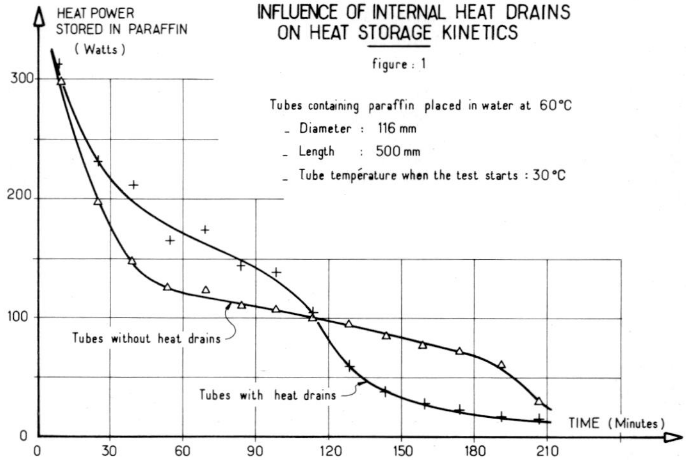

– INFLUENCE OF INTERNAL HEAT DRAINS AND OF HEAT EXCHANGE AREA BETWEEN WATER AND PARAFFIN ON THE VALUES OF STORED HEAT AND C.O.P

figure : 2

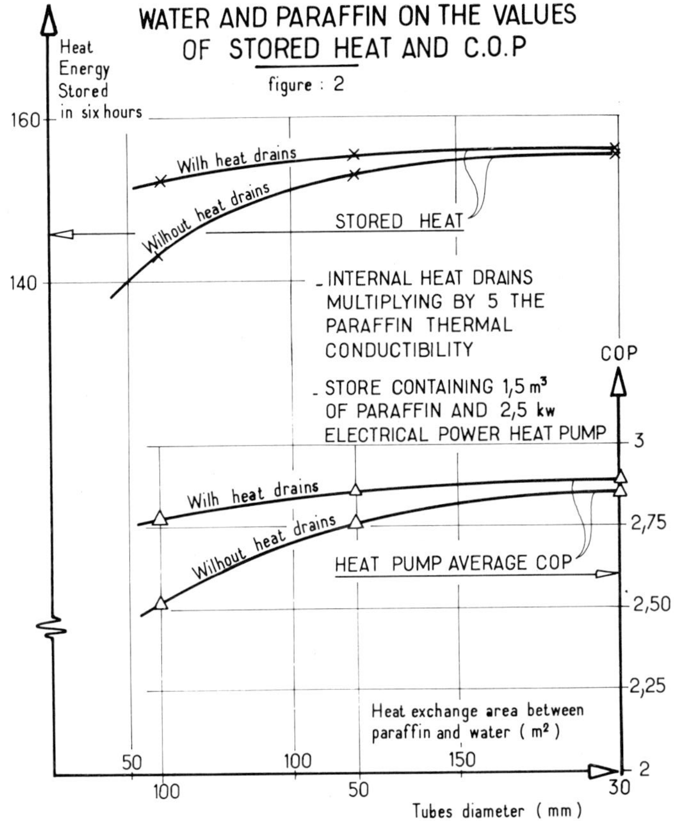

Figure 3

CARACTERISTICS OF PARAFFIN WAX

1) PARAFFIN ARCOVAX 2147 G USED FOR ELEMENTS TESTS

. Supplier : Societe Française des Petroles BP
10 Quai Paul Doumer - 92412 Courbevoie

. Caracteristics

Fusion peak temperature	47°C
Solidification peak temperature	46°C
Fusion enthalpy	32,6 $cal.g^{-1}$
Solid specific heat at 5°C	0,50 cal/°Cg
Liquid specific heat at 60°C	0,54 cal/°Cg
Solid density at 10°C	0,885 g/cm^3
Liquid density at 60°C	0,797 g/cm^3
Thermal conductibility	0,2 W/m°C

2) PARAFFIN INTER B USED FOR THE 1 m3 STORE

. Supplier : Compagnie Française de Raffinage TOTAL
39 à 44 Quai André Citroën - 75739 Paris Cedex 15

. Caracteristics

Fusion peak	53,5°C
Solidification peak	51°C
Fusion enthalpy	39 $cal.g^{-1}$
Solid specific heat at 5°C	0,45 cal/°Cg
Liquid specific heat at 60°C	0,50 cal/°Cg
Density at 10°C (solid)	0,913 g/cm^3
Density at 60°C (liquid)	0,799 g/cm^3
Heat conductibility	0,2 W/m°C

_ INFLUENCE OF THE HEIGHT/THICKNESS RATIO ON HEAT STORAGE KINETICS_

figure : 4

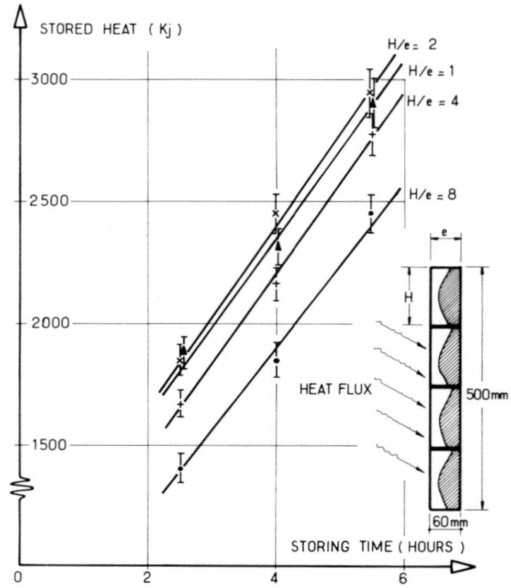

INFLUENCE OF HEAT EXCHANGE AREA ON HEAT STORED AND ELECTRICITY CONSUMPTION _

figure : 5

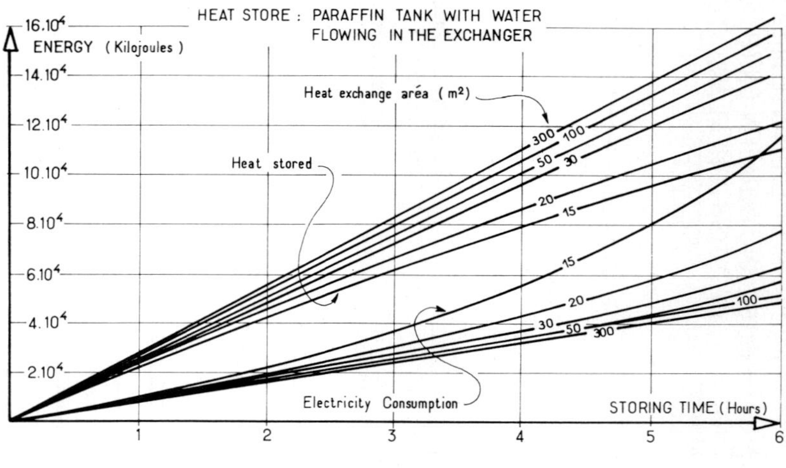

_ SPACE HEATING BY HEAT PUMP AND PARAFFIN HEAT STORE _

OPERATING COST

figure : 6

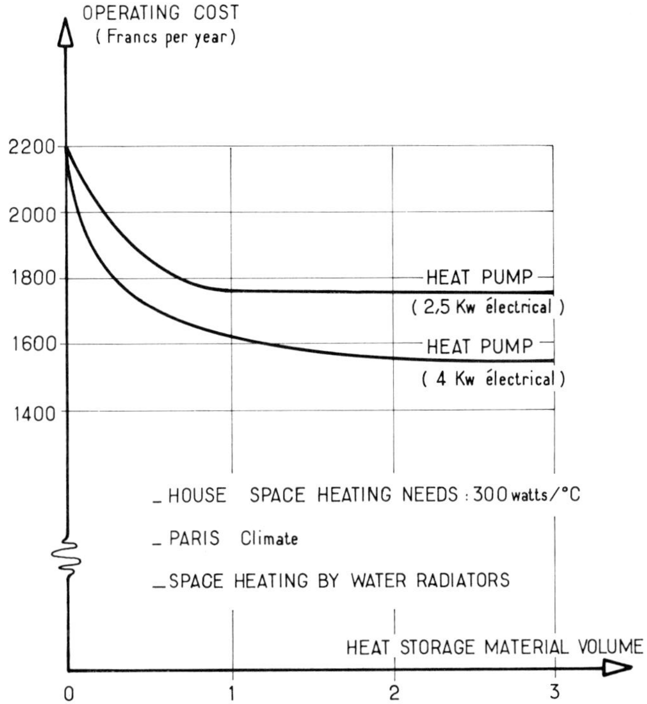

PARAFFIN WATER HEAT STORE WITH ROLL-BOND ALUMINUM WATER HEAT EXCHANGERS _

figure : 7

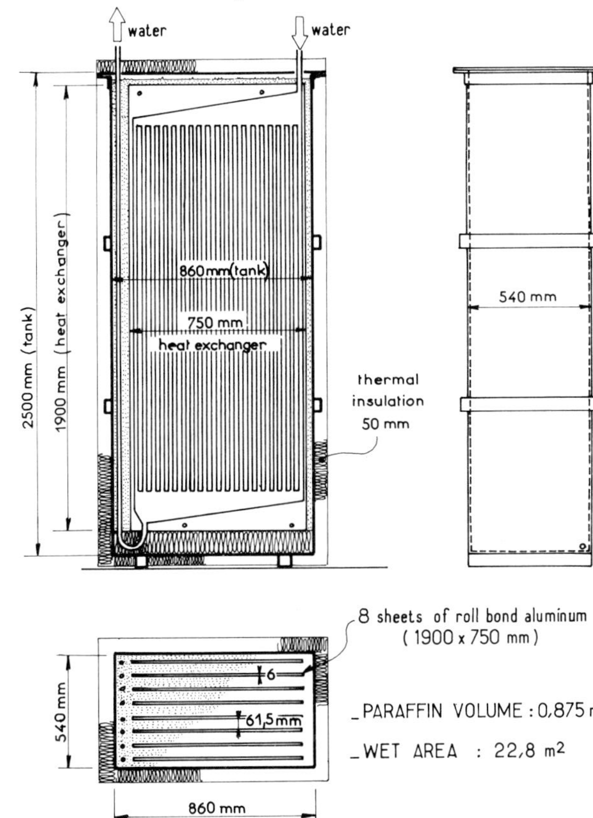

Figure 8 Paraffin store seen from above showing the water heat exchanger made of sheets of roll-bond aluminium

TEST BENCH WITH HEAT PUMP AND WATER PARAFFIN HEAT STORE

HEAT STORAGE

figure: 9

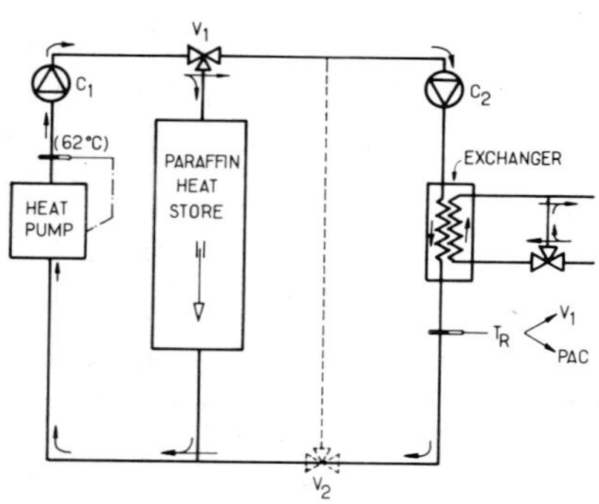

TEST BENCH WITH HEAT PUMP AND WATER PARAFFIN HEAT STORE

HEAT DISCHARGE

figure : 10

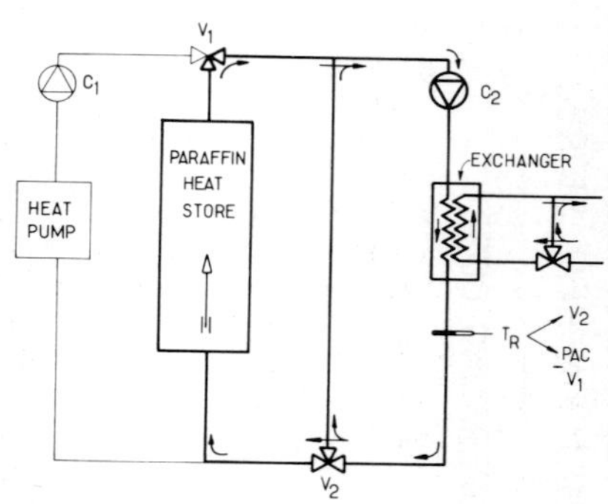

ENERGY SAVING IN THE COMMERCIAL CATERING MARKET

T. G. ENGA
British Gas Corporation, R&D Division, Watson House

Summary

The cost of cooking a meal is comprised of food costs, labour charges, overheads and energy costs, the last item being a very small part of total costs. However, significant savings in energy consumption can be easily made using existing designs. Watson House has carried out a study, co-funded by the Commission of the European Economic Communities, into ways of saving energy in commercial catering. Areas of appliance design, control and usage have been investigated and the potential for waste heat recovery from kitchens assessed. The catering gas consumption could be reduced by 3% if all gas catering appliances were optimised. A further 2% saving could be made by adopting the best currently available designs. Improvements in kitchen management and appliance usage could reduce consumption by another 5% and waste heat recovery equivalent to 3% of gas consumption is feasible for larger kitchens in hotels, restaurants, canteens and institutions. The potential benefits of this study amounts to gas savings of 6500TJ for the UK and 31,000 TJ for the EEC or over 12% of catering gas consumption if all the findings could be implemented.

1. INTRODUCTION

Most catering appliances on sale today have not been designed specifically to be efficient. This situation has arisen because chefs and caterers are more concerned with appliance performance and capital costs than running costs. The rapid rise in fuel prices has been matched by the rise in food and labour costs, leaving the proportion of fuel expenses per meal still negligible. Any controls that are fitted to the appliances are generally to meet safety standards or to improve performance. The traditional caterer still clings to conservative designs and greets new energy efficient concepts with suspicion. Consequently, there has been little acceptance of appliances offering significant energy savings.

This situation need not continue. Unlike most domestic cooking, catering is based on international cooking concepts and training - certainly within Europe and North America. Eating out fashions spread rapidly from one country to another. The necessary appliances and training moving with each fashion. Internationally, kitchens have many common features. Thus energy saving measures, whether based on appliance design, servicing or usage could be applied throughout the EEC. The negligible savings per meal add up to massive potential savings globally, while the market for new designs need not be limited by the specific eating styles of any one country.

To devise measures which might reduce the energy consumed by gas-fuelled catering appliances, the Commission of the European Communities has co-funded a research programme with the British Gas Corporation. The objectives of this research programme have been to investigate areas of appliance design, control and usage so that potential cost effective ways

Open top range   Solid top range

(range = hotplate and natural draught oven)

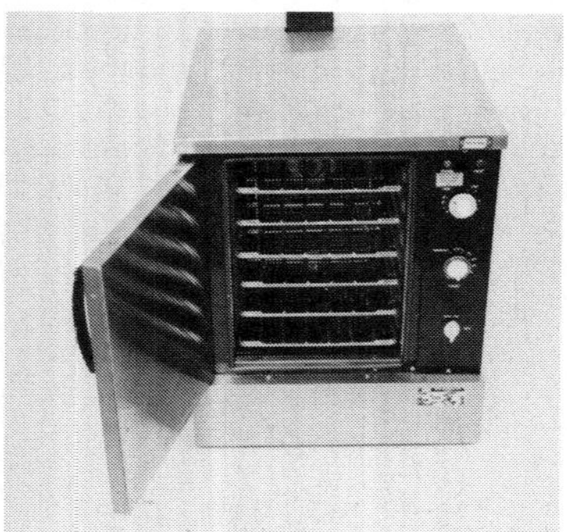

Forced convection oven

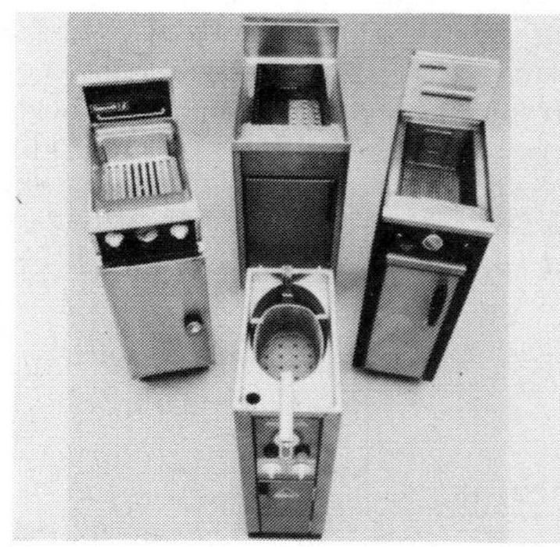

Deep fat fryers

Grill

Griddle

Figure 1: A Selection of catering appliances

of reducing energy consumption can be identified. The data from this study could then be available to stimulate both appliance manufacturers and users into adopting the proposals.

To assess the present levels of appliance efficiency and usage the programme was divided into two parts. Firstly a number of catering appliances were laboratory tested for performance and efficiency and secondly several commercial catering kitchens were monitored to determine usage. This paper presents these results and the economic implications of them.

2. METHODS OF INVESTIGATION

It is very difficult to give a precise definition of efficiency when describing catering appliances. The quantity of heat required to cook food depends on a number of inter-related parameters. Heat transfer takes place via the burner and other parts of the appliance, a cooking vessel and usually a liquid or gaseous medium. The rate at which it proceeds varies according to the cooking method being employed (boiling, frying, baking, grilling etc.) and the thermal properties of the food. It is necessary to maintain the food at the required temperature long enough for the physical and chemical changes involved in the cooking process, but only a small part of the heat supplied might be retained at the end of the cooking period. Indeed if the food were eaten cold it could be argued that the overall efficiency was zero. However, this is clearly not so, because at least part of the heat supplied will have performed a useful function before dissipation.

Inefficiency in catering can occur because of appliance problems such as lack of optimisation or poor design both of which were investigated in the laboratory. It can also occur if the appliances are used badly and this aspect was studied in various monitoring exercises. The overall efficiency of a kitchen might be improved if the waste heat normally expelled through the kitchen air extractors could be recovered. This too was assessed during the kitchen monitoring.

3. LABORATORY INVESTIGATION

Only the most widely used catering appliances were purchased for study in the laboratory. These were hotplates (solid and open top) ovens (roasting and forced convection), fryers, grills and griddles (See Figure 1).

3.1. Appliance Efficiency In order to determine which appliances used the least energy in relation to their cooking performance, beefburgers were cooked on each. The results under full load conditions yielded direct comparisons between hotplate, oven and grill efficiencies. Forced convection ovens were found to be the most efficient, natural convection roasting ovens and hotplates were roughly equal in second place and grills proved least efficient. However the relative positions of ovens and hotplates would be altered in favour of ovens for longer cooking processes (such as casseroles), where the gas used for oven preheating would be less of a disadvantage to the process.

3.2. Open Top Hotplates Hotplates were examined closely because of their popularity and because their gross efficiency can easily be determined. These hotplates were usually found unoptimised on delivery. Figure 2 shows the effect of optimising the burner to pan separation yielding on average an improvement of about 3 to 6 percentage points of efficiency. Redesigning the pan support to limit flame impingement improved hotplate efficiency by up to 4 percentage points (Figure 3).

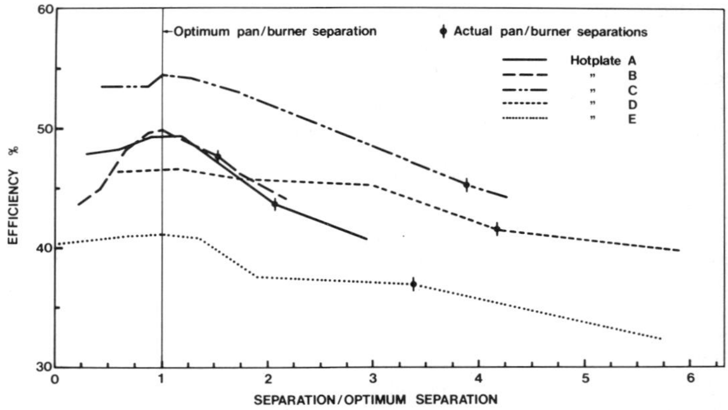

Figure 2: The effect on efficiency of varying burner-to-pan separation on catering hotplates.

Space is not in short supply on catering hotplates. It is primarily needed to accommodate the large pans necessary for catering, but it also allows plenty of ventilation for the burners which also tend to be large. The burners themselves have low flame port loadings (typically 21 kJ/mm^2) which offer little resistance to the entrainment of primary aeration allowing typical primary aerations of about 80%. However these same burners were found to have poor port geometry and the aeration adjusters were frequently incorrectly set. When the burners were fully optimised their efficiency improved by 5 percentage points.

3.3. Solid Top Hotplates

Solid top hotplates examined at Watson House have been optimised but because of their large thermal mass, work has concentrated on trying to reduce the thickness of the solid top itself. Most solid top hotplates are made of grey cast iron. Spheroidal graphite cast iron and stainless steel both offer better strength to weight ratio and are capable of standing the extremely harsh conditions normally experienced by solid tops. Prototypes, all based on the same design, have been made in the three materials and have shown almost identical temperature profiles (See Figure 4) which will permit their being fitted to existing models.

Figure 3: Non Optimised Catering Burner

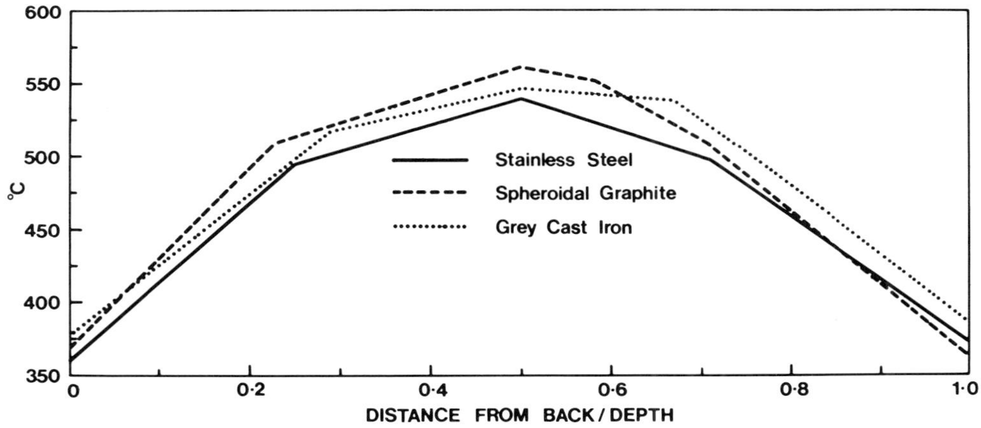

Figure 4: Temperature profile across depth for Solid Top Materials.

However, both the spheroidal graphite and stainless steel prototypes have slighly improved preheat times amounting to a 5% reduction in gas consumption.

3.4. Hotplate Usage Some aspects of hotplate usage have been examined in the laboratory. Saucepan lids can prevent water vapour escaping from the pan. Measurements have shown that at temperatures up to 70°C the rate of evaporation is insufficient to affect efficiency. Above 70°C but below boiling point significant quantities of water evaporate invisibly. The saucepan lid reduces this liquid loss and recovers the latent heat of vaporisation. Improvements of 7 percentage points in efficiency have been recorded with burners having a sufficient turn down ratio.

Catering saucepans come in many sizes and to accommodate them the hotplates have pan supports which allow complete stability wherever pans are placed. Therefore pans can easily be placed off centre and their weight discourages subsequent repositioning. Heat emerging around the side of the pan is then used inefficiently. It has been found that displacing the pan by 25% of its diameter reduces the efficiency by 15 percentage points.

3.5. Pan Sensors It is quite common for burners to be left alight when not in use in catering kitchens, especially in cases where it can be inconvenient to re-light each time they are required. One solution to this problem might be the use of devices which ensure that burners only operate when a cooking vessel is in position. In some cases pan sensors could save up to 25% of gas consumption. Several designs based on purely mechanical or electronic concepts were unsuitable because of their inability to survive the harsh catering environment and in some cases their high capital cost. A design using a mechanical sensor attached to a microswitch (Figure 5) offered the most cost effective solution with a pay-back period of 2-3 years for typical catering hotplates.

Figure 5: Prototype Pan Sensor.

3.6. Ovens Gas fuelled catering ovens can be either natural draught roasting ovens or forced convection ovens while either type may be directly or indirectly heated. In directly heated ovens the heat from the burner directly enters the oven box. This creates a temperature gradient with the hottest part of the oven at the top. Indirectly heated ovens have the heat from the burner absorbed and dissipated by a heat exchanger at the bottom of the oven which becomes the hottest part (See Figure 6).

Taking natural draught roasting ovens first, the best examined in the laboratory consumed 5% less gas than the average under full load cooking tests. Directly heated roasting ovens were found to use 40% less gas to maintain a given oven temperature than indirectly heated ovens (due allowance for differences in oven size being made).

Indirectly Heated Oven

Directly Heated Oven

Figure 6: Typical Natural Draught Gas Ovens

All the ovens examined had ample insulation and no improvements could be made with further additions. The flues were generally oversized. Up to a 75% reduction in flue cross-sectional area was possible without producing combustion products outside normal limits. Reducing the flue cross-section reduced the flue losses and hence saved energy - especially under steady state conditions - typically between 10 and 20% of gas consumption (See Figure 7).

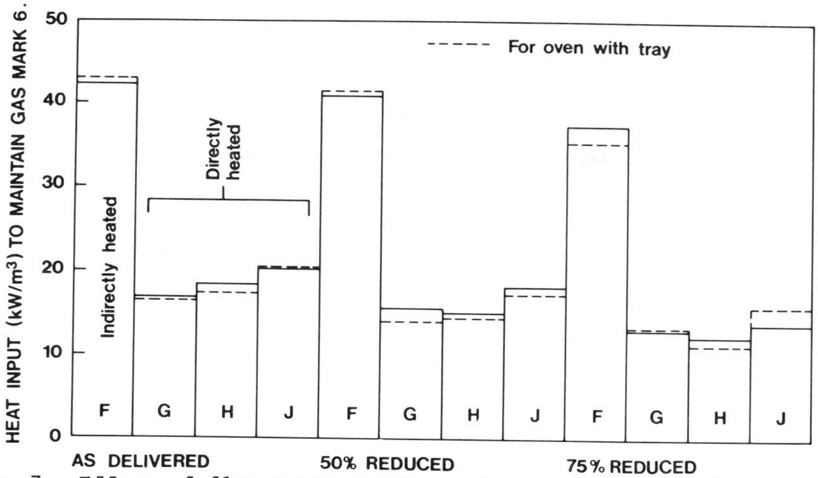

Figure 7: Effect of flue cross-section reduction on oven maintenance rate

The best forced convection ovens examined consumed 5% less gas than the average. As a general rule forced convection ovens have better heat transfer than natural draught roasting ovens, because of the rapid movement of air within the oven box. This can reduce fuel consumption by cooking food more rapidly and/or allowing greater quantities of food within the oven.

The oven temperature may also be reduced in some cases. The extent by which each of these parameters can be improved varies according to the type of food being cooked. Initial results with forced convection ovens indicate that there is some scope for flue size optimisation as with natural draft ovens.

3.7. Grills and Griddles Grills are usually one of the least efficient appliances. Measurements made with a calorimeter and in one case with a radiometer have shown considerable spread of efficiencies for commercial grills and griddles (See Table I).

APPLIANCE	EFFICIENCY	
	(Calorimeter)	(Radiometer)
High Temp. Radiant Grill (Top Shelf)	48.1	-
High Temp. Radiant Grill (Bottom Shelf)	27.0	-
Grill with Ceramic Frets (Top Shelf)	20.24	20.33
Grill with Ceramic Frets (Bottom Shelf)	17.8	-
Griddle Plate	34.4	-
Flare Grill	20.2	-

TABLE I SHOWING GRILL/GRIDDLE EFFICIENCIES

Griddle plates were more efficient than most grills, but the grill with high temperature radiant burners (surface combustion burners) was also found very efficient.

3.8. Fryers Fryers already achieve high efficiencies. However, those designs which possess a deep 'vee' pan heat exchanger were found no better in terms of food quality or efficiency than experimental designs using flat bottomed tanks. The advantage of the flat bottomed designs was that they required less oil to fill and because of this warmed up faster and generally had lower running costs.

4. KITCHEN MONITORING

Catering kitchens have been monitored to study the hot water consumption of dishwashing machines, the energy consumption and usage of gas fuelled appliances and the potential for waste heat recovery through the kitchen air extractor system.

4.1. Dishwashing Monitoring The dishwashing machines monitored were all sized to meet a maximum capacity of 50 racks per hour. In practice, most sites monitored showed the machines were usually under-used (See Figure 8). Two types of machine were encountered - those that were fully automatic with all operations being timed - and those where the operator manually controlled the duration of each process. In the latter case consumption varied from 25 litres to 29,000 litres per week depending on the care of the operator and the ease with which the water valves could be opened or closed.

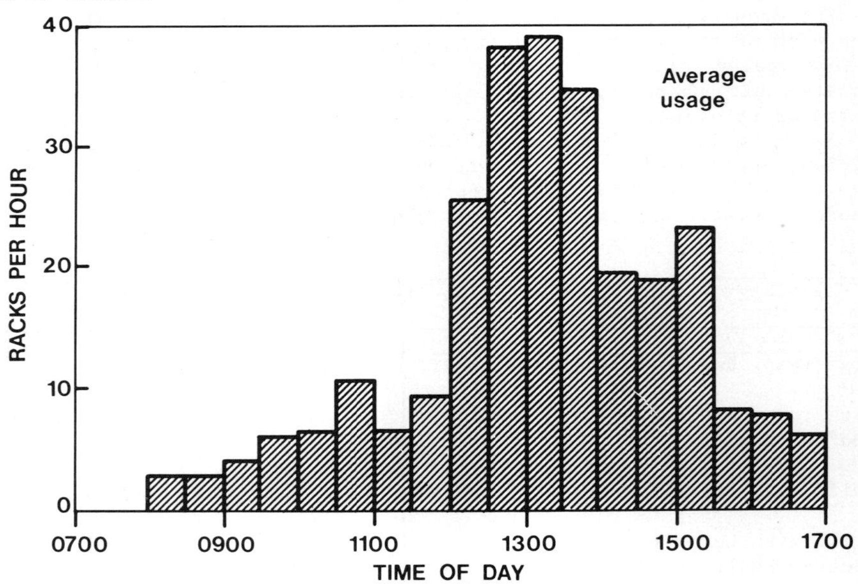

Figure 8: Average Dishwashing machine usage in a canteen

4.2. Catering Appliance Monitoring Schools, hospitals, canteens, hotels, public houses (pubs) and restaurants have had their kitchen appliances monitored to determine typical usage patterns and energy consumptions. Even in well managed kitchens, the general attitude is to leave appliances burning gas throughout the working day irrespective of use. This can be seen with the bains marie which should only be used to keep food hot prior to serving. Figure 9 shows that typically the bains marie was in use from 08.00 to 18.00 every day, yet it is highly unlikely that food was being kept hot for these ten hours. Another site was found to waste 13% of its bains marie gas consumption during overnight and weekend unattended misuse.

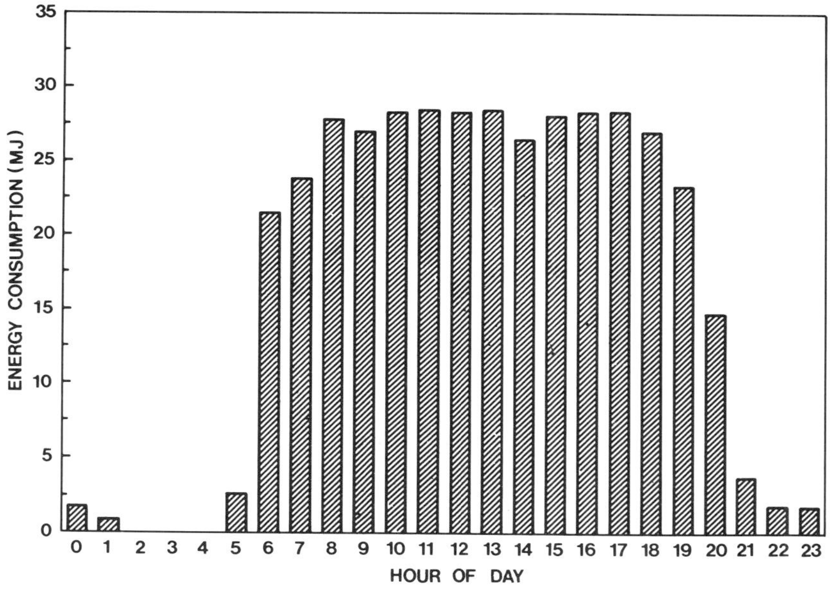

Figure 9: Average bains marie usage

4.3. Waste Heat Recovery Kitchen extractor hoods have been monitored at two sites. In a normal working day from 07.00 to 18.00 the extractor was working for 4.5 hours on average. When working, the hood temperature was between 30°C and 40°C and the extraction rate was approximately 5,400 m^3/hr. This entails a heat flow through the extractor hood in excess of 35kW. This seems compatible with the demands of canteen dishwashing machines for 20kW of power over the same period enabling suitable heat pumps to recover 25% of the extractor hood heat.

5. ECONOMIC ASSESSMENT

The total UK amount of gas consumed in catering for 1982 has been estimated at 52,000TJ. The measures proposed here can be divided into three areas:- appliance design improvements, appliance usage improvements and waste heat recovery. If fully implemented the energy saved would be:-

appliance design improvements	2500TJ
appliance usage improvements	2400TJ
waste heat recovery (canteens, hotels, restaurants and institutions).	1600TJ

adding up to a total of 6500 TJ or over 12% of UK gas consumption for catering. The figure is larger than previously reported because of recent findings reported here regarding ovens and kitchen waste heat recovery. Extrapolating these results to the EEC as a whole gives total catering consumption of 250,000 TJ and potential savings of 31,000 TJ (based on population only and assuming that the distribution of gas between catering and other market sectors is similar throughout the EEC countries).

6. CONCLUSIONS AND PROSPECTS

Significant savings in energy consumption have been found possible in every aspect of catering that has been examined. The main conclusions are as follows:-

- Most catering appliances could benefit from the optimisation of existing designs. The gas consumption of hotplates could be reduced by up to 25% and of ovens by 10% to 20%. Overall consumption could be reduced by 3%.

- Future appliances incorporating all the preferred design options if fully adopted, could reduce gas consumption in catering by a further 2%.

- Improvements in appliance usage by fitting controls and improving kitchen staff management and training could reduce gas consumption by 5%.

- Waste heat recovery utilising existing kitchen extraction systems and heat pumps could recover and use the equivalent of 3% of gas consumption.

Most of the proposed means of reducing gas consumption in catering only require the care and attention of manufacturers and kitchen staff. However, complete implementation will require a change in the traditionally conservative nature of the catering industry. When energy saving appliances are commonplace in catering, the benefits will not only be financial, but will also include a better working environment and in some cases improved appliance performance. It is our hope that these will prove attractive enough to encourage the necessary changes.

RADICAL REDUCTION IN DOMESTIC NEED FOR ELECTRICITY, EXEMPLIFIED BY THE REFRIGERATOR

J.S. NØRGÅRD and J. HEEBØLL
Physics Laboratory 3,
Technical University of Denmark
DK-2800 Lyngby, Denmark

Summary

The technological options for saving electricity have to some extend been neglected. In a study of these options for household appliances we have designed and built a low energy refrigerator. From a users point of view this refrigerator is not much different from a traditional model, but its annual electricity consumption is only 104 kWh under standard test conditions, compared with 230 kWh for the best models on the market today and 420 kWh for the average model in use today. Refrigerators are not unique with respect to savings potentials. Other equipment show similar possibilities and in general the economy of saving electricity is superior to the economy of supplying the electricity. Furthermore we decrease the air pollution and increase the security of supply by lowering our demand for electricity and for energy in general.

1. BACKGROUND

1.1 Theoretical Work

The potentials for saving electricity were rather unnoticed in the early years after the 1973-energy crisis, compared to the attention paid to other energy saving measures such as better thermal insulation of houses. In the period 1974-79 we carried out some theoretical investigations of the possibilities for designing more energy efficient appliances for private households (1)(2). Our criterion was that the new appliances should be able to provide essentially the same comfort and services as the already existing models, but with much lower consumption of electricity. We studied three levels of efforts to reduce the power consumption, namely moderate, strong, and radical measures. On the average we found that through radical measures the annual electricity consumption per appliance could be reduced to around one third of what it was in 1975.

1.2 Technological progres

The actual development since then has indicated that we were too conservative. As an example the most efficient refrigerators on the market today use less than half of the electricity consumed by the average 1975 stock, see Fig. 2. Similar patterns are typical for other appliances. This development was the basis for later analyses which indicated that even though the stock of appliances was improved, it would still be possible to reduce its electricity consumption to around one third (3)(4). Actually it turned out to be easier than first anticipated due to a technical development, triggered by the 1973-oil crisis.

1.3 Three appliances investigated

In 1981 we started a study, the aim of which was to demonstrate the possibilities of building low energy appliances like the ones described in our earlier work. This study was financially supported by the Energy Conservation R&D programme of the EEC, and carried out in cooperation with manufacturers of appliances.

In this study we have focused on three major domestic uses of electricity, namely for refrigeration, for cooking, and for washing. So far we have been most successfull in building a low energy refrigerator. In this paper we will use the refrigerator as an example of how equipment can be designed for more efficient use of electricity.

2. BUILDING A LOW ENERGY REFRIGERATOR

2.1 Partners in the project

With the analyses mentioned above as a background we were in the fall 1981 ready to start designing and building proto types of low energy refrigerators, LER. As industrial partners for this part of our project, we invited the refrigerator manufacturer Bdr. Gram A/S, Vojens, Denmark and the compressor and control system manufacturer Danfoss A/S, Nordborg, Denmark. They both accepted and so did the Danish Home Economic Council (Statens Husholdningsråd) when asked to assist in evaluating the prototypes.

As a reference for comparing our low energy refrigerators, we selected a Gram K215 model, which is among the most electricity efficient models on the market today. K215 has a useful storage volume close to the 200 litres of our LER-prototypes.

2.2 Refrigerator cabinets

We designed and analyzed a better insulated cabinet for the refrigerator. The improved thermal insulation was obtained through increased thickness of the polyurethane foam. The normal 3 to 4 cm of insulation was increased to 7 cm. We wanted to maintain the external standard width of 59,5 cm, which also enabled us to use an outer steel cabinet from a standard refrigerator. The inner box then had to be redesigned as to make space for the extra insulation. For the purpose of making a few prototypes we decided to make the new inner boxes out of fiber glass reinforced plastic, using a special mould made of wood. Seven new inner boxes were made.

The new inner boxes were installed in outer steel boxes by the Gram factory. Next the polyurethane foam was expanded into the cavity between the inner and outer boxes by the usual technique. This process created a rather high pressure, for which reason support is required on both sides. All seven cabinets were insulated satisfactorily, but only four continued to the assembling process.

2.3 Modified compressors

Parallel to the work on the cabinets, two modified compressors were designed and built at Danfoss. The main modification was a refrigerant suction directly into the compressor instead of just into the whole hermetically sealed compressor unit. Furthermore the motor was made more efficient by optimizing its design for the purpose.

Test results for the modified compressors point towards an improvement in the coefficient of performance (COP) of 33 percent. This means a reduction of electricity consumption per unit of cooling capacity around 25 percent for standard cooling processes.

2.4 Installing cooling system

Evaporators for the LER prototypes were choosen 60 percent larger than that of the reference refrigerator K215. As for the condensors the overall area is about the same as that of K215, but the radiator grid for LER has a more efficient design.

Control systems were based on an evaporator thermostat, type 90B from Danfoss, but modified for the purpose.

One prototype, LER1, was equipped with the smallest standard compressor from Danfoss (TL2A). For two others, LER2 and LER3, were used the modified compressors.

3. MAIN RESULTS FOR REFRIGERATORS

3.1 General consumer features

The low energy refrigerators, LER, shown in Fig. 1, are of the cold cabinet type with 200 litres of usable refrigerated volume at +5°C and no freezer compartment. (Most Danish household have a freezer as a separate unit). Automatic defrosting is achieved in the common way in European refrigerators by letting the evaporator temperature rise above 0°C before starting the compressor.

Outer dimensions of LER prototypes are: height 126 cm, width 59.5 cm, and depth 66 cm. The reference, K215, has 206 litres of usable inner volume, and the corresponding outer dimensions are 106 cm, 55 cm, and 63 cm. The extra insulation in LER takes up 110 litres of the kitchen space. This is typically 0.5 percent of the kitchen volume or 0.05 m^2 of floor space.

Compared to K215, the storage space in LER is more accessible as the shelves are less deep, and as there is 62 litres of space available in the door, compared to 44 litres in K215. In LER, like in all other Gram refrigerators, a system of various removable containers is available for utilization of the door space, see Fig. 1.

Figure 1

The low energy refrigerator prototype does not look much different from other refrigerators despite the heavier insulation

3.2 Electrical consumption

For comparison of the power consumption we have used the standard test procedures: ISO/R 824 - DIN 8951. These standards prescribe among other conditions an interior temperature of +5°C and no door openings. To compensate for the unrealistic condition of no door opening, the ambient room temperature is required to be 25°C.

With these conditions we have estimated average electricity consumption per year for the Danish stock of refrigerators in 1975 and 1983 as shown on Fig. 2.

Our LER prototypes have so far been tested in three different places. At Danfoss special efforts were made to adjust the refrigeration system and the control system. For LER3 was measured a power consumption of only 280 Wh per 24 hours, corresponding to 104 kWh/year, see Fig. 2.

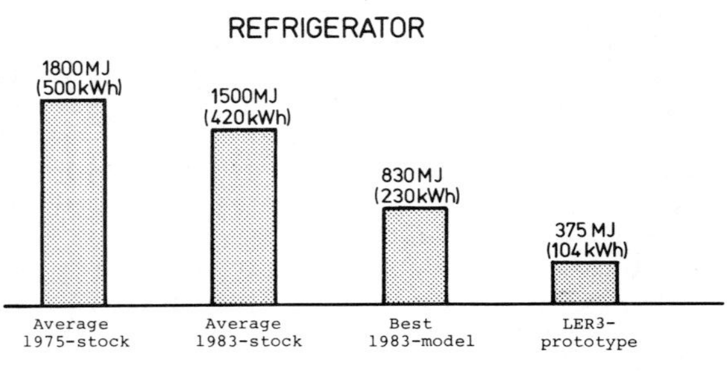

Figure 2

Annual electricity consumption according to standard test conditions (ISO/R 824-DIN 8951) for different versions of a refrigerators with around 200 litres of storage volume. LER refers to the low energy refrigerators described in this paper.

3.3 General observations

As expected even the smallest compressor available was oversized in capacity for LER. This showed up in the running periods of the compressor, which amounted to only 20 percent of the time. Normally it is around 50 percent. The low duty cycle means a low utilization of the evaporator and condensor.

Power consumption under real consumer conditions is difficult to estimate. The deviations from the standard measurements will depend mainly on 1) the number of door openings per day, 2) the amount of food inserted into the refrigerator per day, and 3) the real ambient room temperature.

At Gram a test was run with 51 door openings and 6 kilograms of food at 25°C inserted per day. It showed a 60 percent increase in power consumption, when 25°C is maintained in the test room. At Physics Laboratory 3 we have run some preliminary tests with LER at lower ambient temperature. These experiments indicate that with a room temperature down to 21°C and practically no door openings, the power consumption will be down by more than 40 percent, corresponding to an annual consumption around 60 kWh.

4. CONSEQUENCES OF IMPROVED REFRIGERATORS

4.1 Electricity savings and the environment

The most direct consequence of gradually replacing the present stock of 2 millions refrigerators in Danish households with models like our LER3 of prototype will be an annual saving of 650 millions kWh of electricity. As coal dominates Danish power supply (~95 percent) the savings will reduce air pollution per year by around 5500 tons of sulfur dioxides and 2000 tons of nitrogen oxides. The amount of fly ashes produced per year at power plants will be down by 20000 tons.

Another benefit of lowering energy demand in general is the proportional reduction in coal mine hazards, oil tankers, ship wrecks, and other risks associated with supply of energy.

4.2 Security of supply

Electricity black-outs of a few hours will have only minor effects in well insulated refrigerators and freezers. If the 200 litres LER-refrigerator contains food corresponding to 50 kilogram of water, the temperature will only rise from +5°C to 6°C in 4 hours, if the door is kept closed.

For a nation or a region as a whole the supply of fuel can be secured by stockpiling an amount of coal or other fuels. Such stocks for power plants will obviously be stretched when demand for electricity is decreased or -alternatively - the investment in fuel storage can be reduced.

Finally the general pressure on the world's limited stock of non-renewable energy resources can be relieved through a more efficient use of energy.

4.3 Economy of the conservation

The private economy of choosing a low energy refrigerator like our LER prototype is attractive, when an old refrigerator needs replacement anyway. We have estimated the additional retail cost in Denmark to be around 350 Dkr (45 ECU) including the Danish taxes which account for 32 percent of the price. Also included is the 50 Dkr cost of the extra space taken up in the kitchen. With an electricity price of 0.80 Dkr (0.10 ECU) per kWh the annual saving on the electricity bill will amount to 250 Dkr (30 ECU) when compared to an average refrigerator in use, and 100 Dkr (12 ECU) when compared to choosing the best on the market today. Corresponding payback periods will be 1.5 years and 3.5 years respectively.

Without the Danish taxes on the refrigerator the payback period will be around 2.5 years compared to bying the best presently on the market. With an average lifetime of 13 years for the refrigerator the rate of return is around 35 percent per year. It should be kept in mind that the estimate of the extra cost is rather uncertain, but we believe they are on the safe (high) side. Actually it has not been possible to associate any price increase with the improved efficiency already achieved on the models on the market today.

From a national point of view the economy seems similarly attractive, but more complicated to define and evaluate. On top of the direct economic benefits come, however, the already mentioned external social benefits of cleaner air and more security of supply.

4.4 Future possibilities of refrigerators

As indicated in 3.3, the refrigeration system of LER3 is running inefficiently, due to the overcapacity of the compressor used. A lower yield compressor could further lower the power consumption of the refrigerator. Presently we plan to pursue this path. A couple of low capacity compressor prototypes might be designed and built by Danfoss and installed in some of the remaining LER cabinets at Gram.

5. CONCLUDING REMARKS

5.1 Refrigerators are not unique

It is characteristic for electricity that its use is spread out on many equipments, each of which has a modest power consumption. Analysis, however, indicate that most of those small uses of electricity can be made much more efficient. Refrigerators are not unique in that respect. We will mention a few other dominating uses of electricity, mainly associated with private households.

5.2 Freezers

Obviously the options for domestic freezers are very similar to those of refrigerators. From an average annual consumption of 800 kWh for the 1975 stock of chest freezers, it is now possible to purchase models which use only 320 kWh per year. With improvements similar to those introduced in the LER refrigerator we suggest it is possible to decrease the consumption to around 150 kWh per year.

5.3 Cooking

For many appliances, like those for cooking, it is not possible to distinguish clearly between technological and behavioral savings. Often the technical improvements will require small changes in habits.

Clearly cooking does not require electricity at all, but for oven cooking it is a convinient energy form to use. Within our project we have designed and built a small low energy oven, LEO, which we estimate to be sufficiently large for around 80 to 90 percent of the oven cooking in an average family. Preliminary tests of LEO indicate an overall power consumption around one third of what is normal.

For electrical cooking hot plates our efforts have been directed mainly towards electronic control systems.

5.4 Washing

Many habits are also integrated in the way people wash their clothes, also when using washing machines. If we assume, however, that the purpose of washing is to provide clean clothes, then this goal can be achieved with considerable less electricity than what is typical today.

Within our project we are investigating different options for reducing electricity consumption to a small fraction of the 1975 average of 550 kWh per year. The options involve the use of less water and/or lower temperatures.

5.5 Other small uses of electricity

Many other, if not all, small uses of electricity show similar potentials for reduced consumption. Some of it comes by itself like in the field of electronics, where for instance, a color TV is down from its earlier 400 Watts to around 50 Watts for some models sold today. Other savings require a special effort and attention and maybe an extra cost. Circulation pumps for private central heating systems typically used 300 kWh per year, running continously over the heating season. Today it seems possible through installing a smaller pump and switching it on and off as part of the heat control system to reach an annual consumption of around 30 kWh, or one tenth of the former.

For those technical savings which do require an extra cost, the economy of saving electricity is much better than the economy of supplying electricity. This is analysed to be the case until specific electricity consumption for equipments in homes is reduced by at least 50 percent of present consumptions. In addition we get the external social benefits of better environment and more security of supply.

REFERENCES

1. Nørgård, Jørgen S. : "Husholdninger og Energi", report no. 4 from DEMO-project, Polyteknisk Forlag, DK 2800 Lyngby, An English draft translation, "Households and Energy", from US/DOE is available from Friends of the Earth, 1045 Sansome Street, San Francisco.

2. Nørgård, Jørgen S.: "Improved efficiency in domestic electricity use£, Energy Policy, pg. 43, March 1979.
3. Strabo, F. et al.:"El/forbrugets udvikling i en landkommune", Nysted Rådhus, DK 4880 Nysted, 1981.
4. Nørgård, Jørgen S.: "Low-electricity future households" in IEE conference publication no. 186: Effective Use of Electricity in Buildings, 1980.

IMPROVED CONVENTIONAL HEATING SYSTEMS - R & D PROJECTS

Rapporteur: Dr. S.J. LEACH

As the Chairman has explained I am reporting on 9 to 14 projects and this will be followed by separate 10 minutes presentations on the 5 that remain. I have had te same difficulties as the other rapporteurs and I too have to apologise to the 9 authors whose excellent papers I am about to condense drastically.

I think I have a wider spread of topics than experienced in other sessions and below are the headings which I have used to link the group projects for my presentation. These are :

Heat emitters
Appliances
Control and metering
Heat storage
System performance

These headings are also appropriate for the individual presentations to follow mine. I shall say where these fit in. These 5 headings are all important aspects involved in improving the performance of conventional heating systems.

Below are the projects I have grouped in the first heading - Heat emitters.

Water-air conventor made with plastics
of improved thermal conductivity
Mr. P. Dubois, et al, F

Domestic warm-air distribution with
low-grade heat systems
Mr. H.B. Awbi, UK

Development of a computer program
for boiler plant selection
Dr. A.J. Meijnen, NL

You can see that this covers the development of a radiator for wet heating systems made of plastic and also studies of the design parameters needed for the use of low temperature warm air heat distribution systems, both of which I shall cover. There will later be a very important paper on Boiler Plant Selection from TNO which logically should come just before the emitters. This paper by Mr. Meijnen describes the development of a computer program for boiler selection to give low fuel consumption based on the hourly heat demand. This is to be supported by measurements on existing plant. It provides a tool for designers with prospects of up to 20 % savings in energy consumption. This work will be described by

Mr. Meijnen immediately after my talk. The two papers I shall present are linked to each other and to the review by Mr. Zegers yesterday in that they both concern the use of low temperature heat which is a means of improving the efficiency COP of Heat Pumps.

Heat Emitters

The first of the projects concerns the development of low cost, plastic, water/air radiators for domestic heating systems by Mr. Dubois, and his colleagues from Marcoussis, France.

Radiators made of plastic have two major advantages - low cost and resistance to corrosion. It is not unusual for pressed steel radiators to need replacement after 10 years, or even less when inhibitors have not been correctly used. Plastic radiators however have two important disadvantages, firstly the thermal conductivity of the plastic is less than that of their metallic counterpart which leads to lower surface temperatures and hence a requirement for a larger sized radiator for a given performance, and secondly their performance needs to be established over the repeated temperature cycling typical of domestic operation.

Because of the problem caused by the low thermal conductivity a new plastic material with improved conductivity was developed. This consists of a polyethylene matrix with an ethylene propylene elastomer and the addition of carbon black. A 1 kW output radiator was designed for a water inlet temperature of 50°C, chosen for heat pump operation where low temperatures are of course desirable. A 300 W output version was constructed.

To turn now to the cost of the radiators. Assuming that the retail price is two and a half times the manufacturing price then it is calculated that plastic radiators will sell to the public at a price about 60 % of that of the cost of conventional metal radiators.

The next project under topic of heat emitters is research into domestic warm air distribution with a low grade heat input carried out by Dr Awbi whilst he was working at the Building Services Research and Information Association, UK.

The performance of heating systems using alternative energy sources such as solar energy and heat pumps is improved if the system uses low temperature heat. Using low temperature heat with conventional wet heating systems requires an increase in the area of the radiator as we have seen with the plastic radiators. An alternative is that this requires a large volume of air for a given heat input which in turn can lead to problems with obtaining satisfactory thermal confort in the occupied zone. This means that low temperature heat distribution systems are probably most suitable for heating low energy houses where the insulation of the fabric has led to a low overall requirement for the heat input.

The project was set up to determine the optimum position for the air inlets together with the limitations on input velocity and temperature difference between the incoming air and that in the room.

A test room was developed specifically for these studies. Its dimensions were 5.5 x 3.5 x 2.5 metres and was constructed of timber and slotted steel sections with slightly better insulation properties for the external walls than the current design values of UK dwellings. The thermal properties of a panel in the wall were made equal to that of a double glazed window. Two adjacent walls and the suspended floor of the test room were enclosed by a compartment in which the air could be cooled to - 1°C. This is the normal wintertime design temperature used when sizing heating systems in the UK. This led to a design heat load for the test room of just over 600 W. Air to the test facility was provided by a cold air system supplying the external compartment and a warm air system supplying warm air to the room. It was capable of providing a wide range of internal and external conditions as well as the flexibility to incorporate various configurations of air terminal devices.

The room is used with a two directional traversing system which houses velocity and temperature sensors. The outputs from thermocouples and anemometers are fed to a computer for storage and analysis.

The conditions in the occupied zone were assessed using the PPD method of Prof. Fanger as a thermal comfort criterion - PPD being the "predicted percentage dissatisfied" of the occupants. This requires measurements of air velocity, air temperature and dry resultant temperature.

Figure 1 shows the results of some of the experiments. The vertical axis shows the air temperatures to give comfortable conditions based upon a PPD of less than 10 % plotted against the air change rate. The top curve gives the supply temperature and the lower curve the extract temperature. To give acceptable conditions over the full range of air change rates shown it is necessary to use a low level supply mode for low air change rates, less than round about 10, and a high level supply mode for the higher change rates. If a wall diffuser is used on the inlet, this provides comfortable conditions for a wider air change rate than using a direct discharge.

It is concluded that to maintain a comfortable environment with the optimum utilisation of low grade energy it would be necessary to chose an appropriate supply terminal. In each situation a detailed design assessment would be needed in order to determine the most economic solution. Much will depend upon the temperature output of the low grade heating system and the results in the paper provide a basis for such optimised designs. There is an important next stage of the work. This will investigate the effects of furniture to bring the results closer to a realistic, practical condition.

Appliances

Below are the projects grouped under the next of the main headings - Appliances.

Radical reduction in domestic need for electricity, exemplified by the refrigerator
Mr. J. Norgaard and Mr. J. Heeboll, DK

Energy saving in the commercial catering market
Mr. T. G. Enga, UK

We have two papers, one on the reduction of energy requirements for domestic refrigerators which I shall cover, and the second to be covered later in a 10 minutes presentation by Mr Enga of British Gas on energy savings in the commercial catering market. Mr Enga's work comprehensively covers many ways of increasing the energy efficiency of cooking. This is a neglected aspect of energy use by building services installations and in some building types, e.g. hospitals, can account for a substantial proportion of the energy consumption. Possibilities exist for savings from improved appliance and burner design, better control and more sensible use (better management) of existing appliances.

There is of course one thing we must not forget - the success of the work on appliances will reduce the incidental gains we allow for in system sizing and design. We must all be aware of the future development of efficient appliances when allowing for these gains which are in many buildings today a major part of the heat input.

Dr. Norgaard and his colleagues at the Physics Laboratory of the Technical University, Denmark, are working on domestic electricity using machines, including washing machines, refrigerators, cookers. If you look at electricity use in the home in most European Community countries we find that this is an area of substantial growth. It is also an area where less attention has been given since the energy crisis of 1973 than to most aspects of energy use.

Figure 2 shows the average annual consumption of washing machines in use in Denmark and it can be seen that this was 650 kW hours averages in 1975, falling over 6 years to 550 kW hours but with a target consumption for the future of 100 kW hours supplied by electricity together with about twice as much supplied by alternative heating such as from a district heating network rather than electricity, a considerable improvement overall.

The figure also shows a similar range in performance for refrigerators from 500 kW hours round about 1975 to a target of 100 kW hours for the future.

Finally it shows a similar range for cooking, falling from 950 kW hours to 450 kW hours. This is a less spectacular improvement in performance.

So far the research that has been carried out has concentrated on the domestic refrigerator. It was carried out in collaboration with both the manufacturers of the refrigerator and the manufacturers of the compressor and control system.

The first change was to improve the thermal insulation of the cabinet by increasing the thickness of the polyurethane foam from the normal 3-4 cm to 7-8 cm. This also led to a redesign of the interior to make maximum use of space contained within it for fixed outside dimensions, despite the increase in the thermal insulation.

In parallel to the work on the cabinets, modifications were made to the compressor cooling system. This involved changes in the design and operation of the compressor, its control system and the heat exchangers - both the evaporator and the condensor were made more efficient, in the case of the evaporator by a 60 % increase in size.

The low energy refrigerators of 200 l have been built with different compressor cooling systems. The performance of the best refrigerator (104 kWh/year) was very close to the planned 100 kWh/year. The payback time is two years.

In addition to the reduction in the energy consumption there is a further benefit from using such refrigerators because an electricity failure lasting a few hours would only have a minor affect on the internal temperature.

The economic assessment of the improved refrigerator concludes that the replacement by a low energy refrigerator is attractive when carried out at the time when an old refrigerator needs replacement anyway. The estimated additional retail cost in Denmark is about 350 Danish Kroner, including the taxes. This also includes an estimate for the cost of the extra space taken up in the kitchen. The annual saving amounts to about 250 Danish Kroner when compared to an average refrigerator and about 100 Danish Kroner when compared with choosing the best on the market today. The payback periods are therefore short. Significant savings can be expected in the appliance sector. Later Mr. Enga will give important results in the commercial catering sector. We should all, I think, take note of these developments which will in some cases lead to increased space heating bills but of course with a corresponding reduction in the electricity cost.

Controls and Metering

Below are the papers grouped under the heating controls and metering.

Two applications of microprocessors in climate and lighting in office buildings
Dr. P. Nicolaas, NL

Instrument for automatic heating control
Mr. F. Fonzi and Mr. P. Segalini, I

Heat meter without rotating parts
Mr. W. Gramatte, D

Heat meter with microprocessor; final results of the study
Mr. R. Leopardo, I

Microcomputer network for heating system
Mr. V. Mauro, I

There are three projects in the above group on the application of microelectronic devices to control heating or lighting systems. Two of these I shall present and the one by Mr Mauro which relates to the use of a microcomputer network will be presented later. There are two projects on the development of heat meters. There have always been two problems with heat meters : firstly they tend to be expensive if they are adequately accurate, and secondly a reliance upon moving parts to measure the flow rate leads to unreliability, usually caused by fouling from the rather unclean water normally present in district heating systems. The two projects concern devices being developed which avoid the use of moving parts and both show great promise. One makes use of the Lorenz force. In water normally a small quantity of ions is always present. In a flow of water which is subjected to a magnetic field perpendicular to the water flow direction, these ions experience a force which is perpendicular to both the magnetic field and the flow direction (Lorenz force). With electrodes, one will be able to detect a small current. In this heat meter an electric field opposed to the Lorenz force is regulated in such a way that it compensates this force resulting in a zero current. The electric field is then a measure for the velocity of the mass flow and the mass flow can now be determined if the magnetic and electric field are known. First measurements show a very satisfactory operation in particular at low flow values. This work by Leopardo and Cena will be described later by Dr Leopardo. The other has been developed by Mr Gramatte and his colleagues working at the Battelle Institute in Germany which makes use of an elegant and simple principle which I shall now describe.

Two pipes carrying incoming and outgoing warm water are linked by copper heat conductor. The heat flow through the conductor depends upon the temperature difference between its two ends, the heat transfer coefficient of copper and the dimensions of the copper rod. This in turn equals the heat being extracted by the copper rod from the incoming water flow and its heat flow, is determined by the difference in temperature in the incoming water flow just before and after the copper rod, the mass flow of the water and the specific heat of the water. It can be shown that by measuring the temperatures at three points it is possible to obtain the mass flow rate and therefore the heat flow rate.

There are two basic problems which need to be studied and overcome to make such a heat meter work and these are : firstly that the temperature differences which need to be measured are extremely small which pushes temperature measurement beyond the current state-of-the-art, and secondly the temperature sensor devices must be not significantly affected by deposits which may arise within the system.

The work is proceeding by a mixture of laboratory and theoretical analysis to optimise the design to be followed by field tests of prototype meters. Studies were carried out of the deposits on the surfaces of heating systems and it is broadly concluded that deposit layers less than two microns in thickness have insignificant thermodynamic effects and that thicker deposits can be avoided by taking appropriate steps during the design to arrange for adequately high flow rates.

The method of measurement of temperature being investigated uses platinum thin film temperature sensors. These have been investigated from the point of view of drift and resistance to wear.

Very high measuring accuracy is required of the temperature sensors because to obtain accuracy of a few per cent in the measurement of the flow rate an accuracy of thousandth of a degree Kelvin is needed in the temperature measurement. The work has now reached a stage of prototype development. It can be concluded from the results achieved so far that in the end a prototype meter will be available which will satisfy the requirements.

An economic assessment concludes that if the meter is marketed at a price of 1,300 German Marks the payback time will be less than 4 years, based on an energy saving of about 15 %.

Controls

The three projects on electronic controls cover their application to heating systems in two projects undertaken in Italy together with work in Holland which includes as well the control of lighting and cooling. The work by Mr Mauro on a microcomputer network for controlling heating systems in blocks of flats, schools, offices, etc is completed and has shown energy savings up to 20 %, and will be described later. To turn now to the work on electronic heating control systems by Dr Fonzi and his colleagues from Sulmona, Italy. They are developing micro-computer based devices to improve both the control of the burner in a central heating boiler and the flexibility the user has to programme his requirements.

Three prototype devices have been developed known as HECOS 1, HECOS 2 and HECOS 1 - HECOS standing for Heating Control System.

HECOS 1 controls the operation of the boiler. The objective is to control combustion so that it happens in stoiciometric conditions. It links all the aspects of combustion and safety in a microcomputer.

HECOS 2 provides control of control heating by the addition of external temperature sensors and a provision for presetting the desired internal temperature. Both HECOS 1 and 2 were presented at a previous contractors' conference.

The latest device - HECOS I - has the possibility of being able to be programmed by the user of the heating system. With the built in clock and the programme facilities available the user can set up the desired heating for the week ahead, with different requirements during each day for both temperature in the room and duration. It also includes a digital clock and thermometer and a clear visual display of the set conditions. I will not say any more because you have seen the device itself during the poster session and this is better than any description I can give.

To turn now to the work of Mr Nicolaas from TNO, Holland, who is looking into the applications of microprocessors for the control of climate and lighting in office building.

Microelectronic devices in the form of micro-computers have now become relatively cheap and extremely reliable. Simultaneously there is the development of computer programmes for the simulation of the thermal behaviour of buildings. These programmes make possible to simulate the performance of buildings under real weather conditions. It is therefore possible to analyse the performance of the building and its control system to see if suitable algorithms can be developed for the operation of microprocessor based control systems to achieve significant energy savings.

Microprocessors can also be used to optimise the system performance in relation to air handling, the starting time for the cooling plant, the use of solar controls and the use of mechanical ventilation to cool down the building during the night and weekends.

Two particular aspects are currently being given the most attention. These are the possibility of using mechanical ventilation to cool during the night and the use of measurements of solar radiation on a vertical plane to optimise the lighting control for a specific building.

A lighting control system is being developed for a rectangular office building which has the standard shape with a central corridor and daylight available in all rooms. Each office has the lighting arranged in banks near the window and near the corridor. The artificial lighting is automatically switched on, or half on or off according to the daylight level, the occupants can override the automatic controls and lighting is available at half the maximum level outside working hours for the cleaning and security.

Computer simulation has shown that mechanical ventilation operating at night can reduce the incidence of excessive daytime temperatures. The mechanical ventilation is switched on according to predicted external air temperatures on the basis of weather forecasts and on the level of the indoor air temperature.

Experimental investigations are still being made and the final conclusions of the study have yet to be obtained.

Heat Storage

Below are the papers grouped under the next heading - Heat Storage.

Heat storage by paraffin waxfusion
for space heating by a heat pump
Mr. P. Dubois and Mr. M. Berthet, F

Feasibility of underground heat storage
in captive aquifer
Mr. J. Despois and Mr. B. Conseil, F

You can see that there are two papers. The first is on heat storage by utilisation of the latent heat of melting of paraffin wax which I shall present, and the second is on underground heat storage which will be presented later. We thus have examples of the two extremes of heat storage - small scale within the house and large scale in an aquifer. Mr. Despois has been investigating the feasibility of underground interseasonal hot water heat storage in a suburb of Paris by driving a bore hole to study the geological properties together with computer simulation to determine the thermal capacity. Dubois and Berthet working at their laboratory in Marcoussis have carried out a comprehensive study of the possibility of using paraffin wax as a heat store in combination with a heat pump for domestic space heating.

This draws attention to another way of improving the performance of the heat pumps we discussed yesterday by maintaining its performance more continuously. Also users can take advantage of more advantageous electricity prices during night time operation which reduces costs if not saving energy.

During the course of the study, encapsulation techniques were developed, the behaviour of stores of different designs were modelled and a store of the appropriate size of the house was built and tested in association with a heat pump.

The best encapsulation results were obtained with a triple layer film comprising a high density polyethylene film, a polyester film and a bonding adhesive, and this form of encapsulation is shown in the poster room.

Mathematical and experimentals studies of the design of the heat exchanger led to a design without encapsulation using a roll-bonded aluminium.

The water flows through the aluminium panels which are immersed in a tank of paraffin wax.

A series of laboratory experiments were carried out using a test-rig which included a 1 cm^3 store and an air/water 2.5 kW heat pump. It was concluded that the installation of the heat stores makes possible a 3 % gain in the coefficient of the performance of the heat pump and also a reduction in the operating costs of 30 % arising from the use of the heat pump during the lower tariff hours.

On the assumption of a retail price between two and three times the factory cost the paraffin wax store would retail at about 15 000 French francs which is to be compared with an annual electricity saving worth at French Electricity tariffs of about 500 francs. This leads to the conclusion that such a store is not an economic proposition. However, there are other possibilities for the use of the paraffin encapsulation techniques including, for example, its use as an alternative to a pebble bed heat store in the air loop of a solar collector space heating system. Cost savings come significantly from the major reduction in weight and volume of the paraffin store making it possible to install it without reinforcement of the structure of the house.

System Performance

Below are the papers grouped under the last of the headings - System Performance.

> BRE calibrated house experiments
> Dr. R. Rayment, UK
>
> An investigation of the performance of various heating systems in a matched pair of houses : Initial results
> Mr. T. Fuller, et al, IRL

This is research into what happens in a full-scale system with all the various factors involved including the one often forgotten - the occupants.

There are two papers and they are both strongly related. One is from the UK and the other from Ireland and both use essentially the same experimental technique which I shall describe briefly.

To start with the work of the Building Research Establishment, UK carried out by Dr Rayment and his colleagues. The basis of the study is the use of one house as an experimental control whilst conditions in another house are varied with both houses undergoing a simulated pattern of use of a typical family. Both houses are carefully set up before the experiment itself takes place to ensure that the energy consumption of the control house is identical to the test house when the two houses are operated in the same way.

The pair of houses used and the occupation simulation was described at the last of these conferences and I shall not go through the method involved except to say that there is under computer control simulation of the heat gains from the occupants, of lighting, of television use, and of cooking. There is automatic draw off of hot water, which is left standing in the bath so that the house would experience a heat gain similar to that from a person bathing, and also automatic drawing of the curtains.

Figure 3 shows the two schedules used to drive the occupancy simulation and operation of the heating system. It is based upon observations of the pattern of use of houses from a large number of sources, averaged to give typical patterns of use. The results from the experiments then can be used to define the performance in practice without the problems caused by trying to do experiments in actual occupied houses.

There are three experiments to report now : one concerns the use of secondary windows, the second the use of electronic room thermostats and the third the use of cylinder thermostats to control temperature of stored hot water.

Figure 4 shows the comparison between the test house with secondary window and the control house with single glazing. The line with the solid dots is the calibration experiment where both houses were operated with identical single glazing and you can see that over a very wide of external conditions - heat consumptions per day ranging from about 150 MJ to nearly 600 MJ - that the two houses are performing identically to a very high degree of accuracy. The introduction of the secondary glazing in one of the houses recudes the consumption by 12 %. This is more than can be expected from the change in the U-value alone and the additional improvement results from the reduction in the amount of air infiltration.

To turn now to the use of electronic room thermostats. A conventional room thermostat, which consist of a bimetallic strip with a resistance heated accelerator was used as the basis for comparison. The thermostats are shown in the next slide. Such a thermostat experiences a change in set point according to the heading load. This change in set point is such that house temperatures tend to be two or three degrees Centigrade hotter in mild weather compared with cold weather conditions. Electronic thermostats do not show this characteristics and their performance was compared with the conventional thermostat.

There was a saving of about 2 % of the annual consumption from the use of electronic thermostat. This small level of saving which could not be determined by any other experimental method than that used here because the saving is so small. Nevertheless because the electronic room thermostat cost are small even the saving of 2 % is sufficient to give a payback period of about 8 years.

To turn now to the use of cylinder thermostats. These have not been widely used in the UK, certainly in older central heating systems. In its simplest form a strap-on cylinder stat switches the boiler on or off according to the temperature of the hot water. This method only works in summertime when there is no requirement for space heating. Only by fitting a motorised valva can the full benefit of the cylinder stat be realised for the summertime and wintertime operation.

The results depend upon the basis of comparison. Accordingly the savings obtained range from 5 % to 29 %. This in turn means that there is a wide variety of payback periods depending upon which system is considered as an alternative. The conclusion is that they are generally cost-effective.

Recently eight houses have been commissioned but no results from them are yet available. They replace the single pair of houses whichg were used to obtain the results I have described and will be used for a wide range of studies on controls and on thermal insulation.

To turn now to the work of Mr Finn and his colleagues at the Irish Institute for Construction Research.

Essentially the same experimental technique is being used as that just described. The main objective is to compare the performance of various types of heating systems in terms of overall efficiency and the effectiveness of alternative controls. Studies are being carried out of a high output back boiler from a living room open fire, an oil boiler and a low thermal capacity natural gas boiler.

The houses used are very similar in size to those used in the previous project but differ in the level of thermal insulation. The Irish houses are relatively well insulated with a U-value of 0.6 W/m °C for the walls.

During the calibration processes the houses were heated using oil filled electric radiators to give a specific heating regime. Control of temperature was achieved by having the on-site computer monitor the temperature in each room and which switched on the radiator and gave temperature control to within 0.2 °C. Pressurisation tests by BRE identified air leakage paths and showed that the air change rates were similar - around 1 at 3.5 m/s wind speed - in both houses.

To briefly summarise the initial findings on the three appliances :

Firstly the coal fired back boiler had an efficiency - defined as total daily output to the house divided by the gross calorific value of the fuel used - of between 30 and 40 %. To maintain 40 %, daily cleaning of the back boiler is necessary. The low efficiency arising mainly because the pattern of heat output from the boiler was not well matched to the load pattern and because of the poor combustion air flow control. Detailed suggestions for improving efficiency are given in the paper.

Secondly the oil fired boiler had an efficiency of 73 % at full load but because the smallest size of boiler available on the market was appreciably larger than that needed even on design day or peak load conditions the percentage "on" time was low - averaging 15 % - which led to a significant fall in boiler efficiency to below 60 %.

Thirdly the gas boiler averaged 77 % efficiency and maintained this efficiency even at very low loads - exceeding 70 % with a load of only 11 % of the peak. Thus a low thermal capacity, rapid response gas boiler is well suited to meeting the widely varying demand pattern of these houses.

Mr. Chairman, I should like to mention that the last two projects are an example of detailed collaboration within the European Community programme at the working level with ideas being shared and - in the case of the pressurisation measurements - equipment also being shared. This collaboration means that complementary programmes are carried out and resources saved in developing methodology. This is something quite rare in the research world.

Finally, Mr. Chairman - if I could briefly mention the future - , I see a need for better mathematical models simulating the performance of conventional heating systems and also more information on whole system performance which takes account both of the requirements and behaviour of the user and of the building itself. These studies would bring together the many aspects affecting the efficiency of conventional heating systems.

Acknowledgement

The views expressed are those of the author alone and not necessarily those of DOE.

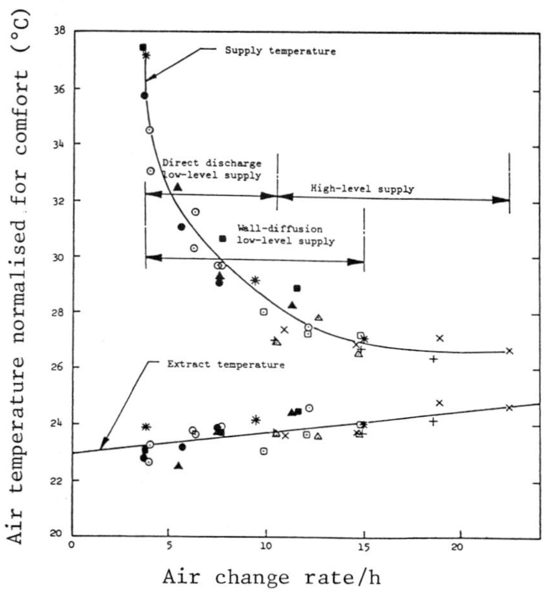

Fig. 1 - Supply and extract air temperatures required for different air change rates

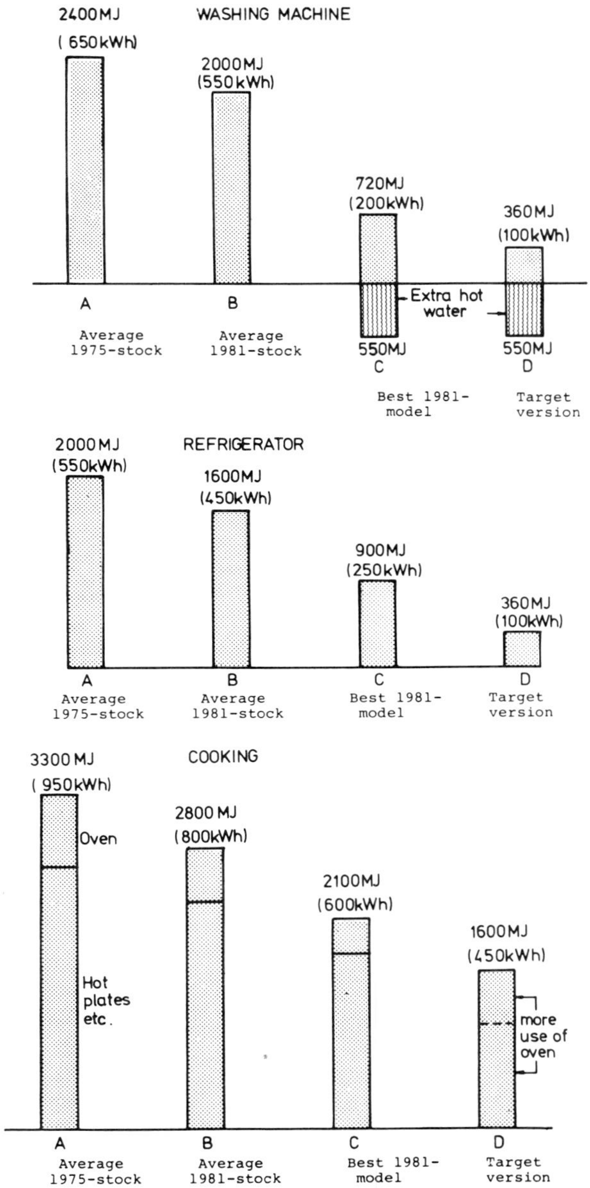

Fig. 2 - Target values in Denmark for the annual electricity consumption of washing machines, refrigerators and cooking equipment and their average values for 1975 and 1981

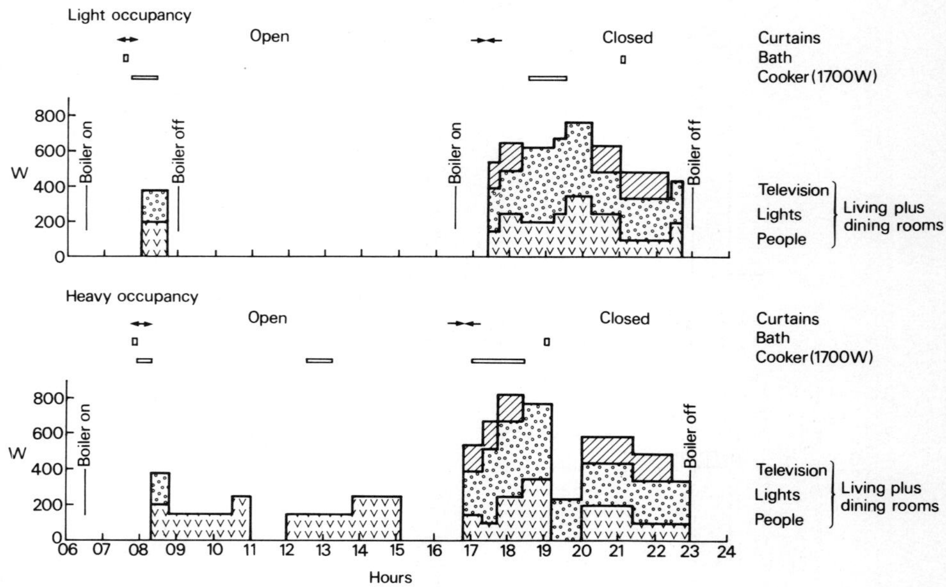

Fig. 3 - Selected occupancy heat gains and times

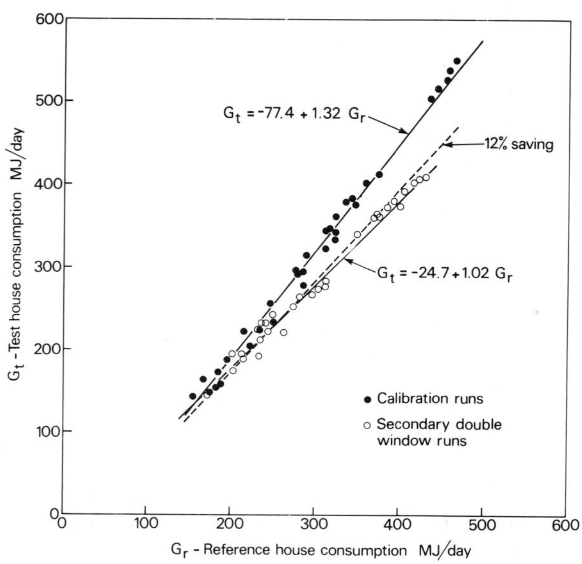

Fig. 4 - Daily consumptions during calibration and secondary double glazing experiments

SESSION III - IMPROVED CONVENTIONAL HEATING SYSTEMS

B - DEMONSTRATION PROJECTS

Chairman: Ir. J. UYTTENBROECK
Rapporteur: Mr. J. DURANDET

CONCLUDING REMARKS by

Dr. M. DAVIS, Directorate General for Energy, Commission of the European Communities, Brussels

FIELD CAMPAIGN OF A COMPLETE SERIES OF A BLUFLAM BURNER BOILER SYSTEM FOR DOMESTIC HEATING.

V. Recchi, E. Triggiani, S. Inserra Imparato.

otb partecipazioni S.p.A.
Zona Industriale - Contrada del Parco
70123 BARI - Italy - tel.: 080/371144

Abstract

The design principles and the main characteristics of a complete serie of Bluflam domestic burner-boiler systems are described, particularly evidencing its peculiarities of fuel saving and pollutant emissions reduction. The aim of the field campaign is presented; it consists in the investigation of the behaviour of this type of system on a large enough statistical basis, in installations distribuited all over the Italian territory to consider different climatic conditions and plant types. The field campaign was based on periodical controls, by which several data on the plants and the state of the system were collected.
The analysis of this data, about 1.000.000 ,collected over 700 installed Bluflam systems and processed by means of a minicomputer, assessed that the system mantains in installation its characteristics of fuel saving and low pollutant emission during all the heating season.
Besides, with some modifications selected in the first year of the campaign, the system resulted reliable and quite easy to install and to maintain.
On the basis of this results the complessive energy saved by the 700 installed in respect of conventional systems has been evaluated by means of a computer program based on plant climatic conditions and cyclic efficiency curves.
In conclusion, the system resulted feasible from the industrial and economics point of view; it is proven

also by more than 4.500 Bluflam already installed in Italy Unfortunately the Italian heating market crisis adversed a larger diffusion of the system, so that the development of a second more economically competitive version of the Bluflam is already under development.

1. OBJECTIVE AND DESCRIPTION OF WORK

From 1st October 1978 to 31st October 1980, a field campaign was carried out about a full range of Bluflam integrated burner-boiler system for domestic heating, ranging from 31 to 407 KW.

The main goal of such campaign, was to test the Bluflam system 91 heat generators burning n°2 oil in normal working conditions and on a large enough statistic basis.

Such a system was designed within an intense research programme aiming at developing a range of domestic heating generators burning n°2 oil being highly efficient and little polluting.

In this system, the boiler is strictly integrates with a burner using combustion with cold flue gas recirculation.

With this technique a percentage of the flue gases (approximately 30%), at a temperature of about 180°C, is mixed with fresh air and conveyed back to the combustion chamber as shown in fig.1.

Thanks to this procedure, the combustion takes place with little excess of air and an extremely low level of pollutant emissions in comparison with traditional generators, as can be seen in fig.2-3-4-5.

It is worth noting that if the percentage of recycled gases in sufficiently high, the flue gases contain no soot; this is also true when the excess air is close to stoichiometric.

Besides, the emissions of nitrogen oxides at about 25% of cold flue gas recirculation is half that of traditional generators.

Finally the output of CO does not exceed 20 p.p.m. and that of HC 2 p.p.m. up to a rate of excess air of about 1.1.

These results were achieved thanks to a high intensity combustion head which was specially designed for combustion with cold flue gas recirculation.

The Bluflam system is also characterized by a pressurized combustion chamber around which highly efficient channels for the convective heat exchange are wrapped (fig.6).

It is therefore possible to have the stack temperature of the flue at about 180°C, with a small-sized compact boiler and a high ratio of heating rate to external surface.

High efficiency is thus obtained both in continuous operating conditions (91-93%) and in cyclic ones, thanks to a process of combustion carried out with a little excess air, low stack temperatures of the flues, and small and pressurized boiler.

In fig.7, the curve of the Bluflam System global efficiency in cycle is compared with those of some traditional systems.

It should be noted that the Bluflam higher performance is not reduced at lower loads.

The field campaign dealt with the full range of Bluflam generators (700 total) which included 13 models, ranging from 31 KW to 407 KW, heating only or heating and hot water generating systems, installed all over Italy (fig.8) so as to cover different working conditions and kinds of dwelling.

2. METHODS OF INVESTIGATION.

First of all 2 cards were prepared for collecting the data.

The 1st one shows the data relating to the installation and the checks which were carried out when the thermal generator was put in operation; the 2nd one shows the data relating to the setting point, to the working conditions of the generator and of its components.

These cards were filled in during periodic checks, 5 per season for heating generators, 8 for heating and hot water ones.

The data collection was carried out by specially trained field engineers; they had suitable instruments, including portable analyzers of CO_2 and O_2 as well as meters of Bacharach index.

The collected data (about 1.000.000) were registered on magnetic tapes and then evaluated by a small computer.

Such recording made it possible to point out the trend of both the performance and the main polluting emissions, for each model of the Bluflam System, during the whole campaign.

In order to evaluate the reliability of the system, the results of the checks on the generator working conditions were classified.

This enable us to find out the causes and kinds of the troubles classified according to the scheme in table 1.

Then the data were statistically processed thus obtaining (for each model) the histogram of the distribution per cent of the causes and kinds of the troubles.

A number of changes in the system to increase its reliability were developped by analyzing the above-mentioned diagrams.

In particular, the components with troubles amounting to over 5% were improved.

In the second part of the field campaign the whole procedure was repeated on the modified generators, in order to test whether the changes were effective.

3. MAIN RESULTS.

The statistical analysis of the data showed that in real and normal working conditions the stack temperature of the flue gases keeps nearly steady throughout the season.

It has also been noticed that the drift of the excess air is almost negligible, as shown by the trends in fig. 9 .

This picture illustrates the temperatures and the percentage of CO_2 in the stack, in the case of a BF 27,000 Kcal/h.

As a consequence, also the efficiency constantly keeps between 91% and 93% as said above.

Once confirmed the laboratory data, we were able to estimate the quantity of fuel that the examined installations had saved in 2 years time.

The estimation was made by studying the cyclic efficiency curve of the Bluflam thermal generators.

Such a curve had been experimentally determined (fig.7) and then suitably inserted into a computer programme which, making use of the climatic data of the various installations and of the allowed working time, calculates the fuel saving of the new generators, as compared to the average traditional ones having the same power rating.

Table I shows the estimated amount of fuel (in litres), saved by the 700 tested installations.

Fig.10 illustrates the saving capability of the Bluflam 350,000 Kcal/h generator in comparison with a traditional one, vs degree-days of the installation size.

Moreover, the analysis of the data showed that the polluting emissions had kept extremely low throught the field campaign.

It was noticed that the Bacharach index kept zero in each generator, which meant there was no soot in the flue gases.

A more complex analysis was necessary to evaluate the reliability of the system and above all to find out the main reasons for troubles.

A distinction was made between the troubles due to the components and those due to the system itself.

The classification and the above-described technique were useful to draw the histograms showing the probability rate of troubles causes and kinds based on the checks carried out on all the installed models.

After analysing the histograms, some changes were deemed to be necessary: particularly in the system of the air-flue mixture device and in that of the flame control, in the combustor can, in the ignition electrodes, in the fuel solenoid valve.

Picture 11a illustrates the probability rate of trouble causes which required servicing. Such a graph concerns all the installed models ranging from 27 to 115,000 Kcal/h and refers to the first and second phases of the field campaign.

In picture 11b the distribution concern 140 to 350,000 Kcal/h generators.

It can be easily noticed that in the second phase, and thanks to the mentioned changes, there was a reduction of the most probable troubles such as the operation of the air-gas mixture system, that of the ignition electrodes (fig.11a) flame control, and fouling deposition in the diffuser, (fig. 11b).

Similarly, fig.12a and 12b shows the rate of probability of the various kinds of troubles. Once again, there was a considerable reduction of all them in the second phase.

Thanks to the introduced changes, it was possible to improve the reliability which reached a good level.

The overall results were therefore very satisying, specially referring to the more complex type of generator.

4. CONCLUSIONS AND PROSPECTS

The field campaign of the Bluflam System 91 generators showed that a substantial fuel saving can be obtained also in real working conditions, and keeping for a long operating period a low rate of polluting emissions (soot in particular), as from the laboratory tests.

Moreover, the reliability of the system, operating with high efficiency, proved to be high enough, as shown by the statistical analysis of the collected data.

Besides, the modifications introduced during the second phase of the campaign gave to the range of generators a good level of industrial and commercial feasibility; provided that the servicing personnel is suitably trained, the system itself proved to be easy to be installed, and to require little maintenance for its operation.

In Italy, in fact, some 4,300 fuel and natural gas Bluflam System 91 generators have already been installed.

A new range of Bluflam, however, is now under development mainly to reduce its cost; improvement of reliability and operating qualities are also expected.

The same design criteria were utilized in order to keep its energy saving and its low polluting emission level.

5. ECONOMIC

The economic aspect of the feasibility of a technologically advanced system for domestic heating was also considered.

It can be noticed, in fact, that the fuel saving rapidly pays back the higher cost of this system if compared to the traditional ones.

Unfortunately, the energy saving systems characterized by higher selling prices are not favoured by the structure of the Italian market: on the one hand, 55% of the Italian population live in rented houses, on the other, the choice of the heat generators is made by installers, who are not directly interested in energy saving systems.

Thus, the above said reasons, together with the crisis of the building sector, prevent the sales of innovative products from being promoted.

Evidently, new laws precisely stating the quality standards of the generators for domestic heating and allocating funds for in the installation of "energy saving" systems, at both national and communitary levels, may help the diffusion of such Systems.

Anyway, the marketing of the new range of Bluflam System 91 generators now under study, as said before, will be better promoted by the reduced production costs and, consequently, selling prices.

RILIEVI RISCONTRATI

Motivazioni

1 . - Locali angusti, poco aerati e difettologia di intervento causata in generale da installazione.

2 . - Tensione di alimentazione troppo bassa.

3 . - Tensione di alimentazione fase-fase.

4 . - Errori di montaggio in generale.

5 . - Cuscinetti rumorosi.

6 . - Problemi connessi { alla funzionalità del miscelatore (1^ Serie) / palette flap (2^ Serie) }

7 . - Accensione a freddo difficoltosa.

8 . - Scarsa funzionalità della testa di combustione.

9 . - Mancanza di rilevazione fiamma.

10. - Scarsa funzionalità elettrodi accensione.

14. - Sostituzione elettrovalvole.

15. - Sostituzione miscelatori.

18. - Formazione di depositi.

19. - Deriva di funzionamento.

26. - Controllo generale.

Tipologia

11. - Sostituzione elettrodi accensione.

12. - Sostituzioni E.C.F.

13. - Sostituzioni raddrizzatori.

16. - Sostituzione ugelli.

17. - Sostituzioni varie di componenti.

20. - Variazione ugello.

21. - Variazione pulegge.

22. - Variazione distanze testa combustione.

23. - Variazione pressione gasolio.

24. - Variazione tempo di flap.

25. - Sostituzione bruciatore.

TAB. I

Working period	Fuel saving (liters)	
24 h	conventional efficiency	reduced efficiency (soot considered)
	2.098.440	2.224.436
continuous working hours *	1.271.254	1.347.529
not continuous working hours *	1.136.242	1.204.416

* according to the Italian Energy Saving Law, April 30, 1976, n° 373.

TAB. II

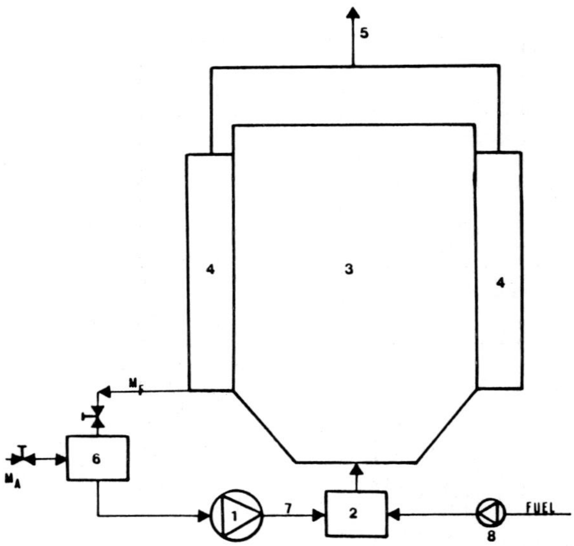

1 - Fan
2 - Combustion head
3 - Combustion chamber
4 - Flue gas collectors/ heat exchanger
5 - Stack
6 - Mixing box
7 - Fan diffuser outlet
8 - Fuel pump
m_a - Fresh air mass flow rate
m_f - CFGR mass flow rate
m - Mixture mass flow rate

Fig. 1: Principal scheme of a system using cold flue gas recirculation, CFGR (1)

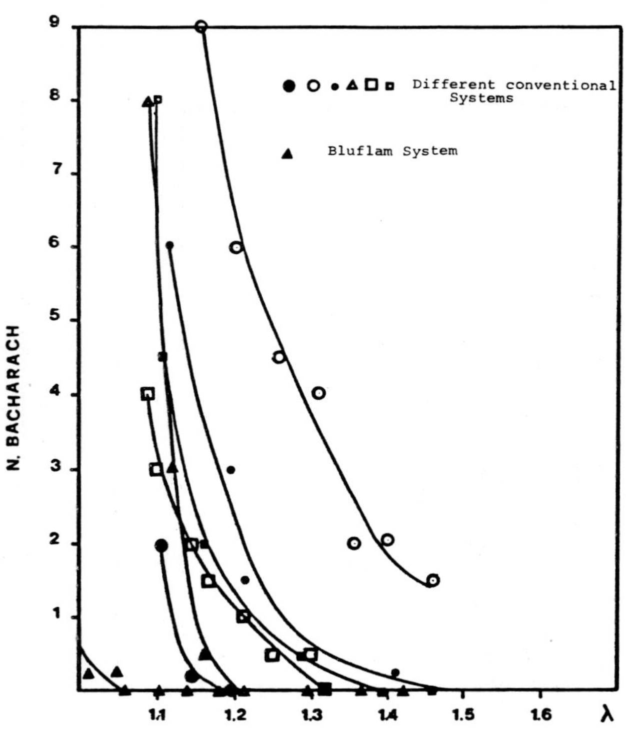

Fig. 2: Soot emission vs excess of air

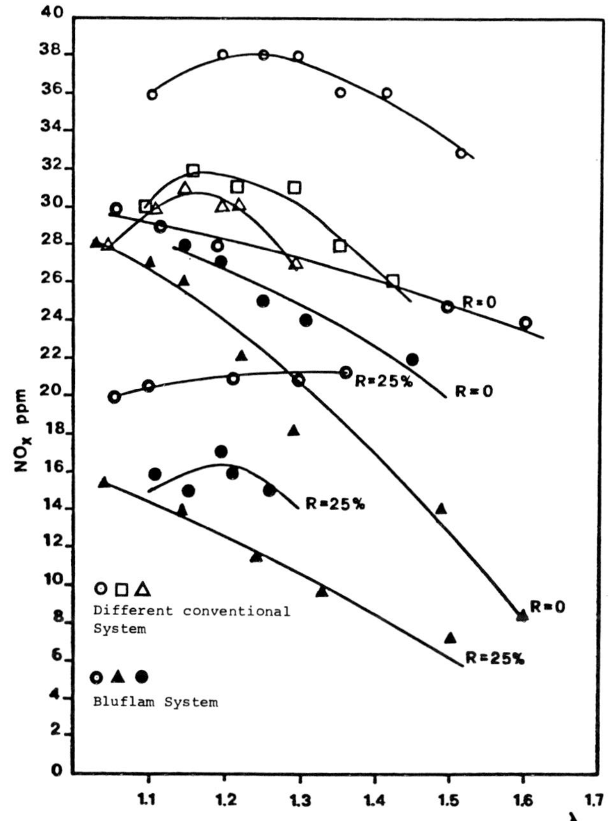

Fig. 3: NO_x emissions vs excess of air

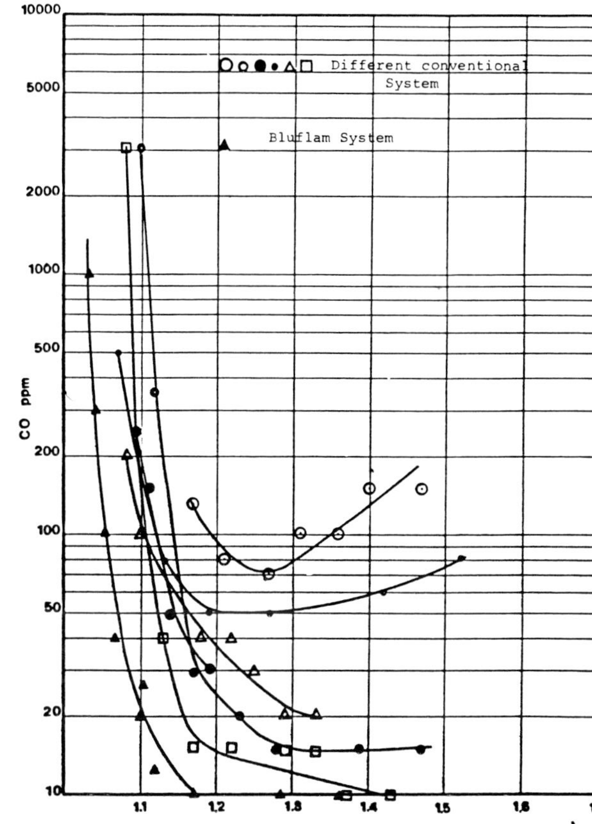

Fig. 4: CO emissions vs excess of air

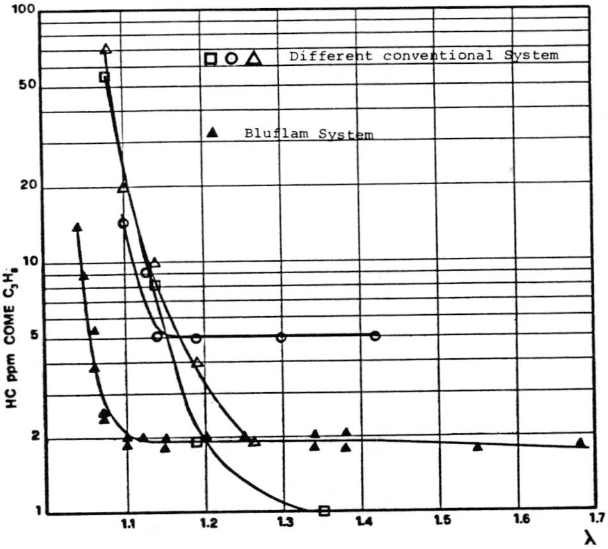

Fig. 5: Unburned hydrocarbon emissions vs excess of air

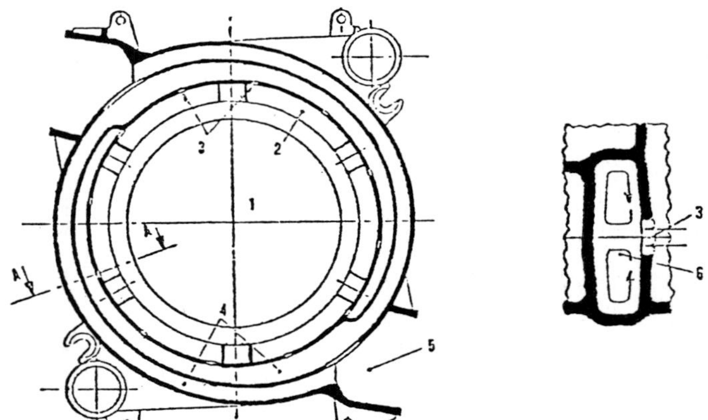

1. Combustion Chamber
2. Distributions Channels
3. Nozzles
4. Convective Channels
5. Flue-gas Exit
6. Secondary movements

Fig. 6: Spiral heat transfer channels arrangements

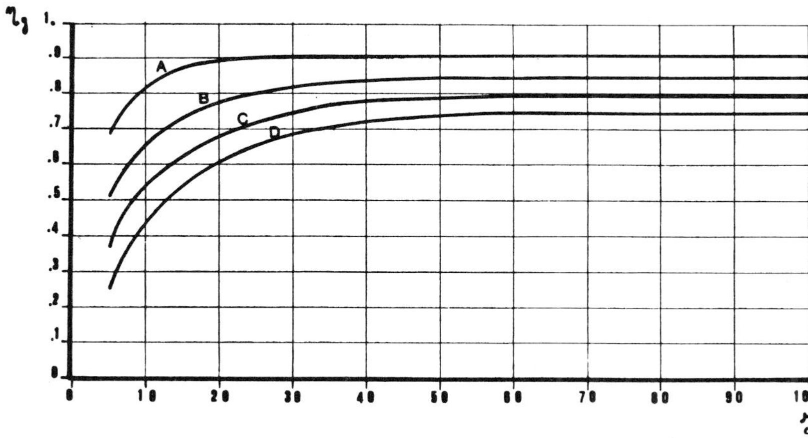

Fig. 7: Global cycling efficiency vs burner fractional "on" time% for different heating generators

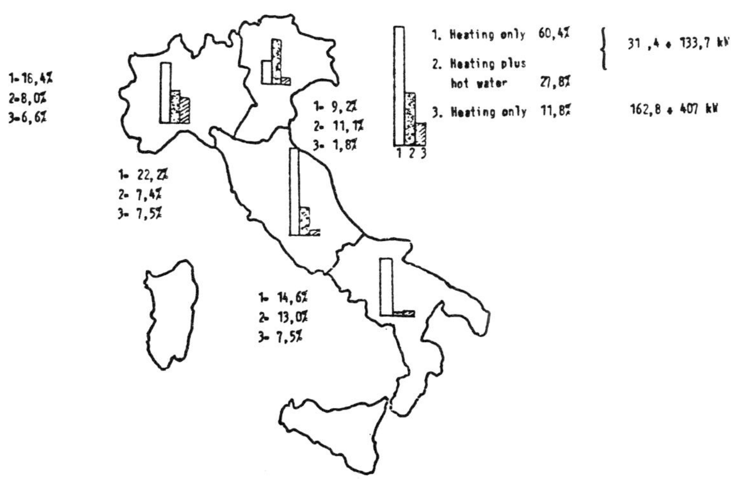

Fig. 8: Heating generators BF System 91 installed in seasons 1978-79, 1979-80: DISTRIBUTION FOR AREAS

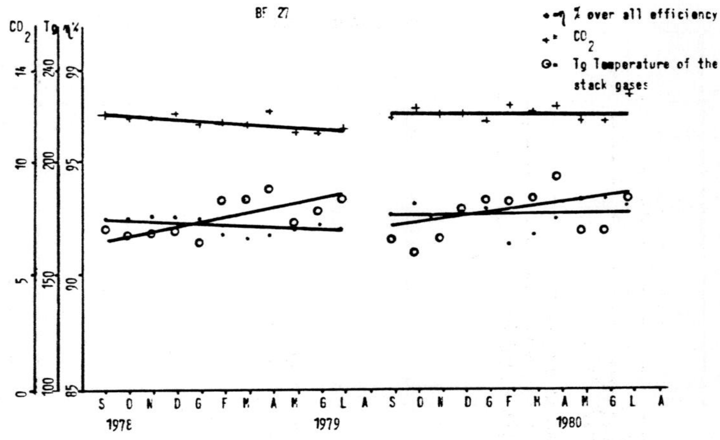

Fig. 9: Relief of measurements of combustion parameters related to BF 27 (31,4 kW)

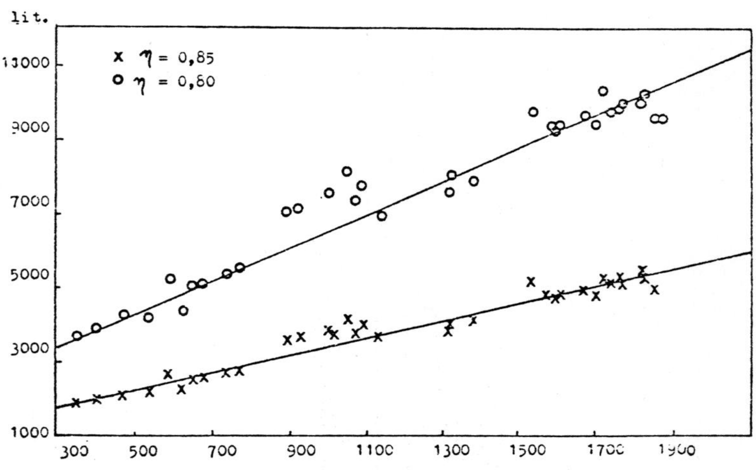

Fig. 10: Saving energy capability of Bluflam vs degree-days. Heating capacity 350.000 kcal/h

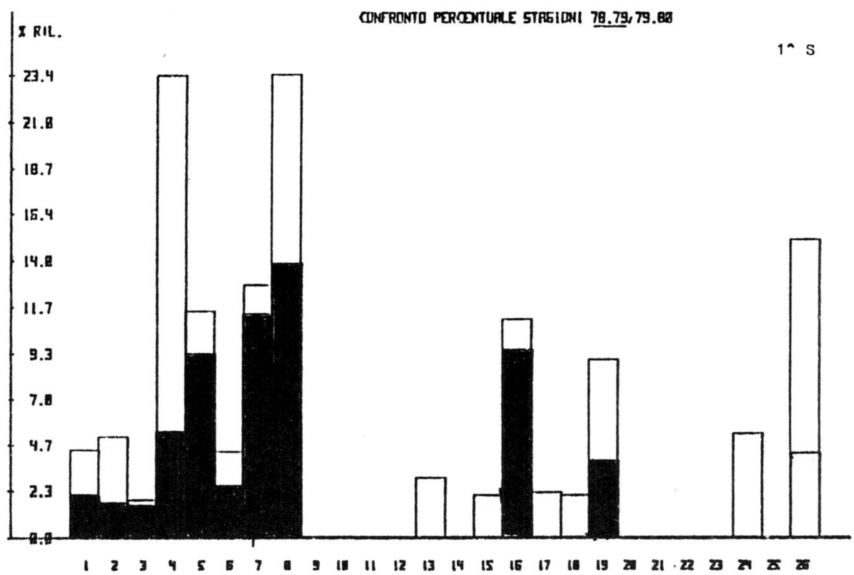

Fig. 11a: "Troubles causes"

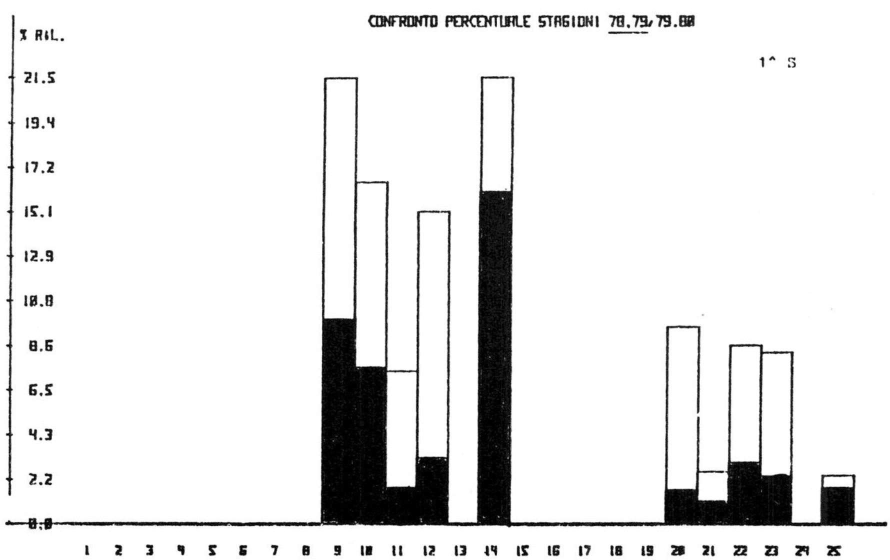

Fig. 11b: "Troubles causes"

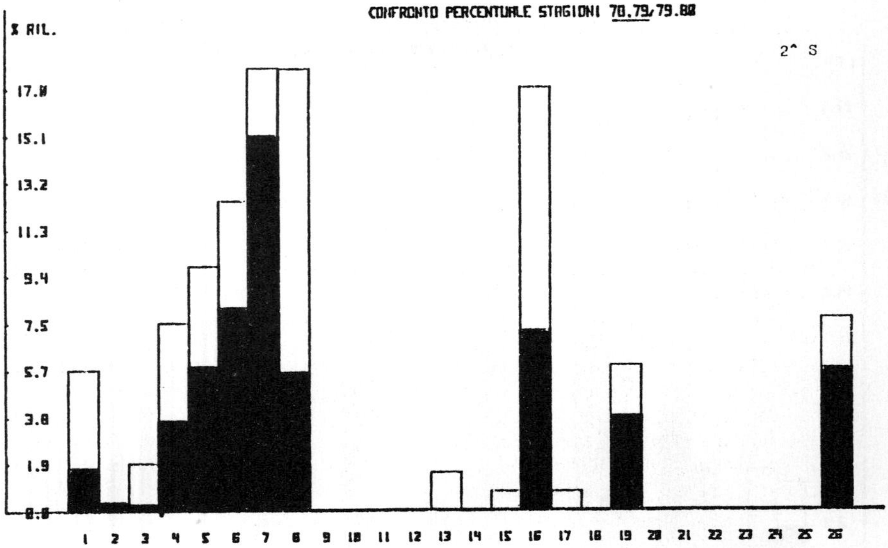

Fig. 12a: "Troubles kinds"

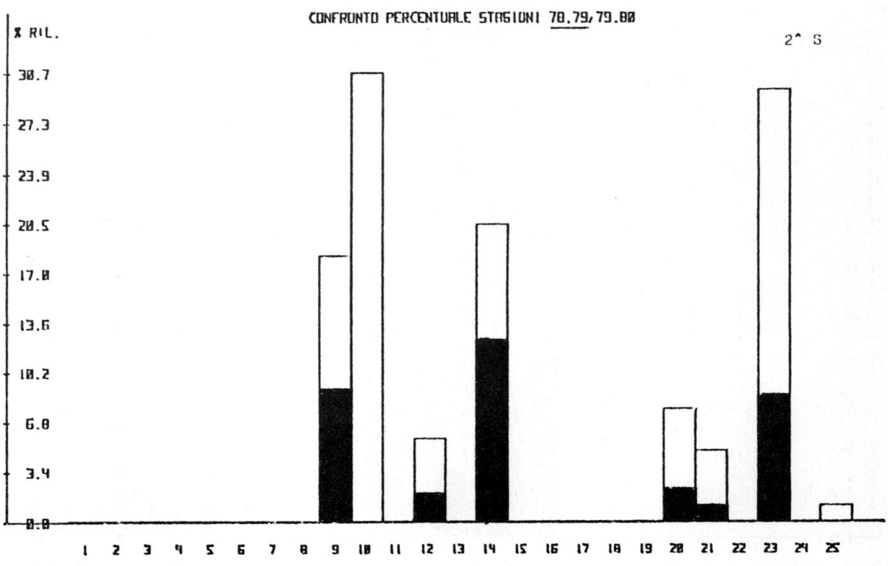

Fig. 12b: "Troubles kinds"

COGENERATION PLANT WITH POLYASSISTED HEAT PUMP FOR A COMMERCIAL BUILDING

P.L. Da Col and M. Gini
SOLARIA Engineering Ltd
Via Gaeta 31-UDINE (Italy)

SUMMARY

This article describes a heating and air conditioning system which puts into practice some of the techniques of reducing energy requirements: thermo-electric cogeneration, heat pumps, collection of "labile" thermal energy, and long term storage. The system is realised in a commercial building, used also for small scale manufacturing, in the suburbs of Udine (Italy).

This is one of the few examples, where the various techniques have been coordinated and used together in a building which has been constructed.

The authors of the project describe the principle characteristics, including the use of heat storage for both winter and summer (in summer heat is collected and stored for the following winter, while in winter the water stored is cooled, ready for use in summer).

1. INTRODUCTION

There has been much discussion in recent years about the saving of energy, and one of the most promising aspects that has evolved is the idea of a more rational and intelligent use of exixsting resources. This is already possible hovever, using present day technology as long as the complete design process is throughly organised and coordinated with this objective in mind.

In the past the topic of energy saving has often been approached on a theoretical level only as it has been difficult to put into practice. This can hovever happen in the case of a specific design project when both the skilled labour and the time are avaiable, and when everyone concerned is totally dedicated to the project.

We are therefore pleased to be able to describe a system, which has recently been realised in Udine, where the above factors were all present.

The design project under consideration is a total conditioning system for summer/winter use. The most important unit of this system is the thermal plant used to provide heat during winter and cold during summer. The project, hovever, has given more importance than usual to the fact that the thermal plant can also produce heat during summer and cold during winter. We feel that the efforts made to conserve this energy in a storage tank (described in greater detail later in this article) represent an original approach to this problem. Any heat generated during the winter can therefore be utilised by the occupants of the building and the tank is left to cool down, while in summer the process is reversed, heat being extracted from the building and stored in the tank.

During the course of a year, due to climatic and operating conditions, the demand for heat is considerably above that for cold and other sources of heat have therefore been included in the desing to maintain the balance between the production of heat and cold. These heat sources are "labile", and in general free as they are drawn from the environment. They include: solar collectors, heat produced during the process of generating electricity and heat from a refuse incinerator.

2. DESCRIPTION OF THE BUILDING

The commercial building under discussion has a internal volume of about 20,000 m^3. The peak thermal demand for winter and summer is about 580 kW and 230 kW respectively.

The building has the headquarters of two commercial companies and therefore contains officies, showrooms, sales areas, and storage facilities.

The thermal plant has been designed to ensure an indoor temperature of 20 °C with a winter outdoor minimum of -6 °C, and a summer indoor temperature of 27 °C with an outdoor maximum of 32 °C. These conditions can, hovever, be relaxed in specific areas of the building as required, and the resulting flexibility of the system is an important factor in the energy saving.

3. COGENERATION

The electrical energy needed for the functioning of the thermal plant is, as already mentioned, generated on the premises. There are five cogeneration units, each of which produces 15 kW_e and 33,000 kcal/h (38 kW_t) from 6 m^3/h of methane. The 15 kW_e can be used to generate, on average, a further 50,000 kcal/h by means of a heat pump. The total production of 83,000 kcal/h from 6 m^3 of methane is in itself an important result.

It should also be noted that the production of electricity on the premises is not only limited to the technical requirements of the plant but it is also used to provide lighting and power for equipment in the shed. The use of 5 generating units is therefore an important aspect, as each unit can be br ought into operation according to requirements. It is more convenient to generate electricity using this system than to buy it as any heat produced is always immediately re-used in the thermal plant in winter, while in summer it is sent to the storage tank for use as required.

4. OTHER HEAT SOURCES

The solar collectors, with a double transparent covering, are over an area of 105 m^2 on one of the roofs (see fig. 2). During the winter months the heat generated can be immediately used in the heating plant, while during the summer months it is sent to the storage tank. The collectors work for the water supply and they are of the "dry" type, that is they empty when no heat is produced.

Another source of free thermal energy is the refuse incinerator. This has a primary function of burning and dispossing of refuse and a secondary one of providing heat. As far as possible the incinerator is programmed to operate during the winter months, howewer heat produced at other times can be collected and either used immediately or sent to the warm sector of the storage tank for later use.

5. STORAGE OF HEAT

This can be considered the most important element of the system. Storage is carried out in a large elongated tank, 60 m. long, containing 1,200 m^3 of water (see fig. 1 for a plan and section).

During the summer, the temperature of the water rises due to the input of heat from sources such as the solar collectors, the refrigeration and cogeneration systems and the incinerator, when used. At the beginning of the winter months there is therefore a surplus of accumulated heat which is first used directly in the conditioning system and then by means of heat pumps. During the winter months the temperature in the tank slowly decreases due to the effect of the heat pumps so that there is a surplus of cooled water ready for use during the first part of summer. As the season progresses, cold is obtained from the compressors and any resulting heat produced is stored in the tank.

From a functional point of view, the processes of heating and cooling, and of storing water in the tank are helped by its elongated shape. From a thermal point of view, the tank has hot water at one end, while the other end has cold or tepid water according to the season.
Heat is always therefore available at the hot end; the opposite end can either give or receive the cold or the heat produced by the refrigeration equipment. Naturally the volume of heat stored increases during the summer months, while in winter the volume of cold water stored increases until the final condition already described is reached at the beginning of the hot season (see fig.3).

The walls of the tank are either in contact with the surrounding soil or exposed to the open air. The part of the tank in contact whith the soil can be considered as providing low heat losses and the soil itself is also used for its capacity to store thermal energy over a period of time .

The part of the tank open to the air, hovever, is due to be thermally protected by an insulated covering, because any heat (of either positive or negative value) reaching the atmosphere is heat lost and this must be reduced to a minimum. See fig . 4 for further details.

6. SUPPLY OF THERMAL REQUIREMENTS TO THE USERS

See fig. 6 for a general description of the system and figs. 5 and 7 for the heating plant.
The sources of heat available during winter are therefore: solar collectors, storage of surplus heat from summer, supplementary heat from the heat pumps, heat produced during the generation of electricity, and the refuse incinerator. There is also a conventional boiler, placed last in the order of heat sources, which is only used in case of emergency.
The sources of cold available during summer are : storage of residual cold from the use of heat pumps during the winter, and heat from the refrigeration plant using freon (R12) compressors.

7. ENVIRONMENTAL DISPLAY PANEL AND MEASUREMENTS

The overall operation of the system can be monitored on an environmental display panel, and near this is an electrical switchboard (see fig. 7).

There are also a series of metering circuits, which are used to optimise the operating programme and to analyse and follow its progress at intervals during the different seasons. These include :

a) remote temperature sensors, placed in key positions, displayed on the main panel;
b) counters relating to the activity of the cogenerators, the R-compressors and the boiler;
c) heat counters connected to: the main user elements of the system, the solar collectors, the incinerator.

The instrumentation is comprehensive, but without any unnecessary extras, as the main emphasis is on practical usage.

8. ECONOMIC ANALYSIS

See table 1 for a breakdown of the cost of execution of the project.

Table 2 compares the estimated running costs for a conventional design and for the one actually adopted, considering an average value for each of the thermal sources available. As can be seen, the two designs have approximately the same running costs, taking into account the capital consumption allowance and the higher figure we have conservatively estimated for the labour costs of the design adopted.

After the first years, hovever, we hope that these costs will be reduced as the workforce gains experience.

We considered it preferable, however in view of similar total running costs to adopt the design which has a high energy saving content (above 68%) and which can offer continuous period of operation, even though this results in higher labour costs.

9. APPENDIX

We include here (in alphabetic order) a brief glossary of some of the terms used in the text, as they may not all be familiar to our readers.

Assisted : a source of heat which is helped in its operation by other devices.
As an example, the storage of heat can be assisted by a heat pump, the temperature available is thus raised. A heat pump can also reuse some of the energy previously stored (autoassisted) or it can draw on other sources (heteroassisted, or in particular helioassisted). In the present design the system is polyassisted.

Bivalent thermal storage : storage of heat or cold according to the season.
It accumulates during one season for reuse during another season.

Cogeneration : a process of simultaneously generating electrical power and heat.
In the design under consideration motors operating from a methane supply are used to drive alternators and thus generate electricity. Any heat produced is used to increase the temperature of the water of the system.

Labile energy : this is energy, which if not collected immediately, will be lost to the external environment.

Thermal energy source : equipment or process which generates useful heat.
The heat can have a positive or negative sign, i.e. it can be used to warm or cool a circulating fluid. In the former situation it is usually referred to as "heat", in the latter as "cold".

Thermal plant : a system of compressors and circuits containing a circulating fluid to transport heat.
In the case of heat pumps (and release of cold) useful heat is generated, while in the case of refrigeration plant (and release of heat) useful cold is generated. In the simplest system there is an attempt to balance the production of useful heat with the corresponding release of energy.

TABLE 1 : breakdown of project costs (in millions of it. £ , 1980)

Preparation for storage tank	85
Piping, valves and various	23
Insulation	18
Fan coils	36
Boiler and accessories	6
Incinerator	10
Refrigeration plant and cold water tank	40
Automatic controls	44,5
Instrumentation, control panel and el. lines	41
Cogeneration units	40
Solar plant	45
Pumps and associated equipment	9
Design and supervision of construction	14.5
Total	**375**

TABLE 2 : estimates of running costs for a conventional design and for the design adopted, using average values for each of the thermal sources avaiable.

		conventional design	design adopted	
		(values in 1,000 it.£)		
1. ELECTRICITY				
1.1 Summer	80 kW x 700 h x 80 £/kWh	4,480	1,493	1/3 recovering of cold water + 1/3 from generating costs (methane) + 1/3 from mains
1.2 Winter	12 kW x 2000 h x 80 £/kWh	1,920	1,680	21,000 kWh x 80 £/kWh :heat pump
1.3 Annual	hot water boiler	195	1,920	24,000 kWh x 80 £/kWh :hot water, pumps and miscella neous equipment.
2. METHANE				
2.1 Summer			2,488	used by totem to produce 56,000 kW_e x 1/3

2.2 Winter	$\frac{350,000 \text{ kcal/h} \times 2,000 \text{ h}}{70 \% \times 8,200 \text{ kc/m}^3} \times 250$ £/ m^3	30,488	22,500	totem, for 2,700 h/each = 13,500 h total at 6 m^3/h
			2,938	supplementary use of boiler as conventional design plus 50 %
3. LABOUR CHARGES				
3.1 Summer	1/3 x 700 h x 12,000 £/h	2,800	4,200	as above
3.2 Winter	1/4 x 2,000 h x 12,000 £/h	6,000	9,000	as above
			3,538	totem maintenace, including spares : 14,744 h x x 240 £/h
			2,400	normal labour charges : 200 h x 12,000 £/h
			1,050	for maintenance and measurements. As above, but skilled : 70 h x 15,000 £/h
			1,000	as above, graduates : 50 h x 20,000 £/h
4. DEPRECIATION				
4.1 Summer	5 % x 90,000,000 £	4,500	4,500	as conventional design
4.2 Winter	10 % x 20,000,000 £	2,000	2,000	as above
4.3 Other parts	5 % x 50,000,000 £	2,500	2,500	as above
4.4 Various		1,000	1,000	as conventional design
			2,200	2% x 110,000,000 £ (storage)
			1,800	4% x 45,000,000 £ (solar plant)
			5,250	15% x 35,000,000 £ (instrumentation)
			1,250	5% x 25,000,000 £ (remaining plant)
5. TOTAL		55,833	74,552	
6. RETURNS (to be deducted)			17,693	electricity generated on the premises : 15 kW x 14,744 h x 80 £/kWh
7. EFFECTIVE ANNUAL RUNNING COSTS		**55,833**	**57,059**	

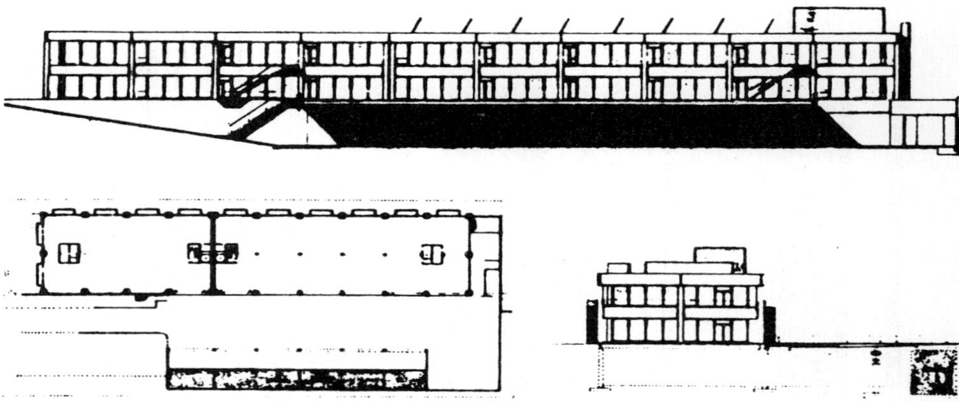

Fig.1: Plan of the building showing storage tank East elevation and cross section.

Fig.2: View of the solar collectors on the roof of the building.

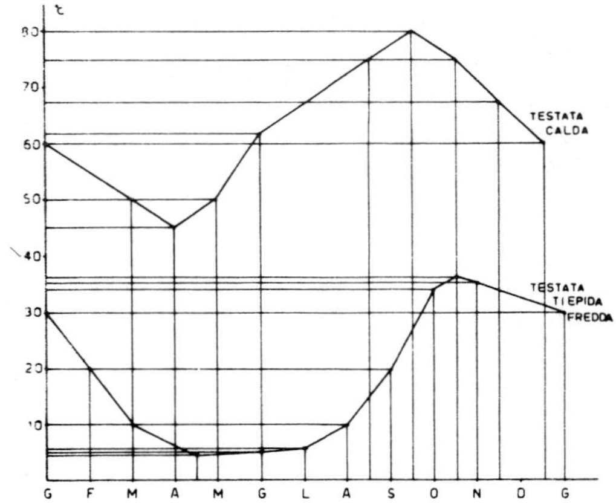

Fig. 3: Theoretical annual average temperature variation in the storage tank.

Fig. 4: The underground garage. The storage tank is behind the end wall. The pipes connecting the various parts of the system and the stock of insulating material (for the end wall) a-re also visible. As can be seen, the pipes have already been placed at the correct distance from the wall.

Fig. 5 : A partial view of the heating plant, with the heat pumps in the left foreground. The boiler is on the right and the thermo-electric cogenerators are in the background.

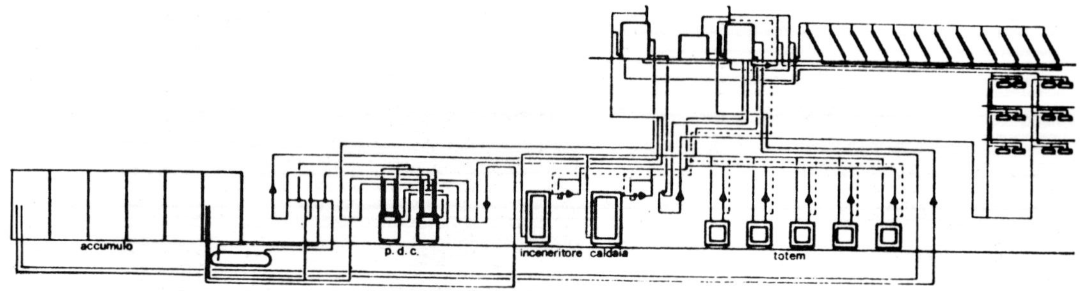

Fig. 6 : General layout of the system.

Fig. 7 : Main electrical switchboard with environmental display panel, heat pump and boiler.

INSTALLATION AND TESTS ON A STATIC EMULSIFIER FOR LIQUID FUELS BURNERS

STATIC FUEL OIL/WATER EMULSIFIER

Ing. M. Carello and Ing. F. Pica

Solar 77 - AgipPetroli S.p.A. - Rome (I)

SUMMARY

It has been known for quite some time that emulsification of liquid fuels with water improves combustion.

In the programme dealt with here, aqueous fuel emulsions have been burnt in heat-production plants representative of the existing range of burners.

Fifty plants rated at between 20 to 4000 kg/h running on liquid fuel, installed in hotels, schools, apartment blocks and factories were converted to operate on emulsion.

A check has been made of the reliability and soundness of the Emulsistem by evaluating the direct and indirect benefits (slight reduction in consumption on the one hand, less fouling and pollution on the other) which may render this application interesting in many cases. An important feature of the system is that it sometimes permits the use of less valuable fuels which means that various petroleum products can be better utilized.

INTRODUCTION

Petroleum derivatives still play a fundamental role in the global energy scenario. Though there is the prospect of new primary sources of energy redimensioning the role of oil, this fuel will certainly remain to the forefront at least until the early years of the second millenium.

It is evident, therefore, that best use must be made of petroleum derivatives by trying to keep consumption low, while endeavouring to use each in the most appropriate manner.

Fuel-oil/water emulsion must be viewed in this context. It has been known for quite some time now that the emulsification of a liquid fuel with water improves combustion; this technique can be used on heating plants and internal combustion engines.

The principle is based on the action of droplets of water contained within the drops of fuel. The droplets evaporate violently when introduced into an ambient at a very high temperature (combustion chamber of an engine or the furnace of a boiler) and these "micro-explosions" cause further atomization of the fuel.

Technical and commercial interest in emulsion equipment has increased in recent years because the method permits better use of the fuel oils involved and enables lower-grade products to be utilized, since the emission of unburnt solid particles is greatly reduced when an emulsion is burnt.

EMULSISTEM operating principles

The EMULSISTEM uses a static emulsifier, so it is very reliable and simple to install. The operation of the emulsifier proper is based on the superimposition of number of physical effects, such as diffraction of the jet, impact, attrition among fluid particles, boundary-layer flow effect and sudden variation of pressure.

The emulsifier used in the EMULSISTEM is a passive device, there being no moving parts and it only requires a small head loss (2 bar) to ensure good operation and a fine emulsion.

Emulsion quality is a decisive factor as regards longterm stability.

EMULSISTEM ensures a very stable fuel oil emulsion that will remain unchanged for several months at room temperature or a week at a constant temperature of 90°C. This stability is provided by the very fine dispersion of the water droplets around one micron in size. This very fine dispersion also guarantees higher efficiency in the combustion phase, since the size of the water droplets is far smaller than that of drops leaving the atomizer nozzle (60 microns).

Let us now examine the operation of a burner running on emulsion. Fig. 1a illustrates the schematized circuit of a conventional single-stage fuel-oil burner, while Fig 1b shows the circuit after modification to operate with EMULSISTEM. Fig. 1c and 1d provide the same comparison for a dual-stage burner.

The system essentially has two main parts: the mixer-metering device and the emulsifier.

- The mixer is the device which ensures the mixing and the addition of the right amount of water for the fuel oil. In the case of enbloc single-stage and dual-stage burners, the mixer consists of a component in which one or two jets are inserted for metering the desired quantity of water (the best oil-water ration is about 20% by weight).

- The emulsifier is the device which actually produces the emulsion. It requires a roughly mixed feed containing the right proportion of water. The emulsifier consists of a conduit of constant cross-section which runs to a cylindrical receiver. The mixture delivered under pressure to the convergent inlet line, passes rapidly through the short lenght of cylindrical line an then enters the receiver where the emulsion is formed. This leaves the device by a discharge line. Operation of the emulsifier is based on the superimposition of several physical phenomena whose effect is to greatly reduce the size of the water droplets thus producing a uniform emulsion.

For industrial plants with burners of the modulating type, the solutions for changing over to emulsion feed require studying case by case. Fig. 2a illustrates the simplified schematic of a modulating F.O. burner, while Fig. 2b shows the modification needed to run on emulsion.

For modulating burners the mixer is replaced by a mixing valve designed to permit a wide range of modulation so as to enable the right quantity of water to be fed into the circuit continuously to suit the fuel oil flow, thus maintaining the F.O.-water ratio virtually constant over virtually the whole of the operating range.

Objective and description of work

The aim of the work reported here has been to check on applicability of the EMULSISTEM on a wide range of heating plants representative of those commonly in use today.

In detail, the objective was to:

- check the long-term operation of the system installed and to ascertain what operating problems may arise;
- check the long-term trend of the combustion values and especially to ascertain whether the reduction in the amount of smoke compared with usual values minimizes soot deposits on the heat-exchange surfaces of the boilers, thus ensuring more constant flue-gas temperatures and hence fuel efficiency;
- compare the results obtained with those achieved prior to the switch to emulsion.

From eighty-six plants originally considered, fifty were chosen for the trial. These included the plants held to be really significant as regards type and validity of the programme or as regards improvement in efficiency or the need to reduce polluting emissions.

It is as well here to mention that there were many reasons which led to the selection of the fifty plants included in the programme. The idea was not solely to select plants where it would be relatively easy to attain marked improvements in efficiency (increases of 15% were achieved on some). Instead, many plants were included which had air-pollution and smoke problems and these were solved by the use of emulsion, sometimes at the expense of a slight loss of efficiency.

The programme was in two stages:

A) Conversion of fifty plants to emulsion. The plants were located in various parts of Italy and ranged in capacity from 20 to 4000 kg/h fuel. They included seven hotels, eleven factories, one school, two sports centers, twenty-four blocks of flats and five hospitals.

B) Campaign of measurements and checks on the modified plants to evaluate reliability and to ascertain combustion values over a lengthy period of time.

Control methods adopted

During the weekly check visits the technicians has to collect two kinds of information:

- Control data regarding the correct operation of the emulsion plant (water pressure, water flow rate, oil-water ratio, emulsion preheating temperature, pressures at emulsifier, operation of electrically-controlled water valve, correct re-ignition after shutdowns, amount of fouling, etc.).
- Measurement data concerning combustion efficiency (flue-gas temperature, percentage carbon dioxide in flue gases, Bacharach smoke number, room temperature and efficiency).

Because of the need to ensure continuity, comparability and reliability of a great number of measurements made by different people on plants with diverse characteristics, it was very important to devise an organized method of data collection which would permit all the objectives to be attained.

Main results

As a result of the measuring campaign the following data were acquired for each plant:

- Efficiency on fuel oil alone.
- Efficiency on emulsion when the plant was converted.
- Average efficiency on emulsion.

To define the difference between efficiency with oil alone and with emulsion, account has to be taken of the fact that while the plants where running on emulsion, weekly measurements were made so the average efficiency is known, but as regards operation on fuel oil alone, only the efficiency at the time of the modification is known.

Boiler fouling, of course, causes quite a marked drop in combustion efficiency. The soot deposit on the heat-exchange surfaces forms rapidly and grows to a stage where it is so thick that further deposition is prevented. The efficiency of the boiler then stabilizes at an almost constant but obviously lower value (see Fig. 3).

As this fouling is avoided when a F.O.-water emulsion is burnt, a meaningful comparison can only be made between the average efficiency values measured over the course of time on boilers operating on straigh F.O. or on F.O.-water emulsions.

Consideration of average emissions with straight F.O. compared with those in the case of F.O.-water emulsions (see Fig. 4) and of the results of long-term tests run in the laboratory on a plant operating on straight F.O. and having a smoke number of 5 on the Bacharach scale, reveals a drop in average efficiency of about 2.5 percentage points after 200 hours operation. Considering an average of 800 hours for each heating season and one boiler-cleaning per year in the case of domestic installations, it can be estimated that there is a 6% drop in their average efficiency over the season. Therefore a figure of 6% has been added for non-factory plants and 3% for factory plants (where the boiler is cleaned more frequently) to take account of the above factor when calculating the differences in efficiency between straight F.O. and emulsion for evaluating the fuel saving achieved.

On the sample plants as a whole the smoke number dropped by an average of 4 points on the Bacharach scale, while there was an average increase of 2 points in the CO_2%, thus signifying a reduction of the excess air, higher efficiency and especially the almost total elimination of soot deposits due to unburnt material, thanks to the reduction in the Bacharach smoke number attainable with emulsion operation even when heavy fuel oil is involved.

Considering the average results measured on the fifty plants and bearing in mind the points already made on efficiency, it can be estimated that the average saving attained with the use of emulsion was around 3,5%.

Table 1 sets forth significant data on some of the plants converted to run on emulsion. It includes the technical characteristics of the plants and the combustion data for straight fuel oil and emulsion operation.

Conclusions and prospects

It emerges from the evaluation of the overall results obtained during the two-year programme that any type of burner can be modified to run on emulsion. The positive effects of the emulsion in improving efficiency or, in other cases, in drastically reducing smoke have emerged clearly from the results obtained, as has the good reliability of the system components.

The advantages of the EMULSISTEM both for domestic and industrial plants can thus be summarized as a slight reduction in fuel consumption, lower maintenance costs and fewer plant stoppages, which is a very important feature where industrial facilities are concerned.

To conclude, the use of F.O.-water emulsions has interesting possibilities, especially as regards improvement of long-term efficiency and reduction in pollution.

Economic assessments

The EMULSISTEM is easy to apply to any heating plant operating on liquid fuel. The small dimensions of the components and the simplicity in installing the system ensure that installation costs are kept within economically viable limits.

The direct and indirect benefits (lower fuel consumption and less fouling on the one hand and lower maintenance and reduced pollution on the other) depend on many factors, but on the whole they provicee a spur for investing in the system, taking account of the points made above.

The time needed to recover the investment has to be worket out case by case, depending on the type of plant, the number of hours operation per year and the possibility of being able to use a cheaper fuel.

Bibliography

ANGLESIO P. — Fundamental characteristics of the combustion of fuel-oil/water emulsions in boilers (Italian text). Turin Polytechnic - ATI LOMBARDA Meeting 1981

MONTI F. — EMULSISTEM AGIP - Characteristics of use and performance in domestic and industrial plants (Italian text). AgipPetroli - ATI LOMBARDA Meeting 1981

CARELLO M. — Benefits, results and prospects of the EMULSISTEM for burning fuel-oil/water emulsion (Italian text). Solar 77 - Meeting on "The energy problem in small and medium-sized factories" ATI LAZIO SECTION, November 1980

ANGLESIO P. — Energy evaluation of the use of fluid-fuel/water emulsion for heating buildings (Italian text). La Rivista dei Combustibili - XXXIII 1979

Data on some plants running on emulsion

CLIENT	APPLICATION	PLANT SPECIFICATIONS	FUEL	Combustion tests					
				Straight fuel oil			Emulsion		
				Bacharach	CO^2%	Fuel gas temp. °C	Bacharach	CO^2%	Fuel gas temp. °C
ISTITUTO GIORGI Genoa	SCHOOL HEATING	ENBLOC BURNER 1.000.000 Kcal/H	FO 3-5°E	7	9,5	280	1	12,5	260
ASSICUR. GENERALI Rome	Apartment-block heating	3 Riello enbloc burners	FO 3-5°E	8	10,5	250	2	13,5	240
				7	9,5	240	0-1	12,5	230
				7	9,5	250	0-1	13	240
JOLLY HOTEL Florence	Hotel heating and air conditioning	2 S.Andrea enbloc burners 2x4.000.000 Kcal/h	FO 3-5°E	7	11	280	1	14	270
				7	10	275	1	14	265
HATU' Bologna	Steam generation	2 Monarch enbloc burners 2x1.900.000 Kcal/h	FO 3-5°E	8	9	290	1-2	13	270
				9	9,5	290	1-2	12,5	275
		1 ICI Modul. burner 4.000.000 Kcal/h		8	9,5	280	1-2	12,5	260
AMMINEX Rome	Steam generation for factory	Ray enbloc burner 9.000.000 Kcal/h	FO 3-5°E	8	10	280	0	14	240
PLASMON Milan	Steam generation	Vamco modulating burner 3.300.000 Kcal/h	FO	4	8	217	1	11,5	187
Hotel "Villa Pamphili" Rome	Hotel heating and air conditioning	3 enbloc burners 3x1.000.000 Kcal/h	FO 3-5°E	7	11	300	0-1	14	270
				7	11	300	0-1	14	260
				7	11	300	0-1	14	260
TENNIS CLUB EUR Rome	Hot-water production	Enbloc burner 250.000 Kcal/h	Gas Oil 1-14°E	2-3	10	280	0	12	270
IMPDAI Rome	Apartment-block heating	Thermonafta burner 500.000 Kcal/h	Heavy FO	3	9	180	1	12	150

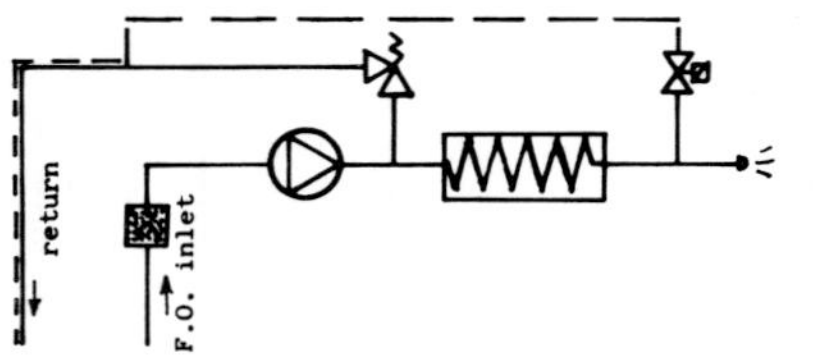

Fig. 1a: One-stage fuel oil burner
- - - pre-purging line
——— working line

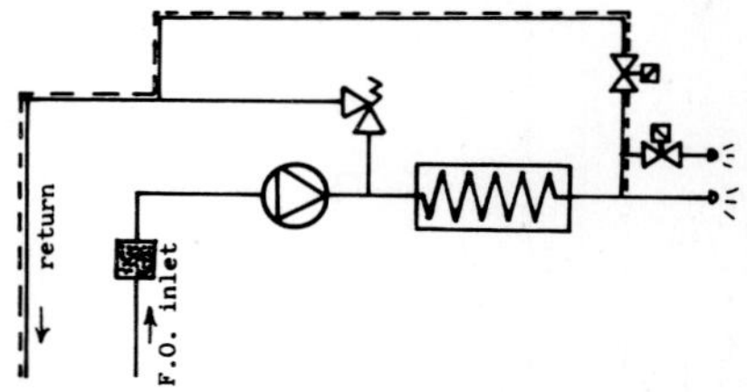

Fig. 1c: Two-stage fuel oil burner

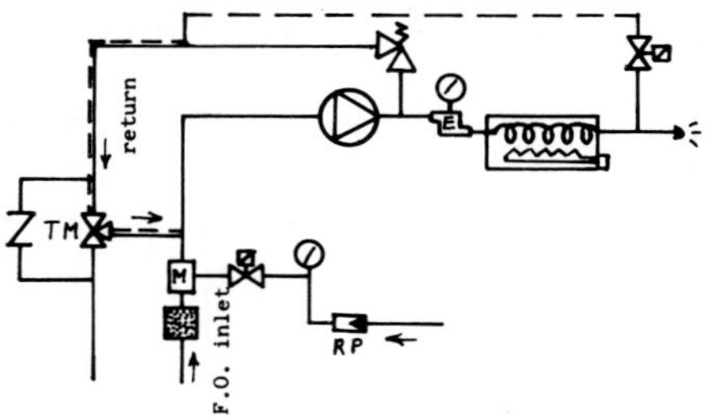

Fig. 1b: One-stage emulsion adapted burner. E: emulsifier M: mixer RP: pressure regulator TM: three way manual valve

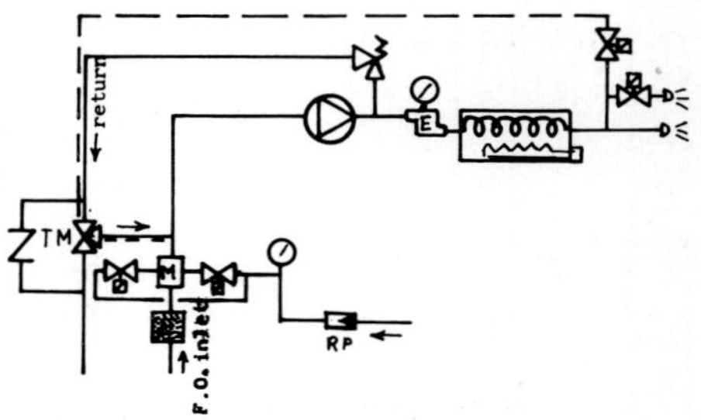

Fig. 1d: Two-stage emulsion adapted burner

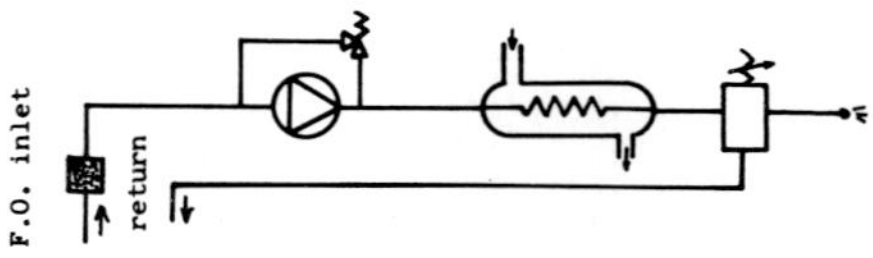

Fig. 2a: Modulating fuel oil burner

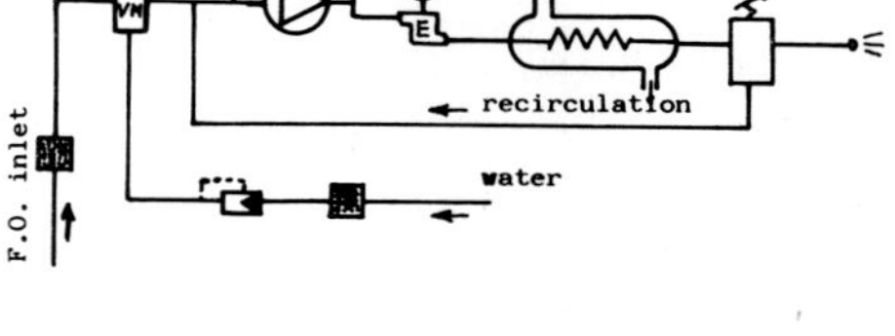

Fig. 2b: Modulating emulsion adapted burner - VM: mixing water - E: emulsifier

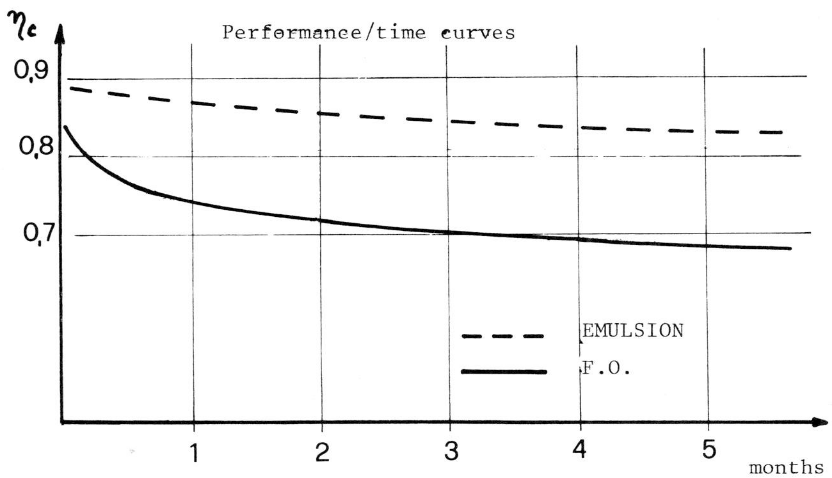

Fig. 3

F.O. EMISSIONS 540°E at 50°C

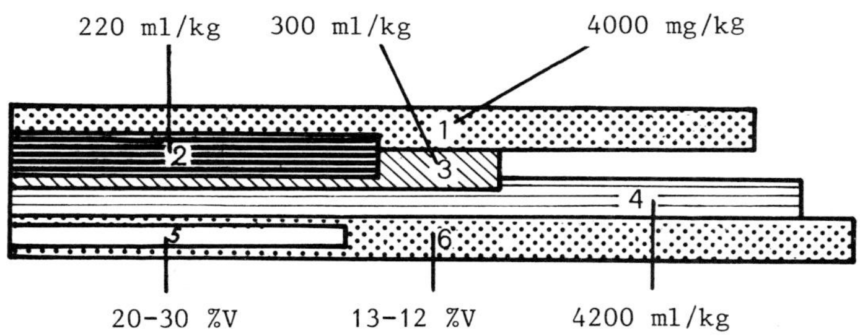

EMULSION EMISSIONS (F.O. 40°E at 50°C + 18% water)

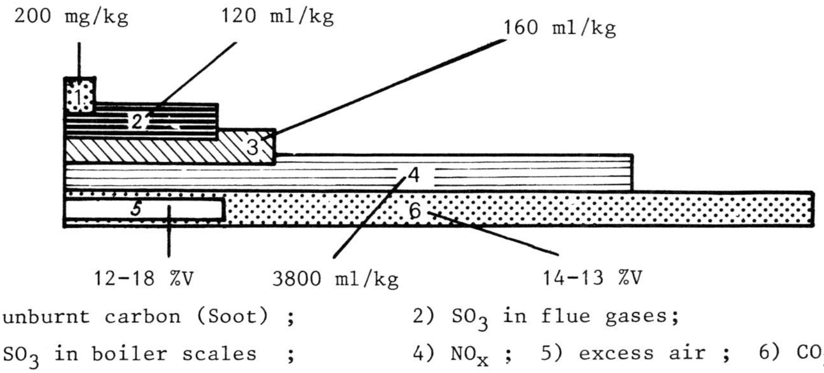

1) unburnt carbon (Soot) ; 2) SO_3 in flue gases;
3) SO_3 in boiler scales ; 4) NO_x ; 5) excess air ; 6) CO_2

Fig. 4

POMPE A CHALEUR ELECTRIQUE UTILISEE POUR RECUPERER LA CHALEUR D'UN LAVE-VAISSELLE

G. SCHOEFFLER
Etablissements BONNET S.A.

Résumé

Les machines à laver la vaisselle pour collectivités consomment une grande quantité d'énergie électrique pour leur fonctionnement, en particulier pour le chauffage et le maintien en température de l'eau servant au lavage et au rinçage de la vaisselle. Une approche de récupération d'énergie a été faite par l'intermédiaire d'un échangeur à circulation d'eau prenant la chaleur des buées produites par la machine. Cette solution permet d'économiser 20 % de la quantité d'énergie nécessaire à la machine, l'eau étant peu chauffée dans ce condenseur de buées. Une solution mettant en jeu une pompe de chaleur comme récupérateur d'énergie semble intéressante si son fonctionnement est adapté aux lave-vaisselle de collectivités. Une économie de 45 à 50 % est envisageable. Son application permet de chauffer l'eau de la ou des cuves de lavage et maintenir la température à une valeur permettant une action efficace, de préchauffer jusqu'à 60° C l'eau de rinçage, et de conditionner l'air ambiant du local de lavage de la vaisselle. La démonstration qui a eu lieu dans la laverie de la cafétaria de l'hyper marché Carrefour Nice (France) a permis, par comparaison des mesures effectuées avant et après installation de la pompe de chaleur, de mettre en évidence une économie d'énergie de 41,5 %.

1. BUT ET DESCRIPTION DU PROJET

Récupération d'énergie par pompe de chaleur sur une machine à laver la vaisselle de collectivités permettant une économie de 45 % sur la consommation d'énergie pour une machine ayant une puissance installée de 95 kW.

12. L'installation

L'installation de démonstration sera réalisée dans la laverie de vaisselle de la cafétaria de l'hypermarché Carrefour à Nice Lingostière. Une machine à laver la vaisselle BONNET, type INO C 50 DP, équipée d'une pompe de chaleur spécialement étudiée pour cette application servira de référence pour le projet de démonstration. Les notices de calcul et l'étude de la pompe de chaleur seront faites par les techniciens de BONNET suivant les caractéristiques de fonctionnement d'une machine à laver la vaisselle de collectivités.

Des relevés faits antérieurement sur la machine montrent que la température moyenne de l'eau froide de remplissage est de 15° C (mini 12° C - maxi 20° C). Cette valeur servira de référence pour les calculs de dimensionnement du système échangeur sur eaux usées qui permettra d'élever la température de l'eau froide jusqu'à 30° C.

Les caractéristiques principales de la notice de calcul sont données dans le tableau suivant.

Caractéristiques	Pompe à chaleur				Réchauffeur instantané
	Chauffage eau de lavage		Chauffage eau de rinçage		
Quantité d'eau à réchauffer	290 l		900 l/heure		900 l/heure
Température eau d'alimentation	15° C		30° C		60° C
Température eau sortie échangeur	60° C		60° C		85° C
Puissance calorifique nécessaire	13.050 kCal		27.000 kCal/h		22.500 kCal/h
	Air				
	Entrée	Sortie	Entrée	Sortie	
Température au : . bulbe humide . bulbe sec	 20° C 20° C	 0° C 0° C	 50° C 49° C	 20° C 20° C	
Humidité relative HR %	100 %	100 %	95 %	100 %	
Enthalpie kCal/kg air sec	13,7	2,25	61,5	13,7	
Teneur en eau g/kg air sec	14,85	3,75	80,65	14,85	
Volume massique m3/kg air sec	0,85	0,778	1,033	0,85	
Coefficient de transfert			1,4		
Puissance à l'évaporateur	5152,5 kCal/h		19285 kCal/h		
Débit d'air	350 m3/h		343,4 m3/h		
Température évaporation	- 27° C		12,5° C		
Température condensation	+ 25° C		65° C		
Puissance	2,7 kW		6,33 kW		

La figure I présente une vue générale de la machine équipée de la pompe de chaleur.

13. La pompe de chaleur

Il est retenu deux compresseurs hermétiques "Maneurop" type MT 80 HT ce choix est guidé par le faible niveau sonore des appareils, leur faible encombrement et la possibilité de pouvoir travailler avec le fluide R 12 à 75° C en condensation. Le R 12 est retenu en tenant compte du fait que les fortes températures de condensation n'entraînent pas des pressions trop élevées : 65° C / 15,87 bars.

La pompe de chaleur est constituée de deux circuits fonctionnant en parallèle, l'un pour le chauffage de l'eau de lavage, l'autre pour le préchauffage de l'eau de rinçage suivant figures II et III .

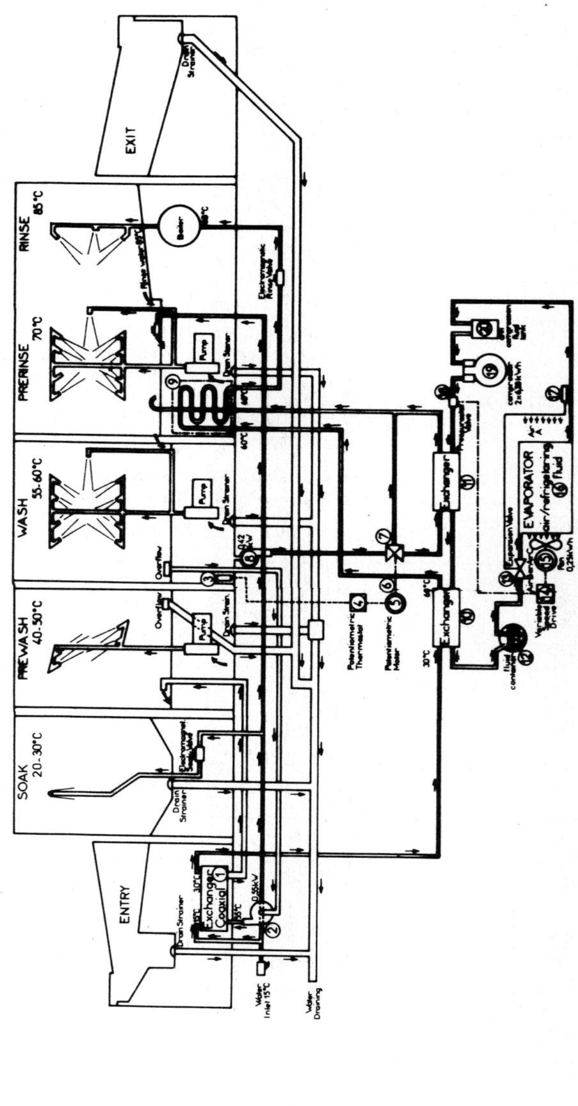
ENTRY
SOAK 20-30°C
PREWASH 40-50°C
WASH 55-60°C
PRERINSE 70°C
RINSE 85°C
EXIT
Boiler
Pump
Pump
Pump
Drain Strainer
Overflow
Overflow
Exchanger Coaxial
Exchanger
Exchanger
EVAPORATOR air/refrigerating fluid
Potentiometric Thermostat
Potentiometric Mixer
Expansion Valve
Variable Speed Drive
Water Inlet 15°C
Water Draining

figure I

figure II

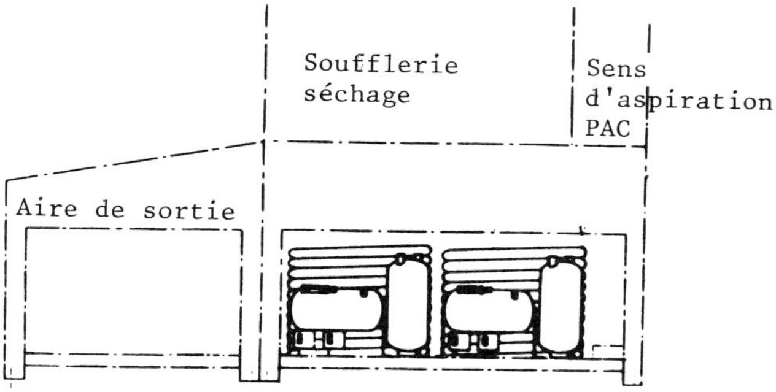

Implantation Pompe à chaleur sous la machine

figure III

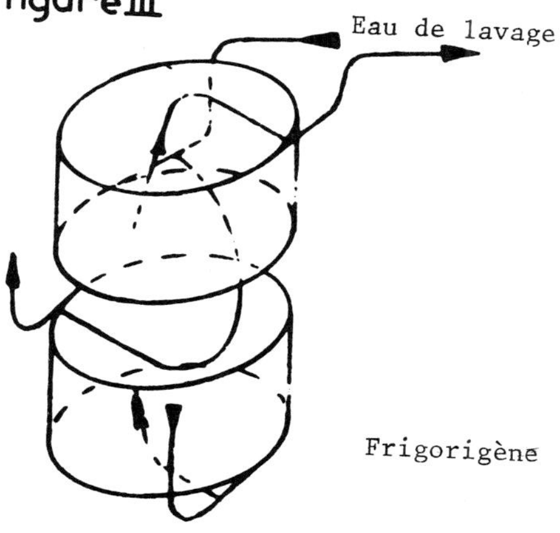

Circuits des fluides

La circulation de l'eau de lavage et celle de l'eau de rinçage ne sont pas simultanées. Priorité est donnée à l'eau de remplissage rinçage, l'eau de lavage est réchauffée durant les périodes où il n'y a pas de circulation d'eau de rinçage.

Nous pouvons retenir deux séquences significatives de fonctionnement du système :

* En fonction chauffage eau de cuve de lavage

. les pompes de lavage ne fonctionnent pas,

. l'eau de rinçage ne circule pas,

. la pompe de cyclage force l'eau des cuves de lavage et prérinçage à travers les condenseurs de la pompe de chaleur réservés à l'eau de lavage pour la réchauffer.

* En fonction lavage-rinçage ou remplissage

. les pompes de lavage fonctionnent (lavage-rinçage) ou ne fonctionnent pas (remplissage),

. l'eau de rinçage ou d'alimentation circule ; elle est préchauffée dans l'échangeur statique sur eaux usées, puis elle passe dans les condenseurs rinçage de la pompe de chaleur où elle est réchauffée jusqu'à 60° C.

Les besoins calorifiques de la machine lors de la phase lavage rinçage sont les plus importants puisqu'il faut porter la température de l'eau d'alimentation de 15° C à 85° C, soit 81,34 kWh.

La consommation horaire de la pompe de chaleur telle que définie et additionnée de la consommation du réchauffeur électrique d'appoint est de 42,38 kWh, soit le rendement de la pompe de chaleur = 1,92.

14. Régulation

Lorsque les cuves de lavage et prérinçage sont en température, les besoins calorifiques sont moindres. Ils sont dus aux dépertitions par les parois et au réchauffage de la vaisselle.

La vaisselle circulant, il y a rinçage donc préchauffage de l'eau de rinçage. Le maintien de la température de l'eau de lavage à 60° C et celui de l'eau de prérinçage à 70° C se fait par l'apport total de l'eau de rinçage à 85° C.

Cette solution, moins coûteuse, a été préférée à celle qui consistait à faire circuler une quantité d'eau de lavage plus ou moins importante dans le condenseur échangeur lavage par l'intermédiaire d'une vanne trois voies motorisée pilotée par un thermostat potentiométrique ; cette dernière solution n'apportait, en effet, que peu d'avantages et pouvait créer des causes de panne dans une profession où les machines sont assez mal entretenues.

La variation de vitesse du ventilateur d'extraction n'a pas été retenue, cette solution était prévue au cas où les échanges calorifiques ne seraient pas suffisants au niveau des condenseurs dans les périodes où nous n'aurions pas de phase de rinçage dans la machine. Des essais en usine ont montré que ce système n'était pas primordial dans la régulation du système.

15. Organisation du programme de mesures

Les paramètres à mesurer sont nombreux :

. températures d'eau

. temps de fonctionnement

. quantité d'eau consommée

. humidité de l'air

. températures d'air

. pressions des fluides

. consommation d'électricité

La plupart de ces mesures pourront être enregistrées par une centrale de mesure automatique à 60 voies AOIP SAM 60, équipée de capteurs judicieusement disposés. Les mesures seront vérifiées ou complétées par des compteurs horaires, des compteurs d'énergie électrique, des compteurs de consommation d'eau.

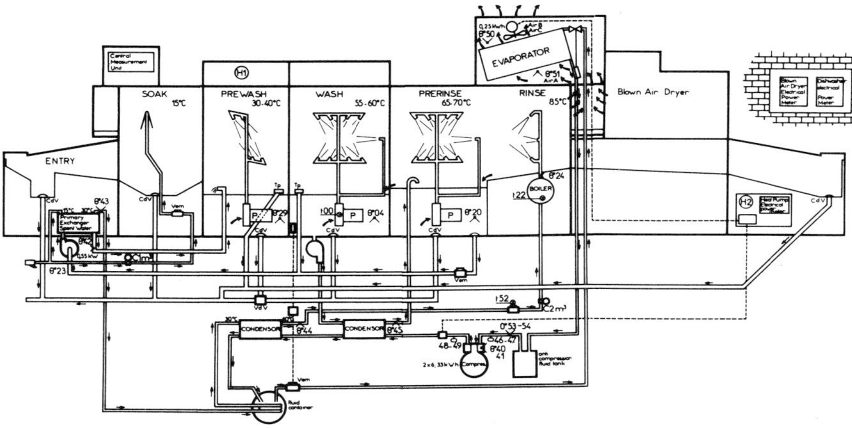

C1m^3 Water meter, water consumption for the Soak
C2m^3 Water meter, water consumption for filling the tanks and for rinsing

H1 Boiler heater time recorder
H2 Heat pump time recorder

t00 Contact 0.1 to measure operating time of pumps
t22 Contact 0.1 to measure operating boiler heating time
t52 Contact 0.1 to measure water filling time

δ46 δ47 Pressure entry of the compressors
δ48 δ49 Pressure discharge of the compressors

θ°04 Thermocouple to measure temperature of washing tank
θ°20 Thermocouple to measure temperature of prerinsing tank
θ°23 Thermocouple to measure temperature of filling water
θ°24 Thermocouple to measure rinsing temperature
θ°29 Thermocouple to measure temperature of prewashing tank
θ°40-θ°41 Thermocouple to measure temperature of compressors
θ°42 Thermocouple to measure temperature of water return to prewashing tank
θ°44 Thermocouple to measure temperature of water discharged from heat pump for the rinsing
θ°45 Thermocouple to measure temperature of water discharged from heat pump for the washing
θ°50 Thermocouple to measure air temperature after the heat pump evaporator
θ°51 Thermocouple to measure steam temperature entry of the heat pump evaporator
θ°53-θ°54 Thermocouple to measure temperature entry of compressors
θ°43 Thermocouple to measure water temperature discharged of the primary exchanger

2. REALISATION DU PROJET

La première opération consiste à trouver un client de la société BONNET susceptible d'accueillir l'installation de démonstration. La Société Carrefour donne son accord en mai 1980 pour que les travaux relatifs au projet puissent être exécutés dans sa laverie de vaisselle de la cafétaria de l'hypermarché de Nice Lingostière.

Le projet se déroulera en 5 phases pour lesquelles la majorité des tâches sont exécutées par le personnel de la société BONNET :

A - Adaptation et définition du système pompe à chaleur,
B - Achat et fabrication du matériel de démonstration,
C - Définition du programme de mesures et achats du matériel de mesures,
D - Adaptation eau froide de la laveuse, mise en place du matériel de

mesures et réalisation du programme de mesures,

E - Installation de la pompe de chaleur et réalisation du programme de mesures sur pompe à chaleur,

F - Période de gestion démonstrative.

Selon le diagramme PERT et la décomposition suivante :

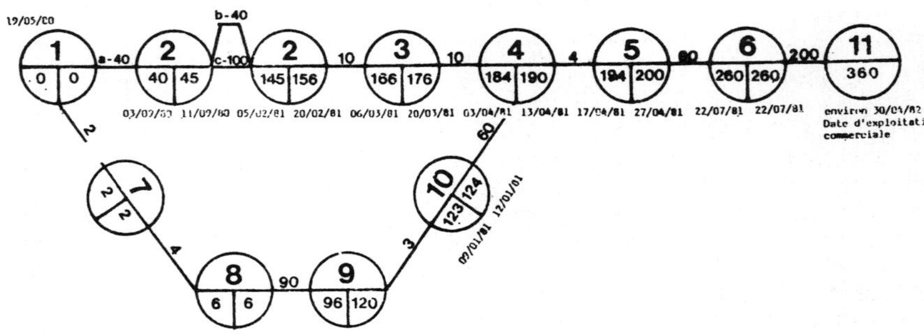

Phase		Désignation
Réf.	Tâche	
A	1	Définition technique de la pompe de chaleur pour l'installation projetée.
	1-2a	Etude de l'adaptation de la pompe de chaleur sur la machine à laver la vaisselle.
	2a-2b	Dessin du prototype.
B	2a-2c	Approvisionnement
	2-3	Montage de la pompe à chaleur.
	3-4	Mise en route. Essais en usine. Simulation du cycle pompe à chaleur / laveuse de vaisselle.
C	1-7	Recensement des paramètres à mesurer.
	7-8	Choix des appareils de mesures et adaptation sur la machine de démonstration.
	8-9	Approvisionnement du matériel, essais et mesures.
D	9-10	Mise en place du matériel de mesures sur l'installation primitive et adaptation en eau froide de la machine à laver.
	10-4	Réalisation du programme de mesures et traitement des informations.

3. MISE EN MARCHE DE L'INSTALLATION

Le programme de travail se poursuit selon le planning défini pour mesurer la consommation globale en énergie électrique d'une machine à laver la vaisselle alimentée en eau froide uniquement et équipée d'une pompe de chaleur.

Phase		Désignation
Réf.	Tâche	
E	4-5	Transformation de l'installation primitive pour mise en place de la pompe de chaleur et du nouveau matériel de mesures.
	5-6	Réalisation du programme de mesures et traitement des informations.

4. PHASE DE MESURES ET RESULTATS

41. Résultats communiqués par la centrale

Les possibilités de mesures, grâce à la centrale de mesures, sont les suivantes : lecture et enregistrement de toutes les voies lors de chaque mise en route de la machine, couplée avec la mise en route de la centrale - lecture et enregistrement de toutes les voies suivant programmation avec un intervalle de lecture de 15 minutes - lecture permanente de chaque voie sur le tableau afficheur de la centrale - lecture et enregistrement lors de chaque franchissement de seuil haut ou bas.

Les résultats donnés par la centrale permettent de faire une analyse statistique et dynamique du fonctionnement de la machine.

42. Les compteurs

Compteurs de consommation d'eau - compteurs horaires de temps de fonctionnement - compteurs d'énergie électrique (compteur machine et compteur compresseur).

Les mesures sont effectuées du 3/7/1981 au 18/1/1982. La période de gestion démonstrative continuera du 19/1/1981 au 4/5/1982.

Des pannes plus ou moins importantes ont affecté les mesures, en particulier panne de deux thermoplongeurs de chaudière impliquant une surconsommation estimée pour la chaudière de rinçage, la consommation d'énergie étant prise à la partir du temps de fonctionnement de la chaudière, ainsi que fuites de fréon au niveau de la pompe de chaleur.

Le tableau suivant donne la comparaison des résultats obtenus selon les 2 phases de mesures. Pour des raisons de facilité, les résultats sont ramenés sur une année de fonctionnement.

Rubrique	Machine initiale	Machine avec pompe de chaleur	Ecart
1. Consommation d'eau fonctionnement rinçage + remplissage	777 m3	761 m3	
2. Consommation d'eau annexes	1109 m3	575 m3	48,2 %
3. Consommation globale énergie système	84844 kWh	49672 kWh	41,5 %
4. Ratio $\frac{\text{Energie}}{\text{m3 eau traitée}}$ $\frac{3}{1}$	109,2 kWh/m3	65,27 kWh/m3	40,2 %
5. Consommation énergie réchauffeur uniquement	77796 kWh	42804kWh	45,0 %

La consommation globale d'énergie de la machine à chuté de 41,5 %, soit 35.172 kWh = 7,8 T.E.P.

5. PLANNING REEL

Le diagramme PERT suivant donne le déroulement du projet.

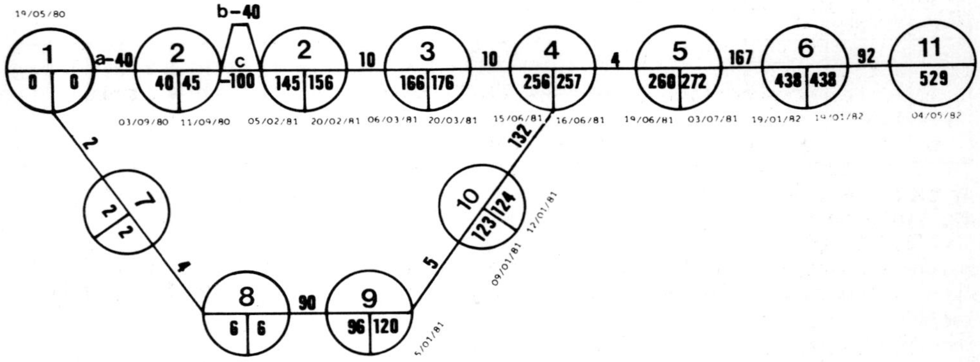

Les travaux ont commencé le 19/5/1980. Le déroulement a été conforme au planning pour les phases : A 1 ; A 1-2a ; A 2a-2b ; B 2a-2c ; B 2-3 ; B 3-4 C 1-7 ; C 7-8 ; C 8-9 ; D 9-10 ; E 4-5. La phase D 9-10 a duré 5 jours au lieu des 4 prévus, en raison de difficultés rencontrées sur le chantier. Les phases suivantes, en particulier celles d'essais et mesures D 10-4 ; E 5-6 ; F 6-11 ont subi des modifications en durée.

. D 10-4 : 132 jours au lieu de 60 jours ouvrables car la machine ne fonctionnait que 2,5 heures au lieu des 5 heures prévues journellement.

. E 5-6 : 167 jours au lieu de 60 jours ouvrables en raison du fonctionnement journalier de 2,5 heures au lieu des 5 heures prévues.

. F 6-11 : 92 jours au lieu de 200 jours ouvrables. Cette période a été réduite en raison de l'augmentation de la durée des périodes précédentes et parce que le client ne désirait pas garder le système pompe de chaleur.

Nous avons néanmoins poursuivi, à titre privé, les essais et mesures durant 5 mois supplémentaires qui ont permis de confirmer les résultats précédents.

Les difficultés rencontrées provenaient principalement du faible taux d'utilisation de la machine, de la non motivation du personnel qui entretenait peu ou pas la machine, du retard dans la déclaration des pannes survenues à la machine, du mauvais repérage des bandes de relevés de mesures délivrées par la centrale et de l'éloignement de BONNET Lyon par rapport au lieu de la démonstration Nice qui ne permettait pas une action rapide et fréquente.

6. COUT REEL DU PROJET - EN FRANCS FRANCAIS

Phase	Frais de personnel		Fournitures diverses		Total	
	Prévu	Réalisé	Prévu	Réalisé	Prévu	Réalisé
A	124.000	127.875,55		5.340,49	124.000	133.216,04
B	76.000	100.436,55	67.000	42.038,00	143.000	142.474,55
C	23.000	26.344,00	62.000	52.198,29	85.000	78.542,29
D	57.000	92.592,75	55.000	18.837,69	112.000	111.430,44
E	56.000	78.284,50	42.000	23.152,06	98.000	98.436,54
F						
Total	336.000	422.533,35	226.000	141.566,53	562.000	564.099,88

7. RENTABILITE - VIABILITE ECONOMIQUE

71. Rentabilité

Selon le tableau comparatif du paragraphe 4, sur une année, les économies réalisées sont :

. au niveau consommation électrique de la machine de 35.172 kWh, soit 7,74 T.E.P.

. au niveau consommation globale d'eau de 550 m3.

Les coûts respectifs de l'électricité et de l'eau étant estimés actuellement à 0,41 F.F. le kWh et 4,50 F.F. le m3, l'économie annuelle réalisée sera de 16.895,52 F.F. dont 14.402,52 F.F. pour l'électricité et 2.475 F.F. pour l'eau.

Le bilan annuel d'une machine sans pompe de chaleur est de 43.273 F.F.

L'économie réalisée au niveau exploitation est de 16.895 F.F., soit 39,04 %.

72. Viabilité économique

Le coût d'une telle pompe de chaleur fabriquée en série est de l'ordre de 120.000 F.F. Le retour sur investissement est de 7 ans pour cette machine. Il faut noter que cette machine ne fonctionnait que 2,5 heures par jour, ce qui est insuffisant.

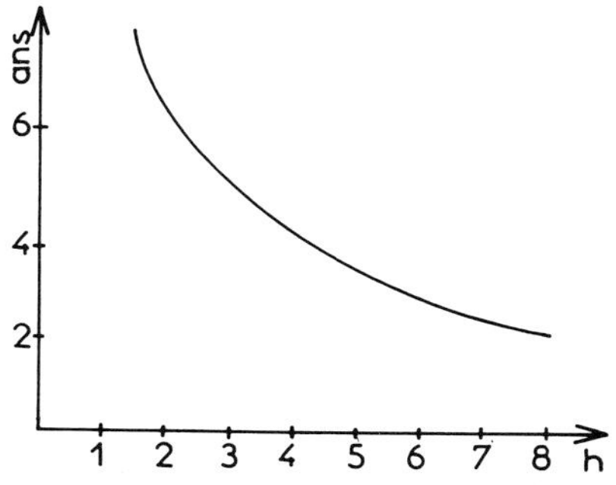

Courbe durée d'amortissement en fonction utilisation journalière

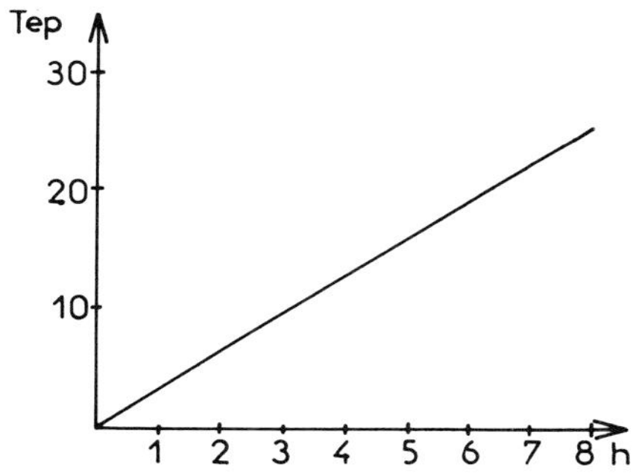

Courbe économie T.E.P. en fonction utilisation journalière

8. PERSPECTIVES COMMERCIALES

Le système semble très intéressant puisqu'il permet de réduire de 42 % la consommation énergétique de la machine à laver ; de plus, au niveau du bilan d'exploitation, il permet de réduire le budget global de l'exploitation de 39 %

Le système utilisé dans la démonstration était complexe, il peut être simplifié en supprimant le système échangeur de chaleur sur eaux usées placé avant la pompe de chaleur et dans ce cas un gain de 15.000 à 20.000 F.F. peut être envisagé sur l'installation globale réduisant la durée du retour sur investissement du système.

Les résultats du projet de démonstration montrent que l'application d'une pompe de chaleur sur une machine à laver la vaisselle permet de réduire à la fois la consommation d'énergie électrique et la consommation d'eau.

Ces deux facteurs ont une importance capitale dans le coût d'exploitation, mais le bilan énergétique du lavage de la vaisselle est souvent mal connu dans le bilan global d'exploitation d'une collectivité publique ou privée (comme c'était le cas où se trouvait l'installation de démonstration).

De plus, il faut tenir compte des conditions réelles de fonctionnement d'une machine, et il ne faut pas placer une pompe de chaleur sur une machine fonctionnant moins de 5 heures par jour.

Il est très important d'informer les gestionnaires et les responsables d'exploitation sur les économies réalisables par un travail profond de marketing, et dans ces conditions, on peut estimer que :

BONNET commercialise environ 40 machines par an pouvant justifier d'investissement de pompe de chaleur, le potentiel de la Communauté Européenne est de 400 machines, ce qui représenter une économie annuelle de 3.096 T.E.P. et atteindre 18.600 T.E.P. en 1990 si la commercialisation de tels systèmes intervient en 1984.

Il faudra, néanmoins, résoudre le problème de l'adaptation d'une telle pompe de chaleur à toutes les sortes de machines existant sur le marché, des différences notables intervenant au niveau des cycles de fonctionnement.

Le projet de démonstration faisait appel à une pompe de chaleur air-eau et limitait la température de réchauffage de l'eau à 65° C, mais une évolution est possible vers les pompes de chaleur eau-eau échangeant sur les eaux usées à 55° C ; dans ce cas, nous pourrions évoluer vers un système réchauffant l'eau à 85° C et ne nécessitant plus de réchauffeur électrique instantané pour l'eau de rinçage. Des problèmes de filtration et de décantation des eaux usées doivent précédemment être résolus, mais c'est vers cette solution ou vers une solution mitigée comprenant un premier étage -air-eau - et un deuxième étage - eau-eau - que peut se trouver l'adaptation des pompes de chaleur sur les machines à laver la vaisselle de collectivités.

9. DIVERS

Jusqu'à ce jour, la société BONNET n'a pas déposé de brevet concernant ce système.

La commercialisation, si elle a lieu, se fera par l'intermédiaire de BONNET et de son réseau de concessionnaires pour l'équipement des machines BONNET. Aucun contact n'a encore été pris avec d'autres fabricants de machines pour une commercialisation extérieure à BONNET.

BONNET n'envisageant pas une commercialisation immédiate en raison d'une durée relativement longue du retour sur investissement, n'a pas encore fait paraître sur les revues spécialisées d'articles concernant le système pompe à chaleur/machine à laver la vaisselle de collectivités.

IMPROVED CONVENTIONAL HEATING SYSTEMS - DEMONSTRATION PROJECTS

Rapporteur: J. DURANDET

There is a somewhat loose connection between the four papers proposed at our session. They all concern the general and very broad topic of the generation and conversion of heat. Research ranges from the preparation of fuels (B.3.3) to the development of efficient burners and furnaces (B.3.1). It also includes the cogeneration of heat and power with the help of solar energy (B.3.2) as well as the recovery and upgrading of waste heat (B.3.4).

- M. Carello and F. Monti (B.3.3) are interested in developing a static system for making fuel-oil/water emulsions. Such emulsions have the advantage of improving the combustion efficiency for fuel oils with a corresponding decrease in the emission of fumes and the fouling of furnaces. The system proposed is a particularly interesting one because it is easy to build and hence not very expensive. Likewise, because of its reduced bulk, it can be adapted to existing burners.

The device mainly consists of two parts : (1) a mixer for making a suitably proportioned water/oil mixture; (2) an emulsifier to transform this mixture, especially under the effect of pressure, into a uniform emulsion of fine water droplets in oil; a pressure drop of 2 bars ensures a good emulsion. Because of the very fine dispersion of water, with droplets measuring about one micron, the emulsion is very stable and remains unchanged for several months.

Such systems have been installed in 50 existing furnaces and differ considerably by the nature of the fuel used (gas oil, light and heavy fuel oils), the fuel-oil flow rates (from 20 to 4000 kg/hr), and the use made of the heat produced (24 for home heating, 11 for industrial power plants and district heating, and 7 for hotels).

The results of two years of testing are quite conclusive. The savings made were about 3.5 %. In addition to these relatively modest savings, there was also a reduction in the maintenance costs of the furnaces. The emulsifier also has the great advantage of considerably reducing pollution by limiting the amounts of fumes given off.

- Paper B.3.1 will be presented directly by its authors, V. Recchi, E. Triggiani and S. Inserra Imparato. It deals with a new type of furnace called the BLUE-FLAM 91 for home heating. A great many such furnaces have been installed. The measurement campaign lasted for two years, and the demonstration was conclusive. The savings made are about 10 %.

- The paper by M. Gini and P.L. Da Col (B.3.2) considers the overall applications of the following techniques to a commercial building : (1) combined production of heat and electricity by five Fiat-Totem generators with a capacity of 38 kWt and 15 kWt apiece; (2) heat pumps (209 kWt) also used with refrigeration plants (174 kWt); (3) incinerator for packing and miscellaneous wastes; (4) heat storage in 1200 m3 of water for both winter and summer (in summer, heat is collected and stored for the following winter, while in winter the water stored is cooled, ready for use in summer); (5) solar collectors, 105 m2.

This is one of the few examples where the various techniques have been coordinated and used together in a building which has been constructed. The project is a total conditioning system for summer/winter use.

The overall investment for the installation has been estimated to be 390 million Italian Lire (1980 value). An economic analysis shows that the design has approximately the same running costs as a conventional design. The project was to have been completed in March 1983.

- G. Schoeffler (B.3.4) presented us the results of operating an electric heat pump using the heat given off by a community dishwasher. A large share of the heat fed to the dishwasher is normally dissipated into the atmosphere, causing a rise in both the heat and the humidity of the ambient air. The heat pump installed, of the air/water type, recovers the heat contained in the mist, thus improving the quality of the air in the room. The system thus conditions the air of the premises while recycling usable heat to the dishwasher. The heat pump includes 2 circuits, one to heat the washing water, the other to preheat the rinsing water. The thermal capacity of the pump is 31 kW. It creates a heat transfer at approximate heat levels of from 30°C to 60°C.

The cost of the unassembled equipment required for building a prototype of the pump is about FF 50,000. The device operated correctly while producing an energy saving of 45 % during the year. The cost of this type of heat pump, when mass produced, is about FF 120,000. Payback time is 7 years under these conditions. These figures could be improved because the moderate commercial success of the "restaurateur" did not enable the installation to be operated under optimum conditions.

CONCLUDING REMARKS by

Dr. M. DAVIS

Commission of the European Communities, Directorate General for Energy, Brussels

Mr. Chairman,
Ladies and Gentlemen,

The building sector with which we have been dealing in these last three days is certainly a very important one : it is the largest energy using sector and it absorbs some 38 % of our global energy consumption. We believe that the realizable potential for saving in this sector alone is about a third of the energy used and this could be realized by the last years of the century. There is of course a much greater potential theoretical saving which, if circumstances changed, could probably also be secured. But just taking a realizable potential : it would require an investment of something like 7 or 8 billion ecus a year, which is a big expenditure. But, on the other hand, this would be equivalent to something like 3/4 of a million jobs, not all new jobs, but certainly including a fair proportion of new jobs. If this were done, in ten years or so, we would be saving energy, and largely energy imports, worth some 30 billion ecus a year.

This conference, Mr. Chairman, I believe, is important because it has brought those concerned from all parts of the Community to The Hague. It is one means of achieving information flow. But, as Prof. Elias said, we must improve information flow in many ways.

This is the first conference in which R & D and Community Demonstration has been brought together in one presentation. The frontier between Research and Development and Demonstration is one in which we must learn how to improve our management methods. The Community Research and Development programme, which you have been hearing about, is virtually all applied and, in so far as it is not ultimately put into use, it is a waste of time. The tendency, as it has been remarked, is for Research and Development to solve some problems but also to raise other questions, and hence to generate the need for more R & D. Now, we must not let this become merely a circular process. We must also adopt a methodology to enable our research projects to be compared usefully one with another. And this is a point which came out of this conference.

Demonstration is by no means the only way in which successful Research and Development can be brought to practical use. But it certainly provides a valuable mechanism to encourage the exploitation of R & D results. I suggest that researchers must frequently ask themselves the question as to when their discoveries are ripe to be demonstrated. The Commission welcomes ideas which can lead to the formulation of demonstration projects. As was remarked by Mr. Britten this morning, Demonstration in meaningless unless it leads to replication. I agree with him. We have a problem of dissemination. It is even greater for demonstration than for R & D. And that is because the people who should know the results of demonstration are more dispersed and more disparate than the scientific community. The Commission is reflecting on how better to do this and we would also welcome ideas.

I have been impressed by the points which both Prof. Boffa and Steimle and, indeed, Prof. Verhoeven and our present Chairman, Ir. Uyttenbroeck have made, that energy saving in buildings depends on three main factors : the building characteristics, the characteristics of the energy consuming equipment and the occupants' requirements and behaviour.

This conference has dealt extensively with the first two and rather less with the third. I am sure we must do more on the human factor, what people want and how to influence them to satisfy their wants and to enable our society to respond to the requirements of our energy strategy and, hence to our economic prosperity. Primitive man was preoccupied with food and comfort. He spent most of his days seeking food, and much of his nights seeking comfort. We have come a long way since then. Without necessarily attributing it to the common agricultural policy, we are no longer preoccupied with the search of food. But we still have to define what we mean by comfort in buildings.

Turning now to some specific points.

Old buildings are far more numerous than new and we must pay great attention to retrofitting.

Thermic diagnostics can be very valuable as Mr. Boncorps said in his keynote speech. But it is also important that, having diagnosed what should be done and, having done it, the effects must be measured. We must know what has been achieved.

Prof. Steimle emphasized and sketched out an inventory of what systems are available to us. This is an important matter. I would like to mention hospitals. We have heard of one project, which is well advanced and which is expected to save 50 to 60 % of energy over a conventional hospital. Hospitals are complex; they have a great variety of needs and can almost be considered as a test bed that tests all kinds of ideas which, subsequently, if successful, will be available for use in simpler systems as well as in hospitals.

On heat pumps, this conference has received a number of papers and presentations on the state of the art. I have the impression that this considerable activity of the Community's programmes anticipates the optimization of a range of heat pumps of many sizes and different working fluids employing the principles of both compression and absorption. Clearly, such heat pumps have a very large potential role in the efficient heating of buildings and, in industry besides. It is necessary to consider heat pumps as part of a system; they cannot be opportunities to agree and define certain features of a suitable range of heat pumps and how to measure their characteristics, so that the Community wide market may soon be established and hence enable costs of heat pumps to be brought down, so that their potential in energy conservation may be realized.

Turning now to some more general considerations.

I believe that this conference has provided clear indications on new methodologies for fining suitable energy saving interventions in various climates and for various building types in the Community countries. A lot of technological know-how has been provided related to new types of interventions for the building shell, the transparent and opaque surfaces of buildings, also in improved heating systems. There are a number of detailed lessons which several of our chairmen, notably Ing. Knobbout and Prof. Knoche, indeed all of them, have suggested to us.

They are too detailed to be mentioned by me now, but I can assure you that these and also all the contributions, including the much valuable contributions of the rapporteurs of the sessions, have been taken note of and will be reflected on.

There is another point I would like to make : consider for a moment the Research and Development programme. What you have been hearing in Research and Development represents work from the second R & D energy programme which is now terminated. The Commission has proposed to the Council a third programme which will take us up to 1987. The Council has already discussed this and we hope very much that they will take a decision before the end of 1983.

Mr. Chairman, I would not like to conclude my remarks without thanking all those responsible, beginning first of all with the Netherlands Ministry of Economics who have been our partners in organizing this conference in The Hague; and they have done all the local organizing. I think you will agree this has been very successful. I would like also to thank all the Chairmen, the Rapporteurs, and those who have written and presented papers to us, and thus made this conference possible, and finally, those of you in the audience for having taken part in our conference and asked interesting questions and made useful comments. Thank you very much, Mr. Chairman.

LIST OF PARTICIPANTS

ALLE, M.
Serv. de Programmation de la
Politique Scientifique
8, rue de la Science
B - 1040 BRUXELLES

ANDREANI, A.-M.
Centre d'Etudes Nucléaires
CEA/CEN-G
B.P. 85 X
F - 38000 GRENOBLE

ANDREASSEN, B.
Jydsk Telefon A/S
Sletvej 30
DK - 8310 AARHUS - Tranbjerg J.

ARACHELIAN, G.
Aifea S.p.A.
Via Biancospini 19
I - 20146 MILANO

ASHFORD, C.
London Borough of Heckney
Honer Road
GB - LONDON E9

AWBI, H.B.
B.S.R.I.A.
Napier College
Colinton Road
GB - EDINBURGH EH10 5DT

BAGLIONE, H.
Leroy-Somer
F - 1601 ANGOULEME Cedex

BAKER, G.J.
The Building Services
Research and Information
Association
Old Bracknell Lane West
GB - BRACKNELL RG12 4AH

BALESTRINI, G.
Amministrazione Provinciale di
Pesaro-Urbino
Via Gramsci
I - 61100 PESARO

BALLADA, R.
Bassani Ticino S.p.A.
Viale Luigi Borri 231
I - 21100 VARESE

BEKKER, J.
Bronswerk Airconditioning
Research Centre
P.O. Box 28
NL - 3800 HC AMERSFOORT

BERNS, W.
Volkshuisvesting Rotterdam
P.O. Box 6623
NL - 3002 AP ROTTERDAM

BERTHET, M.
Laboratoires de Marcoussis
Route de Nozay
F - 91460 MARCOUSSIS

BIASIN, K.
Rheinisch-Westfälisches
Elektrizitätswerk
Kruppstrasse 5
D - 4300 ESSEN 1

BIGOLARO, G.
Consiglio Nazionale delle
Ricerche - Istituto per la
Tecnica del Freddo
Corso Stati Uniti 4
I - 35100 PADOVA

BIRGEN, J.
Instal S.A.
51, rue Haute
L - LUXEMBOURG

BIRGEN, J.
Instal S.A.
51, rue Haute
L - LUXEMBOURG

BLAKENBURG, J.P.M.
Warmtepomp Nederland B.V.
P.O. Box 137
NL - 7300 AC APELDOORN

BOEKWIJT, W.O.
T.N.O. - Department for
Building Physics
P.O. Box 49
NL - 2288 GH RIJSWIJK

BOFFA, C.
Consiglio Nazionale delle Ricerche
Progr. Fin. "Energetica"
C.so Duca degli Abruzzi 24
I - 10129 TORINO

BOHLANDER, G.A.
Delft University of
Technology
Rotterdamseweg 139a
NL - 2628 AL DELFT

BOKELMANN, H.
Universität Essen
Thermodynamik
Universitätsstrasse 15
D - 4300 ESSEN 1

BONCORPS, J. Cl.
Agence Française pour la
Maîtrise de l'Energie
27, rue Louis Vicat
F - 75015 PARIS

BONSEL, A.M.T.G.
Nederlandse Omroep Stichting (NOS)
P.O. Box 10
NL - 1200 JB HILVERSUM

BONTIUS, G.H.
V.D.E.N.
Utrechtseweg 310
NL - 6812 AR ARNHEM

BOONSTOPPEL, F.
Raadgevend Technies Buro
Van Heugten B.V.
St. Annastraat 145
NL - 6524 EP NIJMEGEN

BOUDEWIJN, A.J.
PUEM N.V.
P.O. Box 110
NL - 350 AC UTRECHT

BOUMA, J.
T.N.O. - Division of
Technology for Society
P.O. Box 342
NL - 7300 AH APELDOORN

BOY, E.
Fraunhofer-Institut für
Bauphysik
Nobelstrasse 12
D - 7000 STUTTGART 80

BRANDNER, R.
IEA Heat Pump Center
Analysis Center Graz
Petersgasse 45
A - 8010 GRAZ

BREMMERS, H.H.P.G.
Nederlandse Energie
Ontwikkelingsmaatschappij B.V.
P.O. Box 17
NL - 6130 AA SITTARD

BRITTEN, J.R.
Building Research Establishment
Energy Conservation Support Unit
Bucknalls Lane
GB - GARSTON, Watford

BROEKHUIZEN, M.
Twijnstra Gudde N.V.
Pikeursbaan 15
NL - 7411 GT DEVENTER

BURTON, H.
Economic and Social Committee of
the European Communities
2, rue Ravenstein
B - 1000 BRUXELLES

BYBERG, M.R.
Thermal Insulation Laboratory
Technical University of Denmark
DK - 2800 LYNGBY

CAMPORESE, R.
Consiglio Nazionale delle Ricerche
Istituto per la Tecnica del Freddo
Corso Stati Uniti 4
I - 35100 PADOVA

CARELLO, M.
Solar 77 S.p.A.
Via del Commercio 1/A
I - 00154 ROMA

CASTAGNETTA, V.
Industria Italiana
Petrolia S.p.A.
Piazza della Vittoria 1
I - 16121 GENOVA

CAZZANIGA, E.
Commission of the
European Communities
Joint Research Centre
I - 21020 ISPRA (Varese)

CLARKE, J.A.
Abacus - Department of
Architecture
University of Strathclyde
GB - GLASGOW G4 ONG

COLLET, P.
The Technological Institute
Gregersensvej
DK - 2630 TAASTRUP

CORPELEIJN, B.G.
Erasmus University Rotterdam
Dr. Molewaterplein 50
NL - 3015 GE ROTTERDAM

COTTMAN, G.P.
Loughborough University of
Technology
Department of Physics
GB - LOUGHBOROUGH LE11 3TU

COURT, M.
Centre d'Etudes Nucléaires
de Grenoble
B.P. 85X
F - 38041 GRENOBLE Cedex

CROSSAN, G.
Central Regional Council
Viewforth
GB - STIRLING FK8 2ET

DART, E.K.
Public Works Amsterdam
Wibautstraat 3
NL - 1091 GH AMSTERDAM

DAVID, J.
Building Services Publications
Builder House
1-3 Pemberton Road
GB - LONDON EC4P

DAVIS, M.
Commission of the European
Communities - Directorate General
"Energy"
200, rue de la Loi
B - 1049 BRUXELLES

DE GIDS, W.F.
T.N.O. Research Institute for
Environmental Hygiene
P.O. Box 214
NL - 2600 AE DELFT

DE HOE, J.M.
Laborelec
Section 8
B.P. 11
B - 1640 RHODE ST. GENESE

DE HOOG, M.D.
Nederlandse Aanneming
Maatschappij
NEDAM B.V.
Ln. van Kronenburg 2
NL - 1180 BE AMSTELVEEN

DE JONG, A.
N.V. Prov. Gelderse
Elektriciteits-Mij.
P.O. Box 9039
NL - 6800 EZ ARNHEM

DE JONG, T.
Grontmij.
P.O. Box 153
NL - 3700 AD ZEIST

DE KLEIN, J.A.
VEG-Gasinstitut
Wilmersdorf 50
NL - 7300 AC APELDOORN

DE LEEUW DEN BOUTER, J.A.
D.S.M.
P.O. Box 601
NL - 6160 AP GELEEN

DE WAARD, W. Th.
Crooijmans Raadgevende
Ingenieurs B.V.
P.O. Box 14077
NL - 3508 SC UTRECHT

DEBLIRE, J.
Cabinet du Ministre J. Mayence
Reg. Wallone
35, square de Meeûs
B - 1040 BRUXELLES

DECLERCK, I.
Ebes N.V.
Zone Vlaanderen
Press Rooseveltlaan 1
B - 9000 GENT

DEKKER, K.G.
Delft University of Technology
Mekelweg 2
NL - 2628 CD DELFT

DESPOIS, J.Y.L.
Commissariat à l'Energie Atomique
Centre d'Etudes Nucléaires de
Saclay
F - 91191 GIF-sur-YVETTE

DEVEEN, V.
Commission of the European
Communities
Directorate general "Energy"
200, rue de la Loi
B - 1049 BRUSSELS

DEWSBURY, J.
The Building Services Research
and Information Association
Old Bracknell Lane West
GB - BRACKNELL RG12 4AH

DOGGART, J.
Energy Conscious Design
Emerald Street 11-15
GB - LONDON WC1N 3QL

DUBBELD, M.
T.N.O. Research Institute for
Environmental Hygiene
P.O. Box 214
NL - 2600 AE DELFT

DUBOIS, P.
Centre de Recherches de la C.G.E.
Route de Nozay
F - 91460 MARCOUSSIS

DUIJVESTEIN, C.A.J.
Delft Technical University
Berlageweg 1
NL - 2628 CR DELFT

DUMONT, P.
Ateliers de Constructions
Electriques de Charleroi
B.P. 4
B - 6000 CHARLEROI

DUPAGNE, A.
LPB - Université de Liège
Institut Math.
15, avenue des Tilleuls
B - 4000 LIEGE

DURANDET, J.
Institut Français du Pétrole
F - RUEIL-MALMAISON

EDDI, W.
Ministère de l'Industrie
et de la Recherche
1, rue Descartes
F - PARIS 1

EHMKE, H.J.
Universität Essen
Thermodynmaik
Universitätsstrasse 15
D - 4300 ESSEN 1

EHRINGER, H.
Commission of the European
Communities
Directorate General "Science,
Research and Development"
200, rue de la Loi
B - 1049 BRUXELLES

EISENBEISS, G.
B.M.F.T.
Heinemannstrasse 2
D - 5300 BONN 2

ELIAS, G.
Finalized Energy Project of
National Research Council
Via Nizza 128
I - 00198 ROMA

ELLIS, J.R.J.
Building Design Partnership
St. John Street 24
GB - MANCHESTER M3 4FB

ENGA, T.G.
British Gas Corporation
Watson House
Peterborough Road
GB - LONDON SW6 3HN

ENGLISH, E.C.
Freeman Enercon
Unit S
Science Park, Milton Road
GB - CAMBRIDGE

EUSER, P.
Institute of Applied Physics
T.N.O - TH
P.O. Box 155
NL - 2600 AD DELFT

EUSSEN, L.
Limagas N.V.
Minckelersstraat 2
NL - 6390 ZL NIEUWENHAGEN

FAYET, P.
Novelerg
12, rue de la Baume
F - 75008 PARIS

FILIPPI,
Dipartimento Energetica
del Politecnico
C.so Duca degli Abruzzi 24
I - 10129 TORINO

FINN, M.
An Foras Forbartha
St. Martin's House
Waterloo Road
IRL - DUBLIN 4

FINZI, F.
E.N.E.A.
S.M. Galeria
Via Anguilarese KM 1,3
I - 00197 ROMA

FISCHER, A.
World Association for Element-
Building and Prefabrication
Schlüterstrasse 6
D - 2000 HAMBURG 13

FULVIO, F.
Italenergie SpA
Adriatica Comp. Elett.
Viale della Repubblica 39
I - 67039 SULMONA

GALANIS, I.
Commission of the European
Communities
Directorate General "Information
Market and Innovation"
P.O. Box 1907
L - 2920 LUXEMBOURG

GIJZEN, J.A.A.M.
Apon, v.d. Berg, ter Braak,
Tromp, Arch. B.V.
Rotterdamse Rijweg 48
NL - 3043 BM ROTTERDAM

GONAY, P.
Corr. Revue Architecture
Belgium/Neuf/Decors
116, avenue du Diament
B - 1040 BRUXELLES

GOODMAN, H.
Department for Health and
Social Security
Euston Tower, Euston Road 286
GB - LONDON NW1 3DN

GOSSELIN, F.
Service de Programmation de la
Politique Scientifique
8, rue de la Science
B - 1040 BRUXELLES

GOSSELINK, J.D.
Burgplan B.V.
P.O. Box 360
NL - 3430 AJ NIEUWEGEN

GRABENHENRICH, H.B.
Institut für Technische
Thermodynamik
RWTH Aachen
Schinkelstrasse 8
D - 5100 AACHEN

GRAMATTE, W.
Battelle Institut
Am Römerhof 35
D - 6000 FRANKFURT/MAIN 90

GROENEVELD, G.
Erasmus University Rotterdam
Dr. Molewaterplein 50
NL - 3015 GE ROTTERDAM

GROSSIN, R.
Société Bertin & Cie
B.P. 3
F - 78373 PLAISIR

GRUHLE, W.D.
Technische Hochschule Darmstadt
Institut für Regelungstechnik
Schlossgraben 1
D - 6100 DARMSTADT

GUILLAUME, M.
Centre Scientifique et Technique de la Construction
41, rue du Lombard
B - 1000 BRUXELLES

GUNIA, J.
Vereinigte Elektrizitätswerke Westfalen AG
Rheinlanddam 24
D - 4600 DORTMUND 1

HAARSMA, J.D.
Erasmus University Rotterdam
Burg. Oudlaan 50
NL - 3062 PA ROTTERDAM

HABETS, J.
Eindhoven Technical University
P.O. Box 513
NL - 5600 MB EINDHOVEN

HAINE, J.
Cegedel
29, ave de la Porte Neuve
L - 2089 LUXEMBOURG

HALLGREEN, L.
Teknologisk Institut
P.O. Box 141
DK - 2630 TAASTRUP

HANSEN, E.
The Technological Institute
Gregersensvej
DK - 2630 TAASTRUP

HANSSENS, C.
V.z.w. Het Pand
Het Pand 312
B - 8790 WAREGEM

HASTERT, C.
Ministère de l'Energie
19, bvd Royal
L - LUXEMBOURG

HAUET, J.P.
Novelerg
12, rue de la Baume
F - 75008 PARIS

HAUGLUSTAINE, J.-M.
LPB - Université de Liège
Inst. Math.
15, avenue des Tilleuls
B - 4000 LIEGE

HAVERKAMP, L.J.
Witteveen en Bos Raadgevend Ingenieursbureau
P.O. Box 233
NL - 7400 AE DEVENTER

HEIDWEILLER, D.J.
Tebodin
Raadgevende Ingenieurs B.V.
Ln. v. Nw. Oost-Indie 25
NL - 2593 BJ 's-GRAVENHAGE

HELCKE, G.
Commission of the European Communities
Joint Research Centre
Ispra Establishment
I - 21020 ISPRA (Varese)

HENNESSY, T.D.
Electricity Supply Board
Lr. Fitzwilliam Street 27
IRL - DUBLIN 2

HENRIQUEZ, V.
Caribean Consulting Engineers
Gaitoweg 6
P.O. Box 391
CURACAO

HENSEN, J.
Institute of Applied Physics
T.N.O. - TH
Stieltjesweg 1
NL - 2628 CK DELFT

HOGELING, J.
Stichting ISSO
P.O. Box 20740
NL - 3001 JA ROTTERDAM

HOLMES, P.
Foster Wheeler Power Products Ltd.
Brenda Road
GB - HARTLEPOOL, CLEVELAND TS25 2BU

HOLZAPFEL, K.
Fachinformationszentrum Energie,
Physik - IEA Heat Pump Centre
D - 7514 EGGENSTEIN

HOUGHTON, F.
Foster Wheeler Power Products Ltd.
Brenda Road
GB - HARTLEPOOL, CLEVELAND TS25 2BU

HOWSON, R.P.
Loughborough University of
Technology - Dept. of Physics
GB - LOUGHBOROUGH LE11 3TU

HOYAUX, G.
Commission of the European
Communities
Directorate General "Science,
Research and Development"
200, rue de la Loi
B - 1049 BRUXELLES

HUBERS, H.
Delft University of Technology
Mekelweg 2
NL - 2628 CD DELFT

HUGHES, D.W.
New University of Ulster
Northern Ireland - COLERAINE

IEDEMA, P.D.
Delft University of Technology
Mekelweg 2
NL - 2628 CD DELFT

INCE, I.M.
South London Consortium
125, Camberwell Road
GB - LONDON SE5 OHB

ISERMANN, R.
INFU (Universität Dortmund)
Postfach 500 500
D - 4600 DORTMUND 50

ISINGS, J.
Neth. Organisation for Applied
Scientific Research - T.N.O.
P.O. Box 342
NL - 7300 AH APELDOORN

JACKMAN, P.J.
The Building Services
Research and Information Assoc.
Old Bracknell Lane West
GB - BRACKNELL RG12 4AH

JACKSON, S.
Stephen George and Partners
London Road 170
GB - LEICESTER LE2 1ND

JENSEN, O.
Danish Building
Research Institute
Dr. Neergaardsvej 15
DK - 2970 HORSHOLM

JEREMIASSEN, P.
Technological Institute
Gregersensvej
DK - 2630 TAASTRUP

JUNNE, J.
Marcq & Roba
221, boulevard Leopold II
B - 1080 BRUXELLES

KALLMANN, E.
Omphale S.A.
33, rue Godefroy
F - 92800 PUTEAUX

KAVANAGH, R.
International Energy Agency
2, rue André Pascal
F - 75775 PARIS Cedex 16

KEMPEN, G.
Limagas N.V.
Minckelerstraat 2
NL - 6390 ZL NIEUWENHAGEN

KLAUS, H.
Vereinigte Elektrizitätswerke
Westfalen AG
Rheinlanddamm 24
D - 4600 DORTMUND 1

KLOSTER, O.
Skive Kommune
Ostergade 13
DK - 7800 SKIVE

KNIPSCHEER, H.J.M.
IBK Nederland B.V.
P.O. Box 138
NL - 3800 AC AMERSFOORT

KNOBBOUT, J.
Neth. Organisation for Applied Scientific Research T.N.O.
P.O. Box 342
NL - 7300 AH APELDOORN

KNOCHE, K.F.
Institut für Technische Thermodynamik
RWTH Aachen
Schinkelstr. 8
D - 5100 AACHEN

KOOREMAN, J.
PUEM N.V.
P.O. Box 110
NL - 3500 AC UTRECHT

KOPPENOL, A.
T.N.O. - Division of Technology for Society
P.O. Box 342
NL - 7300 AH APELDOORN

KOUDSTAAL, W.
Van Leer Services
Energieweg 128
NL - 3641 RT MIJDRECHT

KRAINER, A.
University E. Kardelj of Ljubljana
Department of Civil Engineering
P.O. Box 579
YU - 61000 LJUBLJANA

KRAM, T.
Netherlands Energy Research Foundation
P.O. Box 1
NL - 1755 ZG PETTEN

KRISTINSSON, J.
Kristinsson Architectural Engineering
Noordenbergsingel 10
NL - 7411 SE DEVENTER

KRONBORG, M.
Jydsk Telefon a/S
Sletvej 30
DK - 8310 AARHUS - Tranbjerg J.

KULING, D.M.A.
Delft University of Technology
Kanaalweg 2B
NL - 2828 EB DELFT

KVISGAARD, B.
The Technological Institute
Gregersensvej
DK - 2630 TAASTRUP

LAIMER, R.
Stadt Landshut
Postfach
D - 8300 LANDSHUT

LAZZARIN, R.
Università di Padova
Istituto di Fisica Tecnica
Via Marzolo 9
I - 35100 PADOVA

LE VAGUERESE, P.
C.E.T.I.A.T.
B.P. 6089
F - 69604 VILLEURBANNE Cedex

LEACH, S.J.
Building Research Establishment
Garston
GB - WATFORD, Herts. WD2 7JR

LEBRUN, J.
Université de Liège
Inst. Math.
15, ave des Tilleuls
B - 4000 LIEGE

LEMAIRE, T.
Delft Technical University
Stevinweg 1
NL - 2628 CD DELFT

LEMMING, J.
Technological Institute
Gregersensvej
DK - 2630 TAASTRUP

LEOPARDO, R.
Ottico Meccanica Italiana
Omi S.p.A.
Via della Vasca Navale 79-81
I - ROMA

LETELLIER, B.
Novelerg
12, rue de la Baume
F - 75008 PARIS

LITTLER, J.
Building Unit
Polytechnic of Central London
Marylebone Road 35
GB - LONDON NW1

LIVINGSTONE, H.A.
Central Regional Council
Viewforth
GB - STIRLING FK8 2ET

LOESCH, A.
Administration des
bâtiments publics
10, rue du St. Esprit
L - LUXEMBOURG

LOOGMAN, J.D.
Public Works Amsterdam
Wibautstraat 3
NL - 1091 GH AMSTERDAM

LOS, S.
Agip Nucleare S.p.A.
I - MILANO

LOWES, G.
Borough of Darlington
Town Hall
GB - DARLINGTON DL1 5QN

LUECK, K.
Fichtel & Sachs
Energietechnik
Ernst-Sachs-Strasse 62
D - 8720 SCHWEINFURT

LUSCUERE, P.G.
Nederlandse Philips Bedrijven
Cederlaan 4
NL - 5616 SC EINDHOVEN

MAC RANDAL, D.
Abacus, Depart. of Architecture
University of Strathclyde
GB - GLASGOW G4 ONG

MADDALENA, L.
Ministero Lavori Pubblici
Via Nomentana 2
I - 00100 ROMA

MAIER, A.
U.S. Dept. of Energy
Lawrence Berkeley Lab.
University of California
Building 90H
USA - BERKELEY, CA 94720

MAKKAR, L.
South London Consortium
Camberwell Road 125
GB - LONDON SE5 OHB

MALEWSKI, W.
Borsig GmbH
Berliner Strasse 27-37
D - 1000 BERLIN 27

MALLON, W.
N.V. Nederlandse Gasunie
P.O. Box 19
NL - 9700 MA GRONINGEN

MARDORF, L.
Institut für Chemische Antriebe
und Verfahrenstechnik
Lampoldshausen
D - 7101 HARDTHAUSEN

MARSEGAGLIA, A.
San Giorgio Elettro-
domestici S.p.A.
Via Melara 40
I - 19027 LA SPEZIA Sp

MARTIN, D.J.
Energy Technology Support Unit
AERE Harwell
Didcot
GB - OXON OX11 ORA

MARTIN, P.L.
The Oscar Faber Partnership
Upper Marlborough Road
GB - ST. ALBANS, Hertf. AL1 3UT

MATHIEU, Y.
Renault - Techniques
nouvelles appliquées
9-11, avenue du 18 juin 1940
F - 92500 RUEIL-MALMAISON

MAURO, V.
C.S.E.A.
Via Ventimiglia 201
I - 10127 TORINO

MCMULLAN, J.T.
New University of Ulster
Northern Ireland - COLERAINE

MEIJNEN, A.J.
T.N.O. - Division of
Technology for Society
P.O. Box 342
NL - 7300 AH APELDOORN

MENKVELD, H.
Projectbureau Energieonderzoek
P.O. Box 342
NL - 7300 AH APELDOORN

MEYER, L.
Projectbureau Energieonderzoek
P.O. Box 342
NL - 7300 AH APELDOORN

MEYER, T.
Fraunhofer Institut für System-
technik und Innovationsforschung
Breslauer Strasse 48
D - 7500 KARLSRUHE 1

MILBANK, N.O.
Building Research Establishment
Garston
GB - WATFORD WD2 7JR

MINKHORST, J.H.
F.D.O. Technische Adviseurs B.V.
P.O. Box 379
NL - 1000 AJ AMSTERDAM

MOLSNER, H.D.
Ingenieurbüro für
Baugrunduntersuchungen
Hamburger Chaussee 169
D - 2300 KIEL 1

MOSS, R.
Medical Architectural
Research Unit
Holloway Road 383
GB - LONDON N7 8BD

MUISWINKEL, G.M.
European Investment Bank
100, bvd Konrad Adenauer
L - 2950 LUXEMBOURG

MULDERS, R.
Delft University of Technology
P.O. Box 535
NL - 2600 AM DELFT

MULLER, P.
Vereinigte Elektrizitätswerke
Westfalen AG
Rheinlanddamm 24
D - 4600 DORTMUND 1

McADAM, W.A.
Central Regional Council
Viewforth
GB - STIRLING FK8 2ET

NICOLAAS, H.J.
Institute of Applied Physics
T.N.O. - TH
P.O. Box 155
NL - 2600 AD DELFT

NICOLAY, D.
Commission of the European
Communities
Directorate General "Information
Market and Innovation"
P.O. Box 1907
L - 2920 LUXEMBOURG

NIELSEN, O.
Skive Kommune
Ostergade 13
DK - 7800 SKIVE

NOORDZIJ, L.
VEG-Gasinstituut N.V.
P.O. Box 137
NL - 7300 AC APELDOORN

NORGARD, J.S.
Technical University of Denmark
Physics Laboratory III
DK - 2800 LYNGBY

O'FARRELL, F.
An Foras Taluntais
The Agricultrual Institute
Sandymount Avenue
IRL - DUBLIN 4

OLDENGARM, J.
Institute of Applied Physics
T.N.O. - TH
P.O. Box 155
NL - 2600 AD DELFT

OLSEN, A.D.
The Technological Institute
Gregersensvej
DK - 2630 TAASTRUP

OOSTENDORP, P.A.
T.N.O. - Division of Technology
for Society
P.O. Box 342
NL - 7300 AH APELDOORN

OPFERGELT, D.
Université Catholique de Louvain
Centre Recherche et Architecture
Place du Levant 1 (Bât. Vinci)
B - LOUVAIN-LA-NEUVE

OSHIMA, T.
Matsushita Electric Works Ltd.
Kaiser-Friedrich-Ring 59
D - 4000 DUESSELDORF 11

OSSE, H.J.M.
Raadgevend Ingenieursbureau
Ir. H.J.M. Osse BV
Groesbeekseweg 4
NL - 6524 DA NIJMEGEN

OUDELAAR, G.
B.V. GTI-Holding
Department T.R.D.
Kabelweg 25
NL - 1014 BA AMSTERDAM

PAELINCK, G.J.G.
Raadgevend Ingenieursbureau
Aronsohn B.V.
Lijnbaan 131
NL - 3000 CK ROTTERDAM

PATUREAU, J.P.
Bertin & Cie
B.P. 3
F - 78373 PLAISIR Cedex

PAULICK, W.
Fichtel & Sach AG
Energietechnik
Ernst-Sachs-Strasse 62
D - 8720 SCHWEINFURT

PEDERSON, K.
Skive Kommune
Ostergade 13
DK - 7800 SKIVE

PELLEGRINI, G.
Commission of the European
Communities
Joint Research Centre
Ispra Establishment
I - 21020 ISPRA (Varese)

PERERA, M.D.A.E.S.
Building Research Station
Garston
GB - WATFORD, Herts. WD2 7JR

PERNOT, C.
Institute of Applied Physics
T.N.O. - TH
Stieltjesweg 1
NL - 2628 CK DELFT

PERQUIN, G.M.A.
Adviesbureau Peutz & Assoc.
P.O. Box 407
NL - 6500 AK NIJMEGEN

PETERS, H.J.
Jakob Peters Heizung-Klima-
Sanitär
Lindenstrasse 35
D - 5522 SPEICHER

PEZZANI, M.
E.N.E.A.
S.M. Galeria
Via Anguilarese KM 1,3
I - 00197 ROMA

PICA, F.
Agip Petroli
Via Lavrentina 449
I - ROMA 00142

PILAVACHI, P.
Commission of the European
Communities
Directorate general "Science,
Research and Development"
200, rue de la Loi
B - 1049 BRUXELLES

PISKI, E.
VDI
D - 4000 DUESSELDORF 1

PLOEGER, G.
Fraunhofer-Institut für
Bauphysik
Nobelstrasse 12

PLUIJMERS, B.J.
Delft University of Technology
Berlageweg 1
D - 2628 CR DELFT

POST, J.M.
Delft University of Technology
Berlageweg 1
NL - 2628 CR DELFT

PULITZER, N.
Agip Nucleare S.p.A.
I - MILANO

RAES, J.
Gfe-Soc. Belge de Gestion
d'Energie
Holstraat 61-A1
B - 8790 WAREGEM

RAMET, C.
Institut Français du Pétrole
F - RUEIL-MALMAISON

RAUKKO, M.
Embassy of Finland
489, ave Louise
B - 1050 BRUXELLES

RAVELET, R.
Laboratoires de Marcoussis
Route de Nozay
F - 91460 MARCOUSSIS

RAYMENT, R.
Building Research Establishment
Bucknalls Lane
GB - GARSTON, Herts.

RECCHI, V.
O.T.B. Officine Termotecniche
Breda S.p.A.
I - BARI

REEVES, C.
Borough of Darlington
Town Hall
GB - DARLINGTON DL1 5QN

REINAUDO, Ch.
C.G.E. - Laboratoires de
Marcoussis
Route de Nozay
F - 91460 MARCOUSSIS

REPUSSARD, J.P.
Commission of the European
Communities
Joint Interpreting and
Conference Service
200, rue de la Loi
B - 1049 BRUXELLES

RICHELLE, P.
Société d'Isolation Thermique S.A.
29, quai de Brabant
B - 6000 CHARLEROI

RICOSSA, L.
Bassani Ticino S.p.A.
Viale Luigi Borri 231
I - 21100 VARESE

RIENKS, C.
Verhaar-Uniad Technische
Adviesbureaus B.V.
Westersingel 41
Nl - 9718 CD GRONINGEN

ROEL, H.C.
Crooijmans Raadgevend Ing. B.V.
P.O. Box 14077
NL - 3508 SC UTRECHT

ROOS, H.M.
NV Nederlandse Gasunie
Postbus 19
NL - 9700 MA GRONINGEN

ROSSINI, G.
Aifea S.p.A.
Via Biancospini 19
I - 20146 MILANO

ROUSSEAU, G.
ELF Research Centre
B.P. 22
F - 69360 SAINT-SYMPHORIEN D'OZON

ROUX, G.
C.S.T.B.
24, rue Jos. Fourier
F - 38400 ST MARTIN D'HERES

ROYER, D.
CEA
CEN Cadarache
B.P. 1
F - 13115 ST PAUL LES DURANCE

RUBRECH, M.J.A.
Klimaatkontrakt B.V.
Pieter Calandweg 54
NL - 6827 BK ARNHEM

RUYSSEVELT, P.A.
Building Unit
Polytechnic of Central London
35, Marylebone Road
GB - LONDON NW1

SANDGAARD, M.
Rockwool International A/S
Hovegaden 501
DK - 2640 HEDEHUSENE

SCHMITZ, F.
LPB - Université de Liège
Inst. Math.
15, avenue des Tilleuls
B - 4000 LIEGE

SCHNEIDER, F.
Delft University of Technology
Berlageweg 1
NL - 2628 CD DELFT

SCHOEFER, C.
Commission of the European Communities
Directorate general "Science, Research and Development"
200, rue de la Loi
B - 1049 BRUXELLES

SCHOEFFLER, G.
Société Bonnet
117, rue Grenette
F - 69653 VILLEFRANCHE CEDEX

SCHROEDER, H.
COSMIN
p/a Huyghenslaan 28
NL - 6824 JH ARNHEM

SLAGT, G.
Nederlandse Omroep Stichting (NOS)
P.O. Box 10
NL - 1200 JB HILVERSUM

SMITH, G.T.
Ian G. Lindsay & Partners
3, Walker Street
GB - EDINGBURGH EH3 7JY

SMITH, I.E.
Cranfield Institute
GB - BEDFORD

SOLAINI, G.
Luigi-Rossi
A.R.C.O. S.r.L.
I - MILANO

SOUPAULT, O.
Bertin & Cie
B.P. 3
F - 78373 PLAISIR CEDEX

SPACH, Ch.G.
Spach & Fils S.A.
La Claquette
F - 67570 ROTHAU

SPIJKER, F.
Delft University of Technology
Mekelweg 2
NL - 2628 CD DELFT

SPLINTER, M.
Eindhoven University of Technology
P.O. Box 513
NL - 5600 MB EINDHOVEN

STAD, P.
Prov. Electriciteitsbedr. van Noord Holland
P.O. Box 150
NL - 2060 AD BLOEMENDAAL

STANSFIELD, F.
British Rail
Chief Architects Department
Melbury House, Melbury Terrace
GB - LONDON NW1 6JU

STARING, E.
Centre for Appropriate Technology
P.O. Box 5048
NL - 2600 GA DELFT

STEHMEIER, D.
Institut für Technische Thermodynamik
RWTH Aachen
Schinkelstrasse 8
D - 5100 AACHEN

STEIMLE, F.
Universität Essen
Thermodynamik
Universitätsstrasse 15
D - 4300 ESSEN 1

STEPHENS, W.H.
General Technology Systems Ltd.
Forge House
20, Market Place
GB - BRENTFORD, Middlesex TW8 8EQ

STOLK, A.L.
Delft University of Technology
Lab. Refrigerating Eng.
Mekelweg 2
NL - 2628 CD DELFT

STRAETER, M.
T.N.O. - Energy Research
Project Office
P.O. Box 342
NL - 7300 AH APELDOORN

STRUB, A.
Commission of the European
Communities
Directorate general "Science,
Research and Development"
200, rue de la Loi
B - 1049 BRUXELLES

STRUCK, W.
M.A.N. AG
Neue Technologie
Dachauerstrasse 667
D - 8000 MUENCHEN 50

TEMPELMAN, H.S.
GEB Rotterdam
Werkgr. Energiebesparing
P.O. Box 1313
NL - 3000 DE ROTTERDAM

THEYSE, F.H.
Theyse Energieberatung
A. Wipperfürtherstrasse 138
D - 5060 BERGISCH GLADBACH 2

THOMASSEN, A.A.J.
VEG-Gasinstituut N.V.
P.O. Box 137
NL - 7300 AC APELDOORN

TOONK, H.
Adviesbureau P.W. Deerns B.V.
P.O. Box 16102
NL - 2500 BC THE HAGUE

TURNER, P.G.
Van Leer Research Laboratories
Passfield
GB - LIPHOOK, Hampshire GU30 7RN

UYTTENBROECK, J.
W.T.C.B. - C.S.T.C.
41, rue du Lombard
B - 1000 BRUXELLES

VALSTAR, C.B.
Witteveen & Bos Raadgevend
Ingenieursbureau
P.O. Box 233
NL - 7400 AE DEVENTER

VAN BERGEN, F.M.
T.N.O. - Division of
Technology for Society
P.O. Box 342
NL - 7300 AH APELDOORN

VAN DEN AREND, J.
Comprimo B.V.
James Wattstraat 79
NL - 1097 DL AMSTERDAM

VAN DEN BERG, M.
N.V. Prodira S.A.
98-100, chaussée Vilvorde
B - 1120 BRUXELLES

VAN DEN ELST, F.M.N.
F.D.O. Technische Adviseurs B.V.
P.O. Box 194
NL - 7553 LG HENGELO

VAN DER GRINTEN, P.M.C.
Delft University of Technology
Berlageweg 1
NL - 2628 CR DELFT

VAN DER KOOI, J.
Delft University of Technology
Mekelweg 2
NL - 2628 CD DELFT

VAN DER KRIEKE, H.E.
N.V. Kema
P.O. Box 9035
NL - 6800 ET ARNHEM

VAN DER LANS, J.C.F.
Delft University of Technology
Mekelweg 2
NL - 2628 CD DELFT

VAN DER STEK, D.J.
Warmtepomp Nederland B.V.
P.O. Box 137
NL - 7300 AC APELDOORN

VAN DER VLOED, L.
Adv. bureau v. Klimaatbeheersing
Ir. G.G.B. Halmos
Wassenaarseweg 32
NL - 2596 CJ THE HAGUE

VAN DIJK, J.W.
N.V. Nederlandse Gasunie (TP/OE)
Laan Corpus den Hoorn 102
NL - 9700 MA GRONINGEN

VAN EKEREN, W.
Tebodin
Raadgevende Ingenieurs B.V.
Ln. v. Nw. Oost-Indie 25
NL - 2593 BJ s'-GRAVENHAGE

VAN HALTEREN, K.
Public Works Emmen
P.O. Box 242
NL - 7800 AE EMMEN

VAN MEIJGAARD, P.
Nationale Woningraad
P.O. Box 50051
NL - 1305 AB ALMERE

VAN OOSTWAARD, T.
Projectbureau Energieonderzoek
P.O. Box 342
NL - 7300 AH APELDOORN

VAN PAASSEN, A.H.C.
Delft University of Technology
Mekelweg 2
NL - 2628 CD DELFT

VAN RANST, J.
Architektenburo Storme-Van Ranst
38, Van Putlei
B - 2018 ANTWERPEN

VAN TILBURG, C.
Wilma Nederland B.V.
Kasteelsingel 10
NL - 6001 EZ WEERT

VAN ZANTEN, W.
Ned. Energie Ontwikkelings-
maatschappij B.V.
P.O. Box 17
NL - 6130 AA SITTARD

VAN'T HULLENAAR, A.W.J.
Public Works Amsterdam
Wibautstraat 3
NL - 1091 GH AMSTERDAM

VANDERMEULEN, P.
Gfe-Soc. Belge de
Gestion d'Energie
Holstraat 61-A1
B - 8790 WAREGEM

VERHAGEN, J.
Public Works Department
Rotterdam
Galvanistraat 15
NL - 3029 AD ROTTERDAM

VERHOEVEN, A.C.
Delft University of
Technology
Stevinweg 1
NL - 2628 CN DELFT

VERWEIJ, C.G.
Delft University of Technology
Mekelweg 2
NL - 2628 CD DELFT

VIERHOUT, R.M.
Comprimo B.V.
James Wattstraat 79
NL - 1079 DL AMSTERDAM

VOORTER, P.
Journal "Energiebesparing"
P.O. Box 34
Nl - 2501 AG THE HAGUE

VOS, M.
Arch. en Ing. bureau Op ten
Noort Blijdenstein
Beneluxlaan 4
NL - UTRECHT

WARREN, B.
Borough of Darlington
Town Hall
GB - DARLINGTON DL1 5QN

WASSENAAR, R.H.
Delft University of Technology
Mekelweg 2
Nl - 2628 CD DELFT

WETZEL-LIMBOSCH, S.
Neth. Org. for Applied
Scientific Research - T.N.O.
P.O. Box 342
NL - 7300 AH APELDOORN

WILLIAMS, G.J.
G.E.C. Power Engineering Ltd.
Cambridge Road
GB - WHETSTONE, Leicester LE8 3LH

YOUNG, R.
London Borough of Heckney
Honer Road
GB - LONDON E9

ZABOROWSKI, S.
Energy, Mines and Resources
460 O'Connor Street
Canada - OTTAWA, Ontario K1S 5H3

ZACCARIA, G.
Agip Nucleare S.p.A.
Viale Brenta 27-29
I - 20139 MILANO

ZEGERS, P.
Commission of the European
Communities
Directorate general "Science,
Research and Development"
200, rue de la Loi
B - 1049 BRUSSELS

ZITO, U.
Commission of the European
Communities
Directorate general "Energy"
200, rue de la Loi
B - 1049 BRUSSELS

INDEX OF AUTHORS